Lecture Notes in Physics

The Lecture Notes in Physics

The series Lecture Notes in Physics (LNP), founded in 1969, reports new developments in physics research and teaching – quickly and informally, but with a high quality and the explicit aim to summarize and communicate current knowledge in an accessible way. Books published in this series are conceived as bridging material between advanced graduate textbooks and the forefront of research to serve the following purposes:

• to be a compact and modern up-to-date source of reference on a well-defined topic;

• to serve as an accessible introduction to the field to postgraduate students and nonspecialist researchers from related areas;

• to be a source of advanced teaching material for specialized seminars, courses and schools.

Both monographs and multi-author volumes will be considered for publication. Edited volumes should, however, consist of a very limited number of contributions only. Proceedings will not be considered for LNP.

Volumes published in LNP are disseminated both in print and in electronic formats, the electronic archive is available at springerlink.com. The series content is indexed, abstracted and referenced by many abstracting and information services, bibliographic networks, subscription agencies, library networks, and consortia.

Proposals should be sent to a member of the Editorial Board, or directly to the managing editor at Springer:

Dr. Christian Caron
Springer Heidelberg
Physics Editorial Department I
Tiergartenstrasse 17
69121 Heidelberg/Germany
christian.caron@springer.com

Miguel A.L. Marques   Carsten A. Ullrich
Fernando Nogueira   Angel Rubio
Kieron Burke   Eberhard K.U. Gross

Time-Dependent Density Functional Theory

Springer

Editors

Miguel A.L. Marques
Departamento de Física
Universidade de Coimbra
Rua Larga 3004 – 516
Coimbra, Portugal
E-mail: marques@tddft.org

Angel Rubio
Dpto. Fisica de Materiales
Facultad de Quimicas, U. Pais Vasco
Centro Mixto CSIC-UPV/EHU, DIPC
Apdo. 1072, 20018 San Sebastian, Spain
E-mail: arubio@sc.ehu.es

Carsten A. Ullrich
Department of Physics
University of Missouri-Columbia
Columbia, MO 65211, U.S.A.
E-mail: ullrichc@missouri.edu

Kieron Burke
Dept. of Chemistry & Chem. Biology
Rutgers University
610 Taylor Rd, Piscataway
NJ 08854, U.S.A.
E-mail: kieron@dft.rutgers.edu

Fernando Nogueira
Departamento de Física
Universidade de Coimbra
Rua Larga 3004 – 516
Coimbra, Portugal
E-mail: fnog@teor.fis.uc.pt

Eberhard K. U. Gross
Institut für Theoretische Physik
Freie Universität Berlin
Arnimallee 14
14195 Berlin, Germany
E-mail: hardy@physik.fu-berlin.de

M.A.L. Marques et al., *Time-Dependent Density Functional Theory*,
Lect. Notes Phys. 706 (Springer, Berlin Heidelberg 2006), DOI 10.1007/b11767107

ISSN 0075-8450

ISBN 978-3-642-07128-7 e-ISBN 978-3-540-35426-0

Springer is a part of Springer Science+Business Media
springer.com

Printed in The Netherlands

Cover design: *design & production* GmbH, Heidelberg

Preface

The year 2004 was a remarkable one for the growing field of time-dependent density functional theory (TDDFT). Not only did we celebrate the 40th anniversary of the Hohenberg-Kohn paper, which had laid the foundation for ground-state density functional theory (DFT), but it was also the 20th anniversary of the work by Runge and Gross, establishing a firm footing for the time-dependent theory. Because the field has grown to such prominence, and has spread to so many areas of science (from materials to biochemistry), we feel that a volume dedicated to TDDFT is most timely.

TDDFT is based on a set of ideas and theorems quite distinct from those governing ground-state DFT, but employing similar techniques. It is far more than just applying ground-state DFT to time-dependent problems, as it involves its own exact theorems and new and different density functionals. Presently, the most popular application is the extraction of electronic excited-state properties, especially transition frequencies. By applying TDDFT after the ground state of a molecule has been found, we can explore and understand the complexity of its spectrum, thus providing much more information about the species. TDDFT has a especially strong impact in the photochemistry of biological molecules, where the molecules are too large to be handled by traditional quantum chemical methods, and are too complex to be understood with simple empirical frontier orbital theory.

Today, the use of TDDFT is continuously growing in all areas where interactions are important, but direct solution of the Schrödinger equation is too demanding. New and exciting applications are beginning to emerge, from ground-state energies extracted from TDDFT to transport through single molecules, to high-intensity laser and nonequilibrium phenomena, to nonadiabatic excited-state dynamics, to low-energy electron scattering. In each case, the present approximations were applied, and found to work well for some properties, but occasionally fail for others. Thus the search for more accurate, reliable approximations will continue, and over time, should attain the same maturity as present ground-state DFT.

So, whether you're a physicist calculating optical absorption of a metal cluster, or a chemist trying to determine the HOMO and LUMO for a chromophore, we hope you'll try TDDFT, and be pleasantly surprised at the usefulness of the results. And may the force (or at least, a good functional) be with you.

The Editors

User's Guide

This book is not the usual compendium of independent research articles that results from having many contributors. We, the editors, share a common aim of providing an accessible introduction to the subject, a comprehesive review of today's applications of TDDFT, and a survey of some of the most recent work. We have worked hard to do this, and so have the many contributors. Among many other things, there is a common notation and set of abbreviations (see front pages) and a single, comprehensive set of references (see index).

We hope this volume will be useful to a variety of audiences, who can use the book in many different ways. We imagine the following:

Students. In editing this book, we assumed a basic familiarity with the concepts and applications of ground-state DFT. For the student wholly new to DFT, we strongly recommend the book, *A Primer in DFT* [Fiolhais 2003], as background reading *before* this one.

Experienced Users of Ground-State DFT. For these researchers, we hope this book shows how TDDFT goes beyond the ground state, and provides both the conceptual framework and many examples of applications and implementations.

Users of TDDFT. The book provides the conceptual underpinnings of the formal theory and a survery of implementations and applications. For example, those solving the Casida linear response equations might be interested in the various real-time methodologies, and vice versa.

Developers of TDDFT. The book shows the broad scope of present-day applications, and gives a reference for many of the developments up to the current time, and also includes novel applications.

This volume is composed of a general introduction to TDDFT (Basics), followed by 32 refereed contributions divided in six parts. In the following we give a brief overview of these contributions and try to place them in a more general context.

Basics. The first chapter is a brief introduction to the entire subject of TDDFT, and should really be read by anyone using this book. It is simple and straightforward and explains how TDDFT works, including the original proof of Runge and Gross.

Formal Theory. The first part of the book deals with formal theory. This is a collection of chapters on general topics within TDDFT, either fleshing out some of the more esoteric consequences of the theory or generalizing it in some way or another. We start by adding to the original theorem, especially discussing the action functional and the Keldysh formalism as a very natural and useful description of the exact quantum mechanics of many-body systems under time-dependent perturbations, which can then be related to TDDFT. Next the difficult problem of initial-state dependence and its relation to memory in TDDFT is addressed. We then provide the motivation behind sometimes preferring time-dependent current DFT to TDDFT in Chap. 5. There are several cases throughout the book, e.g., for the description of transport or for the calculation of optical response of solids, where this change of variables makes for better approximations. Then we move to Chap. 6, where the first steps towards a theory including both nuclei and electrons on the same footing are given, which is extremely important for, e.g., photodissociation. We conclude this section with Chap. 7, which applies the language of second harmonic response, including super-operators and generalized response functions, to TDDFT.

Approximate Functionals. The heart of every practical DFT treatment of matter is the approximation for the xc functionals. As the theory is exact in principle, the quality of the results depends on the quality of the functional used.

Any ground-state functional yields an adiabatic approximation for use in TDDFT, and most calculations today are performed with such functionals. But many groups are searching for approximations that go beyond this, i.e., functionals with memory. In this context, we show a geometric approach based on ALDA that produces the time-dependent deformation approximation. Then, from a completely different starting point, we consider orbital-dependent functionals, and so we begin with exact exchange and then move to non-universal functional approximations to the kernel based on a correspondence with many-body perturbation theory. Lastly, Chap. 11 reviews the exact conditions of DFT and TDDFT, which should prove useful in guiding construction of approximations or testing them.

Numerical Aspects. For anyone considering performing a TDDFT calculation, there is always the choice of the numerical implementation. For strong fields, one needs to solve the time-dependent Schrödinger equation in real time. If only the optical response is needed, one can choose between real time or frequency space. This is ontop of the choice of basis sets, which can be loosely divided into real-space grids, localized basis functions, or planewaves for periodic codes. Chapter 12 surveys real-time methods for the time-dependent Schrödinger equation, and what happens when they are applied to the time-dependent Kohn-Sham equations. Then we present the intricacies of the linear response formalism in a basis set, and in Chap. 14 we discuss molecular dynamics in a TDDFT excited state. We end this Part with a comparison of time- versus frequency-space techniques.

Applications: Linear Response. This is our largest part, demonstrating the huge variety of systems that linear response TDDFT has been applied to. The highly interesting and particularly difficult case of open-shell molecules is discussed in Chap. 16. There have been many applications to clusters, as described in Chap. 17, and to semiconductor nanostructures, using current density functionals, as discussed in Chap. 18. Progress for extended systems has been more challenging, as local and semilocal density functionals fail in the thermodynamic limit (although current-density and orbital functionals do not). This general feature is discussed in Chapters 19 and 20. This basic difficulty also appears in the polarizability of conjugated polymers, and it is analysed in the following chapter.

The bulk of the applications so far have been to molecules, especially those of biological significance. Chapter 22 discuss in detail how TDDFT can be combined with biochemical methods to extract information on biological chromophores, while Chap. 23 gives a broad survey of these applications.

Applications: Beyond Linear Response. Another area where TDDFT promises to have major impact is in the rapidly-growing field of atoms and molecules in strong laser fields. With the advent of attosecond pulses, there is great demand for efficient methods of solution of the full time-dependent Schrödinger equation. An overview is given in the opening chapter. While some impressive progress has been made here, difficulties abound, such as the calculation of multiple ionization probabilities. Non-linear phenomena in clusters are discussed in Chap. 26. This part closes with a chapter on excited-state dynamics in solids.

New Frontiers. The last part of this volume deals with some recent novel applications of TDDFT. The first three chapters discuss how to use linear-response TDDFT to tackle problems that are difficult for ground-state DFT. For example, these include correct dissociation of molecules, and the inclusion of van der Waals forces between fragments of matter. These ideas are first tested in the simplest possible case, the uniform gas in Chap. 28. Then Chap. 29 shows how to find the exchange-correlation potential from such calculations, and how it leads to improved band-gaps for solids. Finally, Chap. 30 provides a detailed survey of how this formalism produces van der Waals interactions and results for weakly interacting slabs.

The next two chapters deal with the modern problem of how to calculate the transport characteristics of an atomic-sized transistor, such as a single molecule. Two different formalisms are presented, both going beyond the treatment common in present-day calculations. The first one starts from a Master equation for the electrons coupled to the phonons, instead of the pure time-dependent Schrödinger equation. The second one uses the Keldysh formalism described in Chap. 3 to produce a purely (time-dependent) electronic approach to the problem.

Our ultimate chapter focusses on a problem that Gross had in mind when formulating modern TDDFT, namely that of scattering. Linear-response TDDFT is shown to yield information on continuum states as well as bound ones, including elastic scattering of electrons from atoms and molecules.

We expect that these and other new developments will bear fruit over the next several years.

Contents

List of Contributors

Carl-Olof Almbladh
Department of Solid State Theory
Institute of Physics
Lund University, Sölvegatan 14 A
S-223 62 Lund
Sweden
`Carl-Olof.Almbladh@teorfys.lu.se`

André D. Bandrauk
Département de Chimie
Faculté de Sciences
Université de Sherbrooke
Sherbrooke (Québec)
Canada J1K 2R1
`Andre.Bandrauk@USherbrooke.ca`

Kieron Burke
Dept. of Chemistry & Chem. Biology
Rutgers University
610 Taylor Rd, Piscataway
NJ 08854, U.S.A.
`kieron@dft.rutgers.edu`
`http://dft.rutgers.edu/`

Roberto Car
Princeton University
Chemistry Physics
and Materials Institute
Washington Road, Princeton
NJ 08544, U.S.A.
`rcar@Princeton.edu`
`http://www.princeton.edu/~cargroup/`

Mark E. Casida
Laboratoire d'Études Dynamiques
et Structurales de la Selectivité
(LÉDSS)
Institut de Chimie Moléculaire
de Grenoble
Université Joseph Fourier
(Grenoble I)
F-38041 Genoble Cedex 9, France
`mark.casida@ujf-grenoble.fr`

Alberto Castro
Institut für Theoretische Physik
Freie Universität Berlin
Arnimalle 14
D-14195 Berlin, Germany
`alberto@physik.fu-berlin.de`

James R. Chelikowsky
Institute for Computational
Engineering and Sciences (ICES)
(C0200)
ACES Building, Room 4.324
201 East 24th Street ACES
1 University Station
University of Texas at Austin
Austin, Texas 78712, U.S.A.
`jrc@ices.utexas.edu`
`http://tesla.ices.utexas.edu/`

Adam E. Cohen
Department of Physics
Stanford University
382 Via Lomita, Stanford CA 94305
U.S.A.
`aecohen@stanford.edu`

Felipe Cordova
Laboratoire d'Études Dynamiques et Structurales de la Selectivité (LÉDSS)
Institut de Chimie Moléculaire de Grenoble
Université Joseph Fourier (Grenoble I)
F-38041 Genoble Cedcx 9, France
felipe.cordova@ujf-grenoble.fr

Nils E. Dahlen
Theoretical Chemistry
Materials Science Centre
Rijksuniversiteit Groningen
Nijenborgh 4
NL-9747 AG Groningen
The Netherlands
N.E.Dahlen@rug.nl

Paul L. de Boeij
Theoretical Chemistry
Materials Science Centre
Rijksuniversiteit Groningen
Nijenborgh 4
NL-9747 AG Groningen
The Netherlands
p.l.de.boeij@rug.nl
http://theochem.chem.rug.nl/~deboeij

Rodolfo Del Sole
Via della Ricerca Scientifica 1
I-00133 Rome
Italy
Rodolfo.Delsole@roma2.infn.it

John Dobson
School of Science
Griffith University
Nathan Queensland 4111, Australia
j.dobson@griffith.edu.au

Martin Fuchs
Fritz-Haber-Institut der MPG
Faradayweg 4-6
D-14195 Berlin, Germany
fuchs@fhi-berlin.mpg.de

Filipp Furche
Institut für Physikalische Chemie
Lehrstuhl für Theoretische Chemie
Universität Karlsruhe (TH)
Kaiserstrasse 12
D-76128 Karlsruhe, Germany
filipp.furche@chemie.uni-karlsruhe.de

Pablo García-González
Departamento de Física Fundamental
Universidad Nacional de Educación a Distancia
Apto. 60141
E-28080 Madrid
Spain
pgarcia@fisfun.uned.es

Ralph Gebauer
The Abdus Salam International Centre for Theoretical Physics (ICTP)
Condensed Matter Section
Strada Costiera 11
I-34014 Trieste, Italy
rgebauer@ictp.trieste.it

Andreas Görling
Lehrstuhl für Theoretische Chemie
Universität Erlangen-Nürnberg
Egerlandstr. 3
D-91058 Erlangen
Germany
Andreas.Goerling@chemie.uni-erlangen.de

Eberhard K. U. Gross
Institut für Theoretische Physik
Freie Universität Berlin
Arnimallee 14
D-14195 Berlin
Germany
hardy@physik.fu-berlin.de
http://www.physik.fu-berlin.de/~ag-gross/

Upendra Harbola
Department of Chemistry
University of California
Irvine, CA 92697-2025
U.S.A.
uharbola@uci.edu

Jürg Hutter
Physical Chemistry Institute
University of Zurich
Winterthurerstrasse 190
CH-8057 Zurich, Switzerland
hutter@pci.unizh.ch

Andrei Ipatov
Laboratoire d'Études Dynamiques et Structurales de la Selectivité (LÉDSS)
Institut de Chimie Moléculaire de Grenoble
Université Joseph Fourier (Grenoble I)
F-38041 Genoble Cedex 9, France
andrei.ipatov@ujf-grenoble.fr

Stefan Kurth
Institut für Theoretische Physik
Freie Universität Berlin
Arnimalle 14
D-14195 Berlin
Germany
kurth@physik.fu-berlin.de

Manfred Lein
Max-Planck-Institut für Kernphysik
Saupfercheckweg 1
D-69117 Heidelberg
Germany
manfred.lein@mpi-hd.mpg.de
http://www.mpi-hd.mpg.de/keitel/mlein/

Xabier Lopez
Kimika Fakultatea
Euskal Herriko Unibertsitatea
E-20080 Donostia
Spain
poplopex@sq.ehu.es
http://www.sc.ehu.es/powgep99/dcytp/teoricos/xabier/english.html

Neepa T. Maitra
Department of Physics and Astronomy
Hunter College and City University of New York
695 Park Avenue, New York
NY 10021, U.S.A.
nmaitra@hunter.cuny.edu
http://www.ph.hunter.cuny.edu/faculty/Maitra/neepa.html

Andrea Marini
Department of Physics
University of Rome "Tor Vergata"
Via della ricerca scientifica 1
I-0133 Rome, Italy
andrea.marini@roma2.infn.it
http://people.roma2.infn.it/~marini/

Miguel A. L. Marques
Departamento de Física
Universidade de Coimbra
Rua Larga
P-3004 516 Coimbra
Portugal
marques@tddft.org

Yoshiyuki Miyamoto
Fundamental Research Laboratories
NEC Corporation
34, Miyukigaoka
Tsukuba, Ibaraki 305-8501
Japan
y-miyamoto@ce.jp.nec.com

Shaul Mukamel
Department of Chemistry
University of California
Irvine, CA 92697-2025
U.S.A.
smukamel@uci.edu
http://mukamel.ps.uci.edu

Yann M. Niquet
Atomistic Simulation Laboratory
CEA/DRFMC/SP2M/L_Sim
17 avenue des Martyrs
F-38054 Grenoble Cedex 9
France
yniquet@cea.fr

Fernando Nogueira
Departamento de Física
Universidade de Coimbra
Rua Larga
P-3004 516 Coimbra
Portugal
fnog@teor.fis.uc.pt

Dmitrij Rappoport
Institut für Physikalische Chemie
Lehrstuhl für Theoretische Chemie
Universität Karlsruhe (TH)
Kaiserstrasse 12
D-76128 Karlsruhe
Germany
rappoport@chemie.uni-karlsruhe.de

Paul-Gerhard Reinhard
Institut für Theoretische Physik
Universität Erlangen
Staudtstr. 7
D-91058 Erlangen
Germany
reinhard@theorie2.physik.uni-erlangen.de

Angel Rubio
Dpto. Fisica de Materiales
Facultad de Quimicas
U. Pais Vasco
Centro Mixto CSIC-UPV/EHU
DIPC Apdo. 1072
E-20018 San Sebastian
Spain
arubio@sc.ehu.es
http://dipc.ehu.es/arubio/

Yousef Saad
Department of Computer Science
and Engineering
University of Minnesota
4-192 EE/CSci Building
200 Union Street S.E. Minneapolis
MN 55455, U.S.A.
saad@cs.umn.edu
http://www-users.cs.umn.edu/~saad/

Gianluca Stefanucci
Institut für Theoretische Physik
Freie Universität Berlin
Arnimalle 14
D-14195 Berlin
Germany
gianluca.stefanucci@teorfys.lu.se

Osamu Sugino
Fundamental Research Laboratories
NEC Corporation
34, Miyukigaoka
Tsukuba, Ibaraki 305-8501
Japan
sugino@issp.u-tokyo.ac.jp

Eric Suraud
Laboratoire de Physique Theéorique
Université Paul Sabatier
118 Route de Narbonne
F-31062 Toulouse cedex, France
suraud@irsamc.ups-tlse.fr

Ilya V. Tokatly
Lehrstuhl für Theoretische Festkörperphysik
Universität Erlanger
Staudtstr. 7/B2
D-91058 Erlangen
Germany
Ilya.Tokatly@physik.uni-erlangen.de

Carsten A. Ullrich
Department of Physics
University of Missouri-Columbia
Columbia, MO 65211, U.S.A.
ullrichc@missouri.edu
http://www.missouri.edu/~ullrichc/

Robert van Leeuwen
Theoretical Chemistry
Materials Science Centre
Rijksuniversiteit Groningen
Nijenborgh 4
NL-9747 AG Groningen
The Netherlands
R.van.Leeuwen@rug.nl
http://theochem.chem.rug.nl/~leeuwen

Igor Vasiliev
Department of Physics
MSC 3D
New Mexico State University
P.O. Box 30001
Las Cruces, NM 88003-8001, U.S.A.
vasiliev@nmsu.edu

Valérie Véniard
Laboratoire des Solides Irradiés
CNRS-CEA-École Polytechnique
91128 Palaiseau, France
vv@ccr.jussieu.fr

Giovanni Vignale
Department of Physics
University of Missouri-Columbia
Columbia, MO 65211, U.S.A.
vignaleg@missouri.edu
http://www.missouri.edu/~physwww/people/GiovanniVignale.html

Ulf von Barth
Department of Solid State Theory
Institute of Physics
Lund University, Sölvegatan 14 A
S-223 62 Lund, Sweden
Ulf.von_Barth@teorfys.lu.se

Adam Wasserman
Dept. of Chemistry & Chem. Biology
Rutgers University
610 Taylor Rd, Piscataway
NJ 08854, U.S.A.
awasser@rutchem.rutgers.edu

Abbreviations

AA	Adiabatic approximation
ACFD	Adiabatic-connection and fluctuation-dissipation
ALDA	Adiabatic local density approximation (see TDLDA)
AMED	Average magitude of energy denominators
ATD	Above-threshold dissociation
ATI	Above-threshold ionization
B3LYP	Becke's three-parameter hybrid with Lee, Yang, and Parr correlation
BP86	Becke and Perdew 1986
BPG	Burke, Petersilka, and Gross
BSE	Bethe-Salpeter equation
BTE	Boltzmann transport equation
c	Correlation
CEDA	Common energy denominator approximation
CEO	Collective electronic oscillator
CIS	Configuration Interaction Singles
CN	Crank and Nicholson
CREI	Charge Resonance Enhanced Ionization
CVD	Chemical vapor deposition
DFT	Density functional theory
DODS	Different Orbitals for Different Spins
EM	Exponential midpoint
ETRS	Enforced time-reversal symmetry
EXX	Exact exchange
FWHM	Full width at half maximum
GGA	Generalized gradient approximation
GGG	Gonze, Ghosez, and Godby
GK	Gross and Kohn

GRF	Generalized response function
GS	Ground state
GWA	*GW* approximation
H	Hartree
HF	Hartree-Fock
HHG	High-harmonic generation
HOMO	Highest occupied molecular orbital
HPT	Harmonic potential theorem
Hxc	Hartree and exchange-correlation
IR	Infrared
ISB	Intersubband
KLI	Krieger, Li, and Iafrate
KMS	Kubo, Martin, and Schwinger
KS	Kohn-Sham
LB94	van Leeuwen and Baerends 1994
LCAO	Linear combination of atomic orbitals
LDA	Local density approximation
LHF	Localized Hartree-Fock
LIED	Laser Induced Electron Diffraction
LIMP	Laser Induced Molecular Potential
LR-TDDFT	Linear Response Time-Dependent Density-Functional Theory
LRC	Long range correction
l.h.s.	Left hand side
LO	Localized orbital
LUMO	Lowest unoccupied molecular orbital
MBPT	Many-body perturbation theory
MD	Molecular dynamics
MP	Moeller-Plesset
MO	Molecular orbital
MPI	Multi-photon ionization
NEGF	Nonequilibrium Green's function
occ	Occupied
OEP	Optimized effective potential
PBE	Perdew, Burke, and Ernzerhof
PDB	Protein data bank
PES	Photoelectron spectra

PGG	Petersilka, Gossmann, and Gross
QMC	Quantum Monte Carlo
QM/MM	Quantum mechanics / molecular mechanics
QP	Quasiparticle
RA	Richardson and Ashcroft
RG	Runge and Gross
r.h.s.	Right hand side
RPA	Random-phase approximation
SAE	Single active electron
SCF	Self-consistent field (used in ΔSCF)
SIC	Self-interaction correction
SO	Split-operator
SOD	Second order differencing
SODS	Same Orbitals for Different Spins
SOMO	Singly Occupied Molecular Orbital
SSE	Sham-Schlüter equation
ST	Suzuki and Trotter
STM	Scanning tunneling microscope
SPA	Single-pole approximation
TC	Triplet coupled
TDA	Tamm-Dancoff Approximation
TDDFT	Time-dependent density functional theory
TDEHF	Time Dependent Extended Hartree Fock
TDHF	Time-dependent Hartree-Fock
TDLDA	Time-dependent local density approximation (see ALDA)
TDLDefA	Time-dependent local deformation approximation
TDSE	Time-dependent Schrödinger equation
TDKS	Time-dependent Kohn-Sham
unocc	Unoccupied
UV	Ultraviolet
vdW	van der Waals
VIS	Visible
VK	Vignale and Kohn
VUC	Vignale, Ullrich, and Conti
VUU	Vlasov, Ühling, and Uhlenbeck
xc	Exchange-correlation
x	Exchange

Notation

General

$\boldsymbol{r}$	A point in 3-D space, (r_1, r_2, r_3)
t	An instant in time
ω	Frequency (Fourier transform of time)
$\boldsymbol{x}$	Combined space and time coordinates $(\boldsymbol{r}, t)$
$f(\boldsymbol{r})$	f is a *function* of the variable $\boldsymbol{r}$
$f[n]$	f is a *functional* of the function n

DFT

E	Ground-state total energy
E_{xc}	Ground-state exchange-correlation energy functional
e_{xc}	Exchange-correlation energy per electron
$n(\boldsymbol{r})$	Ground-state electronic density
$\rho(\boldsymbol{r}, \boldsymbol{r}')$	Ground-state electronic density matrix
$\varphi_i(\boldsymbol{r})$	Ground-state Kohn-Sham wave-function
ε_i	Ground-state Kohn-Sham eigenvalue
n_i	Occupation number of state i
A	Quantum-mechanical action
$n(\boldsymbol{r}, t)$	TD electronic density
$\boldsymbol{j}(\boldsymbol{r}, t)$	TD electronic current
$\boldsymbol{v}(\boldsymbol{r}, t)$	TD electronic velocity
$v_{\text{KS}}(\boldsymbol{r}, t)$	TD Kohn-Sham potential

$v_{\mathrm{ext}}(\boldsymbol{r}, t)$	TD external potential
$v_{\mathrm{H}}(\boldsymbol{r}, t)$	TD Hartree potential
$v_{\mathrm{xc}}(\boldsymbol{r}, t)$	TD exchange-correlation potential
$\varphi_i(\boldsymbol{r}, t)$	TD Kohn-Sham single-particle wave-function
$\Psi(\boldsymbol{r}_1, \boldsymbol{r}_2, \dots, \boldsymbol{r}_N, t)$	Interacting many-body wave-function
$\Phi(\boldsymbol{r}_1, \boldsymbol{r}_2, \dots, \boldsymbol{r}_N, t)$	Kohn-Sham many-body Slater determinant

Operators

$\hat{H}$	Hamiltonian
$\hat{H}_{\mathrm{KS}}$	Kohn-Sham Hamiltonian
$\hat{T}$	Kinetic energy
$\hat{V}$	Potential
$\hat{V}_{\mathrm{ee}}$	Two-body (Coulomb) interaction
$\hat{\mathcal{T}}$	Time-ordering
$\hat{U}(t, t')$	Evolution operator
$\hat{\mathcal{L}}$	Laplace transform
$\breve{A}$	Super-operator

Many-body and linear response

$G(\boldsymbol{r}, \boldsymbol{r}', \omega)$	Green's function
$\Sigma(\boldsymbol{r}, \boldsymbol{r}', \omega)$	Self energy
$v_{\mathrm{ee}}(\boldsymbol{r}, \boldsymbol{r}')$	Bare Coulomb interaction ($1/\lvert\boldsymbol{r} - \boldsymbol{r}'\rvert$)
$W(\boldsymbol{r}, \boldsymbol{r}', \omega)$	Screened Coulomb interaction
$\chi(\boldsymbol{r}, \boldsymbol{r}', \omega)$	Density-density response function
$f_{\mathrm{xc}}(\boldsymbol{r}, \boldsymbol{r}', \omega)$	Exchange-correlation kernel

Varia

$\mathcal{V}$	Volume
η	Positive infinitesimal
β	Inverse temperature
μ	Chemical potential
r_{s}	Wigner-Seitz radius
ω_{p}	Plasma frequency
I	Ionization potential

1 Basics

E.K.U. Gross and K. Burke

1.1 Introduction

Suppose you are given some piece of matter, such as a molecule, cluster, or solid, and you have already solved the ground-state electronic problem highly accurately. You now ask, how can we best calculate the behavior of the electrons when some time-dependent perturbation, such as a laser field, is applied? The direct approach to this problem is to solve the time-dependent Schrödinger equation. But this can be an even more demanding task than solving for the ground state, and becomes prohibitively expensive as the number of electrons grows, due to their Coulomb repulsion.

We will show in this chapter that, under certain quite general conditions, there is a one-to-one correspondence between time-dependent one-body densities $n(\boldsymbol{r}, t)$ and time-dependent one-body potentials $v_{\text{ext}}(\boldsymbol{r}, t)$, for a given initial state. That is, a given evolution of the density can be generated by at most one time-dependent potential. This statement, first proven by Runge and Gross [Runge 1984] (RG), is the time-dependent analog of the celebrated Hohenberg-Kohn theorem [Hohenberg 1964]. Then one can define a fictitious system of noninteracting electrons moving in a time-dependent effective potential, whose density is precisely that of the real system. This effective potential is known as the time-dependent Kohn-Sham potential. Just as in ground-state density functional theory (DFT), it consists of an external part, the Hartree potential, and the exchange-correlation potential, $v_{\text{xc}}(\boldsymbol{r}, t)$, which is a functional of the entire history of the density, $n(\boldsymbol{r}, t)$, the initial interacting wavefunction, $\Psi(0)$, and the initial Kohn-Sham wavefunction, $\Phi(0)$. This functional is a very complex one, much more so than the ground-state case. Knowledge of it implies solution of all time-dependent Coulomb-interacting problems.

In practice, we always need to approximate unknown functionals. An obvious and simple choice for TDDFT is the adiabatic local density approximation (ALDA), sometimes called time-dependent LDA, in which we use the ground-state potential of the uniform gas with that instantaneous and local density, i.e., $v_{\text{xc}}^{\text{ALDA}}[n](\boldsymbol{r}, t) = v_{\text{xc}}^{\text{unif}}(n(\boldsymbol{r}, t))$. This gives us a working Kohn-Sham scheme, just as in the ground state. We can then apply this DFT technology to every problem involving time-dependent electrons. These applications fall into three general categories: nonperturbative regimes, linear (and

E.K.U. Gross and K. Burke: *Basics*, Lect. Notes Phys. **706**, 1–17 (2006)
DOI 10.1007/3-540-35426-3_1

higher-order) response, and ground-state applications. The rapidly growing number of such applications, and the diversity of systems, ranging from chemistry to biology to materials science, forms the motivation for this book.

The first of these applications involves atoms and molecules in strong laser fields [Marques 2004], in which the field is so intense that perturbation theory does not apply. In these situations, the perturbing electric field is comparable to or much greater than the static electric field due to the nuclei. Experimental aims would be to enhance, e.g., the 27th harmonic, i.e., the response of the system at 27 times the frequency of the perturbing electric field [Christov 1997], or to cause a specific chemical reaction to occur (quantum control) [Rice 2000]. Previously, only one and two electron systems could be handled computationally, as full time-dependent wavefunction calculations are very demanding [Parker 2000]. Crude and unreliable approximations had to be made to tackle larger systems. But with the advent of TDDFT, larger systems with more electrons can now be tackled (see Chaps. 22, 23, 26 and 27).

When the perturbing field is weak, as in typical spectroscopic experiments, perturbation theory applies. Then, instead of needing knowledge of the functional $v_{\rm xc}[n](\boldsymbol{r},t)$ at densities that are changing significantly with time, which might differ substancially from a ground-state density, we only need to know this potential in the vicinity of the initial state, which we take to be a nondegenerate ground-state. These changes are characterized by a new functional, the exchange-correlation kernel [Gross 1985]. The exchange-correlation kernel is much more manageable than the full time-dependent exchange-correlation potential, because it is a functional of the ground-state density alone. Analysis of the linear response then shows [Appel 2003] that the latter is (usually) dominated by the response of the ground-state Kohn-Sham system, but corrected by TDDFT via matrix elements of the exchange-correlation kernel. In the absence of Hartree-exchange-correlation effects, the allowed transitions are exactly those of the ground-state Kohn-Sham potential. But the presence of the kernel shifts the transition frequencies away from the Kohn-Sham values to the true values. The intensities of the optical transitions are also affected by the kernel [Petersilka 1996a, Casida 1996, Rubio 1996].

Several approaches to extracting excitations from TDDFT for atoms, molecules, and clusters are currenty being used. The standard approach in quantum chemistry is to very efficiently convert the search for poles of response functions into a large eigenvalue problem [Casida 1996, Görling 1999a, Furche 2005a], in a space of the single-particle excitations of the system. The eigenvalues yield transition frequencies, while the eigenvectors yield oscillator strengths. This allows use of many existing fast algorithms to extract the lowest few excitations (see Chap. 23). In this way, TDDFT has been programmed into most standard quantum chemical packages [Bauernschmitt 1996a] and, after a molecule's structure has been found, it is usually not too costly to extract its low-lying spectrum [Furche 2005c]. Physicists, on the other hand, tend to solve the time-dependent Kohn-Sham equations by evolving the system in real time in the presence of a weak field [Yabana 1996]. Fourier-transform of the time-dependent dipole matrix element then yields the optical spectrum. Using either methodology, the number of these TDDFT response calculations for transition frequencies is growing exponentially at present [Burke 2005a]. Overall, results tend to be fairly good (0.1 to 0.2 eV errors, typically), but little is understood about their reliability [Furche 2005c]. Challenges remain for the application of TDDFT to solids [Onida 2002], because the present generation of approximate functionals (local and semi-local) lose important effects in the thermodynamic limit, but much work (some reported in this book) is currently in progress.

The last class of application of TDDFT is, perhaps surprisingly, to the ground-state problem. One can extract the ground-state exchange-correlation energy from a response function, in the same fashion as perturbation theory yields expressions for ground-state contributions in terms of sums over excited states, i.e., via the DFT version of the fluctuation-dissipation theorem [Langreth 1975, Gunnarsson 1976]. Thus, any approximation for the exchange-correlation kernel of TDDFT yields an approximation to the exchange-correlation energy, $E_{\rm xc}$, of *ground-state* DFT. Although such calculations are significantly more demanding than regular ground-state DFT calculations [Furche 2001c, Fuchs 2002], they produce a natural method for incorporating time-dependent fluctuations in the exchange-correlation energy. In particular, as a system is pulled apart into fragments, this approach includes correlated fluctuations on the two separated pieces. While in principle all this is included in the exact ground-state functional, in practice TDDFT provides a natural methodology for modeling these fluctuations [van Gisbergen 1995, Kohn 1998, Lein 1999].

1.2 One-to-One Correspondence

The Runge-Gross paper is usually cited as the beginning of modern TDDFT. There were several calculations before this, including those of Ando [Ando 1977a, Ando 1977b], Peuckert [Peuckert 1978], and of Zangwill and Soven [Zangwill 1980a], as well as proofs of the one-to-one correspondence under more limited conditions [Deb 1982]. But the RG paper established this correspondence for a sufficiently general class of problems to make TDDFT rigorous for most of the subsequent applications.

The evolution of the wavefunction is governed by the time-dependent Schrödinger equation:

$$\hat{H}(t)\Psi(t) = \mathrm{i}\frac{\mathrm{d}\Psi(t)}{\mathrm{d}t}\,, \qquad \Psi(0) \text{ given} \tag{1.1}$$

where $\hat{H}(t)$ is the Hamiltonian operator. Because this is a first-order differential equation in time, the initial wavefunction must be specified. We consider

N nonrelativistic electrons, mutually interacting via the Coulomb repulsion, in a time-dependent external potential. We write

$$\hat{T} = -\frac{1}{2}\sum_{i=1}^{N} \nabla_i^2 \tag{1.2}$$

for the kinetic energy, where the label i denotes the particle coordinates $\boldsymbol{r}_i$. We use atomic units throughout this chapter ($e^2 = \hbar = m = 1$) and so all distances arc in Bohr and energies in Hartrees (1 H = 27.21 eV = 627.5 kcal/mol). The electron-electron repulsion is given by

$$\hat{V}_{\text{ee}} = \frac{1}{2}\sum_{i\neq j}^{N} \frac{1}{|\boldsymbol{r}_i - \boldsymbol{r}_j|}, \tag{1.3}$$

where the sum is over all pairs, and the factor of 1/2 avoids double counting. Last, we denote the one-body potential as

$$\hat{V}_{\text{ext}} = \sum_{i=1}^{N} v_{\text{ext}}(\boldsymbol{r}_i, t), \tag{1.4}$$

which differs from problem to problem. For a hydrogenic atom with nuclear charge Z in an alternating electric field of strength $\mathcal{E}$ oriented along the z-axis and of frequency ω, $v_{\text{ext}}(\boldsymbol{r}, t) = -Z/r + \mathcal{E} \cdot z\cos(\omega, t)$. An important point to note is that *only* $v_{\text{ext}}(\boldsymbol{r}, t)$ and the particle number differ in the many problems we address; the interparticle repulsion and statistics never change. As the system evolves in time from some initial point (say $t = 0$), its one-particle density changes. This electron density is given by

$$n(\boldsymbol{r}, t) = N \int \mathrm{d}^3 r_2 \ldots \int \mathrm{d}^3 r_N \, |\Psi(\boldsymbol{r}, \boldsymbol{r}_2, \ldots, \boldsymbol{r}_N, t)|^2 , \tag{1.5}$$

and has the interpretation that $n(\boldsymbol{r}, t)\mathrm{d}^3 r$ is the probability of finding any electron in a region $\mathrm{d}^3 r$ around $\boldsymbol{r}$ at time t. The density is normalized to the number of electrons

$$\int \mathrm{d}^3 r \, n(\boldsymbol{r}, t) = N . \tag{1.6}$$

The analog of the Hohenberg-Kohn theorem for time-dependent problems is the one-to-one correspondence proven by RG [Runge 1984]. We consider N nonrelativistic electrons, mutually interacting via the Coulomb repulsion, in a time-dependent external potential. The theorem states that the densities $n(\boldsymbol{r}, t)$ and $n'(\boldsymbol{r}, t)$ evolving from a common initial state $\Psi(t = 0)$ under the influence of two potentials $v_{\text{ext}}(\boldsymbol{r}, t)$ and $v'_{\text{ext}}(\boldsymbol{r}, t)$ (both Taylor expandable about the initial time 0) eventually differ if the potentials differ by more than a purely time-dependent ($\boldsymbol{r}$-independent) function:

$$\Delta v_{\text{ext}}(\boldsymbol{r}, t) = v_{\text{ext}}(\boldsymbol{r}, t) - v'_{\text{ext}}(\boldsymbol{r}, t) \neq c(t) . \tag{1.7}$$

Under these conditions, there is a one-to-one mapping between densities and potentials, which implies that the potential is a *functional* of the density.

We prove this theorem by first showing that the corresponding current densities must differ. The current density is given by

$$\boldsymbol{j}(\boldsymbol{r},t) = N \int \mathrm{d}^3 r_2 \ldots \int \mathrm{d}^3 r_N \, \Im \left\{ \Psi(\boldsymbol{r},\boldsymbol{r}_2,\ldots,\boldsymbol{r}_N,t) \nabla \Psi^*(\boldsymbol{r},\boldsymbol{r}_2,\ldots,\boldsymbol{r}_N,t) \right\}, \tag{1.8}$$

where $\Im f$ denotes the imaginary part of f. One can easily prove continuity from the time-dependent Schrödinger equation, (1.1):

$$\frac{\partial n(\boldsymbol{r},t)}{\partial t} = -\nabla \cdot \boldsymbol{j}(\boldsymbol{r},t) \tag{1.9}$$

Return now to the problem of two different systems [i.e., $\Delta v_{\mathrm{ext}}(\boldsymbol{r},t) \neq c(t)$]. Because the corresponding Hamiltonians differ only in their one-body potentials, the equation of motion for the difference of the two current densities is, at $t = 0$:

$$\begin{aligned}
\frac{\partial}{\partial t} \left\{ \boldsymbol{j}(\boldsymbol{r},t) - \boldsymbol{j}'(\boldsymbol{r},t) \right\}_{t=0} &= -\mathrm{i} \langle \Psi_0 | \left[\hat{\boldsymbol{j}}(\boldsymbol{r},t), \{ \hat{H}(0) - \hat{H}'(0) \} \right] | \Psi_0 \rangle \\
&= -\mathrm{i} \langle \Psi_0 | \left[\hat{\boldsymbol{j}}(\boldsymbol{r}), \{ v_{\mathrm{ext}}(\boldsymbol{r},0) - v'_{\mathrm{ext}}(\boldsymbol{r},0) \} \right] | \Psi_0 \rangle \\
&= -n_0(\boldsymbol{r}) \nabla \{ v_{\mathrm{ext}}(\boldsymbol{r},0) - v'_{\mathrm{ext}}(\boldsymbol{r},0) \},
\end{aligned} \tag{1.10}$$

where $n_0(\boldsymbol{r}) = n(\boldsymbol{r},0)$ is the initial density. Thus we see that if, at the initial time, the two potentials differ (by more than just a constant), the first derivative of the currents must differ. Then the currents will change infinitesimally soon thereafter. One can go further, by repeatedly using the equation of motion, and considering $t = 0$, to find [Runge 1984]

$$\frac{\partial^{k+1}}{\partial t^{k+1}} \left\{ \boldsymbol{j}(\boldsymbol{r},t) - \boldsymbol{j}'(\boldsymbol{r},t) \right\}_{t=0} = -n_0(\boldsymbol{r}) \nabla \frac{\partial^k}{\partial t^k} \left\{ v(\boldsymbol{r},t) - v'(\boldsymbol{r},t) \right\}_{t=0}. \tag{1.11}$$

If (1.7) holds, and the potentials are Taylor expandable about $t = 0$, then there must be some finite k for which the right hand side of (1.10) does *not* vanish, so that

$$\boldsymbol{j}(\boldsymbol{r},t) \neq \boldsymbol{j}'(\boldsymbol{r},t). \tag{1.12}$$

For two Taylor-expandable potentials that differ by more than just a trivial constant, the corresponding currents must be different. This is the first part of the theorem, which establishes a one-to-one correspondence between current densities and external potentials.

In the second part, we extend the proof to the densities. Taking the gradient of both sides of (1.11), and using continuity, (1.9), we find

$$\frac{\partial^{k+2}}{\partial t^{k+2}} \left\{ n(\boldsymbol{r},t) - n'(\boldsymbol{r},t) \right\}_{t=0} = \nabla \cdot \left[n_0(\boldsymbol{r}) \nabla \frac{\partial^k}{\partial t^k} \{ v_{\mathrm{ext}}(\boldsymbol{r},t) - v'_{\mathrm{ext}}(\boldsymbol{r},t) \}_{t=0} \right]. \tag{1.13}$$

Now, if not for the divergence on the right-hand-side, we would be done, i.e., if $f(\boldsymbol{r}) = \partial^k\{v_{\rm ext}(\boldsymbol{r},t) - v'_{\rm ext}(\boldsymbol{r},t)\}/\partial t^k|_{(t=0)}$ is nonconstant for some k, then the density difference must be nonzero.

Does the divergence allow some escape from this conclusion? The answer is no, for any physical density in a finite system. To see this, write

$$\int \mathrm{d}^3r\, f(\boldsymbol{r})\nabla\cdot[n_0(\boldsymbol{r})\nabla f(\boldsymbol{r})] = \int \mathrm{d}^3r\, \{\nabla\cdot[f(\boldsymbol{r})n_0(\boldsymbol{r})\nabla f(\boldsymbol{r})] - n_0(\boldsymbol{r})|\nabla f(\boldsymbol{r})|^2\}\,. \tag{1.14}$$

The first term on the right may be written as a surface integral at $r = \infty$, and vanishes for all realistic potentials (which fall off at least as fast as $-1/r$). If we imagine that ∇f is nonzero somewhere, then the second term on the right is definitely negative, so that the integral on the left cannot vanish, and its integrand must be nonzero somewhere. Thus there is no way for ∇f to be nonzero, and yet have $\nabla(n_0\nabla f)$ vanish everywhere [Gross 1990].

Since the density determines the potential up to a time-dependent constant, the wavefunction is in turn determined up to a time-dependent phase, that in turn cancels out of the expectation value of any operator. Thus the expectation value of any operator is a functional of the time-dependent density and initial state, completing our proof.

1.3 Time-Dependent Kohn-Sham Equations

Having established that the one-body potential is a functional of the density and initial state, we next define a fictious system of noninteracting electrons that satisfy time-dependent Kohn-Sham equations:

$$\mathrm{i}\frac{\partial\varphi_j(\boldsymbol{r},t)}{\partial t} = \left[-\frac{\nabla^2}{2} + v_{\rm KS}[n](\boldsymbol{r},t)\right]\varphi_j(\boldsymbol{r},t)\,, \tag{1.15}$$

whose density,

$$n(\boldsymbol{r},t) = \sum_{j=1}^{N}|\varphi_j(\boldsymbol{r},t)|^2\,, \tag{1.16}$$

is defined to be precisely that of the real system. By virtue of the one-to-one correspondence proven in the previous section, the potential $v_{\rm KS}(\boldsymbol{r},t)$ yielding this density is unique. We then *define* the exchange-correlation potential via:

$$v_{\rm KS}(\boldsymbol{r},t) = v_{\rm ext}(\boldsymbol{r},t) + v_{\rm H}(\boldsymbol{r},t) + v_{\rm xc}(\boldsymbol{r},t)\,, \tag{1.17}$$

where the Hartree potential has the usual form,

$$v_{\rm H}(\boldsymbol{r},t) = \int \mathrm{d}^3r'\,\frac{n(\boldsymbol{r}',t)}{|\boldsymbol{r}-\boldsymbol{r}'|}\,, \tag{1.18}$$

but for a time-dependent density. The exchange-correlation potential is then a functional of the entire history of the density, $n(\boldsymbol{r}, t)$, the initial interacting wavefunction $\Psi(0)$, and the initial Kohn-Sham wavefunction, $\Phi(0)$. This functional is a very complex one, much more so than the ground-state case. Knowledge of it implies solution of all time-dependent Coulomb interacting problems.

By the arguments of the previous section, if both the interacting and KS initial wavefunctions are nondegenerate ground states, the xc potential is a functional of the time-dependent density alone.

In ground-state DFT, the exchange-correlation potential is the functional derivative of $E_{\text{xc}}[n]$. It would be nice to find a functional of $n(\boldsymbol{r}, t)$ for which $v_{\text{xc}}(\boldsymbol{r}, t)$ was the functional derivative, called the exchange-correlation action. A plausible action was given in the RG paper [Runge 1984]. However, it turned out later [Gross 1995a] that this action leads to a paradox, namely that the resulting exchange-correlation kernel (see next section) violates causality. Two different solutions to this problem are offered in this volume (see Chaps. 2 and 7). In either case, some generalization of the space of densities is required.

Several extensions and further advances on the Runge-Gross paper will appear later in this book, or have appered in the literature. Here we make several points:

- **Construction of Potential:** The one-to-one correspondence tells us only that the potential is a functional of the density and initial state, but not what that functional is. Van Leeuwen gives a constructive procedure for finding such a potential, thereby "solving" the representability problem for the time-dependent case (see Chap. 2).
- **Currents:** The first part of the proof produced a one-to-one correspondence between currents and potentials, while to get a density functional, we needed to invoke a surface condition. This renders the application of TDDFT to extended systems nontrivial. We will see later in the book that the first step is easily generalized to produce a one-to-one correspondence between currents and any vector potential (producing time-dependent current density functional theory) and that some effects are indeed highly nonlocal, when expressed as density functionals (see Chap. 5).
- **Allowed Time-Dependence:** The one-to-one correspondence does not apply to the adiabatic switching often assumed in perturbation theory, $\exp(\gamma t)$, where $\gamma \to 0$. Such potentials are singular around their initial time ($-\infty$ in this case). But one can take the adiabatic limit of well-behaved potentials, as described in Sect. 11.4.1.
- **Initial-State Dependence:** Later in this book, it is shown that, while for one electron the initial state wavefunction is completely fixed by requiring it to yield a given density, for two or more electrons, it is possible to find more than one initial wavefunction, thereby producing two different KS potentials for a given evolution (see Chap. 4).

- **Removal of Initial-State Dependence:** Note that a dependence on the initial wavefunction would render TDDFT useless in practice, as we would need a different density functional for every possible initial wavefunction. However, if the initial state is a nondegenerate ground state, then, by virtue of the *Hohenberg-Kohn* theorem, its wavefunction is a functional of the ground-state density alone. Then the time-dependent potential is a functional of the time-dependent density alone. This is the case for almost all practical applications (see Parts IV and V of this volume).

1.4 Linear Response

In this section, we show how the TDKS equations, with a given approximation for the xc potential, can be used to extract the electronic excitations of a system. This is a powerful tool for studying, e.g., photoluminescence in biological molecules (see Chap. 23), where the quantitative treatment of the absorption spectrum can be key to identifying the underlying mechanism.

In principle, we already have sufficient technology to do these calculations. Simply perturb the system at time $t = 0$ with a weak electric field, and then propagate the TDKS equations for a while, evaluating the dipole of interest as you go. The Fourier transform of that function of time is precisely the optical absorption spectrum. In fact, this procedure is followed in many real-time codes, as described in later chapters. However, it is very enlightening and fruitful to analyze the situation in detail using standard linear response theory.

When the perturbing field is weak, as in normal spectroscopic experiments, perturbation theory applies. Instead of needing knowledge of v_{xc} for densities that are changing significantly with time, we need only know this potential for densities close to that of the initial state, which we take to be a nondegenerate ground-state. Writing $n(\boldsymbol{r},t) = n_{\mathrm{GS}}(\boldsymbol{r},t) + \delta n(\boldsymbol{r},t)$, we have

$$v_{\mathrm{xc}}[n_{\mathrm{GS}}+\delta n](\boldsymbol{r},t) = v_{\mathrm{xc}}[n_{\mathrm{GS}}](\boldsymbol{r}) + \int \mathrm{d}t' \int \mathrm{d}^3r'\, f_{\mathrm{xc}}[n_{\mathrm{GS}}](\boldsymbol{r},\boldsymbol{r}',t-t')\delta n(\boldsymbol{r}',t')\,, \tag{1.19}$$

where f_{xc} is called the exchange-correlation kernel, evaluated on the ground-state density:

$$f_{\mathrm{xc}}[n_{\mathrm{GS}}](\boldsymbol{r},\boldsymbol{r}',t-t') = \left.\frac{\delta v_{\mathrm{xc}}(\boldsymbol{r},t)}{\delta n(\boldsymbol{r}',t')}\right|_{n=n_{\mathrm{GS}}}\,. \tag{1.20}$$

While still more complex than the ground-state exchange-correlation potential, the exchange-correlation kernel is much more manageable than the full time-dependent exchange-correlation potential, because it is a functional of the ground-state density alone. To understand why f_{xc} is important for linear response, define the point-wise susceptibility $\chi[n_{\mathrm{GS}}](\boldsymbol{r},\boldsymbol{r}',t-t')$ as the response of the ground state to a small change in the external potential:

$$\delta n(\boldsymbol{r},t) = \int \mathrm{d}t' \int \mathrm{d}^3r'\, \chi[n_{\mathrm{GS}}](\boldsymbol{r},\boldsymbol{r}',t-t')\, \delta v_{\mathrm{ext}}(\boldsymbol{r}',t')\,, \tag{1.21}$$

i.e., if you make a small change in the external potential at point $\boldsymbol{r}'$ and time t', χ tells you how the density will change at point $\boldsymbol{r}$ and later time t. The ground-state Kohn-Sham system has its own analog of χ, which we denote by χ_{KS}. This function tells you how the noninteracting KS electrons would respond to $\delta v_{\mathrm{KS}}(\boldsymbol{r}',t')$, which is quite different from the interacting case. But both must yield the same density response:

$$\delta n(\boldsymbol{r},t) = \int \mathrm{d}t' \int \mathrm{d}^3r'\, \chi_{\mathrm{KS}}[n_{\mathrm{GS}}](\boldsymbol{r},\boldsymbol{r}',t-t') \times \left\{\delta v_{\mathrm{ext}}(\boldsymbol{r}',t') + \delta v_{\mathrm{H}}(\boldsymbol{r}',t') + \delta v_{\mathrm{xc}}(\boldsymbol{r}',t')\right\}. \tag{1.22}$$

Equating this density change with that of the interacting system, (1.21), and using the definition of the xc kernel, (1.20), we find the central equation of TDDFT linear response (in frequency space):

$$\chi(\boldsymbol{r},\boldsymbol{r}',\omega) = \chi_{\mathrm{KS}}(\boldsymbol{r},\boldsymbol{r}',\omega) + \int \mathrm{d}^3r_1 \int \mathrm{d}^3r_2\, \chi_{\mathrm{KS}}(\boldsymbol{r},\boldsymbol{r}_1,\omega) \left\{\frac{1}{|\boldsymbol{r}_1-\boldsymbol{r}_2|} + f_{\mathrm{xc}}(\boldsymbol{r}_1,\boldsymbol{r}_2,\omega)\right\} \chi(\boldsymbol{r}_2,\boldsymbol{r}',\omega)\,, \tag{1.23}$$

where all objects are functionals of the ground-state density. This is a Dyson-like equation, because it has the same mathematical form as the Dyson equation which relates the one-particle Green's function to its free counterpart via a kernel that is the self-energy.

This equation contains the key to electronic excitations via TDDFT. When ω matches a true transition frequency of the system, the response function χ blows up, i.e., has a pole as a function of ω. Likewise χ_{KS} has a set of such poles, at the single-particle excitations of the KS system:

$$\chi_{\mathrm{KS}}(\boldsymbol{r},\boldsymbol{r}',\omega) = 2 \lim_{\eta\to 0+} \sum_q \left\{\frac{\xi_q(\boldsymbol{r})\,\xi_q^*(\boldsymbol{r}')}{\omega-\omega_q+\mathrm{i}\eta} - \frac{\xi_q^*(\boldsymbol{r})\,\xi_q(\boldsymbol{r}')}{\omega+\omega_q-\mathrm{i}\eta}\right\} \tag{1.24}$$

where q is a double index, representing a transition from occupied KS orbital i to unoccupied KS orbital a,

$$\omega_q = \varepsilon_a - \varepsilon_i\,, \tag{1.25}$$

and

$$\xi_q(\boldsymbol{r}) = \varphi_i^*(\boldsymbol{r})\varphi_a(\boldsymbol{r})\,, \tag{1.26}$$

where ε_j is the eigenenergy of the KS state φ_j. Thus χ_{KS} is purely a product of the ground-state KS calculation. In the absence of Hartree-exchange-correlation effects, $\chi = \chi_{\mathrm{KS}}$, and so the allowed transitions are exactly those of the ground-state KS potential. But the presence of the kernel in (1.23) shifts the transitions away from the KS values to the true values. Moreover, the strengths of the poles can be simply related to optical absorption intensities (oscillator strengths) and so these are also affected by the kernel.

Casida showed that, for frequency-independent kernels, finding the poles of χ is equivalent to solving the eigenvalue problem [Casida 1996]:

$$\sum_{q'} R_{qq'}\, F_{q'} = \Omega_q^2\, F_q\,, \tag{1.27}$$

The matrix is

$$R_{qq'} = \omega_q^4\, \delta_{qq'} + 4\sqrt{\omega_q\, \omega_{q'}}\, K_{q,q'}\,, \tag{1.28}$$

where

$$K_{q,q'} = \int \mathrm{d}^3 r \int \mathrm{d}^3 r'\, \xi_q^*(\boldsymbol{r})\, f_{\mathrm{Hxc}}(\boldsymbol{r},\boldsymbol{r}')\, \xi_{q'}(\boldsymbol{r}')\,. \tag{1.29}$$

In this equation, f_{Hxc} is the Hartree-exchange-correlation kernel:

$$f_{\mathrm{Hxc}}(\boldsymbol{r},\boldsymbol{r}') = \frac{1}{|\boldsymbol{r}-\boldsymbol{r}'|} + f_{\mathrm{xc}}(\boldsymbol{r},\boldsymbol{r}')\,. \tag{1.30}$$

The solution of (1.27) yields the excitation energies Ω, while the oscillator strengths can be obtained from the eigenvectors [Casida 1996]. Efficient algorithms exist for extracting just the lowest transitions. It is in this form that most quantum chemical codes extract TDDFT excitations.

A simple approximate solution to the TDDFT linear response equations is the single-pole approximation (SPA) [Petersilka 1996a], in which the TDDFT corrections to a given transition are included, but not its coupling to the other KS transitions. This amounts to neglecting all other poles in χ_{KS}, or simply taking the diagonal matrix elements in Casida's equations:

$$\Omega^4 \approx \omega_q^4 + 4\,\omega_q\, K_{q,q}\,, \tag{1.31}$$

While not needed numerically, the SPA yields much insight into how and why TDDFT produces the results it does [Appel 2003], and will be used in several occasions in this book.

1.5 Adiabatic Connection Formula

Lastly, we discuss how TDDFT produces sophisticated approximations to the *ground-state* exchange-correlation energy. Almost thirty years ago, the fluctuation-dissipation theorem was applied carefully in DFT. The resulting adiabatic connection fluctuation-dissipation formula is:

$$E_{\rm xc}[n_{\rm GS}] = -\frac{1}{2}\int_0^1 {\rm d}\lambda \int {\rm d}^3r \int {\rm d}^3r' \, \frac{1}{|\boldsymbol{r}-\boldsymbol{r}'|} \times \int_0^\infty \frac{{\rm d}\omega}{\pi} \left\{\chi_\lambda[n_{\rm GS}](\boldsymbol{r},\boldsymbol{r}',\omega) + n_{\rm GS}(\boldsymbol{r})\delta(\boldsymbol{r}-\boldsymbol{r}')\right\} \quad (1.32)$$

where the coupling-constant λ is defined to multiply the electron-electron repulsion in the Hamiltonian, but the external potential is adjusted to keep the density fixed [Langreth 1975, Gunnarsson 1976].

This intriguing formula means that approximations in TDDFT can be used to yield approximations to the ground-state xc energy functional, by constructing χ_λ from the λ-dependent generalization of (1.23). This scheme is much more computationally demanding than a regular ground-state DFT calculation with, e.g., a GGA, but can naturally accomodate effects that are difficult to capture with simple ground-state functionals. For example, van der Waals interactions are very naturally described in this scheme [Kohn 1998], as the correlated fluctuations between distant atoms or molecules are correctly described by the frequency-dependent contributions. Furthermore, ground-state DFT has difficulty describing how bonds break as a function of bond length, because the single Slater determinant of KS DFT is a poor representation of the interacting (Heitler-London) wavefunction of separated atoms [Perdew 1995]. But TDDFT treatments of the adiabatic connection formula can yield the entire dissociation curve [Fuchs 2002].

1.6 Adiabatic Approximation

Although the one-to-one correspondence has given us in principle an exact description of many-electron quantum mechanics in a time-dependent potential, it yields no hint of the missing density functional. By constructing a set of time-dependent KS equations, a significant fraction of that missing functional is included exactly. But the scheme is incomplete without some approximation to the missing xc potential.

As we have noted, the exact exchange-correlation potential depends on the entire history of the density, as well as the initial wavefunctions of both the interacting and the Kohn-Sham systems:

$$v_{\rm xc}[n,\Psi(0),\Phi(0)](\boldsymbol{r},t) = v_{\rm KS}[n,\Psi(0),\Phi(0)](\boldsymbol{r},t) - v_{\rm ext}(\boldsymbol{r},t) - v_{\rm H}[n](\boldsymbol{r},t)\,. \quad (1.33)$$

However, in the special case of starting from a nondegenerate ground state (both interacting and noninteracting), the initial wavefunctions themselves are functionals of the initial density, and so the initial-state dependence disappears. This is the usual case in which TDDFT is used. But even then, the exchange-correlation potential at $\boldsymbol{r}$ and t has a functional dependence not just on $n(\boldsymbol{r},t)$ but on all $n(\boldsymbol{r},t')$ for $0 \le t' \le t$, and for arbitrary points $\boldsymbol{r}'$ in space. Thus the potential *remembers* the density's past, and we say it has memory.

The adiabatic approximation is one in which we ignore all dependence on the past, and allow only a dependence on the instantaneous density:

$$v_{\rm xc}^{\rm adia}[n](\boldsymbol{r},t) = v_{\rm xc}^{\rm approx}[n(t)](\boldsymbol{r}) , \qquad (1.34)$$

i.e., it approximates the functional as being local in time. If the time-dependent potential changes very slowly (adiabatically), this approximation will be valid. But the electrons will remain always in their instantaneous ground state. To make the adiabatic approximation exact for the only systems for which it can be exact, we require

$$v_{\rm xc}^{\rm adia}[n](\boldsymbol{r},t) = v_{\rm xc}^{\rm GS}[n_{\rm GS}](\boldsymbol{r})|_{n_{\rm GS}(\boldsymbol{r}')=n(\boldsymbol{r}',t)} , \qquad (1.35)$$

where $v_{\rm xc}^{\rm GS}[n_{\rm GS}](\boldsymbol{r})$ is the *exact* ground-state exchange-correlation potential of the density $n_{\rm GS}(\boldsymbol{r})$. This is the analog of the argument made to determine the function used in LDA calculations for the ground-state energy: The only sensible choice there is the xc energy of the uniform electron gas. Throughout this book, we will see many successes of this approximation, but some striking failures also.

In practice, the spatial nonlocality of the functional is also approximated, since we do not know the exact xc energy functional, even in the static case. As mentioned above, the ALDA is the mother of all TDDFT density approximations, but any ground-state functional, such as a GGA or hybrid, automatically yields an adiabatic approximation for use in TDDFT calculations.

The adiabatic approximation is particularly simple for linear response, yielding an xc kernel of the form:

$$f_{\rm xc}^{\rm adia}[n_{\rm GS}](\boldsymbol{r},\boldsymbol{r}',t-t') = \frac{\delta v_{\rm xc}^{\rm GS}[n_{\rm GS}](\boldsymbol{r})}{\delta n_{\rm GS}(\boldsymbol{r}')}\delta(t-t') , \qquad (1.36)$$

i.e., it is completely local in time. When Fourier-transformed, this implies that $f_{\rm xc}^{\rm adia}(\omega)$ is *frequency-independent.*

By construction, ALDA should work only for systems with very small density gradients in space and time. It may seem like a drastic approximation to neglect all nonlocality in time. However, the success of modern ground-state density functional theory can be traced to how well LDA works beyond its obvious range of validity, due to satisfaction of sum rules, etc., and how its success was built upon with GGA's and hybrids [Perdew 2005]. We will see in the chapters concerned with applications that some of the same magic applies in TDDFT ALDA calculations.

1.7 Relation to Ground-State DFT

Time-dependent DFT is *not* a simple extension of ground-state DFT. It is an application of DFT philosophy to the world of driven systems, i.e., to the time-dependent Schrödinger equation, a first-order differential equation in time. Thus while many of the statements look similar, the functionals themselves contain greatly different physics.

On the other hand, TDDFT calculations enhance our knowledge and understanding gained from ground-state DFT calculations. One of the most important applications is to produce a spectroscopic signature of a given molecule, not just its thermochemistry and geometric structure. An important example is the interpretation of KS orbital eigenvalues. In ground-state DFT, it was often said that, apart from the highest occupied molecular orbital (HOMO), these had no direct physical significance. With the advent of TDDFT, we can see that they are a zero-order approximation to the optical excitation energies.

Part I

Formal Theory

2 Beyond the Runge-Gross Theorem

R. van Leeuwen

2.1 Introduction

The Runge-Gross theorem [Runge 1984] states that for a given initial state the time-dependent density is a unique functional of the external potential. Let us elaborate a bit further on this point. Suppose we could solve the time-dependent Schrödinger equation (TDSE) for a given many-body system, i.e., we specify an initial state $|\Psi_0\rangle$ at $t = t_0$ and evolve the wave function in time using the Hamiltonian $\hat{H}(t)$. Then, from the wave function, we can calculate the time-dependent density $n(\boldsymbol{r}, t)$. We can then ask the question whether exactly the same density $n(\boldsymbol{r}, t)$ can be reproduced by an external potential $v'_{\text{ext}}(\boldsymbol{r}, t)$ in a system with a different given initial state and a different two-particle interaction, and if so, whether this potential is unique (modulo a purely time-dependent function). The answer to this question is obviously of great importance for the construction of the time-dependent Kohn-Sham equations. The Kohn-Sham system has no two-particle interaction and differs in this respect from the fully interacting system. It has, in general, also a different initial state. This state is usually a Slater determinant rather than a fully interacting initial state. A time-dependent Kohn-Sham system therefore only exists if the question posed above is answered affirmatively. Note that this is a v-representability question: Is a density belonging to an interacting system also noninteracting v-representable? We will show in this chapter that, with some restrictions on the initial states and potentials, this question can indeed be answered affirmatively [van Leeuwen 1999, van Leeuwen 2001, Giuliani 2005]. We stress that we demonstrate here that the interacting-v-representable densities are also noninteracting-v-representable rather than aiming at characterizing the set of v-representable densities. The latter question has inspired much work in ground state density functional theory (for extensive discussion see [van Leeuwen 2003]) and has only been answered satisfactorily for quantum lattice systems [Chayes 1985].

2.2 The Extended Runge-Gross Theorem: Different Interactions and Initial States

We start by considering the Hamiltonian

R. van Leeuwen: *Beyond the Runge-Gross Theorem*, Lect. Notes Phys. **706**, 17–31 (2006)
DOI 10.1007/3-540-35426-3_2

$$\hat{H}(t) = \hat{T} + \hat{V}_{\mathrm{ext}}(t) + \hat{V}_{\mathrm{ee}}\,, \tag{2.1}$$

where $\hat{T}$ is the kinetic energy, $\hat{V}_{\mathrm{ext}}(t)$ the (in general time-dependent) external potential, and $\hat{V}_{\mathrm{ee}}$ the two-particle interaction. In second quantization the constituent terms are, as usual, written as

$$\hat{T} = -\frac{1}{2}\sum_{\sigma}\int \mathrm{d}^3r\, \hat{\psi}^\dagger_\sigma(\boldsymbol{r})\nabla^2\hat{\psi}_\sigma(\boldsymbol{r})\,, \tag{2.2a}$$

$$\hat{V}_{\mathrm{ext}}(t) = \sum_{\sigma}\int \mathrm{d}^3r\, v_{\mathrm{ext}}(\boldsymbol{r},t)\hat{\psi}^\dagger_\sigma(\boldsymbol{r})\hat{\psi}_\sigma(\boldsymbol{r})\,, \tag{2.2b}$$

$$\hat{V}_{\mathrm{ee}} = \frac{1}{2}\sum_{\sigma\sigma'}\int \mathrm{d}^3r\int \mathrm{d}^3r'\, v_{\mathrm{ee}}(|\boldsymbol{r}-\boldsymbol{r}'|)\hat{\psi}^\dagger_\sigma(\boldsymbol{r})\hat{\psi}^\dagger_{\sigma'}(\boldsymbol{r}')\hat{\psi}_{\sigma'}(\boldsymbol{r}')\hat{\psi}_\sigma(\boldsymbol{r})\,. \tag{2.2c}$$

where σ and σ' are spin variables. For the readers not used to second quantization we note that the first few basic steps in this chapter can also be derived in first quantization. For details we refer to [Giuliani 2005, Vignale 2004]. However, good understanding of second quantization is indispensable to understand the next chapter. Good introductions to second quantization are found in [Fetter 1971, Runge 1991].

The two-particle potential $v_{\mathrm{ee}}(|\boldsymbol{r}-\boldsymbol{r}'|)$ in (2.2c) can be arbitrary, but will in practice almost always be equal to the repulsive Coulomb potential. We then consider some basic relations satisfied by the density and the current density. The time-dependent density is given as the expectation value of the density operator

$$\hat{n}(\boldsymbol{r}) = \sum_{\sigma}\hat{\psi}^\dagger_\sigma(\boldsymbol{r})\hat{\psi}_\sigma(\boldsymbol{r})\,, \tag{2.3}$$

with the time-dependent many-body wavefunction, $n(\boldsymbol{r},t) = \langle\Psi(t)|\hat{n}(\boldsymbol{r})|\Psi(t)\rangle$. In the following we consider two continuity equations. If $|\Psi(t)\rangle$ is the state evolving from $|\Psi_0\rangle$ under the influence of Hamiltonian $\hat{H}(t)$ we have the usual continuity equation

$$\frac{\partial}{\partial t}n(\boldsymbol{r},t) = -\mathrm{i}\langle\Psi(t)|\,[\hat{n}(\boldsymbol{r}),\hat{H}(t)]\,|\Psi(t)\rangle = -\nabla\cdot\boldsymbol{j}(\boldsymbol{r},t)\,, \tag{2.4}$$

where the current operator is defined as

$$\hat{\boldsymbol{j}}(\boldsymbol{r}) = \frac{1}{2\mathrm{i}}\sum_{\sigma}\left\{\hat{\psi}^\dagger_\sigma(\boldsymbol{r})\nabla\hat{\psi}_\sigma(\boldsymbol{r}) - [\nabla\hat{\psi}^\dagger_\sigma(\boldsymbol{r})]\hat{\psi}_\sigma(\boldsymbol{r})\right\}\,, \tag{2.5}$$

and has expectation value $\boldsymbol{j}(\boldsymbol{r},t) = \langle\Psi(t)|\hat{\boldsymbol{j}}(\boldsymbol{r})|\Psi(t)\rangle$. This continuity equation expresses, in a local form, the conservation of particle number. Using Gauss' law the continuity equation says that the change of the number of particles within some volume can simply be measured by calculating the flux of the current through the surface of this volume.

As a next step, we can consider an analogous continuity equation for the current itself. We have

$$\frac{\partial}{\partial t}\boldsymbol{j}(\boldsymbol{r},t) = -\mathrm{i}\langle\Psi(t)|\,[\hat{\boldsymbol{j}}(\boldsymbol{r}),\hat{H}(t)]\,|\Psi(t)\rangle\,. \tag{2.6}$$

If we work out the commutator in more detail, we find the expression [Martin 1959]

$$\frac{\partial}{\partial t}j_\alpha(\boldsymbol{r},t) = -n(\boldsymbol{r},t)\frac{\partial}{\partial r_\alpha}v_{\mathrm{ext}}(\boldsymbol{r},t) - \sum_\beta \frac{\partial}{\partial r_\beta}T_{\beta\alpha}(\boldsymbol{r},t) - V_{\mathrm{ee}\,\alpha}(\boldsymbol{r},t)\,. \tag{2.7}$$

Here we have defined the momentum-stress tensor $\hat{T}_{\beta\alpha}$ (part of the energy-momentum tensor)

$$\hat{T}_{\beta\alpha}(\boldsymbol{r}) = \frac{1}{2}\sum_\sigma \Big\{\frac{\partial}{\partial r_\beta}\hat{\psi}^\dagger_\sigma(\boldsymbol{r})\frac{\partial}{\partial r_\alpha}\hat{\psi}_\sigma(\boldsymbol{r}) + \frac{\partial}{\partial r_\alpha}\hat{\psi}^\dagger_\sigma(\boldsymbol{r})\frac{\partial}{\partial r_\beta}\hat{\psi}_\sigma(\boldsymbol{r}) - \frac{1}{2}\frac{\partial^2}{\partial r_\beta \partial r_\alpha}[\hat{\psi}^\dagger_\sigma(\boldsymbol{r})\hat{\psi}_\sigma(\boldsymbol{r})]\Big\}\,, \tag{2.8}$$

and the quantity $\hat{V}_{\mathrm{ee}\,\alpha}$ as

$$\hat{V}_{\mathrm{ee}\,\alpha}(\boldsymbol{r}) = \sum_{\sigma,\sigma'}\int \mathrm{d}^3r'\,\hat{\psi}^\dagger_\sigma(\boldsymbol{r})\hat{\psi}^\dagger_{\sigma'}(\boldsymbol{r}')\frac{\partial}{\partial r_\alpha}v_{\mathrm{ee}}(|\boldsymbol{r}-\boldsymbol{r}'|)\hat{\psi}_{\sigma'}(\boldsymbol{r}')\hat{\psi}_\sigma(\boldsymbol{r})\,. \tag{2.9}$$

The expectation values that appear in (2.7) are defined as $T_{\beta\alpha}(\boldsymbol{r},t) = \langle\Psi(t)|\hat{T}_{\beta\alpha}(\boldsymbol{r})|\Psi(t)\rangle$ and $V_{\mathrm{ee}\,\alpha}(\boldsymbol{r},t) = \langle\Psi(t)|\hat{V}_{\mathrm{ee}\,\alpha}(\boldsymbol{r})|\Psi(t)\rangle$. The continuity equation (2.7) is a local quantum version of Newton's third law. Taking the divergence of (2.7) and using the continuity (2.4) we find

$$\frac{\partial^2}{\partial t^2}n(\boldsymbol{r},t) = \nabla\cdot[n(\boldsymbol{r},t)\nabla v_{\mathrm{ext}}(\boldsymbol{r},t)] + q(\boldsymbol{r},t)\,, \tag{2.10}$$

with $\hat{q}$ and $q(\boldsymbol{r},t)$ being defined as

$$\hat{q}(\boldsymbol{r}) = \sum_{\alpha,\beta}\frac{\partial^2}{\partial r_\beta \partial r_\alpha}\hat{T}_{\beta\alpha}(\boldsymbol{r}) + \sum_\alpha \frac{\partial}{\partial r_\alpha}\hat{V}_{\mathrm{ee}\,\alpha}(\boldsymbol{r})\,, \tag{2.11a}$$

$$q(\boldsymbol{r},t) = \langle\Psi(t)|\hat{q}(\boldsymbol{r})|\Psi(t)\rangle\,. \tag{2.11b}$$

Equation (2.10) will play a central role in our discussion of the relation between the density and the potential. This is because it represents an equation which directly relates the external potential and the electron density. From (2.10) we further see that $q(\boldsymbol{r},t)$ decays exponentially at infinity when $n(\boldsymbol{r},t)$ does, unless $v_{\mathrm{ext}}(\boldsymbol{r},t)$ grows exponentially at infinity. In the following we will, however, only consider finite systems with external potentials that are bounded at infinity (for a discussion of the set of allowed external potentials in ground state DFT we refer to [Lieb 1983, van Leeuwen 2003]).

Let us now assume that we have solved the time-dependent Schrödinger equation for the many-body system described by the Hamiltonian $\hat{H}(t)$ of (2.1) and initial state $|\Psi_0\rangle$ at $t = t_0$. We have thus obtained a many-body wavefunction $|\Psi(t)\rangle$ and density $n(\boldsymbol{r}, t)$. We further assume that $n(\boldsymbol{r}, t)$ is analytic at $t = t_0$. For our system, (2.10) is satisfied. We now consider a second system with Hamiltonian

$$\hat{H}'(t) = \hat{T} + \hat{V}'_{\text{ext}}(t) + \hat{V}'_{\text{ee}} . \tag{2.12}$$

The terms $\hat{V}'_{\text{ext}}(t)$ and $\hat{V}'_{\text{ee}}$ represent again the one- and two-body potentials. We denote the initial state by $|\Psi'_0\rangle$ at $t = t_0$ and the time-evolved state by $|\Psi'(t)\rangle$. The form of $\hat{V}'_{\text{ee}}$ is assumed to be such that its expectation value and its derivatives are finite. For the system described by the Hamiltonian $\hat{H}'$ we have an equation analogous to (2.10).

$$\frac{\partial^2}{\partial t^2} n'(\boldsymbol{r}, t) = \nabla \cdot [n'(\boldsymbol{r}, t) \nabla v'_{\text{ext}}(\boldsymbol{r}, t)] + q'(\boldsymbol{r}, t) , \tag{2.13}$$

where $q'(\boldsymbol{r}, t)$ is the expectation value

$$q'(\boldsymbol{r}, t) = \langle \Psi'(t) | \hat{q}'(\boldsymbol{r}) | \Psi'(t) \rangle , \tag{2.14}$$

for which we defined

$$\hat{q}' = \sum_{\beta,\alpha} \frac{\partial^2}{\partial r_\beta \partial r_\alpha} \hat{T}_{\beta\alpha}(\boldsymbol{r}) + \sum_k \frac{\partial}{\partial r_\alpha} \hat{V}'_{\text{ee}\,\alpha}(\boldsymbol{r}) . \tag{2.15}$$

Our goal is now to choose v'_{ext} in (2.13) so that $n'(\boldsymbol{r}, t) = n(\boldsymbol{r}, t)$. We will do this by constructing v'_{ext} in such a way that for the k-th derivatives of the density at $t = t_0$ we have $\frac{\partial^k}{\partial t^k} n'(\boldsymbol{r}, t)|_{t=t_0} = \frac{\partial^k}{\partial t^k} n(\boldsymbol{r}, t)|_{t=t_0}$. First we need to discuss some initial conditions. As a necessary condition for the potential v'_{ext} to exist, we have to require that the initial states $|\Psi_0\rangle$ and $|\Psi'_0\rangle$ yield the same initial density, i.e.,

$$n'(\boldsymbol{r}, t_0) = \langle \Psi'_0 | \hat{n}(\boldsymbol{r}) | \Psi'_0 \rangle = \langle \Psi_0 | \hat{n}(\boldsymbol{r}) | \Psi_0 \rangle = n(\boldsymbol{r}, t_0) . \tag{2.16}$$

We now note that the basic (2.10) is a second order differential equation in time for $n(\boldsymbol{r}, t)$. This means, as we will see soon, that we still need as additional requirement that $\frac{\partial}{\partial t} n'(\boldsymbol{r}, t) = \frac{\partial}{\partial t} n(\boldsymbol{r}, t)$ at $t = t_0$. With the help of the continuity equation (2.4) this yields the condition

$$\frac{\partial}{\partial t} n'(\boldsymbol{r}, t)\Big|_{t=t_0} = \langle \Psi'_0 | \nabla \cdot \hat{\boldsymbol{j}}(\boldsymbol{r}) | \Psi'_0 \rangle = \langle \Psi_0 | \nabla \cdot \hat{\boldsymbol{j}}(\boldsymbol{r}) | \Psi_0 \rangle = \frac{\partial}{\partial t} n(\boldsymbol{r}, t)\Big|_{t=t_0} . \tag{2.17}$$

This constraint also implies the weaker requirement that the initial state $|\Psi'_0\rangle$ must be chosen such that the initial momenta $\boldsymbol{P}(t_0)$ of both systems are the same. This follows directly from the fact that the momentum of the system is given by

$$\boldsymbol{P}(t) = \int \mathrm{d}^3r\, \boldsymbol{j}(\boldsymbol{r},t) = \int \mathrm{d}^3r\, \boldsymbol{r}\, \frac{\partial}{\partial t} n(\boldsymbol{r},t) \,. \tag{2.18}$$

The equality of the last two terms in this equation follows directly from the continuity equation (2.4) and the fact that we are dealing with finite systems for which, barring pathological examples [van Leeuwen 2001, Maitra 2001], currents and densities are zero at infinity. For notational convenience we first introduce the following notation for the k-th time-derivative at $t = t_0$ of a function f:

$$f^{(k)}(\boldsymbol{r}) = \frac{\partial^k}{\partial t^k} f(\boldsymbol{r},t)\Big|_{t=t_0} \,. \tag{2.19}$$

Then our goal is to choose v'_{ext} in such a way that $n'^{(k)} = n^{(k)}$ for all k. Let us see how we can use (2.13) to do this. If we first evaluate (2.13) at $t = t_0$ we obtain, using the notation of (2.19), the expression

$$n'^{(2)}(\boldsymbol{r}) = \nabla \cdot [n'^{(0)}(\boldsymbol{r}) \nabla v'^{(0)}_{\mathrm{ext}}(\boldsymbol{r})] + q'^{(0)}(\boldsymbol{r}) \,. \tag{2.20}$$

Since we want that $n'^{(2)} = n^{(2)}$ and have chosen the initial state $|\Psi'_0\rangle$ in such a way that $n'^{(0)} = n^{(0)}$ we obtain the following determining equation for $v'^{(0)}_{\mathrm{ext}}$:

$$\nabla \cdot [n^{(0)}(\boldsymbol{r}) \nabla v'^{(0)}_{\mathrm{ext}}(\boldsymbol{r})] = n^{(2)}(\boldsymbol{r}) - q'^{(0)}(\boldsymbol{r}) \,. \tag{2.21}$$

The right hand side is determined since $n^{(0)}$ and $n^{(2)}$ are given and $q'^{(0)}$ is calculated from the given initial state $|\Psi'_0\rangle$ as $q'^{(0)}(\boldsymbol{r}) = \langle \Psi'_0 | \hat{q}'(\boldsymbol{r}) | \Psi'_0 \rangle$. Equation (2.21) is of Sturm-Liouville type and has a unique solution for v'_0 provided we specify a boundary condition. We will specify the boundary condition that $v'^{(0)}_{\mathrm{ext}}(\boldsymbol{r}) \to 0$ for $\boldsymbol{r} \to \infty$. With this boundary condition we also fix the gauge of the potential. Having obtained $v'^{(0)}_{\mathrm{ext}}$ let us now go on to determine $v'^{(1)}_{\mathrm{ext}}$. To do this we differentiate (2.13) with respect to time and evaluate the resulting expression in $t = t_0$. Then we obtain the expression:

$$n'^{(3)}(\boldsymbol{r}) = \nabla \cdot [n'^{(0)}(\boldsymbol{r}) \nabla v'^{(1)}_{\mathrm{ext}}(\boldsymbol{r})] + \nabla \cdot [n'^{(1)}(\boldsymbol{r}) \nabla v'^{(0)}_{\mathrm{ext}}(\boldsymbol{r})] + q'^{(1)}(\boldsymbol{r}) \,. \tag{2.22}$$

Since we want to determine $v'^{(1)}$ such that $n'^{(3)} = n^{(3)}$ and the conditions on the initial states are such that $n'^{(0)} = n^{(0)}$ and $n'^{(1)} = n^{(1)}$, we obtain the following equation for $v'^{(1)}_{\mathrm{ext}}$:

$$\nabla \cdot [n^{(0)}(\boldsymbol{r}) \nabla v'^{(1)}_{\mathrm{ext}}(\boldsymbol{r})] = n^{(3)}(\boldsymbol{r}) - q'^{(1)}(\boldsymbol{r}) - \nabla \cdot [n^{(1)}(\boldsymbol{r}) \nabla v'^{(0)}_{\mathrm{ext}}(\boldsymbol{r})] \,. \tag{2.23}$$

Now all quantities on the right hand side of (2.23) are known. The initial potential $v'^{(0)}_{\mathrm{ext}}$ was already determined from (2.21) whereas the quantity $q'^{(1)}$ can be calculated from

$$q'^{(1)}(\boldsymbol{r}) = \frac{\partial}{\partial t} q'(\boldsymbol{r},t)\Big|_{t=t_0} = -\mathrm{i} \langle \Psi'_0 | \, [\hat{q}'(\boldsymbol{r}), \hat{H}'(t_0)] \, | \Psi'_0 \rangle \,. \tag{2.24}$$

From this expression we see that $q'^{(1)}$ can be calculated from the knowledge of the initial state and the initial potential $v'^{(0)}_{\text{ext}}$ which occurs in $\hat{H}'(t_0)$. Therefore, (2.23) uniquely determines $v'^{(1)}_{\text{ext}}$ (again with boundary conditions $v'^{(1)}_{\text{ext}} \to 0$ for $\boldsymbol{r} \to \infty$). We note that in order to obtain (2.23) from (2.22) we indeed needed both conditions of (2.16) and (2.17). It is now clear how our procedure can be extended. If we take the k-th time-derivative of (2.13) we obtain the expression

$$n'^{(k+2)}(\boldsymbol{r}) = q'^{(k)}(\boldsymbol{r}) + \sum_{l=0}^{k} \binom{k}{l} \nabla \cdot [n'^{(k-l)}(\boldsymbol{r}) \nabla v'^{(l)}_{\text{ext}}(\boldsymbol{r})] \,. \tag{2.25}$$

Demanding that $n'^{(k)} = n^{(k)}$ then yields

$$\nabla \cdot [n^{(0)}(\boldsymbol{r}) \nabla v'^{(k)}_{\text{ext}}(\boldsymbol{r})] = n^{(k+2)}(\boldsymbol{r}) - q'^{(k)}(\boldsymbol{r}) - \sum_{l=0}^{k-1} \binom{k}{l} \nabla \cdot [n^{(k-l)}(\boldsymbol{r}) \nabla v'^{(l)}_{\text{ext}}(\boldsymbol{r})] \,. \tag{2.26}$$

The right hand side of this equation is completely determined since it only involves the potentials $v'^{(l)}_{\text{ext}}$ for $l = 1 \ldots k-1$ which were already determined. Similarly the quantities $q'^{(k)}$ can be calculated from multiple commutators of the operator $\hat{q}'$ and time-derivatives of the Hamiltonian $\hat{H}'(t_0)$ up to order $k-1$ and therefore only involves knowledge of the initial state and $v'^{(l)}_{\text{ext}}$ for $l = 1 \ldots k-1$. We can therefore uniquely determine all functions $v'^{(k)}_{\text{ext}}$ from (2.26) (again taking into account the boundary conditions) and construct the potential $v'_{\text{ext}}(\boldsymbol{r}, t)$ from its Taylor series as

$$v'_{\text{ext}}(\boldsymbol{r}, t) = \sum_{k=0}^{\infty} \frac{1}{k!} \, v'^{(k)}_{\text{ext}}(\boldsymbol{r}) \, (t - t_0)^k \,. \tag{2.27}$$

This determines $v'_{\text{ext}}(\boldsymbol{r}, t)$ completely within the convergence radius of the Taylor expansion. There is, of course, the possibility that the convergence radius is zero. However, this would mean that $v'_{\text{ext}}(\boldsymbol{r}, t)$ and hence $n(\boldsymbol{r}, t)$ and $v_{\text{ext}}(\boldsymbol{r}, t)$ are nonanalytic at $t = t_0$. Since the density of our reference system was supposed to be analytic we can disregard this possibility. If the convergence radius is non-zero but finite, we can propagate $|\Psi'_0\rangle$ to $|\Psi'(t_1)\rangle$ until a finite time $t_1 > t_0$ within the convergence radius and repeat the whole procedure above from $t = t_1$ by regarding $|\Psi'(t_1)\rangle$ as the initial state. This amounts to analytic continuation along the whole real time-axis and the complete determination of $v'_{\text{ext}}(\boldsymbol{r}, t)$ at all times. This completes the constructive proof of $v'_{\text{ext}}(\boldsymbol{r}, t)$.

Let us now summarize what we proved. We specify a given density $n(\boldsymbol{r}, t)$ obtained from a many-particle system with Hamiltonian $\hat{H}$ and initial state $|\Psi_0\rangle$. If one chooses an initial state $|\Psi'_0\rangle$ of a second many-particle system with two-particle interaction $\hat{V}'_{\text{ee}}$ in such a way that it yields the correct initial density and initial time-derivative of the density, then, for this system, there is a unique external potential $v'_{\text{ext}}(\boldsymbol{r},t)$ [determined up to a purely time-dependent function $c(t)$] that reproduces the given density $n(\boldsymbol{r},t)$.

Let us now specify some special cases. If we take $\hat{V}'_{\text{ee}} = 0$ we can conclude that, for a given initial state $|\Psi'_0\rangle = |\Phi_0\rangle$ with the correct initial density and initial time derivative of the density, there is a unique potential $v_{\text{KS}}(\boldsymbol{r},t)$ [modulo $c(t)$] for a noninteracting system that produces the given density $n(\boldsymbol{r},t)$ at all times. This solves the noninteracting v-representability problem, provided we can find an initial state with the required properties. If the many-body system described by the Hamiltonian $\hat{H}$ is stationary for times $t < t_0$, the initial state $|\Psi_0\rangle$ at t_0 leads to a density with zero time-derivative at $t = t_0$. In that case, a noninteracting state with the required initial density and initial time-derivative of the density (namely zero) can be obtained via the so-called Harriman construction [Harriman 1981, Lieb 1983]. Therefore a Kohn-Sham potential always exists for this kind of switch-on processes. The additional question whether this initial state can be chosen as a ground state of a noninteracting system is equivalent to the currently unresolved noninteracting v-representability question for stationary systems [Kohn 1983a, Ullrich 2002a, Dreizler 1990] (for an extensive discussion see [van Leeuwen 2003]).

We now take $\hat{V}'_{\text{ee}} = \hat{V}_{\text{ee}}$. We therefore consider two many-body systems with the same two-particle interaction. Our proof then implies that for a given v-representable density $n(\boldsymbol{r},t)$ that corresponds to an initial state $|\Psi_0\rangle$ and potential $v_{\text{ext}}(\boldsymbol{r},t)$, and for a given initial state $|\Psi'_0\rangle$ with the same initial density and initial time derivative of the density, we find that there is a unique external potential $v'_{\text{ext}}(\boldsymbol{r},t)$ [modulo $c(t)$] that yields this given density $n(\boldsymbol{r},t)$. The case $|\Psi_0\rangle = |\Psi'_0\rangle$ (in which the constraints on the initial state $|\Psi'_0\rangle$ are trivially satisfied) corresponds to the well-known Runge-Gross theorem. Our results in this section therefore provide an extension of this important theorem. As a final note we mention that the proof discussed here has recently been extended in an elegant way by Vignale [Vignale 2004] to time-dependent current-density functional theory. In that work it is shown that currents from an interacting system with some vector potential are also representable by a vector potential in a noninteracting system. This is, however, not true anymore if one considers scalar potentials. Interacting-v-representable currents are in general not noninteracting-v-representable [D'Agosta 2005a].

2.3 Invertibility of the Linear Density Response Function

In this section we will address the question if we can recover the potential variation $\delta v_{\text{ext}}(\boldsymbol{r},t)$ from a given density variation $\delta n(\boldsymbol{r},t)$ that was produced by it. There is, of course, an obvious non-uniqueness since both $\delta v_{\text{ext}}(\boldsymbol{r},t)$ and $\delta v_{\text{ext}}(\boldsymbol{r},t) + c(t)$, where $c(t)$ is an arbitrary time-dependent function, produce the same density variation. However, this is simply a gauge of the potential and is easily taken care of. Thus, by an inverse we will always mean an inverse modulo a purely time-dependent function $c(t)$ and by different potentials we will always mean that they differ more than a gauge $c(t)$.

From the work of Mearns and Kohn [Mearns 1987] we know that different potentials can yield the same density variations. However, in their examples these potentials are always potentials that exist at all times, i.e. there is no t_0 such that $\delta v_{\text{ext}} = 0$ for times $t < t_0$. On the other hand, we know from the Runge-Gross proof that a potential $\delta v_{\text{ext}}(\boldsymbol{r}, t)$ (not purely time-dependent) that is switched on at $t = t_0$ and is analytic at t_0 always causes a nonzero density variation $\delta n(\boldsymbol{r}, t)$. In this proof, the first nonvanishing time-derivative of δn at t_0 is found to be linear in the corresponding derivative of δv_{ext} and therefore the linear response function is invertible. Note that this conclusion holds even for an arbitrary initial state. The conclusion is therefore true for linear response to an already time-dependent system for which the linear response function depends on both t and t' separately, rather than on the time-difference $t - t'$. In the following we give an explicit proof for the invertibility of the linear response function for which the system is initially in its ground state. However, we will relax the condition that δv_{ext} is an analytic function in time, and we therefore allow for a larger class of external potentials than assumed in the Runge-Gross theorem. For clarification we further mention that it is sometimes assumed that the Dyson-type response equations of TDDFT are based on an adiabatic switch-on of the potential at all times. This is, however, not the case. The response functions can simply be derived by first order perturbation theory on the TDSE using a sudden switch-on of the external time-dependent potential [Fetter 1971]. The typical imaginary infinitesimals that occur in the denominator of the response functions result from the Fourier-representation of the causal Heaviside function (written as a complex contour integral) in the retarded density response function rather than from an adiabatically switched-on potential. The linear response equations of TDDFT are therefore in perfect agreement with a sudden switch-on of the potential.

We consider a many-body system in its ground state. At $t = 0$ (since the system is initially described by a time-independent Hamiltonian we can, without loss of generality, put the initial time $t_0 = 0$) we switch on an external field $\delta v_{\text{ext}}(\boldsymbol{r}, t)$ which causes a density response δn. We want to show that the linear response function is invertible for these switch-on processes. From simple first order perturbation theory on the TDSE we know that the linear density response is given by [Fetter 1971]

$$\delta n(\boldsymbol{r}_1, t_1) = \int_0^{t_1} \mathrm{d}t_2 \int \mathrm{d}^3 r_2 \, \chi_R(\boldsymbol{r}_1 t_1, \boldsymbol{r}_2 t_2) \delta v_{\text{ext}}(\boldsymbol{r}_2, t_2) \, , \tag{2.28}$$

where

$$\chi_R(\boldsymbol{r}_1 t_1, \boldsymbol{r}_2 t_2) = -\mathrm{i}\,\theta(t_1 - t_2) \langle \Psi_0 | \left[\Delta \hat{n}_H(\boldsymbol{r}_1, t_1), \Delta \hat{n}_H(\boldsymbol{r}_2, t_2) \right] | \Psi_0 \rangle \, , \tag{2.29}$$

is the retarded density response function. Note that here, instead of the density operator $\hat{n}_H$ (in the Heisenberg picture with respect to the ground state Hamiltonian $\hat{H}$), we prefer to use the density fluctuation operator $\Delta\hat{n}_H = \hat{n}_H - \langle\hat{n}_H\rangle$ in the response function, where we use that the commutator of the density operators is equal to the commutator of the density fluctuation operators. Now we go over to a Lehmann representation of the response function and we insert a complete set of eigenstates of $\hat{H}$:

$$\delta n(\boldsymbol{r}_1, t_1) = \mathrm{i}\sum_n \int_0^{t_1} \mathrm{d}t_2 \int \mathrm{d}^3 r_2\, \mathrm{e}^{\mathrm{i}\Omega_n(t_1-t_2)} f_n^*(\boldsymbol{r}_1) f_n(\boldsymbol{r}_2)\delta v_{\mathrm{ext}}(\boldsymbol{r}_2, t_2) + \mathrm{c.c.}\,, \tag{2.30}$$

where $\Omega_n = E_n - E_{\mathrm{GS}} > 0$ are the excitation energies of the unperturbed system (we assume the ground state to be nondegenerate) and the functions f_n are defined as

$$f_n(\boldsymbol{r}) = \langle\Psi_{\mathrm{GS}}|\Delta\hat{n}(\boldsymbol{r})|\Psi_n\rangle\,. \tag{2.31}$$

The density response can then be rewritten as

$$\delta n(\boldsymbol{r}_1, t_1) = \mathrm{i}\sum_n f_n^*(\boldsymbol{r}_1) \int_0^{t_1} \mathrm{d}t_2\, a_n(t_2)\mathrm{e}^{\mathrm{i}\Omega_n(t_1-t_2)} + \mathrm{c.c.}\,, \tag{2.32}$$

where we defined

$$a_n(t) = \int \mathrm{d}^3 r\, f_n(\boldsymbol{r})\delta v_{\mathrm{ext}}(\boldsymbol{r}, t)\,. \tag{2.33}$$

Now note that the time integral in (2.32) has the form of a convolution. This means that we can simplify this equation using Laplace transforms. The Laplace transform and its deconvolution property are given by

$$\hat{\mathcal{L}}f(s) = \int_0^{\infty} \mathrm{d}t\, \mathrm{e}^{-st} f(t)\,, \tag{2.34a}$$

$$\hat{\mathcal{L}}(f * g)(s) = \hat{\mathcal{L}}f(s)\hat{\mathcal{L}}g(s)\,. \tag{2.34b}$$

where the convolution product is defined as

$$(f * g)(t) = \int_0^t \mathrm{d}\tau f(\tau)g(t-\tau)\,. \tag{2.35}$$

If we now take the Laplace transform of δn in (2.32) we obtain the equation:

$$\hat{\mathcal{L}}(\delta n)(\boldsymbol{r}_1, s) = \mathrm{i}\sum_n f_n^*(\boldsymbol{r}_1)\frac{1}{s - \mathrm{i}\Omega_n}\hat{\mathcal{L}}a_n(s) + \mathrm{c.c.} \tag{2.36}$$

If we multiply both sides with the Laplace transform $\hat{\mathcal{L}}(\delta v_{\mathrm{ext}})$ of δv_{ext} and integrate over $\boldsymbol{r}_1$ we obtain

$$\int \mathrm{d}^3 r_1 \, \hat{\mathcal{L}}(\delta v_{\mathrm{ext}})(\boldsymbol{r}_1, s) \hat{\mathcal{L}}(\delta n)(\boldsymbol{r}_1, s) = \mathrm{i} \sum_n \frac{1}{s - \mathrm{i}\Omega_n} |\hat{\mathcal{L}} a_n(s)|^2 + \text{c.c.}$$
$$= -2 \sum_n \frac{\Omega_n}{s^2 + \Omega_n^2} |\hat{\mathcal{L}} a_n(s)|^2 \,. \qquad (2.37)$$

This is the basic relation that we use to prove invertibility. If we assume that $\delta n = 0$ then also $\hat{\mathcal{L}}(\delta n) = 0$ and we obtain

$$0 = \sum_n \frac{\Omega_n}{s^2 + \Omega_n^2} |\hat{\mathcal{L}} a_n(s)|^2 \,. \qquad (2.38)$$

However, since each prefactor of $|\hat{\mathcal{L}} a_n|^2$ in the summation is positive the sum can only be zero if $\hat{\mathcal{L}} a_n = 0$ for all n. This in its turn implies that $a_n(t)$ must be zero for all n. This means also that

$$\int \mathrm{d}^3 r \, \Delta\hat{n}(r) \delta v_{\mathrm{ext}}(\boldsymbol{r}, t) |\Psi_0\rangle = \sum_n |\Psi_n\rangle \int \mathrm{d}^3 r \, \langle \Psi_n | \Delta\hat{n}(\boldsymbol{r}) | \Psi_0 \rangle \delta v_{\mathrm{ext}}(\boldsymbol{r}, t)$$
$$= \sum_n a_n(t) |\Psi_n\rangle = 0 \,. \qquad (2.39)$$

Note that $a_0(t)$ is automatically zero since obviously $\langle \Psi_{\mathrm{GS}} | \Delta\hat{n}(x) | \Psi_{\mathrm{GS}} \rangle = 0$. If we write out the above equation in first quantization again we have

$$\sum_{k=1}^{N} \Delta v_{\mathrm{ext}}(\boldsymbol{r}_k, t) |\Psi_{\mathrm{GS}}\rangle = 0 \,, \qquad (2.40)$$

where N is the number of electrons in the system and $\Delta v_{\mathrm{ext}}(\boldsymbol{r}, t)$ is defined as

$$\Delta v_{\mathrm{ext}}(\boldsymbol{r}, t) = \delta v_{\mathrm{ext}}(\boldsymbol{r}, t) - \frac{1}{N} \int \mathrm{d}^3 r \, n_{\mathrm{GS}}(\boldsymbol{r}) \delta v_{\mathrm{ext}}(\boldsymbol{r}, t) \,, \qquad (2.41)$$

where n_{GS} is the density of the unperturbed system. Now (2.40) immediately implies that $\Delta v_{\mathrm{ext}} = 0$ and, since the second term on the right hand side of (2.41) is a purely time-dependent function, we obtain

$$\delta v_{\mathrm{ext}}(\boldsymbol{r}, t) = c(t) \,. \qquad (2.42)$$

We have therefore proven that only purely time-dependent potentials yield zero density response. In other words, the response function is, modulo a trivial gauge, invertible for switch-on processes. Note that the only restriction we put on the potential $\delta v_{\mathrm{ext}}(\boldsymbol{r}, t)$ is that it is Laplace-transformable. This is a much weaker restriction on the potential than the constraint that it be an analytic function at $t = t_0$, as required in the Runge-Gross proof. One should, however, be careful with what one means with an inverse response function. The response function defines a mapping $\chi : \delta\mathcal{V}_{\mathrm{ext}} \to \delta\mathcal{N}$ from the set of potential variations from a nondegenerate ground state, which we call

$\delta\mathcal{V}_{\text{ext}}$, to the set of first order density variations $\delta\mathcal{N}$ that are reproduced by it. We have shown that the inverse $\chi^{-1} : \delta\mathcal{N} \to \delta\mathcal{V}_{\text{ext}}$ is well-defined modulo a purely time-dependent function. However, there are density variations that can never be produced by a finite potential variation and are therefore not in the set $\delta\mathcal{N}$. An example of such a density variation is one which is identically zero on some finite volume.

Another consequence of the above analysis is the following. Suppose the linear response kernel has eigenfunctions, i.e. there is a λ such that

$$\int \mathrm{d}t_2 \int \mathrm{d}\boldsymbol{r}_2 \, \chi_R(\boldsymbol{r}_1 t_1, \boldsymbol{r}_2 t_2) \zeta(\boldsymbol{r}_2, t_2) = \lambda \zeta(\boldsymbol{r}_1, t_1) \, . \tag{2.43}$$

Laplace transforming this equation yields

$$\int \mathrm{d}^3 r_2 \, \Xi(\boldsymbol{r}_1, \boldsymbol{r}_2, s) \hat{\mathcal{L}}\zeta(\boldsymbol{r}_2, s) = \lambda \hat{\mathcal{L}}\zeta(\boldsymbol{r}_1, s) \, , \tag{2.44}$$

where Ξ is the Laplace transform of χ explicitly given by

$$\Xi(\boldsymbol{r}_1, \boldsymbol{r}_2, s) = \mathrm{i} \sum_n \frac{f_n^*(\boldsymbol{r}_1) f_n(\boldsymbol{r}_2)}{s - \mathrm{i}\Omega_n} + \text{c.c.} \tag{2.45}$$

Since Ξ is a real Hermitian operator, its eigenvalues λ are real and its eigenfunctions $\hat{\mathcal{L}}\zeta$ can be chosen to be real. Then ζ is real as well and (2.37) implies (if we take $\delta v_{\text{ext}} = \zeta$ and $\delta n = \lambda\zeta$)

$$\lambda \int \mathrm{d}^3 r \, [\hat{\mathcal{L}}\zeta(\boldsymbol{r}, s)]^2 < 0 \, , \tag{2.46}$$

which implies $\lambda < 0$. We have therefore proven that if there are density variations that are proportional to the applied potential, then this constant of proportionality is negative. In other words, the eigenvalues of the density response function are negative. In this derivation we made again explicit use of Laplace transforms and therefore of the condition that $\zeta = 0$ for $t < 0$. The work of Mearns and Kohn shows that positive eigenvalues are possible when this restriction is not made. The same is true when one considers response functions for excited states [Gaudoin 2004]. We finally note that similar results are readily obtained for the static density response function [van Leeuwen 2003] in which case the negative eigenvalues of the response function are an immediate consequence of the Hohenberg-Kohn theorem.

Let us now see what our result implies. We considered the density $n[v_{\text{ext}}]$ as a functional of v_{ext} and established that the response kernel $\chi[v_{\text{GS}}] = \delta n/\delta v_{\text{ext}}[v_{\text{GS}}]$ is invertible where v_{GS} is the potential in the ground state and that $\delta n/\delta v_{\text{ext}}[v_{\text{GS}}] < 0$ in the sense that its eigenvalues are all negative definite. We can now apply a fundamental theorem of calculus, the *inverse function theorem.* For functions of real numbers the theorem states that if a continuous function $y(x)$ is differentiable at x_0 and if $\mathrm{d}y/\mathrm{d}x(x_0) \neq 0$ then locally there exists an inverse $x(y)$ for y close enough to y_0, where $y(x_0) = y_0$. The theorem can be extended to functionals on function spaces (to be precise Banach spaces, for details see [Choquet-Bruhat 1991]). For our case, this theorem implies that if the functional $n[v_{\text{ext}}]$ is differentiable at the ground state potential v_{GS} and the derivative $\chi[v_{\text{GS}}] = \delta n/\delta v_{\text{ext}}[v_{\text{GS}}]$ is an invertible kernel then for potentials v_{ext} close enough to v_{GS} (in Banach norm sense) the inverse map $v_{\text{ext}}[n]$ exists. Since we have shown that the linear response function $\chi[v_{\text{GS}}]$ is invertible this then proves the Runge-Gross theorem for Laplace transformable switch-on potentials for systems initially in the ground state.

2.4 Consequences of v-Representability for the Quantum Mechanical Action

The role that is played by the energy functional in stationary density-functional theory is played by the action functional in time-dependent density functional theory. The correct form of the action appears naturally within the framework of Keldysh theory and is discussed in detail in Chap. 3. However, historically the first action within time-dependent density-functional theory context was defined by Peuckert [Peuckert 1978] (who already made a connection to Keldysh theory) and later in the Runge-Gross paper [Runge 1984]. However, as was discovered later [Gross 1996, Burke 1998b] this form of the action leads to paradoxical results. Rajagopal [Rajagopal 1996] attempted to introduce an action principle in TDDFT using the formalism of Jackiw and Kerman [Jackiw 1979] for deriving time-ordered n-point functions in quantum field theory. However, due to the time-ordering inherent in the work of Jackiw and Kerman the basic variable of Ragagopal's formalism is not the time-dependent density but an transition element of the density operator between a wavefunction evolving from the past and a wavefunction evolving from the future to a certain time t. Moreover, the action functional in this formalism suffers from the same difficulties as the action introduced by Runge and Gross. In this section we will show that these difficulties arise due to a restriction of the variational freedom as a consequence of v-representability constraints. For two other recent discussions of these points we refer to [van Leeuwen 2001, Maitra 2002c].

We start with the following time-dependent action functional

$$A[\Psi] = \int_{t_0}^{t_1} \mathrm{d}t \; \langle \Psi | \mathrm{i}\frac{\partial}{\partial t} - \hat{H}(t) | \Psi \rangle \; . \tag{2.47}$$

The usual approach is to require the action to be stationary under variations $\delta\Psi$ that satisfy $\delta\Psi(t_0) = \delta\Psi(t_1) = 0$. We then find after a partial integration

$$\delta A = \int_{t_0}^{t_1} \mathrm{d}t \; \langle \delta\Psi | \mathrm{i}\frac{\partial}{\partial t} - \hat{H}(t) | \Psi \rangle + \text{c.c.} + \mathrm{i}\langle \Psi | \delta\Psi \rangle \Big|_{t_0}^{t_1} \; . \tag{2.48}$$

With the boundary conditions and the fact that the real and imaginary part of $\delta\Psi$ can be varied independently we obtain the result that

$$\left[\mathrm{i}\frac{\partial}{\partial t} - \hat{H}(t)\right] |\Psi\rangle = 0 \,, \tag{2.49}$$

which is just the time-dependent Schrödinger equation. We see that the variational requirement $\delta A = 0$, together with the boundary conditions is equivalent to the time-dependent Schrödinger equation.

A different derivation [Löwdin 1972] which does not put any constraints on the variations at the endpoints of the time interval is the following. We consider again a first order change in the action due to changes in the wavefunction and require that the action is stationary. We have the general relation

$$0 = \delta A = \int_{t_0}^{t_1} \mathrm{d}t \, \langle \delta\Psi | \mathrm{i}\frac{\partial}{\partial t} - \hat{H}(t) | \Psi \rangle + \int_{t_0}^{t_1} \mathrm{d}t \, \langle \Psi | \mathrm{i}\frac{\partial}{\partial t} - \hat{H}(t) | \delta\Psi \rangle \,. \tag{2.50}$$

We now choose the variations $\delta\Psi = \delta\tilde{\Psi}$ and $\delta\Psi = \mathrm{i}\delta\tilde{\Psi}$ where $\delta\tilde{\Psi}$ is arbitrary. We thus obtain

$$0 = \delta A = \int_{t_0}^{t_1} \mathrm{d}t \, \langle \delta\tilde{\Psi} | \mathrm{i}\frac{\partial}{\partial t} - \hat{H}(t) | \Psi \rangle + \int_{t_0}^{t_1} \mathrm{d}t \, \langle \Psi | \mathrm{i}\frac{\partial}{\partial t} - \hat{H}(t) | \delta\tilde{\Psi} \rangle \tag{2.51a}$$

$$0 = \delta A = -\mathrm{i} \int_{t_0}^{t_1} \mathrm{d}t \, \langle \delta\tilde{\Psi} | \mathrm{i}\frac{\partial}{\partial t} - \hat{H}(t) | \Psi \rangle + \mathrm{i} \int_{t_0}^{t_1} \mathrm{d}t \, \langle \Psi | \mathrm{i}\frac{\partial}{\partial t} - \hat{H}(t) | \delta\tilde{\Psi} \rangle \,. \tag{2.51b}$$

From (2.51a) and (2.51b) we obtain

$$0 = \int_{t_0}^{t_1} \mathrm{d}t \, \langle \delta\tilde{\Psi} | \mathrm{i}\frac{\partial}{\partial t} - \hat{H}(t) | \Psi \rangle \,. \tag{2.52}$$

Since this must be true for arbitrary $\delta\tilde{\Psi}$ we again obtain the time-dependent Schrödinger equation

$$\left[\mathrm{i}\frac{\partial}{\partial t} - \hat{H}(t)\right] |\Psi\rangle = 0 \,. \tag{2.53}$$

We did not need to put any boundary conditions on the variations at all. We only required that if $\delta\tilde{\Psi}$ is an allowed variation that then also $\mathrm{i}\delta\tilde{\Psi}$ is an allowed variation.

Let us now discuss the problems with the variational principle when one attempts to construct a time-dependent density-functional theory. The obvious definition of a density functional would be [Runge 1984]

$$A[n] = \int_{t_0}^{t_1} \mathrm{d}t \, \langle \Psi[n] | \mathrm{i}\frac{\partial}{\partial t} - \hat{H}(t) | \Psi[n] \rangle \,, \tag{2.54}$$

where $|\Psi[n]\rangle$ is a wavefunction which yields the density $n(\boldsymbol{r},t)$ and evolves from a given initial state $|\Psi_0\rangle$ with initial density $n_0(\boldsymbol{r})$. By the Runge-Gross theorem such a wave function is determined up to a phase factor. In order to define the action uniquely we have to make a choice for this phase factor. An obvious choice would be to choose the $|\Psi[n]\rangle$ that evolves in the external potential $v_{\mathrm{ext}}(\boldsymbol{r},t)$ that vanishes at infinity and yields the density $n(\boldsymbol{r},t)$. This corresponds to choosing a particular kind of gauge. There are of course many more phase conventions possible. The trouble obviously arises from the fact that the density only determines the wavefunction up to an arbitrary time-dependent phase. However, there are more problems. Suppose we avoid the phase problem by defining a functional of the external potential rather than the density

$$A[v] = \int_{t_0}^{t_1} \mathrm{d}t \, \langle \Psi[v] | \mathrm{i}\frac{\partial}{\partial t} - \hat{H}(t) | \Psi[v] \rangle \, . \tag{2.55}$$

Note that the potential v in the argument of the action is only used to parametrize the set of wavefunctions used in the action principle. This potential v is therefore not the same as the external potential in the Hamiltonian $\hat{H}(t)$ of (2.55) as this Hamiltonian is fixed. The state $|\Psi[v]\rangle$ is a state that evolves from a given initial state $|\Psi_0\rangle$ by solution of a time-dependent Schrödinger equation with potential v as its external potential. As the potential obviously defines $|\Psi[v]\rangle$ uniquely, including its phase, the action is well-defined. The question is now whether one can recover the time-dependent Schrödinger equation by making the action stationary with respect to potential variations δv. It is readily seen that this is not the case. The reason for this is that all variations $\delta\Psi$ of the wave function must now be caused by potential variations δv which leads to variations over a restricted set of wave functions. In other words, the variations $\delta\Psi$ must be v-representable. For instance, when deriving the Schrödinger equation from the variational principle one can not assume the boundary conditions $\delta\Psi(t_0) = \delta\Psi(t_1) = 0$. Since the time-dependent Schrödinger equation is first order in time, the variation $\delta\Psi(t)$ at times $t > t_0$ is completely determined by the boundary condition for $\delta\Psi(t_0)$. We are thus no longer free to specify a second boundary condition at a later time t_1. Moreover, we are not allowed to treat the real and imaginary part of $\delta\Psi$ as independent variations since both are determined simultaneously by the potential variation δv. This means that the first derivation of the TDSE that we presented in this section can not be carried out. It is readily seen that also the second derivation based on (2.51a) and (2.51b) fails. If $\delta\Psi$ is a variation generated by some $\delta\hat{V}(t) = \int \mathrm{d}^3 r \, \hat{n}(\boldsymbol{r}) \delta v(\boldsymbol{r},t)$, then $\delta\Psi$ satisfies

$$\left[\mathrm{i}\frac{\partial}{\partial t} - \hat{H}_v(t) \right] |\delta\Psi\rangle = \delta\hat{V}(t) |\Psi\rangle \, , \tag{2.56}$$

where $\hat{H}_v$ is a Hamiltonian with potential v and we neglected terms of higher order. Multiplication by the imaginary number "i" yields that the variation $\mathrm{i}\delta\Psi$ must be generated by potential $\mathrm{i}\delta v$. This potential variation is however imaginary and therefore not an allowed variation since all potential variations must be real.

We therefore conclude that time-dependent density-functional theory can not be based on the usual variational principle, and indeed attempts to do so have led to paradoxes. In Chap. 3 we will discuss how an extended type of action functional defined on a Keldysh time-contour [van Leeuwen 2001, van Leeuwen 1998] can be used as a basis from which the time-dependent Kohn-Sham equations can be derived. This also has the immediate advantage that the action functional can then be directly related to the elegant formalism of nonequilibrium Green function theory which offers a systematic way of constructing time-dependent density functionals. Some examples of such functionals can be found in reference [von Barth 2005]. With hindsight it is interesting to see that already the work of Peuckert [Peuckert 1978], which is one of the very first papers in TDDFT, makes a connection to Keldysh Green functions, and in fact several of his results (such as the adiabatic connection formula) are perfectly valid when interpreted in terms of the action formalism of the next chapter.

3 Introduction to the Keldysh Formalism

R. van Leeuwen, N.E. Dahlen, G. Stefanucci, C.-O. Almbladh, and U. von Barth

3.1 Introduction

In this chapter we give an introduction to the Keldysh formalism, which is an extremely useful tool for first-principles studies of nonequilibrium many-particle systems. Of particular interest for TDDFT is the relation to non-equilibrium Green functions (NEGF), which allows us to construct exchange-correlation potentials with memory by using diagrammatic techniques. For many problems, such as quantum transport or atoms in intense laser pulses, one needs exchange-correlation functionals with memory, and Green function techniques offer a systematic method for developing these. The Keldysh formalism is also necessary for defining response functions in TDDFT and for defining an action functional needed for deriving TDDFT from a variational principle. In this chapter, we give an introduction to the nonequilibrium Green function formalism, intended to illustrate the usefulness of the theory. The formalism does not differ much from ordinary equilibrum theory, the main difference being that all time-dependent functions are defined for time-arguments on a contour, known as the Keldysh contour.

The Green function $G(\boldsymbol{r}t, \boldsymbol{r}'t')$ is a function of two space- and time-coordinates, and is obviously more complicated than the one-particle density $n(\boldsymbol{r}, t)$, which is the main ingredient of TDDFT. However, the advantage of NEGF methods is that we can systematically improve the approximations by taking into account particular physical processes (represented in the form of Feynman diagrams) that we believe to be important. The Green function provides us directly with all expectation values of one-body operators (such as the density and the current), and also the total energy, ionization potentials, response functions, spectral functions, etc. In relation to TDDFT, this is useful not only for developing orbital functionals and exchange-correlation functionals with memory, but also for providing insight in the exact properties of the noninteracting Kohn-Sham system.

In the following, we shall focus on systems that are initially in thermal equilibrium. We will start by introducing the Keldysh contour and the nonequilbrium Green function, which is one particular example of a function defined on the contour. In Sect. 3.4 we will explain how to combine and manipulate functions of time variables on the contour. These results, that are summarized in Table 3.1, are highly important, since the class of functions

R. van Leeuwen et al.: *Introduction to the Keldysh Formalism*, Lect. Notes Phys. **706**, 33–59 (2006)
DOI 10.1007/3-540-35426-3_3

also include response functions and self-energies. The results derived in this section are essential for defining action functionals and response functions, as we will do in Sect. 3.9, and are also used extensively in Chap. 32. The equations of motion for the Green function, known as the Kadanoff-Baym equations, are explained in Sect. 3.5. While in TDDFT we take exchange and correlation effects into account through $v_{\mathrm{xc}}[n]$, the corresponding quantity in Green function theory is the self-energy $\Sigma[G]$. Just like v_{xc}, the self-energy functional must be approximated. For a given functional $\Sigma[G]$, it is important that the resulting observables obey the macroscopic conservation laws, such as the continuity equation. These approximations are known as *conserving*, and will be discussed briefly in Sect. 3.7. In the last part of this chapter we discuss the applications of the Keldysh formalism in TDDFT, including the relation between Σ and v_{xc}, the derivation of the Kohn-Sham equations from an action functional, and the derivation of an f_{xc} functional. As an illustrative example, we discuss the time-dependent exchange-only optimized effective potential approximation.

3.2 The Keldysh Contour

In quantum mechanics we associate with any observable quantity O a Hermitian operator $\hat{O}$. The expectation value $\mathrm{Tr}\{\hat{\rho}_0\hat{O}\}$ gives the value of O when the system is described by the density operator $\hat{\rho}_0$ and the trace denotes a sum over a complete set of states in Hilbert space. For an isolated system the Hamiltonian $\hat{H}_0$ does not depend on time, and the expectation value of *any* observable quantity is constant, provided that $[\hat{\rho}_0, \hat{H}_0] = 0$. In the following we want to discuss how to describe systems that are isolated for times $t < 0$, such that $\hat{H}(t<0) = \hat{H}_0$, but disturbed by an external time-dependent field at $t > 0$. The expectation value of $\hat{O}$ at $t > 0$ is then given by the average on the initial density operator $\hat{\rho}_0$ of the operator $\hat{O}$ in the Heisenberg representation,

$$O(t) = \langle \hat{O}_H(t) \rangle \equiv \mathrm{Tr}\{\hat{\rho}_0 \hat{O}_H(t)\} = \mathrm{Tr}\{\hat{\rho}_0 \hat{U}(0,t)\hat{O}\hat{U}(t,0)\} \,, \tag{3.1}$$

where the operator in the Heisenberg picture has a time-dependence according to $\hat{O}_H(t) = \hat{U}(0,t)\hat{O}\hat{U}(t,0)$. The evolution operator $\hat{U}(t,t')$ is the solution of the equations

$$\mathrm{i}\frac{\mathrm{d}}{\mathrm{d}t}\hat{U}(t,t') = \hat{H}(t)\hat{U}(t,t') \qquad \text{and} \qquad \mathrm{i}\frac{\mathrm{d}}{\mathrm{d}t'}\hat{U}(t,t') = -\hat{U}(t,t')\hat{H}(t') \,, \tag{3.2}$$

with the boundary condition $\hat{U}(t,t) = 1$. It can be formally written as

$$\hat{U}(t,t') = \begin{cases} \hat{T} \exp[-\mathrm{i}\int_{t'}^{t} \mathrm{d}\bar{t}\, \hat{H}(\bar{t})] & t > t' \\ \hat{\bar{T}} \exp[-\mathrm{i}\int_{t'}^{t} \mathrm{d}\bar{t}\, \hat{H}(\bar{t})] & t < t' \end{cases} . \tag{3.3}$$

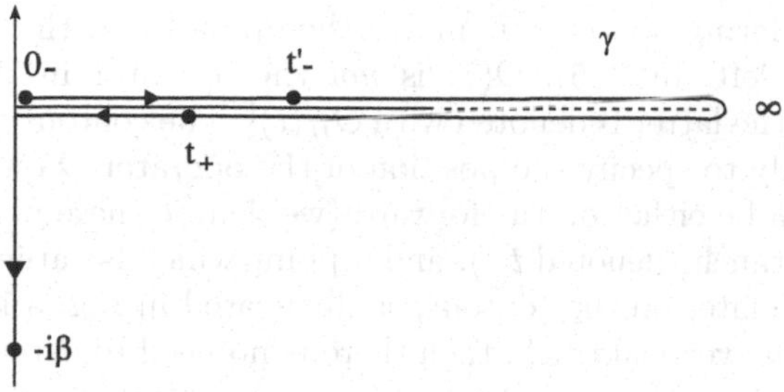

Fig. 3.1. The Keldysh contour in the complex time-plane, starting at $t = 0$, and ending at $t = -\mathrm{i}\beta$, with t on the backward branch and t' on the forward branch. By definition, any point lying on the vertical track is later than a point lying on the forward or backward branch

In (3.3), $\hat{\mathcal{T}}$ is the time-ordering operator that rearranges the operators in chronological order with later times to the left; $\hat{\tilde{\mathcal{T}}}$ is the anti-chronological time-ordering operator. The evolution operator satisfies the group property $\hat{U}(t,t_1)\,\hat{U}(t_1,t') = \hat{U}(t,t')$ for any t_1. Notice that if the Hamiltonian is time-independent in the interval between t and t', then the evolution operator becomes $\hat{U}(t,t') = \exp[-\mathrm{i}\hat{H}(t-t')]$. If we now let the system be initially in thermal equilibrium, with an inverse temperature $\beta \equiv 1/k_B T$ and chemical potential μ, the initial density matrix is $\hat{\rho}_0 = \exp[-\beta(\hat{H}_0 - \mu\hat{N})]/\,\mathrm{Tr}\{\exp[-\beta(\hat{H}_0 - \mu\hat{N})]\}$. Assuming that $\hat{H}_0$ and $\hat{N}$ commute, $\hat{\rho}_0$ can be rewritten using the evolution operator $\hat{U}$ with a complex time-argument, $t = -\mathrm{i}\beta$, according to $\hat{\rho}_0 = \exp[\beta\mu\hat{N}]\hat{U}(-\mathrm{i}\beta,0)/\,\mathrm{Tr}\{\exp[\beta\mu\hat{N}]\hat{U}(-\mathrm{i}\beta,0)\}$. Inserting this expression in (3.1), we find

$$O(t) = \frac{\mathrm{Tr}\left\{\mathrm{e}^{\beta\mu\hat{N}}\hat{U}(-\mathrm{i}\beta,0)\hat{U}(0,t)\hat{O}\hat{U}(t,0)\right\}}{\mathrm{Tr}\left\{\mathrm{e}^{\beta\mu\hat{N}}\hat{U}(-\mathrm{i}\beta,0)\right\}}\,. \tag{3.4}$$

Reading the arguments in the numerator from the right to the left, we see that we can design a time-contour γ with a forward branch going from 0 to t, a backward branch coming back from t and ending in 0, and a branch along the imaginary time-axis from 0 to $-\mathrm{i}\beta$. This contour is illustrated in Fig. 3.1. Note that the group property of $\hat{U}$ means that we are free to extend this contour up to infinity. We can now generalize (3.4), and let z be a time-contour variable on γ. We will in the following stick to the notation that the time-variable on the contour is denoted z unless we specify on which branch of the contour it is located. This time-variable can therefore be real or complex. Letting the variable $\bar{z}$ run along this same contour, (3.4) can be formally recast as

$$O(z) = \frac{\mathrm{Tr}\left\{\mathrm{e}^{\beta\mu\hat{N}}\hat{\mathcal{T}}_\mathrm{c}\,\mathrm{e}^{-\mathrm{i}\int_\gamma \mathrm{d}\bar{z}\,\hat{H}(\bar{z})}\,\hat{O}(z)\right\}}{\mathrm{Tr}\left\{\mathrm{e}^{\beta\mu\hat{N}}\hat{\mathcal{T}}_\mathrm{c}\,\mathrm{e}^{-\mathrm{i}\int_\gamma \mathrm{d}\bar{z}\,\hat{H}(\bar{z})}\right\}}\,. \tag{3.5}$$

The contour ordering operator $\hat{\mathcal{T}}_c$ moves the operators with "later" contour variable to the left. In (3.5), $\hat{O}(z)$ is *not* the operator in the Heisenberg representation [the latter is denoted with $\hat{O}_H(t)$]. The contour-time argument in $\hat{O}$ is there only to specify the position of the operator $\hat{O}$ on γ. A point on the real axis can be either on the forward (we denote these points t_-), or on the backward branch (denoted t_+), and a point which is earlier in real time, can therefore be later on the contour, as illustrated in Fig. 3.1.

If z lies on the vertical track, then there is no need to extend the contour along the real axis. Instead, we have

$$O(z) = \frac{\mathrm{Tr}\left\{e^{\beta\mu\hat{N}} e^{-i\int_z^{-i\beta} d\bar{z}\hat{H}_0}\, \hat{O}\, e^{-i\int_0^z d\bar{z}\hat{H}_0}\right\}}{\mathrm{Tr}\left\{e^{-\beta(\hat{H}_0-\mu\hat{N})}\right\}} = \frac{\mathrm{Tr}\left\{e^{-\beta(\hat{H}_0-\mu\hat{N})}\,\hat{O}\right\}}{\mathrm{Tr}\left\{e^{-\beta(\hat{H}_0-\mu\hat{N})}\right\}}, \tag{3.6}$$

where the cyclic property of the trace has been used. The right hand side is independent of z and coincides with the thermal average $\mathrm{Tr}\{\hat{\rho}_0\hat{O}\}$. It is easy to verify that (3.5) would give exactly the same result for $O(t)$, where t is real, if the Hamiltonian was time-independent, i.e., $\hat{H}(t) = \hat{H}_0$ also for $t > 0$.

To summarize, in (3.5) the variable z lies on the contour of Fig. 3.1; the r.h.s. gives the time-dependent statistical average of the observable O when z lies on the forward or backward branch, and the statistical average before the system is disturbed when z lies on the vertical track.

3.3 Nonequilibrium Green Functions

We now introduce the NEGF, which is a function of two contour time-variables. In order to keep the notation as light as possible, we here discard the spin degrees of freedom; the spin index may be restored later as needed. The field operators $\hat{\psi}(\boldsymbol{r})$ and $\hat{\psi}^\dagger(\boldsymbol{r})$ destroy and create an electron in $\boldsymbol{r}$ and obey the anticommutation relations $\{\hat{\psi}(\boldsymbol{r}), \hat{\psi}^\dagger(\boldsymbol{r}')\} = \delta(\boldsymbol{r}-\boldsymbol{r}')$. We write the Hamiltonian $\hat{H}(t)$ as the sum of a quadratic term

$$\hat{h}(t) = \int d^3r \int d^3r'\, \hat{\psi}^\dagger(\boldsymbol{r})\langle \boldsymbol{r}|\boldsymbol{h}(t)|\boldsymbol{r}'\rangle\hat{\psi}(\boldsymbol{r}')\,, \tag{3.7}$$

and the interaction operator

$$\hat{V}_{ee} = \frac{1}{2}\int d^3r \int d^3r'\, \hat{\psi}^\dagger(\boldsymbol{r})\hat{\psi}^\dagger(\boldsymbol{r}')v_{ee}(\boldsymbol{r},\boldsymbol{r}')\hat{\psi}(\boldsymbol{r}')\hat{\psi}(\boldsymbol{r})\,. \tag{3.8}$$

We use boldface to indicate matrices in one-electron labels, e.g., $\boldsymbol{h}$ is a matrix and $\langle \boldsymbol{r}|\boldsymbol{h}|\boldsymbol{r}'\rangle$ is the $(\boldsymbol{r},\boldsymbol{r}')$ matrix element of $\boldsymbol{h}$. When describing electrons in an electro-magnetic field, the quadratic term is given by $\langle \boldsymbol{r}|\boldsymbol{h}(t)|\boldsymbol{r}'\rangle = \delta(\boldsymbol{r}-\boldsymbol{r}')\left\{[-i\nabla + \boldsymbol{A}_{\mathrm{ext}}(\boldsymbol{r},t)]^2/2 + v_{\mathrm{ext}}(\boldsymbol{r},t)\right\}$.

The definition of an expectation value in (3.1) can be generalized to the expectation value of two operators. The Green function is defined as

$$G(\boldsymbol{r}z, \boldsymbol{r}'z') = \langle \boldsymbol{r}|\boldsymbol{G}(z, z')|\boldsymbol{r}'\rangle \equiv -\mathrm{i}\langle \mathcal{T}_c\, \hat{\psi}_H(\boldsymbol{r}, z)\hat{\psi}_H^\dagger(\boldsymbol{r}', z')\rangle , \qquad (3.9)$$

where the contour variable in the field operators specifies the position in the contour ordering. The operators have a time-dependence according to the definition of the Heisenberg picture, e.g., $\hat{\psi}_H^\dagger(\boldsymbol{r}, z) = \hat{U}(0, z)\hat{\psi}^\dagger(\boldsymbol{r})\hat{U}(z, 0)$. Notice that if the time-argument z is located on the real axis, then $\hat{\psi}_H(\boldsymbol{r}, t_+) = \hat{\psi}_H(\boldsymbol{r}, t_-)$. If the time-argument is on the imaginary axis, then $\hat{\psi}(\boldsymbol{r}, -\mathrm{i}\tau)$ is *not* the adjoint of $\hat{\psi}(\boldsymbol{r}, -\mathrm{i}\tau)$ since $\hat{U}^\dagger(-\mathrm{i}\tau, 0) \neq \hat{U}(0, -\mathrm{i}\tau)$. The Green function can be written

$$\boldsymbol{G}(z, z') = \theta(z, z')\boldsymbol{G}^>(z, z') + \theta(z', z)\boldsymbol{G}^<(z, z') . \qquad (3.10)$$

The function $\theta(z, z')$ is defined to be 1 if z is later on the contour than z', and 0 otherwise.[1] From the definition of the time-dependent expectation value in (3.4), it follows that the greater Green function $\boldsymbol{G}^>(z, z')$, where z is later on the contour than z', is

$$\mathrm{i}G^>(\boldsymbol{r}z, \boldsymbol{r}'z') = \frac{\mathrm{Tr}\left\{\mathrm{e}^{\beta\mu\hat{N}}\hat{U}(-\mathrm{i}\beta, 0)\hat{\psi}_H(\boldsymbol{r}, z)\hat{\psi}_H^\dagger(\boldsymbol{r}', z')\right\}}{\mathrm{Tr}\left\{\mathrm{e}^{\beta\mu\hat{N}}\hat{U}(-\mathrm{i}\beta, 0)\right\}} . \qquad (3.11)$$

If z' is later on the contour than z, then the Green function equals

$$\mathrm{i}G^<(\boldsymbol{r}z, \boldsymbol{r}'z') = -\frac{\mathrm{Tr}\left\{\mathrm{e}^{\beta\mu\hat{N}}\hat{U}(-\mathrm{i}\beta, 0)\hat{\psi}_H^\dagger(\boldsymbol{r}', z')\hat{\psi}_H(\boldsymbol{r}, z)\right\}}{\mathrm{Tr}\left\{\mathrm{e}^{\beta\mu\hat{N}}\hat{U}(-\mathrm{i}\beta, 0)\right\}} . \qquad (3.12)$$

The extra minus sign on the right hand side comes from the contour ordering. More generally, rearranging the field operators $\hat{\psi}$ and $\hat{\psi}^\dagger$ (later arguments to the left), we also have to multiply by $(-1)^P$, where P is the parity of the permutation. From the definition of the Green function, it is easily seen that the electron density, $n(\boldsymbol{r}, z) = \langle\hat{\psi}_H^\dagger(\boldsymbol{r}, z)\hat{\psi}_H(\boldsymbol{r}, z)\rangle$ and current are obtained according to

$$n(\boldsymbol{r}, z) = -\mathrm{i}G(\boldsymbol{r}z, \boldsymbol{r}z^+) , \qquad (3.13)$$

and

$$\boldsymbol{j}(\boldsymbol{r}, z) = -\left\{\left[-\mathrm{i}\frac{\nabla}{2} + \mathrm{i}\frac{\nabla'}{2} + \boldsymbol{A}_{\mathrm{ext}}(\boldsymbol{r}, z)\right]\mathrm{i}G(\boldsymbol{r}z, \boldsymbol{r}'z')\right\}_{z'=z^+} . \qquad (3.14)$$

where z^+ indicates that this time-argument is infinitesimally later on the contour.

The Green function $\boldsymbol{G}(z, z')$ obeys an important cyclic relation on the Keldysh contour. Choosing $z = 0_-$, which is the earliest time on the contour, we find $\boldsymbol{G}(0_-, z') = \boldsymbol{G}^<(0, z')$, given by (3.12) with $\hat{\psi}_H(\boldsymbol{r}, 0) = \hat{\psi}(\boldsymbol{r})$.

[1] This means that if z is parametrized according to $z(s)$, where the parameter s runs from linearly from s_i to s_f, then $\theta(z, z') = \theta(s - s')$.

Inside the trace we can move $\hat{\psi}(\boldsymbol{r})$ to the left. Furthermore, we can exchange the position of $\hat{\psi}(\boldsymbol{r})$ and $\exp\{\beta\mu\hat{N}\}$ by noting that $\hat{\psi}(\boldsymbol{r})\exp\{\beta\mu\hat{N}\} = \exp\{\beta\mu(\hat{N}+1)\}\hat{\psi}(\boldsymbol{r})$. Using the group identity $\hat{U}(-\mathrm{i}\beta,0)\hat{U}(0,-\mathrm{i}\beta) = 1$, we obtain

$$\begin{aligned} \mathrm{i}G(\boldsymbol{r}\,0_-,\boldsymbol{r}'z') &= -\frac{\mathrm{Tr}\left\{\hat{\psi}_H(\boldsymbol{r})\mathrm{e}^{\beta\mu\hat{N}}\hat{U}(-\mathrm{i}\beta,0)\hat{\psi}^\dagger_H(\boldsymbol{r}',z')\right\}}{\mathrm{Tr}\left\{\mathrm{e}^{\beta\mu\hat{N}}\hat{U}(-\mathrm{i}\beta,0)\right\}} \\ &= -\mathrm{e}^{\beta\mu}\frac{\mathrm{Tr}\left\{\mathrm{e}^{\beta\mu\hat{N}}\hat{U}(-\mathrm{i}\beta,0)\hat{\psi}_H(\boldsymbol{r},-\mathrm{i}\beta)\hat{\psi}^\dagger_H(\boldsymbol{r}',z')\right\}}{\mathrm{Tr}\left[\mathrm{e}^{\beta\mu\hat{N}}\hat{U}(-\mathrm{i}\beta,0)\right]} . \end{aligned} \tag{3.15}$$

The r.h.s. equals $-\mathrm{e}^{\beta\mu}\langle\boldsymbol{r}|\mathrm{i}\boldsymbol{G}(-\mathrm{i}\beta,z')|\boldsymbol{r}'\rangle$. Together with a similar analysis for $\boldsymbol{G}(z,0_-)$, we conclude that

$$\boldsymbol{G}(0_-,z') = -\mathrm{e}^{\beta\mu}\boldsymbol{G}(-\mathrm{i}\beta,z') \quad \text{and} \quad \boldsymbol{G}(z,0_-) = -\mathrm{e}^{-\beta\mu}\boldsymbol{G}(z,-\mathrm{i}\beta) . \tag{3.16}$$

These equations constitute the so called Kubo-Martin-Schwinger (KMS) boundary conditions [Kubo 1957, Martin 1959]. From the definition of the Green function in (3.9), it is easily seen that the $\boldsymbol{G}(z,z)$ has a discontinuity at $z = z'$,

$$\boldsymbol{G}^>(z,z) = \boldsymbol{G}^<(z,z) - \mathrm{i}\mathbf{1} . \tag{3.17}$$

Furthermore, for both time-arguments on the real axis we have the important symmetry $\left[\boldsymbol{G}^{\lessgtr}(t',t)\right]^\dagger = -\boldsymbol{G}^{\lessgtr}(t,t')$. As we shall see, these relations play a crucial role in solving the equation of motion.

3.4 The Keldysh Book-Keeping

The Green function belongs to a larger class of functions of two time-contour variables that we will refer to as Keldysh space. These are functions that can be written on the form

$$k(z,z') = \delta(z,z')k^\delta(z) + \theta(z,z')k^>(z,z') + \theta(z',z)k^<(z,z') , \tag{3.18}$$

where the δ-function on the contour is defined as $\delta(z,z') = \mathrm{d}\theta(z,z')/\mathrm{d}z$.[2] The Green function, as defined in (3.10), has no such singular part. Another example of a function belonging to the Keldysh space, is the self-energy Σ, which will be discussed below. The singular part, Σ^δ, of the self-energy is the Hartree-Fock self-energy, while the terms $\Sigma^{\lessgtr}$ represent the correlation part.

[2] In general, functions containing singularities of the form $\mathrm{d}^n\delta(z,z')/\mathrm{d}z^n$ belong to the Keldysh space (see [Danielewicz 1984]). Notice that if the contour variable z is parametrized according to $z(s)$, where the parameter s runs linearly from some value s_i to s_f, we have $\delta(z,z') = [\mathrm{d}z/\mathrm{d}s]^{-1}\mathrm{d}\Theta(s-s')/\mathrm{d}s = [\mathrm{d}z/\mathrm{d}s]^{-1}\delta(s-s')$.

The functions in Keldysh space are somewhat complicated due to the fact that each of the time-arguments can be located on three different branches of the contour, as illustrated in Fig. 3.1. Below we systematically derive a set of identities that are commonly used for dealing with such functions and will be used extensively in the following sections. Most of the relations are well known [Langreth 1976], while others, equally important [Wagner 1991], are not. Our aim is to provide a self-contained derivation of all of them. A table at the end of the section summarizes the main results. For those who are not familiar with the Keldysh contour, we strongly recommend to scan what follows with pencil and paper.

It is straightforward to show that if $a(z, z')$ and $b(z, z')$ belong to the Keldysh space, then

$$c(z, z') = \int_\gamma \mathrm{d}\bar{z}\; a(z, \bar{z}) b(\bar{z}, z') \tag{3.19}$$

also belongs to the Keldysh space. For any $k(z, z')$ in the Keldysh space we define the *greater* and *lesser* functions on the physical time axis

$$k^>(t, t') \equiv k(t_+, t'_-), \quad k^<(t, t') \equiv k(t_-, t'_+) \,. \tag{3.20}$$

We also define the following two-point functions with one argument t on the physical time axis and the other τ on the vertical track

$$k^\rceil(t, \tau) \equiv k(t_\pm, \tau), \quad k^\lceil(\tau, t) \equiv k(\tau, t_\pm) \,. \tag{3.21}$$

In the definition of $k^\rceil$ and $k^\lceil$ we can arbitrarily choose t_+ or t_- since τ is later than both of them. The symbols "$\rceil$" and "$\lceil$" have been chosen in order to help the visualization of the time arguments. For instance, "$\rceil$" has a horizontal segment followed by a vertical one; correspondingly, $k^\rceil$ has a first argument which is real (and thus lies on the horizontal axis) and a second argument which is imaginary (and thus lies on the vertical axis). We will also use the convention of denoting the real time with latin letters and the imaginary time with greek letters.

If we write out the contour integral in (3.19) in detail, we see – with the help of Fig. 3.1 – that the integral consists of four main parts. First, we must integrate along the real axis from $\bar{z} = 0_-$ to $\bar{z} = t'_-$, for which $a = a^>$ and $b = b^<$. Then, the integral goes from $\bar{z} = t'_-$ to $\bar{z} = t_+$, where $a = a^>$ and $b = b^>$. The third part of the integral goes along the real axis from $\bar{z} = t_+$ to $\bar{z} = 0_+$, with $a = a^<$ and $b = b^>$. The last integral is along the imaginary track, from 0_+ to $-\mathrm{i}\beta$, where $a = a^\rceil$ and $b = b^\lceil$. In addition, we have the contribution from the singular parts, a^δ and b^δ, which is trivial since these integrals involve a δ-function. With these specifications, we can drop the "$\pm$" subscripts on the time-arguments and write

$$
\begin{aligned}
c^>(t,t') = {} & a^>(t,t')b^\delta(t') + a^\delta(t)b^>(t,t') \\
& + \int_0^{t'} \mathrm{d}\bar{t}\; a^>(t,\bar{t})b^<(\bar{t},t') + \int_{t'}^{t} \mathrm{d}\bar{t}\; a^>(t,\bar{t})b^>(\bar{t},t') \\
& + \int_t^{0} \mathrm{d}\bar{t}\; a^<(t,\bar{t})b^>(\bar{t},t') + \int_0^{-\mathrm{i}\beta} \mathrm{d}\bar{\tau}\; a^\rceil(t,\bar{\tau})b^\lceil(\bar{\tau},t') \,. \qquad (3.22)
\end{aligned}
$$

The second integral on the r.h.s. is an ordinary integral on the real axis of two well defined functions and may be rewritten as

$$
\int_{t'}^{t} \mathrm{d}\bar{t}\; a^>(t,\bar{t})b^>(\bar{t},t') = \int_{t'}^{0} \mathrm{d}\bar{t}\; a^>(t,\bar{t})b^>(\bar{t},t') + \int_0^{t} \mathrm{d}\bar{t}\; a^>(t,\bar{t})b^>(\bar{t},t') \,. \quad (3.23)
$$

Using this relation, the expression for $c^>$ becomes

$$
\begin{aligned}
c^>(t,t') = {} & a^>(t,t')b^\delta(t') + a^\delta(t)b^>(t,t') - \int_0^{t'} \mathrm{d}\bar{t}\; a^>(t,\bar{t})[b^>(\bar{t},t') - b^<(\bar{t},t')] \\
& + \int_0^{t} \mathrm{d}\bar{t}\; [a^>(t,\bar{t}) - a^<(t,\bar{t})]b^>(\bar{t},t') + \int_0^{-\mathrm{i}\beta} \mathrm{d}\bar{\tau}\; a^\rceil(t,\bar{\tau})b^\lceil(\bar{\tau},t') \,. \qquad (3.24)
\end{aligned}
$$

Next, we introduce two other functions on the physical time axis

$$
k^{\mathrm{R}}(t,t') \equiv \delta(t,t')k^\delta + \theta(t-t')[k^>(t,t') - k^<(t,t')] \,, \qquad (3.25a)
$$

$$
k^{\mathrm{A}}(t,t') \equiv \delta(t,t')k^\delta - \theta(t'-t)[k^>(t,t') - k^<(t,t')] \,. \qquad (3.25b)
$$

The *retarded* function $k^{\mathrm{R}}(t,t')$ vanishes for $t < t'$, while the *advanced* function $k^{\mathrm{A}}(t,t')$ vanishes for $t > t'$. The retarded and advanced functions can be used to rewrite (3.24) in a more compact form

$$
c^>(t,t') = \int_0^{\infty} \mathrm{d}\bar{t}\; [a^>(t,\bar{t})b^{\mathrm{A}}(\bar{t},t') + a^{\mathrm{R}}(t,\bar{t})b^>(\bar{t},t')] + \int_0^{-\mathrm{i}\beta} \mathrm{d}\bar{\tau}\; a^\rceil(t,\bar{\tau})b^\lceil(\bar{\tau},t') \,. \qquad (3.26)
$$

It is convenient to introduce a short hand notation for integrals along the physical time axis and for those between 0 and $-\mathrm{i}\beta$. The symbol "$\cdot$" will be used to write $\int_0^\infty \mathrm{d}\bar{t}\; f(\bar{t})g(\bar{t})$ as $f \cdot g$, while the symbol "$\star$" will be used to write $\int_0^{-\mathrm{i}\beta} \mathrm{d}\bar{\tau}\; f(\bar{\tau})g(\bar{\tau})$ as $f \star g$. Then

$$
c^> = a^> \cdot b^{\mathrm{A}} + a^{\mathrm{R}} \cdot b^> + a^\rceil \star b^\lceil \,. \qquad (3.27)
$$

Similarly, one can prove that

$$
c^< = a^< \cdot b^{\mathrm{A}} + a^{\mathrm{R}} \cdot b^< + a^\rceil \star b^\lceil \,. \qquad (3.28)
$$

Equations (3.27)–(3.28) can be used to extract the retarded and advanced component of c. By definition

$$
\begin{aligned}
c^{\mathrm{R}}(t,t') &= \delta(t-t')c^{\delta}(t) + \theta(t-t')[c^{>}(t,t') - c^{<}(t,t')] \\
&= a^{\delta}(t)b^{\delta}(t')\delta(t-t') + \theta(t-t')\int_0^{\infty} \mathrm{d}\bar{t}\, a^{\mathrm{R}}(t,\bar{t})[b^{>}(\bar{t},t') - b^{<}(\bar{t},t')] \\
&\quad + \theta(t-t')\int_0^{\infty} \mathrm{d}\bar{t}\, [a^{>}(t,\bar{t}) - a^{<}(t,\bar{t})]b^{\mathrm{A}}(\bar{t},t') \,. \qquad (3.29)
\end{aligned}
$$

Using the definitions (3.25a) and (3.25b) to expand the integrals on the r.h.s. of this equation, it is straightforward to show that

$$
c^{\mathrm{R}} = a^{\mathrm{R}} \cdot b^{\mathrm{R}} \,. \qquad (3.30)
$$

Proceeding along the same lines, one can show that the advanced component is given by $c^{\mathrm{A}} = a^{\mathrm{A}} \cdot b^{\mathrm{A}}$. It is worth noting that in the expressions for c^{R} and c^{A} no integration along the imaginary track is required.

Next, we show how to extract the components $c^{\rceil}$ and $c^{\lceil}$. We first define the Matsubara function $k^{\mathrm{M}}(\tau,\tau')$ with both arguments in the interval $(0,-\mathrm{i}\beta)$:

$$
k^{\mathrm{M}}(\tau,\tau') \equiv k(z=\tau, z'=\tau') \,. \qquad (3.31)
$$

Let us focus on $k^{\rceil}$. Without any restrictions we may take t_- as the first argument in (3.21). In this case, we find

$$
\begin{aligned}
c^{\rceil}(t,\tau) &= a^{\delta}(t)b^{\rceil}(t,\tau) + \int_{0_-}^{t_-} \mathrm{d}\bar{z}\, a^{>}(t_-,\bar{z})b^{<}(\bar{z},\tau) \\
&\quad + \int_{t_+}^{0_+} \mathrm{d}\bar{z}\, a^{<}(t_-,\bar{z})b^{<}(\bar{z},\tau) + \int_{0_+}^{-\mathrm{i}\beta} \mathrm{d}\bar{z}\, a^{<}(t_-,\bar{z})b(\bar{z},\tau) \,. \qquad (3.32)
\end{aligned}
$$

Converting the contour integrals in integrals along the real time axis and along the imaginary track, and taking into account the definition (3.25a)

$$
c^{\rceil} = a^{\mathrm{R}} \cdot b^{\rceil} + a^{\rceil} \star b^{\mathrm{M}} \,. \qquad (3.33)
$$

The relation for $c^{\lceil}$ can be obtained in a similar way and reads $c^{\lceil} = a^{\lceil} \cdot b^{\mathrm{A}} + a^{\mathrm{M}} \star b^{\lceil}$. Finally, it is straightforward to prove that the Matsubara component of c is simply given by $c^{\mathrm{M}} = a^{\mathrm{M}} \star b^{\mathrm{M}}$.

There is another class of identities we want to discuss for completeness. We have seen that the convolution (3.19) of two functions belonging to the Keldysh space also belongs to the Keldysh space. The same holds true for the product

$$
c(z,z') = a(z,z')b(z',z) \,. \qquad (3.34)
$$

Omitting the arguments of the functions, one readily finds (for $z \neq z'$)

$$
c^{>} = a^{>}b^{<}, \quad c^{<} = a^{<}b^{>}, \quad c^{\rceil} = a^{\rceil}b^{\lceil}, \quad c^{\lceil} = a^{\lceil}b^{\rceil}, \quad c^{\mathrm{M}} = a^{\mathrm{M}}b^{\mathrm{M}} \,. \qquad (3.35)
$$

The retarded function is then obtained exploiting the identities (3.35). We have (for $t \neq t'$)

Table 3.1. Table of definitions of Keldysh functions and identities for the convolution and the product of two functions in the Keldysh space

Definition	$c(z,z') = \int_\gamma \mathrm{d}\bar{z}\, a(z,\bar{z})b(\bar{z},z')$	$c(z,z') = a(z,z')b(z',z)$
$k^>(t,t') = k(t_+,t'_-)$	$c^> = a^> \cdot b^A + a^R \cdot b^> + a^\rceil \star b^\lceil$	$c^> = a^> b^<$
$k^<(t,t') = k(t_-,t'_+)$	$c^< = a^< \cdot b^A + a^R \cdot b^< + a^\rceil \star b^\lceil$	$c^< = a^< b^>$
$k^R(t,t') = \delta(t-t')k^\delta(t) + \theta(t-t')[k^>(t,t') - k^<(t,t')]$	$c^R = a^R \cdot b^R$	$c^R = \begin{cases} a^R b^< + a^< b^A \\ a^R b^> + a^> b^A \end{cases}$
$k^A(t,t') = \delta(t-t')k^\delta(t) - \theta(t'-t)[k^>(t,t') - k^<(t,t')]$	$c^A = a^A \cdot b^A$	$c^A = \begin{cases} a^A b^< + a^< b^R \\ a^A b^> + a^> b^R \end{cases}$
$k^\rceil(t,\tau) = k(t_\pm,\tau)$	$c^\rceil = a^R \cdot b^\rceil + a^\rceil \star b^M$	$c^\rceil = a^\rceil b^\lceil$
$k^\lceil(\tau,t) = k(\tau,t_\pm)$	$c^\lceil = a^\lceil \cdot b^A + a^M \star b^\lceil$	$c^\lceil = a^\lceil b^\rceil$
$k^M(\tau,\tau') = k(z=\tau, z'=\tau')$	$c^M = a^M \star b^M$	$c^M = a^M b^M$

$$c^R(t,t') = \theta(t-t')[a^>(t,t')b^<(t',t) - a^<(t,t')b^>(t',t)] \,. \tag{3.36}$$

We may get rid of the θ-function by adding and subtracting $a^< b^<$ or $a^> b^>$ to the above relation and rearranging the terms. The final result is

$$c^R = a^R b^< + a^< b^A = a^R b^> + a^> b^A \,. \tag{3.37}$$

Similarly one finds $c^A = a^A b^< + a^< b^R = a^A b^> + a^> b^R$. The time-ordered and anti-time-ordered functions can be obtained in a similar way and the Reader can look at Table 3.1 for the complete list of definitions and identities.

For later purposes, we also consider the case of a Keldysh function $k(z,z')$ multiplied on the left by a scalar function $l(z)$. The scalar function is equivalent to the singular part of a function belonging to Keldysh space, $\tilde{l}(z,z') = l(z)\delta(z,z')$, meaning that $\tilde{l}^{R/A} = \tilde{l}^M = \tilde{l}$ and $\tilde{l}^\lessgtr = \tilde{l}^\rceil = \tilde{l}^\lceil = 0$. Using Table 3.1, one immediately realizes that the function l is simply a prefactor: $\int_\gamma \mathrm{d}\bar{z}\, \tilde{l}(z,\bar{z})k^x(\bar{z},z') = l(z)k^x(z,z')$, where x is one of the Keldysh components ($\lessgtr$, R, A, $\rceil$, $\lceil$, M). The same is true for $\int_\gamma \mathrm{d}\bar{z}\, k^x(z,\bar{z})\tilde{r}(\bar{z},z') = k^x(z,z')r(z')$, where $\tilde{r}(z,z') = r(z)\delta(z,z')$ and $r(z)$ is a scalar function.

3.5 The Kadanoff-Baym Equations

The Green function, as defined in (3.10), satisfies the equation of motion

$$\mathrm{i}\frac{\mathrm{d}}{\mathrm{d}z}\boldsymbol{G}(z,z') = \mathbf{1}\delta(z,z') + \boldsymbol{h}(z)\boldsymbol{G}(z,z') + \int_\gamma \mathrm{d}\bar{z}\, \boldsymbol{\Sigma}(z,\bar{z})\boldsymbol{G}(\bar{z},z') \,, \tag{3.38}$$

as well as the adjoint equation

$$-\mathrm{i}\frac{\mathrm{d}}{\mathrm{d}z'}\boldsymbol{G}(z,z') = \mathbf{1}\delta(z,z') + \boldsymbol{G}(z,z')\boldsymbol{h}(z') + \int_\gamma \mathrm{d}\bar{z}\, \boldsymbol{G}(z,\bar{z})\boldsymbol{\Sigma}(\bar{z},z') \,. \tag{3.39}$$

The external potential is included in $\boldsymbol{h}$, while the self-energy $\boldsymbol{\Sigma}$ is a functional of the Green function, and describes the effects of the electron interaction. The self-energy belongs to Keldysh space and can therefore be written in the form $\boldsymbol{\Sigma}(z,z') = \delta(z,z')\boldsymbol{\Sigma}^{\delta}(z) + \theta(z,z')\boldsymbol{\Sigma}^{>}(z,z') + \theta(z',z)\boldsymbol{\Sigma}^{<}(z,z')$. The singular part of the self-energy can be identified with the Hartree–Fock potential, $\boldsymbol{\Sigma}^{\delta}(z) = \mathbf{v}_{\mathrm{H}}(z) + \boldsymbol{\Sigma}_{\mathrm{x}}(z)$. The self-energy obeys the same anti-periodic boundary conditions at $z = 0_-$ and $z = -\mathrm{i}\beta$ as $\boldsymbol{G}$. We will discuss self-energy approximations in more detail below.

Calculating the Green function on the time-contour now consists of two steps: (i) First one has to find the Green function for imaginary times, which is equivalent to finding the equilibrium Matsubara Green function $\boldsymbol{G}^{\mathrm{M}}(\tau,\tau')$. This Green function depends only on the difference between the time-coordinates, and satisfies the KMS boundary conditions according to $\boldsymbol{G}^{\mathrm{M}}(\tau+\mathrm{i}\beta,\tau') = -\mathrm{e}^{\beta\mu N}\boldsymbol{G}^{\mathrm{M}}(\tau,\tau')$. Since the self-energy depends on the Green function, this amounts to solving the finite-temperature Dyson equation to self-consistency. (ii) The Green function with one or two time-variables on the real axis can now be found by propagating according to (3.38) and (3.39). Starting from $t = 0$, this procedure corresponds to extending the time-contour along the real time-axis. The process is illustrated in Fig. 3.2. Writing out the equations for the components of $\boldsymbol{G}$ using Table 3.1, we obtain the equations known as the Kadanoff-Baym equations [Kadanoff 1962],

$$\begin{aligned}\mathrm{i}\frac{\mathrm{d}}{\mathrm{d}t}\boldsymbol{G}^{\lessgtr}(t,t') &= \boldsymbol{h}(t)\boldsymbol{G}^{\lessgtr}(t,t') + [\boldsymbol{\Sigma}^{\mathrm{R}}\cdot\boldsymbol{G}^{\lessgtr}](t,t') + [\boldsymbol{\Sigma}^{\lessgtr}\cdot\boldsymbol{G}^{\mathrm{A}}](t,t') \\ &\quad + [\boldsymbol{\Sigma}^{\rceil}\star\boldsymbol{G}^{\lceil}](t,t')\,, \qquad (3.40\mathrm{a}) \\ -\mathrm{i}\frac{\mathrm{d}}{\mathrm{d}t'}\boldsymbol{G}^{\lessgtr}(t,t') &= \boldsymbol{G}^{\lessgtr}(t,t')\boldsymbol{h}(t') + [\boldsymbol{G}^{\lessgtr}\cdot\boldsymbol{\Sigma}^{\mathrm{A}}](t,t') + [\boldsymbol{G}^{\mathrm{R}}\cdot\boldsymbol{\Sigma}^{\lessgtr}](t,t') \\ &\quad + [\boldsymbol{G}^{\rceil}\star\boldsymbol{\Sigma}^{\lceil}](t,t')\,, \qquad (3.40\mathrm{b})\end{aligned}$$

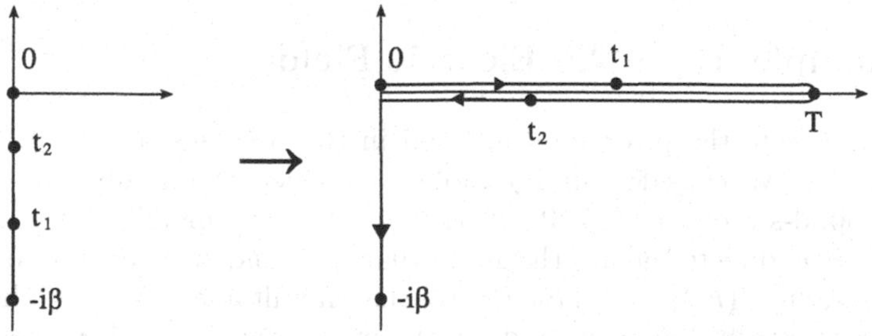

Fig. 3.2. Propagating the Kadanoff-Baym equations means that one first determines the Green function for time-variables along the imaginary track. One then calculates the Green function with one or two variables on an expanding time-contour

$$\mathrm{i}\frac{\mathrm{d}}{\mathrm{d}t}\boldsymbol{G}^{\rceil}(t,\tau) = \boldsymbol{h}(t)\boldsymbol{G}^{\rceil}(t,\tau) + [\boldsymbol{\Sigma}^{\mathrm{R}} \cdot \boldsymbol{G}^{\rceil}](t,\tau) + [\boldsymbol{\Sigma}^{\rceil} \star \boldsymbol{G}^{\mathrm{M}}](t,\tau) \,, \tag{3.40c}$$

$$-\mathrm{i}\frac{\mathrm{d}}{\mathrm{d}t}\boldsymbol{G}^{\lceil}(\tau,t) = \boldsymbol{G}^{\lceil}(\tau,t)\boldsymbol{h}(t) + [\boldsymbol{\Sigma}^{\lceil} \cdot \boldsymbol{G}^{\mathrm{A}}](\tau,t) + [\boldsymbol{\Sigma}^{\mathrm{M}} \star \boldsymbol{G}^{\lceil}](\tau,t) \,. \tag{3.40d}$$

The equations (3.40a) and (3.40c) can both be written on the form

$$\mathrm{i}\frac{\mathrm{d}}{\mathrm{d}t}\boldsymbol{G}^{\times}(t,z') = \boldsymbol{h}^{\mathrm{HF}}(t)\boldsymbol{G}^{\times}(t,z') + \boldsymbol{I}^{\times}(t,z') \,, \tag{3.41}$$

while (3.40b) and (3.40d) can be written as the adjoint equations. The term proportional to $\boldsymbol{h}^{\mathrm{HF}} \equiv \boldsymbol{h} + \boldsymbol{\Sigma}^{\delta}$ describes a free-particle propagation, while $\boldsymbol{I}^{\times}$ is a collision term, which accounts for electron correlation and introduces memory effects and dissipation. Considering the function $\boldsymbol{G}^{\lessgtr}(t,t')$, it is easily seen that if we denote by T the largest of the two time-arguments t and t', then the collision terms $\boldsymbol{I}^{\lessgtr}(t,t')$ depend on $\boldsymbol{G}^{\lessgtr}(t_1,t_2)$, $\boldsymbol{G}^{\lceil}(\tau_1,t_2)$ and $\boldsymbol{G}^{\rceil}(t_1,\tau_2)$ for $t_1, t_2 \leq T$. In other words, given the functions $\boldsymbol{G}^{x}(t,t')$ for time arguments up to T, we can calculate $\boldsymbol{I}^{x}(t,t')$, and consequently find $\boldsymbol{G}^{x}$ for time-arguments $t+\Delta$ and $t'+\Delta$, by a simple time-stepping procedure based on (3.41). The Green function $\boldsymbol{G}^{\lessgtr}(t,t')$ is thus obtained for time-arguments within the expanding square given by $t,t' \leq T$. Simultaneously, one calculates $\boldsymbol{G}^{\rceil}(t,\tau)$ and $\boldsymbol{G}^{\lceil}(\tau,t)$ for $t \leq T$. The resulting $\boldsymbol{G}$ then automatically satisfies the KMS boundary conditions.

When propagating the Kadanoff-Baym equations one therefore starts at $t = t' = 0$, with the initial conditions given by $\boldsymbol{G}^{<}(0,0) = \lim_{\eta\to 0} \boldsymbol{G}^{\mathrm{M}}(0,-\mathrm{i}\eta)$, $\boldsymbol{G}^{>}(0,0) = \lim_{\eta\to 0} \boldsymbol{G}^{\mathrm{M}}(-\mathrm{i}\eta,0)$, $\boldsymbol{G}^{\lceil}(\tau,0) = \boldsymbol{G}^{M}(\tau,0)$ and $\boldsymbol{G}^{\rceil}(0,\tau) = \boldsymbol{G}^{M}(0,\tau)$. As can be seen from (3.40a)–(3.40d), the only contribution to $\boldsymbol{I}^{\times}(0,0)$ comes from terms containing time-arguments on the imaginary axis. These terms therefore contain the effect of initial correlations, since the time-derivative of $\boldsymbol{G}$ would otherwise correspond to that of an uncorrelated system, i.e., $\boldsymbol{I}^{\times}(0,0) = 0$.

3.6 Example: H_2 in An Electric Field

We can illustrate the procedure outlined in the previous section by a simple example. We consider an H_2 molecule, which is initially (at $t = 0$) in its ground-state. At $t = 0$ we then switch on an additional electric field, which is directed along the molecular axis and will remain constant, adding a term $v'(\boldsymbol{r},t) = -zE_0\theta(t)$ to the Hamiltonian. We will here focus on the electron dynamics, and let the nuclei remain fixed in their equilibrium positions. The functions $\boldsymbol{G}$, $\boldsymbol{\Sigma}$, $\boldsymbol{h}$ and $\boldsymbol{I}$ defined in the previous section are all expanded in a molecular orbital basis, and the first step therefore consists of choosing these orbitals, e.g. by performing a Hartree-Fock calculation. The resulting Green function is independent of this choice of orbitals. Given this basis, the Green function is represented on matrix form,

$\langle \boldsymbol{r}|\boldsymbol{G}(z,z')|\boldsymbol{r}'\rangle = \sum_{ij} \varphi_i(\boldsymbol{r}) G_{ij}(z,z')\varphi_j^*(\boldsymbol{r}')$, where the indices i refer to the molecular orbitals $\varphi_i(\boldsymbol{r}) = \langle \boldsymbol{r}|\varphi_i\rangle$. We then solve the Dyson equation for the ground state, when the Hamiltonian (without the additional electric field) is time-independent. The Matsubara Green function only depends on the difference between the two imaginary time-coordinates, and we consequently have to solve the equation[3]

$$\mathrm{i}\frac{\mathrm{d}}{\mathrm{d}\tau}\boldsymbol{G}^M(\tau-\tau') = \mathbf{1}\delta(\tau,\tau') + \boldsymbol{h}\boldsymbol{G}^M(\tau) + \left[\boldsymbol{\Sigma}^M \star \boldsymbol{G}^M\right](\tau-\tau') \tag{3.42}$$

with the anti-periodic boundary condition $\boldsymbol{G}^M(\tau+\mathrm{i}\beta) = -\mathrm{e}^{-\beta\mu}\boldsymbol{G}^M(\tau)$. In this example, we have use the second-order approximation to the self-energy $\boldsymbol{\Sigma}$, as illustrated in Fig. 3.4(b). Since the self-energy depends on the Green function, the Dyson equation should be solved to self-consistency, which can be done with an iterative procedure [Dahlen 2005b, Ku 2002]. The Matsubara Green function itself contains a wealth of information about the ground state system, and quantities such as the energy, ionization potential and the density matrix are readily given.

The time-propagation of the time-dependent matrix equations (3.40a)–(3.40d) is relatively straightforward, the main difficulty rising from the fact that the Green function $G^{\lessgtr}$ has to be stored for all times $t, t' \leq T$. The self-energy approximation used here, is given by the same second-order diagrams that was used for the ground-state calculation. The plots in Fig. 3.3 show the imaginary part of the matrix element $G^<_{\sigma_g,\sigma_g}(t,t')$ calculated for time-variables within the square $t, t' \leq T = 20.0$ a.u., i.e., we have extended the contour in Fig. 3.2 to $T = 20$ a.u. The time-variables are here represented on an even-spaced grid. In the plot to the left, there is no added external potential and the molecule remains in equilibrium. This means that the Green function depends only on the difference $t_2 - t_1$ (for $t_1, t_2 \geq 0$) precisely like the ordinary equilibrium Green functions. Time-propagation without any added time-dependent field can in this way provide us with information about the ground state of the system. For instance, the Fourier transformed Green function $\boldsymbol{G}(\omega) = \int \mathrm{d}(t_1-t_2)\mathrm{e}^{\mathrm{i}\omega(t_1-t_2)}\boldsymbol{G}(t_1-t_2)$ has poles at the ionization potentials and electron affinities of the system [Fetter 1971]. The density matrix at a time t is given by the time-diagonal, $-\mathrm{i}\boldsymbol{G}^<(t,t)$, and one can therefore define time-dependent natural orbitals (and corresponding natural orbital occupation numbers) by diagonalizing the time-dependent density matrix. As the Green function illustrated in Fig. 3.3(a) is largely diagonal in the HF orbital indices, the frequency of the oscillations in the matrix element $G_{\sigma_g,\sigma_g}(t_1,t_2)$ is for this reason practically identical to the first ionization potential of the molecule. Also the value of $-iG^<_{\sigma_g,\sigma_g}(t_1,t_1) = \mathrm{Im}G^<_{\sigma_g,\sigma_g}(t_1,t_1)$ (the Green function is imaginary on the diagonal), which is constant along the diagonal

[3] This equation looks slightly different from how it usually appears in textbooks, e.g. in [Fetter 1971]. The conventional form is obtained by redefining $\boldsymbol{G}^M \to -i\boldsymbol{G}^M$, $\boldsymbol{\Sigma}^M \to -i\boldsymbol{\Sigma}^M$ and $\tau \to -i\tau$. The new quantities are then all real.

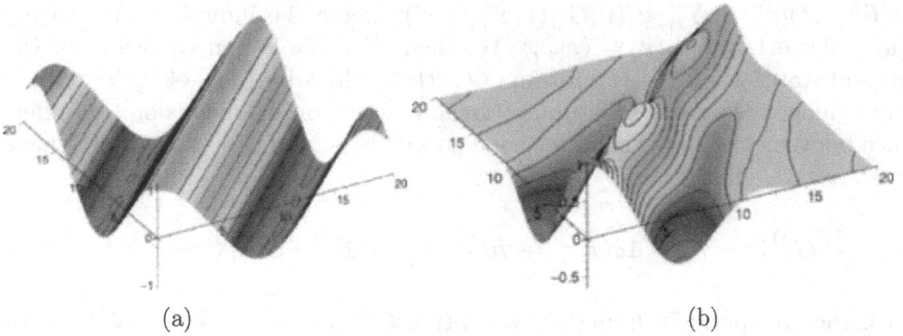

Fig. 3.3. The figures show the Green function $\text{Im}\, G^{<}_{\sigma_g,\sigma_g}(t_1,t_2)$ in the double time-plane, where the matrix indices refer to the groundstate σ_g Hartree–Fock orbital of the molecule. The figure on the *left* shows the system in equilibrium, while the system on the *right* has an additional electric field, $\theta(t)E_0$ along the molecular axis. The times t_1 and t_2 on the axes are given in atomic units

ridge in Fig. 3.3, is almost identical to the occupation number of the $1\sigma_g$ natural orbital.

The figure on the right shows the same matrix element, but now in the presence of an additional electric field which is switched on at $t = 0$. The oscillations along the ridge $t_1 = t_2$ can be interpreted as oscillations in the occupation number. We emphasize that Fig. 3.3 only shows the evolution of one matrix element. To calculate observables from the Green function we must of course take all matrix elements into account.

3.7 Conserving Approximations

In the Dyson-Schwinger equations (3.38) and (3.39), we introduced the electronic self-energy functional $\boldsymbol{\Sigma}$, which accounts for the effects of the electron interaction. The self-energy is a functional of the Green function, and will have to be approximated in practical calculations. Diagrammatic techniques provide a natural scheme for generating approximate self-energies and for systematically improving these approximations. There are no general prescriptions for how to select the relevant diagrams, which means that this selection must be guided by physical intuition. There are, however, important conservation laws, like the number conservation law or the energy conservation law, that should always be obeyed. We will in the following discuss an exact framework for generating such *conserving approximations*.

Let us first discuss the conservation laws obeyed by a system of interacting electrons in an external field given by the electrostatic potential $v_{\text{ext}}(\boldsymbol{r},t)$ and vector potential $\boldsymbol{A}_{\text{ext}}(\boldsymbol{r},t)$. An important relation is provided by the continuity equation

$$\frac{\mathrm{d}}{\mathrm{d}t} n(\boldsymbol{r}, t) + \nabla \cdot \boldsymbol{j}(\boldsymbol{r}, t) = 0 . \tag{3.43}$$

The density and the current density can be calculated from the Green function using (3.13) and (3.14). Whether these quantities will obey the continuity equation will depend on whether the Green function is obtained from a conserving self-energy approximation. If we know the current density we can also calculate the total momentum and angular momentum expectation values in the system from the equations

$$\boldsymbol{P}(t) = \int \mathrm{d}^3 r \, \boldsymbol{j}(\boldsymbol{r}, t) \qquad \text{and} \qquad \boldsymbol{L}(t) = \int \mathrm{d}^3 r \, \boldsymbol{r} \times \boldsymbol{j}(\boldsymbol{r}, t) . \tag{3.44}$$

For these two quantities the following relations should be satisfied

$$\frac{\mathrm{d}}{\mathrm{d}t} \boldsymbol{P}(t) = - \int \mathrm{d}^3 r \, \{ n(\boldsymbol{r}, t) \, \boldsymbol{E}(\boldsymbol{r}, t) + \boldsymbol{j}(\boldsymbol{r}, t) \times \boldsymbol{B}(\boldsymbol{r}, t) \} \tag{3.45a}$$

$$\frac{\mathrm{d}}{\mathrm{d}t} \boldsymbol{L}(t) = - \int \mathrm{d}^3 r \, \{ n(\boldsymbol{r}, t) \boldsymbol{r} \times \boldsymbol{E}(\boldsymbol{r}, t) + \boldsymbol{r} \times [\boldsymbol{j}(\boldsymbol{r}, t) \times \boldsymbol{B}(\boldsymbol{r}, t)] \} . \tag{3.45b}$$

where $\boldsymbol{E}$ and $\boldsymbol{B}$ are the electric and magnetic fields calculated from

$$\boldsymbol{E}(\boldsymbol{r}, t) = \nabla v_{\mathrm{ext}}(\boldsymbol{r}, t) - \frac{\mathrm{d}}{\mathrm{d}t} \boldsymbol{A}_{\mathrm{ext}}(\boldsymbol{r}, t) \qquad \text{and} \qquad \boldsymbol{B}(\boldsymbol{r}, t) = \nabla \times \boldsymbol{A}_{\mathrm{ext}}(\boldsymbol{r}, t) . \tag{3.46}$$

The (3.45a) and (3.45b) tell us that the change in momentum and angular momentum is equal to the total force and total torque on the system. In the absence of external fields these equations express momentum and angular momentum conservation. Since the right hand sides of (3.45a) and (3.45b) can also directly be calculated from the density and the current and therefore from the Green function, we may wonder whether they are satisfied for a given approximation to the Green function.

Finally we consider the case of energy conservation. Let $E(t) = \langle \hat{H}(t) \rangle$ be the energy expectation value of the system, then we have

$$\frac{\mathrm{d}}{\mathrm{d}t} E(t) = - \int \mathrm{d}^3 r \, \boldsymbol{j}(\boldsymbol{r}, t) \cdot \boldsymbol{E}(\boldsymbol{r}, t) . \tag{3.47}$$

This equation tells us that the energy change of the system is equal to the work done on the system. The total energy is calculated from the Green function using the expression

$$E(t) = -\frac{\mathrm{i}}{2} \int \mathrm{d}^3 r \, \langle \boldsymbol{r} | \left[\mathrm{i} \frac{\mathrm{d}}{\mathrm{d}t} + \boldsymbol{h}(t) \right] \boldsymbol{G}^{<}(t, t') | \boldsymbol{r} \rangle \Big|_{t'=t} . \tag{3.48}$$

The question is now whether the energy and the current calculated from an approximate Green function satisfy the relation in (3.47).

Baym and Kadanoff [Baym 1961, Baym 1962] showed that conserving approximations follow immediately if the self-energy is obtained as the functional derivative,

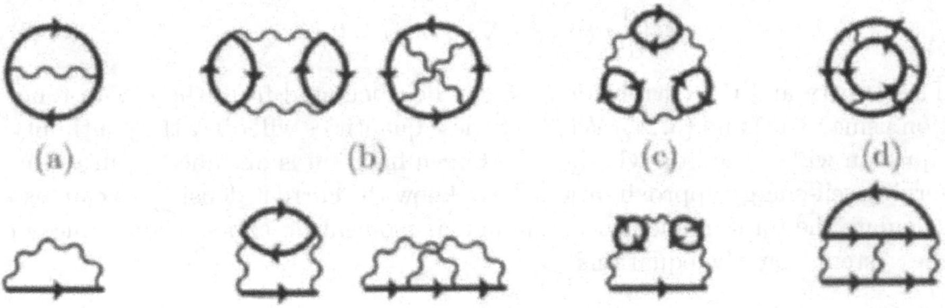

Fig. 3.4. Diagrams for the generating functional $\Phi[G]$, and the corresponding self-energy diagrams. In (**a**) we have the exchange diagram, and (**b**) the second order approximation. The diagrams in (**c**) and (**d**) belong to the GW approximation and the T-matrix approximation respectively

$$\Sigma(1,2) = \frac{\delta\Phi[G]}{\delta G(2,1)} . \tag{3.49}$$

Here, and in the following discussion, we use numbers to denote the contour coordinates, such that $1 = (\boldsymbol{r}_1, z_1)$. A functional $\Phi[G]$ can be constructed, as first shown in a seminal paper by Luttinger and Ward [Luttinger 1960], by summing over irreducible self-energy diagrams closed with an additional Green function line and multiplied by appropriate numerical prefactors,

$$\Phi[G] = \sum_{n,k} \frac{1}{2n} \int \mathrm{d}\bar{1} \int \mathrm{d}\bar{2}\, \Sigma_n^{(k)}(\bar{1},\bar{2}) G(\bar{2},\bar{1}) . \tag{3.50}$$

In this summation, $\Sigma_n^{(k)}$ denotes a self-energy diagram of n-th order, i.e., containing n interaction lines. The time-integrals go along the contour, but the rules for constructing Feynman diagrams are otherwise exactly the same as those in the ground-state formalism [Fetter 1971]. Notice that the functional derivative in (3.49) may generate other self-energy diagrams in addition to those used in the construction of $\Phi[G]$ in (3.50). In Fig. 3.4 we show some examples of typical $\Phi[G]$ diagrams. Examples of Φ-derivable approximations include Hartree–Fock, the second order approximation (also known as the second Born approximation), the GW approximation and the T-matrix approximation.

When the Green function is calculated from a conserving approximation, the resulting observables agree with the conservation laws of the underlying Hamiltonian, as given in (3.43), (3.45a), (3.45b), and (3.47). This guarantees the conservation of particles, energy, momentum, and angular momentum. All these conservation laws follow from the invariance of $\Phi[G]$ under specific changes in G. We will here only outline the principles of the proofs, without going into the details, which can be found in [Baym 1961, Baym 1962].

- *Number conservation* follows from the gauge invariance of $\Phi[G]$. A gauge transformation $\boldsymbol{A}_{\mathrm{ext}}(1) \to \boldsymbol{A}_{\mathrm{ext}}(1) + \nabla\Lambda(1)$, where $\Lambda(\boldsymbol{r}, 0_-) = \Lambda(\boldsymbol{r}, -\mathrm{i}\beta)$

leaves $\Phi[G]$ unchanged. A consequence of the gauge invariance is that a pure gauge cannot induce a change in the density or current. The invariance is therefore closely related to the Ward-identities and to the f-sum rule for the density response function [van Leeuwen 2004a].

- *Momentum conservation* follows from the invariance of $\Phi[G]$ under spatial translations, $\boldsymbol{r} \to \boldsymbol{r} + \boldsymbol{R}(z)$. The invariance is a consequence of the electron interaction $v(1,2) = \delta(z_1, z_2)/|\boldsymbol{r}_1 - \boldsymbol{r}_2|$ being instantaneous and only depending on the difference between the spatial coordinates.
- *Angular momentum conservation* follows from the invariance of $\Phi[G]$ under a rotation of the spatial coordinates.
- *Energy conservation* follows from the invariance of $\Phi[G]$ when described by an observer using a "rubbery clock", measuring time according to the function $s(z)$. The invariance relies on the electron interaction being instantaneous.

3.8 Noninteracting Electrons

In this section we focus on noninteracting electrons. This is particularly relevant for TDDFT, where the electrons are described by the noninteracting Kohn-Sham system. While the Kohn-Sham Green function differs from the true Green function, they both produce the same time-dependent density. This is important since the density is not only an important observable in, e.g., quantum transport, but also since the density is the central ingredient in TDDFT. The use of NEGFs in TDDFT is therefore important due to the relation between v_{xc} and the self-energy.

For a system of noninteracting electrons $\hat{V}_{\mathrm{ee}} = 0$ and it is straightforward to show that the Green function obeys the equations of motion (3.38) and (3.39), with $\boldsymbol{\Sigma} = 0$. For any $z \neq z'$, the equations of motion can be solved by using the evolution operator on the contour,

$$\boldsymbol{U}(z, z') = \hat{\mathcal{T}}_{\mathrm{c}} \left\{ \mathrm{e}^{-\mathrm{i}\int_{z'}^{z} \mathrm{d}\bar{z}\; \boldsymbol{h}(\bar{z})} \right\} , \tag{3.51}$$

which solves $\mathrm{i}\frac{\mathrm{d}}{\mathrm{d}z}\boldsymbol{U}(z,z') = \boldsymbol{h}(z)\boldsymbol{U}(z,z')$ and $-\mathrm{i}\frac{\mathrm{d}}{\mathrm{d}z'}\boldsymbol{U}(z,z') = \boldsymbol{U}(z,z')\boldsymbol{h}(z')$. Therefore, any Green function

$$\boldsymbol{G}(z,z') = \theta(z,z')\boldsymbol{U}(z,0_-)\boldsymbol{f}^{>}\boldsymbol{U}(0_-,z') + \theta(z',z)\boldsymbol{U}(z,0_-)\boldsymbol{f}^{<}\boldsymbol{U}(0_-,z') , \tag{3.52}$$

satisfying the constraint (3.17) on the form

$$\boldsymbol{f}^{>} - \boldsymbol{f}^{<} = -\mathrm{i}\mathbf{1} , \tag{3.53}$$

is a solution of (3.38)–(3.39). In order to fix the matrix $\boldsymbol{f}^{>}$ or $\boldsymbol{f}^{<}$ we impose the KMS boundary conditions. The matrix $\boldsymbol{h}(z) = \boldsymbol{h}_0$ for any z on the vertical track, meaning that $\boldsymbol{U}(-\mathrm{i}\beta, 0_-) = \mathrm{e}^{-\beta \boldsymbol{h}_0}$. Equation (3.16) then implies

$\boldsymbol{f}^< = -\mathrm{e}^{-\beta(\boldsymbol{h}_0-\mu)}\boldsymbol{f}^>$, and taking into account the constraint (3.53) we conclude that

$$\boldsymbol{f}^< = \frac{\mathrm{i}}{\mathrm{e}^{\beta(\boldsymbol{h}_0-\mu)}+1} = \mathrm{i}f(\boldsymbol{h}_0)\,, \tag{3.54}$$

where $f(\omega) = 1/[\mathrm{e}^{\beta(\omega-\mu)}+1]$ is the Fermi distribution function. The matrix $\boldsymbol{f}^>$ takes the form $\boldsymbol{f}^> = \mathrm{i}[f(\boldsymbol{h}_0) - \mathbf{1}]$.

The Green function $\boldsymbol{G}(z, z')$ for a system of noninteracting electrons is now completely fixed. Both $\boldsymbol{G}^>$ and $\boldsymbol{G}^<$ depend on the initial distribution function $f(\boldsymbol{h}_0)$, as it should according to the discussion of Sect. 3.3. Another way of writing $-\mathrm{i}\boldsymbol{G}^<$ is in terms of the eigenstates $|\varphi_n\rangle \equiv |\varphi_n(0)\rangle$ of $\boldsymbol{h}_0$ with eigenvalues ε_n. From the time-evolved eigenstate $|\varphi_n(t)\rangle = \boldsymbol{U}(t,0)|\varphi_n\rangle$ we can calculate the time-dependent wavefunction $\varphi_n(\boldsymbol{r},t) = \langle\boldsymbol{r}|\varphi_n(t)\rangle$. Inserting $\sum_n |\varphi_n(0)\rangle\langle\varphi_n(0)|$ in the expression for $\boldsymbol{G}^<$ we find

$$\begin{aligned} -\mathrm{i}G^<(\boldsymbol{r}t,\boldsymbol{r}'t') &= -\mathrm{i}\sum_{m,n}\langle\boldsymbol{r}|\boldsymbol{U}(t,0)|\varphi_m\rangle\langle\varphi_m|\boldsymbol{f}^<|\varphi_n\rangle\langle\varphi_n|\boldsymbol{U}(0,t')|\boldsymbol{r}'\rangle \\ &= \sum_n f(\varepsilon_n)\varphi_n(\boldsymbol{r},t)\varphi_n^*(\boldsymbol{r}',t')\,, \end{aligned} \tag{3.55}$$

which for $t = t'$ reduces to the time-dependent density matrix. The Green function $\boldsymbol{G}^>$ becomes

$$-\mathrm{i}G^>(\boldsymbol{r}t,\boldsymbol{r}'t') = -\sum_n[1 - f(\varepsilon_n)]\varphi_n(\boldsymbol{r},t)\varphi_n^*(\boldsymbol{r}',t')\,. \tag{3.56}$$

Knowing the greater and lesser Green functions we can also calculate $\boldsymbol{G}^{\mathrm{R,A}}$. By definition we have

$$\boldsymbol{G}^{\mathrm{R}}(t,t') = \theta(t-t')[\boldsymbol{G}^>(t,t') - \boldsymbol{G}^<(t,t')] = -\mathrm{i}\theta(t-t')\boldsymbol{U}(t,t')\,, \tag{3.57}$$

and similarly

$$\boldsymbol{G}^{\mathrm{A}}(t,t') = \mathrm{i}\theta(t'-t)\boldsymbol{U}(t,t') = [\boldsymbol{G}^{\mathrm{R}}(t',t)]^\dagger\,. \tag{3.58}$$

In the above expressions the Fermi distribution function has disappeared. The information carried by $\boldsymbol{G}^{\mathrm{R,A}}$ is the same contained in the one-particle evolution operator. There is no information on how the system is prepared (how many particles, how they are distributed, etc). We use this observation to rewrite $\boldsymbol{G}^{\lessgtr}$ in terms of $\boldsymbol{G}^{\mathrm{R,A}}$

$$\boldsymbol{G}^{\lessgtr}(t,t') = \boldsymbol{G}^{\mathrm{R}}(t,0)\boldsymbol{G}^{\lessgtr}(0,0)\boldsymbol{G}^{\mathrm{A}}(0,t')\,. \tag{3.59}$$

Thus, $\boldsymbol{G}^{\lessgtr}$ is completely known once we know how to propagate the one-electron orbitals in time and how they are populated before the system is perturbed [Blandin 1976, Cini 1980, Stefanucci 2004a]. We also observe that an analogous relation holds for $\boldsymbol{G}^{\rceil,\lceil}$

$$G^{\rceil}(t,\tau) = \mathrm{i}G^{\mathrm{R}}(t,0)G^{\rceil}(0,\tau), \quad G^{\lceil}(\tau,t) = -\mathrm{i}G^{\lceil}(\tau,0)G^{\mathrm{A}}(0,\tau)\,. \tag{3.60}$$

Let us now focus on a special kind of disturbance, namely $h(t) = h_0 + \theta(t)h_1$. In this case

$$G^{\mathrm{R}}(t,t') = -\mathrm{i}\theta(t-t')\mathrm{e}^{-\mathrm{i}(h_0+h_1)(t-t')} \tag{3.61}$$

depends only on the difference between the time arguments. Let us define the Fourier transform of $G^{\mathrm{R,A}}$ from

$$G^{\mathrm{R,A}}(t,t') = \int \frac{\mathrm{d}\omega}{2\pi}\,\mathrm{e}^{-\mathrm{i}\omega(t-t')}G^{\mathrm{R,A}}(\omega)\,. \tag{3.62}$$

The step function can be written as $\theta(t-t') = \int \frac{\mathrm{d}\omega}{2\pi\mathrm{i}} \frac{\exp\{\mathrm{i}\omega(t-t')\}}{\omega-\mathrm{i}\eta}$, with η an infinitesimally small positive constant. Substituting this representation of the θ-function into (3.61) and shifting the ω variable one readily finds

$$G^{\mathrm{R}}(\omega) = \frac{1}{\omega - h_0 - h_1 + \mathrm{i}\eta}\,, \tag{3.63}$$

and therefore $G^{\mathrm{R}}(\omega)$ is analytic in the upper half plane. On the other hand, from (3.58) it follows that $G^{\mathrm{A}}(\omega) = [G^{\mathrm{R}}(\omega)]^{\dagger}$ is analytic in the lower half plane. What can we say about the greater and lesser components? Do they also depend only on the difference $t-t'$? The answer to the latter question is negative. Indeed, we recall that they contain information on how the system is prepared before h_1 is switched on. In particular the original eigenstates are eigenstates of h_0 and in general are not eigenstates of the Hamiltonian $h_0 + h_1$ at positive times. From (3.59) one can see that $G^{\lessgtr}(t,t')$ cannot be expressed only in terms of the time difference $t-t'$. For instance

$$G^{<}(t,t') = \mathrm{e}^{-\mathrm{i}(h_0+h_1)t}\;\mathrm{i}f(h_0)\;\mathrm{e}^{\mathrm{i}(h_0+h_1)t'}\,, \tag{3.64}$$

and, unless h_0 and h_1 commute, it is a function of t and t' separately.

It is sometimes useful to split $h(t)$ in two parts[4] and treat one of them perturbatively. Let us think, for instance, of a system composed of two connected subsystems $A+B$. In case we know how to calculate the Green function of the isolated subsystems A and B, it is convenient to treat the connecting part as a perturbation. Thus, we write $h(t) = \mathcal{E}(t) + \mathcal{V}(t)$, and we define g as the Green function when $\mathcal{V} = 0$. The function g is a solution of

$$\left\{\mathrm{i}\frac{\mathrm{d}}{\mathrm{d}z} - \mathcal{E}(z)\right\} g(z,z') = \mathbf{1}\delta(z,z')\,, \tag{3.65}$$

and of the corresponding adjoint equation of motion. Furthermore, the Green function g obeys the KMS boundary conditions. With these we can use g to convert the equations of motion for G into integral equations

[4] This can be done using projection operators. See [Stefanucci 2004a].

$$\boldsymbol{G}(z,z') = \boldsymbol{g}(z,z') + \int_\gamma \mathrm{d}\bar{z}\, \boldsymbol{g}(z,\bar{z})\boldsymbol{\mathcal{V}}(\bar{z})\boldsymbol{G}(\bar{z},z')$$
$$= \boldsymbol{g}(z,z') + \int_\gamma \mathrm{d}\bar{z}\, \boldsymbol{G}(z,\bar{z})\boldsymbol{\mathcal{V}}(\bar{z})\boldsymbol{g}(\bar{z},z') ; \tag{3.66}$$

the integral on $\bar{z}$ is along the generalized Keldysh contour of Fig. 3.1. One can easily check that this $\boldsymbol{G}$ satisfies both (3.38) and (3.39). $\boldsymbol{G}$ also obeys the KMS boundary conditions since the integral equation is defined on the contour of Fig. 3.1.

In order to get some familiarity with the above perturbation scheme, we consider explicitly the system $A + B$ already mentioned. We partition the one-electron Hilbert space in states of the subsystem A and states of the subsystem B. The "unperturbed" system is described by $\boldsymbol{\mathcal{E}}$, while the connecting part by $\boldsymbol{\mathcal{V}}$ and

$$\boldsymbol{\mathcal{E}} = \begin{bmatrix} \boldsymbol{\mathcal{E}}_{AA} & 0 \\ 0 & \boldsymbol{\mathcal{E}}_{BB} \end{bmatrix}, \qquad \boldsymbol{\mathcal{V}} = \begin{bmatrix} 0 & \boldsymbol{\mathcal{V}}_{AB} \\ \boldsymbol{\mathcal{V}}_{BA} & 0 \end{bmatrix}. \tag{3.67}$$

Taking into account that $\boldsymbol{g}$ has no off-diagonal matrix elements, the Green function projected on one of the two subsystems, e.g., $\boldsymbol{G}_{BB}$, is

$$\boldsymbol{G}_{BB}(z,z') = \boldsymbol{g}_{BB}(z,z') + \int_\gamma \mathrm{d}\bar{z}\, \boldsymbol{g}_{BB}(z,\bar{z})\boldsymbol{\mathcal{V}}_{BA}(\bar{z})\boldsymbol{G}_{AB}(\bar{z},z') \tag{3.68}$$

and

$$\boldsymbol{G}_{AB}(z,z') = \int_\gamma \mathrm{d}\bar{z}\, \boldsymbol{g}_{AA}(z,\bar{z})\boldsymbol{\mathcal{V}}_{AB}(\bar{z})\boldsymbol{G}_{BB}(\bar{z},z') . \tag{3.69}$$

Substituting this latter equation into the first one, we obtain a closed equation for $\boldsymbol{G}_{BB}$:

$$\boldsymbol{G}_{BB}(z,z') = \boldsymbol{g}_{BB}(z,z') + \int_\gamma \mathrm{d}\bar{z} \int \mathrm{d}\bar{z}' \boldsymbol{g}_{BB}(z,\bar{z})\boldsymbol{\Sigma}_{BB}(\bar{z},\bar{z}')\boldsymbol{G}_{BB}(\bar{z}',z') , \tag{3.70}$$

with

$$\boldsymbol{\Sigma}_{BB}(\bar{z},\bar{z}') = \boldsymbol{\mathcal{V}}_{BA}(\bar{z})\boldsymbol{g}_{AA}(\bar{z},\bar{z}')\boldsymbol{\mathcal{V}}_{AB}(\bar{z}') \tag{3.71}$$

the embedding self-energy. The retarded and advanced component can now be easily computed. With the help of Table 3.1 one finds

$$\boldsymbol{G}_{BB}^{\mathrm{R,A}} = \boldsymbol{g}_{BB}^{\mathrm{R,A}} + \boldsymbol{g}_{BB}^{\mathrm{R,A}} \cdot \boldsymbol{\Sigma}_{BB}^{\mathrm{R,A}} \cdot \boldsymbol{G}_{BB}^{\mathrm{R,A}} . \tag{3.72}$$

Next, we have to compute the lesser or greater component. As for the retarded and advanced components, this can be done starting from (3.70). The reader can soon realize that the calculation is rather complicated, due to the mixing of pure real-time functions with functions having one real time argument and one imaginary time argument, see Table 3.1. Below, we use (3.59) as a feasible short-cut. A closed equation for the retarded and advanced component has been already obtained. Thus, we simply need an equation for $\boldsymbol{G}^{\lessgtr}(0,0)$. Let us focus on the lesser component $\boldsymbol{G}^{<}(0,0) = \mathrm{i}\boldsymbol{f}^{<}$. Assuming that the Hamiltonian $\boldsymbol{h}_0$ is hermitian, the matrix $(\omega - \boldsymbol{h}_0)^{-1}$ has poles at frequencies equal to the eigenvalues of $\boldsymbol{h}_0$. These poles are all on the real frequency axis, and we can therefore write

$$\boldsymbol{G}^{<}(0,0) = \mathrm{i}f(\boldsymbol{h}_0) = \int_\gamma \frac{\mathrm{d}\zeta}{2\pi} f(\zeta)\frac{1}{\zeta - \boldsymbol{h}_0} , \tag{3.73}$$

where the contour γ encloses the real frequency axis.

3.9 Action Functional and TDDFT

We define the action functional

$$\tilde{A} = \mathrm{i} \ln \mathrm{Tr}\left\{ \mathrm{e}^{\beta\mu\hat{N}} \hat{U}(-\mathrm{i}\beta, 0) \right\} , \tag{3.74}$$

where the evolution operator $\hat{U}$ is the same as defined in (3.3). The action functional is a tool for generating equations of motion, and is not interesting *per se*. Nevertheless, one should notice that the action, as defined in (3.74) has a numerical value equal to $\tilde{A} = \mathrm{i}\ln Z$, where Z is the thermodynamic partition function. In the zero temperature limit, we then have $\lim_{\beta\to\infty} i\tilde{A}/\beta = E - \mu N$.

It is easy to show that if we make a perturbation $\delta\hat{V}(z)$ in the Hamiltonian, the change in the evolution operator is given by

$$\mathrm{i}\frac{\mathrm{d}}{\mathrm{d}z}\delta\hat{U}(z,z') = \delta\hat{V}(z)\hat{U}(z,z') + \hat{H}(z)\delta U(z,z') . \tag{3.75}$$

A similar equation for the dependence on z', and the boundary condition $\delta\hat{U}(z,z) = 0$ gives

$$\delta\hat{U}(z,z') = -\mathrm{i}\int_{z'}^{z} \mathrm{d}\bar{z}\, \hat{U}(z,\bar{z})\delta\hat{V}(\bar{z})\hat{U}(\bar{z},z') . \tag{3.76}$$

We stress that the time-coordinates are on a contour going from 0 to $-\mathrm{i}\beta$. The variation in, e.g., $V(t_+)$ is therefore independent of the variation in $V(t_-)$. If we let $\delta\hat{V}(z) = \int \mathrm{d}^3 r\, \delta v(\boldsymbol{r},z)\hat{n}(\boldsymbol{r})$, a combination of (3.74) and (3.76) yields [compare to (3.4)] the expectation values of the density,

$$\begin{aligned}\frac{\delta\tilde{A}}{\delta v(\boldsymbol{r},z)} &= \frac{\mathrm{i}}{\mathrm{Tr}\left\{\mathrm{e}^{\beta\mu\hat{N}}\hat{U}(-\mathrm{i}\beta,0)\right\}} \frac{\delta}{\delta v(\boldsymbol{r},z)} \mathrm{Tr}\left\{\mathrm{e}^{\beta\mu N}\hat{U}(-\mathrm{i}\beta,0)\right\} \\ &= \frac{\mathrm{Tr}\left\{\mathrm{e}^{\beta\mu\hat{N}}\hat{U}(-\mathrm{i}\beta,0)\hat{U}(0,z)\hat{n}(\boldsymbol{r})\hat{U}(z,0)\right\}}{\mathrm{Tr}\left\{\mathrm{e}^{\beta\mu\hat{N}}\hat{U}(-\mathrm{i}\beta,0)\right\}} = n(\boldsymbol{r},z) . \end{aligned} \tag{3.77}$$

A physical potential is the same on the positive and on the negative branch of the contour, and the same is true for the corresponding time-dependent density, $n(\boldsymbol{r},t) = n(\boldsymbol{r},t_\pm)$. A density response function defined for time-arguments on the contour is found by taking the functional derivative of the density with respect to the external potential. Using the compact notation $1 = (\boldsymbol{r}_1, z_1)$, the response function is written

$$\chi(1,2) = \frac{\delta n(1)}{\delta v(2)} = \frac{\delta^2 \tilde{A}}{\delta v(1)\delta v(2)} = \chi(2,1)\,. \tag{3.78}$$

This response function is symmetric in the space and time-contour coordinates. We again stress that the variations in the potentials at t_+ and t_- are independent. If, however, one uses this response function to calculate the density response to an actual physical perturbing electric field, we obtain

$$\delta n(\boldsymbol{r},t) = \delta n(\boldsymbol{r},t_\pm) = \int_\gamma \mathrm{d}z' \int \mathrm{d}^3r'\, \chi(\boldsymbol{r}t_\pm, \boldsymbol{r}'z')\delta v(\boldsymbol{r}', z')\,, \tag{3.79}$$

where γ indicates an integral along the contour. In this expression, the perturbing potential (as well as the induced density response) is independent of whether it is located on the positive or negative branch, i.e., $\delta v(\boldsymbol{r}', t'_\pm) = \delta v(\boldsymbol{r}', t')$. We consider a perturbation of a system initially in equilibrium, which means that $\delta v(\boldsymbol{r}', t') \neq 0$ only for $t' > 0$, and we can therefore ignore the integral along the imaginary track of the time-contour. The contour integral then consists of two parts: (i) First an integral from $t' = 0$ to $t' = t$, in which $\chi = \chi^>$, and (ii) an integral from $t' = t$ to $t' = 0$, where $\chi = \chi^<$. Writing out the contour integral in (3.79) explicitly then gives

$$\begin{aligned}\delta n(\boldsymbol{r},t) &= \int_0^t \mathrm{d}t' \int \mathrm{d}^3r' \left[\chi^>(\boldsymbol{r}t, \boldsymbol{r}'t') - \chi^<(\boldsymbol{r}t, \boldsymbol{r}'t')\right] \delta v(\boldsymbol{r}', t') \\ &= \int_0^\infty \mathrm{d}t' \int \mathrm{d}^3r'\, \chi^{\mathrm{R}}(\boldsymbol{r}t, \boldsymbol{r}'t')\delta v(\boldsymbol{r}', t')\,.\end{aligned} \tag{3.80}$$

The response to a perturbing field is therefore given by the retarded response function, while $\chi(1,2)$ defined on the contour is symmetric in $(1 \leftrightarrow 2)$.

If we now consider a system of noninteracting electrons in some external potential v_{KS}, we can similarly define a noninteracting action-functional $\tilde{A}_{\mathrm{KS}}$. The steps above can be repeated to calculate the noninteracting response function. The derivation is straightforward, and gives

$$\chi_{\mathrm{KS}}(1,2) = \frac{\delta^2 \tilde{A}_{\mathrm{KS}}}{\delta v_{\mathrm{KS}}(1)\delta v_{\mathrm{KS}}(2)} = -\mathrm{i}G_{\mathrm{KS}}(1,2)G_{\mathrm{KS}}(2,1)\,. \tag{3.81}$$

The noninteracting Green function G_{KS} has the form given in (3.52), (3.55) and (3.56). The retarded response-function is

$$\begin{aligned}\chi^{\mathrm{R}}_{\mathrm{KS}}(\boldsymbol{r}_1t_1,\boldsymbol{r}_2t_2) &= -\mathrm{i}\theta(t_1-t_2)\left[G^{>}_{\mathrm{KS}}(\boldsymbol{r}_1t_1,\boldsymbol{r}_2t_2)G^{<}_{\mathrm{KS}}(\boldsymbol{r}_2t_2,\boldsymbol{r}_1t_1)\right.\\ &\qquad\left.-G^{<}_{\mathrm{KS}}(\boldsymbol{r}_1t_1,\boldsymbol{r}_2t_2)G^{>}_{\mathrm{KS}}(\boldsymbol{r}_2t_2,\boldsymbol{r}_1t_1)\right]\\ &= \mathrm{i}\theta(t_1-t_2)\sum_{n,m}[f(\varepsilon_m)-f(\varepsilon_n)]\\ &\qquad\times\varphi_n(\boldsymbol{r}_1,t_1)\varphi^*_m(\boldsymbol{r}_1,t_1)\varphi_m(\boldsymbol{r}_2,t_2)\varphi^*_n(\boldsymbol{r}_2,t_2)\,,\end{aligned} \tag{3.82}$$

where we have used (3.55) and (3.56) in the last step.

Having defined the action functional for both the interacting and the noninteracting systems, we now make a Legendre transform, and define

$$A[n] = -\tilde{A}[v] + \int \mathrm{d}1\, n(1)v(1)\,, \tag{3.83}$$

which has the property that $\delta A[n]/\delta n(1) = v(1)$. We also observe that the functional $A_{v_0}[n] = A[n] - \int \mathrm{d}1\, n(1)v_0(1)$, where v_0 is a fixed potential, is variational in the sense that

$$\frac{\delta A_{v_0}[n]}{\delta n(1)} = v(1) - v_0(1) = 0 \tag{3.84}$$

when $v = v_0$. This equation can be used as a basis for a variational principle in TDDFT [von Barth 2005]. Similar to the Legendre transform in (3.83), we define the action functional

$$A_{\mathrm{KS}}[n] = -\tilde{A}_{\mathrm{KS}}[v_{\mathrm{KS}}] + \int \mathrm{d}1\, n(1)v_{\mathrm{KS}}(1)\,. \tag{3.85}$$

with the property $\delta A_{\mathrm{KS}}[n]/\delta n(1) = v_{\mathrm{KS}}(1)$. The Legendre transforms assume the existence of a one-to-one correspondence between the density and the potential. From these action functionals, we now define the exchange-correlation part to be

$$A_{\mathrm{xc}}[n] = A_{\mathrm{KS}}[n] - A[n] - \frac{1}{2}\int \mathrm{d}1 \int \mathrm{d}2\, \delta(z_1,z_2)\frac{n(1)n(2)}{|\boldsymbol{r}_1-\boldsymbol{r}_2|}\,. \tag{3.86}$$

Taking the functional derivative with respect to the density gives

$$v_{\mathrm{KS}}[n](1) = v(1) + v_{\mathrm{H}}(1) + v_{\mathrm{xc}}[n](1) \tag{3.87}$$

where $v_{\mathrm{H}}(1)$ is the Hartree potential and $v_{\mathrm{xc}}(1) = \delta A_{\mathrm{xc}}/\delta n(1)$. Again, for time-arguments on the real axis, these potentials are independent of whether the time is on the positive or the negative branch. If we, however, want to calculate the response function from the action functional, then it is indeed important which part of the contour the time-arguments are located on.

As mentioned in the beginning in the section, we can make a connection to ground state DFT if we restrict ourselves to a time-independent Hamiltonian. In that case, the Kohn-Sham action takes the numerical value

$\lim_{\beta\to\infty} \mathrm{i}\tilde{A}_{\mathrm{KS}}/\beta = \sum_{i=1}^{N}(\epsilon_i - \mu) = T_{\mathrm{KS}}[n] + \int \mathrm{d}^3r\, n(\boldsymbol{r})v_{\mathrm{KS}}(\boldsymbol{r}) - \mu N$. Using $i/\beta \int \mathrm{d}1\, n(1)v_{\mathrm{KS}}(1) = \int \mathrm{d}^3r\, n(\boldsymbol{r})v_{\mathrm{KS}}(\boldsymbol{r})$, we can obtain, for a fixed potential v_0,

$$E_{v_0}[n] - \mu N = -\lim_{\beta\to\infty} \frac{\mathrm{i}}{\beta} A_{v_0}[n] = T_{\mathrm{KS}}[n] + \int \mathrm{d}^3r\, n(\boldsymbol{r})v_0(\boldsymbol{r}) + \frac{1}{2}\int \mathrm{d}^3r \int \mathrm{d}^3r'\, \frac{n(\boldsymbol{r})n(\boldsymbol{r}')}{|\boldsymbol{r}-\boldsymbol{r}'|} + \lim_{\beta\to\infty} \frac{\mathrm{i}}{\beta} A_{\mathrm{xc}} - \mu N \quad (3.88)$$

from which we identify the relation

$$E_{\mathrm{xc}}[n] = \lim_{\beta\to\infty} \frac{\mathrm{i}}{\beta} A_{\mathrm{xc}}[n] \,. \quad (3.89)$$

As an example, we can consider the ALDA action functional defined according to

$$A_{\mathrm{xc}}^{\mathrm{ALDA}}[n] = \int \mathrm{d}1\, n(1)e_{\mathrm{xc}}(n(1)) \quad (3.90)$$

where e_{xc} is the exchange-correlation energy density. The resulting energy expression is

$$E_{\mathrm{xc}}^{\mathrm{LDA}}[n] = \lim_{\beta\to\infty} \frac{\mathrm{i}}{\beta} \int_0^{-\mathrm{i}\beta} d\tau \int \mathrm{d}^3r\, n(\boldsymbol{r})e_{\mathrm{xc}}(n(\boldsymbol{r})) = \int \mathrm{d}^3r\, n(\boldsymbol{r})e_{\mathrm{xc}}(n(\boldsymbol{r})) \,. \quad (3.91)$$

We mention that much more sophisticated approximations to the exchange-correlation action functional can be derived using Green function techniques [von Barth 2005].

We already described how to define response function on the contour, both in the interacting (3.78) and the noninteracting (3.81) case. Given the exact Kohn-Sham potential, the TDDFT response function should give exactly the same density change as the exact response function,

$$\delta n(1) = \int \mathrm{d}2\, \chi(1,2)\delta v(2) = \int \mathrm{d}2\, \chi_{\mathrm{KS}}(1,2)\delta v_{\mathrm{KS}}(2) \,. \quad (3.92)$$

The change in the Kohn-Sham potential is given by

$$\begin{aligned}\delta v_{\mathrm{KS}}(1) &= \delta v(1) + \int \mathrm{d}2\, \frac{\delta v_{\mathrm{H}}(1)}{\delta n(2)}\delta n(2) + \int \mathrm{d}2\, \frac{\delta v_{\mathrm{xc}}(1)}{\delta n(2)}\delta n(2) \\ &= \delta v(1) + \int \mathrm{d}2\, f_{\mathrm{Hxc}}(1,2)\delta n(2) \\ &= \delta v(1) + \int \mathrm{d}2 \int \mathrm{d}3\, f_{\mathrm{Hxc}}(1,2)\chi(2,3)\delta v(3) \,, \end{aligned} \quad (3.93)$$

where $f_{\mathrm{Hxc}}(1,2) = \delta(z_1,z_2)/|\boldsymbol{r}_1 - \boldsymbol{r}_2| + \delta v_{\mathrm{xc}}(1)/\delta n(2)$. Inserted in (3.92), we obtain

$$\chi(1,2) = \chi_{\mathrm{KS}}(1,2) + \int \mathrm{d}3 \int \mathrm{d}4\, \chi_{\mathrm{KS}}(1,3) f_{\mathrm{Hxc}}(3,4) \chi(4,2) \,. \tag{3.94}$$

This is the response function defined for time-arguments on the contour. If we want to calculate the response induced by a perturbing potential, the density change will be given by the retarded response function. Using Table 3.1, we can just write down

$$\chi^{\mathrm{R}}(\boldsymbol{r}_1 t_1, \boldsymbol{r}_2 t_2) = \chi^{\mathrm{R}}_{\mathrm{KS}}(\boldsymbol{r}_1 t_1, \boldsymbol{r}_2 t_2) + \int \mathrm{d}t_3 \int \mathrm{d}t_4 \int \mathrm{d}^3 r_3 \int \mathrm{d}^3 r_4$$
$$\chi^{\mathrm{R}}_{\mathrm{KS}}(\boldsymbol{r}_1 t_1, \boldsymbol{r}_3 t_3) f^{\mathrm{R}}_{\mathrm{Hxc}}(\boldsymbol{r}_3 t_3, \boldsymbol{r}_4 t_4) \chi^{\mathrm{R}}(\boldsymbol{r}_4 t_4, \boldsymbol{r}_2 t_2) \,. \tag{3.95}$$

The time-integrals in the last expression go from 0 to ∞. As expected, only the retarded functions are involved in this expression. We stress the important result that while the function $f_{\mathrm{Hxc}}(1,2)$ is symmetric under the coordinate-permutation $(1 \leftrightarrow 2)$, it is the retarded function

$$f^{\mathrm{R}}_{\mathrm{Hxc}}(\boldsymbol{r}_1 t_1, \boldsymbol{r}_2 t_2) = \frac{\delta(t_1, t_2)}{|\boldsymbol{r}_1 - \boldsymbol{r}_2|} + f^{\mathrm{R}}_{\mathrm{xc}}(\boldsymbol{r}_1 t_1, \boldsymbol{r}_2 t_2) \,, \tag{3.96}$$

which is used when calculating the response to a perturbing potential.

3.10 Example: Time-Dependent OEP

We will close this section by discussing the time-dependent optimized effective potential (TDOEP) method in the exchange-only approximation. This is a useful example of how to use functions on the Keldysh contour. While the TDOEP equations can be derived from an action functional, we use here the time-dependent Sham-Schlüter equations as starting point [van Leeuwen 1996]. This equation is derived by employing a Kohn-Sham Green function, $G_{\mathrm{KS}}(1,2)$ which satisfies the equation of motion

$$\left\{ \mathrm{i}\frac{\mathrm{d}}{\mathrm{d}z_1} + \frac{\nabla_1^2}{2} - v_{\mathrm{KS}}(\boldsymbol{r}_1, z_1) \right\} G_{\mathrm{KS}}(\boldsymbol{r}_1 z_1, \boldsymbol{r}_2 z_2) = \delta(z_1, z_2) \delta(\boldsymbol{r}_1 - \boldsymbol{r}_2) \,, \tag{3.97}$$

as well as the adjoint equation. The Kohn-Sham Green function is given by (3.55) and (3.56) in terms of the time-dependent Kohn-Sham orbitals. Comparing (3.97) to the Dyson-Schwinger (3.38), we see that we can write an integral equation for the interacting Green function in terms of the Kohn-Sham quantities,

$$G(1,2) = G_{\mathrm{KS}}(1,2) +$$
$$\int \mathrm{d}\bar{1} \int \mathrm{d}\bar{2}\, G_{\mathrm{KS}}(1,\bar{1}) \left\{ \Sigma(\bar{1},\bar{2}) + \delta(\bar{1},\bar{2}) [v_{\mathrm{ext}}(\bar{1}) - v_{\mathrm{KS}}(\bar{1})] \right\} G(\bar{2},2) \,. \tag{3.98}$$

It is important to keep in mind that this integral equation for $G(1,2)$ differs from the differential equations (3.38) and (3.39) in the sense that we have imposed the boundary conditions of G_{KS} on G in (3.98). This means that if $G_{\mathrm{KS}}(1,2)$ satisfies the KMS boundary conditions (3.16), then so will $G(1,2)$.

If we now assume that for any density $n(1) = -\mathrm{i}G(1,1^+)$ there is a potential $v_{\mathrm{KS}}(1)$ such that $n(1) = -\mathrm{i}G_{\mathrm{KS}}(1,1^+)$, we obtain the time-dependent Sham-Schlüter equation,

$$\int \mathrm{d}\bar{1} \int \mathrm{d}\bar{2}\, G_{\mathrm{KS}}(1,\bar{1})\Sigma(\bar{1},\bar{2})G(\bar{2},1) = \int \mathrm{d}\bar{1}\, G_{\mathrm{KS}}(1,\bar{1})[v_{\mathrm{KS}}(\bar{1}) - v_{\mathrm{ext}}(\bar{1})]G(\bar{1},1)\,. \tag{3.99}$$

This equation is formally correct, but not useful in practice since solving it would involve first calculating the nonequilibrium Green function. Instead, one sets $G = G_{\mathrm{KS}}$ and $\Sigma[G] = \Sigma[G_{\mathrm{KS}}]$. For a given self-energy functional, we then have an integral equation for the Kohn-Sham equation. This equation is known as the time-dependent OEP equation. Defining $\Sigma = v_H + \Sigma_{\mathrm{xc}}$ and $v_{\mathrm{KS}} = v_{\mathrm{ext}} + v_{\mathrm{H}} + v_{\mathrm{xc}}$, the TDOEP equation can be written

$$\int \mathrm{d}\bar{1} \int \mathrm{d}\bar{2}\, G_{\mathrm{KS}}(1,\bar{1})\Sigma_{\mathrm{xc}}[G_{\mathrm{KS}}](\bar{1},\bar{2})G_{\mathrm{KS}}(\bar{2},1) = \int \mathrm{d}\bar{1}\, G_{\mathrm{KS}}(1,\bar{1})v_{\mathrm{xc}}(\bar{1})G_{\mathrm{KS}}(\bar{1},1)\,. \tag{3.100}$$

In the simplest approximation, Σ_{xc} is given by the exchange-only self-energy of Fig. 3.4a,

$$\Sigma_{\mathrm{x}}(1,2) = \mathrm{i}G^<_{\mathrm{KS}}(1,2)v_{\mathrm{ee}}(1,2) = -\sum_j n_j \varphi_j(1)\varphi_j^*(2)v_{\mathrm{ee}}(1,2) \tag{3.101}$$

where n_j is the occupation number. This approximation leads to what is known as the exchange-only TDOEP equations [Ullrich 1995a, Ullrich 1995b, Görling 1997] (see Chap. 9). Since the exchange self-energy Σ_{x} is local in time, there is only one time-integration in (3.100). The x-only solution for the potential will be denoted v_{x}. With the notation $\tilde{\Sigma}(3,4) = \Sigma_{\mathrm{x}}(\boldsymbol{r}_3 t_3, \boldsymbol{r}_4 t_3) - \delta(\boldsymbol{r}_3 - \boldsymbol{r}_4)v_{\mathrm{x}}(\boldsymbol{r}_3 t_3)$ we obtain from (3.100)

$$\begin{aligned} 0 = &\int_0^{t_1} \mathrm{d}t_3 \int \mathrm{d}^3 r_3 \int \mathrm{d}^3 r_4 \Big[G^<_{\mathrm{KS}}(1,3)\tilde{\Sigma}(3,4)G^>_{\mathrm{KS}}(4,1) - G^>_{\mathrm{KS}}(1,3)\tilde{\Sigma}(3,4)G^<_{\mathrm{KS}}(4,1)\Big] \\ &+ \int_0^{-\mathrm{i}\beta} \mathrm{d}t_3 \int \mathrm{d}^3 r_3 \int \mathrm{d}^3 r_4\, G^{\rceil}_{\mathrm{KS}}(1,3)\tilde{\Sigma}(3,4)G^{\lceil}_{\mathrm{KS}}(4,1)\,. \end{aligned} \tag{3.102}$$

Let us first work out the last term which describes a time-integral from 0 to $-\mathrm{i}\beta$. On this part of the contour, the Kohn-Sham Hamiltonian is time-independent, with $v_{\mathrm{x}}(\boldsymbol{r},0) \equiv v_{\mathrm{x}}(\boldsymbol{r})$, and $\varphi_i(\boldsymbol{r},t) = \varphi_i(\boldsymbol{r})\exp(-\mathrm{i}\varepsilon_i t)$. Since Σ_{x} is time-independent on this part of the contour, we can integrate

$$\int_0^{-\mathrm{i}\beta} \mathrm{d}t_3\, G^{\rceil}_{\mathrm{KS}}(1, \boldsymbol{r}_3 t_3) G^{\lceil}_{\mathrm{KS}}(\boldsymbol{r}_4 t_3, 1)$$
$$= -\mathrm{i} \sum_{i,k} n_i(1 - n_k) \varphi_i(1) \varphi_i^*(\boldsymbol{r}_3) \varphi_k(\boldsymbol{r}_4) \varphi_k^*(1) \frac{\mathrm{e}^{\beta(\varepsilon_i - \varepsilon_k)} - 1}{\varepsilon_i - \varepsilon_k} . \quad (3.103)$$

If we then use $n_i(1 - n_k)(\mathrm{e}^{\beta(\varepsilon_i - \varepsilon_k)} - 1) = n_k - n_i$ and define the function $u_{\mathrm{x},j}$ by

$$u_{\mathrm{x},j}(1) = -\frac{1}{\varphi_j^*(1)} \sum_k n_k \int \mathrm{d}2\, \varphi_j^*(2) \varphi_k(2) \varphi_k^*(1) v_{\mathrm{ee}}(1, 2) , \quad (3.104)$$

we obtain from (3.103) and (3.101)

$$i \int_0^{-\mathrm{i}\beta} \mathrm{d}t_3 \int \mathrm{d}^3 r_3 \int \mathrm{d}^3 r_4\, G^{\rceil}_{\mathrm{KS}}(1, 3) \tilde{\Sigma}(3, 4) G^{\lceil}_{\mathrm{KS}}(4, 1)$$
$$= -\int \mathrm{d}^3 r_2 \sum_j n_j \sum_{k \neq j} \frac{\varphi_j^*(\boldsymbol{r}_2) \varphi_k(\boldsymbol{r}_2)}{\varepsilon_j - \varepsilon_k} \varphi_j(1) \varphi_k^*(1) \left[u_{\mathrm{x},j}(\boldsymbol{r}_2) - v_{\mathrm{x}}(\boldsymbol{r}_2) \right] + \text{c.c.}$$
$$(3.105)$$

The integral along the real axis on the l.h.s. of (3.102) can similarly be evaluated. Collecting our results we obtain the OEP equations on the same form as in [Görling 1997],

$$0 = \mathrm{i} \sum_j \sum_{k \neq j} n_j \int_0^{t_1} \mathrm{d}t_2 \int \mathrm{d}^3 r_2 \left[v_{\mathrm{x}}(2) - u_{\mathrm{x},j}(2) \right] \varphi_j(1) \varphi_j^*(2) \varphi_k^*(1) \varphi_k(2) + \text{c.c.}$$
$$+ \sum_j \sum_{k \neq j} n_j \frac{\varphi_j(1) \varphi_k^*(1)}{\varepsilon_j - \varepsilon_k} \int \mathrm{d}^3 r_2\, \varphi_j^*(\boldsymbol{r}_2) \left[v_{\mathrm{x}}(\boldsymbol{r}_2) - u_{\mathrm{x},j}(\boldsymbol{r}_2) \right] \varphi_k(\boldsymbol{r}_2) + \text{c.c.} .$$
$$(3.106)$$

The last term represents the initial conditions, expressing that the system is in thermal equilibrium at $t = 0$. The equations have exactly the same form if the initial condition is specified at some other initial time t_0. The second term on the r.h.s. can be written as a time-integral from $-\infty$ to 0 if one introduces a convergence factor. In that case the remaining expression equals the one given in [van Leeuwen 1996, Ullrich 1995a, Ullrich 1995b]. The OEP equation (3.106) in the so-called KLI approximation have been successfully used by Ullrich et al. [Ullrich 1995b] to calculate properties of atoms in strong laser fields (see Chap. 24).

4 Initial-State Dependence and Memory

N.T. Maitra

4.1 Introduction

In ground-state DFT, the fact that the xc potential is a functional of the density is a direct consequence of the one-to-one mapping between ground-state densities and potentials. In TDDFT, the one-to-one mapping is between densities and potentials for a given initial-state. This means that the potentials, most generally, are functionals of the initial state of the system, as well as of the density; and, not just of the instantaneous density, but of its entire history. These dependences are explicitly displayed in (1.33) of Chap. 1. Of particular interest is the xc potential, as that is the quantity that must be approximated. This is the difference between the KS potential and the sum of the external and Hartree potentials. The Hartree potential has no memory, as the classical Coulomb interaction depends on the instantaneous density only, but since both the interacting and non-interacting mappings can depend on the initial state, the xc potential must be a functional of both the initial states and density.

We use the term memory to refer to this dependence on quantities at earlier times: there is memory due to initial-state dependence, and memory due to the history-dependence of the density.

In a sense, memory arises in TDDFT because of the reduced nature of the density as a basic variable: If the wavefunction of the system was known, there would be no memory dependence, since the wavefunction at time t contains the complete information about the system at time t, from which we can determine any observable. The density however traces out much of the information, desirably reducing the description in terms of $3N$ spatial variables plus time to a description in terms of 3 variables plus time. Analogously to the theories of open systems, out of this tracing out of degrees of freedom emerges memory dependence. In the treatment of open systems one traces out the bath degrees of freedom in order to get a reduced description in terms of system variables only: the effect of the bath is embodied in an influence functional that is non-local in time. Much like in open system theory with a low-dimensional bath, the TDDFT memory of early history persists at long times: time does not wash it away (as it would if we were tracing out a bath of a continuous spectrum).

N.T. Maitra: *Initial-State Dependence and Memory*, Lect. Notes Phys. **706**, 61–74 (2006)
DOI 10.1007/3-540-35426-3_4

Even before Runge and Gross (RG) formally established their theory [Runge 1984], there were calculations of optical spectra that plugged the instantaneous density into the LDA [Ando 1977a, Ando 1977b, Zangwill 1980a, Zangwill 1981]. This is an adiabatic approximation, as discussed in Sect. 1.6, and neglects any memory dependence. Since the inception of the RG theorem, there have been attempts to develop functionals with some memory dependence, with varying degrees of success and applicability. The earliest and simplest was the Gross-Kohn approximation (GK) for the xc kernel [Gross 1985, Gross 1985, Gross 1990]. Non-local dependence in the time domain translates into non-constant frequency dependence when a time-frequency Fourier transform is done. Considering densities that are slowly varying in space, GK bootstraps the local density approximation to finite frequencies, i.e., the frequency-dependent kernel is approximated as the homogeneous electron gas response at finite frequency. In the mid-nineties, it was realized, however, that a theory that depends on the density non-locally in time must also depend on it non-locally in space; otherwise, exact conditions are violated [Vignale 1995a, Dobson 1994a]. In particular, the GK approximation violates the harmonic potential theorem (see also Chap. 5). Vignale and Kohn, in 1996, showed that a theory local in space and non-local in time, is possible in terms of the current density [Vignale 1996, Vignale 1997]. Their functional, discussed in Chap. 5, has begun to be tested on a variety of systems; some of these are discussed in Part IV of this book. Most of these applications involve response properties, but the extension to the non-linear regime by Vignale, Ullrich and Conti [Vignale 1997], has recently been studied in quantum wells [Wijewardane 2005]. The idea that memory is locally carried by the electron "fluid", in a Lagrangian framework, was also exploited by Dobson, Bünner, and Gross [Dobson 1997]. In 2003, Tokatly and Pankratov [Tokatly 2003] extended the hydrodynamic formulation to the fully spatially- and time-nonlocal regime, using Landau Fermi-liquid theory (see Chap. 8). Most recently, Tokatly [Tokatly 2005a, Tokatly 2005b] further developed this, formulating TDDFT based on exact quantum many-body dynamics in the co-moving Lagrangian frame (see also Chap. 8). Also, Kurzweil and Baer [Kurzweil 2004] have formulated a theory based on a "memory action functional".

Today, almost all applications of TDDFT utilize an adiabatic approximation, where the instantaneous density is used as input for the "ground-state" density. For example, in the ubiquitous ALDA, $v_{\rm xc}^{\rm ALDA}(\boldsymbol{r},t) = v_{\rm xc}^{\rm unif}[n(\boldsymbol{r},t)]$. Certainly, an adiabatic approximation will work well if the system is slowly-varying enough that the system remains in a slowly-evolving ground-state, but this is hardly the typical case in dynamics.

Yet, excitation energies are generally well approximated (to tenths of an eV) when an adiabatic approximation is used in the linear response formalism. An exception are states of double-excitation character, where indeed it is the lack of memory dependence that dooms adiabatic TDDFT. Part of the reason for the success of adiabatic TDDFT in many excitations is that the KS eigenvalues are themselves reasonably close to the true excitations. Species in intense fields are a different story (see Part V), where memory dependence can play a vital role.

The remainder of this chapter will be investigating memory properties of the *exact* functional. Understanding how the exact functional behaves should prove a useful tool in constructing accurate approximations.

4.2 History Dependence: An Example

Consider beginning a system in its ground state, which we will assume to be non-degenerate. By virtue of the Hohenberg-Kohn theorem, a non-degenerate ground-state is a functional of its own density: initial-state dependence need not be explicitly included in the functional, provided that the functional space is reduced to that where the initial state is a non-degenerate ground state.

So, putting aside initial-state dependence, we may ask how far back does the system remember its past? How far back in time do observables at the present time depend on the density in the past?

A useful tool to study this question is a time-dependent problem with at least two electrons, for which both the Kohn-Sham system and the interacting system are exactly, or exactly numerically, solvable. Two electrons in a Mathieu oscillator provides a good case [Hessler 2002]; the external potential has the form:

$$v(\boldsymbol{r}) = k(t)r^2/2, \text{ with } k(t) = \bar{k} - \epsilon\cos(\omega t)\,. \tag{4.1}$$

The static version is often called the Hooke's atom; a paradigm for studies of exchange and correlation in the ground state [Taut 1993, Frydel 2000], largely because, for some parameters, the interacting problem can be solved analytically. For the exact interacting solution of the time-dependent problem, (4.1), we transform to center-of-mass and relative coordinates as this renders the Hamiltonian separable. Due to the spherical symmetry, one needs only to solve numerically two uncoupled one-dimensional time-dependent Schrödinger equations. From the evolving wavefunction, which begins in the ground state, the exact evolving density is obtained. The KS wavefunction involves just one doubly-occupied spatial orbital, evolving in time. By requiring its density to yield half the density of the interacting wavefunction, one can invert the KS equation to obtain the KS potential in terms of the evolving density [Hessler 2002].

One may then compare the exact calculation with that of an adiabatic (albeit not self-consistent) one, where one inputs the instantaneous density into a ground-state functional. The exact interacting dynamics in the Mathieu oscillator has the useful property that the evolving density retains a Gaussian-like profile, such that at each time t, it is very close to the density of a *ground state* of a certain Hooke's atom of spring constant k_{eff}. This spring constant is not equal to the actual spring constant in (4.1) at time t, except when the latter is modulated slowly enough such that the state remains an instantaneous ground state. In the general case, the state is *not* a ground state, but its *density* is that of a ground state of some Hooke's atom.

Many interesting phenomena arise [Hessler 2002]; one typically finds the differences between the adiabatic approximation and the exact KS case to be very large (except for very slow modulations). For example, the instantaneous correlation energy can become positive, which is impossible in any adiabatic approximation, since for ground states E_{c} is tied down below zero by the variational principle.

Of particular interest for this chapter is that the correlation potential displays severe non-locality in time due to history dependence. Before demonstrating this, we first note a simple way to track down the behavior in time of the correlation potential. The first time-derivative of the correlation energy, $\dot{E}_{\text{c}}(t)$, is intimately related to the correlation potential, $v_{\text{c}}(t)$, via [Hessler 1999]

$$\dot{E}_{\text{c}}(t) = \int \mathrm{d}^3r\, v_{\text{c}}(\boldsymbol{r},t)\dot{n}(\boldsymbol{r},t)\,. \tag{4.2}$$

This shows that if $\dot{E}_{\text{c}}(t)$ depends not just on the density at and near time t, but also on its earlier history, then $v_{\text{c}}(t)$ must too. That is, non-locality in $\dot{E}_{\text{c}}$ directly implies non-locality in the correlation potential $v_{\text{c}}(t)$. We turn now to Fig. 4.1. In the top panel is a plot of the value of $k_{\text{eff}}(t)$, which, as discussed earlier, completely identifies the density profile. The density profiles for a time range centered near $t = 4.8$ and centered near $t = 28.9$ are almost the same, yet the values of $\dot{E}_{\text{c}}(t)$ near those times are significantly different. The density at times near t is not enough to specify $\dot{E}_{\text{c}}$: it depends on the entire history and this is what we mean by non-locality in time. Likewise, the exact correlation potential $v_{\text{c}}(t)$ is a highly non-local functional of the density. Any adiabatic approximation has no history dependence and will fail to capture this effect.

4.3 Initial-State Dependence

For a given time-dependent density $n(\boldsymbol{r},t)$, how does the potential that yields that density depend on the choice of the initial wave-function? Initial-state dependence has only begun to be explored [Maitra 2001, Maitra 2002a, Holas 2002]; unlike the dependence on the density, there is no precedent for initial-state dependence in ground-state DFT. For example, there is no analogue of the adiabatic approximation that could be used as a starting point for exploration.

One may wonder whether initial-state dependence actually exists. That is, if we are constraining the density to evolve in a certain way, are the implicit

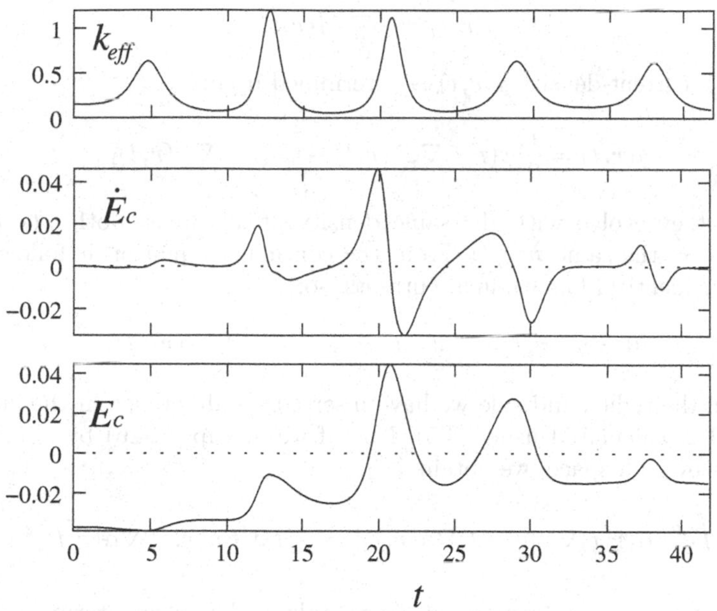

Fig. 4.1. Non-locality in time: the *top panel* shows $k_{\text{eff}}(t)$, *middle panel* $\dot{E}_c$, *bottom panel* E_c. The parameters in (4.1) are $\bar{k} = 0.25$, $\omega = 0.75$ and $\epsilon = 0.1$. Atomic units are used here, and throughout this chapter: energies are measured in Hartrees, distances in Bohr radii, and times are in units of 2.419×10^{-17}s

constraints on the initial-state enough to completely determine it? If this were the case, then there would be no initial-state dependence: knowing the history of the density would be enough to determine the functionals. We shall argue shortly that this is in fact the case for one electron, but not for more than one.

Let us first rephrase the question: Consider a many-electron density $n(\boldsymbol{r}, t)$ evolving in time under an external time-dependent potential $v_{\text{ext}}(\boldsymbol{r}, t)$. Can we obtain the same density evolution by propagating a different initial state in a different potential?

One Electron Case

Consider one electron, evolving with density $n(\boldsymbol{r}, t)$. Let the electron's wavefunction be $\varphi(\boldsymbol{r}, t)$, where $n(\boldsymbol{r}, t) = |\varphi(\boldsymbol{r}t)|^2$. An alternate candidate wavefunction $\tilde{\varphi}(\boldsymbol{r}, t)$ that evolves with identical density (in a different potential) must then be related to $\varphi(\boldsymbol{r}, t)$, by a (real) phase $\alpha(\boldsymbol{r}, t)$:

$$\tilde{\varphi}(\boldsymbol{r}, t) = \varphi(\boldsymbol{r}, t)\mathrm{e}^{\mathrm{i}\alpha(\boldsymbol{r},t)} . \tag{4.3}$$

The wavefunction at time t determines not just the density at time t but also its first time-derivative through the continuity equation:

$$\dot{n}(\boldsymbol{r},t) = -\nabla \cdot \boldsymbol{j}(\boldsymbol{r},t) \,, \tag{4.4}$$

where the current-density $\boldsymbol{j}(\boldsymbol{r},t)$ is determined from:

$$\boldsymbol{j}(\boldsymbol{r},t) = \frac{\mathrm{i}}{2}[\varphi(\boldsymbol{r},t)\nabla\varphi^*(\boldsymbol{r},t) - \varphi^*(\boldsymbol{r},t)\nabla\varphi(\boldsymbol{r},t)] \,. \tag{4.5}$$

Because they evolve with the same density at all times, both $\varphi(\boldsymbol{r},t)$ and $\tilde{\varphi}(\boldsymbol{r},t)$ share the same $\dot{n}(\boldsymbol{r},t)$. From the continuity equation it follows that they have identical longitudinal currents, so:

$$0 = \dot{n}_{\varphi}(\boldsymbol{r},t) - \dot{n}_{\tilde{\varphi}}(\boldsymbol{r},t) = \nabla \cdot [n(\boldsymbol{r},t)\nabla\alpha(\boldsymbol{r},t)] \,, \tag{4.6}$$

where on the right-hand side we have inserted the difference in the currents of $\tilde{\varphi}$ and φ, calculated using (4.5). Now if we multiply (4.6) by $\alpha(\boldsymbol{r},t)$ and integrate over all space, we obtain

$$0 = \int \mathrm{d}^3r\, \alpha(\boldsymbol{r},t)\nabla \cdot [n(\boldsymbol{r},t)\nabla\alpha(\boldsymbol{r},t)] = -\int \mathrm{d}^3r\, n(\boldsymbol{r},t)|\nabla\alpha(\boldsymbol{r},t)|^2 \tag{4.7}$$

In the last step, we integrated by parts, taking the surface term $\int_S \mathrm{d}\sigma\, \boldsymbol{e_n} \cdot (\alpha n \nabla\alpha)$, evaluated on a closed surface at infinity, to be zero. This will be true for any finite system, where the electron density decays at infinity, while any physical potential remains finite (or, if the potential grows, the density decays still faster).

Because the integrand in (4.7) cannot be negative, yet it integrates to zero, the integrand must be identically zero. Thus $\nabla\alpha(\boldsymbol{r},t) = 0$ everywhere. This is true even at nodes of the wavefunction, where $n(\boldsymbol{r}_0,t) = 0$: If $\nabla\alpha$ was zero everywhere except at the nodes, then as a distribution it is equivalent to being zero everywhere, unless it was a delta-function at the node – but in that case the potential would be highly singular, and therefore unphysical. So, for physical potentials, $\alpha(\boldsymbol{r},t)$ must be constant in space, i.e., the wavefunctions $\varphi(\boldsymbol{r},t)$ and $\tilde{\varphi}(\boldsymbol{r},t)$ differ only by an irrelevant time-dependent phase. Thus, only one initial-state (and one potential) can give rise to a particular density: the evolving density is enough to completely determine the potential and the initial states.

A note about the vanishing of the surface term in (4.7): This can be compared with the requirement on the potential in the Runge-Gross theorem, as discussed after (1.14). In [Maitra 2001], an example of a pathological initial state is given, where the surface term does not vanish, even though the density decays exponentially at large distances: the potential in which it lives plummets to minus infinity at large distances, yielding wildly oscillatory behavior in the tails of the decaying wavefunction, embodying infinite kinetic energy and momentum. Such unphysical states are beyond consideration!

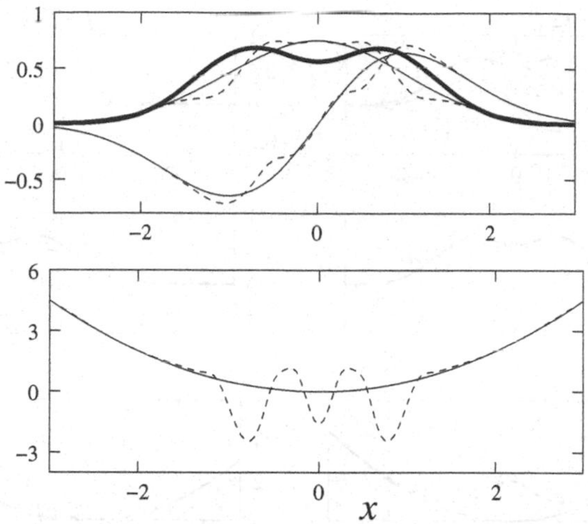

Fig. 4.2. An example of initial-state dependence for two non-interacting electrons: two different wavefunctions may evolve with the same density in different potentials. In the top plot, the *solid lines* are the two occupied orbitals of one wavefunction, which happens to be a stationary state of the harmonic oscillator potential, shown in the lower plot as a *solid line*. The density is shown as the *thick solid line* in the top figure. The *dashed lines* are the two orbitals of an alternative initial wavefunction, that evolves with the same density in the potential which, at the initial time, is shown as the *dashed line* in the lower figure

Many Electrons

For many electrons, initial-state dependence is real and alive: one can find two or more more different initial-states which evolve with identical density for all time in different external potentials.

A few simple examples of this are shown in Figs. 4.2 and 4.3. In Fig. 4.2, the density (thick solid line) of two non-interacting electrons in one dimension in an eigenstate (thin solid line) of the harmonic potential is considered [Maitra 2001, Holas 2002]. The two orbitals are the thin solid lines. If we keep this potential constant, the density will remain constant. We then ask, can we find another potential in which another non-interacting wavefunction evolves with this same, constant density for all time? There are in fact an infinite number of them, and one is shown here (dashed lines). The alternate potential was constructed using van Leeuwen's prescription [van Leeuwen 1999], and is shown here at the initial time. It is not constant in time: both the alternate potential and the alternate orbitals evolve in time, in such a way as to keep the density constant at all times.

The significance of this for TDDFT comes to light when we imagine the density as the density of some *interacting* electronic system. For a KS calculation, we are free to choose any initial KS state which has this initial

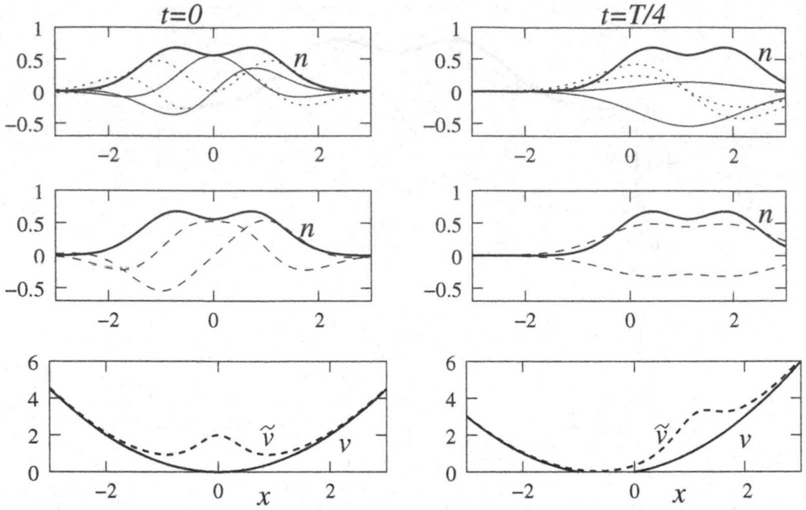

Fig. 4.3. *Top left panel*: the real and imaginary parts of the driven harmonic oscillator Floquet orbitals $\varphi_0(x,0)$ (*solid*) and $\varphi_1(x,0)$ (*dotted*) at time $= 0$, together with their density (*thick line*). *Middle left panel*: The real and imaginary parts of the alternative doubly-occupied Floquet orbital $\tilde{\varphi}(x,0)$ (*dashed*), which has the same density shown (*thick line*). *Bottom left panel*: the two potentials, v is the *solid*, and $\tilde{v}$ is *dashed*. The *right hand side* shows the same quantities at $t = T/4$

density: that is, both the potentials shown in the lower panel of Fig. 4.2, along with their respective orbitals, are fair game. The difference between these two KS potentials is exactly the difference in the xc potential, since the Hartree and external potentials are the same [see (1.33)]. So depending on this choice, the xc potentials are very different. Any functional without initial-state dependence would predict the same potential in both cases.

A note about the construction of the potential: In [van Leeuwen 1999], van Leeuwen answered the question of non-interacting v-representability in TDDFT. That is, given a density evolution of an interacting system, does a non-interacting system exist that can reproduce this? Given some restrictions on the initial state, the answer is in the affirmative, and it is also shown there how to construct such a potential for an appropriate choice of the initial state.

Figure 4.3 is another example of two different initial states that evolve with the same density for all time. This example, again of two non-interacting electrons, demonstrates that there is no one-to-one mapping between time-periodic densities of Floquet states and time-periodic potentials [Maitra 2002a]. Consider a periodically driven harmonic oscillator, containing two non-interacting electrons in a spin-singlet occupying two distinct quasi-energy orbitals. One can show that the density then periodically sloshes back and forth in the well. This is illustrated in the top panels of Fig. 4.3. The middle panel of Fig. 4.3 shows a doubly-occupied Floquet orbital (real and imaginary parts are the dashed lines) whose density (solid line) evolves identically to the density of the Floquet state in the top panel. This orbital sloshes back and forth in its potential, in a similar way to the orbitals of the driven oscillator. The lowest panel shows the potentials: the solid is the periodically driven harmonic potential corresponding to the Floquet state of the top panel, and the dashed is the periodically driven potential corresponding to that of the middle panel. Now, assuming there corresponds an interacting electron system whose density evolves exactly as shown, then both the Floquet state in the top panel and the middle panel are possible Kohn-Sham wavefunctions, and both the solid and dashed potentials in the lower panels are possible Kohn-Sham potentials; again, their difference (the sloshing "bump" in the figure) is the difference in the xc potential.

Not only any adiabatic approximation, but any density-functional approximation that lacks initial-state dependence – even with history-dependence – would incorrectly predict the same potential for all choices of KS initial states that propagate with the same density. In the next section, we discuss how, in many cases, one can eliminate the need for initial-state dependence altogether, by transforming it into a history-dependence.

4.4 Memory: An Exact Condition

Part of what makes the memory dependence complex, is the intricate entanglement of initial-state and history effects. This has consequences even when the initial state is a ground state. On the other hand, we shall see that due to the entanglement, memory dependence can often be reduced to history dependence alone.

Consider an interacting system, beginning with wavefunction $\Psi(0)$ at time 0, and evolving in time, with density $n(\boldsymbol{r},t)$. The xc potential at time t is determined by the density at all previous times, the initial interacting wavefunction, and the choice of the initial KS wavefunction $\Phi(0)$ for the KS calculation. Now say we can calculate the interacting wavefunction at a later time t', where $0 < t' < t$. Then, we may think of t' as the "initial" time for the inputs into the functional arguments of v_{xc}: that is,

$$v_{\text{xc}}[n_{t'},\Psi(t'),\Phi(t')](\boldsymbol{r},t) = v_{\text{xc}}[n,\Psi(0),\Phi(0)](\boldsymbol{r},t) \quad \text{for} \quad t > t' \,, \tag{4.8}$$

Here, $\Psi(t') = \hat{U}(t')\Psi(0)$, where $\hat{U}(t)$ is the unitary evolution operator, and $\Phi(t') = \hat{U}_{\text{KS}}(t')\Phi(0)$ where $\hat{U}_{\text{KS}}(t')$ is the KS evolution operator. The subscript on the density means that the density is undefined for times earlier than the subscript, and it equals the evolving density $n(\boldsymbol{r},t)$ for times t greater than the subscript.

Equation (4.8) displays the entanglement of the memory effects: Any dependence of the xc potential on the density at prior times may be transformed into an initial-state dependence and vice versa.

Like other exact conditions (discussed in Chap. 11), (4.8) may be used as a test for approximate functionals, but it is a very difficult condition to satisfy. For example, any of the recent attempts to include history-dependence, while ignoring initial-state dependence, must fail. If we restrict their application to systems beginning in the ground state, then (4.8) still produces a strict test of such functionals: Imagine an exact time-dependent calculation beginning in the ground state of some system. Later, when $\Psi(t')$ is no longer a ground state, we evolve *backwards* in time in a different external potential, that leads us backward to a different ground state at a different initial time. The history during the time before t' is different from the original history, but the xc potential for all times greater than t' should be the same for both the original evolution and the evolution along the alternative path. The extent to which these two differ is a measure of the error in a given history-dependent approximation, even applied only to initial ground states. Note that any adiabatic approximation ignoring initial-state dependence produces no difference. By ignoring both history dependence *and* initial-state dependence, the ALDA trivially satisfies (4.8).

A technical note: Although the RG theorem was proven only for analytic potentials (i.e., those that equal their Taylor series expansions at t in a neighbourhood of t in $[0, \infty)$ for all $t \geq 0$) [Runge 1984], it holds also for piecewise analytic potentials, i.e., potentials that are analytic in each of a finite number of intervals [Maitra 2002b]. This means that alternative allowed "pseudo-prehistories" can connect to the same wavefunction at some later time.

This raises the possibility of eliminating the initial-state dependence altogether: If we can evolve an initial interacting wavefunction that is not a ground state, backwards in time to a non-degenerate ground state, then the initial-state dependence may be completely absorbed into a history-dependence along this pseudo-prehistory.

In the usual formulation of TDDFT, one may choose any initial KS state which reproduces the density and divergence of the current of the interacting initial state [van Leeuwen 1999]. Here, this choice is translated into the choice of which ground state the interacting wavefunction $\Psi(0)$ evolves back to, together with the pseudo-prehistory of the density thus generated. One can imagine that for a given wavefunction $\Psi(0)$ there may be many paths which evolve back to some ground state, each path generating a different pseudo-prehistory. Only for those which result in the same KS wavefunction $\Phi(0)$ [and of course interacting wavefunction $\Psi(0)$] will the xc potentials be identical after time 0.

In the linear response regime, the memory formula (4.8) yields an exact condition relating the xc kernel to initial-state variations [Maitra 2005a]. We consider applying (4.8) in the perturbative regime, with the initial states at time 0 (on the right-hand-side) being ground-states. The initial states on the left-hand-side (i.e., the states at t') are not ground-states. We wish to express deviations from the ground-state values through functional derivatives with respect to the density and with respect to the initial states. Because the initial state determines the initial density and its first time-derivative, and puts constraints on higher-order time-derivatives of the density, the definition of a partial derivative with respect to the initial state is not trivial: what should be held fixed in the variation? The partial derivative with respect to the density, holding the initial state fixed, is simpler; for example, for the external potential this is a generalized inverse susceptibility, generalized to initial states which are not ground-states [cf. 1.21]. Variations of the density at times greater than zero are included. In order to define an initial-state derivative, one considers an extension of the functionals to a higher space in which the initial-state variable and density variable are independent: one drops (1.5) and (1.9). Then one can show that

$$\sum_\alpha \int \mathrm{d}^3 r_1 \left. \frac{\delta v_{\mathrm{KS}}[n_{t'}, \Phi_{t'}](\boldsymbol{r},t)}{\delta \varphi_{t',\alpha}(\boldsymbol{r}_1)} \right|_{(n_{\mathrm{GS}}, \Phi_{\mathrm{GS}}[n_{\mathrm{GS}}])} \delta\varphi_{t',\alpha}(\boldsymbol{r}_1)$$
$$- \int \mathrm{d}x_1 \ldots \int \mathrm{d}x_N \left. \frac{\delta v_{\mathrm{ext}}[n_{t'}, \Psi_{t'}](\boldsymbol{r},t)}{\delta \Psi_{t'}(\boldsymbol{x}_1, \ldots, \boldsymbol{x}_N)} \right|_{(n_{\mathrm{GS}}, \Psi_{\mathrm{GS}}[n_{\mathrm{GS}}])} \delta\Psi_{t'}(\boldsymbol{x}_1, \ldots, \boldsymbol{x}_N) + \text{c.c.}$$
$$= \int \mathrm{d}^3 r_1 \int_0^{t'} \mathrm{d}t_1 \, f_{\mathrm{xc}}[n_{\mathrm{GS}}](\boldsymbol{r}, \boldsymbol{r}_1, t-t_1) \delta n(\boldsymbol{r}_1, t_1) \,, \quad 0 < t' < t \,, \tag{4.9}$$

where the variables $\boldsymbol{x}_i = (\boldsymbol{r}_i, \sigma_i)$ represent spatial and spin coordinates, $\delta\varphi_{t',\alpha} = \delta\varphi_\alpha(t') = \varphi_\alpha(t') - \varphi_{\alpha,\mathrm{GS}}[n_{\mathrm{GS}}]$ represent the deviations at time t' of the spin orbitals of the KS Slater determinant away from the ground-state values and $\delta\Psi$ is similarly the deviation of the interacting state away from its ground-state. This equation demonstrates the entanglement of initial-state dependence and history-dependence in the linear response regime: the expression for the xc kernel on the right is entirely expressed in terms of initial-state dependence on the left.

4.5 Role of Memory in Quantum Control Phenomena

The challenges memory dependence presents to TDDFT emerge immediately when we attempt to describe quantum control phenomena. In recent years there have been huge advances in the control of chemical reactions, where nuclei are manipulated. The development of attosecond laser pulses opens the door to the possibility of manipulating electronic processes as well.

Let us say we are interested in driving a molecule from its ground state Ψ_0 in potential $v_{\mathrm{ext},0}$ to its mth excited state Ψ_m. Let us say we are lucky enough to know the external time-dependent field that achieves this after a time $\tilde{t}$. The field is then turned off at time $\tilde{t}$ so that the molecule remains in the excited eigenstate. We now ask how this process is described in the corresponding KS system, i.e., what is the KS potential? Initially, this is the ground-state potential $v_{\mathrm{KS},0}$ whose ground-state Φ_{KS}, has density n_{GS}, the density of the interacting ground state of the molecule. The first observation is that the KS potential after time $\tilde{t}$ does *not* typically return to the initial KS potential, in contrast to the case of the interacting system. This is because, by definition, the density of the KS state equals the interacting density at all times; in particular, after time $\tilde{t}$ it is the density of the interacting excited state of potential $v_{\mathrm{ext},0}$, but this is *not* guaranteed to be the density of the KS excited state of potential $v_{\mathrm{KS},0}$. Only the ground-state density of an interacting system is shared by its KS counterpart, not the higher excited states; the final KS state of the molecule will not typically be an eigenstate of $v_{\mathrm{KS},0}$.

There are two possibilities for the KS potential after time $\tilde{t}$. The first is that it does becomes static, and the static final density, call it $\tilde{n}$, is that of an eigenstate of it. From the above argument, the KS potential is however different from the initial, and does not equal the ground-state KS counterpart of the interacting case. For example, if we are exciting from the ground state of the helium atom to an excited state, the external potential of the interacting system is both initially and finally $-2/r$. The initial KS potential is the ground-state KS potential of helium, but the final is not; the final KS eigenstate has the same density as the excited state of interest in helium. Now we will argue that any adiabatic approximation, or indeed any potential that is not ultranonlocal in time, is unlikely to do well. Consider a time t beyond $\tilde{t}$. The density for times near t is constant, so any semi-local approximation for the potential will be any one of the potentials for which the density $\tilde{n}$ is the density of some eigenstate of it. In particular, for an adiabatic approximation, the potential is that for which the excited state density $\tilde{n}$ is the density of its *ground state*. There is no way for an approximate semi-local KS system to know that it should be the potential corresponding to the interacting system that has an mth excited state density of $\tilde{n}$. This information is encoded in the early history of the density, from times $0 < t < \tilde{t}$: the exact KS potential must be ultranonlocal in time, since, as time gets very large, it never forgets the early history. Alternatively, taking the "initial" time to be $\tilde{t}$ in the memory formula (4.8), this effect is an initial-state effect where the initial interacting (and non-interacting state) is not a ground-state.

The other possibility is that the KS (and xc) potential never becomes constant: it continues to change in time, with KS orbitals and orbital-densities changing in time in such a way that the total KS density remains static and equal to $\tilde{n}$. It is clear that any semi-local approximation will fail here, because for times beyond $\tilde{t}$, it will predict a constant potential since the density is constant. The exact xc (and KS) potential will be ultranonlocal in time; as time gets very large, one has to go way back in time, to times less than $\tilde{t}$, in order to capture any time dependence in the density.

This extreme non-locality is very difficult for a density functional approximation to capture: it may be that orbital functionals, which are implicit

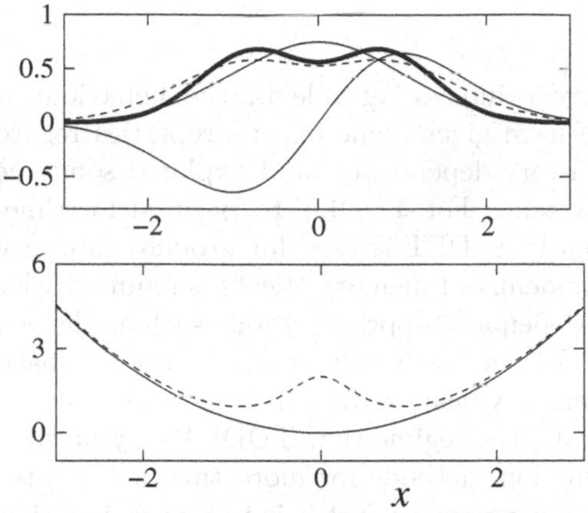

Fig. 4.4. In the *top panel* is the density (*thick solid line*) of the two electron singlet excited state (*solid lines*) of the harmonic oscillator (*solid line* in the *lower panel*). The *dashed line* is the doubly-occupied orbital resulting from evolving the singlet ground-state to a state of the same density as the excited state. The potential in which this is an eigenstate – the ground-state – is shown as the *dashed line* in the *lower panel*

density functionals, provide a promising approach. Even so, there are cases where TDDFT faces a formidable challenge. Consider two electrons, beginning in a spin-singlet ground state (e.g. the ground state of helium). Imagine now evolving the interacting state to a singlet singly-excited state (e.g. 1s2p of helium). Now, the KS ground state is a single Slater determinant with a doubly-occupied spatial orbital. This evolves under a one-body evolution operator, the KS Hamiltonian, so must remain a single Slater determinant. But a single excitation is a *double* Slater determinant, so can never be attained even with an orbital-dependent functional. This is a case of severe static correlation, where a single Slater determinant is inadequate to describe a truly multi-determinantal state. A simplified model of this is shown in Fig. 4.4. Here the density of the first excited state of two electrons in a harmonic oscillator, considered to be the final KS potential, is shown. Attempting to evolve to this density from the ground state of the harmonic oscillator, which is a doubly-occupied orbital, the best we can do is reach another doubly-occupied orbital (dashed), whose potential is shown in the lower panel (dashed).

We note that such problems do not arise in linear response regime, where we do not need to drive the system entirely into a single excited state: only perturbations of the ground-state are needed which have a small, non-zero projection on to the various excited states of the system.

4.6 Outlook

Memory plays a very interesting role in the behaviour of functionals in TDDFT. Here we have given some exact properties regarding initial-state dependence and history-dependence, and explored some memory effects on exactly solvable systems. For TDDFT to be used for time-dependent phenomena as confidently as DFT is used for ground-state problems, more understanding and modeling of memory effects is required, along with developments of memory-dependent approximations such as those mentioned in the introduction Sect. 4.1. In linear response, why the adiabatic approximation works so well for many typical excitations is not well understood. Strong field dynamics is especially the regime that TDDFT may be the only feasible approach, as wavefunction methods for more than a few interacting electrons becomes prohibitively expensive. Yet it is in this regime that memory effects appear to be significant, and fundamentally so for quantum control problems.

5 Current Density Functional Theory

G. Vignale

5.1 Introduction

The nonlocality of the exchange-correlation (xc) potential, i.e., the fact that the xc potential at a certain position depends on the global distribution of the particle density in space, is the curse of density functional theory. It is mainly because of this fact that, even after years of intensive studies, the exact form of the xc potential as a functional of the density remains unknown. Nevertheless, it is true that many accurate and useful results can be obtained from the use of an approximation – the local density approximation (LDA) – which ignores the problem altogether. Apparently, the nonlocal dependence of the Kohn-Sham orbitals on the density is sufficient in many cases to give the right quantum chemistry. Furthermore, a number of successful strategies have been designed to go beyond the LDA when needed: in one such approach (the generalized gradient approximation – GGA) one goes beyond the LDA by including the dependence of the xc potential on the *gradient* of the local density; in another one expresses the xc potential as a functional of the Kohn-Sham orbitals, and, finally, in the "meta-GGA" approach one fights the problem by including additional local variables, such as the kinetic energy density.

In this chapter we are going to see that the nonlocality problem affects in a more severe form the time-dependent density functional theory. This complication arises as a consequence of *memory* (see Chap. 4). The xc potential at a time t (now) depends on the density at earlier times t'. But at these earlier times a small volume element of the system, which is now located at $\boldsymbol{r}$, was located at a different position $\boldsymbol{r}'$. Retardation in time thus implies nonlocality in space (see Chap. 8). We will see that, when retardation is taken into account, the local density approximation breaks down *even in the limit of slowly varying density.* We will refer to this feature of the time-dependent theory as *ultranonlocality*, to distinguish it from the ordinary nonlocality, which becomes harmless in the limit of slowly varying density. Furthermore, we will see that "ultranonlocality" is not related to the presence of a long-range interaction between the particles, but, more in depth, implies that the particle density is not a well-chosen variable (although, in principle a legitimate one) for the description of effects that involve retardation in time. It is also evident that the inclusion of additional variables

G. Vignale: *Current Density Functional Theory*, Lect. Notes Phys. **706**, 75–91 (2006)
DOI 10.1007/3-540-35426-3_5

might cure the "disease": for example, by looking at the current density of an infinitesimal volume element of the fluid at position $\boldsymbol{r}$ at a certain time we might be able to estimate its position at an earlier time, and thus arrive at a local or quasi-local expression for the retarded xc potential. These general ideas will be explored in some detail in the following sections.

5.2 First Hints of Ultranonlocality: The Harmonic Potential Theorem

Historically, the first hint of ultranonlocality in TDDFT came from the work of John Dobson [Dobson 1994a] on the collective dynamics of electrons in parabolic quantum wells. Under the action of a uniform time-dependent electric field the density of such system oscillates back and forth without changing its shape, i.e., one has $n(\boldsymbol{r},t) = n_{\mathrm{GS}}(\boldsymbol{r} - \boldsymbol{R}(t))$ where $n_{\mathrm{GS}}(\boldsymbol{r})$ is the ground-state density and $\boldsymbol{R}(t)$ is the position of the center of mass of the system. The latter moves exactly as a single classical particle of mass m and charge $-e$ under the action of the external electric field: this is the content of the "harmonic potential theorem" (HPT). It is easy to see that the exact TDDFT satisfies the HPT for, according to the translational invariance condition (11.19) (see also Appendix to this chapter), the xc potential created by the oscillating density $n_{\mathrm{GS}}(\boldsymbol{r} - \boldsymbol{R}(t))$ is given by $v_{\mathrm{xc}}(\boldsymbol{r},t) = v_{\mathrm{xc,\,GS}}(\boldsymbol{r} - \boldsymbol{R}(t))$, where $v_{\mathrm{xc,\,GS}}(\boldsymbol{r})$ is the xc potential in the ground-state. In an accelerated frame of reference that moves together with the center of mass of the system, the external electric field is cancelled by the inertial force while the xc potential has exactly the form that is needed to preserve the ground-state density distribution.

Dobson observed that a naïve application of the local density approximation, including a local but retarded xc potential [Gross 1985], leads to results that are in conflict with the HPT. For example, one finds a density-dependent shift in the frequency of the oscillatory motion of the center of mass, and this motion becomes "damped". The reason for this difficulty is that the local density approximation is unable to distinguish between a situation in which the density variation is due to local compression/rarefaction of the electron liquid (as in the case of a long-wavelength plasmon) and the present one, in which this variation is due to a global translation of a system, without compression or rarefaction. The "obvious" choice of [Gross 1985] amounts to choosing the first option in both cases: thus introducing fictitious dissipative processes when none is present.

From a mathematical point of view, the link between ultranonlocality and the HPT can be seen as follows [Vignale 1998]. First of all, notice that the translational invariance identity (11.19) (which ensures satisfaction of the HPT) is intimately related to the zero-force theorem, (11.11a), (and 5.25 in the Appendix), which in turn implies that the exact xc kernel of any system must satisfy the equation

$$\int d^3r' \, f_{xc}(\boldsymbol{r}, \boldsymbol{r}', \omega) \nabla n_{GS}(\boldsymbol{r}') = \nabla v_{xc,\,GS}(\boldsymbol{r}) \,, \tag{5.1}$$

where $n_{GS}(\boldsymbol{r})$ and $v_{xc,\,GS}(\boldsymbol{r})$ are the density and the xc potential in the ground-state. Notice that the quantity on the right hand side of this equation is frequency-independent, implying that the integral over $\boldsymbol{r}'$ on the left hand side must somehow "wash out" the frequency dependence of the integrand. Now assume that f_{xc} has a finite range in the sense that the integral

$$\int d^3r' \, f_{xc}(\boldsymbol{r}, \boldsymbol{r}', \omega) \tag{5.2}$$

is finite. Indeed, this condition is satisfied by the xc kernel of a strictly homogeneous electron liquid, since it is known that the Fourier transform $f_{xc}(\boldsymbol{k}, \omega)$ of the homogeneous xc kernel has finite limit for $\boldsymbol{k} \to 0$ (see discussion in Sect. 5.5). Suppose now that $n_{GS}(\boldsymbol{r})$ is very slowly varying on the scale of the range of $f_{xc}(\boldsymbol{r}, \boldsymbol{r}', \omega)$. Then we can pull $\nabla n_{GS}(\boldsymbol{r}')$ out of the integral of (5.1) and get

$$\nabla n_{GS}(\boldsymbol{r}) \int d^3r' \, f_{xc}(\boldsymbol{r}, \boldsymbol{r}', \omega) = \nabla v_{xc,\,GS}(\boldsymbol{r}) \,. \tag{5.3}$$

In the limit that the density approaches uniformity, the integral on the left hand of this expression ought to converge (if it converges at all) to the $\boldsymbol{k} \to 0$ limit of the homogeneous electron gas kernel $f_{xc}(\boldsymbol{k}, \omega)$, which, as we have just stated, is a function of frequency. Since the right hand side of the expression is still frequency-independent we have arrived at a contradiction. This proves the fallacy of the initial assumption, namely, the finiteness of the integral (5.2) in a weakly inhomogeneous system. Indeed, the divergence of the integral (5.2) is the mathematical signature of what we have dubbed "the ultranonlocality problem". Notice that, unlike ordinary nonlocality, this problem is present in systems that are arbitrarily close to a homogenous electron liquid, and has nothing to do with the range of the interaction.

5.3 TDDFT and Hydrodynamics

Looking at TDDFT with hindsight one can easily understand why the description of many-body forces as gradients of a density-dependent potential led to the difficulties described in the previous section. It is not by accident that the two major classical theories of the dynamics of continuous media, namely, elasticity, for solids, and hydrodynamics, for fluids, express the many-body forces not as gradients of a scalar potential, but as divergences of a stress tensor, which is a local functional of the displacement or the velocity field. For example, in hydrodynamics the current density, $\boldsymbol{j}(\boldsymbol{r}, t)$, satisfies the Navier-Stokes equation [Landau 1987]

$$m \left(\frac{\partial}{\partial t} + \boldsymbol{v} \cdot \nabla \right) \boldsymbol{j}(\boldsymbol{r}, t) = \boldsymbol{F}(\boldsymbol{r}, t) + \nabla \cdot \overleftrightarrow{\sigma}(\boldsymbol{r}, t) \,, \tag{5.4}$$

where the velocity field, $\boldsymbol{v}(\boldsymbol{r},t)$, is defined as the ratio of the current density to the particle density, i.e., $\boldsymbol{v}(\boldsymbol{r},t) \equiv \boldsymbol{j}(\boldsymbol{r},t)/n(\boldsymbol{r},t)$. Here $\boldsymbol{F}(\boldsymbol{r},t)$ is the external volume force density (in which we also include the force density generated by the Hartree potential) while the second term on the right hand-side is the contact force exerted on the volume element by the surrounding medium. The stress tensor is given by

$$\sigma_{ij}(\boldsymbol{r},t) = -p\delta_{ij} + \eta\left(\frac{\partial v_i}{\partial r_j} + \frac{\partial v_j}{\partial r_i} - \frac{2}{d}\nabla\cdot\boldsymbol{v}\delta_{ij}\right) + \zeta\nabla\cdot\boldsymbol{v}\,\delta_{ij}\,, \tag{5.5}$$

where p is the equilibrium pressure associated with the local instantaneous density (and temperature), d is the number of spatial dimensions, and the coefficients η and ζ are, respectively, the shear viscosity and the bulk viscosity of the fluid.

Notice that, by expressing the force density as the divergence of a stress tensor, the Navier-Stokes equation is guaranteed to satisfy Newton's third law. Indeed, the net contact force exerted on a certain volume $\mathcal{V}$ of the fluid by the fluid that surrounds it is given by

$$\int_{\mathcal{V}} \mathrm{d}^3r\, \nabla\cdot \overleftrightarrow{\sigma}(\boldsymbol{r},t) = \int_S \mathrm{d}\Omega\, \boldsymbol{e_n}\cdot \overleftrightarrow{\sigma}(\boldsymbol{r},t)\,, \tag{5.6}$$

where the integral on the right hand side is done over the boundary surface S of the volume under consideration and $\boldsymbol{e_n}$ is the unit vector normal to this surface. The surface integral vanishes when $\mathcal{V}$ encloses the whole system because $\overleftrightarrow{\sigma}$ vanishes at infinity: this is another way of saying that the system, or each of its parts, does not exert a net force upon itself (the corresponding result for the torque can be shown to follow from the symmetry relation $\sigma_{ij} = \sigma_{ji}$, see [Landau 1987]).

In comparison to hydrodynamics, TDDFT seems to take a "back step" since it attempts to represent the contact force density $\boldsymbol{F}_{\mathrm{xc}}$ in terms of a scalar exchange-correlation potential, i.e., $\boldsymbol{F}_{\mathrm{xc}} = -n(\boldsymbol{r},t)\nabla v_{\mathrm{xc}}(\boldsymbol{r},t)$, where v_{xc} is a functional of the density. This is admissible in principle, but may lead to a violation of Newton's third law when an approximate form of v_{xc} is used.

To understand why it might be better to formulate the theory in terms of the current density (and its conjugate field, the vector potential), consider the continuity equation

$$\frac{\partial n(\boldsymbol{r},t)}{\partial t} = -\nabla\cdot\boldsymbol{j}(\boldsymbol{r},t)\,, \tag{5.7}$$

which expresses the local conservation of particle number. Starting from this equation it is relatively easy to get $n(\boldsymbol{r},t)$ from $\boldsymbol{j}(\boldsymbol{r},t)$ by taking the divergence of the latter and integrating over time. But the inverse problem, getting $\boldsymbol{j}(\boldsymbol{r},t)$ from $n(\boldsymbol{r},t)$, is far more difficult and does not possess a unique solution (we can always add to the solution the curl of an arbitrary vector

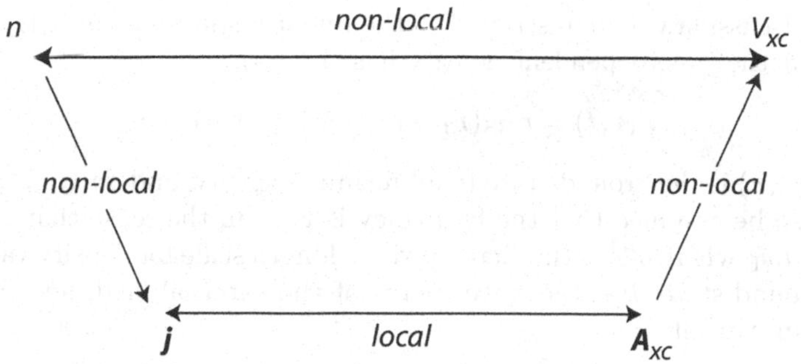

Fig. 5.1. Diagram showing the relation between current density functional theory and density functional theory. The nonlocal relation between n and $v_{\rm xc}$ is transformed into a local relation between $\boldsymbol{j}$ and $\boldsymbol{A}_{\rm xc}$ by means of two non-local transformations from n to $\boldsymbol{j}$ and from $v_{\rm xc}$ to $\boldsymbol{A}_{\rm xc}$

field, i.e., a purely transverse vector field). Only the longitudinal part of $\boldsymbol{j}$ is determined by the continuity equation and even this involves an integration over the whole space, i.e., one has

$$\boldsymbol{j}_L(\boldsymbol{r},t) = \int \mathrm{d}^3r' \, \frac{\partial n(\boldsymbol{r}',t)}{\partial t} \nabla_{\boldsymbol{r}} \frac{1}{4\pi|\boldsymbol{r}-\boldsymbol{r}'|} \, . \tag{5.8}$$

This relation is highly nonlocal (as is the corresponding relation between a scalar potential and the equivalent longitudinal vector potential, see (5.11) below), and this is why we expect to be able to drastically reduce the nonlocality of the density-dependent scalar potential by re-expressing it as a current-dependent vector potential. The idea is schematically depicted in Fig. 5.1. As a byproduct, the formulation in terms of $\boldsymbol{j}$ gives also direct access to the transverse component of the current density. This is important because the transverse current cannot be reliably extracted from the Kohn-Sham orbitals.

The path in front of us is now clear. In the next three sections we will develop a time-dependent current-density functional theory (TD-CDFT) in which the basic variable is the current density and the ordinary xc potential is replaced by an xc vector potential. We will see that in this theory the exchange-correlation force density has the form of the contact force density of hydrodynamics, i.e., it is the divergence of a stress tensor, and that this stress tensor can be safely approximated as a local functional of the current density.

This theory will enable us to calculate not only the density but also the current density from an effective Kohn-Sham equation. In order to accomplish this we will have to generalize (5.5) by endowing the viscosity constants η and ζ with both real and imaginary parts (the latter representing the dynamical bulk and shear moduli of the liquid) and making them functions of the frequency as well as the local particle density.

Our discussion will be restricted to the linear response regime. By this we mean that the time dependent density has the form

$$n(\boldsymbol{r},t) = n_{\mathrm{GS}}(\boldsymbol{r}) + \delta n(\boldsymbol{r})\mathrm{e}^{-i\omega t} + \mathrm{c.c.}\,, \tag{5.9}$$

where $n_{\mathrm{GS}}(\boldsymbol{r})$ is the ground-state (equilibrium) density, and $\delta n(\boldsymbol{r}) \ll n_{\mathrm{GS}}(\boldsymbol{r})$. It will also be assumed that the frequency is high in the sense that $\omega \gg qv_{\mathrm{F}}$ and $\omega \gg kv_{\mathrm{F}}$ where q^{-1} is the characteristic length scale for density variations in the ground-state, $\boldsymbol{k}$ is the wave vector of the external field, and v_{F} is the local Fermi velocity.

5.4 Current Density Functional Theory

In time-dependent CDFT we consider a broader class of Hamiltonians than those considered in the original Runge-Gross formulation, namely Hamiltonians of the form

$$\hat{H} = \sum_i \left\{ \frac{1}{2m}\left(\boldsymbol{p}_i + \frac{e}{c}\boldsymbol{A}_{\mathrm{ext},\,i}\right)^2 + v_{\mathrm{ext},\,i} \right\} + \hat{V}_{\mathrm{ee}}\,, \tag{5.10}$$

where $\boldsymbol{A}_{\mathrm{ext},\,i}$ is the external vector potential evaluated at the position $\boldsymbol{r}_i$ of the ith particle, $v_{\mathrm{ext},\,i}$ is the scalar potential at the position of the ith particle, and $\hat{V}_{\mathrm{ee}}$ represents the electron-electron interaction. The reason why this is a proper generalization of the Runge-Gross (RG) Hamiltonian is that every scalar potential $v(\boldsymbol{r})$ can be represented as a *longitudinal* vector potential $\boldsymbol{A}_v(\boldsymbol{r})$ by choosing the latter as the solution of the equation

$$\frac{e}{c}\frac{\partial \boldsymbol{A}_v(\boldsymbol{r},t)}{\partial t} = \nabla v(\boldsymbol{r},t)\,. \tag{5.11}$$

Of course, *transverse* vector potential represents different physics (magnetic fields).

It can be easily proved that for Hamiltonians of the form (5.10) the time-dependent current density, together with the initial state, uniquely determine the scalar and the vector potential, up to a gauge transformation that leaves the initial state unchanged. A first proof of this generalized RG theorem was provided by Ghosh and Dhara [Ghosh 1988], and I have recently found a simpler proof [Vignale 2004]. Therefore, following the usual arguments, one expects to be able to construct, uniquely, a Kohn-Sham hamiltonian, $\hat{H}_{\mathrm{KS}}$, that produces the correct current of the many-body sytem. This hamiltonian will have the form

$$\hat{H}_{\mathrm{KS}} = \sum_i \left\{ \frac{1}{2m}\left(\boldsymbol{p}_i + \frac{e}{c}\boldsymbol{A}_{\mathrm{KS},\,i}\right)^2 + v_{\mathrm{KS},\,i} \right\}\,, \tag{5.12}$$

and notice that the effective vector potential $\boldsymbol{A}_{\mathrm{KS}}$ will have in general longitudinal and transverse component even though the original external vector potential $\boldsymbol{A}_{\rm ext}$ was purely longitudinal. This equation (unlike the Kohn-Sham equation of ordinary TDDFT) determines in principle the whole current – not just its longitudinal component. The particle density is, of course, an immediate by-product of the longitudinal current.

So far goes the formalism. Now in order to find a concrete expression for $\boldsymbol{A}_{\rm KS} = \boldsymbol{A}_{\rm ext} + \boldsymbol{A}_{\rm xc}$ we resort to linear response theory; namely we assume that we are close to equilibrium and therefore $\boldsymbol{A}_{\rm xc}$ can be approximated as a linear functional of the current, with coefficients that depend on the equilibrium density. In other words we assume that $\boldsymbol{A}_{\rm xc}$ has the form

$$\boldsymbol{A}_{\rm xc}(\boldsymbol{r},\omega) = \int {\rm d}^3r' \; \overleftrightarrow{\boldsymbol{f}}_{\rm xc}(\boldsymbol{r},\boldsymbol{r}',\omega)\cdot \boldsymbol{j}(\boldsymbol{r}',\omega)\;, \tag{5.13}$$

where the tensor kernel $\overleftrightarrow{\boldsymbol{f}}_{\rm xc}$ is a generalization of the scalar xc kernel of TDFT. We will discuss its structure in the next section. It must be borne in mind, however, that after doing the linear response approximation on $\boldsymbol{A}_{\rm xc}$, we lose control on the terms proportional to $A^2_{\rm xc}$, which arise from the expansion of the kinetic energy operator in the Kohn-Sham equation.

5.5 The xc Vector Potential for The Homogeneous Electron Liquid

Let us first consider the tensor exchange-correlation kernel $\overleftrightarrow{\boldsymbol{f}}_{\rm xc}(\boldsymbol{r},\boldsymbol{r}',\omega)$ in a homogeneous electron liquid of density n. Translational invariance makes $\overleftrightarrow{\boldsymbol{f}}_{\rm xc}(\boldsymbol{r},\boldsymbol{r}',\omega)$ a function of $\boldsymbol{r}-\boldsymbol{r}'$ and we will therefore focus on its Fourier transform

$$\overleftrightarrow{\boldsymbol{f}}_{\rm xc}(\boldsymbol{k},\omega) = \int {\rm d}^3r \; \overleftrightarrow{\boldsymbol{f}}_{\rm xc}(\boldsymbol{r},\omega)\, {\rm e}^{{\rm i}\boldsymbol{k}\cdot\boldsymbol{r}}\;. \tag{5.14}$$

Furthermore, we make use of rotational invariance to express the full kernel in terms of just two independent scalar functions of $k = |\boldsymbol{k}|$, the longitudinal component $f_{\rm xc,\,L}(k,\omega)$ and the transversal component $f_{\rm xc,\,T}(k,\omega)$:

$$[\overleftrightarrow{\boldsymbol{f}}_{\rm xc}(\boldsymbol{k},\omega)]_{\alpha\beta} = \left[f_{\rm xc,\,L}(k,\omega)\hat{\boldsymbol{k}}_\alpha\hat{\boldsymbol{k}}_\beta + f_{\rm xc,\,T}(k,\omega)(\delta_{\alpha\beta}-\hat{\boldsymbol{k}}_\alpha\hat{\boldsymbol{k}}_\beta)\right]\frac{ck^2}{e\omega^2}\;, \tag{5.15}$$

where $\hat{\boldsymbol{k}}$ is the unit vector in the direction of $\boldsymbol{k}$. The factor $ck^2/e\omega^2$ is introduced here as a matter of convenience, in order to make $f_{\rm xc,\,L}(k,\omega)$ coincide with the xc kernel of the ordinary density functional theory.

Given the kernels $f_{\rm xc,\,L}$ and $f_{\rm xc,\,T}$ it is easy to construct the linear response of the homogeneous electron liquid to an external vector potential $\boldsymbol{A}_{\rm ext}(\boldsymbol{k},\omega)$. As discussed in the previous section, this is just the response of the non-interacting electron gas (at the same density n) to the effective field $\boldsymbol{A}_{\rm ext}(\boldsymbol{k},\omega)+\boldsymbol{A}_{\rm xc}(\boldsymbol{k},\omega)$. (Once again, the "external" field $\boldsymbol{A}_{\rm ext}$ is assumed to include the Hartree potential $\boldsymbol{A}_{\rm H}(\boldsymbol{k},\omega) = 4\pi e^2\hat{\boldsymbol{k}}/\omega^2$ in three dimensions.)

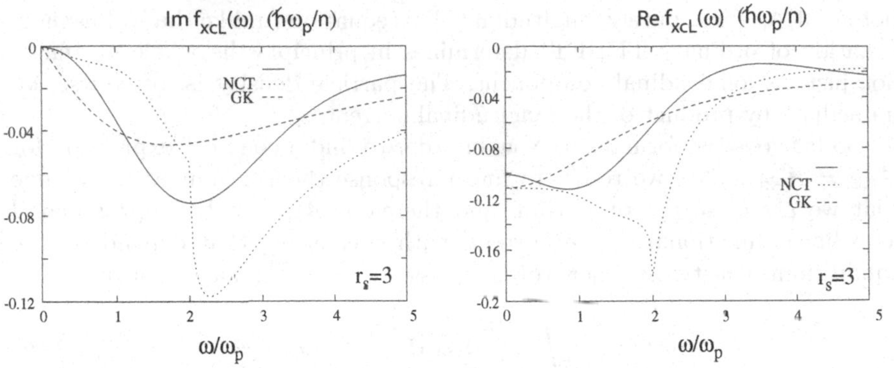

Fig. 5.2. The imaginary and the real parts of $f_{\mathrm{xc,\,L}}(\omega)$ (in units of $\hbar\omega_p/n$) in a homogeneous electron liquid at $r_s = 3$. The *short-dashed line* (NCT) is the result of a mode-coupling calculation [Nifosì 1998]. The *long-dashed line* (GK) [Gross 1985] and the *solid line* (QV) [Qian 2002] are interpolation formulas based on exact limiting forms

The connection between the xc kernels and the linear response functions of the electron liquid is the basis of the microscopic calculation of $f_{\mathrm{xc,\,L}}$ and $f_{\mathrm{xc,\,T}}$ [Nifosì 1998, Qian 2002, Qian 2003]. These calculations are too technical to be described here, but the following features should be noted:

(i) Both $f_{\mathrm{xc,\,L}}(k,\omega)$ and $f_{\mathrm{xc,\,T}}(k,\omega)$ tend to finite limits, denoted by $f_{\mathrm{xc,\,L}}(\omega)$ and $f_{\mathrm{xc,\,T}}(\omega)$, when $k \to 0$ at finite ω (this is a consequence of translational invariance, as it implies that the electron liquid accelerates uniformly in response to a uniform electric field)

(ii) The $k = 0$ kernels $f_{\mathrm{xc,\,L}}(\omega)$ and $f_{\mathrm{xc,\,T}}(\omega)$ have both real and imaginary parts. The real parts have finite limiting values at $\omega = 0$ and $\omega = \infty$ and may have either sign; the imaginary parts are negative at all frequencies and tends to zero linearly for $\omega \to 0$ and as $\omega^{-d/2}$ for $\omega \to \infty$: the coefficients of these asymptotic behaviors are known analytically.

Representative plots of the longitudinal kernel $f_{\mathrm{xc\,L}}(\omega)$ and of the transverse kernel $f_{\mathrm{xc,\,T}}(\omega)$ *vs* ω are shown in Fig. 5.2 and Fig. 5.3.

Let us now return to the full xc vector potential. Combining the longitudinal and transverse components, and making use of the existence of the $k \to 0$ limit of $f_{\mathrm{xc,\,L(T)}}(k,\omega)$ we see that, up to order k^2, the xc vector potential can be written as

$$\frac{e}{c}\boldsymbol{A}_{\mathrm{xc}}(\boldsymbol{k},\omega) = [f_{\mathrm{xc,\,L}}(\omega)\hat{\boldsymbol{k}}\cdot\boldsymbol{j} - f_{\mathrm{xc,\,T}}(\omega)\hat{\boldsymbol{k}}\times(\hat{\boldsymbol{k}}\times\boldsymbol{j})]\frac{k^2}{\omega^2}\,. \tag{5.16}$$

From this we want to separate the adiabatic LDA contribution. Recall that in the adiabatic LDA the xc potential is just the xc component of the chemical potential, μ_{xc}, evaluated at the intantaneous local density

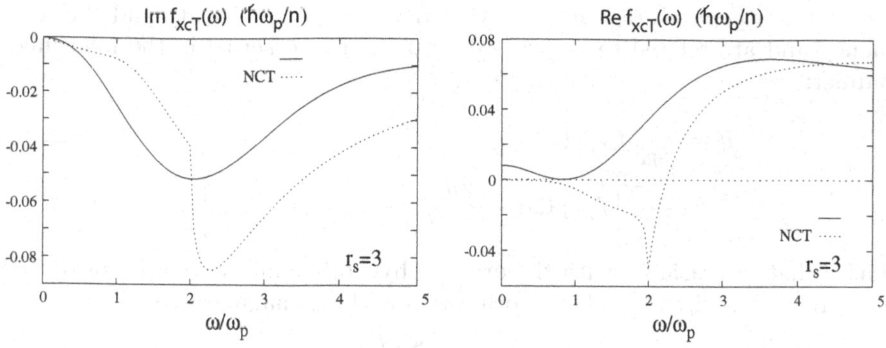

Fig. 5.3. Same as the previous figure for the imaginary and real parts of $f_{\mathrm{xc,\,T}}(\omega)$ at $r_{\mathrm{s}} = 3$

$n = n_{GS} + n_1 \exp\{\mathrm{i}(\boldsymbol{k}\cdot\boldsymbol{r} - \omega t)\} + \mathrm{c.c.}$ Thus, in the linear response approximation we have

$$v_{\mathrm{xc}}^{\mathrm{ALDA}}(\boldsymbol{k},\omega) = \mu'_{\mathrm{xc}}(n_{GS})n(\boldsymbol{k},\omega) , \tag{5.17}$$

where the prime denotes a derivative with respect to n. Making use of the continuity equation $n(\boldsymbol{k},\omega) = \boldsymbol{k}\cdot\boldsymbol{j}(\boldsymbol{k},\omega)/\omega$, and recasting $v_{\mathrm{xc}}^{\mathrm{ALDA}}$ as a longitudinal vector potential according to the formula

$$\frac{e}{c}\boldsymbol{A}_{\mathrm{xc}}^{\mathrm{ALDA}}(\boldsymbol{k},\omega) = \frac{\boldsymbol{k}\, v_{\mathrm{xc}}^{\mathrm{ALDA}}(\boldsymbol{k},\omega)}{\omega} , \tag{5.18}$$

we arrive at

$$\begin{aligned}\frac{e}{c}\boldsymbol{A}_{\mathrm{xc}}(\boldsymbol{k},\omega) &= \frac{e}{c}\boldsymbol{A}_{\mathrm{xc}}^{\mathrm{ALDA}}(\boldsymbol{k},\omega)\\ &\quad + \left\{[f_{\mathrm{xc,\,L}}(\omega) - \mu'_{\mathrm{xc}}]\hat{\boldsymbol{k}}\cdot\boldsymbol{j} - f_{\mathrm{xc,\,T}}(\omega)\hat{\boldsymbol{k}}\times(\hat{\boldsymbol{k}}\times\boldsymbol{j})\right\}\frac{k^2}{\omega^2} .\end{aligned} \tag{5.19}$$

We are now very close to the promised hydrodynamic form. All that remains to be done is to Fourier-transform the expression for $\boldsymbol{A}_{\mathrm{xc}}$ back to real space, keeping in mind that under this transformation $\mathrm{i}\boldsymbol{k}$ becomes the ∇ operator. It is also convenient to focus on the force exerted by the vector potential on the volume element rather than the vector potential itself: this is given by $\boldsymbol{F}_{\mathrm{xc}}(\boldsymbol{k},\omega) = -ne\boldsymbol{E}_{\mathrm{xc}}(\boldsymbol{k},\omega) = -\mathrm{i}\omega n\frac{e}{c}\boldsymbol{A}_{\mathrm{xc}}(\boldsymbol{k},\omega)$. (This is strictly speaking only the electric force. The magnetic Lorentz force is neglected, being of higher order in both $\boldsymbol{j}$ and $\boldsymbol{k}$). Thus, after some straightforward algebra we arrive at the following expression for the force density:

$$\boldsymbol{F}_{\mathrm{xc}}(\boldsymbol{r},\omega) = \boldsymbol{F}_{\mathrm{xc}}^{\mathrm{ALDA}}(\boldsymbol{r},\omega) - \nabla\cdot \overleftrightarrow{\sigma}_{\mathrm{xc}}(\boldsymbol{r},t) , \tag{5.20}$$

where

$$[\overleftrightarrow{\sigma}_{\mathrm{xc}}(\boldsymbol{r},t)]_{\alpha\beta} = \tilde{\eta}\left(\frac{\partial v_\alpha}{\partial r_\beta} + \frac{\partial v_\beta}{\partial r_\alpha} - \frac{2}{d}\nabla\cdot\boldsymbol{v}\delta_{\alpha\beta}\right) + \tilde{\zeta}\nabla\cdot\boldsymbol{v}\delta_{\alpha\beta} , \tag{5.21}$$

and $\tilde{\eta}$, $\tilde{\zeta}$ are *generalized viscosities* that depend on the density and the frequency and are related to the $k \to 0$ limit of the xc kernel in the following manner:

$$\tilde{\eta} = -\frac{n^2}{\mathrm{i}\omega} f_{\mathrm{xc,\,T}}(\omega) , \tag{5.22a}$$

$$\tilde{\zeta} = -\frac{n^2}{\mathrm{i}\omega} \left[f_{\mathrm{xc,\,L}}(\omega) - \frac{2(d-1)}{d} f_{\mathrm{xc,\,T}}(\omega) - \mu'_{\mathrm{xc}} \right] . \tag{5.22b}$$

Notice that, at variance with the original hydrodynamic viscosities of (5.5), the generalized viscosities have both a real and an imaginary part:

$$\tilde{\eta}(\omega) = \eta(\omega) - \frac{S_{\mathrm{xc}}(\omega)}{\mathrm{i}\omega} \tag{5.23a}$$

$$\tilde{\zeta}(\omega) = \zeta(\omega) - \frac{B_{\mathrm{xc}}^{\mathrm{dyn}}(\omega)}{\mathrm{i}\omega} , \tag{5.23b}$$

where η, ζ, S_{xc}, and $B_{\mathrm{xc}}^{\mathrm{dyn}}$ are all real quantities. Clearly $\eta(\omega)$ and $\zeta(\omega)$ describe the physical viscosity of the liquid. On the other hand, $S_{\mathrm{xc}}(\omega)$ and $B_{\mathrm{xc}}^{\mathrm{dyn}}(\omega)$ describe the *elasticity of the electron liquid* [Conti 1999]: they are the dynamical shear modulus and the dynamical bulk modulus, respectively. Notice that there is no static shear modulus in a liquid, hence the superscript "dyn" is not needed for S_{xc}; on the other hand $B_{\mathrm{xc}}^{\mathrm{dyn}}(\omega)$ denotes the difference between the frequency-dependent bulk modulus, $B_{\mathrm{xc}}(\omega)$ and its static value $B_{xc}(0)$. These elastic constants are absent in hydrodynamics because hydrodynamics deals with the collisional regime $\omega\tau \ll 1$ in which frequent collisions between the particles, (occurring at a rate $1/\tau$) create a situation of local equilibrium in a time that is short compared to the period of the oscillations. But at frequencies higher than $1/\tau$ the system is out of equilibrium, the Fermi surface is deformed, and there is a energy cost to pay for such a deformation. The elastic constants of the electron liquid are precisely the stiffnesses of the Fermi surface against deformations.

Based on the above discussion one could expect that S_{xc} and $B_{\mathrm{xc}}^{\mathrm{dyn}}$ vanish in the $\omega \to 0$ limit. This expectation is borne out for the dynamical bulk modulus, but, surprisingly, not for the shear modulus. The reason is that even a very small frequency $\omega \ll \epsilon_F$ (where ϵ_F is the Fermi energy) is large in comparison with the inverse relaxation time from electron-electron collisions, which vanishes as T^2 when the temperature, T, tends to zero. Indeed, since the real part of $f_{\mathrm{xc,\,T}}(\omega)$ tends to a finite limit for $\omega \to 0$, while the imaginary part tends to zero, it turns out that the shear modulus is the dominant contribution to the xc field in the $\omega \to 0$ limit. It is precisely this term that makes the difference between CDFT and ordinary adiabatic LDA in the calculation of the polarizability of long polymer chains, which is described later in this book (Chap. 21). In any case, the lesson to be learned is that the $\omega \to 0$ limit of the time-dependent CDFT is not the same as the adiabatic ALDA, provided that the limit is taken in such a way that local equilibrium is *not* reached.

5.6 The xc Vector Potential for the Inhomogeneous Electron Liquid

The main result of the previous section (5.20), is written so that it can immediately be turned into a local density approximation for the xc electric field of an inhomogeneous electron liquid through the replacement $n \to n_{\rm GS}(\boldsymbol{r})$, where $n_{\rm GS}(\boldsymbol{r})$ is the ground-state density of the inhomogeneous liquid. Of course, the xc kernels must also be evaluated at the local density. The correctness of this procedure is confirmed by a careful study of the structure of the tensor exchange-correlation kernel of a weakly inhomogeneous electron liquid. This study was carried out by VK [Vignale 1996] and is reviewed in [Vignale 1998]. In [Vignale 1996] VK considered an electron liquid modulated by a charge-density wave of small amplitude γ and small wave vector $\boldsymbol{q}$. The wave vector k of the external field and q were assumed to be small not only in comparison to the Fermi wave vector $k_{\rm F}$ but also in comparison to $\omega/v_{\rm F}$ ($v_{\rm F}$ being the Fermi velocity). The second condition ensures that the phase velocity of the density disturbance is higher than the Fermi velocity, so that no form of static screening can occur. Under these assumptions, all the components of the tensorial kernel $\overleftrightarrow{\boldsymbol{f}}_{\rm xc}$ could be calculated, up to first order in the amplitude of the charge density wave, and to second order in the wave vectors k and q. The calculations were greatly facilitated by a set of sum rules which are mathematically equivalent to the zero-force and zero-torque requirements discussed in Chap. 11 and in the Appendix of this chapter. The result of the analysis was a rather complicated but regular expression for the various components of $f_{\rm xc}$ in the limit of small k and q. Finally, this expression could be rearranged [Vignale 1997, Ullrich 2002b] in the elegant form

$$\frac{e}{c}\boldsymbol{A}_{\rm xc}(\boldsymbol{r},\omega) = \frac{e}{c}\boldsymbol{A}_{\rm xc}^{\rm ALDA}(\boldsymbol{r},\omega) - \frac{1}{{\rm i}\omega n_{\rm GS}(\boldsymbol{r})}\nabla\cdot \overleftrightarrow{\sigma}_{\rm xc}(\boldsymbol{r},t)\,, \tag{5.24}$$

which is of course equivalent to (5.20).

As the occurrence of two spatial derivatives of the velocity field in the second term on the right hand side of this equation is dictated by general principles, (5.20), with the xc stress tensor given by (5.21), is expected to remain valid even for large values of the velocity, i.e., in the nonlinear regime, provided v and n are slowly varying. The argument goes as follows: Suppose we tried to extend (5.20) into the nonlinear regime by including terms of order v^2. Because the stress tensor must depend on first derivatives of v, such corrections would have to go as $(\nabla v)^2$. But then the force density, given by the derivative of the stress tensor, would have to involve at least three derivatives. Thus, for sufficiently small spatial variation of the density and velocity fields, the nonlinear terms can be neglected. Since the ALDA is an intrinsically nonlinear approximation, VUC proposed that (5.20), written in the time domain, could provide an appropriate description of both linear and nonlinear response properties. A nonlinear, retarded expression for $\delta v_{\rm xc}$ was also proposed by Dobson et al. [Dobson 1997]. The two approximations coincide in "one-dimensional systems" (i.e., when one has a unidirectional current density field that depends only on one coordinate), but differ in the general case. More recently, Ilya Tokatly has developed a more general and beautiful theory, based on the use of Lagrangian coordinates, which is applicable also when the gradient of the velocity field is not small. An accessible presentation of his theory can be found in Chap. 8 of this book.

5.7 Applications

Several applications of time-dependent CDFT have appeared so far in the literature and are described in Chaps. 18 and 19. Here I limit myself to a very brief summary, illustrating how new effects can be described which were not accessible by the standard TDDFT.

Collective Excitations in Semiconductor Quantum Wells – This is an ideal application, because the electronic density profile of semiconductor quantum wells is defined by electrostatic gating and is therefore slowly varying in space. A particularly important problem is the characterization of intersubband transitions, which take place between two different levels of quantized motion in the growth direction. These transitions are highly collective (i.e., they consists of a complex superposition of electron-hole pairs) and cannot be described within a single-particle picture. It turns out that the ordinary TDDFT, in the adiabatic approximation, does a reasonably good job of predicting the energy of these excitations. However, the calculation of the linewidth is completely beyond the power of ordinary TDFT. Chapter 18 explains how the linewidth is calculated in TDCDFT, and how it can been brought to excellent agreement with experiment by the inclusion of extrinsic effects such as impurity scattering and surface roughness. Unfortunately, the excellent agreement found in simple quantum wells does not extend to more complex structures, such as the double quantum well (a single well split into two parts by a potential barrier). The non classical, low-density barrier region is problematic because the velocity field has a large gradient there. Thus double-well systems expose some of the limitations of the local current density approximation.

Atoms – Ullrich and Burke[Ullrich 2004] have applied TDCDFT to the calculation of the excitation energies of divalent atoms. The results are mixed. For $^1S \rightarrow ^1S$ transitions they find a small improvement upon the ordinary DFT (in local density approximation), but for $^1S \rightarrow ^1P$ transitions things get worse. Furthermore, CDFT predicts a small but finite linewidth for all atomic excitations. This is an artifact of the theory, since the linewidth arises from the continuum excitation spectrum of the homogeneous electron gas to which the electronic cloud of the atom is locally assimilated.

Conjugated Polymers – TDCDFT has been applied, in the zero-frequency limit, to the study of the static dielectric response of long insulating polymer chains (polyacetylene and other Pi-conjugated polymers). The interest and importance of the problem arises from the fact that the ordinary LDA-based DFT leads to a serious overestimation of the dielectric polarizability – an overestimation that grows with the length of the chain, and is due to the ultra-nonlocality of the xc functional. Indeed, Hartree-Fock calculations, in which the nonlocality of the potential is preserved, do not suffer from this problem (at least not in such a serious form). In applying the TD-CDFT one must keep in mind that the zero-frequency limit of this theory does not coincide with the naive LDA. There is an additional term, which arises from the elasticity of the electron liquid in the insulator, and this term is (within the local current density approximation) taken to be equal to the dynamical shear modulus of a uniform electron gas of the same density. The results are quite spectacular for polyacetylene and other polymer chains which have well extended electronic states at the top of the valence band: the new term corrects the overestimation of the dielectric constant and brings the result in excellent agreement with state-of-the art wave function methods (MP2 perturbation theory). The method is not equally successful for other polymer chain (e.g., long H_2 chains) which are characterized by much more localized electronic states. In this case the improvement upon the LDA is marginal. It is of course expected that the local approximation should be less effective in dealing with systems that are further away from the homogeneous electron liquid. A more detailed discussion of these results is provided in Chap. 19.

Optical Spectra of Semiconductors – The problem of calculating the optical spectra of semiconductors (e.g. Si) is historically very important, for it exposes one of the main weakness of DFT: its lack of accuracy in predicting the value of the optical gap (LDA typically underestimates it by a significant fraction). Can TDCDFT solve the band-gap problem? I think the jury is still out on this important question. At the fundamental level the answer is probably no. Although TDCDFT does possess the two features which are known to be necessary for a renormalization of the band gap, namely, a finite imaginary part and an infinite range in space of the xc kernel f_{xc}, it appears that the uniform electron-gas approximation is not sufficient to produce a sharp change in the optical gap (in fact the present approximations predict a finite absorption coefficient at all frequencies). In practice, however, one could still obtain improved optical spectra, since the xc potential may alter the oscillator strength of transitions near the gap in such a way as to simulate an increase in the band gap. This is what seems to be happening in recent calculations of the optical spectra carried out by the Groningen group. These calculations, however, are still based on a frequency-independent exchange-correlation kernel. More details will be found in Chap. 19.

Irreversible Dynamics and Transport – Finally, I mention that the inclusion of the current in the xc potential makes the latter non-invariant under time reversal, and thus opens the way to a first principle treatment of irreversible effects, such as the relaxation of an excited state [Wijewardane 2005, D'Agosta 2005b]. It is remarkable that the irreversible behavior of a global variable (the current) can be computed without leaving the framework of the Kohn-Sham equation, i.e., within a pure-state description of the system. TDCDFT can also be applied to electrical transport problems, since it gives direct access to the current and therefore to the resistance. Recently it has been shown that the low-frequency viscosity gives a substantial contribution to the resistance of molecular junctions and quantum point contacts [Sai 2005]. Remarkably, the term in the xc potential which causes this effect in a conductor is the same that modifies the dielectric polarizability in an insulator.

In conclusion, the time-dependent CDFT is a powerful and versatile tool for the study of optical, dielectric, transport, and relaxational properties of electronic systems. However, the local approximation upon which xc functionals are based does have severe shortcomings, particularly in atomic systems. More research is needed to produce functionals that work in difficult situations.

Acknowledgments

Work supported by NSF grant No. DMR-0313681.

Appendix – The Zero Force Theorem and Generalized Translational Invariance

It is important to become familiar with the few known exact properties of the exchange-correlation potential in TDDFT since they provide insight into the structure of the theory and are useful constraints on approximations. The simplest exact condition probably is the *zero-force theorem*, which states that

$$\int \mathrm{d}^3r\, n(\boldsymbol{r},t)\nabla v_{\mathrm{xc}}(\boldsymbol{r},t) = 0\,, \tag{5.25}$$

i.e., the net force exerted by the xc potential on the system is zero [Vignale 1995a]. This is a fairly obvious statement, following from Newton's third law, and can be proved in a completely elementary manner [Gross 1996]. One simply notices that the average center of mass coordinate of a system of N identical particles of mass m, defined as

$$\boldsymbol{R} \equiv \frac{1}{N}\int \mathrm{d}^3r\, \boldsymbol{r}\, n(\boldsymbol{r},t)\,, \tag{5.26}$$

satisfies the equation of motion

$$Nm\ddot{\boldsymbol{R}}(t) = -\int \mathrm{d}^3r\, n(\boldsymbol{r},t)\nabla v_{\mathrm{ext}}(\boldsymbol{r},t)\ , \tag{5.27}$$

where $v_{\mathrm{ext}}(\boldsymbol{r},t)$ is the external potential and $\ddot{\boldsymbol{R}}(t)$ denotes the second derivative of $\boldsymbol{R}(t)$ with respect to time. The internal particle-particle interactions cancel out in pairs. Recalling that the Kohn-Sham system has precisely the same density (and therefore precisely the same center of mass coordinate) as the true many-particle system, we can also write

$$Nm\ddot{\boldsymbol{R}}(t) = -\int \mathrm{d}^3r\, n(\boldsymbol{r},t)\nabla[v_{\mathrm{ext}}(\boldsymbol{r},t) + v_{\mathrm{H}}(\boldsymbol{r},t) + v_{\mathrm{xc}}(\boldsymbol{r},t)]\ . \tag{5.28}$$

Subtracting (5.27) from (5.28), and noting that the Hartree potential explicitly satisfies the zero-force theorem, we immediately arrive at (5.25).

The zero-force theorem can also be cast in a differential form which turns out to be useful in the analysis of linear response. To this end consider the linear response regime $n(\boldsymbol{r},t) = n_{\mathrm{GS}}(\boldsymbol{r}) + \delta n(\boldsymbol{r},t)$ and $v_{\mathrm{xc,\,GS}}(\boldsymbol{r}) + \delta v_{\mathrm{xc}}(\boldsymbol{r},t)$, with $\delta n(\boldsymbol{r},t) \ll n_{\mathrm{GS}}(\boldsymbol{r})$ and $v_{\mathrm{xc,\,GS}}(\boldsymbol{r}) \ll \delta v_{\mathrm{xc}}(\boldsymbol{r},t)$, where $n_{\mathrm{GS}}(\boldsymbol{r})$ is the ground-state density and $v_{\mathrm{xc,\,GS}}(\boldsymbol{r})$ is the xc potential associated with it. Expanding (5.25) to first order in δn we get

$$\int \mathrm{d}^3r\, n_{\mathrm{GS}}(\boldsymbol{r})\nabla\delta v_{\mathrm{xc}}(\boldsymbol{r},t) + \int \mathrm{d}^3r\, \delta n(\boldsymbol{r},t)\nabla v_{\mathrm{xc,\,GS}}(\boldsymbol{r}) = 0\ . \tag{5.29}$$

The first term on the left hand side can be integrated by parts yielding

$$\int \mathrm{d}^3r\, \delta v_{\mathrm{xc}}(\boldsymbol{r},t)\nabla n_{\mathrm{GS}}(\boldsymbol{r}) = \int \mathrm{d}^3r\, \delta n(\boldsymbol{r},t)\nabla v_{\mathrm{xc,\,GS}}(\boldsymbol{r})\ . \tag{5.30}$$

We now express δv_{xc} in terms of δn according to the linear relation

$$\delta v_{\mathrm{xc}}(\boldsymbol{r},t) = \int \mathrm{d}t' \int \mathrm{d}^3r'\, f_{\mathrm{xc}}(\boldsymbol{r},\boldsymbol{r}',t-t')\delta n(\boldsymbol{r}',t') \tag{5.31}$$

where f_{xc} is the xc kernel of the ground-state, which depends only on the difference $t-t'$ and vanishes for $t<t'$. Substituting this into (5.30) and Fourier-transforming both sides with respect to time we arrive at

$$\int \mathrm{d}^3r \int \mathrm{d}^3r'\, f_{\mathrm{xc}}(\boldsymbol{r},\boldsymbol{r}',\omega)\delta n(\boldsymbol{r}',\omega)\nabla n_{\mathrm{GS}}(\boldsymbol{r}) = \int \mathrm{d}r'\, \delta n(\boldsymbol{r}',\omega)\nabla v_{\mathrm{xc,\,GS}}(\boldsymbol{r}')\ . \tag{5.32}$$

Finally, taking into account the arbitrariness of $\delta n(\boldsymbol{r},\omega)$ we see that the above equation implies

$$\int \mathrm{d}^3r\, f_{\mathrm{xc}}(\boldsymbol{r},\boldsymbol{r}',\omega)\nabla n_{\mathrm{GS}}(\boldsymbol{r}) = \nabla v_{\mathrm{xc,\,GS}}(\boldsymbol{r}')\ . \tag{5.33}$$

Furthermore, since $f_{\mathrm{xc}}(\boldsymbol{r},\boldsymbol{r}',\omega)$ is symmetric under interchange of $\boldsymbol{r}$ and $\boldsymbol{r}'$ [Vignale 1998] we also have

$$\int \mathrm{d}r' \, f_{\mathrm{xc}}(\boldsymbol{r}, \boldsymbol{r}', \omega) \nabla n_{\mathrm{GS}}(\boldsymbol{r}') = \nabla v_{\mathrm{xc,\,GS}}(\boldsymbol{r}) \, . \tag{5.34}$$

We make use of this form of the identity in the discussion of "ultranonlocality" in Chap. 5.

Deep down, the above conditions are consequences of the fact that we are free to choose an arbitrary reference frame to describe the time evolution of a many-particle system [Vignale 1995a]. Of course, the dynamics in some reference frames will be more complicated than in others. However, the transformation rules for the particle density and for the xc potential are easily worked out, and since the xc potential in one frame must be the same universal functional of the density (and initial state) that it is in any other frame, the knowledge of these transformation rules leads to exact constraints on the form of the functional.

For example consider the transformation

$$\boldsymbol{r} = \boldsymbol{r}' + \boldsymbol{R}(t) \, , \tag{5.35}$$

where $\boldsymbol{R}(t)$ is an arbitrary time-dependent vector. Evidently, the particle density in the transformed frame is related to the particle density in the original frame by

$$n'(\boldsymbol{r}', t) = n(\boldsymbol{r}' + \boldsymbol{R}(t), t) \, . \tag{5.36}$$

and the initial state is transformed according the unitary transformation

$$|\psi'\rangle = \exp\left\{ -\mathrm{i} \sum_{i=1}^{N} m\boldsymbol{r}_i \cdot \dot{\boldsymbol{R}}(0) \right\} \exp\left\{ \mathrm{i} \sum_{i=1}^{N} \boldsymbol{p}_i \cdot \boldsymbol{R}(0) \right\} |\psi\rangle \, , \tag{5.37}$$

where $\boldsymbol{r}_i$ and $\boldsymbol{p}_i$ are the position and momentum operators of the i-th particle. Then it can be shown [Vignale 1995a] that

$$v_{\mathrm{xc}}[|\psi'\rangle, |\phi'\rangle, n'](\boldsymbol{r}', t) = v_{\mathrm{xc}}[|\psi\rangle, |\phi\rangle, n](\boldsymbol{r}' + \boldsymbol{R}(t), t) \, , \tag{5.38}$$

where $v_{\mathrm{xc}}[|\psi\rangle, |\phi\rangle, n](\boldsymbol{r}, t)$ is the xc potential at position $\boldsymbol{r}$ and time t produced by the density n in a system that starts its evolution in the initial many-body state $|\psi\rangle$, with initial Kohn-Sham state $|\phi\rangle$ (see Sect. 11.4.4). $|\phi'\rangle$ and $|\psi'\rangle$ are related to $|\phi\rangle$ and $|\psi\rangle$ respectively by the transformation (5.37). Equation (5.38) can be interpreted as a "generalized translational invariance condition", telling us that the exchange-correlation potential "rides on top" of a globally accelerating density, remaining a constant functional of the instantaneous density. It can be shown that this leads to the zero-force theorem [Vignale 1995a].

A similar set of conditions is arrived at by considering global rotations of the system by an arbitrary time-dependent angle about an arbitrary axis. An example is the "zero torque theorem" given by (11.11b). All these theorems carry over to CDFT with the appropriate modifications. For example the tensor xc kernel of CDFT for a system in the ground-state withdensity $n_{\mathrm{GS}}(\boldsymbol{r})$ satisfies the identity

$$\int \mathrm{d}^3r' \, f_{\mathrm{xc},\,\alpha\beta}(\boldsymbol{r},\boldsymbol{r}',\omega) n_{\mathrm{GS}}(\boldsymbol{r}') = -\frac{\nabla_\alpha \nabla_\beta v_{\mathrm{xc},\,\mathrm{GS}}(\boldsymbol{r})}{\omega^2}\,, \tag{5.39}$$

where α and β denote cartesian components. This is the generalization of (5.34) in CDFT. We refer the reader to [Vignale 1998] for a detailed derivation and discussion of these identities.

6 Multicomponent Density-Functional Theory

R. van Leeuwen and E.K.U. Gross

6.1 Introduction

The coupling between electronic and nuclear motion plays an essential role in a wide range of physical phenomena. A few important research fields in which this is the case are superconductivity in solids, quantum transport where one needs to take into account couplings between electrons and phonons, the polaronic motion in polymer chains, and the ionization-dissociation dynamics of molecules in strong laser fields. Our goal is to set up a time-dependent multicomponent density-functional theory (TDMCDFT) to provide a general framework to describe these diverse phenomena. In TDMCDFT the electrons and nuclei are treated completely quantum mechanically from the outset. The basic variables of the theory are the electron density n, which will be defined in a body-fixed frame attached to the nuclear framework, and the diagonal of the nuclear N-body density matrix Γ, which will depend on all the nuclear coordinates. The chapter is organized as follows: We start out by defining the coordinate transformations to obtain a suitable Hamiltonian for defining our densities to be used as basic variables in the theory. We then discuss the basic one-to-one correspondence between TD potentials and TD densities, and subsequently, the resulting TD Kohn-Sham equations, the action functional, and linear response theory. As an example we discuss a diatomic molecule in a strong laser field.

6.2 Fundamentals

We consider a system composed of $N_{\rm e}$ electrons with coordinates $\{\boldsymbol{r}\}$ and $N_{\rm n}$ nuclei with masses $M_1 \dots M_{N_{\rm n}}$, charges $Z_1 \dots Z_{N_{\rm n}}$, and coordinates denoted by $\{\boldsymbol{R}\}$. By convention, the subscripts "e" and "n" refer to electrons and nuclei, respectively, and atomic units are employed throughout this chapter. In non-relativistic quantum mechanics, the system described above is characterized by the Hamiltonian

$$\hat{H}(t) = \hat{T}_{\rm n}(\{\boldsymbol{R}\}) + \hat{V}_{\rm nn}(\{\boldsymbol{R}\}) + \hat{U}_{\rm ext,\,n}(\{\boldsymbol{R}\},t) + \hat{T}_{\rm e}(\{\boldsymbol{r}\}) + \hat{V}_{\rm ee}(\{\boldsymbol{r}\}) + \hat{U}_{\rm ext,\,e}(\{\boldsymbol{r}\},t) + \hat{V}_{\rm en}(\{\boldsymbol{r}\},\{\boldsymbol{R}\}) , \quad (6.1)$$

R. van Leeuwen and E.K.U. Gross: *Multicomponent Density-Functional Theory*, Lect. Notes Phys. **706**, 93–106 (2006)
DOI 10.1007/3-540-35426-3_6

where

$$\hat{T}_{\mathrm{n}} = -\frac{1}{2}\sum_{\alpha=1}^{N_{\mathrm{n}}} \frac{\nabla_\alpha^2}{M_\alpha} \qquad \text{and} \qquad \hat{T}_{\mathrm{e}} = -\frac{1}{2}\sum_{j=1}^{N_{\mathrm{e}}} \nabla_j^2 \tag{6.2}$$

denote the kinetic-energy operators of the nuclei and electrons, respectively,

$$\hat{V}_{\mathrm{nn}} = \frac{1}{2}\sum_{\substack{\alpha,\beta=1\\ \alpha\neq\beta}}^{N_{\mathrm{n}}} \frac{Z_\alpha Z_\beta}{|\boldsymbol{R}_\alpha - \boldsymbol{R}_\beta|}\,, \qquad \hat{V}_{\mathrm{ee}} = \frac{1}{2}\sum_{\substack{i,j=1\\ i\neq j}}^{N_{\mathrm{e}}} \frac{1}{|\boldsymbol{r}_i - \boldsymbol{r}_j|}\,, \tag{6.3}$$

and

$$\hat{V}_{\mathrm{en}} = -\sum_{j=1}^{N_{\mathrm{e}}}\sum_{\alpha=1}^{N_{\mathrm{n}}} \frac{Z_\alpha}{|\boldsymbol{r}_j - \boldsymbol{R}_\alpha|} \tag{6.4}$$

represent the interparticle Coulomb interactions. Truly external potentials representing, e.g., a laser pulse applied to the system, are contained in

$$\hat{U}_{\mathrm{ext,\,n}}(t) = \sum_{\alpha=1}^{N_{\mathrm{n}}} u_{\mathrm{ext,\,n}}(\boldsymbol{R}_\alpha, t) \tag{6.5a}$$

$$\hat{U}_{\mathrm{ext,\,e}}(t) = \sum_{j=1}^{N_{\mathrm{e}}} u_{\mathrm{ext,\,e}}(\boldsymbol{r}_j, t)\,. \tag{6.5b}$$

Defining electronic and nuclear single-particle densities conjugated to the true external potentials (6.5a) and (6.5b), a multicomponent density-functional theory (MCDFT) formalism can readily be formulated on the basis of the above Hamiltonian [Capitani 1982, Gidopoulos 1998]. However, as discussed in [Kreibich 2000, Kreibich 2001a, Kreibich 2005], such a MCDFT is not useful in practice because the single-particle densities necessarily reflect the symmetry of the true external potentials and are therefore not characteristic of the internal properties of the system. In particular, for all isolated systems where the external potentials (6.5a) and (6.5b) vanish, these densities are constant, as a consequence of the translational invariance of the respective Hamiltonian. A suitable MCDFT is obtained by defining the densities with respect to internal coordinates of the system [Kreibich 2001a, van Leeuwen 2004b]. To this end, new electronic coordinates are introduced according to

$$\boldsymbol{r}'_j = \mathcal{R}(\alpha,\beta,\gamma)\;(\boldsymbol{r}_j - \boldsymbol{R}_{\mathrm{CMN}}) \qquad j = 1\ldots N_{\mathrm{e}}\,, \tag{6.6}$$

where the nuclear center-of-mass is defined as

$$\boldsymbol{R}_{\mathrm{CMN}} := \frac{1}{M_{\mathrm{nuc}}}\sum_{\alpha=1}^{N_{\mathrm{n}}} M_\alpha \boldsymbol{R}_\alpha\,, \quad \text{where} \quad M_{\mathrm{nuc}} = \sum_{\alpha=1}^{N_{\mathrm{n}}} M_\alpha\,. \tag{6.7}$$

The quantity $\mathcal{R}$ is a three-dimensional orthogonal matrix representing the Euler rotations. The Euler angles (α, β, γ) are functions of the nuclear coordinates $\{\boldsymbol{R}\}$ and specify the orientation of the body-fixed coordinate frame.

They can be determined in various ways. One way is by requiring the inertial tensor of the nuclei to be diagonal in the body-fixed frame. The conditions that the off-diagonal elements of the inertia tensor are zero in terms of the rotated coordinates $\mathcal{R}(\boldsymbol{R}_\alpha - \boldsymbol{R}_{\text{CMN}})$ then give three determining equations for the three Euler angles in terms of the nuclear coordinates $\{\boldsymbol{R}\}$ [Villars 1970]. A common alternative to determine the orientation of the body-fixed system is provided by the so-called Eckart conditions [Eckart 1935, Louck 1976, Bunker 1998] (for recent reviews see [Sutcliffe 2000, Meyer 2002]) which are suitable to describe small vibrations in molecules and phonons in solids [van Leeuwen 2004b]. A general and very elegant discussion on the various ways the body-fixed frame can be chosen is given in reference [Littlejohn 1997] . In this work we will not make a specific choice, as our derivations are independent of such choice. The most important point is that, by virtue of (6.6), the electronic coordinates are defined with respect to a coordinate frame that is attached to the nuclear framework and rotates as the nuclear framework rotates. The nuclear coordinates themselves are not transformed any further at this point, i.e.,

$$\boldsymbol{R}'_\alpha = \boldsymbol{R}_\alpha \qquad \alpha = 1 \dots N_{\text{n}} \,. \tag{6.8}$$

Of course, introducing internal nuclear coordinates is also desirable. However, the choice of such coordinates depends strongly on the specific situation to be described: If near-equilibrium situations in systems with well-defined geometries are considered, normal or – for a solid – phonon coordinates are most appropriate, whereas fragmentation processes of molecules are better described in terms of Jacobi coordinates [Meyer 2002, Schinke 1993]. Therefore, to keep a high degree of flexibility, the nuclear coordinates are left unchanged for the time being and are transformed to internal coordinates only prior to actual applications of the final equations that we will derive.

As a result of the coordinate changes of (6.6), the Hamiltonian (6.1) transforms into

$$\begin{aligned}\hat{H}(t) = \hat{T}_{\text{n}}(\{\boldsymbol{R}\}) + \hat{V}_{\text{nn}}(\{\boldsymbol{R}\}) + \hat{U}_{\text{ext, n}}(\{\boldsymbol{R}\}, t) + \hat{T}_{\text{e}}(\{\boldsymbol{r}'\}) + \hat{V}_{\text{ee}}(\{\boldsymbol{r}'\})\\ + \hat{T}_{\text{MPC}}(\{\boldsymbol{r}'\}, \{\boldsymbol{R}\}) + \hat{V}_{\text{en}}(\{\boldsymbol{r}'\}, \{\boldsymbol{R}\}) + \hat{U}_{\text{ext, e}}(\{\boldsymbol{r}'\}, \{\boldsymbol{R}\}, t) \,.\end{aligned} \tag{6.9}$$

Since we transformed to a noninertial coordinate frame a mass-polarization and Coriolis (MPC) term

$$\hat{T}_{\text{MPC}} := \sum_{\alpha=1}^{N_{\text{n}}} -\frac{1}{2M_\alpha}\left[\nabla_{\boldsymbol{R}_\alpha} + \sum_{j=1}^{N_{\text{e}}} \frac{\partial \boldsymbol{r}'_j}{\partial \boldsymbol{R}_\alpha} \nabla_{\boldsymbol{r}'_j}\right]^2 - \hat{T}_{\text{n}}(\{\boldsymbol{R}\}) \tag{6.10}$$

appears. Obviously, this MPC term is not symmetric in the electronic and nuclear coordinates. However, this was not expected since only the electrons refer to a noninertial coordinate frame, whereas the nuclei are still defined with respect to the inertial frame. Therefore, all MPC terms arise solely from the electronic coordinates, representing fictious forces due to the electronic motion in noninertial systems (for a detailed form of these terms within the current coordinate transformation see [van Leeuwen 2004b]). The kinetic-energy operators $\hat{T}_{\rm n}$ and $\hat{T}_{\rm e}$, the electron-electron and nuclear-nuclear interactions, as well as the true external potential $\hat{U}_{\rm ext,\,n}$ acting on the nuclei are formally unchanged in (6.9) and therefore given by (6.2) and (6.3) with the new coordinates replacing the old ones, whereas the electron-nuclear interaction now reads

$$\hat{V}_{\rm en}(\{\boldsymbol{r}'\},\{\boldsymbol{R}\}) = -\sum_{j=1}^{N_{\rm e}}\sum_{\alpha=1}^{N_{\rm n}} \frac{Z_\alpha}{|\boldsymbol{r}'_j - \mathcal{R}(\alpha,\beta,\gamma)(\boldsymbol{R}_\alpha - \boldsymbol{R}_{\rm CMN})|} \, . \tag{6.11}$$

The quantity

$$\boldsymbol{R}''_\alpha = \mathcal{R}(\alpha,\beta,\gamma)(\boldsymbol{R}_\alpha - \boldsymbol{R}_{\rm CMN}) \tag{6.12}$$

that appears in (6.11) is a so-called shape coordinate [Littlejohn 1997, van Leeuwen 2004b], i.e., it is invariant under rotations and translations of the nuclear framework. This is, of course, precisely the purpose of introducing a body-fixed frame: The attractive nuclear Coulomb potential (6.11) that the electrons in the body-fixed frame experience is invariant under rotations or translations of the nuclear framework. As a further consequence of the coordinate transformation (6.6), the true external potential acting on the electrons now not only depends on the electronic coordinates, but also on all the nuclear coordinates:

$$\hat{U}_{\rm ext,\,e}(\{\boldsymbol{r}'\},\{\boldsymbol{R}\},t) = \sum_{j=1}^{N_{\rm e}} u_{\rm ext,\,e}(\mathcal{R}^{-1}\boldsymbol{r}'_j + \boldsymbol{R}_{\rm CMN}, t) \, . \tag{6.13}$$

Therefore, in the chosen coordinate system, the electronic external potential is not a one-body operator anymore, but acts as an effective interaction. Consequences of this fact are discussed later.

6.2.1 Definition of the Densities

As already mentioned above, it is not useful to define electronic and nuclear single-particle densities in terms of the inertial coordinates $\boldsymbol{r}$ and $\boldsymbol{R}$, since such densities necessarily reflect the symmetry of the corresponding true external potentials, e.g., they are constant for vanishing external potentials. Instead, we use the diagonal of the nuclear $N_{\rm n}$-body density matrix

$$\Gamma(\{\boldsymbol{R}\},t) = \sum_{\{s\},\{\sigma\}} \int {\rm d}^3r'_1 \cdots \int {\rm d}^3r'_{N_{\rm e}} \, \left|\Psi_{\{s\}\{\sigma\}}(\{\boldsymbol{R}\},\{\boldsymbol{r}'\},t)\right|^2 \, , \tag{6.14}$$

and the electronic single-particle density referring to the body-fixed frame:

$$n(\boldsymbol{r}',t) = N_{\mathrm{e}} \sum_{\{s\},\{\sigma\}} \int \mathrm{d}^3R_1' \cdots \int \mathrm{d}^3R_{N_{\mathrm{n}}}' \int \mathrm{d}^3r_1' \cdots \int \mathrm{d}^3r_{N_{\mathrm{e}}-1}' \left|\Psi_{\{s\}\{\sigma\}}(\{\boldsymbol{R}\},\{\boldsymbol{r}'\},t)\right|^2 . \quad (6.15)$$

Here $\Psi_{\{s\}\{\sigma\}}(\{\boldsymbol{R}\},\{\boldsymbol{r}'\},t)$ represents the full solution of the TD Schrödinger equation with the Hamiltonian (6.9). The quantities $\{s\}$ and $\{\sigma\}$ denote the nuclear and electronic spin coordinates. The electronic density (6.15) represents a conditional density. It is proportional to the probability density of finding an electron at postion $\boldsymbol{r}'$ as measured from the nuclear center-of-mass, given a certain orientation of the nuclear framework. Therefore the electronic density calculated through (6.15) reflects the internal symmetries of the system, e.g., the cylindrical symmetry of a diatomic molecule, instead of the Galilean symmetry of the underlying space.

6.3 The Runge-Gross Theorem for Multicomponent Systems

In order to set up a density-functional framework, our next task is to prove the analogue of the Runge-Gross theorem [Runge 1984] for multicomponent systems. To this end, we slightly modify the Hamiltonian (6.9) to take the form

$$\hat{H}(t) = \hat{T}_{\mathrm{n}}(\{\boldsymbol{R}\}) + \hat{V}_{\mathrm{nn}}(\{\boldsymbol{R}\}) + \hat{T}_{\mathrm{e}}(\{\boldsymbol{r}'\}) + \hat{V}_{\mathrm{ee}}(\{\boldsymbol{r}'\}) + \hat{T}_{\mathrm{MPC}}(\{\boldsymbol{r}'\},\{\boldsymbol{R}\}) + \hat{V}_{\mathrm{en}}(\{\boldsymbol{r}'\},\{\boldsymbol{R}\}) + \hat{U}_{\mathrm{ext,\,e}}(\{\boldsymbol{r}'\},\{\boldsymbol{R}\},t) + \hat{V}_{\mathrm{ext,\,n}}(\{\boldsymbol{R}\},t) + \hat{V}_{\mathrm{ext,\,e}}(\{\boldsymbol{r}'\},t). \quad (6.16)$$

The potentials $\hat{V}_{\mathrm{ext,\,n}}(\{\boldsymbol{R}\},t)$ and $\hat{V}_{\mathrm{ext,\,e}}(\{\boldsymbol{r}'\},t)$, given by

$$\hat{V}_{\mathrm{ext,\,n}}(\{\boldsymbol{R}\},t) = v_{\mathrm{ext,\,n}}(\{\boldsymbol{R}\},t) \quad \text{and} \quad \hat{V}_{\mathrm{ext,\,e}}(\{\boldsymbol{r}'\},t) = \sum_{j=1}^{N_{\mathrm{e}}} v_{\mathrm{ext,\,e}}(\boldsymbol{r}_j',t)\,, \quad (6.17)$$

are potentials conjugate to the densities $\Gamma(\{\boldsymbol{R}\},t)$ and $n(\{\boldsymbol{r}'\},t)$ and are introduced to provide the necessary mappings between potentials and densities. In the special case $\hat{V}_{\mathrm{ext,\,n}}(\{\boldsymbol{R}\},t) = \hat{U}_{\mathrm{ext,\,n}}(\{\boldsymbol{R}\},t)$ and $\hat{V}_{\mathrm{ext,\,e}}(\{\boldsymbol{r}'\},t) = 0$, the external potentials reduce to those of the Hamiltonian (6.9). It is important to note that the potential $\hat{U}_{\mathrm{ext,\,e}}(\{\boldsymbol{r}'\},\{\boldsymbol{R}\},t)$ depends on both the electronic and nuclear coordinates and is therefore treated as a fixed many-body term in Hamiltonian (6.16). The mass-polarization and Coriolis terms in $\hat{T}_{\mathrm{MPC}}$ are complicated many-body operators. They are treated here as additional electron-nuclear interactions which ultimately enter the exchange-correlation functional. For Hamiltonians of the form (6.16) we can apply the proof of the basic $1-1$ correspondence along the same lines as Li and Tong [Li 1986]. Two sets of densities $\{\Gamma(\{\boldsymbol{R}\},t), n(\boldsymbol{r}',t)\}$ and $\{\Gamma'(\{\boldsymbol{R}\},t), n'(\boldsymbol{r}',t)\}$, which evolve from a common initial state Ψ_0 at $t = t_0$ under the influence of two sets of potentials $\{v_{\mathrm{ext,\,n}}(\{\boldsymbol{R}\},t), v_{\mathrm{ext,\,e}}(\boldsymbol{r}',t)\}$ and $\{v'_{\mathrm{ext,\,n}}(\{\boldsymbol{R}\},t), v'_{\mathrm{ext,\,e}}(\boldsymbol{r}',t)\}$ always become different infinitesimally after t_0 provided that at least one component of the potentials differs by more than a purely time-dependent function:

$$v_{\mathrm{ext,\,n}}(\{\boldsymbol{R}\},t) \neq v'_{\mathrm{ext,\,n}}(\{\boldsymbol{R}\},t) + C(t) \quad \text{or} \quad v_{\mathrm{ext,\,e}}(\boldsymbol{r}',t) \neq v'_{\mathrm{ext,\,e}}(\boldsymbol{r}',t) + C(t) \tag{6.18}$$

Consequently a one-to-one mapping between time-dependent densities and external potentials,

$$\{v_{\mathrm{ext,\,n}}(\{\boldsymbol{R}\},t), v_{\mathrm{ext,\,e}}(\boldsymbol{r}',t)\} \leftrightarrow \{\Gamma(\{\boldsymbol{R}\},t), n(\boldsymbol{r}',t)\} \tag{6.19}$$

is established for a given initial state Ψ_0. We again stress that since the external potential acting on the electrons $\hat{U}_{\mathrm{ext,\,e}}(\{\boldsymbol{r}'\},\{\boldsymbol{R}\},t)$ in the body-fixed frame attains the form of an electron-nuclear interaction, the $1-1$ mapping is still functionally dependent on $u_{\mathrm{ext,\,e}}(\boldsymbol{r}',\{\boldsymbol{R}\},t)$.

6.4 The Kohn-Sham Scheme for Multicomponent Systems

On the basis of the multi-component Runge-Gross (MCRG) theorem we can set up the Kohn-Sham equations. For this we consider an auxiliary system with Hamiltonian

$$\hat{H}_{\mathrm{KS}}(t) = \hat{T}_{\mathrm{n}}(\{\boldsymbol{R}\}) + \hat{T}_{\mathrm{e}}(\{\boldsymbol{r}'\}) + \hat{V}_{\mathrm{KS,\,n}}(\{\boldsymbol{R}\},t) + \hat{V}_{\mathrm{KS,\,e}}(\{\boldsymbol{r}'\},t)\;, \tag{6.20}$$

where we introduced the potentials

$$\hat{V}_{\mathrm{KS,\,n}}(\{\boldsymbol{R}\},t) = v_{\mathrm{KS,\,n}}(\{\boldsymbol{R}\},t) \quad \text{and} \quad \hat{V}_{\mathrm{KS,\,e}}(\{\boldsymbol{r}'\},t) = \sum_{j=1}^{N_{\mathrm{e}}} v_{\mathrm{KS,\,e}}(\boldsymbol{r}'_j,t)\;. \tag{6.21}$$

This represents a system in which the interelectronic interaction as well as the interaction between the nuclei and the electrons has been switched off. According to the MCRG theorem there is at most one set of potentials $\left\{\hat{V}_{\mathrm{KS,\,n}}(\{\boldsymbol{R}\},t), \hat{V}_{\mathrm{KS,\,e}}(\{\boldsymbol{r}'\},t)\right\}$ (up to a purely time-dependent function) that reproduces a given set of densities $\{\Gamma(\{\boldsymbol{R}\},t), n(\boldsymbol{r}',t)\}$. The potentials determined in this way are therefore functionals of the densities n and Γ and will henceforth be denoted as the Kohn-Sham potentials for the multicomponent system. The corresponding Hamiltonian of (6.20) will be denoted as the multicomponent Kohn-Sham Hamiltonian. In the Kohn-Sham Hamiltonian the electronic and nuclear motion have become separated. If we therefore choose the initial Kohn-Sham wavefunction $\Phi_{\mathrm{KS,\,0}}$ to be a product of a nuclear and an electronic wavefunction then the time-dependent Kohn-Sham wavefunction will also be such a product, i.e.,

$$\Phi_{KS,\{s\}\{\sigma\}}(\{\boldsymbol{R}\},\{\boldsymbol{r}'\},t) = \Phi_{\mathrm{e},\{\sigma\}}(\{\boldsymbol{r}'\},t)\,\Phi_{\mathrm{n},\{\mathrm{s}\}}(\{\boldsymbol{R}\},t) \tag{6.22}$$

and the corresponding densities are given by

$$\Gamma(\{\boldsymbol{R}\},t) = \sum_{\{s\}} \left|\Phi_{\mathrm{n},\{\mathrm{s}\}}(\{\boldsymbol{R}\},t)\right|^2 \tag{6.23a}$$

$$n(\boldsymbol{r}',t) = N_{\mathrm{e}} \sum_{\{\sigma\}} \int \mathrm{d}^3 r_1' \cdots \int \mathrm{d}^3 r_{N_{\mathrm{e}}-1}' \left|\Phi_{\mathrm{e},\{\sigma\}}(\{\boldsymbol{r}'\},t)\right|^2 . \tag{6.23b}$$

The electronic and nuclear Kohn-Sham wavefunctions satisfy the equations

$$\left\{\mathrm{i}\frac{\partial}{\partial t} - \hat{T}_{\mathrm{n}}(\{\boldsymbol{R}\}) - \hat{V}_{\mathrm{KS,\,n}}[n,\Gamma](\{\boldsymbol{R}\},t)\right\} \Phi_{\mathrm{n},\{\mathrm{s}\}}(\{\boldsymbol{R}\},t) = 0 \tag{6.24a}$$

$$\left\{\mathrm{i}\frac{\partial}{\partial t} - \hat{T}_{\mathrm{e}}(\{\boldsymbol{r}'\}) - \hat{V}_{\mathrm{KS,\,e}}[n,\Gamma](\{\boldsymbol{r}'\},t)\right\} \Phi_{\mathrm{e},\{\sigma\}}(\{\boldsymbol{r}'\},t) = 0\,. \tag{6.24b}$$

Note that the potential $\hat{V}_{\mathrm{KS,\,n}}$ in the nuclear Kohn-Sham equation (6.24a) is an N_{n}-body interaction, whereas the electronic Kohn-Sham potential $\hat{V}_{\mathrm{KS,\,e}}$ is a one-body operator. Hence, by choosing the initial electronic Kohn-Sham wavefunction as a Slater determinant consisting of orbitals φ_j , the electronic Kohn-Sham (6.24b) attains the usual form

$$\left\{\mathrm{i}\frac{\partial}{\partial t} - \left[-\frac{1}{2}\nabla'^2 + v_{\mathrm{KS,\,e}}[n,\Gamma](\boldsymbol{r}',t)\right]\right\} \varphi_j(\boldsymbol{r}',t) = 0 \tag{6.25a}$$

$$n(\boldsymbol{r}',t) = \sum_{j}^{N_{\mathrm{e}}} |\varphi_j(\boldsymbol{r}',t)|^2 . \tag{6.25b}$$

The nuclear (6.23a) and (6.24a), together with the electronic (6.25a) and (6.25b), provide a formally exact scheme to calculate the electronic density n and N_{n}-body nuclear density Γ. For practical applications it remains to obtain good approximations for the potentials $v_{\mathrm{KS,\,n}}[n,\Gamma]$ and $v_{\mathrm{KS,\,e}}[n,\Gamma]$. More insight into this question is obtained from the multicomponent action functional to be discussed in the next paragraph.

6.5 The Multicomponent Action

We start by defining a multicomponent action functional

$$\tilde{A}[v_{\mathrm{ext,\,e}}, v_{\mathrm{ext,\,n}}] = \mathrm{i}\ln\langle\Psi_0|\hat{\mathcal{T}}_{\mathrm{C}} \exp\left\{-\mathrm{i}\int_C \mathrm{d}t\, \hat{H}(t)\right\}|\Psi_0\rangle \tag{6.26}$$

The Hamiltonian in this expression is the one of (6.16). Furthermore Ψ_0 is the initial state of the system and $\hat{\mathcal{T}}_C$ denotes time-ordering along the Keldysh time contour C running along the real time-axis from t_0 to t and back to t_0. The time-dependent potentials $v_{\mathrm{ext,\,e}}$ and $v_{\mathrm{ext,\,n}}$ are correspondingly defined on this contour. The case discussed here is for an initial pure state. In case the initial system is in thermodynamic equilibrium the expectation value with respect to Ψ_0 can be replaced by a thermodynamic trace and the contour can be extended to include a final vertical stretch running from t_0 to $t_0 - \mathrm{i}\beta$, where β is the inverse temperature of the initial ensemble. In that case the functional is closely related to the grand potential as is extensively discussed in Chap. 3. The main property of the action (6.26) which is important for multicomponent density-functional theory is that

$$\frac{\delta \tilde{A}}{\delta v_{\mathrm{ext,\,e}}(\boldsymbol{r},t)} = n(\boldsymbol{r},t) \quad \text{and} \quad \frac{\delta \tilde{A}}{\delta v_{\mathrm{ext,\,n}}(\{\boldsymbol{R}\},t)} = \Gamma(\{\boldsymbol{R}\},t)\,. \tag{6.27}$$

(From now on, for ease of notation, we will remove the prime from the electronic coordinate.) We now do a Legendre transform to obtain a functional of n and Γ and we define

$$\begin{aligned} A[n,\Gamma] = &-\tilde{A}[v_{\mathrm{ext,\,e}}, v_{\mathrm{ext,\,n}}] + \int_C \mathrm{d}t \int \mathrm{d}^3r\, n(\boldsymbol{r},t)\, v_{\mathrm{ext,\,e}}(\boldsymbol{r},t) \\ &+ \int_C \mathrm{d}t \int \mathrm{d}^3R_1 \cdots \int \mathrm{d}^3R_{N_\mathrm{n}}\, \Gamma(\{\boldsymbol{R}\},t)\, v_{\mathrm{ext,\,n}}(\{\boldsymbol{R}\},t)\,, \end{aligned} \tag{6.28}$$

where in this equation $v_{\mathrm{ext,\,e}}$ and $v_{\mathrm{ext,\,n}}$ (by virtue of the MCRG theorem) are now regarded as functionals of n and Γ. From the chain rule of differentiation we then easily obtain

$$\frac{\delta A}{\delta n(\boldsymbol{r},t)} = v_{\mathrm{ext,\,e}}(\boldsymbol{r},t) \quad \text{and} \quad \frac{\delta A}{\delta \Gamma(\{\boldsymbol{R}\},t)} = v_{\mathrm{ext,\,n}}(\{\boldsymbol{R}\},t)\,. \tag{6.29}$$

For the Hamiltonian $\hat{H}_{\mathrm{KS}}(t)$ of (6.20) we can now further define an action functional analogous to (6.26)

$$\tilde{A}_{\mathrm{KS}}[v_{\mathrm{KS,\,e}}, v_{\mathrm{KS,\,n}}] = \mathrm{i}\ln\langle\Phi_0|\hat{\mathcal{T}}_C \exp\left\{-\mathrm{i}\int_C \mathrm{d}t\, \hat{H}_{\mathrm{KS}}(t)\right\}|\Phi_0\rangle \tag{6.30}$$

where Φ_0 is the initial state of the auxiliary system. By a Legendre transform we then obtain the functional $A_{\mathrm{KS}}[n,\Gamma]$. With $A[n,\Gamma]$ and $A_{\mathrm{KS}}[n,\Gamma]$ well-defined we now can define the exchange-correlation part $A_{\mathrm{xc}}[n,\Gamma]$ of the action through the equation

$$\begin{aligned}
A[n,\Gamma] = A_{\mathrm{KS}}[n,\Gamma] &- \frac{1}{2}\int_C \mathrm{d}t \int \mathrm{d}^3 r_1 \int \mathrm{d}^3 r_2\, v_{\mathrm{ee}}(\boldsymbol{r}_1,\boldsymbol{r}_2) n(\boldsymbol{r}_1,t) n(\boldsymbol{r}_2,t) \\
&- \int_C \mathrm{d}t \int \mathrm{d}^3 r \int \mathrm{d}^3 R_1 \cdots \int \mathrm{d}^3 R_{N_\mathrm{n}} \left[v_{\mathrm{en}}(\boldsymbol{r},\{\boldsymbol{R}\}) \right. \\
&\left. + u_{\mathrm{ext,\,e}}(\boldsymbol{r},\{\boldsymbol{R}\},t) \right] n(\boldsymbol{r},t)\Gamma(\{\boldsymbol{R}\},t) \\
&- \int_C \mathrm{d}t \int \mathrm{d}^3 R_1 \cdots \int \mathrm{d}^3 R_{N_\mathrm{n}}\, v_{\mathrm{nn}}(\{\boldsymbol{R}\})\Gamma(\{\boldsymbol{R}\},t) - A_{\mathrm{xc}}[n,\Gamma] \,, \quad (6.31)
\end{aligned}$$

where we subtracted the Hartree-like parts of the electron-electron and electron-nuclear interaction and the internuclear repulsion, using the definitions

$$v_{\mathrm{en}}(\boldsymbol{r},\{\boldsymbol{R}\}) = -\sum_{\alpha=1}^{N_\mathrm{n}} \frac{Z_\alpha}{|\boldsymbol{r} - \mathcal{R}(\boldsymbol{R} - \boldsymbol{R}_{\mathrm{CMN}})|} \quad (6.32a)$$

$$u_{\mathrm{ext,\,e}}(\boldsymbol{r},\{\boldsymbol{R}\},t) = u_{\mathrm{ext,\,e}}(\mathcal{R}^{-1}\boldsymbol{r} + \boldsymbol{R}_{\mathrm{CMN}},t)\,. \quad (6.32b)$$

These Hartree terms are treated separately because they are expected to be the dominant potential-energy contributions whereas the remainder, $A_{\mathrm{xc}}[n,\Gamma]$, is expected to be smaller. No such dominant contributions arise from the mass-polarization and Coriolis terms which are usually rather small. The contributions coming from $\hat{T}_{\mathrm{MPC}}$ are therefore completely retained in $A_{\mathrm{xc}}[n,\Gamma]$. Differentiation of (6.31) with respect to n and Γ then yields

$$\begin{aligned}
v_{\mathrm{KS,\,e}}(\boldsymbol{r},t) = v_{\mathrm{ext,\,e}}(\boldsymbol{r},t) &+ \int \mathrm{d}^3 r'\, v_{\mathrm{ee}}(\boldsymbol{r},\boldsymbol{r}') n(\boldsymbol{r}',t) \\
&+ \int \mathrm{d}^3 R_1 \cdots \int \mathrm{d}^3 R_{N_\mathrm{n}} \left[v_{\mathrm{en}}(\boldsymbol{r},\{\boldsymbol{R}\}) \right. \\
&\left. + u_{\mathrm{ext,\,e}}(\boldsymbol{r},\{\boldsymbol{R}\},t) \right] \Gamma(\{\boldsymbol{R}\},t) + v_{\mathrm{xc,\,e}}(\boldsymbol{r},t)\,, \quad (6.33)
\end{aligned}$$

and

$$\begin{aligned}
v_{\mathrm{KS,\,n}}(\{\boldsymbol{R}\},t) &= v_{\mathrm{ext,\,n}}(\{\boldsymbol{R}\},t) + v_{\mathrm{nn}}(\{\boldsymbol{R}\}) \\
&+ \int \mathrm{d}^3 r \left[v_{\mathrm{en}}(\boldsymbol{r},\{\boldsymbol{R}\}) + u_{\mathrm{ext,\,e}}(\boldsymbol{r},\{\boldsymbol{R}\},t) \right] n(\boldsymbol{r},t) + v_{\mathrm{xc,\,n}}(\{\boldsymbol{R}\},t)\,, \quad (6.34)
\end{aligned}$$

where we have defined the electronic and nuclear exchange-correlation potentials as

$$v_{\mathrm{xc,\,e}}(\boldsymbol{r},t) = \frac{\delta A_{\mathrm{xc}}[n,\Gamma]}{\delta n(\boldsymbol{r},t)} \quad \text{and} \quad v_{\mathrm{xc,\,n}}(\{\boldsymbol{R}\},t) = \frac{\delta A_{\mathrm{xc}}[n,\Gamma]}{\delta \Gamma(\{\boldsymbol{R}\},t)}. \quad (6.35)$$

The main question is now how to obtain explicit functionals for the exchange-correlation potentials. One of the most promising ways of obtaining these may be the development of orbital functionals as in the OEP approach. Such functionals can be deduced by a diagrammatic expansion of the action functionals.

6.6 Linear Response and Multicomponent Systems

We will now consider the important case of linear response in the multicomponent formalism. Such approach will, for instance, be very useful in the weak field problems such as the electron-phonon coupling in solids. For convenience we first introduce the notation $i = (\boldsymbol{r}_i, t_i)$ and $\underline{i} = (\{\boldsymbol{R}\}, t_i)$. Let us then define the set of response functions:

$$\boldsymbol{\chi}_{12} = \begin{pmatrix} \chi_{\mathrm{ee}}(1,2) & \chi_{\mathrm{en}}(1,\underline{2}) \\ \chi_{\mathrm{ne}}(\underline{1},2) & \chi_{\mathrm{nn}}(\underline{1},\underline{2}) \end{pmatrix} = \begin{pmatrix} \frac{\delta n(1)}{\delta v_{\mathrm{e}}(2)} & \frac{\delta n(1)}{\delta v_{\mathrm{n}}(\underline{2})} \\ \frac{\delta \Gamma(\underline{1})}{\delta v_{\mathrm{e}}(2)} & \frac{\delta \Gamma(\underline{1})}{\delta v_{\mathrm{n}}(\underline{2})} \end{pmatrix} . \tag{6.36}$$

Similarly for the Kohn-Sham system we have

$$\boldsymbol{\chi}_{\mathrm{KS},\,12} = \begin{pmatrix} \chi_{\mathrm{KS,\,ee}}(1,2) & 0 \\ 0 & \chi_{\mathrm{KS,\,nn}}(\underline{1},\underline{2}) \end{pmatrix} = \begin{pmatrix} \frac{\delta n(1)}{\delta v_{\mathrm{KS,\,e}}(2)} & \frac{\delta n(1)}{\delta v_{\mathrm{KS,\,n}}(\underline{2})} \\ \frac{\delta \Gamma(\underline{1})}{\delta v_{\mathrm{KS,\,e}}(2)} & \frac{\delta \Gamma(\underline{1})}{\delta v_{\mathrm{KS,\,n}}(\underline{2})} \end{pmatrix} , \tag{6.37}$$

in which the mixed response functions $\chi_{\mathrm{KS,\,en}} = \chi_{\mathrm{KS,\,ne}} = 0$ since in the Kohn-Sham system the nuclear and electronic systems are decoupled. The two sets of response functions are related by an equation that is very similar to that of ordinary TDDFT

$$\boldsymbol{\chi}_{12} = \boldsymbol{\chi}_{\mathrm{KS},\,12} + \boldsymbol{\chi}_{\mathrm{KS},\,13} \cdot (\boldsymbol{v}_{34} + \boldsymbol{f}_{\mathrm{xc},\,34}) \cdot \boldsymbol{\chi}_{42} \tag{6.38}$$

where "·" denotes a matrix product and integration over the variables 3 and 4, respectively. The matrices $\boldsymbol{f}_{\mathrm{xc}}$ and $\boldsymbol{v}$ are defined as

$$\boldsymbol{f}_{\mathrm{xc},\,12} = \begin{pmatrix} f_{\mathrm{xc,\,ee}}(1,2) & f_{\mathrm{xc,\,en}}(1,\underline{2}) \\ f_{\mathrm{xc,\,ne}}(\underline{1},2) & f_{\mathrm{xc,\,nn}}(\underline{1},\underline{2}) \end{pmatrix} = \begin{pmatrix} \frac{\delta v_{\mathrm{xc,\,e}}(1)}{\delta n(2)} & \frac{\delta v_{\mathrm{xc,\,e}}(1)}{\delta \Gamma(\underline{2})} \\ \frac{\delta v_{\mathrm{xc,\,n}}(\underline{1})}{\delta n(2)} & \frac{\delta v_{\mathrm{xc,\,n}}(\underline{1})}{\delta \Gamma(\underline{2})} \end{pmatrix} \tag{6.39a}$$

$$\boldsymbol{v}_{12} = \begin{pmatrix} v_{\mathrm{ee}}(1,2) & v_{\mathrm{en}}(1,\underline{2}) + u_{\mathrm{ext,\,e}}(1,\underline{2}) \\ v_{\mathrm{en}}(2,\underline{1}) + u_{\mathrm{ext,\,e}}(2,\underline{1}) & 0 \end{pmatrix} . \tag{6.39b}$$

The (6.38) is the central equation of the multicomponent response theory and is readily derived by application of the chain rule of differentiation. As an example we calculate

$$\begin{aligned} \chi_{\mathrm{ee}}(1,2) &= \frac{\delta n(1)}{\delta v_{\mathrm{e}}(2)} \\ &= \int \mathrm{d}3\, \frac{\delta n(1)}{\delta v_{\mathrm{KS,\,e}}(3)} \frac{\delta v_{\mathrm{KS,\,e}}(3)}{\delta v_{\mathrm{ext,\,e}}(2)} + \int \mathrm{d}\underline{3}\, \frac{\delta n(1)}{\delta v_{\mathrm{KS,\,n}}(\underline{3})} \frac{\delta v_{\mathrm{KS,\,n}}(\underline{3})}{\delta v_{\mathrm{ext,\,e}}(2)} \\ &= \frac{\delta n(1)}{\delta v_{\mathrm{KS,\,e}}(2)} + \int \mathrm{d}3\, \frac{\delta n(1)}{\delta v_{\mathrm{KS,\,e}}(3)} \frac{\delta v_{\mathrm{Hxc,\,e}}(3)}{\delta v_{\mathrm{ext,\,e}}(2)} , \end{aligned} \tag{6.40}$$

From which readily follows

$$\chi_{\rm ee}(1,2) = \chi_{\rm KS,\,ee}(1,2) + \int {\rm d}3\, \chi_{\rm KS,\,ee}(1,3) \times \left\{ \int {\rm d}4\, \frac{\delta v_{\rm Hxc,\,e}(3)}{\delta n(4)} \chi_{\rm ee}(4,2) + \int {\rm d}\underline{4}\, \frac{\delta v_{\rm Hxc,\,e}(3)}{\delta \Gamma(\underline{4})} \chi_{\rm ne}(\underline{4},2) \right\}, \tag{6.41}$$

where $v_{\rm Hxc,\,e} = v_{\rm KS,\,e} - v_{\rm e}$. We further have

$$\frac{\delta v_{\rm Hxc,\,e}(3)}{\delta n(4)} = v_{\rm ee}(3,4) + f_{\rm xc,\,ee}(1,2) \tag{6.42a}$$

$$\frac{\delta v_{\rm Hxc,\,e}(3)}{\delta \Gamma(\underline{4})} = v_{\rm en}(3,\underline{4}) + u_{\rm ext,\,e}(3,\underline{4}) + f_{\rm xc,\,en}(3,\underline{4})\,. \tag{6.42b}$$

Inserting these expressions into (6.41) we have established one entry in the matrix (6.38). The other entries can be verified analogously. We finally note that (6.39b) still contains the term $u_{\rm ext,\,e}$, which is inconvenient in practice. However, to calculate the linear response to the true external field we anyway need to expand further in powers of $u_{\rm ext,\,e}$. If we do this we obtain (6.38) with $u_{\rm ext,\,e} = 0$ in (6.39b) and two additional equations for the response functions $\delta n/\delta u_{\rm ext,\,e}$ and $\delta\Gamma/\delta u_{\rm ext,\,e}$ which will not be discussed here [Butriy 2005]. From the structure of the linear response equation (6.38) it is readily seen that electronic Kohn-Sham excitations (poles of $\chi_{\rm ee,\,KS}$) and nuclear vibrational Kohn-Sham excitations (poles of $\chi_{\rm nn,\,KS}$) will in general mix. The exchange-correlation kernels in $\boldsymbol{f}_{\rm xc}$ will then have to provide the additional shift such that the true response functions in $\boldsymbol{\chi}$ will contain the true excitations of the coupled electron-nuclear system.

6.7 Example

As an application of the formalism we discuss the case of a diatomic molecule in a strong laser field. The Hamiltonian of this system in laboratory frame coordinates is given by

$$\hat{H}(t) = -\frac{1}{2M_1}\nabla^2_{\boldsymbol{R}_1} - \frac{1}{2M_2}\nabla^2_{\boldsymbol{R}_2} - \frac{1}{2}\sum_{i=1}^{N_{\rm e}} \nabla_i^2 + v_{\rm en} + v_{\rm ee} + v_{\rm nn} + v_{\rm laser}(t) \tag{6.43}$$

where

$$v_{\rm nn}(\{\boldsymbol{R}\}) = \frac{Z_1 Z_2}{|\boldsymbol{R}_1 - \boldsymbol{R}_2|} \tag{6.44a}$$

$$v_{\rm en}(\{\boldsymbol{r}\},\{\boldsymbol{R}\}) = -\sum_{i=1}^{N_{\rm e}} \left\{ \frac{Z_1}{|\boldsymbol{r}_i - \boldsymbol{R}_1|} + \frac{Z_2}{|\boldsymbol{r}_i - \boldsymbol{R}_2|} \right\} \tag{6.44b}$$

$$v_{\rm laser}(\{\boldsymbol{r}\},\{\boldsymbol{R}\},t) = \left\{ \sum_{i=1}^{N_{\rm e}} \boldsymbol{r}_i - Z_1\boldsymbol{R}_1 - Z_2\boldsymbol{R}_2 \right\} \cdot \boldsymbol{E}(t) \tag{6.44c}$$

and where $\boldsymbol{E}(t)$ represents the electric field of the laser.

We now have to perform a suitable body-fixed frame transformation to refer the electron coordinates to a nuclear frame. For the diatomic molecule a natural choice presents itself: We determine the Euler angles by the requirement that the internuclear axis be parallel to the z-axis in the body-fixed frame, i.e., $\mathcal{R}(\boldsymbol{R}) = R\boldsymbol{e}_z$, where $\boldsymbol{R} = \boldsymbol{R}_1 - \boldsymbol{R}_2$ and $R = |\boldsymbol{R}|$. For the special case of the diatomic molecule only two Euler angles are needed to specify the rotation matrix $\mathcal{R}$. From (6.6) and (6.7) we see that the electron-nuclear interaction and the external laser field transform to

$$v_{\mathrm{en}}(\{\boldsymbol{r}'\},\{\boldsymbol{R}\}) = -\sum_{i=1}^{N_{\mathrm{e}}} \left\{ \frac{Z_1}{|\boldsymbol{r}'_i - \frac{M_2}{M_1+M_2} R\boldsymbol{e}_z|} + \frac{Z_2}{|\boldsymbol{r}'_i + \frac{M_1}{M_1+M_2} R\boldsymbol{e}_z|} \right\} \tag{6.45a}$$

$$v_{\mathrm{laser}}(t) = \left\{ N_{\mathrm{e}}\boldsymbol{R}_{\mathrm{CMN}} - Z_1\boldsymbol{R}_1 - Z_2\boldsymbol{R}_2 + \sum_{i=1}^{N_{\mathrm{e}}} \mathcal{R}^{-1}\boldsymbol{r}'_i \right\} \cdot \boldsymbol{E}(t)\,. \tag{6.45b}$$

With these expressions the Kohn-Sham potentials of (6.33) and (6.34) attain the form

$$v_{\mathrm{KS,\,e}}(\boldsymbol{r},t) = \int \mathrm{d}^3 r'\, v_{\mathrm{ee}}(\boldsymbol{r},\boldsymbol{r}') n(\boldsymbol{r}',t)$$
$$+ \int \mathrm{d}^3 R_1 \cdots \int \mathrm{d}^3 R_{N_{\mathrm{n}}} \left[v_{\mathrm{en}}(\boldsymbol{r},\{\boldsymbol{R}\}) + \mathcal{R}^{-1}\boldsymbol{r}\cdot\boldsymbol{E}(t)\right] \Gamma(\{\boldsymbol{R}\},t) + v_{\mathrm{xc,\,e}}(\boldsymbol{r},t) \tag{6.46}$$

and

$$v_{\mathrm{KS,\,n}}(\{\boldsymbol{R}\},t) = \left[N_{\mathrm{e}}\boldsymbol{R}_{\mathrm{CMN}} - Z_1\boldsymbol{R}_1 - Z_2\boldsymbol{R}_2\right]\cdot\boldsymbol{E}(t) + v_{\mathrm{nn}}(\{\boldsymbol{R}\})$$
$$+ \int \mathrm{d}^3 r \left[v_{\mathrm{en}}(\boldsymbol{r},\{\boldsymbol{R}\}) + \mathcal{R}^{-1}\boldsymbol{r}\cdot\boldsymbol{E}(t)\right] n(\boldsymbol{r},t) + v_{\mathrm{xc,\,n}}(\{\boldsymbol{R}\},t)\,. \tag{6.47}$$

Since the rotation matrix $\mathcal{R}$ only depends on $\boldsymbol{R}$ the nuclear Kohn-Sham potential is readily seen to be separable in terms of the coordinates $\boldsymbol{R}$ and $\boldsymbol{R}_{\mathrm{CMN}}$. The nuclear Kohn-Sham wavefunction can then be written as

$$\Phi_{\mathrm{n},\,s_1,s_2}(\boldsymbol{R}_1,\boldsymbol{R}_2,t) = \Upsilon(\boldsymbol{R}_{\mathrm{CMN}},t)\xi(\boldsymbol{R},t)\theta(s_1,s_2) \tag{6.48}$$

where θ is a nuclear spin function of the nuclear spin coordinates s_1 and s_2 and Υ and ξ satisfy the equations

$$\left\{ \mathrm{i}\partial_t - \left[-\frac{1}{M_{\mathrm{nuc}}}\nabla^2_{\boldsymbol{R}_{\mathrm{CMN}}} + Q_{\mathrm{tot}}\boldsymbol{R}_{\mathrm{CMN}}\cdot\boldsymbol{E}(t)\right]\right\}\Upsilon(\boldsymbol{R}_{\mathrm{CMN}},t) = 0 \tag{6.49a}$$

$$\left\{ i\partial_t - \left[-\frac{1}{2\mu}\nabla^2_{\boldsymbol{R}} + \bar{v}_{\mathrm{KS,\,n}}[n,\Gamma](\boldsymbol{R},t)\right]\right\}\xi(\boldsymbol{R},t) = 0\,, \tag{6.49b}$$

where we defined the total nuclear mass $M_{\mathrm{nuc}} = M_1 + M_2$, the total charge $Q_{\mathrm{tot}} = N_{\mathrm{e}} - Z_1 - Z_2$ and the reduced mass $\mu = M_1 M_2/(M_1 + M_2)$. The potential $\bar{v}_{\mathrm{KS,\,n}}$ has the form

$$\tilde{v}_{\mathrm{KS,\,n}}(\boldsymbol{R},t) = [-q_{\mathrm{n}}\boldsymbol{R} + \boldsymbol{d}(\boldsymbol{R},t)] \cdot \boldsymbol{E}(t) + \frac{Z_1 Z_2}{R}$$
$$- \int \mathrm{d}^3 r\, n(\boldsymbol{r},t) \left\{ \frac{Z_1}{|\boldsymbol{r} - \frac{M_2}{M_1+M_2} R\boldsymbol{e}_z|} + \frac{Z_2}{|\boldsymbol{r} + \frac{M_1}{M_1+M_2} R\boldsymbol{e}_z|} \right\} + v_{\mathrm{xc,\,n}}(\boldsymbol{R},t)\,, \tag{6.50}$$

where we have defined

$$q_{\mathrm{n}} = \frac{M_2 Z_1 - M_1 Z_2}{M_1 + M_2} \tag{6.51a}$$

$$\boldsymbol{d}(\boldsymbol{R},t) = \mathcal{R}^{-1} \int \mathrm{d}^3 r\, n(\boldsymbol{r},t)\boldsymbol{r}\,. \tag{6.51b}$$

We see that the nuclear center-of-mass motion has been decoupled from the nuclear relative motion. The nuclear center-of-mass wavefunction corresponds to a so-called Volkov plane wave. If it is normalized to a volume $\mathcal{V}$ then the nuclear density matrix can be written as

$$\Gamma(\boldsymbol{R}_1, \boldsymbol{R}_2, t) = \frac{1}{\mathcal{V}} N(\boldsymbol{R},t)\,, \tag{6.52}$$

where we defined the density of the relative nuclear "particle" as

$$N(\boldsymbol{R},t) = |\xi(\boldsymbol{R},t)|^2. \tag{6.53}$$

In terms of this quantity, the electronic Kohn-Sham potential (6.46) attains the form

$$v_{\mathrm{KS,\,e}}(\boldsymbol{r},t) = \mathcal{D}^{-1}\boldsymbol{r} \cdot \boldsymbol{E}(t) + \int \mathrm{d}^3 r'\, \frac{n(\boldsymbol{r}',t)}{|\boldsymbol{r} - \boldsymbol{r}'|}$$
$$- \int \mathrm{d}^3 R\, N(\boldsymbol{R},t) \left\{ \frac{Z_1}{|\boldsymbol{r} - \frac{M_2}{M_1+M_2} R\boldsymbol{e}_z|} + \frac{Z_2}{|\boldsymbol{r} + \frac{M_1}{M_1+M_2} R\boldsymbol{e}_z|} \right\} + v_{\mathrm{xc,\,e}}(\boldsymbol{R},t)\,, \tag{6.54}$$

where

$$\mathcal{D}^{-1} = \int \mathrm{d}^3 R\, N(\boldsymbol{R},t)\mathcal{R}^{-1}\,. \tag{6.55}$$

We have now completely defined the multicomponent Kohn-Sham equations for a diatomic molecule in a laser field in the dipole approximation. The next task is to develop appropriate functionals for the exchange-correlaton potentials, particularly for the electron-nuclear correlation. The simplest approach is to treat the electron-nuclear correlation in the Hartree approach where we put $v_{\mathrm{xc,\,n}} = 0$ in (6.50). This approach has been tested [Kreibich 2004] in a one-dimensional model system for H_2^+ which is a suitable testcase since (i) it can be compared to the exact solution of the Schrödinger equation and, (ii) there are no electron-electron correlations.

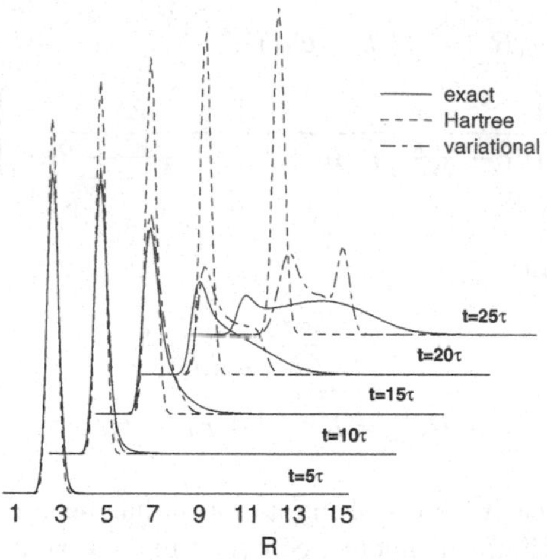

Fig. 6.1. Time-evolution (in units of optical cycles τ) of the nuclear density $N(R,t)$ obtained for a one-dimensional model H_2^+-molecule in a $\lambda = 228\,\mathrm{nm}$, $I = 5 \times 10^{13}\,\mathrm{W/cm^2}$ laser field from the exact solution, the time-dependent Hartree approximation and a time-dependent correlated variational approach

When the model molecule is exposed to a strong laser field, the nuclear density $N(R,t)$ shows a time-dependence as shown in Fig. 6.1. In this plot we can clearly see that the exact nuclear wavepacket splits, and part of the nuclear wavepacket moves away and decribes a dissociating molecule. However, within the Hartree approach to the electron-nuclear correlation, the nuclear wavepacket remains sharply peaked around the molecular equilibrium bond distance. This means that electron-nuclear correlation beyond the Hartree approximation is very important (for a more extensive discussion see [Kreibich 2001a, Kreibich 2004]). This is corroborated by the fact that a variational ansatz for the time-dependent wavefunction in terms of correlated orbitals (denoted as "variational" in Fig. 6.1) does yield the qualitatively correct splitting of the nuclear wavepacket.

6.8 Conclusions

We showed how to set up a multicomponent density-functional scheme for general systems of electrons and nuclei in time-dependent external fields. The basic quantities in this theory are the electron density referred to a body-fixed frame and the nuclear density matrix. Important for future applications will be the development of functionals for electron-nuclear correlations. The first steps in this direction have already been taken in the MCDFT for stationary systems [Kreibich 2001a, Kreibich 2004]. The development of such functionals for the time-dependent case is an important goal for the future.

7 Intermolecular Forces and Generalized Response Functions in Liouville Space

S. Mukamel, A.E. Cohen, and U. Harbola

7.1 Introduction

Consider two interacting sub-systems with nonoverlapping charge distributions. How can the properties of the combined system be expressed in terms of properties of the individual sub-systems alone? This general problem appears in a wide variety of physical, chemical, and biological systems [Joester 2003, Stone 1996, Moore 2001]. In this chapter we provide a prescription for addressing this issue by the computation of (i) response functions and (ii) correlation functions of spontaneous fluctuations of relevant degrees of freedom in the individual sub-systems.

The computation of response and correlation functions is greatly simplified by using the density matrix in Liouville space [Mukamel 1995]. Hilbert and Liouville spaces offer very different languages for the description of nonlinear response. Computing dynamical observables in terms of the wavefunction in Hilbert space requires both forward and backward propagations in time. In contrast, the density matrix calculated in Liouville space should only be propagated forward. The choice is between following only the ket, moving it both forward and backward, or following the joint forward dynamics of the ket and the bra. Artificial time variables (Keldysh loops) commonly used in many-body theory [Haug 1996] are connected with the wavefunction. The density matrix which uses the real laboratory time throughout the calculation offers a more intuitive picture. Wavefunction-based theories calculate transition amplitudes, which by themselves are not observable. The density matrix, on the other hand directly calculates physical observables. Moreover, dephasing processes (damping of off-diagonal elements of the density matrix caused by phase fluctuations) which are naturally included into the Liouville space formulation may not be described in terms of the wavefunction. The "causality paradox" of TDDFT [Gross 1996] can be clearly resolved in Liouville space [Mukamel 2005].

In this chapter we present a method for expressing the joint response of two interacting sub-systems in terms of their correlation and response functions. This factorization appears quite naturally in Liouville space. The pth order response of the individual systems is a linear combination of 2^p distinct $(p+1)$-point correlation functions known as *Liouville space pathways* [Mukamel 1995], which differ by whether the interaction at each time

S. Mukamel et al.: *Intermolecular Forces and Generalized Response Functions in Liouville Space*, Lect. Notes Phys. **706**, 107–123 (2006)
DOI 10.1007/3-540-35426-3_7

is with the bra or the ket. The pth order contribution to the intermolecular interaction requires a different linear combination of these same Liouville space pathways of both molecules. The 2^p Liouville space pathways are conveniently combined into $p+1$ *generalized response functions* (GRFs) [Cohen 2003, Cohen 2003b, Cohen 2005, Chernyak 1995]. One of the GRFs is the ordinary (causal) response function. The other GRFs represent spontaneous fluctuations, and the response of such fluctuations to a perturbation, and are therefore non-causal. The complete set of GRFs is calculated using generalized TDDFT equations in Liouville space.

A direct DFT simulation of molecular complexes by treating them as supermolecules is complicated because it requires nonlocal energy functionals [Hult 1999, Lein 1999, Kohn 1998]. The response approach makes good use of the perturbative nature of the coupling and recasts the energies in terms of properties of individual molecules which, in turn, may be calculated using local functionals [Dobson 1994b, Misquitta 2003].

7.2 Quantum Dynamics in Liouville Space; Superoperators

In this section we introduce the notion of Liouville space superoperators and review some of their useful properties. A detailed discussion of superoperators is given in [Mukamel 2003]. The elements of an $N \times N$ density matrix in Hilbert space are arranged as a vector of length N^2 in Liouville space. An operator in Liouville space then becomes a matrix of dimension $N^2 \times N^2$, and is called a *superoperator*. Two special superoperators, $\breve{A}_L$ and $\breve{A}_R$, are associated with every Hilbert space operator, $\hat{A}$, and implement "left" and "right" multiplication on another operator $\hat{X}$: $\breve{A}_L\hat{X} \Leftrightarrow \hat{A}\hat{X}$, $\breve{A}_R\hat{X} \Leftrightarrow \hat{X}\hat{A}$. These relations are not written as equalities because $\hat{X}$ is a vector in Liouville space and a matrix in Hilbert space.

It will be useful to define the symmetric $\breve{A}^+ \equiv \frac{1}{2}(\breve{A}_L + \breve{A}_R)$ and antisymmetric $\breve{A}^- \equiv \breve{A}_L - \breve{A}_R$ combinations. Hereafter we shall use Greek indices to denote superoperators $\breve{A}^\nu$ with $\nu = +, -$. Using ordinary Hilbert space operators we get

$$\breve{A}^+\hat{X} \Leftrightarrow \frac{1}{2}(\hat{A}\hat{X} + \hat{X}\hat{A}), \quad \breve{A}^-\hat{X} \Leftrightarrow \hat{A}\hat{X} - \hat{X}\hat{A}\,. \tag{7.1}$$

A product of + and − superoperators constitutes a series of nested commutators and anticommutators in Hilbert space. It is easy to verify that

$$(\breve{AB})^- = \breve{A}^+\breve{B}^- + \breve{A}^-\breve{B}^+, \quad (\breve{AB})^+ = \breve{A}^+\breve{B}^+ + \frac{1}{4}\breve{A}^-\breve{B}^-\,. \tag{7.2}$$

We now consider products of superoperators that depend parametrically on time. We introduce a *time ordering operator* $\hat{\mathcal{T}}$ in Liouville space, which orders all superoperators to its right such that time decreases from left to right. This natural time-ordering follows chronologically the various interactions with the density matrix. We can freely commute operators following a $\hat{\mathcal{T}}$ operation without worrying about commutations because in the end the order will be determined by $\hat{\mathcal{T}}$.

The expectation value of any superoperator, $\breve{A}_\nu$, is defined as,

$$\langle \breve{A}^\nu(t)\rangle = \mathrm{Tr}\{\breve{A}^\nu(t)\hat{\rho}_{\mathrm{eq}}\}\,, \tag{7.3}$$

where $\hat{\rho}_{\mathrm{eq}}$ is the equilibrium density matrix. For any two operators $\hat{A}$ and $\hat{B}$, we have

$$\langle \hat{\mathcal{T}}\breve{A}^+(t)\breve{B}^-(t')\rangle = 0 \quad \text{if} \quad t' > t\,. \tag{7.4}$$

$\langle \hat{\mathcal{T}}\breve{A}^+(t)\breve{B}^-(t')\rangle$ is thus a retarded (i.e., causal) function. Equation (7.4) follows from the definitions (7.1): A "−" superoperator corresponds to a commutator in Hilbert space, so for $t < t'$, $\langle \hat{\mathcal{T}}\breve{A}^+(t)\breve{B}^-(t')\rangle$ becomes a trace of a commutator which vanishes. Similarly, the trace of two "−" operators vanishes:

$$\langle \hat{\mathcal{T}}\breve{A}^-(t)\breve{B}^-(t')\rangle = 0 \quad \text{for all } t \text{ and } t'\,. \tag{7.5}$$

We next introduce the interaction picture for superoperators. To that end, we partition the Hamiltonian, $\hat{H} = \hat{H}_0 + \hat{H}_1$, into a reference part, $\hat{H}_0$, which can be diagonalized, and the remainder, interaction part, $\hat{H}_1$. We define the corresponding superoperators, $\breve{\mathcal{H}}^-$, as $\breve{\mathcal{H}}^- = \breve{\mathcal{H}}_0^- + \breve{\mathcal{H}}_1^-$. The time evolution of the density matrix is given by the Liouville equation:

$$\frac{\partial\hat{\rho}}{\partial t} = -\frac{\mathrm{i}}{\hbar}\breve{\mathcal{H}}^-\hat{\rho}\,. \tag{7.6}$$

The formal Green function solution of (7.6) is $\hat{\rho}(t) = \breve{\mathcal{U}}(t,t_0)\hat{\rho}(t_0)$. Note that the time evolution operator, $\breve{\mathcal{U}}$, acts only from the left, implying forward evolution of the density matrix. The total time evolution operator

$$\breve{\mathcal{U}}(t,t_0) = \hat{\mathcal{T}}\exp\left\{-\frac{\mathrm{i}}{\hbar}\int_{t_0}^{t}\mathrm{d}\tau\,\breve{\mathcal{H}}^-(\tau)\right\}\,, \tag{7.7}$$

can be partitioned as:

$$\breve{\mathcal{U}}(t,t_0) = \breve{\mathcal{U}}_0(t,t_0)\breve{\mathcal{U}}_1(t,t_0) \tag{7.8}$$

where $\breve{\mathcal{U}}_0$ describes the time evolution due to $\hat{H}_0$

$$\breve{\mathcal{U}}_0(t,t_0) = \theta(t-t_0)\exp\left\{-\frac{\mathrm{i}}{\hbar}\breve{\mathcal{H}}_0^-(t-t_0)\right\}\,, \tag{7.9}$$

and $\breve{\mathcal{U}}_1$ is the time evolution operator in the *interaction picture*

$$\breve{\mathcal{U}}_1(t,t_0) = \hat{\mathcal{T}}\exp\left\{-\frac{\mathrm{i}}{\hbar}\int_{t_0}^{t}\mathrm{d}\tau\,\breve{\mathcal{H}}_1^{I-}(\tau)\right\}\,. \tag{7.10}$$

The time dependent superoperator $\breve{A}(t)$ in the interaction picture, denoted by a $\breve{A}^I$, is defined as

$$\breve{A}^{I\nu}(t) \equiv \breve{\mathcal{U}}_0^\dagger(t,t_0)\breve{A}^\nu(t_0)\breve{\mathcal{U}}_0(t,t_0) \,. \tag{7.11}$$

The equilibrium density matrix of the interacting system can be generated from the noninteracting density matrix $\hat{\rho}_0$ by adiabatically switching the interaction H_1, starting at time $t = -\infty$: $\hat{\rho}^I_{\text{eq}} = \breve{\mathcal{U}}_I(0,-\infty)\hat{\rho}_0$. In the wavefunction (Gell-Mann-Low) formulation of adiabatic switching, the wavefunction acquires a singular phase which must be cancelled by a denominator given by the closed loop S matrix [Negele 1988]; this unphysical phase never shows up in Liouville space.

For a set of operators $\{\hat{A}_i\}$, the pth order *generalized response functions* (GRF) are defined as

$$R^{\nu_{p+1}\dots\nu_1}_{i_{p+1}\dots i_1}(t_{p+1}\dots t_1) = \left(\frac{-\mathrm{i}}{\hbar}\right)^{p'} \langle \hat{\mathcal{T}}\breve{A}^{\nu_{p+1}}_{i_{p+1}}(t_{p+1})\dots\breve{A}_{i_1\nu_1}(t_1)\rangle_0 \,, \tag{7.12}$$

where $\langle\dots\rangle_0$ represents a trace with respect to $\hat{\rho}_0$, and the indices $\nu_n = +$ or $-$, and p' denote the number of '$-$' indices in the set $\{\nu_{p+1}\dots\nu_1\}$. There are $p+1$, p^{th} order GRFs, having different number of '$-$' indices. Each member of the pth order GRF represents to a different physical process. For example, there are two first order GRFs,

$$R^{++}_{i_2i_1}(t_2,t_1) = \langle\hat{\mathcal{T}}\breve{A}^+_{i_2}(t_2)\breve{A}^+_{i_1}(t_1)\rangle_0 \tag{7.13a}$$

$$R^{+-}_{i_2i_1}(t_2,t_1) = \frac{-\mathrm{i}}{\hbar}\langle\hat{\mathcal{T}}\breve{A}^+_{i_2}(t_2)\breve{A}^-_{i_1}(t_1)\rangle_0 \,. \tag{7.13b}$$

Recasting them in Hilbert space we have

$$\begin{aligned} R^{+-}_{i_2i_1}(t_2,t_1) &= \frac{-\mathrm{i}}{\hbar}\theta(t_2-t_1)\left[\mathrm{Tr}\{\hat{A}_{i_2}(t_2)\hat{A}_{i_1}(t_1)\hat{\rho}_0\} - \mathrm{Tr}\{\hat{A}_{i_2}(t_2)\hat{A}_{i_1}(t_1)\hat{\rho}_0\}\right] \\ &= \hbar^{-1}\theta(t_2-t_1)\mathrm{Im}J(t_2,t_1) \qquad (7.14a) \\ R^{++}_{i_2i_1}(t_2,t_1) &= \mathrm{Tr}\{\hat{A}_{i_1}(t_1)\hat{A}_{i_2}(t_2)\hat{\rho}_0\} + \mathrm{Tr}\{\hat{A}_{i_2}(t_2)\hat{A}_{i_1}(t_1)\hat{\rho}_0\} \\ &= \mathrm{Re}J(t_2,t_1)\,, \qquad (7.14b) \end{aligned}$$

where $J(t_2,t_1) = \mathrm{Tr}\{\hat{A}_{i_2}(t_2)\hat{A}_{i_1}(t_1)\hat{\rho}_0\}$. With the factor $\hbar^{-1}$, the GRF R^{+-} has a well defined classical limit [Mukamel 2003]. R^{+-} is causal [see (7.5)] and represents the response of the system at time t_2 to an external perturbation acting at an earlier time t_1. On the other hand, R^{++} is non-causal and denotes the correlation of A at two times. Each '$-$' index corresponds to the interaction with an external perturbation while a '$+$' index denotes an observation. In general, time-ordered Liouville space correlation functions with one '$+$' and several '$-$' indices, R^{+-}, R^{+--}, R^{+---}, etc., give response functions; all '$+$' correlation functions of the form R^{++}, R^{+++}, R^{++++}, etc. give ground state fluctuations, wheras R^{++-}, R^{++--}, R^{+++-}, etc. represent changes in the fluctuations caused by an external perturbation.

7.3 TDDFT Equations of Motion for the GRFs

Time dependent density functional theory is based on the effective one-body Kohn-Sham (KS) Hamiltonian [Gross 1996],

$$\hat{H}_{\mathrm{KS}}[n] = -\frac{\hbar^2\nabla^2}{2m} + v_{\mathrm{ext}}(\boldsymbol{r}_1) + v_{\mathrm{H}}[n](\boldsymbol{r}_1,t) + v_{\mathrm{xc}}[n](\boldsymbol{r}_1,t) \; , \tag{7.15}$$

where the four terms represent the kinetic energy, the nuclear potential, the Hartree, and the exchange correlation potential, respectively.

We now introduce the reduced single electron density matrix $\hat{\rho}$ [Tretiak 2002, Chernyak 1996, Ring 1980, Blaizot 1986, Berman 2003, Coleman 2000] whose diagonal elements give the charge distribution, $n(\boldsymbol{r}_1,t) = \rho(\boldsymbol{r}_1,\boldsymbol{r}_1,t)$ and the off-diagonal elements, $\rho(\boldsymbol{r}_1,\boldsymbol{r}_2)$ with $\boldsymbol{r}_1 \neq \boldsymbol{r}_2$, represent electronic coherences. We further denote the ground state density matrix by $\hat{\rho}^{\mathrm{GS}}$.

The GRF corresponding to the charge density may be calculated by solving the time dependent generalized KS equation of motion for $\hat{\rho}$ [Mukamel 2005, Harbola 2004],

$$\mathrm{i}\hbar\frac{\partial}{\partial t}\delta\hat{\rho} = [\hat{H}_{\mathrm{KS}},\hat{\rho}(t)] + v_{\mathrm{ket}}(t)\hat{\rho}(t) - \hat{\rho}(t)v_{\mathrm{bra}}(t) \; , \tag{7.16}$$

where $\delta\rho(\boldsymbol{r}_1,\boldsymbol{r}_2,t) \equiv \rho(\boldsymbol{r}_1,\boldsymbol{r}_2,t) - \rho^{\mathrm{GS}}(\boldsymbol{r}_1,\boldsymbol{r}_2)$ is the change in the density matrix induced by the external potentials v_{ket} and v_{bra}. Equation (7.16) differs from the standard TDDFT equations in that the system is coupled to two external potentials, a “left” one v_{ket} acting on the ket and a “right” one v_{bra} acting on the bra.

We next define the new variables,

$$v_{-}(\boldsymbol{r},t) \equiv \frac{1}{2}\left[v_{\mathrm{ket}}(\boldsymbol{r},t) + v_{\mathrm{bra}}(\boldsymbol{r},t)\right] \tag{7.17a}$$

$$v_{+}(\boldsymbol{r},t) \equiv \frac{1}{2}\left[v_{\mathrm{bra}}(\boldsymbol{r},t) - v_{\mathrm{ket}}(\boldsymbol{r},t)\right] \; , \tag{7.17b}$$

and the diagonal matrices

$$v_{-}(\boldsymbol{r}_1,\boldsymbol{r}_2) = v_{-}(\boldsymbol{r}_1)\delta(\boldsymbol{r}_1-\boldsymbol{r}_2) \quad , \quad v_{+}(\boldsymbol{r}_1,\boldsymbol{r}_2) = v_{+}(\boldsymbol{r}_1)\delta(\boldsymbol{r}_1-\boldsymbol{r}_2) \; . \tag{7.18}$$

We further introduce two matrices (operators) in real space $\hat{V}_{-}$ and $\hat{V}_{+}$ with elements $v_{-}(\boldsymbol{r}_1,\boldsymbol{r}_2)$ and $v_{+}(\boldsymbol{r}_1,\boldsymbol{r}_2)$, respectively. $\delta\hat{\rho}$ serves as a *generating function* for GRFs, which are obtained by a perturbative solution of (7.16) in $v_{-}(\boldsymbol{r},t)$ and $v_{+}(\boldsymbol{r},t)$ using $\hat{H}^{(0)} = \hat{H}_{\mathrm{KS}}$ as a reference, as we shall shortly see. The reason for assigning the $+$ and the $-$ subscripts to v in (7.17a) and (7.17b) is to keep track of the perturbative terms that arise from the external potential that couples to the density matrix through commutators and anticommutators.

Equation (7.16) can be recast in terms of superoperators,

$$\mathrm{i}\hbar\frac{\partial}{\partial t}\delta\hat{\rho} = \check{\mathcal{H}}_{\mathrm{KS}}^{-}\hat{\rho}(t) + \check{\mathcal{V}}^{-}(t)\hat{\rho}(t) - \check{\mathcal{V}}^{+}(t)\hat{\rho}(t) \; , \tag{7.19}$$

where

$$\check{\mathcal{V}}^{-}\hat{\rho} \equiv [\hat{V}_{-},\hat{\rho}], \quad \check{\mathcal{V}}^{+}\hat{\rho} \equiv [\hat{V}_{+},\hat{\rho}]_{+}, \quad \check{\mathcal{H}}_{\mathrm{KS}}^{-}\hat{\rho} \equiv [\hat{H}_{\mathrm{KS}},\hat{\rho}] \; . \tag{7.20}$$

The pth order GRFs $\chi^{\nu_{p+1}\dots\nu_1}$, are computed as the kernels in a perturbative expansion of the charge density fluctuation, $\delta n(\boldsymbol{r}_1,t) = \delta\rho(\boldsymbol{r}_1,\boldsymbol{r}_1,t)$, in the applied potentials, v_+ and v . Adopting the abbreviated space-time notation $\boldsymbol{x}_n \equiv (\boldsymbol{r}_n, t_n)$, we get

$$\langle \delta n^{+}(\boldsymbol{x}_{p+1})\rangle^{(p)} \equiv \int \mathrm{d}^3 r_p \int \mathrm{d}t_p \dots \int \mathrm{d}^3 r_1 \int \mathrm{d}t_1$$
$$\chi^{\nu_{p+1}\nu_p\dots\nu_1}(\boldsymbol{x}_{p+1},\boldsymbol{x}_p,\dots,\boldsymbol{x}_1)\, v_{\nu_p}(\boldsymbol{x}_p) v_{\nu_{p-1}}(\boldsymbol{x}_{p-1})\dots v_{\nu_1}(\boldsymbol{x}_1) \; . \tag{7.21}$$

It follows from (7.16) and (7.21) that

$$\chi^{\nu_{p+1}\dots\nu_1}(\boldsymbol{x}_{p+1}\dots\boldsymbol{x}_1) = \left(\frac{-\mathrm{i}}{\hbar}\right)^{p'} \left\langle \hat{\mathcal{T}}\delta n^{\nu_{p+1}}(\boldsymbol{x}_{p+1})\dots\delta n^{\nu_1}(\boldsymbol{x}_1)\right\rangle \tag{7.22}$$

where p' denotes the number of "minus" indices in the set $\{\nu_{p+1}\dots\nu_1\}$. To first order ($p = 1$), we have

$$\chi^{++}(\boldsymbol{x}_1,\boldsymbol{x}_2) = \langle\hat{\mathcal{T}}\delta n^{+}(\boldsymbol{x}_1)\delta n^{+}(\boldsymbol{x}_2)\rangle \tag{7.23a}$$
$$= \theta(t_1 - t_2)\langle\delta n^{+}(\boldsymbol{x}_1)\delta n^{+}(\boldsymbol{x}_2)\rangle + \theta(t_2 - t_1)\langle\delta n^{+}(\boldsymbol{x}_2)\delta n^{+}(\boldsymbol{x}_1)\rangle$$
$$\chi^{+-}(\boldsymbol{x}_1,\boldsymbol{x}_2) = \frac{-\mathrm{i}}{\hbar}\langle\hat{\mathcal{T}}\delta n^{+}(\boldsymbol{x}_1)\delta n^{-}(\boldsymbol{x}_2)\rangle \tag{7.23b}$$
$$= \frac{-\mathrm{i}}{\hbar}\theta(t_1 - t_2)\langle\delta n^{+}(\boldsymbol{x}_1)\delta n^{-}(\boldsymbol{x}_2)\rangle \; .$$

The standard TDDFT equations which only yield ordinary response functions are obtained by setting $v_{\mathrm{ket}} = v_{\mathrm{bra}}$ so that $\check{\mathcal{V}}^{+} = 0$ in (7.19). By allowing v_{ket} to be different from v_{bra} we can generate the complete set of GRF. The ordinary response function χ^{+-} represents the response of the density to an applied potential v_- [Mukamel 1995]. Similarly, χ^{++} can be formally obtained as the response to the artificial external potential, v_+, that couples to the charge density through an anticommutator. χ^{++} represents equilibrium charge fluctuations and is therefore non-retarded.

7.4 Collective Electronic Oscillator Representation of the GRF

Since the TDDFT density matrix, $\hat{\rho}(t)$, corresponds to a many-electron wavefunction given by a single Slater determinant at all times, it can be separated into its electron-hole (interband) part $\hat{\xi}$ and the electron-electron and hole-hole (intraband) components, $\hat{K}(\hat{\xi})$ [Chernyak 1996, Thouless 1961].

$$\delta\hat{\rho}(t) = \hat{\xi}(t) + \hat{K}(\hat{\xi}(t)) . \tag{7.24}$$

Note that $\hat{\rho}(t)$ is the KS density matrix which corresponds to a noninteracting system. It follows from the idempotent property, $\hat{\rho}^2 = \hat{\rho}$, that $\hat{K}$ (and $\delta\hat{\rho}$) is uniquely determined by the interband part $\hat{\xi}$ [Ring 1980, Chernyak 1996].

$$\hat{K}(\hat{\xi}) = \frac{1}{2}(2\hat{\rho}^{\mathrm{GS}} - \hat{I})\left(\hat{I} - \sqrt{\hat{I} - 4\hat{\xi}\hat{\xi}}\right) . \tag{7.25}$$

The elements of $\hat{\xi}$ (but not of $\delta\hat{\rho}$) can thus be considered as independent oscillator coordinates for describing the electronic structure.

We next expand H_{KS} in powers of $\delta n(\boldsymbol{r},t)$:

$$\hat{H}_{\mathrm{KS}} = \hat{H}^{(0)}_{\mathrm{KS}} + \hat{H}^{(1)}_{\mathrm{KS}} + \hat{H}^{(2)}_{\mathrm{KS}} + \dots \tag{7.26a}$$

$$\hat{H}^{(0)}_{\mathrm{KS}}[\bar{n}] = -\frac{\hbar^2}{2m}\nabla^2 + v_{\mathrm{ext}}(\boldsymbol{r}_1) + v_{\mathrm{H}}[\bar{n}](\boldsymbol{r}_1,t) + v_{\mathrm{xc}}[\bar{n}](\boldsymbol{r}_1,t) \tag{7.26b}$$

$$\hat{H}^{(1)}_{\mathrm{KS}}[\delta n] = \int \mathrm{d}^3 r_2 \left\{ \frac{e^2}{|\boldsymbol{r}_1 - \boldsymbol{r}_2|} + f_{\mathrm{xc}}(\boldsymbol{r}_1,\boldsymbol{r}_2) \right\} \delta n(\boldsymbol{r}_2,t) , \tag{7.26c}$$

with f_{xc} the first order exchange correlation kernel in the adiabatic approximation where it is assumed to be frequency independent (see Chap. 1)

$$f_{\mathrm{xc}}(\boldsymbol{r}_1,\boldsymbol{r}_2) = \left.\frac{\delta v_{\mathrm{xc}}[n](\boldsymbol{r}_1)}{\delta n(\boldsymbol{r}_2)}\right|_{\bar{n}} . \tag{7.27}$$

The second order term in density fluctuations is,

$$\hat{H}^{(2)}_{\mathrm{KS}}[\delta n] = \int \mathrm{d}^3 r_2 \int \mathrm{d}^3 r_3 \, g_{\mathrm{xc}}(\boldsymbol{r}_1,\boldsymbol{r}_2,\boldsymbol{r}_3)\delta n(\boldsymbol{r}_2,t)\delta n(\boldsymbol{r}_3,t) , \tag{7.28}$$

with the kernel (in the adiabatic approximation),

$$g_{xc}(\boldsymbol{r}_1,\boldsymbol{r}_2,\boldsymbol{r}_3) = \left.\frac{\delta^2 v_{\mathrm{xc}}[n](\boldsymbol{r}_1)}{\delta n(\boldsymbol{r}_2)\delta n(\boldsymbol{r}_3)}\right|_{\bar{n}} . \tag{7.29}$$

A quasiparticle algebra can be developed for $\hat{\xi}$ by expanding it in the basis set of *collective electronic oscillator* (CEO) modes, $\hat{\xi}_\alpha$, which are the eigenvectors of the linearized TDDFT eigenvalue equation with eigenvalues Ω_α [Chernyak 1996, Tretiak 2002].

$$\breve{L}\hat{\xi}_\alpha = \Omega_\alpha \hat{\xi}_\alpha , \tag{7.30}$$

The linearized Liouville space operator, $\breve{L}$ is obtained by substituting (7.26) into (7.16),

$$\check{L}\hat{\xi}_\alpha = [\hat{H}^{(0)}_{\rm KS}[\bar{n}], \hat{\xi}_\alpha] + [\hat{H}^{(1)}_{\rm KS}[\xi_\alpha], \hat{\rho}^{\rm GS}] \,. \tag{7.31}$$

$\hat{H}^{(0)}_{\rm KS}$ and $\hat{H}^{(1)}_{\rm KS}$ are diagonal matrices with matrix elements

$$H^{(0)}_{\rm KS}[\bar{n}](\boldsymbol{r}_1, \boldsymbol{r}_2) = \delta(\boldsymbol{r}_1 - \boldsymbol{r}_2) H^{(0)}_{\rm KS}[\bar{n}](\boldsymbol{r}_1) \tag{7.32a}$$

$$H^{(1)}_{\rm KS}[\xi_\alpha](\boldsymbol{r}_1, \boldsymbol{r}_2) = \delta(\boldsymbol{r}_1 - \boldsymbol{r}_2) \times \int {\rm d}^3 r_3 \left\{ \frac{e^2}{|\boldsymbol{r}_2 - \boldsymbol{r}_3|} + f_{\rm xc}(\boldsymbol{r}_2, \boldsymbol{r}_3) \right\} \xi_\alpha(\boldsymbol{r}_3, \boldsymbol{r}_3) \tag{7.32b}$$

The eigenmodes $\hat{\xi}_\alpha$ come in pairs represented by positive and negative values of α, and we adopt the notation, $\Omega_{-\alpha} = -\Omega_\alpha$ and $\hat{\xi}_{-\alpha} = \hat{\xi}^\dagger_\alpha$. Each pair of modes represents a CEO and the complete set of modes $\hat{\xi}_\alpha$ may be used to describe all response and spontaneous charge fluctuation properties of the system.

By expanding $\hat{\xi}(t)$ of the externally driven system in the CEO eigenmodes, $\hat{\xi}(t) = \sum_\alpha \bar{z}_\alpha(t)\hat{\xi}_\alpha$, where α runs over all modes (positive and negative) and $\bar{z}_\alpha$ are numerical coefficients, and substituting in (7.25) and (7.24), we obtain the following expansion for the density matrix

$$\delta\rho(\boldsymbol{r}_1, \boldsymbol{r}_2, t) = \sum_\alpha \mu_\alpha(\boldsymbol{r}_1, \boldsymbol{r}_2)\bar{z}_\alpha(t) + \frac{1}{2}\sum_{\alpha,\beta} \mu_{\alpha,\beta}(\boldsymbol{r}_1, \boldsymbol{r}_2)\bar{z}_\alpha(t)\bar{z}_\beta(t) + \ldots \tag{7.33}$$

where we have introduced the auxiliary quantities, $\hat{\mu}_\alpha = \hat{\xi}_\alpha$ and $\hat{\mu}_{\alpha\beta} = (2\hat{\rho}^{\rm GS} - I)(\hat{\xi}_\alpha\hat{\xi}_\beta + \hat{\xi}_\beta\hat{\xi}_\alpha)$.

Upon the substitution of (7.24) and (7.25) into (7.16) we can derive equations of motion for the CEO amplitudes $\bar{z}_\alpha$ which can then be solved successively order by order in the external potentials, v_{ν_1}. To second order we get

$$i\hbar\frac{{\rm d}\bar{z}_\alpha(t)}{{\rm d}t} = \Omega_\alpha\bar{z}_\alpha(t) + K_{-\alpha}(t) + \sum_\beta K_{-\alpha\beta}(t)\bar{z}_\beta(t) \,, \tag{7.34}$$

with the coefficients,

$$K_{-\alpha}(t) = \sum_\nu \int {\rm d}^3 r_1\, v_\nu(\boldsymbol{r}_1, t)\mu^\nu_{-\alpha}(\boldsymbol{r}_1) \tag{7.35a}$$

$$K_{-\alpha\beta}(t) = \sum_\nu \int {\rm d}^3 r_1\, v_\nu(\boldsymbol{r}_1, t)\mu^\nu_{-\alpha\beta}(\boldsymbol{r}_1) \,. \tag{7.35b}$$

Here $\mu^-_\alpha(\boldsymbol{r}_1) \equiv \mu_\alpha(\boldsymbol{r}_1, \boldsymbol{r}_1)$, $\mu^-_{\alpha\beta}(\boldsymbol{r}_1) \equiv \hat{\mu}_{\alpha\beta}(\boldsymbol{r}_1, \boldsymbol{r}_1)$, $\mu^+_\alpha(\boldsymbol{r}_1) \equiv \bar{\mu}_{-\alpha}(\boldsymbol{r}_1, \boldsymbol{r}_1) = \frac{1}{2}(2\hat{\rho}^{\rm GS} - \hat{I})\hat{\xi}_\alpha(\boldsymbol{r}_1, \boldsymbol{r}_1)$, and $\mu^+_{\alpha\beta}(\boldsymbol{r}_1) \equiv \bar{\mu}_{\alpha\beta}(\boldsymbol{r}_1) = \frac{1}{2}(2\hat{\rho}^{\rm GS} - \hat{I})(\hat{\xi}_\alpha\hat{\xi}_\beta - \hat{\xi}_\beta\hat{\xi}_\alpha)(\boldsymbol{r}_1, \boldsymbol{r}_1)$.

We further expand $\bar{z}_\alpha = z^{\nu_1}_\alpha + z^{\nu_1\nu_2}_\alpha + \ldots$, in powers of the external potentials, where $z^{\nu_1\nu_2\ldots\nu_p}$ denotes the $n^{\rm th}$ order term in the potentials, $v_{\nu_1}v_{\nu_2}\ldots v_{\nu_p}$. By comparing the terms in both sides, we obtain equations of motion for $z_\alpha^{\nu_1 \ldots \nu_p}$ for each order in the external potential. This gives to first order,

$$i\hbar \frac{dz_\alpha^{\nu_1}(t)}{dt} = \Omega_\alpha z_\alpha^{\nu_1}(t) + K_{-\alpha}(t) . \tag{7.36}$$

The solution of (7.36) gives the generalized linear response functions

$$\chi^{++}(\boldsymbol{r}_1 t_1, \boldsymbol{r}_2 t_2) = \theta(t_1 - t_2) \sum_\alpha \mu_\alpha(\boldsymbol{r}_1)\bar{\mu}_{-\alpha}(\boldsymbol{r}_2) e^{-\frac{i}{\hbar}\Omega_\alpha (t_1 - t_2)} + \theta(t_2 - t_1) \sum_\alpha \mu_\alpha(\boldsymbol{r}_2)\bar{\mu}_{-\alpha}(\boldsymbol{r}_1) e^{\frac{i}{\hbar}\Omega_\alpha (t_1 - t_2)} \tag{7.37a}$$

$$\chi^{+-}(\boldsymbol{r}_1 t_1, \boldsymbol{r}_2 t_2) = -\frac{i}{\hbar}\theta(t_1 - t_2) \sum_\alpha s_\alpha \mu_\alpha(\boldsymbol{r}_1)\mu_{-\alpha}(\boldsymbol{r}_2) e^{-\frac{i}{\hbar}\Omega_\alpha (t_1 - t_2)} , \tag{7.37b}$$

with $s_\alpha = \text{sign}(\alpha)$. Higher order GRF can be obtained by repeating this procedure [Harbola 2004].

We further consider the generalized susceptibilities defined by the Fourier transform of the response functions to the frequency domain,

$$\langle \delta n^+(\boldsymbol{r}_{p+1}, \omega_{p+1}) \rangle^{(p)} = \int d^3 r_p \int_{-\infty}^{\infty} d\omega_p \ldots \int d^3 r_1 \int_{-\infty}^{\infty} d\omega_1 \, v_{\nu_p}(\boldsymbol{r}_p, \omega_p) \ldots v_{\nu_1}(\boldsymbol{r}_1, \omega_1) \chi^{\nu_{p+1} \ldots \nu_1}(\boldsymbol{r}_{p+1}\omega_{p+1}, \boldsymbol{r}_p \omega_p, \ldots, \boldsymbol{r}_1 \omega_1) , \tag{7.38}$$

where the frequency transform is defined as

$$\chi^{\nu_1 \nu_2}(\boldsymbol{r}_1 \omega_1, \boldsymbol{r}_2 \omega_2) = \int_{-\infty}^{\infty} dt_1 \int_{-\infty}^{\infty} dt_2 \, e^{i(\omega_1 t_1 + \omega_2 t_2)} \chi^{\nu_1 \nu_2}(\boldsymbol{x}_1, \boldsymbol{x}_2) . \tag{7.39}$$

Equation (7.39) together with (7.37) gives

$$\chi^{++}(\boldsymbol{r}_1 \omega_1, \boldsymbol{r}_2 \omega_2) = i\hbar \delta(\omega_1 + \omega_2) \sum_\alpha \left[\frac{\mu_\alpha(\boldsymbol{r}_1)\bar{\mu}_{-\alpha}(\boldsymbol{r}_2)}{\omega_2 - \Omega_\alpha + i\eta} - \frac{\mu_\alpha(\boldsymbol{r}_2)\bar{\mu}_{-\alpha}(\boldsymbol{r}_1)}{\omega_2 + \Omega_\alpha - i\eta} \right] \tag{7.40a}$$

$$\chi^{+-}(\boldsymbol{r}_1 \omega_1, \boldsymbol{r}_2 \omega_2) = \delta(\omega_1 + \omega_2) \sum_\alpha \frac{s_\alpha \mu_\alpha(\boldsymbol{r}_1)\mu_{-\alpha}(\boldsymbol{r}_2)}{\omega_2 + \Omega_\alpha - i\eta} . \tag{7.40b}$$

The CEO representations of the ordinary response functions to third order were given in [Tretiak 2002] and the GRF to second order were given in [Harbola 2004].

The linear GRFs, χ^{++} and χ^{+-}, are connected by the *fluctuation-dissipation* relation,

$$\chi^{++}(\boldsymbol{r}_1 \omega; \boldsymbol{r}_2 - \omega) = \coth\left(\frac{\beta\hbar\omega}{2}\right) \chi^{+-}(\boldsymbol{r}_1 \omega, \boldsymbol{r}_2 - \omega) . \tag{7.41}$$

To linear order, the ordinary response function χ^{+-} provides the complete information. However such fluctuation-dissipation relations are not that obvious for the higher order response functions [Wang 2002] and the complete set of GRF are required to describe all possible fluctuations and response functions of the charge density.

7.5 GRF Expressions for Intermolecular Interaction Energies

We now show how the GRF may be used to compute the energy of two interacting systems a and b with nonoverlapping charge distributions. Wavefunction based theories for intermolecular forces (polarization theory, symmetry-adapted perturbation theory and many-body symmetry-adapted theory) are well developed [Jeziorski 1994]. The response function based formulation presented here can be applied to study non-equilibrium effects (e.g., when the two sub-systems are at different temperaturs) as well as coupling to nuclear degrees of freedom [Cohen 2003]. At time $t = -\infty$ we take the density matrix to be a direct-product of the density matrices of the individual molecules (sub-systems), $\hat{\rho}_0^a$ and $\hat{\rho}_0^b$, $\hat{\rho}_0^{\mathrm{GS}} = \hat{\rho}_0^a \hat{\rho}_0^b$. The Liouville space time-evolution operator transforms this initial state into a correlated state. The GRF allow us to factorize the time-evolution operator into a sum of terms that individually preserve the purity of the direct-product form.

We start with the total hamiltonian of two interacting molecules $\hat{H}_\lambda = \hat{H}_a + \hat{H}_b + \lambda \hat{H}_{ab}$, where H_a and H_b represent the Hamiltonians for the individual molecules and their coupling $\hat{H}_{ab}$ is multiplied by the control parameter λ, $0 \leq \lambda \leq 1$, where $\lambda = 1$ corresponds to the physical Hamiltonian. Primed and unprimed indices will correspond to molecules a and b, respectively. The charge densities of molecules a and b at space points $\boldsymbol{r}$ and $\boldsymbol{r}'$ will be denoted by $n_a(\boldsymbol{r})$ and $n_b(\boldsymbol{r}')$, respectively. $\hat{H}_{ab}$ is the Coulomb interaction

$$
\begin{aligned}
\hat{H}_{ab} = & -\int \mathrm{d}^3 r \int \mathrm{d}^3 r' \, v_{\mathrm{ee}}(\boldsymbol{r} - \boldsymbol{r}') n_a(\boldsymbol{r}) n_b(\boldsymbol{r}') - \sum_{k,k'} v_{\mathrm{ee}}(\boldsymbol{R}_k - \boldsymbol{R}_{k'}) Z_k Z_{k'} \\
& + \sum_{P,k} \int \mathrm{d}^3 r' \, v_{\mathrm{ee}}(\boldsymbol{R}_k - \boldsymbol{r}') Z_k n_b(\boldsymbol{r}') \, , \qquad (7.42)
\end{aligned}
$$

where $v_{\mathrm{ee}}(\boldsymbol{r} - \boldsymbol{r}') \equiv 1/|\boldsymbol{r} - \boldsymbol{r}'|$ and $\boldsymbol{R}_k$ ($\boldsymbol{R}_{k'}$) represents the position of kth (k'th) nucleus in molecule a (b) with charge Z_k ($Z_{k'}$). The symbol $\sum_P$ represents the sum over single permutation of primed and unprimed quantities together with indices a and b. The interaction energy of the coupled system is obtained by switching the parameter λ from 0 to 1 [Kohn 1998]

$$
E_{ab} = \int_0^1 \mathrm{d}\lambda \, \langle \hat{H}_{ab} \rangle_\lambda \, . \qquad (7.43)
$$

Here $\langle \ldots \rangle_\lambda$ denotes the expectation value with respect to the λ-dependent ground state many-electron density matrix of the system, $\hat{\rho}_\lambda$. We next partition the charge densities of both molecules as, $n_a(\boldsymbol{r}) = \bar{n}_a(\boldsymbol{r}) + \delta n_a(\boldsymbol{r})$, $n_b(\boldsymbol{r}') = \bar{n}_b(\boldsymbol{r}') + \delta n_b(\boldsymbol{r}')$, where $\bar{n}$ is the average density, $\bar{n}_a(\boldsymbol{r}) = \rho_a^0(\boldsymbol{r}, \boldsymbol{r})$, and $\bar{n}_b(\boldsymbol{r}') = \rho_b^0(\boldsymbol{r}', \boldsymbol{r}')$. Thus the total interaction energy can be written as, $E_{ab} = E_{ab}^{(0)} + E_{ab}^{(I)} + E_{ab}^{(II)}$, where

$$E_{ab}^{(0)} = -\int \mathrm{d}^3r \int \mathrm{d}^3r'\, v_{\mathrm{ee}}(\boldsymbol{r}-\boldsymbol{r}')\bar{n}_a(\boldsymbol{r})\bar{n}_b(\boldsymbol{r}') - \sum_{k,k'} v_{\mathrm{ee}}(\boldsymbol{R}_k - \boldsymbol{R}_{k'})Z_k Z_{k'} + \sum_{P,k} \int \mathrm{d}^3r'\, v_{\mathrm{ee}}(\boldsymbol{R}_k - \boldsymbol{r}')Z_k \bar{n}_b(\boldsymbol{r}')\,, \tag{7.44}$$

is the average electrostatic energy, and the remaining two terms represent the effects of correlated fluctuations.

$$E_{ab}^{(I)} = -\int_0^1 \mathrm{d}\lambda \int \mathrm{d}^3r \int \mathrm{d}^3r'\, v_{\mathrm{ee}}(\boldsymbol{r}-\boldsymbol{r}')\left[\bar{n}_a(\boldsymbol{r})\langle\delta\hat{n}_b(\boldsymbol{r}')\rangle_\lambda + \bar{n}_b(\boldsymbol{r}')\langle\delta\hat{n}_a(\boldsymbol{r})\rangle_\lambda\right] + \sum_{P,k}\int \mathrm{d}^3r'\, v_{\mathrm{ee}}(\boldsymbol{R}_k - \boldsymbol{r}')Z_k(\boldsymbol{R}_k)\langle\delta\hat{n}_b(\boldsymbol{r}')\rangle_\lambda \tag{7.45a}$$

$$E_{ab}^{(II)} = -\int_0^1 \mathrm{d}\lambda \int \mathrm{d}^3r \int \mathrm{d}^3r'\, v_{\mathrm{ee}}(\boldsymbol{r}-\boldsymbol{r}')\langle\delta\hat{n}_a(\boldsymbol{r})\delta\hat{n}_b(\boldsymbol{r}')\rangle_\lambda\,. \tag{7.45b}$$

The expectation values $\langle\delta\hat{n}_a(\boldsymbol{r}_1)\rangle_\lambda$ and $\langle\delta\hat{n}_a(\boldsymbol{r}_1)\delta\hat{n}_b(\boldsymbol{r}_2')\rangle_\lambda$ can be computed perturbatively in $\lambda\hat{H}_{ab}$ in the interaction picture. The interaction $\lambda\hat{H}_{ab}$ is switched on adiabatically to generate the interacting ground state density matrix in terms of the non-interacting one. Substituting for $\hat{H}_{ab}$ from (7.42), and expanding in powers of λ yields a perturbation series in terms of the pth order joint response function, using $\boldsymbol{x}_n' = (\boldsymbol{r}_n', t_n)$,

$$R_a^{(p)}(\boldsymbol{x}, \boldsymbol{x}_p, \boldsymbol{x}_p' \ldots \boldsymbol{x}_1, \boldsymbol{x}_1') = \langle \hat{T}\delta\hat{n}_a^{I+}(\boldsymbol{x})[\hat{n}_a^I(\boldsymbol{x}_p)\hat{n}_b^I(\boldsymbol{x}_p')]^- \ldots [\hat{n}_a^I(\boldsymbol{x}_1)\hat{n}_b^I(\boldsymbol{x}_1')]^-\rangle_0\,. \tag{7.46}$$

Making use of (7.2) and the fact that the initial density matrix is a direct product of the density matrices of the individual molecules, $R^{(p)}$ can be factorized in terms of GRFs of the individual molecules. For example, the first order joint response function is:

$$\begin{aligned} R_a^{(1)}(\boldsymbol{x}, \boldsymbol{x}_1, \boldsymbol{x}_1') &= \left\langle \hat{T}\delta\hat{n}_a^{I+}(\boldsymbol{x})[\hat{n}_a^I(\boldsymbol{x}_1)\hat{n}_b^I(\boldsymbol{x}_1')]^-\right\rangle_0 \\ &= \mathrm{Tr}\left\{\hat{T}\delta\hat{n}_a^{I+}(\boldsymbol{x})[\hat{n}_a^I(\boldsymbol{x}_1)\hat{n}_b^I(\boldsymbol{x}_1')]^-\hat{\rho}_a^0\hat{\rho}_b^0\right\}\,. \end{aligned} \tag{7.47}$$

Substituting $n_a^{I\,\nu}(\boldsymbol{x}) = \bar{n}_a^\nu(\boldsymbol{x}) + \delta n_a^{I\,\nu}(\boldsymbol{x})$, $n_b^{I\,\nu}(\boldsymbol{x}') = \bar{n}_b^\nu(\boldsymbol{x}') + \delta n_b^{I\,\nu}(\boldsymbol{x}')$, and using the identities, $\langle\delta n_a^{I\,\nu}(\boldsymbol{x})\rangle_{0a} = 0$ and $\langle\bar{n}_a^+(\boldsymbol{x})\rangle_{0a} = \bar{n}_a(\boldsymbol{x})$, we obtain

$$R_a^{(1)}(\boldsymbol{x}, \boldsymbol{x}_1, \boldsymbol{x}_1') = \mathrm{i}\bar{n}_b(\boldsymbol{r}_1')\chi_a^{+-}(\boldsymbol{x}, \boldsymbol{x}_1)\,, \tag{7.48}$$

where χ_a^{+-} represents the linear order GRF for molecule a [see (7.22)]. Similarly, second and higher order joint response functions can be expressed in terms of the GRFs of the individuals molecules.

In the present work we have ignored the contributions due to the interactions with nuclei in (7.42). The quantities $\langle\delta\hat{n}_a(\boldsymbol{r}_1)\rangle_\lambda$ and $\langle\delta\hat{n}_a(\boldsymbol{r}_1)\delta\hat{n}_b(\boldsymbol{r}_1')\rangle_\lambda$, and consequently the interaction energies $E_{ab}^{(I)}$ and $E_{ab}^{(II)}$ can be expanded perturbatively in terms of the GRFs of the individual molecules [Harbola 2004]. We shall collect terms in $E_{ab}^{(I)}$ and $E_{ab}^{(II)}$ by their order with respect to charge fluctuations. The total energy is then, $E_{ab} = \sum_j E_{ab}^{(j)}$, where $E_{ab}^{(j)}$ represents contribution from jth order charge fluctuation. $W^{(0)}$ was given in (7.44) and $E_{ab}^{(1)} = 0$. $E_{ab}^{(j)}$ to sixth order are given in [Harbola 2004].

$$E_{ab}^{(2)} = -\frac{1}{2}\sum_P \int_{-\infty}^{t_1} \mathrm{d}t_2 \int \mathrm{d}^3\underline{r}_1 \int \mathrm{d}^3\underline{r}_2\, \bar{n}_b(\boldsymbol{r}_1')\bar{n}_b(\boldsymbol{r}_2')\chi_a^{+-}(\boldsymbol{x}_1,\boldsymbol{x}_2)v_{\mathrm{ee}}(\boldsymbol{s}_1)v_{\mathrm{ee}}(\boldsymbol{s}_2) \tag{7.49a}$$

$$\begin{aligned} E_{ab}^{(3)} &= \frac{1}{6}\sum_P \int_{-\infty}^{t_1} \mathrm{d}t_2 \int_{-\infty}^{t_1} \mathrm{d}t_3 \int \mathrm{d}^3\underline{r}_1 \int \mathrm{d}^3\underline{r}_2 \int \mathrm{d}^3\underline{r}_3\, v_{\mathrm{ee}}(\boldsymbol{s}_1)v_{\mathrm{ee}}(\boldsymbol{s}_2)v_{\mathrm{ee}}(\boldsymbol{s}_3) \\ &\quad \times \bar{n}_b(\boldsymbol{r}_1')\chi_a^{+--}(\boldsymbol{x}_1,\boldsymbol{x}_2,\boldsymbol{x}_3)\bar{n}_b(\boldsymbol{r}_2')\bar{n}_b(\boldsymbol{r}_)'3 \end{aligned} \tag{7.49b}$$

$$\begin{aligned} E_{ab}^{(4)} &= \frac{1}{6}\sum_P \int_{-\infty}^{t_1} \mathrm{d}t_2 \int_{-\infty}^{t_1} \mathrm{d}t_3 \int \mathrm{d}^3\underline{r}_1 \int \mathrm{d}^3\underline{r}_2 \int \mathrm{d}^3\underline{r}_3\, v_{\mathrm{ee}}(\boldsymbol{s}_1)v_{\mathrm{ee}}(\boldsymbol{s}_2)v_{\mathrm{ee}}(\boldsymbol{s}_3) \\ &\quad \times \left[\bar{n}_a(\boldsymbol{r}_2)\bar{n}_b(\boldsymbol{r}_3')\chi_a^{+-}(\boldsymbol{x}_1,\boldsymbol{x}_3)\chi_b^{+-}(\boldsymbol{x}_1',\boldsymbol{x}_2')\right. \\ &\quad \left. + \bar{n}_b(\boldsymbol{r}_1')\bar{n}_a(\boldsymbol{r}_3)\chi_a^{+-}(\boldsymbol{x}_1,\boldsymbol{x}_2)\chi_b^{+-}(\boldsymbol{x}_2',\boldsymbol{x}_3')\right] \\ &\quad - \frac{1}{2}\sum_P \int_{-\infty}^{t_1} \mathrm{d}t_2 \int \mathrm{d}^3\underline{r}_1 \int \mathrm{d}^3\underline{r}_2\, v_{\mathrm{ee}}(\boldsymbol{s}_1)v_{\mathrm{ee}}(\boldsymbol{s}_2)\chi_a^{++}(\boldsymbol{x}_1,\boldsymbol{x}_2)\chi_b^{+-}(\boldsymbol{x}_1',\boldsymbol{x}_2') \end{aligned} \tag{7.49c}$$

$$\begin{aligned} E_{ab}^{(5)} &= \frac{1}{6}\sum_P \int_{-\infty}^{t_1} \mathrm{d}t_2 \int_{-\infty}^{t_1} \mathrm{d}t_3 \int \mathrm{d}^3\underline{r}_1 \int \mathrm{d}^3\underline{r}_2 \int \mathrm{d}^3\underline{r}_3\, v_{\mathrm{ee}}(\boldsymbol{s}_1)v_{\mathrm{ee}}(\boldsymbol{s}_2)v_{\mathrm{ee}}(\boldsymbol{s}_3) \\ &\quad \times \left\{\chi_b^{+--}(\boldsymbol{x}_1',\boldsymbol{x}_2',\boldsymbol{x}_3')\left[\bar{n}_a(\boldsymbol{r}_1)\chi_a^{++}(\boldsymbol{x}_2,\boldsymbol{x}_3) + \bar{n}_a(\boldsymbol{r}_2)\chi_a^{++}(\boldsymbol{x}_1,\boldsymbol{x}_3)\right.\right. \\ &\qquad \left. + \bar{n}_a(\boldsymbol{r}_3)\chi_a^{++}(\boldsymbol{x}_1,\boldsymbol{x}_2) + \bar{n}_a(\boldsymbol{r}_1)\chi_a^{+-}(\boldsymbol{x}_1,\boldsymbol{x}_3)\right] \\ &\quad \left. + \chi_b^{++-}(\boldsymbol{x}_1',\boldsymbol{x}_2',\boldsymbol{x}_3')[\bar{n}_a(\boldsymbol{r}_2)\chi_a^{+-}(\boldsymbol{x}_1,\boldsymbol{x}_3) + \bar{n}_b(\boldsymbol{r}_3')\chi_b^{+-}(\boldsymbol{x}_1',\boldsymbol{x}_2')]\right\} \end{aligned} \tag{7.49d}$$

$$\begin{aligned} E_{ab}^{(6)} &= \frac{1}{6}\sum_P \int_{-\infty}^{t_1} \mathrm{d}t_2 \int_{-\infty}^{t_1} \mathrm{d}t_3 \int \mathrm{d}^3\underline{r}_1 \int \mathrm{d}^3\underline{r}_2 \int \mathrm{d}^3\underline{r}_3\, v_{\mathrm{ee}}(\boldsymbol{s}_1)v_{\mathrm{ee}}(\boldsymbol{s}_2)v_{\mathrm{ee}}(\boldsymbol{s}_3) \\ &\quad \times \left[\chi_a^{+++}(\boldsymbol{x}_1,\boldsymbol{x}_2,\boldsymbol{x}_3)\chi_b^{+--}(\boldsymbol{x}_1',\boldsymbol{x}_2',\boldsymbol{x}_3')\right. \\ &\quad \left. + \chi_a^{++-}(\boldsymbol{x}_1,\boldsymbol{x}_2,\boldsymbol{x}_3)\chi_b^{++-}(\boldsymbol{x}_1',\boldsymbol{x}_2',\boldsymbol{x}_3')\right] , \end{aligned} \tag{7.49e}$$

where for brevity we have used the notations, $\int \mathrm{d}^3\underline{r}_n = \int \mathrm{d}^3 r_n \int \mathrm{d}^3 r_n'$ and $v_{\mathrm{ee}}(\boldsymbol{s}_n) = v_{\mathrm{ee}}(\boldsymbol{r}_n - \boldsymbol{r}_n')$.

We have now at hand all the ingredients necessary for computing the intermolecular energies using the GRF of the individual molecules calculated at the TDDFT level.

The second term in (7.49c) reproduces McLachlan's expression for the van der Waals intermolecular energy [Misquitta 2003, McLachlan 1963a, McLachlan 1963b]. Since χ^{+-} and χ^{++} are related by the fluctuation-dissipation theorem, the McLachlan expression may be recast solely in terms of the ordinary response of two molecules, χ_a^{+-} and χ_b^{+-}.

$$E_{\mathrm{vdW}}^{(4)} = -\frac{1}{2\hbar}\int_{-\infty}^{\infty}\mathrm{d}\omega \int \mathrm{d}^3\underline{r}_1 \int \mathrm{d}^3\underline{r}_2\, v_{\mathrm{ee}}(\boldsymbol{s}_1)v_{\mathrm{ee}}(\boldsymbol{s}_2)$$
$$\coth\left(\frac{\beta\hbar\omega}{2}\right)\alpha_a^{+-}(\boldsymbol{r}_1,\boldsymbol{r}_2,\omega)\alpha_b^{+-}(\boldsymbol{r}_1',\boldsymbol{r}_2',\omega)\,, \quad (7.50)$$

where $\chi^{\nu_2\nu_1}(\boldsymbol{r}_2\omega_2,\boldsymbol{r}_1\omega_1) = (\hbar^{-1})^{p'}\alpha_a^{\nu_2\nu_1}(\boldsymbol{r}_1,\boldsymbol{r}_2,\omega_1)\delta(\omega_1+\omega_2)$. Equation (7.50) gives

$$E_{\mathrm{vdW}}^{(4)} = -k_{\mathrm{B}}T\sum_{n=0}^{\infty}\int \mathrm{d}^3\int \mathrm{d}^3\underline{r}_1\, \underline{r}_2 v_{\mathrm{ee}}(\boldsymbol{s}_1)v_{\mathrm{ee}}(\boldsymbol{s}_2)\alpha_a^{+-}(\boldsymbol{r}_1,\boldsymbol{r}_2,\mathrm{i}\omega_n)\alpha_b^{+-}(\boldsymbol{r}_1',\boldsymbol{r}_2',\mathrm{i}\omega_n)\,, \quad (7.51)$$

where $\omega_n = (2\pi n k_{\mathrm{B}}T/\hbar)$ are the Matsubara frequencies. However, life is not as simple for the higher order responses. The $(p+1)$, p^{th} order generalized response functions, $\chi^{\nu_{p+1}\nu_p\cdots\nu_1}$, cannot all be related to the fully retarded ordinary response, $\chi^{+-\cdots-}$. The complete set of generalized response functions is thus required to represent the intermolecular forces.

By combining (7.49a)–(7.49e) with the CEO expansion, we can finally express the intermolecular energies in terms of CEO modes of the separate molecules. For example, substituting for χ^{+-} and χ^{++} from (7.37) in (7.49c), the fourth order term is obtained in terms of CEO modes as

$$\begin{aligned}E_{ab}^{(4)} &= -\frac{1}{2\hbar}\sum_P \int \mathrm{d}^3\underline{r}_1 \int \mathrm{d}^3\underline{r}_2 \frac{1}{|\boldsymbol{r}_1-\boldsymbol{r}_1'||\boldsymbol{r}_2-\boldsymbol{r}_2'|}\\
&\quad\times\sum_{\alpha\alpha'}\frac{s_{\alpha'}\bar{\mu}_{-\alpha}(\boldsymbol{r}_1)\mu_\alpha(\boldsymbol{r}_2)\mu_{-\alpha'}(\boldsymbol{r}_1')\mu_{\alpha'}(\boldsymbol{r}_2')}{(\Omega_\alpha+\Omega_{\alpha'})}\\
&\quad-\frac{1}{6\hbar^2}\sum_P\int \mathrm{d}\underline{r}_1\int \mathrm{d}\underline{r}_2\int \mathrm{d}^3\underline{r}_3\frac{1}{|\boldsymbol{r}_1-\boldsymbol{r}_1'||\boldsymbol{r}_2-\boldsymbol{r}_2'||\boldsymbol{r}_3-\boldsymbol{r}_3'|}\bar{n}_b(\boldsymbol{r}_3')\\
&\quad\times\sum_{\alpha\alpha'}\frac{s_\alpha s_{\alpha'}\mu_\alpha(\boldsymbol{r}_3)\mu_{\alpha'}(\boldsymbol{r}_1')\mu_{-\alpha'}(\boldsymbol{r}_2')}{\Omega_\alpha\Omega_{\alpha'}}\left[\bar{n}_a(\boldsymbol{r}_1)\mu_\alpha(\boldsymbol{r}_2)+\bar{n}_a(\boldsymbol{r}_2)\mu_\alpha(\boldsymbol{r}_1)\right]\,.\end{aligned} \quad (7.52)$$

Expansions of higher order terms in CEO modes are given in [Harbola 2004]. The GRF therefore provide a compact and complete description of intermolecular interactions. Both response and correlation functions can be described in terms of Liouville space pathways and can thus be treated along the same footing.

Acknowledgments

We wish to thank Prof. Vladimir Chernyak and Dr. Sergei Tretiak who were instrumental in developing the CEO approach. The support of the National Science Foundation (CHE 0446555), NIRT (EEC 0406750) and CBC (CHE 0533162) is gratefully acknowledged.

Part II

Approximate Functionals

8 Time-Dependent Deformation Approximation

I.V. Tokatly

8.1 Introduction

One of the most important conceptual achievements of DFT is the possibility to formulate the many-body problem in a form of a closed theory that contains only a restricted set of basic variables, such as the density in the static DFT, or the density and the current in TDDFT. In classical physics, a theory of this type is known for more than two centuries. This is classical hydrodynamics. In fact, the Runge-Gross mapping theorem in TDDFT [Runge 1984] proves the existence of the exact quantum hydrodynamics. In this respect, static equilibrium DFT should be viewed as the exact quantum hydrostatics. It is indeed known that the condition of the energy minimum is equivalent to the condition for a local compensation of the external and the internal stress forces exerted on every infinitesimal volume element of the system in equilibrium [Bartolotti 1980]. Interestingly, the equations of TDDFT in the hydrodynamic formulation can be also considered as a force balance condition, but in a local noninertial reference frame moving with the flow. In the time-dependent case there is a local compensation of the external, inertial, and internal stress forces. This demonstrates a close similarity of static DFT and TDDFT in the co-moving frame. The above similarity was the main motivation to reconsider the formulation of TDDFT from the point of view of a local observer in the co-moving Lagrangian reference frame [Tokatly 2005a, Tokatly 2005b]. One of the goals of this work is to present the Lagrangian formulation of TDDFT in a compact and physically transparent form. Simple numerical illustrations of this approach for one-dimensional dynamics can be found in a recent paper [Ullrich 2006].

Practical applications of any DFT rely on the Kohn-Sham (KS) construction [Kohn 1965, Dreizler 1990], which maps the calculation of basic observables in the interacting system to the solution of an auxiliary noninteracting KS problem. Noninteracting KS particles move in a self-consistent exchange-correlation (xc) potential that is adjusted to reproduce the correct values of the basic variables, i.e., the density and the current in TDDFT. From the hydrodynamical point of view, the KS construction allows one to compute exactly the kinetic part of the internal stress force, while treating the xc contribution to the stress in an approximate fashion. Thus, the central problem of any practical DFT reduces to the construction of adequate

I.V. Tokatly: *Time-Dependent Deformation Approximation*, Lect. Notes Phys. **706**, 123–136 (2006)
DOI 10.1007/3-540-35426-3_8

approximations for the xc potentials. In the static DFT a good starting point is provided by the local density approximation (LDA). On one hand, the static LDA, by itself, gives quite reasonable results, and, on the other hand, it allows for further modifications and refinements. The construction of a similar basic local approximation in TDDFT turns out to be problematic. The reason for these problems is the inherent nonlocality of the nonadiabatic nonequilibrium DFT. General arguments [Vignale 1995a, Vignale 1995b], based on the harmonic potential theorem [Dobson 1994a], require that any consistent nonadiabatic xc potential must be a strongly nonlocal functional of the density. Otherwise the harmonic potential theorem is violated.

An important step in resolving the nonlocality problem was made by Vignale and Kohn (VK) [Vignale 1996] who showed that at least in the linear response regime switching the basic variable from the density n to the current $\boldsymbol{j}$ allows one to construct a consistent local nonadiabatic approximation. Shortly afterwards, Vignale, Ullrich, and Conti [Vignale 1997] found that the velocity $\boldsymbol{v} = \boldsymbol{j}/n$ is a more natural variable, as it allows to represent the complicated VK expression in a physically transparent viscoelastic form. The reformulation of TDDFT in the co-moving Lagrangian frame, which is presented in this work, makes the next step in that direction.

We show that the most natural complete set of basic variables in a general nonlinear TDDFT consists of the Lagrangian coordinate $\boldsymbol{\xi}$, the symmetric Green's deformation tensor $g_{\mu\nu}$, and the skew-symmetric vorticity tensor $\widetilde{F}_{\mu\nu}$. These three quantities, one vector, one symmetric and one skew-symmetric tensor, contain twelve numbers that are required for the complete local characterization of a deformed state of any continuous medium [Masson 1964]. Namely, $\boldsymbol{\xi}$, $g_{\mu\nu}$ and $\widetilde{F}_{\mu\nu}$ describe respectively the translation, deformation and rotation of an infinitesimal fluid element. On the other hand, they describe generalized inertia forces in the Lagrangian frame. All three quantities are functionals of the velocity in accordance with the Runge-Gross mapping theorem [Runge 1984]. The new formulation of TDDFT relates the local stress in the system, and thus the xc potential, to the dynamic deformations in the quantum many-body system. Therefore, the functional dependence of the xc potential on the basic variables acquires a clear physical meaning. It corresponds to the stress-deformation relation, which is very natural from the point of view of continuum mechanics. If spatial derivatives of the deformation tensor are small, the stress-deformation relation becomes local and we get the local approximation for the xc potential in TDDFT. In the linear response regime the general stress-deformation relation reduces to the linear Hooke's law, which exactly coincides with the visco-elastic VK approximation [Vignale 1996, Vignale 1997].

8.2 DFT as Exact Quantum Continuum Mechanics

In this section we discuss a hydrodynamic formulation of DFT and introduce a definition of the xc potential in terms of local stress forces.

8.2.1 Conservation Laws and the Hydrodynamic Formulation of DFT

Let us consider a system of N interacting fermions in the presence of a time-dependent external potential $v_{\text{ext}}(\boldsymbol{r},t)$. The hydrodynamic formulation of DFT follows from the equations of motion for the density $n(\boldsymbol{r},t)$ and for the current $\boldsymbol{j}(\boldsymbol{r},t)$,

$$\frac{\partial}{\partial t}n(\boldsymbol{r},t) - \mathrm{i}\langle[\hat{H},\hat{n}(\boldsymbol{r},t)]\rangle = 0 \tag{8.1a}$$

$$\frac{\partial}{\partial t}\boldsymbol{j}(\boldsymbol{r},t) - \mathrm{i}\langle[\hat{H},\hat{\boldsymbol{j}}(\boldsymbol{r},t)]\rangle = 0\,, \tag{8.1b}$$

where $\hat{n}(\boldsymbol{r},t)$ and $\hat{\boldsymbol{j}}(\boldsymbol{r},t)$ are the corresponding Heisenberg operators, and $\hat{H}$ is the Hamiltonian of the system. Angle brackets in the above formulas stand for the averaging with the exact density matrix. Equations (8.1a) and (8.1b) can be represented in the form of hydrodynamics balance equations (see [Tokatly 2005a] for a detailed derivation),

$$D_t n + n\frac{\partial}{\partial r^\mu}v_\mu = 0 \tag{8.2a}$$

$$mnD_t v_\mu + \frac{\partial}{\partial r^\nu}P_{\mu\nu} + n\frac{\partial}{\partial r^\mu}\left(v_{\text{ext}} + v_{\text{H}}\right) = 0\,, \tag{8.2b}$$

where r^ν are the components of the vector $\boldsymbol{r}$. Equation (8.2a) is the usual continuity equation, while (8.2b) corresponds to the local momentum conservation law. In these equations, $\boldsymbol{v} = \boldsymbol{j}/n$ is the velocity of the flow, and $D_t = \partial_t + \boldsymbol{v}\nabla$ is the convective derivative. The exact stress tensor in (8.2b),

$$P_{\mu\nu}(\boldsymbol{r},t) = T_{\mu\nu}(\boldsymbol{r},t) + V^{\text{ee}}_{\mu\nu}(\boldsymbol{r},t)\,, \tag{8.3}$$

contains the kinetic, $T_{\mu\nu}$, and the interaction, $V^{\text{ee}}_{\mu\nu}$, contributions. In [Tokatly 2005a] we derived the following explicit representations for the stress tensors (see also [Martin 1959, Kadanoff 1962, Puff 1968, Zubarev 1974])

$$T_{\mu\nu}(\boldsymbol{r}) = \frac{1}{2m}\left[\lim_{\boldsymbol{r}'\to\boldsymbol{r}}\left(\hat{q}^*_\mu\hat{q}'_\nu + \hat{q}^*_\nu\hat{q}'_\mu\right)\rho(\boldsymbol{r},\boldsymbol{r}') - \frac{\delta_{\mu\nu}}{2}\nabla^2 n(\boldsymbol{r})\right] \tag{8.4a}$$

$$V^{\text{ee}}_{\mu\nu}(\boldsymbol{r}) = -\frac{1}{2}\int \mathrm{d}^3r'\,\frac{r'^\mu r'^\nu}{r'}\frac{\partial v_{\text{ee}}(r')}{\partial r'}\int_0^1 \mathrm{d}\lambda\, G_2(\boldsymbol{r}+\lambda\boldsymbol{r}', \boldsymbol{r}-(1-\lambda)\boldsymbol{r}') \tag{8.4b}$$

where $\hat{\boldsymbol{q}} = -\mathrm{i}\nabla - m\boldsymbol{v}$ is the operator of "relative" momentum, $v_{\mathrm{ee}}(r)$ is the potential of a pairwise interparticle interaction, $\rho(\boldsymbol{r}, \boldsymbol{r}')$ is the one-particle density matrix, and $G_2(\boldsymbol{r}, \boldsymbol{r}')$ is the pair correlation function.

Equations (8.2a) and (8.2b) represent the exact local conservation laws which must be satisfied for an arbitrary evolution of the system. Let us apply them to TDDFT. The first, less restrictive part of the TDDFT mapping theorem [Runge 1984, Li 1985] states the existence of a unique and invertible map: $\boldsymbol{j} \rightarrow v_{\mathrm{ext}}$ or, equivalently, $\boldsymbol{v} \rightarrow v_{\mathrm{ext}}$. This implies that the exact many-body density matrix $\hat{\rho}(t)$ for a given initial condition, $\hat{\rho}(0) = \hat{\rho}_0$, is a functional of the velocity $\boldsymbol{v}$. Hence the stress tensor of (8.3) is a functional of $\boldsymbol{v}$ and of the initial density matrix: $P_{\mu\nu}[\hat{\rho}_0, \boldsymbol{v}]$. Therefore (8.2a) and (8.2b) constitute a formally closed set of exact quantum hydrodynamics equations with the memory of the initial many-body correlations. It is interesting to note that the usual classical hydrodynamics can be viewed as a particular limiting form of TDDFT. In this limiting case the stress tensor functional is known explicitly – it takes the usual Navier-Stokes form [Landau 1987].

In the equilibrium system, (8.2b) reduces to the static force balance equation

$$\frac{\partial}{\partial r^\nu} P_{\mu\nu} + n \frac{\partial}{\partial r^\mu} (v_{\mathrm{ext}} + v_{\mathrm{H}}) = 0\,. \tag{8.5}$$

This equation shows that the force produced by the external and the Hartree potentials is compensated by the force of internal stresses. The net force exerted on every infinitesimal fluid element is zero, which results in zero current density and a stationary particle density distribution. According to the Hohenberg-Kohn theorem [Hohenberg 1964, Dreizler 1990] any equilibrium observable, in particular the stress tensor, is a functional of the density: $P_{\mu\nu} = P_{\mu\nu}[n]$. Hence (8.5) is, in fact, the equation of the exact quantum hydrostatics that uniquely determines the density distribution in the equilibrium system. In the semiclassical limit, (8.5) reduces to the usual hydrostatics equation [Landau 1987]. (For a degenerate high density Fermi gas we recover the Thomas-Fermi theory.)

8.2.2 Definition of the xc Potentials

An important part of any DFT is the KS construction. The current and the density in the interacting system can be reproduced in a system of noninteracting KS particles, moving in a properly chosen self-consistent potential $v_{\mathrm{KS}} = v_{\mathrm{ext}} + v_{\mathrm{H}} + v_{\mathrm{xc}}$. The hydrodynamic formulation of TDDFT allows us to relate the xc potential v_{xc} to the stress density. The hydrodynamics balance equations for the KS system take the form

$$D_t n + n \frac{\partial}{\partial r^\mu} v_\mu = 0 \tag{8.6a}$$

$$mnD_t v_\mu + \frac{\partial}{\partial r^\nu} T^{\mathrm{KS}}_{\mu\nu} + n \frac{\partial}{\partial r^\mu} (v_{\mathrm{ext}} + v_{\mathrm{H}} + v_{\mathrm{xc}}) = 0\,, \tag{8.6b}$$

where the kinetic stress tensor of the KS system, $T_{\mu\nu}^{\mathrm{KS}}$, is given by (8.4a), but with the averaging over the state of noninteracting particles. Comparing (8.2) with (8.6) we find that the velocity $\boldsymbol{v}$ and the density n of the noninteracting and the interacting systems coincide if the xc potential $v_{\mathrm{xc}}(\boldsymbol{r}, t)$ satisfies the equation

$$\frac{\partial v_{\mathrm{xc}}}{\partial r^{\mu}} = \frac{1}{n}\frac{\partial}{\partial r^{\nu}}\left(P_{\mu\nu} - T_{\mu\nu}^{\mathrm{KS}}\right) \equiv \frac{1}{n}\frac{\partial}{\partial r^{\nu}} P_{\mu\nu}^{\mathrm{xc}}, \tag{8.7}$$

where $P_{\mu\nu}^{\mathrm{xc}}$ is the xc stress tensor. Equation (8.7) demonstrates the physical significance of v_{xc}. The xc potential should produce a force that compensates for the difference of the internal stress forces in the real interacting system and in the auxiliary noninteracting KS system. By the continuity equation, the density n is a functional of the velocity. Therefore (8.7) defines v_{xc} (up to an inessential constant) as a functional of $\boldsymbol{v}$.

Apparently the above stress definition of the xc potential apply equally well both to TDDFT and to the static/equilibrium DFT. For practical applications it is possibly more convenient to represent the force definition of v_{xc}, (8.7), in the familiar form of a Poisson equation

$$\nabla^2 v_{\mathrm{xc}}(\boldsymbol{r}, t) = 4\pi \rho_{\mathrm{xc}}(\boldsymbol{r}, t), \tag{8.8}$$

where the quantity $\rho_{\mathrm{xc}}(\boldsymbol{r}, t)$,

$$\rho_{\mathrm{xc}} = \frac{1}{4\pi}\frac{\partial}{\partial r^{\mu}}\left(\frac{1}{n}\frac{\partial}{\partial r^{\nu}} P_{\mu\nu}^{\mathrm{xc}}\right), \tag{8.9}$$

can be interpreted as an xc "charge" density. In this context the xc stress force, $n^{-1}\partial_{\nu} P_{\mu\nu}^{\mathrm{xc}}$, has the clear meaning of an xc "polarization" density. The additional differentiation in (8.8) requires an additional boundary condition. The most natural condition, which we should impose on the solution of (8.8), is the requirement of being bounded at infinity.

Equation (8.7) or, equivalently, (8.8) and (8.9) reduce the problem of approximating v_{xc} to the construction of approximations for the xc stress tensor $P_{\mu\nu}^{\mathrm{xc}}$. Since the stress density has a clear physical and microscopic meaning, there is hope that the latter problem is more tractable.

8.3 Geometric Formulation of TDDFT

8.3.1 Preliminaries: Static LDA vs. Time-Dependent LDA

Let us first derive the standard static LDA from the force definition of v_{xc}, (8.7). Formally, the static $v_{\mathrm{xc}}^{\mathrm{LDA}}(\boldsymbol{r})$ is the solution of (8.7) in lowest order of the gradient expansion. If the density distribution is a semiclassically slowly varying function in space, a small volume element, located at some point $\boldsymbol{r}$, can be considered as an independent homogeneous many-body system. The density in this homogeneous system equals the density $n(\boldsymbol{r})$ at the location of the element. By solving the homogeneous many-body problem for the interacting and noninteracting systems we find the corresponding stress tensors, $P_{\mu\nu}[n] = \delta_{\mu\nu} P(n)$ and $T^{\mathrm{KS}}_{\mu\nu}[n] = \delta_{\mu\nu}\, P_0(n)$, where P and P_0 are the pressure of the interacting system and of an ideal gas, respectively. Therefore, to lowest order in the density gradients we get

$$P^{\mathrm{xc}}_{\mu\nu}[n](\boldsymbol{r}) = \delta_{\mu\nu}\left[P(n(\boldsymbol{r})) - P_0(n(\boldsymbol{r}))\right] = \delta_{\mu\nu} P_{\mathrm{xc}}(n(\boldsymbol{r}))\,. \tag{8.10}$$

Substituting (8.10) into (8.7), and using the common thermodynamic relations, $\mathrm{d}P = n\mathrm{d}\mu$, $\mu = \partial E/\partial n$, we immediately recover the standard static LDA [Dreizler 1990].

The situation in the time-dependent theory is much more complicated. Even if at any instant t the density distribution $n(\boldsymbol{r}, t)$ is a slow function in space, a small volume element, located at some point $\boldsymbol{r}$, can not be considered as a system independent of the surrounding space. In the case of nonadiabatic dynamics, particles arriving at the point $\boldsymbol{r}$ from other regions bring information about other spatial points. This is the physical reason for the well known ultranonlocality of TDDFT [Vignale 1995a, Vignale 1995b]. Clearly, any naïve attempt to extend the above derivation of the static LDA to the time-dependent case fails. Indeed the homogeneous many-body problem, which we get by formally separating a small volume element, corresponds to an infinite system with strongly nonconserved number of particles. Obviously, this problem is meaningless.

Below we show that the problem of nonlocality can be resolved by changing the "point of view" of the nonequilibrium many-body system. Any flow in the system can be considered as a collection of small fluid elements moving along their own trajectories. It is possible to divide the system into elements in such a way that the number of particles in every element is conserved. Indeed, by proper deformation and rotation of a fluid element one can always adjust its shape to the motion of particles and thus prevent the flow through its surface. Let us look at the system from a local reference frame attached to one of those moving elements. The motion of the origin compensates the translational motion of the fluid element. By properly changing scales and directions of the coordinate axes we can compensate both the deformations and the rotation. This means that an observer in the new frame will see no currents in the system, but a stationary density distribution. Thus, from the point of view of the co-moving observer, the nonequilibrium system looks very similar to the equilibrium system viewed by a stationary observer in the laboratory reference frame. This similarity is of course not complete since the particles in the co-moving frame should experience inertia forces. However, these inertia forces are determined only by the local geometric characteristics of the frame. The locality of inertia forces and the stationarity of the density allow us to consider a small volume element in the co-moving frame as an independent many-body system. Therefore we can extend the derivation of the static LDA to the time-dependent case.

8.3.2 TDDFT in the Lagrangian Frame

Formally, the transformation to the co-moving Lagrangian frame is defined as follows. Let $\boldsymbol{v}(\boldsymbol{r},t) = \boldsymbol{j}(\boldsymbol{r},t)/n(\boldsymbol{r},t)$ be the velocity of the flow. By solving the following initial value problem

$$\frac{\partial \boldsymbol{r}(\boldsymbol{\xi},t)}{\partial t} = \boldsymbol{v}(\boldsymbol{r}(\boldsymbol{\xi},t),t), \qquad \boldsymbol{r}(\boldsymbol{\xi},0) = \boldsymbol{\xi}, \tag{8.11}$$

we find the function $\boldsymbol{r}(\boldsymbol{\xi},t)$ that determines the trajectory of a fluid element. The initial point, $\boldsymbol{\xi}$, of the trajectory can be used as a unique label of the element. This initial position of an infinitesimal fluid element is called the Lagrangian coordinate. The transformation from the original $\boldsymbol{r}$-space to the $\boldsymbol{\xi}$-space of initial positions corresponds to the transformation from the Eulerian to the Lagrangian description of a fluid [Masson 1964]. On the other hand, since the equation $\boldsymbol{r} = \boldsymbol{r}(\boldsymbol{\xi},t)$ – which maps $\boldsymbol{r}$ to $\boldsymbol{\xi}$ – describes the trajectory of a fluid element, the transformation from $\boldsymbol{r}$ to $\boldsymbol{\xi}$ is exactly the transformation to the frame attached to this element.

Nonlinear transformation of the coordinates induces a change of metric. The new metric in $\boldsymbol{\xi}$-space is described by Green's deformation tensor $g_{\mu\nu}(\boldsymbol{\xi},t)$ [Tokatly 2005a]

$$g_{\mu\nu} = \frac{\partial r^\alpha}{\partial \xi^\mu}\frac{\partial r^\alpha}{\partial \xi^\nu}, \qquad g^{\mu\nu} = \frac{\partial \xi^\mu}{\partial r^\alpha}\frac{\partial \xi^\nu}{\partial r^\alpha}. \tag{8.12}$$

The determinant, g, of $g_{\mu\nu}$ determines the ratio of unit volumes in the laboratory and in the Lagrangian frames: $\mathrm{d}^3r = \sqrt{g}\,\mathrm{d}^3\xi$. Therefore the density (the number of particles per unit volume) $\widetilde{n}(\boldsymbol{\xi},t)$ in the Lagrangian frame is related to the density $n(\boldsymbol{r},t)$ in the laboratory frame as follows:

$$\widetilde{n}(\boldsymbol{\xi},t) = \sqrt{g(\boldsymbol{\xi},t)}\, n(\boldsymbol{r}(\boldsymbol{\xi},t),t). \tag{8.13}$$

Another important quantity is the velocity vector transformed to the Lagrangian frame $\widetilde{v}^\mu(\boldsymbol{\xi},t)$,

$$\widetilde{v}^\mu(\boldsymbol{\xi},t) = \frac{\partial \xi^\mu}{\partial r^\nu}\, v^\nu(\boldsymbol{r}(\boldsymbol{\xi},t),t). \tag{8.14}$$

In the equations of many-body theory in the Lagrangian frame, the vector $m\widetilde{v}_\nu = mg_{\mu\nu}\widetilde{v}^\mu$ plays the role of an effective "vector potential", while the quantity $-m\widetilde{v}^\mu\widetilde{v}_\mu/2$ enters the theory as an effective "scalar potential" [Tokatly 2005a]. The effective potentials and a nontrivial metric tensor describe generalized inertia forces in the local noninertial frame. The tensor $g_{\mu\nu}$ produces the "geodesic" force, which is responsible for the motion of a free particle along the geodesic in $\boldsymbol{\xi}$-space. The "vector potential" $m\widetilde{v}_\nu$ generates the Coriolis force (an effective Lorentz force) and the linear acceleration force (an effective electric field). The Coriolis force is determined by the skew-symmetric vorticity tensor $\widetilde{F}_{\mu\nu} = \partial_\nu \widetilde{v}_\mu - \partial_\mu \widetilde{v}_\nu$ (local angular velocity of the frame or an effective magnetic field). The "scalar potential" $-m\widetilde{v}^\mu \widetilde{v}_\mu/2$ produces the generalized centrifugal force.

The main property of the Lagrangian frame is that the inertia forces exactly compensate the difference of the external force and the force of internal stress. The local force balance equation can be obtained by transforming the local momentum conservation law (8.2b) to the Lagrangian frame

$$m\frac{\partial \widetilde{v}_\mu}{\partial t} - \frac{m}{2}\frac{\partial \widetilde{v}_\nu \widetilde{v}^\nu}{\partial \xi^\mu} + \frac{\partial}{\partial \xi^\mu}(v_{\rm ext} + v_{\rm H}) + \frac{1}{\widetilde{n}}\frac{\partial \sqrt{g}\widetilde{P}^\nu_\mu}{\partial \xi^\nu} - \frac{\sqrt{g}}{2\widetilde{n}}\frac{\partial g_{\alpha\beta}}{\partial \xi^\mu}\widetilde{P}^{\alpha\beta} = 0\,, \quad (8.15)$$

where $\widetilde{P}_{\mu\nu}$ is the stress tensor in the Lagrangian frame (the last two terms form the covariant divergence of the stress tensor [Dubrovin 1984]). The first two terms in (8.15) are the linear acceleration force and the generalized centrifugal force, while the last two terms correspond to the force of internal stress. Equation (8.15) has precisely the same physical significance as the static force balance equation (8.5) – it shows that the net force exerted on every infinitesimal fluid element in the Lagrangian frame is exactly zero. Therefore the current density in every point of the Lagrangian space vanishes, and the density of particles remains stationary during the whole evolution of the system: $\widetilde{n}(\boldsymbol{\xi}, t) = n_0(\boldsymbol{\xi})$, where $n_0(\boldsymbol{r})$ is the initial density distribution.

Since the linear acceleration force and the generalized centrifugal force are the same for all particles in a given fluid element, these two forces explicitly enter the balance (8.15). In contrast, the "geodesic" force and the Coriolis force depend on the velocity of a particular particle. Therefore they are present in (8.15) only implicitly via the stress tensor. In fact [Tokatly 2005b], the stress tensor $\widetilde{P}_{\mu\nu}$ is a universal functional of the deformation tensor $g_{\mu\nu}$ and the vorticity tensor $\widetilde{F}_{\mu\nu}$ (which determine the "geodesic" and the Coriolis force respectively):

$$\widetilde{P}_{\mu\nu} = \widetilde{P}_{\mu\nu}[g_{\mu\nu}, \widetilde{F}_{\mu\nu}](\boldsymbol{\xi}, t)\,. \quad (8.16)$$

Since the stress tensor is uniquely related to the xc potential, the latter is also a universal functional of $g_{\mu\nu}$ and $\widetilde{F}_{\mu\nu}$. Therefore (8.16) can be viewed as a reformulation of TDDFT in terms of new basic variables. Both $g_{\mu\nu}$ and $\widetilde{F}_{\mu\nu}$ are functionals of the velocity in agreement with the Runge-Gross theorem. However, the physical interpretation of the new formulation of TDDFT becomes much more transparent. The symmetric Green's deformation tensor and the skew-symmetric vorticity tensor completely characterize the deformed state of any system in the Lagrangian description. Hence (8.16) can be interpreted as the exact nonequilibrium "equation of state", which relates the stress tensor to the dynamical deformation.

The main formal advantage of the reformulation of TDDFT in the Lagrangian frame is the stationarity of the density, which allows to derive in a consistent manner the time-dependent local approximation.

8.4 Time-Dependent Local Deformation Approximation

8.4.1 General Formulation of the TDLDefA

By construction, the current density is zero at every point of Lagrangian space. Hence, in the limit of vanishingly small deformation gradients, an infinitesimal volume located at some point $\boldsymbol{\xi}$ can be considered as an independent homogeneously deformed many-body system. The density and the metric (deformation) tensor in this homogeneous system are given by $n_0(\boldsymbol{\xi})$ and $g_{\mu\nu}(\boldsymbol{\xi},t)$, respectively. The corresponding Hamiltonian takes the form

$$\widetilde{H} = \sum_{\boldsymbol{k}} g^{\mu\nu}(t)\frac{k_\mu k_\nu}{2m}\widetilde{a}^\dagger_{\boldsymbol{k}}\widetilde{a}_{\boldsymbol{k}} + \frac{1}{2\sqrt{g(t)}}\sum_{\boldsymbol{p},\boldsymbol{k},\boldsymbol{q}} v_{\mathrm{ee}}(\|\boldsymbol{q}\|)\widetilde{a}^\dagger_{\boldsymbol{k}}\widetilde{a}^\dagger_{\boldsymbol{p}}\widetilde{a}_{\boldsymbol{p}+\boldsymbol{q}}\widetilde{a}_{\boldsymbol{k}-\boldsymbol{q}}\,, \quad (8.17)$$

where $\|\boldsymbol{q}\| = \sqrt{g^{\mu\nu}(t)q_\mu q_\nu}$ is the wave vector norm in the "deformed" momentum space.

By definition, the stress tensor is proportional to the derivative of the energy with respect to the metric tensor: $\sqrt{g}\,\widetilde{P}_{\mu\nu} = 2\langle\delta\widetilde{H}/\delta g^{\mu\nu}\rangle$ [Rogers 2002, Tokatly 2005a]. Calculating the derivative of (8.17) we arrive at the following microscopic expression for the stress tensor:

$$\widetilde{P}_{\mu\nu} = \sum_{\boldsymbol{k}}\left\{\frac{1}{\sqrt{g}}\frac{k_\mu k_\nu}{m}\widetilde{f}(\boldsymbol{k}) + \frac{1}{2g}\left[\frac{k_\mu k_\nu}{\|\boldsymbol{k}\|}v'_{\mathrm{ee}}(\|\boldsymbol{k}\|) + g_{\mu\nu}v_{\mathrm{ee}}(\|\boldsymbol{k}\|)\right]\widetilde{G}_2(\boldsymbol{k})\right\}, \quad (8.18)$$

where $\widetilde{f}(\boldsymbol{k}) = \langle\widetilde{a}^\dagger_{\boldsymbol{k}}\widetilde{a}_{\boldsymbol{k}}\rangle$ is the Wigner function, $\widetilde{G}_2(\boldsymbol{k})$ is the Fourier component of the pair correlation function, and $v'_{\mathrm{ee}}(x) = \mathrm{d}v_{\mathrm{ee}}(x)/\mathrm{d}x$.

By solving the many-body problem defined by (8.17) we find the Wigner function $\widetilde{f}(\boldsymbol{k},t)$, and the pair correlation function $\widetilde{G}_2(\boldsymbol{k},t)$. Substitution of $\widetilde{f}(\boldsymbol{k},t)$ and $\widetilde{G}_2(\boldsymbol{k},t)$ into (8.18) yields the stress tensor functional $\widetilde{P}_{\mu\nu}[g_{\mu\nu}, n_0]$. By repetition of the above procedure for the noninteracting system [i.e., using (8.17), (8.18) with $v_{\mathrm{ee}} = 0$] we find the KS stress tensor, $\widetilde{T}^{\mathrm{KS}}_{\mu\nu}[g_{\mu\nu}, n_0]$, and, finally, the xc stress tensor

$$\widetilde{P}^{\mathrm{xc}}_{\mu\nu}[g_{\mu\nu}(\boldsymbol{\xi},t), n_0(\boldsymbol{\xi})] = \widetilde{P}_{\mu\nu} - \widetilde{T}^{\mathrm{KS}}_{\mu\nu}\,. \quad (8.19)$$

The last step is to transform $\widetilde{P}^{\mathrm{xc}}_{\mu\nu}$ of (8.19) back to the laboratory frame and to substitute it into the force definition of the xc potential (8.8).

The xc stress tensor $\widetilde{P}^{\mathrm{xc}}_{\mu\nu}$, given by (8.19), is a spatially local functional of the deformation tensor. Therefore it is natural to name the approximation (8.19) *time-dependent local deformation approximation* (TDLDefA).

Stress Tensor of the Noninteracting System

A necessary step in the derivation of the TDLDefA is to compute the stress tensor, $\widetilde{T}^{\mathrm{KS}}_{\mu\nu}$, in the noninteracting system. This problem can be solved exactly. In general, the equation of motion for the Wigner function takes the form

$$\frac{\partial \widetilde{f}(\boldsymbol{k})}{\partial t} = -\mathrm{i}\sum_{\boldsymbol{p},\boldsymbol{q}} \frac{v_{\mathrm{ee}}(\|\boldsymbol{q}\|)}{\sqrt{g(t)}} \langle \widetilde{a}^{\dagger}_{\boldsymbol{p}}(\widetilde{a}^{\dagger}_{\boldsymbol{k}}\widetilde{a}_{\boldsymbol{k}-b\boldsymbol{q}} - \widetilde{a}^{\dagger}_{\boldsymbol{k}+\boldsymbol{q}}\widetilde{a}_{\boldsymbol{k}})\widetilde{a}_{\boldsymbol{p}+\boldsymbol{q}}\rangle \,. \tag{8.20}$$

In the noninteracting case ($v_{\mathrm{ee}} = 0$) the right-hand side of (8.20) vanishes. Therefore the distribution function of noninteracting particles in the Lagrangian frame is time-independent:

$$\widetilde{f}(\boldsymbol{k}, t) = \widetilde{f}(\boldsymbol{k}, 0) = n^F_{\boldsymbol{k}} \,, \tag{8.21}$$

where we assumed for definiteness that the system evolves from the equilibrium state ($n^F_{\boldsymbol{k}}$ is the Fermi function). Substituting (8.21) into (8.18) (with $v_{\mathrm{ee}} = 0$) we get the kinetic stress tensor of the noninteracting system in the Lagrangian frame,

$$\widetilde{T}^{\mathrm{KS}}_{\mu\nu}(\boldsymbol{\xi}, t) = \frac{\delta_{\mu\nu}}{\sqrt{g(\boldsymbol{\xi}, t)}} P_0(n_0(\boldsymbol{\xi})) \,, \tag{8.22}$$

where the function $P_0(n)$ is the equilibrium pressure of a noninteracting homogeneous Fermi gas.

For the practical calculation of the xc potential in TDLDefA we need the stress tensor $T^{\mathrm{KS}}_{\mu\nu}(\boldsymbol{r}, t)$ in the laboratory frame. Application of the usual tensor transformation rules [Dubrovin 1984],

$$P_{\mu\nu}(\boldsymbol{r}, t) = \frac{\partial \xi^{\alpha}}{\partial r^{\mu}} \frac{\partial \xi^{\beta}}{\partial r^{\nu}} \widetilde{P}_{\alpha\beta}(\boldsymbol{\xi}(\boldsymbol{r}, t), t) \,, \tag{8.23}$$

to the stress tensor (8.22) yields the result

$$T^{\mathrm{KS}}_{\mu\nu}(\boldsymbol{r}, t) = \bar{g}_{\mu\nu}(\boldsymbol{r}, t)\sqrt{\bar{g}(\boldsymbol{r}, t)}\, P_0\left(\frac{n(\boldsymbol{r}, t)}{\sqrt{\bar{g}(\boldsymbol{r}, t)}}\right) \,, \tag{8.24}$$

where $\bar{g}_{\mu\nu}(\boldsymbol{r}, t)$ is the Cauchy's deformation tensor [Masson 1964]

$$\bar{g}_{\mu\nu}(\boldsymbol{r}, t) = \frac{\partial \xi^{\alpha}}{\partial r^{\mu}} \frac{\partial \xi^{\alpha}}{\partial r^{\nu}} \,. \tag{8.25}$$

It is quite natural that the stress tensor in the laboratory frame depends on $\bar{g}_{\mu\nu}(\boldsymbol{r}, t)$ since the Cauchy's deformation is the usual way of characterizing deformations in the Eulerian formulation of continuum mechanics.

8.4.2 Exchange-Only TDLDefA

The most difficult part in the derivation of an explicit TDLDefA is the solution of the interacting problem defined by the Hamiltonian (8.17). In this subsection we find the exact solution of this problem in the exchange-only approximation. In the x-only case the pair correlation function $\widetilde{G}_2(\boldsymbol{k}, t)$ is completely determined by the one-particle distribution function $\widetilde{f}(\boldsymbol{k}, t)$

$$\widetilde{G}_2(\boldsymbol{k},t) = -\sum_{\boldsymbol{p}} \widetilde{f}(\boldsymbol{k}+\boldsymbol{p},t)\widetilde{f}(\boldsymbol{p},t)\,. \tag{8.26}$$

Performing the mean field decoupling of the four-fermion terms in (8.20) we find that the right hand side of this equation cancels out. Hence in the x-only approximation both the Wigner function and the pair correlation function in the Lagrangian frame preserve their initial (equilibrium) form

$$\widetilde{f}(\boldsymbol{k},t) = n_{\boldsymbol{k}}^F\,, \qquad \widetilde{G}_2(\boldsymbol{k},t) = G_2^{\mathrm{x}}(n_0;k) = -\sum_{\boldsymbol{p}} n_{\boldsymbol{k}+\boldsymbol{p}}^F n_{\boldsymbol{p}}^F\,. \tag{8.27}$$

Here, $k = |\boldsymbol{k}| = \sqrt{k_\mu k_\mu}$ is the usual modulus of $\boldsymbol{k}$, and $G_2^{\mathrm{x}}(n;k)$ is the exchange pair correlation function for an equilibrium Fermi gas of density n. Substituting (8.27) into (8.18) and subtracting the noninteracting stress tensor (8.22) we obtain the exchange contribution to the local stress density in the Lagrangian frame:

$$\widetilde{P}_{\mu\nu}^{\mathrm{x}} = \frac{1}{2g}\sum_{\boldsymbol{k}}\left[\frac{k_\mu k_\nu}{\|\boldsymbol{k}\|}v'_{\mathrm{ee}}(\|\boldsymbol{k}\|) + g_{\mu\nu}v_{\mathrm{ee}}(\|\boldsymbol{k}\|)\right]G_2^{\mathrm{x}}(n_0;k)\,. \tag{8.28}$$

Using the transformation rule (8.23) we get the exchange stress tensor in the laboratory frame:

$$P_{\mu\nu}^{\mathrm{x}}(n,\bar{g}_{\alpha\beta}) = \frac{\sqrt{\bar{g}}}{2}\sum_{\boldsymbol{p}}\left[\frac{p_\mu p_\nu}{p}v'_{\mathrm{ee},\,p} + \delta_{\mu\nu}v_{\mathrm{ee},\,p}\right]G_2^{\mathrm{x}}\left(\frac{n}{\sqrt{\bar{g}}};\sqrt{\bar{g}^{\alpha\beta}p_\alpha p_\beta}\right)\,. \tag{8.29}$$

Equations (8.29), (8.9) and (8.8) uniquely determine the local potential $v_{\mathrm{x}}(\boldsymbol{r},t)$ in x-only TDLDefA. Clearly the exchange potential $v_{\mathrm{x}}(\boldsymbol{r},t)$ is a local (both in space and in time) functional of the density $n(\boldsymbol{r},t)$ and Cauchy's deformation tensor $\bar{g}_{\mu\nu}(\boldsymbol{r},t)$. In the equilibrium system ($\bar{g}_{\mu\nu} = \delta_{\mu\nu}$) the potential, defined by (8.29), (8.9), and (8.8), reduces to that of the usual static local exchange approximation.

8.4.3 Inclusion of Correlations: Elastic TDLDefA

Exchange-Correlation Stress Tensor in the Elastic TDLDefA

The elastic TDLDefA is based on the assumption that both the Wigner function and the pair correlation function in the Lagrangian frame preserve their initial form. This corresponds to dynamics with extremely pronounced memory that is not destroyed by effects of collisional relaxation. To get the stress tensor for the system evolving from the equilibrium state we have to substitute the equilibrium values $\widetilde{f}(\boldsymbol{k},t) = f^{\mathrm{eq}}(n_0;k)$ and $\widetilde{G}_2(\boldsymbol{k},t) = G_2^{\mathrm{eq}}(n_0;k)$ into (8.18). Transforming the result of this substitution to the laboratory frame and subtracting the KS stress tensor (8.24) we obtain the following xc stress tensor in the physical $\boldsymbol{r}$-space:

$$P^{\mathrm{xc}}_{\mu\nu} = \frac{2\bar{g}_{\mu\nu}}{d}\sqrt{\bar{g}}E^{\mathrm{xc}}_{\mathrm{kin}}\left(\frac{n}{\sqrt{\bar{g}}}\right) + \frac{\sqrt{\bar{g}}}{2}\sum_{\boldsymbol{p}}\left[\frac{p_\mu p_\nu}{p}\bar{v}'_{\mathrm{ee},\,p} + \delta_{\mu\nu}\bar{v}_{\mathrm{ee},\,p}\right]$$
$$\times\, G^{\mathrm{eq}}_2\left(\frac{n}{\sqrt{\bar{g}}};\sqrt{\bar{g}^{\alpha\beta}p_\alpha p_\beta}\right), \qquad (8.30)$$

where $E^{\mathrm{xc}}_{\mathrm{kin}}(n)$ is the xc kinetic energy of the interacting homogeneous Fermi system. Equation (8.30) determines the xc stress tensor as a function of the time-dependent density $n(\boldsymbol{r},t)$, and Cauchy's deformation tensor $\bar{g}_{\mu\nu}(\boldsymbol{r},t)$. The tensor $\bar{g}_{\mu\nu}(\boldsymbol{r},t)$ is responsible for the memory related ultranonlocality of $P^{\mathrm{xc}}_{\mu\nu}$. The "elastic" xc potential is the solution to the Poisson equation (8.8), where the xc "charge density" is defined after (8.9) and (8.30).

In the exchange approximation the stress tensor $P^{\mathrm{xc}}_{\mu\nu}$, given by (8.30), reduces to the x-only tensor $P^{\mathrm{x}}_{\mu\nu}$ of (8.29) that is exact in the weak coupling limit. In the linear response regime, the corrections to the density and to Cauchy's deformation tensor are proportional to the displacement vector $\boldsymbol{u} = \boldsymbol{r} - \boldsymbol{\xi}$,

$$n = n_0 - \nabla n_0 \boldsymbol{u}, \qquad \bar{g}_{\mu\nu} = \delta_{\mu\nu} - \frac{\partial u_\mu}{\partial r^\nu} - \frac{\partial u_\nu}{\partial r^\mu}\,. \qquad (8.31)$$

Linearizing the stress tensor (8.30) and using (8.31) we recover the VK approximation [Vignale 1996, Vignale 1997]:

$$\delta P^{\mathrm{xc}}_{\mu\nu} = -\delta_{\mu\nu}\frac{\partial P}{\partial n_0}\boldsymbol{u}\nabla n_0 + \delta_{\mu\nu}K^{\infty}_{\mathrm{xc}}\frac{1}{2}\delta g^{\alpha\alpha} + \mu^{\infty}_{\mathrm{xc}}\left(\delta g^{\mu\nu} - \frac{\delta_{\mu\nu}}{d}\delta g^{\alpha\alpha}\right), \qquad (8.32)$$

where K^{∞}_{xc} and $\mu^{\infty}_{\mathrm{xc}}$ are the high frequency limits of the xc bulk and shear moduli, respectively [Conti 1999, Nifosì 1998, Qian 2002]. Thus, the VK approximation is nothing but Hooke's law in our quantum continuum mechanics.

Self-Consistent Kohn-Sham Equations

Let us formulate the complete set of self-consistent KS equations in the elastic TDLDefA. The Kohn-Sham formulation of TDDFT allows to calculate the density $n(\boldsymbol{r},t)$ and the velocity $\boldsymbol{v}(\boldsymbol{r},t)$ in the interacting N-particle system using the ideal gas formulas

$$n(\boldsymbol{r},t) = \sum_{j=1}^{N}|\varphi_j(\boldsymbol{r},t)|^2, \qquad \boldsymbol{v}(\boldsymbol{r},t) = \frac{\mathrm{i}}{2nm}\sum_{j=1}^{N}\left[\varphi_j\nabla\varphi^*_j - \varphi^*_j\nabla\varphi_j\right]. \qquad (8.33)$$

The single particle orbitals $\varphi_j(\boldsymbol{r},t)$ satisfy the time-dependent KS equation

$$\mathrm{i}\frac{\partial\varphi_j}{\partial t} = -\frac{\nabla^2}{2m}\varphi_j + \left\{v_{\mathrm{ext}} + v_{\mathrm{eff}}[n,\bar{g}_{\mu\nu}]\right\}\varphi_j\,, \qquad (8.34)$$

where $v_{\mathrm{ext}}(\boldsymbol{r},t)$ is the external potential. For the practically important case of a 3D system with Coulomb interaction the effective potential $v_{\mathrm{eff}}[n,\bar{g}_{\mu\nu}](\boldsymbol{r},t)$ is the solution of the following Poisson equation:

$$\nabla^2 v_{\text{eff}} = 4\pi\{e^2 n + \rho_{\text{xc}}[n, \bar{g}_{\mu\nu}]\}. \tag{8.35}$$

The first term in the brackets in (8.35) generates the Hartree potential, v_{H}, while the second term is responsible for the xc potential. The xc "charge density", ρ_{xc}, is a local functional of n and $\bar{g}_{\mu\nu}$,

$$\rho_{\text{xc}} = \frac{1}{4\pi}\frac{\partial}{\partial r^\mu}\left[\frac{1}{n}\frac{\partial}{\partial r^\nu}P^{\text{xc}}_{\mu\nu}(n, \bar{g}_{\mu\nu})\right], \tag{8.36}$$

where $P^{\text{xc}}_{\mu\nu}(n, \bar{g}_{\mu\nu})$ is a *function* of $n(\boldsymbol{r}, t)$ and $\bar{g}_{\mu\nu}(\boldsymbol{r}, t)$, which is defined by (8.30). One can show [Tokatly 2005b] that for a Coulomb system (8.30) simplifies as follows:

$$P^{\text{xc}}_{\mu\nu} = \frac{2}{3}\bar{g}_{\mu\nu}\sqrt{\bar{g}}E^{\text{xc}}_{\text{kin}}\left(\frac{n}{\sqrt{\bar{g}}}\right) + L_{\mu\nu}(\bar{g}_{\alpha\beta})E_{\text{pot}}\left(\frac{n}{\sqrt{\bar{g}}}\right), \tag{8.37}$$

where E_{pot} is the potential energy of a homogeneous electron gas, and $L_{\mu\nu}(\bar{g}_{\alpha\beta})$ is a purely geometric factor that involves only angle integrations. Therefore, the dependence of $P^{\text{xc}}_{\mu\nu}(n, \bar{g}_{\mu\nu})$ on $\bar{g}_{\mu\nu}$ and on $n/\sqrt{\bar{g}}$ is completely factorized, which should significantly simplify practical applications. The kinetic and potential energies of the homogeneous electron gas, $E^{\text{xc}}_{\text{kin}}(n)$ and $E_{\text{pot}}(n)$, can be expressed in terms of the xc energy per particle, $\epsilon_{\text{xc}}(n)$ (see, for example, [Conti 1999]). Hence, our nonadiabatic TDLDefA requires only knowledge of the function $\epsilon_{\text{xc}}(n)$ for the homogeneous electron gas, exactly as the static LDA.

The new basic variable, Cauchy's deformation tensor $\bar{g}_{\mu\nu}$, is uniquely determined by the velocity $\boldsymbol{v}(\boldsymbol{r}, t)$. Using the definition of $\bar{g}_{\mu\nu}$ (8.25) and the trajectory equation (8.11) we find the following equation of motion for Cauchy's deformation tensor $\bar{g}_{\mu\nu}(\boldsymbol{r}, t)$:

$$\left(\frac{\partial}{\partial t} + v^\alpha\frac{\partial}{\partial r^\alpha}\right)\bar{g}_{\mu\nu} = -\frac{\partial v^\alpha}{\partial r^\mu}\bar{g}_{\alpha\nu} - \frac{\partial v^\alpha}{\partial r^\nu}\bar{g}_{\alpha\mu}, \qquad \bar{g}_{\mu\nu}(\boldsymbol{r}, 0) = \delta_{\mu\nu}. \tag{8.38}$$

The system (8.33–8.38) constitutes the complete set of self-consistent KS equations in the nonlinear elastic TDLDefA. In the equilibrium situation ($\bar{g}_{\mu\nu} = \delta_{\mu\nu}$) this system reduces to the usual static KS equations with the LDA xc potential, while in the linear regime it recovers the results of the VK approximation. The nonadiabatic memory effects are described by Cauchy's deformation tensor, which satisfies (8.38). It should be noted that from a computational point of view, the solution of this equation should not introduce any additional difficulties. Formally (8.38) has the same structure as the time-dependent KS equation (8.34). Hence (8.34) and (8.38) can be solved simultaneously by the same method.

Very recently, the VK approximation has been successfully applied to the description of optical and polarization properties of many different systems, such as atoms, molecules, semiconductors and polymers [de Boeij 2001, van Faassen 2002, van Faassen 2003a, Ullrich 2004, van Faassen 2003b]. Since the VK approximation is a linearized version of our theory, we hope that the general TDLDefA also will become a useful tool for studying nonlinear time-dependent phenomena.

9 Exact-Exchange Methods and Perturbation Theory along the Adiabatic Connection

A. Görling

9.1 Preliminary Remarks

In this chapter we consider time-dependent density functional methods that are based on exact exchange (EXX) functionals, i.e., EXX potentials and EXX kernels. Moreover a systematic route to the construction of correlation potentials and kernels is discussed.

The EXX functionals are derived by perturbation theory along the adiabatic connection [Görling 1994, Görling 1995b, Görling 1996, Görling 1997, Görling 1998a, Görling 2005]. This perturbation theory represents a quite general formal framework. In first order it yields EXX functionals, while higher orders give rise to correlation functionals. In infinite order, the exact exchange-correlation (xc) functionals and exact density functional methods are obtained, provided that the perturbative series converges. However, as it often happens in perturbation theory approaches, convergence is an open question [Seidl 2000]. Moreover, the expressions for the higher order terms become increasingly complicated and unmanageable in practical applications. As a result, so far, apart from a few exceptions, only first order perturbation theory was performed and the resulting EXX methods were implemented and applied.

Density functionals of time-dependent density functional theory (TDDFT) were originally defined in analogy to static density functional theory (DFT) as functional derivatives of the quantum mechanical action with respect to the time-dependent electron density [Ullrich 1995a, Gross 1996, Petersilka 1996a, Colwell 1996, Petersilka 1998]. In these definitions, the quantum mechanical action corresponds to the energy in static DFT. TDDFT is most often applied within the response regime, in particular within the linear response regime which yields excitation energies and dynamic linear polarizabilities [Stott 1980, Zangwill 1980b, Mahan 1980, Mahan 1990, Casida 1995a, van Gisbergen 1995, Jamorski 1996, Casida 1996, Bauernschmitt 1996a, Stratmann 1998, Görling 1999b, Heinze 2000, Hirata 1999a, Tozer 1999, Yabana 1999a]. The working equation underlying most linear density functional response methods, the density-matrix based coupled Kohn-Sham (KS) equation [Görling 1999b], was also originally derived within the quantum mechanical action formalism [Casida 1995a, Jamorski 1996]. Later it was shown that the xc kernel, the time-dependent functional derivative of the xc

A. Görling: *Exact-Exchange Methods and Perturbation Theory along the Adiabatic Connection*, Lect. Notes Phys. **706**, 137–159 (2006)
DOI 10.1007/3-540-35426-3_9

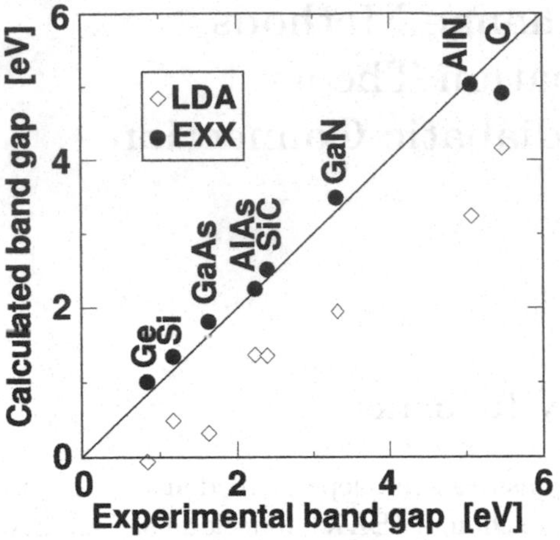

Fig. 9.1. Comparison of LDA and EXX band gaps (in eV) of various semiconductors with experimental data [Städele 1999]. (EXX band gaps were obtained with exact exchange potentials plus LDA correlation potentials)

potential with respect to the electron density, within the action formalism violates basic causality requirements in time [Gross 1996, van Leeuwen 1998] because in the action formalism it is obtained as the second functional derivative of the xc part of the quantum mechanical action. Definitions and derivations within the action formalism are thus questionable from a formal point of view. A derivation of the density-matrix based coupled KS equations which does not rely on the quantum mechanical action formalism can be found in [Görling 1999b, Görling 1998b, Furche 2001a], while definitions of linear and higher order xc kernels not invoking the quantum mechanical action are given in [Görling 1998a, Görling 1998b, Kim 2002b, Heinze 2002]. The perturbation theory approach considered here also does not rely on the quantum mechanical action formalism and is therefore not affected by any causality problems in time.

Density functional response methods consist of two steps. First, a KS calculation for the ground state of the electronic system under consideration has to be performed. This yields the KS orbitals or one-particle functions and their eigenvalues. These are the input for the second step, the actual response calculation, which, in first order response, leads to polarizabilities, excitation energies, and oscillator strengths, i.e., to the linear optical properties. Higher order response yields non-linear optical properties, e.g., hyperpolarizabilities. The density functionals needed for the response calculations are, for the first step – the ground-state calculation – the xc potential; for the second step, in the case of linear response, the xc kernel. For higher order response properties, higher order xc kernels are required. While the xc potential does not depend on time or frequency, the xc kernels do.

Static exact exchange KS methods [Görling 1994, Görling 1995b, Sharp 1953, Talman 1976, Sahni 1982, Shaginyan 1993, Görling 1996, Grabo 1998, Engel 1999, Görling 1999a, Ivanov 1999, Hamel 2001, Veseth 2001, Kotani 1995, Städele 1997, Städele 1999, Magyar 2004, Rinke 2005, Qteisch 2005, Sharma 2005, Kresse 2005, Yang 2002, Kümmel 2003] have a number of advantages compared to conventional KS methods based on the local density approximation or generalized gradient approximations. First, of course, the exchange energy and potential do not need to be approximated. As a result, unphysical self-interactions of electrons present in the Coulomb energy are exactly canceled. In LDA or GGA methods this cancelation is not complete and, as a consequence, the effective KS potential is too repulsive and does not exhibit the correct asymptotic behavior. Accordingly, LDA and GGA orbital and eigenvalue spectra are qualitatively wrong. They, e.g., do not have a Rydberg series and, in the case of semiconducting or insulating solids, too small band gaps are obtained. With EXX methods, on the other hand, physically reasonable and qualitatively correct KS orbitals and eigenvalues are obtained [Görling 1999a, Ivanov 1999]. Furthermore, the EXX eigenvalues and eigenvalue differences are well-defined zeroth order approximations for ionization [Chong 2002, Gritsenko 2002, Gritsenko 2003] and excitation [Görling 1996, Filippi 1997] energies, respectively.

Figure 9.1, moreover, shows that EXX methods yield band gaps close to the experimental ones for semiconductors. Also for organic polymers, in particular for trans-polyacetylene, band gaps are obtained which are much closer to the experimental values than those from LDA or GGA methods [Rohra 2005]. However, for solid noble gases [Magyar 2004] and also for diamond, see Fig. 9.1, EXX band gaps are larger than LDA- or GGA-band gaps but still clearly smaller than experimental band gaps. In this context it should be noted that even the exact KS band gap does not equal the experimental band gap but differs from it by the derivative discontinuity of the xc potential at integer particle numbers [Perdew 1983, Sham 1983]. The agreement between EXX and experimental band gaps displayed in Fig. 9.1 shows that for simple semiconductors the effect of neglecting the derivative discontinuity plus the effect of approximating the correlation potential within the LDA, in sum, is very small. Whether both effects cancel or are individually small remains to be investigated. It should, however, be noted that a complete neglect of the correlation potential or a GGA treatment of the correlation potential changes the band gaps only marginally. This suggests that at least presently available approximate correlation potentials have little effect on the band gap [Städele 1999].

Because KS orbitals and eigenvalues are the input quantities for DFT response methods, the accuracy of the latter depends crucially on the quality of the former. On the basis of LDA or GGA orbitals and eigenvalues, e.g., it is not possible to treat excitations into states with Rydberg character

[Casida 1998a, Della Sala 2003a]. This problem is fundamentally solved by employing EXX orbitals and eigenvalues [Della Sala 2003a].

So far the experience with time-dependent EXX methods is quite limited. It results from two EXX response calculations, one considering solids the other considering molecules, which are reported in [Kim 2002a] and [Hirata 2002]. Full EXX-TDDFT methods beyond the linear regime so far have not been implemented. However, steps towards such methods including numerical results can be found in [Ullrich 1996] and [Marques 2001]. These first experiences for the most part are positive. Moreover, the basic formulas of EXX-TDDFT [Görling 1997, Görling 1998a, Ullrich 1995a, Gross 1996, Petersilka 1996a, Görling 1998b, Kim 2002b] show that EXX-TDDFT goes distinctively beyond LDA- or GGA-TDDFT. EXX-TDDFT, in contrast to LDA- or GGA-TDDFT, contains memory effects in time. Thus, EXX-TDDFT represents a highly promising new line of development.

In the following section we will derive an integral equation, the static EXX equation, for the static exchange potential, and we will consider how this equation and approximations to it can be used in practice. In the subsequent section the derivation of time- and frequency-dependent EXX equations for the time-dependent exchange potential and for the frequency-dependent exchange kernel is presented.

9.2 Exact Exchange Methods and Static Perturbation Theory

9.2.1 Perturbation Theory along the Adiabatic Connection and the Static Exact Exchange Equation

The starting point of a perturbation theory along the adiabatic connection [Görling 1994, Görling 1995b, Görling 1996, Görling 1997, Görling 1998a] is the many-electron Schrödinger equation

$$\left[\hat{T} + \hat{V}^{\lambda} + \lambda\,\hat{V}_{\rm ee}\right] \Psi_0^{\lambda} = E_0^{\lambda}\,\Psi_0^{\lambda}\,. \tag{9.1}$$

In (9.1), $\hat{T}$ is the operator of the kinetic energy, $\hat{V}_{\rm ee}$ is the operator of the electron-electron interactions, $\hat{V}^{\lambda}$ is the operator generated by the local multiplicative potential $v^{\lambda}(\boldsymbol{r})$, Ψ_0^{λ} is the ground-state electronic wave function, E_0^{λ} is the corresponding ground-state energy, and λ is the coupling constant (with values between zero and one), which switches on and off the electron-electron interaction. This means that (9.1) only describes real electrons for $\lambda = 1$, while for $\lambda < 1$ it describes model electrons with a reduced interaction. The potential $v^{\lambda}(\boldsymbol{r})$ is determined by the conditions that (i) the ground-state electronic density, the electronic density associated to the wave functions Ψ_0^{λ}, is independent of the coupling constant λ, and that (ii) $v^{\lambda}(\boldsymbol{r})$ for $\lambda = 1$ equals the external potential $v_{\rm ext}(\boldsymbol{r})$ of the real electron system under consideration.

This is, usually, the potential created by the nuclei of the atom, molecule or solid being studied. This means that (9.1) turns into the Schrödinger equation of the real electronic system for $\lambda = 1$ and that the ground-state electronic density $n(\boldsymbol{r})$ equals the ground-state density of the real electronic system, independently of λ. For $\lambda = 0$, (9.1) becomes the corresponding many-electron KS equation

$$\left[\hat{T} + \hat{V}_{\rm KS}\right] \Phi_0 = E_{KS,0}\, \Phi_0 \tag{9.2}$$

with $v_{\rm KS}(\boldsymbol{r}) = v^{\lambda=0}(\boldsymbol{r})$, $\Phi_0 = \Psi_0^{\lambda=0}$, and $E_{\rm KS,0} = E_0^{\lambda=0}$. The Hohenberg-Kohn theorem [Hohenberg 1964] guarantees the uniqueness of the KS potential $v_{\rm KS}(\boldsymbol{r})$ and generally of the potential $v^{\lambda}(\boldsymbol{r})$.

Next we expand the potential $v^{\lambda}(\boldsymbol{r})$ in a Taylor series with respect to the coupling constant λ

$$\begin{aligned} v^{\lambda}(\boldsymbol{r}) &= \sum_{p=0}^{\infty} \lambda^p v^{(p)}(\boldsymbol{r}) \\ &= v_{\rm KS}(\boldsymbol{r}) + \lambda\, v^{(1)}(\boldsymbol{r}) + \lambda^2 v^{(2)}(\boldsymbol{r}) + \ldots \end{aligned} \tag{9.3}$$

In order to define the exchange potential, we write the first order potential $v^{(1)}(\boldsymbol{r})$ as

$$v^{(1)}(\boldsymbol{r}) = -v_{\rm x}(\boldsymbol{r}) - v_{\rm H}(\boldsymbol{r})\,, \tag{9.4}$$

i.e., as the negative of the exchange potential $v_{\rm x}(\boldsymbol{r})$ plus the negative of the Hartree potential $v_{\rm H}(\boldsymbol{r}) = \int {\rm d}^3 r'\, n(\boldsymbol{r}')/|\boldsymbol{r} - \boldsymbol{r}'|$. This defines the exchange potential as

$$v_{\rm x}(\boldsymbol{r}) = -v^{(1)}(\boldsymbol{r}) - v_{\rm H}(\boldsymbol{r})\,. \tag{9.5}$$

The sum $\sum_{p=2}^{\infty} v^{(p)}(\boldsymbol{r})$ represents the negative of the correlation potential.

With the Taylor expansion (9.3), the Hamiltonian operator of the adiabatic connection Schrödinger (9.1) turns into

$$\hat{T} + \hat{V}_{\rm KS} + \lambda \left[\hat{V}_{\rm ee} - \hat{V}_{\rm x} - \hat{V}_{\rm H}\right] + \lambda^2 \hat{V}^{(2)} + \ldots \tag{9.6}$$

In this adiabatic connection perturbative expansion, the KS Hamiltonian operator $(\hat{T} + \hat{V}_{\rm KS})$ in (9.6) describes the unperturbed system, whereas the perturbation is given by the terms depending on the coupling constant λ. Perturbation theory along the adiabatic connection exhibits some peculiarities: (i) The Hamiltonian operator of the unperturbed system is unknown because, for a given real system, the KS potential is, initially, not known. (ii) The perturbation is also unknown because neither the exchange potential $v_{\rm x}$ nor the higher order potentials $v^{(p)}$ are known in the beginning. (iii) The perturbation is not linear in the coupling constant λ.

Concerning point (i), the KS potential $v_{\rm KS}$ is given by

$$v_{\rm KS}(\boldsymbol{r}) = v_{\rm ext}(\boldsymbol{r}) + v_{\rm x}(\boldsymbol{r}) + v_{\rm H}(\boldsymbol{r}) - v^{(2)}(\boldsymbol{r}) - \ldots \tag{9.7}$$

and thus is the sum of the known external potential v_{ext} with the potentials representing the perturbation. Therefore, these perturbation potentials are required right from the beginning, in collision with point (ii) above. To solve this problem we formally expand the ground-state electronic density into a Taylor series with respect to λ

$$n^{\lambda}(\boldsymbol{r}) = \sum_{p=0}^{\infty} \lambda^{p} n^{(p)}(\boldsymbol{r}) \,. \tag{9.8}$$

However, by construction, the ground-state density $n^{\lambda}(\boldsymbol{r})$ is independent of the coupling constant λ and equals the ground-state density $n(\boldsymbol{r})$ of the real electronic system. For the contributions $n^{\lambda}(\boldsymbol{r})$ hence follows

$$n^{(0)}(\boldsymbol{r}) = n(\boldsymbol{r}) \,, \tag{9.9}$$

and

$$n^{(p)}(\boldsymbol{r}) = 0 \quad \text{for} \quad p \geq 1 \,. \tag{9.10}$$

Equations (9.10) are central for the perturbative scheme we are trying to establish. For each order, the corresponding equation (9.10) leads to an integral equation determining the potential $v^{(p)}$. Here we will consider only the first order.

Textbook perturbation theory yields the expression

$$\sum_{j}{}' \Phi_j \frac{\langle \Phi_j | \hat{V}_{\text{ee}} - \hat{V}_{\text{x}} - \hat{V}_{\text{H}} | \Phi_0 \rangle}{E_{\text{KS},0} - E_{\text{KS},j}} + \text{c.c.} \tag{9.11}$$

for the first-order contribution to the wave function Ψ_0^{λ} due to the perturbation $\hat{V}_{\text{ee}} - \hat{V}_{\text{x}} - \hat{V}_{\text{H}}$. In (9.11) Φ_j stands for excited states of the many-electron KS (9.2) with energies $E_{\text{KS},j}$. The summation in (9.12) runs over all excited states of the KS (9.2), i.e., all eigenstates of (9.2) except Φ_0. The prime in the summation indicates that Φ_0 is not included. The first order contribution (9.11) to the wave function yields the first order contribution $n^{(1)}(\boldsymbol{r})$ to the density and, after substitution into (9.10), we get

$$0 = \sum_{j}{}' \frac{\langle \Phi_0 | \hat{n}(\boldsymbol{r}) | \Phi_j \rangle \langle \Phi_j | \hat{V}_{\text{ee}} - \hat{V}_{\text{x}} - \hat{V}_{\text{H}} | \Phi_0 \rangle}{E_{\text{KS},0} - E_{\text{KS},j}} + \text{c.c.} \,. \tag{9.12}$$

In (9.12) $\hat{n}(\boldsymbol{r})$ denotes the density operator. The many-electron KS equation (9.2) leads to one-particle KS equations

$$\left[\hat{T} + \hat{V}_{\text{KS}} \right] \varphi_i = \varepsilon_i \varphi_i \,, \tag{9.13}$$

for the KS orbitals φ_i from which the many-electron KS wave functions Φ_j are constructed. If the Φ_j are expressed by their orbitals, after rearrangement, (9.12) turns into the EXX equation

$$\int \mathrm{d}^3r' \, \chi_{\mathrm{KS}}(\boldsymbol{r},\boldsymbol{r}')v_{\mathrm{x}}(\boldsymbol{r}') = \Lambda(\boldsymbol{r}) \,, \tag{9.14}$$

where

$$\chi_{\mathrm{KS}}(\boldsymbol{r},\boldsymbol{r}') = 2\sum_i^{\mathrm{occ}}\sum_a^{\mathrm{unocc}} \frac{\varphi_i^*(\boldsymbol{r})\varphi_a(\boldsymbol{r})\varphi_a^*(\boldsymbol{r}')\varphi_i(\boldsymbol{r}')}{\varepsilon_i - \varepsilon_a} + \mathrm{c.c.} \tag{9.15}$$

is the response function of the KS system and

$$\Lambda(\boldsymbol{r}) = 2\sum_i^{\mathrm{occ}}\sum_a^{\mathrm{unocc}} \frac{\varphi_i^*(\boldsymbol{r})\varphi_a(\boldsymbol{r})\langle\varphi_a|\hat{V}_{\mathrm{x}}^{\mathrm{NL}}|\varphi_i\rangle}{\varepsilon_i - \varepsilon_a} + \mathrm{c.c.} \,. \tag{9.16}$$

The nonlocal exchange operator $\hat{V}_{\mathrm{x}}^{\mathrm{NL}}$ is of the form of the Hartree-Fock exchange operator but is constructed from KS orbitals, i.e., it is an integral operator defined by

$$\left[\hat{V}_{\mathrm{x}}^{\mathrm{NL}}\varphi_k\right](\boldsymbol{r}) = -\int \mathrm{d}^3r' \sum_j^{\mathrm{occ}} \frac{\varphi_j(\boldsymbol{r})\varphi_j^{\dagger}(\boldsymbol{r}')\varphi_k(\boldsymbol{r}')}{|\boldsymbol{r}-\boldsymbol{r}'|} \,. \tag{9.17}$$

In (9.15) and (9.16) φ_i and φ_a denote occupied and unoccupied KS orbitals, with eigenvalues ε_i and ε_a, and φ_k denotes an arbitrary, i.e., occupied or unoccupied, orbital. The factor 2 appears due to the spin degrees of freedom.

The EXX equation (9.14) determines the exchange potential. Integral equations determining the contributions to the correlation potential, i.e., the higher order potentials $v^{(p)}$, have the same form as the EXX equation (9.14), merely the right hand side differs and is becoming increasingly complicated for higher orders.

In an exact exchange KS calculation, instead of evaluating an approximate expression for the exchange potential in terms of the electronic density and, in the case of generalized gradient approximations, of the derivatives of the electronic density, we have to solve the EXX integral (9.14).

9.2.2 Implementations of Static Exact Exchange KS Methods

The first EXX method was implemented in the 1970s by Talman and Shadwick as an approximation to the Hartree-Fock method [Talman 1976]. The implementation solves the exact exchange KS equations including the EXX equation (9.14) on a numerical grid and is restricted to spherical systems, i.e., atoms. Much later, EXX implementations for solids [Kotani 1995, Görling 1996, Städele 1997, Städele 1999] as well as molecules [Görling 1999a, Ivanov 1999] were introduced, which expand the KS orbitals as well as the exchange potential in basis sets. In fact, two different basis sets are used, one for representing the KS orbitals, and an auxiliary basis set for representing the quantities in the EXX equation (9.14), i.e., the exchange potential v_{x}, the response function χ_{KS}, and the right hand side Λ.

Within the auxiliary basis set $\{f_k\}$, the EXX equation (9.14) turns into the matrix equation [Görling 1996]

$$\boldsymbol{\chi}_{\mathrm{KS}} \boldsymbol{v}_{\mathrm{x}} = \boldsymbol{\Lambda} \tag{9.18}$$

with matrix elements

$$\chi_{\mathrm{KS},\, k\ell} = \int \mathrm{d}^3 r \int \mathrm{d}^3 r' \, f_k^*(\boldsymbol{r}) \chi_{\mathrm{KS}}(\boldsymbol{r}, \boldsymbol{r}') f_\ell(\boldsymbol{r}') \tag{9.19}$$

and

$$\Lambda_k = \int \mathrm{d}^3 r \, f_k^*(\boldsymbol{r}) \Lambda(\boldsymbol{r}) \,. \tag{9.20}$$

The components $v_{\mathrm{x},\, k}$ of $\boldsymbol{v}_{\mathrm{x}}$ are the expansion coefficients of the exchange potential in the auxiliary basis:

$$v_{\mathrm{x}}(\boldsymbol{r}) = \sum_k v_{\mathrm{x},\, k} f_k(\boldsymbol{r}) \,. \tag{9.21}$$

Alternatively, an exchange charge density $n_{\mathrm{x}}(\boldsymbol{r})$ can be defined [Görling 1999a] as the charge distribution that yields as its electrostatic potential the exchange potential

$$v_{\mathrm{x}}(\boldsymbol{r}) = \int \mathrm{d}^3 r' \, \frac{n_{\mathrm{x}}(\boldsymbol{r}')}{|\boldsymbol{r} - \boldsymbol{r}'|} \,. \tag{9.22}$$

The exchange charge density $n_{\mathrm{x}}(\boldsymbol{r})$ is then related by a Poisson equation to the exchange potential,

$$n_{\mathrm{x}}(\boldsymbol{r}) = (-1/4\pi) \nabla^2 v_{\mathrm{x}}(\boldsymbol{r}) \,. \tag{9.23}$$

After expanding the exchange charge density in the auxiliary basis set

$$n_{\mathrm{x}}(\boldsymbol{r}) = \sum_k n_{\mathrm{x},\, k} f_k(\boldsymbol{r}) \,, \tag{9.24}$$

it is straightforward to obtain the EXX matrix equation

$$\tilde{\boldsymbol{\chi}}_{\mathrm{KS}} \boldsymbol{n}_{\mathrm{x}} = \tilde{\boldsymbol{\Lambda}} \,, \tag{9.25}$$

with matrix elements

$$\tilde{\chi}_{\mathrm{KS},\, k\ell} = \int \mathrm{d}^3 r_1 \int \mathrm{d}^3 r_2 \int \mathrm{d}^3 r_3 \int \mathrm{d}^3 r_4 \, \frac{f_k^*(\boldsymbol{r}_1) \chi_{\mathrm{KS}}(\boldsymbol{r}_2, \boldsymbol{r}_3) f_\ell(\boldsymbol{r}_4)}{|\boldsymbol{r}_1 - \boldsymbol{r}_2||\boldsymbol{r}_3 - \boldsymbol{r}_4|} \tag{9.26}$$

and

$$\tilde{\Lambda}_k = \int \mathrm{d}^3 r_1 \int \mathrm{d}^3 r_2 \, \frac{f_k^*(\boldsymbol{r}_1) \Lambda(\boldsymbol{r}_2)}{|\boldsymbol{r}_1 - \boldsymbol{r}_2|} \,. \tag{9.27}$$

In the matrix elements (9.26) and (9.27) the Coulomb norm $1/|\boldsymbol{r}_1 - \boldsymbol{r}_2|$ was used throughout in integrations in order to obtain an EXX matrix (9.25) with a symmetric matrix $\tilde{\chi}_{\mathrm{KS}}$. The components $n_{\mathrm{x},k}$ of $\boldsymbol{n}_{\mathrm{x}}$ are the expansion coefficients of the exchange charge density in the auxiliary basis [see (9.24)].

The determination of the exchange potential via the exchange charge density, for molecules, has the advantage that the exchange potential exhibits a $1/r$-asymptotic, provided that the exchange charge density integrates to minus one. In molecular programs, moreover, integrals with respect to the Coulomb norm, like the integrals in (9.26) and (9.27), are usually available. This facilitates the implementation of EXX approaches for molecules relying on the exchange charge density.

Similarly to the exchange charge density, also a correlation charge density $n_{\mathrm{c}}(\boldsymbol{r})$ can be defined as the charge distribution which yields as electrostatic potential the correlation potential. The sum of the electron density, the exchange charge density, and the correlation charge density gives the effective density distribution which is, literally speaking, the density seen by an electron of the system due to the presence of the other electrons, and which takes into account many-body effects. This effective density might be of interest in the interpretation of electronic structures.

For solids a natural choice for both the orbital basis set and the auxiliary basis set are plane waves. However, the cutoff of the auxiliary basis set has to be chosen much smaller than that of the orbital basis set in order to avoid numerical instabilities due to unphysical small eigenvalues in the KS response matrix χ_{KS}. For molecules, different Gaussian basis sets are used to represent the orbitals on the one hand, and the exchange potential or the exchange charge on the other. By now, several plane wave EXX KS implementations for solids have been presented [Magyar 2004, Rinke 2005, Qteisch 2005] in addition to the one of [Städele 1997, Städele 1999]. For solids, an EXX KS implementation within the atomic sphere approximation has also been introduced already several years ago [Kotani 1995]. Very recently, EXX KS implementions within the fully linearized augmented plane wave approach [Sharma 2005] and within the projector augmented wave approach [Kresse 2005] were presented. For molecules, a number of Gaussian basis set EXX KS implementations were developed in recent years [Hamel 2001, Veseth 2001, Yang 2002] besides those of [Görling 1999a] and [Ivanov 1999]. Furthermore, a numerical grid EXX KS approach for molecules was recently introduced [Kümmel 2003].

9.2.3 Static Effective Exact Exchange KS Methods

While plane wave EXX KS methods are numerically stable, EXX KS methods employing Gaussian basis sets suffer from numerical instabilities [Hamel 2001, Görling 1995a, Schipper 1997, Hirata 2001, Della Sala 2003b]. The resulting exchange energies and thus, also the total energies and the occupied orbitals, are numerically stable. Exchange potentials, however, exhibit highly oscillatory unphysical features which affect the unoccupied orbitals and their eigenvalues. Because the latter are among the input quantities of density functional response methods for the calculation of, e.g., excitation energies, these numerical problems limit severely the scope and usefulness of EXX KS methods based on Gaussian functions.

There are two known potential sources of the numerical problems. First, the EXX (9.14) is known to be inherently unstable from a numerical point of view [Schipper 1997]. Small changes on the right hand side Λ can lead to huge changes in the resulting exchange potentials v_x. If the right hand side Λ of the EXX (9.14) is expanded in Gaussian functions, it will exhibit small oscillations in space around the correct values. Such oscillations are known to be strongly magnified in the solution of the EXX equation, i.e., in v_x. Secondly, it can happen that linear combinations of auxiliary basis functions couple only poorly to the response function [Görling 1995a]. In the limit that a linear combination of auxiliary functions does not couple at all to the response function, i.e., it is an eigenfunction of the response function with eigenvalue zero, the solution of the EXX equation is undefined with respect to addition of this linear combination. In this context, it should be noted that the validity of the Hohenberg-Kohn theorem is limited if orbitals are expanded in a finite basis set. This is an often overlooked fact. As an example let us consider a KS calculation of a molecule using a Gaussian basis set. The Gaussian basis set is usually localized in the vicinity of the molecule. This, however, means that the KS potential can be arbitrarily changed outside the region in the vicinity of the molecule where the basis set is localized without changing the resulting KS orbitals. Similary, certain combinations of auxiliary basis functions might be added to the KS potential without changing the KS orbitals if the orbital and the auxiliary basis set are not balanced. Such a balancing is easily possible for plane wave basis sets by chosing for the auxiliary basis set a much smaller cutoff than for the orbital basis set. For Gaussian basis sets, however, a simple balancing scheme, at present, is not known.

So far, a convincing solution for the numerical problems of the EXX KS implementations using Gaussian basis sets could not be found. Note that the problems do not result from the method by which the EXX equation is solved. The unphysical oscillatory features are exhibited by mathematically correct solutions of the EXX equation for given combinations of auxiliary and orbital basis sets. Thus, any approach which yields the full solution of the EXX equation should suffer from the numerical problems, be it direct solutions of the EXX equation or iterative ones.

In order to avoid such numerical problems and to decrease the numerical effort, *effective* exact exchange KS schemes have been devised [Della Sala 2003b, Krieger 1992a, Della Sala 2001a, Della Sala 2002a, Della Sala 2002b, Gritsenko 2001, Vitale 2005]. One such effective EXX method was originally derived by making the approximation that the exact exchange-only KS determinant and Hartree-Fock determinant were identical [Della Sala 2001a]. The resulting expression for the effective EXX potential, the localized Hartree-Fock (LHF) exchange potential v_x^{LHF}, is given in the case of real valued orbitals by

$$v_{\mathrm{x}}^{\mathrm{LHF}}(\boldsymbol{r}) = v_{\mathrm{x}}^{\mathrm{Slater}}(\boldsymbol{r}) + 2 \sum_{(i,j)\neq(N,N)}^{\mathrm{occ}} \frac{\varphi_i(\boldsymbol{r})\varphi_j(\boldsymbol{r})}{n(\boldsymbol{r})} \langle\varphi_j|\hat{V}_{\mathrm{x}}^{\mathrm{LHF}} - \hat{V}_{\mathrm{x}}^{\mathrm{NL}}|\varphi_i\rangle \,. \tag{9.28}$$

with the Slater potential

$$v_{\mathrm{x}}^{\mathrm{Slater}}(\boldsymbol{r}) = 2\sum_{ij}^{\mathrm{occ}} \frac{\varphi_i(\boldsymbol{r})\varphi_j(\boldsymbol{r})}{n(\boldsymbol{r})} \int \mathrm{d}^3 r \, \frac{\varphi_j(\boldsymbol{r}')\varphi_i(\boldsymbol{r}')}{|\boldsymbol{r}-\boldsymbol{r}'|} \,. \tag{9.29}$$

In (9.28), $\hat{V}_{\mathrm{x}}^{\mathrm{LHF}}$ denotes the operator generated by the LHF exchange potential $v_{\mathrm{x}}^{\mathrm{LHF}}$. The potential $v_{\mathrm{x}}^{\mathrm{LHF}}$ can be considered as a localization of the nonlocal Hartree-Fock exchange potential. Therefore the method is called localized Hartree-Fock method. When (9.28) is derived, at first, the summation runs over all pairs of occupied orbitals including the pair consisting of twice the HOMO, which is denoted by $(i,j) = (N,N)$, with N being the number of occupied orbitals. This original equation, however, determines the LHF exchange potential only up to an additive constant because both $v_{\mathrm{x}}^{\mathrm{LHF}}(\boldsymbol{r})$ and $v_{\mathrm{x}}^{\mathrm{LHF}}(\boldsymbol{r}) + C$, with C being an arbitrary constant, obey (9.28). By removing the pair $(i,j) = (N,N)$ from the summation, the additive constant is fixed and a unique exchange potential is determined by (9.28). This potential exhibits a $1/r$-asymptotic far from the system, except on the nodal planes of the HOMO.

The LHF exchange potential $v_{\mathrm{x}}^{\mathrm{LHF}}$ occurs on both sides of (9.28). This suggests an iterative solution. Alternatively, (9.28) can be cast into a linear equation which can be solved, for example, by a conjugate gradient approach. The solution of (9.28) not only leads to numerically stable exchange potentials but, moreover, can be carried out quite efficiently. An advantage of the LHF approach is that it requires only occupied orbitals, whereas the full EXX approaches require the use of unoccupied orbitals either by explicit summation over unoccupied orbitals or by the construction of Green functions through the solution of linear equations.

An independent alternative derivation of (9.28) makes the approximation of setting all energy denominators in the original EXX (9.14), or more precisely in the response function χ_{KS}, (9.15), and in the right hand side Λ, (9.16), to a common value and was therefore named common energy denominator approximation (CEDA) [Gritsenko 2001]. It is quite amazing and, up to now, not well understood that two very different approximations lead to the same approximate exchange potential $v_{\mathrm{x}}^{\mathrm{LHF}}$.

The response function χ_{KS} can be alternatively written as

$$\chi_{\mathrm{KS}}(\boldsymbol{r},\boldsymbol{r}') = 2\sum_{i}^{\mathrm{occ}}\sum_{k}{}' \frac{\varphi_i^*(\boldsymbol{r})\varphi_k(\boldsymbol{r})\varphi_k^*(\boldsymbol{r}')\varphi_i(\boldsymbol{r}')}{\varepsilon_i - \varepsilon_k} + \mathrm{c.c.} \tag{9.30}$$

In contrast to the original expressions (9.15), the second summation now runs over all orbitals except orbital i. Both expressions for the response function are identical because the additional terms in expression (9.30) cancel each other. The sum over unoccupied orbitals in the right hand side, Λ, (9.16), can be similarly expanded. If, however, a common energy denominator approximation is applied to the EXX equation with the expanded range for the second summation, it no longer results in the LHF expression (9.28) for the exchange potential, but in another approximation, named KLI approximation after its authors [Krieger 1992a]. The KLI approximation for the exchange potential can be obtained from the LHF approximation by omitting in the summation in expression (9.28) all terms with orbital indices $i \neq j$. A fundamental flaw of the KLI approximation, however, is that it is not invariant with respect to unitary transformations of the occupied orbitals among themselves. Because the density is invariant with respect to such transformations, also all density functionals should exhibit this invariance. The basic LHF expression, i.e., the expression before removing the $(i, j) = (N, N)$ pair in the summation of (9.28), is invariant with respect to unitary transformations of the occupied orbitals.

The reason why the common energy denominator approximation yields different results, the LHF and the KLI exchange potential, when applied to equivalent forms of the EXX equation, the forms with or without an expanded summation, is that it sets energy denominators of the same magnitude but different sign to the same common value and thus prevents the cancelations of those terms which result from an expansion of the second summations in (9.15) and (9.16). This problem can be avoided by assigning an average value not to the eigenvalue differences themselves but only to their magnitudes while retaining their signs. Such an approximation of an average magnitude of energy denominators (AMED) [Della Sala 2003b] yields the LHF exchange no matter to which form of the EXX equation it is applied to. Moreover, it can be applied to the case of open-shell systems which contain, besides fully occupied and empty orbitals, also partially occupied ones. With the help of the AMED approximation, an open-shell LHF method was recently derived [Della Sala 2003b] on the basis of a symmetrized Kohn-Sham formalism [Görling 1993b] that enables a proper treatment of the energetically lowest state of each symmetry of open-shell systems. Within the framework of the generalized adiabatic connection KS approach [Görling 1999c, Görling 2000] also a self-consistent treatment of excited states became possible at the effective EXX level within the open-shell LHF scheme [Vitale 2005]. This generalized adiabatic connection open-shell LHF approach is no longer based on the Hohenberg-Kohn theorem but on a more general theorem relating electronic densities and potentials [Görling 1999c], and represents an alternative to TDDFT for the treatment of excited states.

9.3 Time-Dependent Perturbation Theory and the Exact Exchange Kernel

9.3.1 Time-Dependent Perturbation Theory along the Adiabatic Connection and the Time Dependent Exact Exchange Equation

The starting point of a perturbation theory along the adiabatic connection in the time-dependent case [Görling 1997, Görling 1998a] is the time-dependent many-electron Schrödinger equation

$$\mathrm{i}\frac{\mathrm{d}}{\mathrm{d}t}\Psi^\lambda(t) = \left[\hat{T} + \hat{V}^\lambda(t) + \lambda\hat{V}_{\mathrm{ee}}\right]\Psi^\lambda(t) \, , \tag{9.31}$$

together with the specification of initial conditions

$$\Psi^\lambda(t_0) = \Psi_0^\lambda \tag{9.32}$$

for the wave functions $\Psi^\lambda(t)$ at some, for the moment arbitrary, time t_0. The wave functions Ψ_0^λ have to be chosen in such a way that all, independently of the value of the coupling constant λ, yield the same electronic density $n_0(\boldsymbol{r})$. Apart from this condition, the Ψ_0^λ can be chosen arbitrarily at this point. The subscript 0 of the initial state wave functions Ψ_0^λ and their electronic density $n_0(\boldsymbol{r})$ refers to the initial time t_0 and does not indicate that Ψ_0^λ is a ground-state wave function of any Schrödinger equation or that $n_0(\boldsymbol{r})$ is a ground-state electronic density. For a given initial condition (9.32) the potential $v^\lambda(\boldsymbol{r},t)$ generating the operator $\hat{V}^\lambda(t)$ is determined up to a time-dependent constant through the Runge-Gross theorem [Runge 1984] by the conditions that (i) the time-dependent electronic density $n(\boldsymbol{r},t)$ of the wave functions $\Psi^\lambda(t)$ is independent of the coupling constant λ, and that (ii) $v^\lambda(\boldsymbol{r},t)$ for $\lambda = 1$ equals a given external potential $v_{\mathrm{ext}}(\boldsymbol{r},t)$ of an electronic system with real, i.e., fully interacting, electrons. That is, for $\lambda = 1$, (9.31) represents a Schrödinger equation for real electrons. For $\lambda = 0$, (9.31) turns into the corresponding time-dependent many-electron KS equation

$$\mathrm{i}\frac{\mathrm{d}}{\mathrm{d}t}\Phi(t) = \left[\hat{T} + \hat{V}_{\mathrm{KS}}(t)\right]\Phi(t) \tag{9.33}$$

together with the initial condition

$$\Phi(t_0) = \Psi^{\lambda=0}(t_0) = \Psi_0^{\lambda=0} = \Phi_0 \tag{9.34}$$

and with $v_{\mathrm{KS}}(\boldsymbol{r},t) = v^{\lambda=0}(\boldsymbol{r},t)$ and $\Phi(t) = \Psi^{\lambda=0}(t)$.

An important difference to the static case discussed in the previous section is that, in the time-dependent case, the adiabatic connection for a given system of interacting electrons depends on the choice of initial conditions (9.32). Thus, the wave functions $\Psi^\lambda(t)$ and potentials $v^\lambda(\boldsymbol{r},t)$ are not defined by only specifying $v_{\mathrm{ext}}(\boldsymbol{r},t) = v^{\lambda=1}(\boldsymbol{r},t)$ and the value of the coupling constant λ; additionaly, also the initial conditions have to be specified. On the other hand, the wave functions $\Psi^\lambda(t)$ are not required to be ground-state wave functions or to evolve from an eigenstate or from the ground state.

Like in the static case, we expand the potential $v^\lambda(\boldsymbol{r},t)$ in a Taylor series with respect to the coupling constant λ

$$\begin{aligned} v^\lambda(\boldsymbol{r},t) &= \sum_{p=0}^{\infty} \lambda^p v^{(p)}(\boldsymbol{r},t) \\ &= v_{\mathrm{KS}}(\boldsymbol{r},t) + \lambda\, v^{(1)}(\boldsymbol{r},t) + \lambda^2 v^{(2)}(\boldsymbol{r},t) + \ldots \end{aligned} \tag{9.35}$$

By writing the first order potential $v^{(1)}(\boldsymbol{r},t)$ as

$$v^{(1)}(\boldsymbol{r},t) = -v_{\mathrm{x}}(\boldsymbol{r},t) - v_{\mathrm{H}}(\boldsymbol{r},t)\ , \tag{9.36}$$

i.e., as the negative of the exchange potential $v_{\mathrm{x}}(\boldsymbol{r},t)$ plus the negative of the Hartree potential $v_{\mathrm{H}}(\boldsymbol{r},t) = \int \mathrm{d}^3r'\, n(\boldsymbol{r}',t)/|\boldsymbol{r}-\boldsymbol{r}'|$, the exchange potential

$$v_{\mathrm{x}}(\boldsymbol{r},t) = -v^{(1)}(\boldsymbol{r},t) - v_{\mathrm{H}}(\boldsymbol{r},t) \tag{9.37}$$

is defined in analogy to the static case. The sum $\sum_{p=2}^{\infty} v^{(p)}(\boldsymbol{r},t)$ again represents the negative of the correlation potential.

Additionally, we expand the initial state wave function Ψ_0^λ in a Taylor series with respect to the coupling constant λ

$$\Psi_0^\lambda = \sum_{p=0}^{\infty} \lambda^{(p)} \Psi_0^{(p)} = \Phi_0 + \lambda\, \Psi_0^{(1)} + \lambda^2 \Psi_0^{(2)} + \ldots \tag{9.38}$$

With the Taylor expansion (9.35), the Hamiltonian operator of the adiabatic connection Schrödinger (9.31) turns into

$$\hat{T} + \hat{V}_{\mathrm{KS}}(t) + \lambda \left[\hat{V}_{\mathrm{ee}} - \hat{V}_{\mathrm{x}}(t) - \hat{V}_{\mathrm{H}}(t)\right] + \lambda^2 \hat{V}^{(2)}(t) + \ldots \tag{9.39}$$

Analogously to the static case, the KS potential $v_{\mathrm{KS}}(t)$ is given by

$$v_{\mathrm{KS}}(\boldsymbol{r},t) = v_{\mathrm{ext}}(\boldsymbol{r},t) + v_{\mathrm{x}}(\boldsymbol{r},t) + v_{\mathrm{H}}(\boldsymbol{r},t) - v^{(2)}(\boldsymbol{r},t) - \ldots \tag{9.40}$$

In order to derive a time-dependent EXX equation for the exchange potential $v_{\mathrm{x}}(\boldsymbol{r},t)$, and corresponding integral equations for the higher order potentials $v^{(2)}(\boldsymbol{r},t)$, we again expand the electronic density in a Taylor series with respect to the coupling constant λ

$$n^\lambda(\boldsymbol{r},t) = \sum_{p=0}^{\infty} \lambda^p n^{(p)}(\boldsymbol{r},t)\ . \tag{9.41}$$

and exploit the fact that all but the zeroth order terms in λ have to vanish, i.e.,

$$n^{(p)}(\boldsymbol{r},t) = 0 \quad \text{for} \quad p \geq 1\,, \tag{9.42}$$

as the electronic density, by construction, is independent of the coupling constant λ. In the time-dependent case, however, the terms $n^{(p)}(\boldsymbol{r},t)$ of such Taylor series depend on the initial conditions and are therefore more complicated to derive than in the static case. As a first step, the choice of the initial wave functions Ψ_0^λ, (9.32), has to be made.

From now on we choose initial wave functions Ψ_0^λ which are the ground states of a static coupling constant Schrödinger equation, i.e., of an equation similar to (9.1). In order to avoid confusion between static and time-dependent potentials, from now on, we denote the potential of the static coupling constant Schrödinger equation by $u^\lambda(\boldsymbol{r})$ instead of $v^\lambda(\boldsymbol{r})$. Thus the initial wave functions Ψ_0^λ are eigenstates of the Schrödinger equation

$$\left[\hat{T} + \hat{u}^\lambda + \lambda \hat{V}_{\mathrm{ee}}\right] \Psi_0^\lambda = E_0^\lambda \Psi_0^\lambda\,. \tag{9.43}$$

This means that the subscript 0 of the wave functions Ψ_0^λ can be interpreted both in the sense that it refers to the initial time t_0 and in the sense that Ψ_0^λ are ground-state wave functions of (9.43). The initial KS wave function $\Phi_0 = \Psi_0^\lambda$ then obeys the many-electron KS equation

$$\left[\hat{T} + \hat{U}_{\mathrm{KS}}\right] \Phi_0 = E_{\mathrm{KS},\,0} \Phi_0 \tag{9.44}$$

with the operator $\hat{U}_{\mathrm{KS}}$ generated by the KS potential $u_{\mathrm{KS}}(\boldsymbol{r})$. Equation (9.44) equals (9.2) except that the KS potential and the corresponding operator are denoted by $\hat{U}_{\mathrm{KS}}$ and $u_{\mathrm{KS}}(\boldsymbol{r})$, respectively, instead of $\hat{V}_{\mathrm{KS}}$ and $v_{\mathrm{KS}}(\boldsymbol{r})$, in order to avoid confusion with the corresponding time-dependent quantities. Note that the static potentials $u^\lambda(\boldsymbol{r})$ and $u_{\mathrm{KS}}(\boldsymbol{r})$ and the time-dependent potentials $v^\lambda(\boldsymbol{r},t)$ and $v_{\mathrm{KS}}(\boldsymbol{r},t)$, at this point, are independent of each other. In particular, it is not required, that $v^\lambda(\boldsymbol{r},t_0)$ equals $u^\lambda(\boldsymbol{r})$ or that $v_{\mathrm{KS}}(\boldsymbol{r},t_0)$ equals $u_{\mathrm{KS}}(\boldsymbol{r})$.

Time-dependent perturbation theory then yields the first order contribution

$$(-\mathrm{i})\sum_j \int_{t_0}^{t} \mathrm{d}t'\; \Phi_j(t') \langle \Phi_j(t') | \hat{V}_{\mathrm{ee}} - \hat{V}_{\mathrm{x}}(t') - \hat{V}_{\mathrm{H}}(t') | \Phi(t') \rangle$$

$$+ \sum_j{}' \Phi_j(t) \frac{\langle \Phi_j | \hat{V}_{\mathrm{ee}} - \hat{U}_{\mathrm{x}} - \hat{U}_{\mathrm{H}} | \Phi_0 \rangle}{E_{\mathrm{KS},\,0} - E_{\mathrm{KS},\,j}} \tag{9.45}$$

to the wave function $\Psi^\lambda(t)$ due to the first order perturbation $\hat{V}_{\mathrm{ee}} - \hat{V}_{\mathrm{x}}(t) - \hat{V}_{\mathrm{H}}(t)$ and to the first order term of the initial state $\Psi_0^{(1)}$. The latter is given by (9.45), and the wave functions $\Phi_j(t)$ are those solutions of the time-dependent KS equation (9.33) that evolve from the excited states Φ_j of the static KS (9.44). This means that, depending on whether the wave functions are written as functions $\Phi_j(t)$ of time or simply as Φ_j without time-dependence, they are solutions of the time-dependent KS (9.33) or of the static KS (9.44). The relation $\Phi_j(t_0) = \Phi_j$ holds. Strictly, the time-dependent KS wave function $\Phi(t)$ that evolves from Φ_0 should then be written $\Phi_0(t)$. However, the subscript 0 is omitted in order to stick to the notation used when (9.33) was introduced. The prime in the second summation of expression (9.45) indicates that the KS ground state Φ_0 is omitted from the sum. The potential $u_{\mathrm{x}}(\boldsymbol{r})$ generating the operator $\hat{U}_{\mathrm{x}}$ is the exchange potential corresponding to the static adiabatic connection defined by (9.43), while the operator $\hat{U}_{\mathrm{H}}$ is generated by the corresponding Hartree potential. The first term in (9.45) arises because the potential $v^{\lambda}(\boldsymbol{r},t)$, which equals the KS potential $v_{\mathrm{KS}}(\boldsymbol{r},t)$ at $\lambda = 0$, changes if the coupling constant $\lambda = 0$ is turned on; the second term arises because the initial wave function Ψ_0^{λ}, which equals the KS wave function Φ_0 at $\lambda = 0$, changes if $\lambda = 0$ is turned on. The first-order contribution $n^{(1)}(\boldsymbol{r},t)$ to the density resulting from the first-order contribution (9.45) of the wave function equals zero according to (9.42), i.e.,

$$0 = (-\mathrm{i})\sum_j{}' \int_{t_0}^{t} \mathrm{d}t' \, \langle \Phi(t')|\hat{n}(\boldsymbol{r})|\Phi_j(t')\rangle \langle \Phi_j(t')|\hat{V}_{\mathrm{ee}} - \hat{V}_{\mathrm{x}}(t') - \hat{V}_{\mathrm{H}}(t')|\Phi(t')\rangle + \mathrm{c.c.}$$

$$+ \sum_j{}' \langle \Phi(t)|\hat{n}(\boldsymbol{r})|\Phi_j(t)\rangle \frac{\langle \Phi_j|\hat{V}_{\mathrm{ee}} - \hat{U}_{\mathrm{x}} - \hat{U}_{\mathrm{H}}|\Phi_0\rangle}{E_{\mathrm{KS},0} - E_{\mathrm{KS},j}} + \mathrm{c.c.} \quad (9.46)$$

The time-dependent KS wave functions $\Phi_j(t)$, like the static KS wave functions Φ_j, are Slater determinants constructed from orbitals $\varphi_i(t)$ and φ_i, respectively, because the KS (9.33) and (9.44) decouple in corresponding one-particle equations for the orbitals, i.e., in (9.13) for the static KS orbitals φ_i and in

$$\mathrm{i}\frac{\mathrm{d}}{\mathrm{d}t}\varphi_i(t) = \left[\hat{T} + \hat{V}_{\mathrm{KS}}(t)\right] \varphi_i(t) \quad (9.47)$$

for the time-dependent KS orbitals $\varphi_i(t)$. If these orbitals are substituted into (9.46) and the terms are rearranged, we obtain the time-dependent EXX equation

$$\int_{t_0}^{t} \mathrm{d}t' \int \mathrm{d}^3 r' \, \chi_{\mathrm{KS}}(\boldsymbol{r}t, \boldsymbol{r}'t') v_{\mathrm{x}}(\boldsymbol{r}', t') = \Lambda(\boldsymbol{r}, t) \, . \quad (9.48)$$

The time-dependent EXX equation contains the time-dependent response function of the KS system,

$$\chi_{\mathrm{KS}}(\boldsymbol{r}t, \boldsymbol{r}'t') = 2(-\mathrm{i}) \sum_i^{\mathrm{occ}} \sum_a^{\mathrm{unocc}} \varphi_i^*(\boldsymbol{r},t)\varphi_a(\boldsymbol{r},t)\varphi_a^*(\boldsymbol{r}',t')\varphi_i(\boldsymbol{r}',t') + \mathrm{c.c.} \quad (9.49)$$

The right hand side is given by

$$\Lambda(\boldsymbol{r},t) = 2(-\mathrm{i})\sum_i^{\mathrm{occ}}\sum_a^{\mathrm{unocc}} \varphi_i^*(\boldsymbol{r},t)\varphi_a(\boldsymbol{r},t)\int_{t_0}^{t}\mathrm{d}t'\ \langle\varphi_a(t')|\hat{V}_{\mathrm{x}}^{\mathrm{NL}}(t')|\varphi_i(t')\rangle + \mathrm{c.c.}$$
$$+\,2\sum_i^{\mathrm{occ}}\sum_a^{\mathrm{unocc}} \varphi_i^*(\boldsymbol{r},t)\varphi_a(\boldsymbol{r},t)\,\frac{\langle\varphi_a|\hat{U}_{\mathrm{x}}^{\mathrm{NL}}|\varphi_i\rangle}{\varepsilon_i-\varepsilon_a} + \mathrm{c.c.} \quad (9.50)$$

and the time-dependent nonlocal exchange operator $\hat{V}_{\mathrm{x}}^{\mathrm{NL}}(t)$, is an integral operator defined by

$$\left[\hat{V}_{\mathrm{x}}^{\mathrm{NL}}(t)\varphi_k(t)\right](\boldsymbol{r}) = -2\int\mathrm{d}^3r'\sum_j^{\mathrm{occ}}\frac{\varphi_j(\boldsymbol{r},t)\varphi_j^*(\boldsymbol{r}',t)\varphi_k(\boldsymbol{r}',t)}{|\boldsymbol{r}-\boldsymbol{r}'|}\,. \quad (9.51)$$

The corresponding static nonlocal exchange operator $\hat{U}_{\mathrm{x}}^{\mathrm{NL}}$ is defined according to (9.16). The EXX (9.48) takes into account memory effects through the time integrations contained in it. Moreover, it explicitly depends on the choice of the initial state through the second term on the right hand side of (9.50). Time-dependent LDA or GGA approximations, on the other hand, employ the adiabatic approximation, i.e., they construct the xc potential at a given time taking into account the KS orbitals or the electronic density at this time only, and therefore neglect memory effects and exhibit no explicit initial state dependence. The EXX potential thus differs from common approximations for the xc potential with respect to the behavior in both space and time.

Equation (9.48) yields the exact time-dependent exchange potential, as required in time-dependent KS methods that carry out real time propagation of KS orbitals. In practice, full EXX real time KS propagations so far could not be carried out because of its high complexity. This complexity arises, in particular, from the time integrations in the EXX (9.48) responsible for the accounting of memory effects. An approach which can be considered as a first step towards EXX real time KS propagations is presented in [Ullrich 1996] and [Marques 2001] and employs an adiabatic KLI exchange potential for the propagation in time.

9.3.2 The Exact Exchange Kernel

An EXX equation for the exchange kernel [Gross 1996, Görling 1997, Görling 1998a, Görling 1998b, Kim 2002b] can be derived by taking the functional derivative of the time-dependent EXX (9.48) with respect to the KS potential $v_{\mathrm{KS}}(\boldsymbol{r},t)$ [Görling 1997, Görling 1998a, Görling 1998b, Kim 2002b]. To that end, we note that the orbitals $\varphi_i^*(\boldsymbol{r},t)$ and $\varphi_a^*(\boldsymbol{r},t)$ occuring in the EXX (9.48) are determined by the initial condition at t_0, i.e., by the static one-particle KS (9.13), and by the time-dependent KS potential $v_{\mathrm{KS}}(\boldsymbol{r},t)$ via the time-dependent one-particle KS (9.47). The initial conditions will be kept fixed when taking the functional derivative of the EXX (9.48) with respect to the KS potential $v_{\mathrm{KS}}(\boldsymbol{r},t)$.

Time-dependent perturbation theory yields the functional derivative of the KS orbitals with respect to the KS potential as

$$\frac{\delta \varphi_k(\boldsymbol{r},t)}{\delta v_{\mathrm{KS}}(\boldsymbol{r}',t')} = \sum_{\ell} \varphi_\ell(\boldsymbol{r},t)\varphi_\ell^*(\boldsymbol{r}',t')\varphi_k(\boldsymbol{r}',t')\theta(t-t') \, . \tag{9.52}$$

In (9.52) the index k labels a KS orbital which can be either occupied or unoccupied, while the summation over ℓ runs over all KS orbitals. By $\theta(t-t')$ we denote the usual step function with $\theta(t-t') = 1$ for $t \geq t'$ and $\theta(t-t') = 0$ for $t < t'$.

Next, we consider the point where the functional derivative with respect to the KS potential is taken. This means we consider the time-dependent EXX equation (9.48) for the KS orbitals corresponding to a certain choice of the KS potential $v_{\mathrm{KS}}(\boldsymbol{r},t)$, and then take the functional derivative at this point, i.e., at this KS potential. We choose

$$v_{\mathrm{KS}}(\boldsymbol{r},t) = u_{\mathrm{KS}}(\boldsymbol{r}) \, . \tag{9.53}$$

Thus we take the functional derivative at the special point where the KS potential $v_{\mathrm{KS}}(\boldsymbol{r},t)$ equals the KS potential of (9.44) defining the initial condition. In this special case the KS potential $v_{\mathrm{KS}}(\boldsymbol{r},t)$ is no longer time-dependent. The time-dependent EXX (9.48) is, of course, also valid for this special case; in fact, it reduces to the static EXX (9.14). We, however, consider the time-dependent EXX equation (9.48) for this special case and take the time-dependent functional derivative with respect to the KS potential. For this special case, i.e., for the time-dependent KS potential chosen according to (9.53), the time-dependent KS orbitals are simply given by

$$\varphi_k(\boldsymbol{r},t) = \varphi_k(\boldsymbol{r})\mathrm{e}^{-\mathrm{i}\varepsilon_k(t-t_0)} \, . \tag{9.54}$$

The exchange kernel is required in density functional response methods, i.e., in methods using TDDFT in the response regime in order to calculate polarizabilities and excitation energies. Density functional response methods usually do not work in the time domain but in the frequency domain. Therefore, we will derive here the exact frequency-dependent, and not the exact time-dependent exchange kernel. Switching from time to frequency domain is not trivial due to the causality requirements in time that have to be obeyed by the time-dependent exchange kernel. In order to obtain the frequency-dependent exchange kernel, the functional derivative of the time-dependent EXX (9.48) has to be taken with respect to variations of the time-dependent KS potential given by

$$\delta v_{\mathrm{KS}}(\boldsymbol{r},t) = \int \mathrm{d}\omega \; \delta v_{\mathrm{KS}}(\boldsymbol{r},\omega) \; \mathrm{e}^{-\mathrm{i}\omega t} \; \mathrm{e}^{\eta t} \, , \tag{9.55}$$

with

$$v_{\mathrm{KS}}(\boldsymbol{r},-\omega) = v_{\mathrm{KS}}^*(\boldsymbol{r},\omega) \, . \tag{9.56}$$

In (9.55), ω stands for the frequency and $e^{\eta t}$ with $\eta \to 0^+$ is a convergence factor that guarantees that the perturbation is switched on adiabatically if the initial time t_0 is chosen as $t_0 \to -\infty$. The condition (9.56) guarantees that the variations $\delta v_{\mathrm{KS}}(\boldsymbol{r},t)$ are real valued. Equations (9.55) and (9.56) together with the choice $t_0 \to -\infty$ lead to variations $\delta v_{\mathrm{KS}}(\boldsymbol{r},t)$ compatible with density functional response theory. Equation (9.52) together with

$$\frac{\delta v_{\mathrm{KS}}(\boldsymbol{r},t)}{\delta v_{\mathrm{KS}}(\boldsymbol{r}',\omega)} = \mathrm{e}^{-\mathrm{i}\omega t}\, \mathrm{e}^{\eta t} \delta(\boldsymbol{r}-\boldsymbol{r}') \tag{9.57}$$

following from (9.55), give the functional derivative of the KS orbitals with respect to variations $\delta v_{\mathrm{KS}}(\boldsymbol{r}',\omega)$ of the Fourier components of the KS potential

$$\frac{\delta \varphi_k(\boldsymbol{r},t)}{\delta v_{\mathrm{KS}}(\boldsymbol{r}',\omega)} = \mathrm{e}^{-\mathrm{i}\omega t}\, \mathrm{e}^{\eta t} \sum_{\ell} \varphi_\ell(\boldsymbol{r}) \mathrm{e}^{-\mathrm{i}\varepsilon_k(t-t_0)} \frac{\varphi_\ell^*(\boldsymbol{r}')\varphi_k(\boldsymbol{r}')}{\varepsilon_k - \varepsilon_\ell + \omega + \mathrm{i}\eta} . \tag{9.58}$$

If now the functional derivative of the time-dependent EXX equation (9.48) with respect to variations $\delta v_{\mathrm{KS}}(\boldsymbol{r}',\omega)$ of the Fourier components of the KS potential is taken, then (9.58) gives the derivative of all terms which depend on the time-dependent KS orbitals, i.e., of the orbitals themselves, of the response function $\chi_{\mathrm{KS}}(\boldsymbol{r}t,\boldsymbol{r}'t')$, and of the nonlocal exchange potential $\hat{V}_{\mathrm{x}}^{\mathrm{NL}}(t')$. The only derivative not obtained is the derivative

$$\frac{\delta v_{\mathrm{x}}(\boldsymbol{r},t)}{\delta v_{\mathrm{KS}}(\boldsymbol{r}',\omega)} = \frac{\delta v_{\mathrm{x}}(\boldsymbol{r}',\omega)}{\delta v_{\mathrm{KS}}(\boldsymbol{r}',\omega)} \mathrm{e}^{-\mathrm{i}\omega t}\, \mathrm{e}^{\eta t} \tag{9.59}$$

of the exchange potential with respect to the KS potential. The right hand side of (9.59) follows because in the linear response regime variations of Fourier components of potentials or densities are only coupled if they belong to the same frequency.

Taking the functional derivative of the time-dependent EXX equation (9.48) with respect to variations $\delta v_{\mathrm{KS}}(\boldsymbol{r}',\omega)$, carrying out all time integrations in (9.48), and rearranging the terms, leads to

$$\int \mathrm{d}^3 r''\, \chi_{\mathrm{KS}}(\boldsymbol{r},\boldsymbol{r}'',\omega) \frac{\delta v_{\mathrm{x}}(\boldsymbol{r}'',\omega)}{\delta v_{\mathrm{KS}}(\boldsymbol{r}',\omega)} = h_{\mathrm{x}}(\boldsymbol{r},\boldsymbol{r}',\omega) , \tag{9.60}$$

where the frequency-dependent response function is given by

$$\begin{aligned}\chi_{\mathrm{KS}}(\boldsymbol{r},\boldsymbol{r}',\omega) = &\sum_i^{\mathrm{occ}} \sum_a^{\mathrm{unocc}} \frac{\varphi_i^*(\boldsymbol{r})\varphi_a(\boldsymbol{r})\varphi_a^*(\boldsymbol{r}')\varphi_i(\boldsymbol{r}')}{\varepsilon_i - \varepsilon_a + \omega + \mathrm{i}\eta} \\ &+ \sum_i^{\mathrm{occ}} \sum_a^{\mathrm{unocc}} \frac{\varphi_a^*(\boldsymbol{r})\varphi_i(\boldsymbol{r})\varphi_i^*(\boldsymbol{r}')\varphi_a(\boldsymbol{r}')}{\varepsilon_i - \varepsilon_a - \omega - \mathrm{i}\eta} .\end{aligned} \tag{9.61}$$

The right hand side, $h_{\mathrm{x}}(\boldsymbol{r},\boldsymbol{r}',\omega)$, collects all terms with known functional derivative, i.e., all terms except the one on the left hand side containing the exchange potential. For obtaining (9.60) we took into account that according to (9.56) a change $\delta v_{\mathrm{KS}}(\boldsymbol{r}',\omega)$ is accompanied by a change $\delta v_{\mathrm{KS}}(\boldsymbol{r}',-\omega) = \delta v^*_{\mathrm{KS}}(\boldsymbol{r}',\omega)$.

The right hand side $h_{\mathrm{x}}(\boldsymbol{r},\boldsymbol{r}',\omega)$ of (9.60) is quite complex. It can be decomposed into four terms $h^p_{\mathrm{x}}(\boldsymbol{r},\boldsymbol{r}',\omega)$ with $p = 1,2,3,4$:

$$h_{\mathrm{x}}(\boldsymbol{r},\boldsymbol{r}',\omega) = \sum_{p=1,4} h^p_{\mathrm{x}}(\boldsymbol{r},\boldsymbol{r}',\omega) \,. \tag{9.62}$$

The four terms are given by:

$$\begin{aligned} h^1_{\mathrm{x}}(\boldsymbol{r},\boldsymbol{r}',\omega) = -2\sum_{ij}^{\mathrm{occ}}\sum_{ab}^{\mathrm{unocc}} \Bigg[& \frac{\varphi_i^*(\boldsymbol{r})\varphi_a(\boldsymbol{r})\langle aj|bi\rangle\varphi_t^*(\boldsymbol{r}')\varphi_b(\boldsymbol{r})}{(\varepsilon_i-\varepsilon_a+\omega+\mathrm{i}\eta)(\varepsilon_j-\varepsilon_b+\omega+\mathrm{i}\eta)} \\ & + \frac{\varphi_a^*(\boldsymbol{r})\varphi_i(\boldsymbol{r})\langle ib|ja\rangle\varphi_b^*(\boldsymbol{r}')\varphi_t(\boldsymbol{r})}{(\varepsilon_i-\varepsilon_a-\omega-\mathrm{i}\eta)(\varepsilon_j-\varepsilon_b-\omega-\mathrm{i}\eta)}\Bigg] \,, \end{aligned} \tag{9.63}$$

$$\begin{aligned} h^2_{\mathrm{x}}(\boldsymbol{r},\boldsymbol{r}',\omega) = -2\sum_{ij}^{\mathrm{occ}}\sum_{ab}^{\mathrm{unocc}} \Bigg[& \frac{\varphi_i^*(\boldsymbol{r})\varphi_a(\boldsymbol{r})\langle ab|ji\rangle\varphi_j^*(\boldsymbol{r}')\varphi_b(\boldsymbol{r})}{(\varepsilon_i-\varepsilon_a+\omega+\mathrm{i}\eta)(\varepsilon_j-\varepsilon_b-\omega-\mathrm{i}\eta)} \\ & + \frac{\varphi_a^*(\boldsymbol{r})\varphi_i(\boldsymbol{r})\langle ij|ba\rangle\varphi_b^*(\boldsymbol{r}')\varphi_j(\boldsymbol{r})}{(\varepsilon_i-\varepsilon_a-\omega-\mathrm{i}\eta)(\varepsilon_j-\varepsilon_b+\omega+\mathrm{i}\eta)}\Bigg] \,, \end{aligned} \tag{9.64}$$

$$\begin{aligned} h^3_{\mathrm{x}}(\boldsymbol{r},\boldsymbol{r}',\omega) = -2\sum_{ij}^{\mathrm{occ}}\sum_{a}^{\mathrm{unocc}} \Bigg[& \frac{\varphi_i^*(\boldsymbol{r})\varphi_a(\boldsymbol{r})\langle j|\hat{V}_{\mathrm{x}}^{\mathrm{NL}}-\hat{V}_{\mathrm{x}}|i\rangle\varphi_a^*(\boldsymbol{r}')\varphi_j(\boldsymbol{r})}{(\varepsilon_i-\varepsilon_a+\omega+\mathrm{i}\eta)(\varepsilon_j-\varepsilon_a+\omega+\mathrm{i}\eta)} \\ & + \frac{\varphi_a^*(\boldsymbol{r})\varphi_i(\boldsymbol{r})\langle i|\hat{V}_{\mathrm{x}}^{\mathrm{NL}}-\hat{V}_{\mathrm{x}}|j\rangle\varphi_j^*(\boldsymbol{r}')\varphi_a(\boldsymbol{r})}{(\varepsilon_i-\varepsilon_a-\omega-\mathrm{i}\eta)(\varepsilon_j-\varepsilon_a-\omega-\mathrm{i}\eta)}\Bigg] \\ +2\sum_{i}^{\mathrm{occ}}\sum_{ab}^{\mathrm{unocc}} \Bigg[& \frac{\varphi_i^*(\boldsymbol{r})\varphi_a(\boldsymbol{r})\langle a|\hat{V}_{\mathrm{x}}^{\mathrm{NL}}-\hat{V}_{\mathrm{x}}|b\rangle\varphi_b^*(\boldsymbol{r}')\varphi_i(\boldsymbol{r})}{(\varepsilon_i-\varepsilon_a+\omega+\mathrm{i}\eta)(\varepsilon_i-\varepsilon_b+\omega+\mathrm{i}\eta)} \\ & + \frac{\varphi_a^*(\boldsymbol{r})\varphi_i(\boldsymbol{r})\langle b|\hat{V}_{\mathrm{x}}^{\mathrm{NL}}-\hat{V}_{\mathrm{x}}|a\rangle\varphi_i^*(\boldsymbol{r}')\varphi_b(\boldsymbol{r})}{(\varepsilon_i-\varepsilon_a-\omega-\mathrm{i}\eta)(\varepsilon_i-\varepsilon_b-\omega-\mathrm{i}\eta)}\Bigg] \,, \end{aligned} \tag{9.65}$$

and

$$h_{\mathrm{x}}^{4}(\boldsymbol{r},\boldsymbol{r}',\omega) = -2\sum_{ij}^{\mathrm{occ}}\sum_{a}^{\mathrm{unocc}}\left[\frac{\varphi_j^*(\boldsymbol{r})\varphi_a(\boldsymbol{r})\langle a|\hat{V}_{\mathrm{x}}^{\mathrm{NL}}-\hat{V}_{\mathrm{x}}|i\rangle\varphi_i^*(\boldsymbol{r}')\varphi_j(\boldsymbol{r})}{(\varepsilon_j-\varepsilon_a+\omega+\mathrm{i}\eta)(\varepsilon_i-\varepsilon_a)}\right.$$
$$+\frac{\varphi_j^*(\boldsymbol{r})\varphi_i(\boldsymbol{r})\langle i|\hat{V}_{\mathrm{x}}^{\mathrm{NL}}-\hat{V}_{\mathrm{x}}|a\rangle\varphi_a^*(\boldsymbol{r}')\varphi_j(\boldsymbol{r})}{(\varepsilon_i-\varepsilon_a)(\varepsilon_j-\varepsilon_a+\omega+\mathrm{i}\eta)}$$
$$+\frac{\varphi_a^*(\boldsymbol{r})\varphi_j(\boldsymbol{r})\langle i|\hat{V}_{\mathrm{x}}^{\mathrm{NL}}-\hat{V}_{\mathrm{x}}|a\rangle\varphi_j^*(\boldsymbol{r}')\varphi_i(\boldsymbol{r})}{(\varepsilon_j-\varepsilon_a-\omega-\mathrm{i}\eta)(\varepsilon_i-\varepsilon_a)}$$
$$\left.+\frac{\varphi_i^*(\boldsymbol{r})\varphi_j(\boldsymbol{r})\langle a|\hat{V}_{\mathrm{x}}^{\mathrm{NL}}-\hat{V}_{\mathrm{x}}|i\rangle\varphi_j^*(\boldsymbol{r}')\varphi_a(\boldsymbol{r})}{(\varepsilon_i-\varepsilon_a)\,(\varepsilon_j-\varepsilon_a-\omega-\mathrm{i}\eta)}\right]$$

$$-2\sum_{i}^{\mathrm{occ}}\sum_{ab}^{\mathrm{unocc}}\left[\frac{\varphi_i^*(\boldsymbol{r})\varphi_b(\boldsymbol{r})\langle a|\hat{V}_{\mathrm{x}}^{\mathrm{NL}}-\hat{V}_{\mathrm{x}}|i\rangle\varphi_b^*(\boldsymbol{r}')\varphi_a(\boldsymbol{r})}{(\varepsilon_i-\varepsilon_b+\omega+\mathrm{i}\eta)(\varepsilon_i-\varepsilon_a)}\right.$$
$$+\frac{\varphi_a^*(\boldsymbol{r})\varphi_b(\boldsymbol{r})\langle i|\hat{V}_{\mathrm{x}}^{\mathrm{NL}}-\hat{V}_{\mathrm{x}}|a\rangle\varphi_b^*(\boldsymbol{r}')\varphi_i(\boldsymbol{r})}{(\varepsilon_i-\varepsilon_a)(\varepsilon_i-\varepsilon_b+\omega+\mathrm{i}\eta)}$$
$$+\frac{\varphi_b^*(\boldsymbol{r})\varphi_i(\boldsymbol{r})\langle i|\hat{V}_{\mathrm{x}}^{\mathrm{NL}}-\hat{V}_{\mathrm{x}}|a\rangle\varphi_a^*(\boldsymbol{r}')\varphi_b(\boldsymbol{r})}{(\varepsilon_i-\varepsilon_b-\omega-\mathrm{i}\eta)(\varepsilon_i-\varepsilon_a)}$$
$$\left.+\frac{\varphi_b^*(\boldsymbol{r})\varphi_a(\boldsymbol{r})\langle a|\hat{V}_{\mathrm{x}}^{\mathrm{NL}}-\hat{V}_{\mathrm{x}}|i\rangle\varphi_i^*(\boldsymbol{r}')\varphi_b(\boldsymbol{r})}{(\varepsilon_i-\varepsilon_a)(\varepsilon_i-\varepsilon_b-\omega-\mathrm{i}\eta)}\right]. \tag{9.66}$$

Matrix elements of the type $\langle aj|bi\rangle$ are defined by

$$\langle aj|bi\rangle = \int \mathrm{d}^3r \int \mathrm{d}^3r'\, \varphi_a^*(\boldsymbol{r})\varphi_j^*(\boldsymbol{r}')\varphi_b(\boldsymbol{r})\varphi_i^*(\boldsymbol{r}')/|\boldsymbol{r}-\boldsymbol{r}'|\,. \tag{9.67}$$

The functional derivative of the exchange potential with respect to the KS potential can be expressed with the help of the chain rule as

$$\begin{aligned}\frac{\delta v_{\mathrm{x}}(\boldsymbol{r},\omega)}{\delta v_{\mathrm{KS}}(\boldsymbol{r}',\omega)} &= \int \mathrm{d}^3r''\,\frac{\delta v_{\mathrm{x}}(\boldsymbol{r},\omega)}{\delta n(\boldsymbol{r}'',\omega)}\frac{\delta n(\boldsymbol{r}'',\omega)}{\delta v_{\mathrm{KS}}(\boldsymbol{r}',\omega)}\\ &= \int \mathrm{d}^3r''\, f_{\mathrm{x}}(\boldsymbol{r},\boldsymbol{r}'',\omega)\chi_{\mathrm{KS}}(\boldsymbol{r}'',\boldsymbol{r}',\omega)\,.\end{aligned} \tag{9.68}$$

The second line of (9.68) follows because the functional derivative of the exchange potential with respect to the electronic density is, by definition, the exchange kernel

$$f_{\mathrm{x}}(\boldsymbol{r},\boldsymbol{r}',\omega) = \frac{\delta v_{\mathrm{x}}(\boldsymbol{r},\omega)}{\delta n(\boldsymbol{r}',\omega)}\,, \tag{9.69}$$

the quantity we are interested in. If (9.68) is inserted into (9.60), we obtain the EXX equation for the frequency-dependent exchange kernel

$$\int d^3r_2 \int d^3r_3\, \chi_{\text{KS}}(\boldsymbol{r}_1,\boldsymbol{r}_2,\omega)\, f_{\text{x}}(\boldsymbol{r}_2,\boldsymbol{r}_3,\omega)\, \chi_{\text{KS}}(\boldsymbol{r}_3,\boldsymbol{r}_4,\omega) = h_{\text{x}}(\boldsymbol{r}_1,\boldsymbol{r}_4,\omega) \,. \tag{9.70}$$

Compared to the EXX equations (9.14) and (9.48) for the the static and time-dependent exchange potential, respectively, the EXX (9.70) for the frequency-dependent kernel is more complicated: the response functions occurs twice in the integral equation and the right hand side, $h_{\text{x}}(\boldsymbol{r}_1,\boldsymbol{r}_4,\omega)$, is quite complicated, see above.

In order to demonstrate why the EXX kernel is qualitatively superior to LDA or GGA kernels we consider the density-matrix based coupled KS equation [Casida 1995a, Heinze 2000] solved in most density functional response methods to calculate excitation energies

$$\left[\boldsymbol{\omega}^2 + 4\boldsymbol{\omega}^{1/2}\boldsymbol{K}(\omega_q)\boldsymbol{\omega}^{1/2}\right]\boldsymbol{F}_q = \Omega_q^2 \boldsymbol{F}_q \,, \tag{9.71}$$

with

$$K_{ia,jb}(\omega) = \int d^3r \int d^3r'\, \varphi_i^*(\boldsymbol{r})\varphi_a^*(\boldsymbol{r}) \left[\frac{1}{|\boldsymbol{r}-\boldsymbol{r}'|} + f_{\text{xc}}(\boldsymbol{r},\boldsymbol{r}',\omega)\right] \varphi_j(\boldsymbol{r}')\varphi_b(\boldsymbol{r}') \,, \tag{9.72}$$

and

$$\omega_{ia,jb} = \delta_{ia,jb}(\varepsilon_a - \varepsilon_i) \,. \tag{9.73}$$

Above, Ω_k denotes the q-th excitation frequency, and $\boldsymbol{F}_q$ stands for the corresponding eigenvector that yields the oscillator strengths and characterizes the excitation. Equation (9.71) is a matrix equation of a dimensionality given by the number of occupied times unoccupied orbitals; the matrix and vector elements are labeled by the superindices ia and jb. Within the adiabatic LDA or GGA the xc kernel assumes the simplified form

$$f_{\text{xc}}(\boldsymbol{r},\boldsymbol{r}',\omega) \approx f_{\text{xc}}^{\text{LDA/GGA}}(\boldsymbol{r})\delta(\boldsymbol{r}-\boldsymbol{r}') \,. \tag{9.74}$$

This means that LDA or GGA kernels effectively depend on only one and not two spatial variables and thus have an oversimplified structure. Moreover, due to the adiabatic approximation, LDA or GGA kernels do not depend on the frequency. As a result, the LDA or GGA coupled KS (9.71) is a linear equation, a standard eigenvalue equation, while the true coupled KS equation is not linear in frequency. The adiabatic approximation and the accompanying linearity means that the number of excitations which can be obtained by the LDA or GGA coupled KS (9.71) is given by the number of occupied times unoccupied orbitals. This equals the number of excitations which, in an independent particle picture, are described by single-particle excitations. The true number of excitations, due to the double- and multiple-particle excitations is, however, much higher. An exact density functional response treatment would yield all excitations. This, however, is only possible due to the nonlinearity of the exact coupled KS equation. With the EXX kernel, the coupled KS (9.71) is nonlinear as required, and can thus yield more than just single-particle excitations.

The nonlinearity of the coupled KS (9.71) also constitutes an important difference between EXX density-functional response and time-dependent Hartree-Fock approaches. One could assume that both approaches should be somehow similar because both treat exchange exactly, even though a differently defined exchange, and neglect correlation. The time-dependent Hartree-Fock equation is, however, linear and, due to the same dimensionality reasons as those discussed above, can only describe single-particle excitations or linear combinations of the latter, like excitations with plasmon character, whereas multiple-particle excitations can not be described. The nonlinearity of EXX TDDFT with the possibility to obtain a number of excitation higher than the number of single-particle excitations therefore is a fundamental advantage over time-dependent Hartree-Fock.

So far the EXX kernel was employed in practice only in two cases [Kim 2002a, Hirata 2002]. Within a plane wave implementation, the absorption spectrum of silicon was calculated with the adiabatic EXX kernel [Kim 2002a], i.e., the zero-frequency limit of the full EXX kernel. The results were encouraging because the excitonic structures of the silicon spectrum, which are missing in LDA or GGA spectra, were obtained. It was, however, necessary to omit certain long-range Coulomb terms in the construction of the kernel in order to achieve good results, a point which requires further investigations. Within a Gaussian basis set implementation, the use of the adiabatic EXX kernel leads to results for excitation energies of small molecules which were not better than those obtained with standard TDDFT methods [Hirata 2002]. The Gaussian basis set implementation, however, was most likely strongly impaired by numerical problems and therefore the results obtained are probably of limited significance. In any case, further investigations of the performance of the exact-exchange TDDFT method seem to be highly desirable, in particular with the full frequency-dependent exchange kernel.

10 Approximate Functionals from Many-Body Perturbation Theory

A. Marini, R.D. Sole, and A. Rubio

10.1 Motivations

As discussed in previous chapters, one of the main ingredients in TDDFT is the exchange-correlation (xc) kernel, $f_{xc}(\boldsymbol{r}, \boldsymbol{r}', \omega)$, that includes all the many-body effects beyond the Hartree approximation.

In contrast with the original static derivation of density functional theory (DFT), that is not applicable to excited state properties, TDDFT has become a promising and appealing approach to the study of linear response properties. TDDFT is appealing because f_{xc} is a functional of the ground-state density only. Moreover, in TDDFT there exists an exact and simple relation between f_{xc} and the polarization function χ. TDDFT is promising because it gives excellent results in several cases even using simple approximations for f_{xc} [Onida 2002]. In particular, the photoabsorption cross section and polarizabilities of simple metal clusters and biomolecules is well reproduced by the standard time-dependent local-density approximation (TDLDA) [Onida 2002, Marques 2003a]. For these systems, like in the case of atoms and molecules, the Hartree term is dominant and the TDLDA only modifies slightly the result of a simpler calculation performed within the random-phase approximation (RPA). However, the scenario changes rapidly if we increase the size of the system towards a periodic structure in one, two, and three dimensions (i.e., polymers, slabs, surfaces, or solids). Difficulties arise, for example, in long conjugated molecular chains, where the strong non-locality of the exact functional is not well reproduced in the local and semi-local approximations. A related problem appears for semiconductors and insulators where these functionals fail to describe the optical absorption experiments. As we will discuss in the next section, the reason for this has been traced to the fact that the xc kernel f_{xc} should behave asymptotically, in momentum space, as $1/q^2$ when $|\boldsymbol{q}| \to 0$ [Onida 2002], which is not the case for the adiabatic LDA or GGA [Perdew 1996b].

An alternative, more traditional, approach to the study of correlation in many-body systems is given by many-body perturbation theory (MBPT) [Abrikosov 1975] where the response function χ is expanded in powers of the screened electron-electron interaction W. At variance with TDDFT, W, in MBPT, is a well defined quantity, but there is not a simple relation between W and χ. In practical applications, the calculation of χ within MBPT

A. Marini et al.: *Approximate Functionals from Many-Body Perturbation Theory*, Lect. Notes Phys. **706**, 161–180 (2006)
DOI 10.1007/3-540-35426-3_10

can be cumbersome, but the results are often in very good agreement with experiment, both for finite and infinite systems [Onida 2002].

Thus the question is: can we benefit in some way of the good performance of MBPT to derive a more efficient approximation to f_{xc}? In what follows we will discuss different approaches to link TDDFT and MBPT: from the Bethe-Salpeter kernel (Sect. 10.3.1) or using the fully interacting response function (Sect. 10.3.2). Both approaches will be revisited in Sect. 10.4 in the spirit of a more general link between TDDFT and MBPT based on the many-body vertex function and Hedin's equations. We also discuss some possible ideas for further developments and establish contact with other approaches based on the EXX or the Sham-Schlüter equation.

10.2 Hedin's Equations and the Vertex Function

MBPT is a rigorous approach to describe the excited-state properties of condensed matter based on the Green's function method [Abrikosov 1975], and provides a proper framework for accurately computing excited state properties. For example, knowledge of the one-particle and two-particle Green's functions yields information, respectively, on the quasiparticle (QP) spectrum and optical response of a system.

For details of the Green's function formalism and many-body techniques applied to condensed matter, we refer the reader to several comprehensive papers in the literature [Onida 2002, Abrikosov 1975, Hedin 1965, Hedin 1999, Hedin 1969, Aryasetiawan 1998, Aulbur 2000, Strinati 1988]. Here we shall just present some of the main equations used for the quasiparticle and optical spectra calculations. (To simplify the presentation, we use in the following atomic units, $e = \hbar = m = 1$.)

The basic brick in a perturbative expansion is the reference, non-interacting system whose Green's functions G_0, that are known, enter in the terms of the perturbative expansion of G.

For the sake of simplicity, we consider here a non relativistic N-electrons system whose Hamiltonian $\hat{H}$ is decomposed into a non-interacting part plus a term containing the remaining electron-electron interaction. The system is assumed to interact with an external scalar potential through the operator

$$\hat{J}(t) = \int \mathrm{d}^3 r\, \hat{\psi}^\dagger(\boldsymbol{r}) J(\boldsymbol{r}, t) \hat{\psi}(\boldsymbol{r}) \,. \tag{10.1}$$

Following the equation of motion approach [Strinati 1988], or alternatively the standard diagrammatic technique [Abrikosov 1975] the exact Green's function G is found to satisfy the Dyson equation

$$G^{-1}(1,2) = G_0^{-1}(1,2) - J(1)\delta(1,2) - \Sigma(1,2) \,, \tag{10.2}$$

that, connecting G to G_0, defines the self-energy operator Σ (numbers stands for space, time, and spin coordinates)

$$\Sigma(1,2) = -\mathrm{i} \int \mathrm{d}3 \int \mathrm{d}4\, v_{\mathrm{ee}}(1^+,3) G_2(13,43^+) G^{-1}(4,2)\,. \tag{10.3}$$

All operators are in the interaction representation [Abrikosov 1975] ($\hat{\psi}(1) = \exp\{\mathrm{i}\hat{H}t_1\}\hat{\psi}(\boldsymbol{r}_1)\exp\{-\mathrm{i}\hat{H}t_1\}$, $\hat{J}_{\mathrm{I}}(t) = \exp\{\mathrm{i}\hat{H}t\}\hat{J}(t)\exp\{-\mathrm{i}\hat{H}t\}$) and G_2 is the two-particle Green's function

$$G_2(12,34) = (-\mathrm{i})^2 \langle \Psi | \hat{\mathcal{T}}\{\hat{U}\hat{\psi}(1)\hat{\psi}(2)\hat{\psi}^\dagger(4)\hat{\psi}^\dagger(3)\} | \Psi \rangle\,, \tag{10.4}$$

with $\hat{U} = \exp\{-\mathrm{i}\int_{-\infty}^{\infty}\mathrm{d}t\; \hat{J}_{\mathrm{I}}(t)\}$. It is interesting to note that the self-energy Σ, even if directly connected to the single-particle Green's function (10.2) contains a reference to the two-particle Green's function. This is the source of the difficulties in developing simple approximations to Σ, as this corresponds to finding simple approximations to the complicate two-body operator G_2.

An important step forward in the derivation of simple and efficient approximations to Σ has been done in the Hedin's equations [Hedin 1965, Hedin 1969] where G_2 in is rewritten in terms of more "physical" quantities related to microscopic polarization effects. We derive shortly this set of equations using the identity

$$G_2(13,23^+) = G(1,2)G(3,3^+) - \frac{\delta G(1,2)}{\delta J(3)}\,, \tag{10.5}$$

and introducing the total potential $V(1) = J(1) - \mathrm{i}\int \mathrm{d}3\, v_{\mathrm{ee}}(1,3)G(3,3^+)$. After some mathematical manipulations of (10.3), Σ can be rewritten as

$$\Sigma(1,2) \equiv \Sigma_{\mathrm{H}}(1,2) - \mathrm{i}\int \mathrm{d}3 \int \mathrm{d}4 \int \mathrm{d}5\, v_{\mathrm{ee}}(1^+,3) G(1,4) \frac{\delta G^{-1}(4,2)}{\delta V(5)} \frac{\delta V(5)}{\delta J(3)}\,. \tag{10.6}$$

Here $\Sigma_{\mathrm{H}}(1,2)$ is the Hartree self-energy

$$\Sigma_{\mathrm{H}}(1,2) = \delta(1,2)\left[-\mathrm{i}\int \mathrm{d}3\, v_{\mathrm{ee}}(1,3)G(3,3^+)\right]\,. \tag{10.7}$$

Equation (10.6) introduces two important quantities: (i) the scalar (irreducible) vertex function $\tilde{\Gamma}(12,3)$

$$\tilde{\Gamma}(12,3) \equiv -\frac{\delta G^{-1}(1,2)}{\delta V(3)}\,; \tag{10.8}$$

(ii) the inverse (longitudinal) dielectric function ϵ^{-1}

$$\epsilon^{-1}(1,2) = \frac{\delta V(1)}{\delta J(2)}\,. \tag{10.9}$$

The self-energy can be rewritten in terms of $\tilde{\Gamma}$ and ϵ^{-1} as

$$\Sigma(1,2) \equiv \Sigma_{\mathrm{H}}(1,2) + \mathrm{i}\int \mathrm{d}3 \int \mathrm{d}4\, W(1^+,3) G(1,4) \tilde{\Gamma}(42,3)\,, \tag{10.10}$$

with $W(1,2) = \int \mathrm{d}3\, \epsilon^{-1}(1,3) v_{\mathrm{ee}}(3,2)$ the screened electron-electron interaction. $\tilde{\Gamma}$ is a key quantity in Hedin's equations. The relation between ϵ^{-1} and $\tilde{\Gamma}$ follows from the introduction of the reducible polarization function χ

$$\epsilon^{-1}(1,2) = \delta(1,2) + \int \mathrm{d}3\, v_{\mathrm{ee}}(1,3) \chi(3,2)\,, \tag{10.11}$$

with $\chi(1,2) = \tilde{\chi}(1,2) + \int \mathrm{d}3 \int \mathrm{d}4\, \tilde{\chi}(1,3) v_{\mathrm{ee}}(3,4) \chi(4,2)$ and $\tilde{\chi}$ the irreducible response function

$$\tilde{\chi}(1,2) \equiv \frac{\delta \langle \hat{n}(1) \rangle}{\delta V(2)} = -\mathrm{i} \int \mathrm{d}3 \int \mathrm{d}4\, G(1,3) G(4,1) \tilde{\Gamma}(34,2)\,. \tag{10.12}$$

Within the linear response theory formalism, the inverse dielectric function $\epsilon^{-1}(\omega)$ (10.9) relates the total effective potential to the external perturbation v_{ext} applied on an electronic system. χ describes the dynamical properties of $\epsilon^{-1}(\omega)$ relating the charge response (δn) to the external potential: $\delta n = \chi \delta v_{\mathrm{ext}}$.

Now the scheme is clear: once an approximated expression for $\tilde{\Gamma}$ is given the response function $\chi(1,2)$ and the self-energy Σ are fully defined.

In practical implementations, the so-called *GW* approximation (GWA), the self-energy operator Σ is taken to be the first order term of a series expansion in terms of the screened Coulomb interaction W and the dressed Green function G

$$\Sigma(1,2) = \Sigma_{\mathrm{H}}(1,2) + \mathrm{i} G(1,2) W(1^+,2)\,. \tag{10.13}$$

Vertex corrections are not included in this approximation that corresponds to the simplest approximation for $\tilde{\Gamma}$: the vertex is assumed to be diagonal in space and time coordinates $\tilde{\Gamma}(12,3) \sim \delta(1,2)\delta(1,3)$. Most ab-initio *GW* applications solve the Dyson equation at the *GW* level non self-consistently, finding the poles of G corresponding to the quasiparticle energies, while keeping fixed the wavefunctions at the DFT level. This corresponds to the $G_0 W_0$ scheme for the calculation of quasiparticle energies as a first-order perturbation to the Kohn-Sham energy [Aryasetiawan 1998].

10.2.1 The Bethe-Salpeter Equation

As we have already discussed, the GWA to the self-energy uses a rather rough approximation to the vertex function. More importantly the corresponding expression for χ, the random-phase approximation (RPA), evaluated from (10.12) with $\tilde{\Gamma}(12,3) \sim \delta(1,2)\delta(1,3)$, does not yield optical absorption spectra in good agreement with experiments for several insulating and metallic systems [Onida 2002, Albrecht 1998, Benedict 1998a, Rohlfing 1998a, Rohlfing 2000a, Marini 2003a]. The reason is that the absorption intensity is given by the $\Im[\epsilon(\omega)]$, that is related to the response function via (10.11). The response function χ measures the change in the electronic density induced by the external applied potential. In a non-interacting system the RPA for χ is exact, but when self-energy corrections are included the electronic density $n(1) \equiv -\mathrm{i}G(1,1^+)$ changes and, consequently, the RPA approximation is not valid anymore. This change is reflected in the vertex function, as can be devised from (10.8)

$$\tilde{\Gamma}(12,3) = -\frac{\delta G^{-1}(1,2)}{\delta V(3)} = \delta(1,2)\delta(1,3) + \mathrm{i}\frac{\delta[G(1,2)W(1^+,2)]}{\delta V(3)} \tag{10.14}$$

Thus, the vertex function can be viewed as the linear response of the self-energy to a change in the total potential of the system. The vertex corrections account for xc effects between an electron and the other electrons in the screening density cloud. In particular this includes the electron-hole attraction (excitonic effects) in the dielectric response. Indeed, neglecting terms proportional to $\delta W/\delta V$ in (10.14) and using the chain rule we obtain

$$\tilde{\Sigma}(12,3) = \delta(1,2)\delta(1,3) + \mathrm{i}W(1,2)\frac{\delta G(1,2)}{\delta V(3)} . \tag{10.15}$$

Then using the identity

$$\frac{\delta G(1,2)}{\delta V(3)} = -\int \mathrm{d}4 \int \mathrm{d}5\, G(1,4)\frac{\delta G^{-1}(4,5)}{\delta V(3)}G(5,2) , \tag{10.16}$$

to get the final closed equation for $\tilde{\Gamma}$:

$$\tilde{\Gamma}(12,3) = \delta(1,2)\delta(1,3) + \mathrm{i}W(1,2) \int \mathrm{d}6 \int \mathrm{d}7\, G(1,6)G(7,2)\tilde{\Gamma}(67,3) . \tag{10.17}$$

This is the Bethe-Salpeter equation (BSE) for the irreducible vertex function. The BSE defines a consistent method to go beyond the RPA. When the vertex $\tilde{\Gamma}$, solution of (10.2.1), is inserted in (10.12), the corresponding expression for χ (and thus for ϵ) describes correctly the experimental spectra for a wide range of materials, including wide-gap insulators characterized by the presence of excitons (bound electron-hole states) [Onida 2002, Albrecht 1998, Marini 2003a]. The Bethe-Salpeter approach to the calculation of two-particle excited states is a natural extension of the GWA for the calculation of one-particle excited states, within the same theoretical framework and set of approximations (the GW-BSE scheme). The GW-BSE approach has successfully yielded the absorption spectra for a wide range of systems from molecules to nanostructures, from bulk semiconductors to surfaces and one-dimensional polymers and nanotubes [Onida 2002].

10.3 The xc Kernel: Different Schemes Based on MBPT

After the short review on MBPT, the aim of this section is to show how the knowledge gleaned in MBPT can be translated into a practical xc kernel, i.e, the challenge is now to construct $f_{\rm xc}$ using MBPT.

In TDDFT the exact, irreducible, response function is given by [Gross 1996, Petersilka 1996a, Petersilka 1996b]

$$\tilde{\chi}(\boldsymbol{r}_1, \boldsymbol{r}_2, \omega) = \chi_{\rm KS}(\boldsymbol{r}_1, \boldsymbol{r}_2, \omega) + \int \mathrm{d}^3 r_3 \int \mathrm{d}^3 r_4 \, \chi_{\rm KS}(\boldsymbol{r}_1, \boldsymbol{r}_3, \omega) f_{\rm xc}(\boldsymbol{r}_3, \boldsymbol{r}_4, \omega) \tilde{\chi}(\boldsymbol{r}_4, \boldsymbol{r}_2, \omega), \quad (10.18)$$

with $f_{\rm xc} \equiv \delta v_{\rm xc}/\delta n$, and $\chi_{\rm KS}$ the Kohn-Sham response function. The challenge is now how to link (10.18) with the BSE and construct $f_{\rm xc}$ using MBPT.

A first difficulty is $\chi_{\rm KS}$ and the xc potential $v_{\rm xc}$. In MBPT there is a clear distinction between the self-energy and the electron-electron interaction effects. To obtain a final response function in agreement with experiment, the single-particle Green's functions entering the BSE *must* have poles corresponding to the physical quasiparticle states (10.2). This is not true in TDDFT, as the exact $v_{\rm xc}$ is not supposed to give the measurable quasiparticle energies. However, while $\chi_{\rm KS}$ differs from $\chi_{\rm QP}$ in MBPT, the final response function is exact in both frameworks. This means that looking at TDDFT from the MBPT point of view, any discrepancy between the KS and quasiparticle eigenvalues must be accounted for by $f_{\rm xc}$.

This point had a strong impact on the development of many-body based $f_{\rm xc}$'s. Two different classes of approaches can be identified: (i) methods based on approximated xc functionals [Städele 1997, Ullrich 1995a] or on the Sham-Schlüter equation (that impose the many-body density to be equal to the DFT one [Tokatly 2002, Stubner 2004]), (ii) methods that assume that the DFT Green's functions to be the same as the QP ones [Marini 2003b, Adragna 2003, Reining 2002, Sottile 2003, Del Sole 2003, Sottile 2005]. In the first class of methods, the xc potential and the kernel are derived consistently. As a consequence, $f_{\rm xc}$ contains terms that account for the difference between the bare and the KS states. In the second group of methods, the attention is concentrated on the the effect of the electron-hole interaction in order to prove that $f_{\rm xc}$ can account for the excitonic effects provided by the BSE. To this end, the KS and the QP states are assumed to be the same.

In the next two sections we discuss two methods (contained in the second group of approaches) where the idea is to benefit from the good performance of MBPT response functions. We will analyze, in particular, the many-body properties of $f_{\rm xc}$ and the relation between its perturbative character and the inclusion of self-energy terms coming from $v_{\rm xc}$.

10.3.1 Static Long-Range Kernels

A pioneering study of the link between MBPT and TDDFT was done by Reining et al. [Reining 2002], analyzing the structure of the BSE and of (10.18) when both are rewritten in the same electron-hole basis. The point is that the TDDFT equation for $\tilde{\chi}$ can be expanded in the transition space, i.e., transformed using a basis of pairs of single-particle states. This allowed Reining et al. [Reining 2002] to compare directly (10.19) to the BSE scheme, for which this formulation is naturally adopted [Albrecht 1998, Benedict 1998a, Rohlfing 1998a, Rohlfing 2000a].

To start, we rewrite the Dyson-like matrix equation for $\tilde{\chi}$ as

$$\hat{\tilde{\chi}}(\boldsymbol{q},\omega) = \hat{\chi}_{\mathrm{KS}}(\boldsymbol{q},\omega) + \hat{\chi}_{\mathrm{KS}}(\boldsymbol{q},\omega)\hat{f}_{\mathrm{xc}}(\boldsymbol{q},\omega)\hat{\tilde{\chi}}(\boldsymbol{q},\omega)\,, \tag{10.19}$$

with $\boldsymbol{q}$ the transferred momentum and $\hat{\chi}_{\mathrm{KS}}(\boldsymbol{q},\omega)$ the matrix of the reciprocal space components of χ_{KS}:

$$[\chi_{\mathrm{KS}}(\boldsymbol{q},\omega)]_{\boldsymbol{G},\boldsymbol{G}'} = -\frac{2}{\Omega N_k}\sum_{cv\boldsymbol{k}} \frac{\xi^*_{cv\boldsymbol{k}}(\boldsymbol{q},\boldsymbol{G})\xi_{cv\boldsymbol{k}}(\boldsymbol{q},\boldsymbol{G}')}{\omega - \varepsilon^{\mathrm{KS}}_{c\boldsymbol{k}} + \varepsilon^{\mathrm{KS}}_{v\boldsymbol{k}-\boldsymbol{q}} + \mathrm{i}0^+}\,. \tag{10.20}$$

Here we have considered only the resonant part (positive energy poles). The oscillators ξ are given by $\xi_{cv\boldsymbol{k}}(\boldsymbol{q},\boldsymbol{G}) = \langle c\boldsymbol{k}|\exp\{\mathrm{i}(\boldsymbol{q}+\boldsymbol{G})\cdot\boldsymbol{r}\}|v\boldsymbol{k}-\boldsymbol{q}\rangle$, in terms of the conduction and valence Kohn-Sham states with energies $\varepsilon^{\mathrm{KS}}_{c\boldsymbol{k}}$ and $\varepsilon^{\mathrm{KS}}_{v\boldsymbol{k}-\boldsymbol{q}}$.

If we introduce the *interacting*, resonant, and irreducible electron-hole Green's function $\tilde{L}$ as

$$[\tilde{\chi}(\boldsymbol{q},\omega)]_{\boldsymbol{G},\boldsymbol{G}'} = -\frac{2}{\Omega N_k}\sum_{\substack{cv\boldsymbol{k}\\ c'v'\boldsymbol{k}'}} \xi^*_{cv\boldsymbol{k}}(\boldsymbol{q},\boldsymbol{G})\xi_{c'v'\boldsymbol{k}'}(\boldsymbol{q},\boldsymbol{G}')\tilde{L}_{cv\boldsymbol{k},c'v'\boldsymbol{k}'}(\boldsymbol{q},\omega) \tag{10.21}$$

the BSE can be rewritten as an equation for $\tilde{L}$

$$\hat{\tilde{L}}(\boldsymbol{q},\omega) = \hat{L}_{\mathrm{QP}}(\boldsymbol{q},\omega) + \hat{L}_{\mathrm{QP}}(\boldsymbol{q},\omega)\hat{W}(\boldsymbol{q})\hat{\tilde{L}}(\boldsymbol{q},\omega)\,, \tag{10.22}$$

with $[L_{\mathrm{QP}}]cv\boldsymbol{k},c'v'\boldsymbol{k}'(\boldsymbol{q},\omega) = \mathrm{i}\delta_{cc'}\delta_{vv'}\delta_{\boldsymbol{k},\boldsymbol{k}'}(\omega - \varepsilon^{\mathrm{QP}}_{c\boldsymbol{k}} - \varepsilon^{\mathrm{QP}}_{v\boldsymbol{k}-\boldsymbol{q}} + \mathrm{i}0^+)^{-1}$, in terms of the QP energies ($\varepsilon^{\mathrm{QP}}$). All quantities in (10.22) are matrices in the electron-hole basis. $\hat{W}(\boldsymbol{q})$ is the Coulombic part of the Bethe-Salpeter kernel,

$$W_{cv\boldsymbol{k},c'v'\boldsymbol{k}'}(\boldsymbol{q}) = \mathrm{i}\langle c\boldsymbol{k}, v\boldsymbol{k}-\boldsymbol{q}|W(\boldsymbol{r}_1,\boldsymbol{r}_2)|c'\boldsymbol{k}', v'\boldsymbol{k}'-\boldsymbol{q}\rangle\,, \tag{10.23}$$

with $W(\boldsymbol{r}_1,\boldsymbol{r}_2)$ the statically screened electron-hole interaction, $W(\boldsymbol{r}_1,\boldsymbol{r}_2) = \int \mathrm{d}^3r_3\,\epsilon^{-1}(\boldsymbol{r}_1,\boldsymbol{r}_3,\omega=0)v_{\mathrm{ee}}(\boldsymbol{r}_3,\boldsymbol{r}_2)$.

Following the same approach, (10.19) can be expanded in the $|cv\boldsymbol{k}\rangle$ basis using the expression for $\tilde{\chi}$ given in (10.21)

$$\hat{\tilde{L}}(\boldsymbol{q},\omega) = \hat{L}_{\mathrm{KS}}(\boldsymbol{q},\omega) + \hat{L}_{\mathrm{KS}}(\boldsymbol{q},\omega)\hat{F}(\boldsymbol{q})\hat{\tilde{L}}(\boldsymbol{q},\omega)\,, \tag{10.24}$$

with

$$[\hat{F}(\boldsymbol{q},\omega)]_{cv\boldsymbol{k},c'v'\boldsymbol{k}'} \equiv \sum_{\boldsymbol{G},\boldsymbol{G}'} \xi_{cv\boldsymbol{k}}(\boldsymbol{q},\boldsymbol{G})[f_{\mathrm{xc}}(\boldsymbol{q},\omega)]_{\boldsymbol{G},\boldsymbol{G}'}\xi^*_{c'v'\boldsymbol{k}'}(\boldsymbol{q},\boldsymbol{G}') . \quad (10.25)$$

A comparison of the TDDFT equation (10.24) and the BSE (10.22) tells us immediately that the BSE and TDDFT equations yield the same spectrum if

$$[F(\boldsymbol{q})]_{cv\boldsymbol{k},c'v'\boldsymbol{k}'} = -I\delta_{cc'}\delta_{vv'}\delta_{\boldsymbol{k}\boldsymbol{k}'}\left[\varepsilon^{\mathrm{QP}}_{c\boldsymbol{k}} - \varepsilon^{\mathrm{QP}}_{v\boldsymbol{k}} - \varepsilon^{\mathrm{KS}}_{c\boldsymbol{k}} + \varepsilon^{\mathrm{KS}}_{v\boldsymbol{k}}\right] + W_{cv\boldsymbol{k},c'v'\boldsymbol{k}'}(\boldsymbol{q}) . \quad (10.26)$$

It is important to note that a static f_{xc} in reciprocal (or real) space follows exactly from (10.26) only if the oscillators ξ can be inverted in (10.25) generating a one to one relation between $[f_{\mathrm{xc}}]_{\boldsymbol{G}\boldsymbol{G}'} \leftrightarrow [F]_{cv\boldsymbol{k},c'v'\boldsymbol{k}'}$. In practice this has been proved to hold only for a model case [Sottile 2003], while a more general approach presented in [Sottile 2003] has shown how to obtain a frequency-dependent f_{xc} imposing (10.26) at the level of the response function.

Nevertheless, the static kernel derived by Reining et al. [Reining 2002], given in (10.26) has been of great importance to discuss some of its features, and in particular its long-range behavior. In fact, for valence $(v,\boldsymbol{k})$ and conduction $(c,\boldsymbol{k}+\boldsymbol{q})$ states, $\xi_{cv\boldsymbol{k}}(\boldsymbol{q},\boldsymbol{G})$ goes to zero like q for small q. Since $W_{vc\boldsymbol{k},v'c'\boldsymbol{k}'}$ in this limit behaves as a constant, an $f_{\mathrm{xc}}(\mathbf{q},\mathbf{G}=\mathbf{G}'=0)$ obtained from (10.26) must behave as $1/q^2$. There is in fact a positive long-range contribution stemming from the QP shift of eigenvalues (as also predicted in [Gonze 1997a]), and a negative one resulting from the electron-hole interaction.

To show the crucial importance of the long-range tail of f_{xc}, a simple calculation using a model $f_{\mathrm{xc}}(\mathbf{r},\mathbf{r}') \sim -\alpha/|\mathbf{r}-\mathbf{r}'|$, with the empirical value $\alpha = 0.2$ works pretty well in the case of silicon (see Fig. 10.1 and the extended discussion in Chap. 20 of this book).

10.3.2 A Perturbative Scheme

In the past section we have seen that it is possible to link f_{xc} with the BSE using the electron-hole basis and comparing directly the TDDFT and Bethe-Salpeter kernel [(10.26)]. An important property of the static kernel of (10.26) and of its dynamical extension [Sottile 2003] is that it is to first order in W. In this section, we introduce a more general approach for deriving a perturbative expansion of f_{xc}, without the assumption of linearity between f_{xc} and W.

The idea is to benefit from the good performance of MBPT *response functions* and build an f_{xc} that mimics those results. The kernel f_{xc} is derived by imposing TDDFT to reproduce the perturbative expansion of the BSE in terms of the screened Coulomb interaction at any order. In practice this

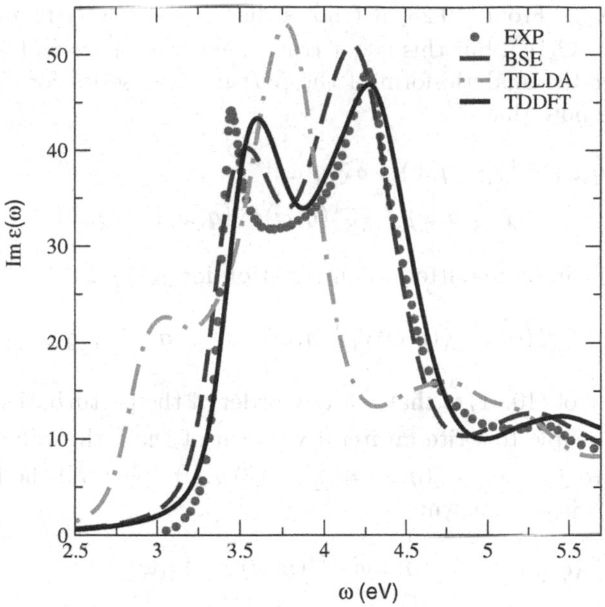

Fig. 10.1. Silicon, optical absorption spectra. *Dots*: experiment. *Dot-dashed curve*: TDLDA result. *Dashed curve*: result obtained through the Bethe-Salpeter method. *Continuous curve*: TDDFT result using the long-range model f_{xc} described in the text. (Adapted from [Reining 2002])

means that we assume $\hat{\chi}^{\mathrm{KS}}(\boldsymbol{q},\omega) = \hat{\chi}^{\mathrm{QP}}(\boldsymbol{q},\omega)$, i.e, the "exact" Kohn-Sham DFT response function is approximated by the independent-quasiparticle response. Second, we *assume* that there exists an $\hat{f}_{\mathrm{xc}}(\boldsymbol{q},\omega)$ that reproduces the BSE spectra [Del Sole 2003], i.e., we impose

$$\hat{\tilde{\chi}}_{\mathrm{BSE}}(\boldsymbol{q},\omega) \equiv \hat{\tilde{\chi}}_{\mathrm{TDDFT}}(\boldsymbol{q},\omega)\,. \tag{10.27}$$

From (10.19) we obtain an equation for the reciprocal-space matrix components of f_{xc} [Del Sole 2003]

$$\hat{f}_{\mathrm{xc}}(\boldsymbol{q},\omega) = \hat{\chi}_{\mathrm{KS}}^{-1}(\boldsymbol{q},\omega)\delta\hat{\tilde{\chi}}(\boldsymbol{q},\omega)\hat{\tilde{\chi}}^{-1}(\boldsymbol{q},\omega)\,, \tag{10.28}$$

with $\delta\hat{\tilde{\chi}}$ given by $\hat{\tilde{\chi}} = \hat{\chi}_{\mathrm{KS}} + \delta\hat{\tilde{\chi}}$. More explicitly, from (10.21) and (10.22) we have

$$[\delta\tilde{\chi}(\boldsymbol{q},\omega)]_{\boldsymbol{G},\boldsymbol{G}'} = -\frac{2}{\Omega N_k}\sum_{\substack{c_1v_1\boldsymbol{k}_1,c_2v_2\boldsymbol{k}_2\\ c_3v_3\boldsymbol{k}_3}} \xi^*_{c_1v_1\boldsymbol{k}_1}(\boldsymbol{q},\boldsymbol{G})\xi_{c_3v_3\boldsymbol{k}_3}(\boldsymbol{q},\boldsymbol{G}') \tag{10.29}$$
$$\times[L_{\mathrm{KS}}(\boldsymbol{q},\omega)]_{c_1v_1\boldsymbol{k}_1}W_{c_1v_1\boldsymbol{k}_1,c_2v_2\boldsymbol{k}_2}[\tilde{L}(\boldsymbol{q},\omega)]_{c_2v_2\boldsymbol{k}_2,c_3v_3\boldsymbol{k}_3}\,,$$

with $L_{\mathrm{KS}} = L_{\mathrm{QP}}$. From (10.28) it follows that f_{xc} is exactly first-order in W only if $\delta\hat{\hat{\chi}}/\hat{\hat{\chi}} = \mathcal{O}(W)$, but this is not true in general (as we will see shortly).

To inspect the analytic form of the perturbative series for f_{xc} obtained by (10.28) we note that

$$\begin{aligned}\hat{\hat{\chi}}^{-1}(\boldsymbol{q},\omega) &= [\hat{\chi}_{\mathrm{KS}}(\boldsymbol{q},\omega) + \delta\hat{\hat{\chi}}(\boldsymbol{q},\omega)]^{-1} \\ &= \hat{\chi}_{\mathrm{KS}}^{-1}(\boldsymbol{q},\omega) - \hat{\chi}_{\mathrm{KS}}^{-1}(\boldsymbol{q},\omega)\delta\hat{\hat{\chi}}(\boldsymbol{q},\omega)\hat{\hat{\chi}}^{-1}(\boldsymbol{q},\omega) .\end{aligned} \tag{10.30}$$

Thus, (10.28) can be rewritten as an equation for $f_{\mathrm{xc}}(\boldsymbol{q},\omega)$

$$\hat{f}_{\mathrm{xc}}(\boldsymbol{q},\omega) = \hat{\chi}_{\mathrm{KS}}^{-1}(\boldsymbol{q},\omega)\delta\hat{\hat{\chi}}(\boldsymbol{q},\omega)\hat{\chi}_{\mathrm{KS}}^{-1}(\boldsymbol{q},\omega) - \hat{\chi}_{\mathrm{KS}}^{-1}(\boldsymbol{q},\omega)\delta\hat{\hat{\chi}}(\boldsymbol{q},\omega)\hat{f}_{\mathrm{xc}}(\boldsymbol{q},\omega) . \tag{10.31}$$

The advantage of (10.31) is that, for any order of the perturbative expansion in W, it is possible to write an iterative form of the p-th order $[\hat{f}_{\mathrm{xc}}^{(p)}(\boldsymbol{q},\omega)]$ contribution to f_{xc}, i.e., $\hat{f}_{\mathrm{xc}}(\boldsymbol{q},\omega) = \sum_p \hat{f}_{\mathrm{xc}}^{(p)}(\boldsymbol{q},\omega)$, when all the lower order terms of the BSE are known:

$$\begin{aligned}\hat{f}_{\mathrm{xc}}^{(p)}(\boldsymbol{q},\omega) = \; & \hat{\chi}_{\mathrm{KS}}^{-1}(\boldsymbol{q},\omega)\delta\hat{\hat{\chi}}^{(p)}(\boldsymbol{q},\omega)\hat{\chi}_{\mathrm{KS}}^{-1}(\boldsymbol{q},\omega) \\ & - \sum_{m=1}^{p-1} \hat{\chi}_{\mathrm{KS}}^{-1}(\boldsymbol{q},\omega)\delta\hat{\hat{\chi}}^{(m)}(\boldsymbol{q},\omega)f_{\mathrm{xc}}^{(p-m)}(\boldsymbol{q},\omega) ,\end{aligned} \tag{10.32}$$

when $p > 1$, while the 1st order is simply given by

$$\hat{f}_{\mathrm{xc}}^{(1)}(\boldsymbol{q},\omega) = \hat{\chi}_{\mathrm{KS}}^{-1}(\boldsymbol{q},\omega)\delta\hat{\hat{\chi}}^{(1)}(\boldsymbol{q},\omega)\hat{\chi}_{\mathrm{KS}}^{-1}(\boldsymbol{q},\omega) . \tag{10.33}$$

The scheme proposed in [Reining 2002, Del Sole 2003, Sottile 2003] appears naturally as an approximate solution of (10.32). Moreover in the case of a fully invertible relation between f_{xc} and F, in (10.25), it can be easily shown that only the first order survives in (10.32). This shows the equivalence of the first-order version of (10.31) with the approach of Reining et al. [Reining 2002]. Such an equivalence was rather unexpected, in view of the different assumptions underlying the two methods.

At this stage, it is important to note the work done by Sottile et al. [Sottile 2005] in seeking alternative expressions for $f_{\mathrm{xc}}^{(1)}$, based on the flexibility in the definition of χ_{KS} in (10.33). The idea is to use, instead of χ_{KS}, a less singular function that, yielding the same optical spectra, reduces the numerical instabilities and complexity in evaluating $f_{\mathrm{xc}}^{(1)}$.

Another, correlated, many-body based approach to f_{xc} is through the Sham-Schlüter equation [Sham 1983] (SSE). We refer the reader to the original paper and subsequent extensions [van Leeuwen 1996] for mode details. Here we want to stress that this method links MBPT and TDDFT through the density operator (exact in both schemes). Thus, starting from an expression for the xc energy functional, a perturbative series for v_{xc} and f_{xc} is consistently derived. As showed by Stubner et al. [Tokatly 2002, Stubner 2004], the expression for $f_{\mathrm{xc}}^{(p)}$ given in (10.32) is consistent with the SSE when only irreducible single-particle propagators are considered in the perturbative expansion. More importantly, Stubner et al. derive an integral equation for f_{xc} (derived from the vertex function) that is equivalent to the iterative procedure of (10.32).

As noted by in [Sottile 2003], the first order $f_{\mathrm{xc}}^{(1)}$ allows for a straightforward comparison with the EXX kernel of Kim and Görling [Kim 2002a, Kim 2002b]. Assuming that the EXX-KS states are equal to the Hartree-Fock states, the EXX kernel f_{x} corresponds to $f_{\mathrm{xc}}^{(1)}$ with, however, *W replaced by the bare Coulomb interaction.* The empirical cutoff in [Kim 2002a, Kim 2002b] can therefore be seen as a way to simulate the missing screening, which would be contained in further correlation terms of their expansion.

10.3.3 The f_{xc} Perturbative Series: Convergence and Cancellations

The theoretical scheme presented in the last section allows to construct f_{xc} order by order. If the corresponding series converges, the polarization function calculated through the BSE is, by hypothesis, exactly reproduced by TDDFT, (10.19).

However, f_{xc}, to be useful, should be well described with only a few orders of the perturbative expansion, (10.32). Unfortunately, possible cancellations that can enhance the lower orders terms are difficult to prove from (10.32). We may note that, with the exception of the first order, all other contributions to f_{xc} $[\hat{f}_{\mathrm{xc}}^{(2)}(\omega), \hat{f}_{\mathrm{xc}}^{(3)}(\omega), \ldots]$ contain an even number of terms that, as we will see shortly, tend to cancel each other. But this is not enough, as the direct application of (10.32) leads to spurious oscillations in the calculated optical spectra. Those oscillations are moderate in the case of weakly interacting systems (like diamond) but they tend to destroy the spectra for the case of materials with strong electron-hole attraction, like LiF (showed in Fig. 10.2).

The intensity of the oscillations in $\epsilon(\omega)$ increases with the order of $\hat{f}_{\mathrm{xc}}^{(p)}$ and eventually gives rise to non-physical regions of negative absorption. The reason for this numerical pathology stems from the way the Bethe-Salpeter kernel acts on the spectra: (i) redistributing the optical oscillator strength, (ii) shifting rigidly the spectra to account for the diagonal of the Bethe-Salpeter kernel, $\Delta_{cv\boldsymbol{k}}(\boldsymbol{q}) = W_{cv\boldsymbol{k},cv\boldsymbol{k}}(\boldsymbol{q})$, that should vanish in the limit of infinite $\boldsymbol{k}$-point sampling. However, with a finite $\boldsymbol{k}$-point grid corresponding to a fully converged BSE spectrum, we get $\Delta_{\boldsymbol{q}\to\boldsymbol{0}} \sim -0.9\,\mathrm{eV}$ in LiF. The diagonal of $\hat{W}$ appears in f_{xc} through a series expansion in $\Delta_{\boldsymbol{q}}$, $\hat{f}_{\mathrm{xc}}^{\Delta}$ that can be isolated from (10.32). If we note that $\Delta_{cv\boldsymbol{k}}(\boldsymbol{q}) \sim \Delta_{\boldsymbol{q}}$ for all states $(cv\boldsymbol{k})$ we obtain

$$\hat{f}_{\mathrm{xc}}^{\Delta}(\boldsymbol{q},\omega) = -\sum_p \Delta_{\boldsymbol{q}}^p \frac{\partial^p}{\partial\omega^p}[\hat{\chi}^{\mathrm{KS}}(\boldsymbol{q},\omega)]^{-1} . \tag{10.34}$$

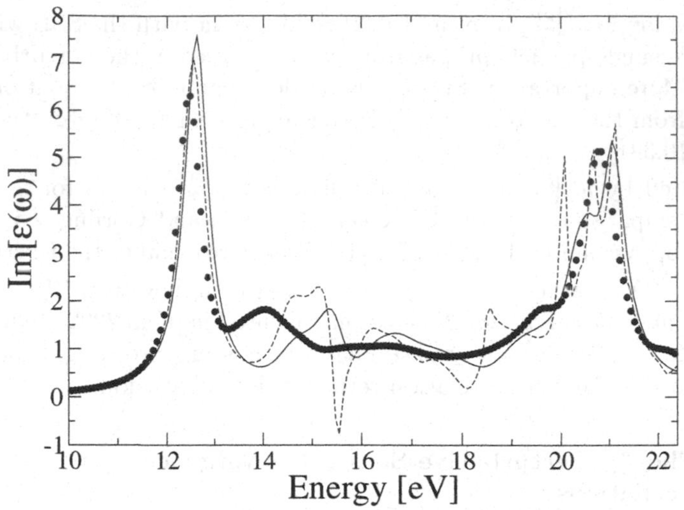

Fig. 10.2. Optical absorption spectra of LiF. *Dots*: BSE. Continuous curve: TDDFT using $f_{\mathrm{xc}}^{(1)}$. *Dashed curve*: TDDFT using $f_{\mathrm{xc}}^{(1)} + f_{\mathrm{xc}}^{(2)}$. While the first order f_{xc} reproduces quite well the main features of the BSE spectra, the second (and higher) order approximations to f_{xc} give unphysical negative absorption intensities

that is meaningful only when $\Delta_{\boldsymbol{q}}$ is sufficiently small. This series, in general, oscillates and cannot be treated perturbatively. As this is not the case for SiO_2 and LiF, natural oscillations are found in the naïve application of the perturbative expansion of (10.32). To circumvent this issue we included the diagonal part of $\hat{W}(\boldsymbol{q})$ into the independent QP response function $\hat{\chi}^{\mathrm{KS}}(\boldsymbol{q},\omega)$ and let $f_{\mathrm{xc}}^{(1)}$ account explicitly for the off-diagonal contributions to the Bethe-Salpeter kernel. Using this idea, the higher order corrections to $f_{\mathrm{xc}}^{(1)}$ are not only well defined, but numerically stable for all orders with the same $\boldsymbol{k}$-point sampling used in a standard BSE calculation.

The same holds for the description of the QP-shifts. Indeed, if we do not assume that the QP levels are equal to the KS ones, an additional contribution to $\Delta(\boldsymbol{q})$ comes from the difference $\chi_{\mathrm{QP}} - \chi_{\mathrm{KS}}$ in the form of a positive $\Delta_{\boldsymbol{q}}^{\mathrm{QP}}$, that is small if and only if the QP and the KS energies are very similar. The interpretation of the above result is straightforward: An important gap correction due to a remarkable difference between the QP and the KS energy levels *cannot* be described with a perturbative (read finite order) f_{xc}.

The clear instability of the f_{xc} series when gap correction terms are included is extremely meaningful. Even if we consider approaches like the SSE or EXX, where v_{xc} and f_{xc} are consistently derived, f_{xc} will always contain terms that renormalize the electronic gap. In the EXX, for example, the f_{x} includes a $\Delta(\boldsymbol{q})$ that is related to the difference between the Hartree-Fock self-energy and the EXX potential.

The last point we want to discuss in this paragraph is the numerical evaluation of $\delta\hat{\tilde{\chi}}^{(1)}(\omega)$ that constitutes the most cumbersome part of $f_{\mathrm{xc}}^{(1)}$:

$$[\delta\tilde{\chi}^{(1)}(\boldsymbol{q},\omega)]_{\boldsymbol{G},\boldsymbol{G}'} = -\frac{2}{\Omega N_k}\sum_{c_1v_1\boldsymbol{k}_1,c_2v_2\boldsymbol{k}_2} \xi^*_{c_1v_1\boldsymbol{k}_1}(\boldsymbol{q},\boldsymbol{G})\xi_{c_3v_3\boldsymbol{k}_3}(\boldsymbol{q},\boldsymbol{G}') \times[L_{\mathrm{KS}}(\boldsymbol{q},\omega)]_{c_1v_1\boldsymbol{k}_1} W_{c_1v_1\boldsymbol{k}_1,c_2v_2\boldsymbol{k}_2}[L_{\mathrm{KS}}(\boldsymbol{q},\omega)]_{c_2v_2\boldsymbol{k}_2}\,, \tag{10.35}$$

So we have to perform two square matrix, and two rectangular matrix multiplications for every frequency. However, by looking at the analytic properties of $\delta\hat{\tilde{\chi}}^{(1)}(\boldsymbol{q},\omega)$ (for an explicit expression see [Del Sole 2003]) we can single out the contribution of degenerate non-interacting electron-hole states in (10.35) and write a general expression for $f_{\mathrm{xc}}^{(1)}$ [Marini 2003b]:

$$\hat{f}_{\mathrm{xc}}^{(1)}(\boldsymbol{q},\omega) = \frac{2}{\Omega}\hat{\chi}_{\mathrm{KS}}^{-1}(\boldsymbol{q},\omega')\sum_{cv\boldsymbol{k}}\left[\frac{\hat{R}^{(\boldsymbol{q})}_{cv\boldsymbol{k}}+\hat{R}^{(\boldsymbol{q})\dagger}_{cv\boldsymbol{k}}}{\omega'-E^{(\boldsymbol{q})}_{cv\boldsymbol{k}}+\mathrm{i}0^+} + \frac{\hat{Q}^{(\boldsymbol{q})}_{cv\boldsymbol{k}}}{(\omega'-E^{(\boldsymbol{q})}_{cv\boldsymbol{k}}+\mathrm{i}0^+)^2}\right]\hat{\chi}_{\mathrm{KS}}^{-1}(\boldsymbol{q},\omega')\,. \tag{10.36}$$

Here, $\omega' = \omega + \Delta_{\boldsymbol{q}}$, $E^{(\boldsymbol{q})}_{cv\boldsymbol{k}} = \varepsilon^{\mathrm{QP}}_{c_1\boldsymbol{k}_1} - \varepsilon^{\mathrm{QP}}_{v_1\boldsymbol{k}_1-\boldsymbol{q}}$, and the sum runs through all independent electron-hole states with residual

$$[R^{(\boldsymbol{q})}_{cv\boldsymbol{k}}]_{\boldsymbol{G}_1,\boldsymbol{G}_2} = \sum_{\substack{c'v'\boldsymbol{k}'\\ E^{(\boldsymbol{q})}_{c'v'\boldsymbol{k}'}\neq E^{(\boldsymbol{q})}_{cv\boldsymbol{k}}}} \frac{\xi^*_{cv\boldsymbol{k}}(\boldsymbol{q},\boldsymbol{G}_1)W_{cv\boldsymbol{k},c'v'\boldsymbol{k}'}(\boldsymbol{q})\xi_{c'v'\boldsymbol{k}'}(\boldsymbol{q},\boldsymbol{G}_2)}{E^{(\boldsymbol{q})}_{cv\boldsymbol{k}}-E^{(\boldsymbol{q})}_{c'v'\boldsymbol{k}'}}\,, \tag{10.37}$$

for non-degenerate states, and

$$[Q^{(\boldsymbol{q})}_{cv\boldsymbol{k}}]_{\boldsymbol{G}_1,\boldsymbol{G}_2} = \sum_{\substack{c'v'\boldsymbol{k}'\\ E^{(\boldsymbol{q})}_{c'v'\boldsymbol{k}'}= E^{(\boldsymbol{q})}_{cv\boldsymbol{k}}}} \xi^*_{cv\boldsymbol{k}}(\boldsymbol{q},\boldsymbol{G}_1)W_{cv\boldsymbol{k},c'v'\boldsymbol{k}'}(\boldsymbol{q})\xi_{c'v'\boldsymbol{k}'}(\boldsymbol{q},\boldsymbol{G}_2)\,, \tag{10.38}$$

for degenerate states. Equation (10.36) is very fast to compute as it has the form of a non-interacting polarization function with modified residuals (Q, R) that are evaluated only once as a result of two simple matrix-vector multiplications. Also, (10.36) can be made causal and be extended to higher orders of the perturbative expansion of f_{xc}.

10.4 The Vertex Function $\tilde{\Gamma}$: a TDDFT-Based Approach

The validity of the GWA is based on the physical argument that the long-range collective excitations dominate the screening process. Consequently, the screened interaction is assumed to be at the RPA level and any effect due to the direct electron-hole interaction (short-range effects) is neglected.

However, we have seen how strong W can be, for example, in insulators, where $\tilde{\chi}$ largely deviates from $\tilde{\chi}_{\mathrm{RPA}}$. This corresponds to a non negligible kernel in the BSE, (10.17), that modifies both the polarization function, (10.12) and the self-energy operator, (10.10). The interplay between those two effects has been strongly debated in the last years, using different approximations for $\tilde{\Gamma}$, and different levels of self-consistency in the solution of the Dyson equation. Regarding the vertex function, two main classes of approximations can be identified: the first reduces the spatial non-locality of $\tilde{\Gamma}$ (the vertex is a three-point function) using two point DFT-based expressions [Mahan 1989, Hindgren 1997, Del Sole 1994], while the second takes only the first order terms in the vertex expansion [Bobbert 1994, Ummels 1994, Shirley 1996]. While those approximations can be justified in the case of the homogeneous electron gas or simple semiconductors, they are inadequate in the case of wide-gap insulators (e.g., LiF). To discuss more extensively this point, we will present recent approaches to $\tilde{\Gamma}$ that, based on a "cooperation" of MBPT and TDDFT, respect the spatial and dynamical properties of the vertex function, without truncating its perturbative expansion.

We start the derivation recalling the definition of the irreducible response function in terms of $\tilde{\Gamma}$, (10.12)

$$\tilde{\chi}(1,2) = -\mathrm{i}\int \mathrm{d}3 \int \mathrm{d}4\, G(1,3)G(4,1)\tilde{\Gamma}(34,2)\,. \tag{10.39}$$

Even if $\tilde{\Gamma}$ is an highly non-local, three-point function, we have seen in Sect. 10.3.2 that, as long as we are interested in the two-point polarization function $\tilde{\chi}$, (10.39) can be cast in terms of f_{xc}

$$\tilde{\chi}(1,2) = \chi_{\mathrm{KS}}(1,2) + \int \mathrm{d}3 \int \mathrm{d}4\, \chi_{\mathrm{KS}}(1,3) f_{\mathrm{xc}}(3,4)\tilde{\chi}(4,2)\,. \tag{10.40}$$

At this point if we take the xc potential corresponding to f_{xc} as a local approximation to the self-energy, then the vertex function can be easily contracted into a two-point function: $\tilde{\Gamma}(67,3) \equiv \tilde{\Gamma}_{\mathrm{loc}}(6,3)\delta(6,7)$ [Mahan 1989, Hindgren 1997, Del Sole 1994], with

$$\tilde{\Gamma}_{\mathrm{loc}}(1,2) = \left[\delta(1,2) - \int \mathrm{d}3\, f_{\mathrm{xc}}(1,3)\chi_{\mathrm{KS}}(3,2)\right]^{-1}\,. \tag{10.41}$$

Thus (10.10) gives

$$\Sigma(1,2) = \mathrm{i}W_{\mathrm{TDDFT}}(1^+,2)G_{\mathrm{KS}}(1,2)\,, \tag{10.42}$$

in terms of the TDDFT effective potential

$$W_{\mathrm{TDDFT}}(1,2) = \int \mathrm{d}3\, v_{\mathrm{ee}}(1,3) \times \left\{\delta(3,2) - \int \mathrm{d}4\, \left[v_{\mathrm{ee}}(3,4) + f_{\mathrm{xc}}(3,4)\right]\chi_{\mathrm{KS}}(4,2)\right\}^{-1} \tag{10.43}$$

This expression for the self-energy is extremely appealing and simple to implement in actual calculations [Del Sole 1994]. As an example we take the QP lifetimes, that, for a generic conduction state c with momentum $\boldsymbol{k}$ is given by

$$\tau_{c\boldsymbol{k}}^{-1} = -2\Omega^{-1} \sum_{\boldsymbol{G}_1,\boldsymbol{G}_2} \sum_{\boldsymbol{q},c'} \xi_{cc'\boldsymbol{k}}(\boldsymbol{q}\boldsymbol{G}_1)\xi^*_{cc'\boldsymbol{k}}(\boldsymbol{q}\boldsymbol{G}_2) \times \Im[W_{\mathrm{TDDFT}}(\boldsymbol{q}, \varepsilon_{c\boldsymbol{k}} - \varepsilon_{c'\boldsymbol{k}-\boldsymbol{q}})]_{\boldsymbol{G}_1\boldsymbol{G}_2}\,, \tag{10.44}$$

Thus, using the f_{xc} given in (10.33), we would have a simple expression for Σ that also corresponds to an accurate response function. This is an important difference compared to previous expressions for $\tilde{\Gamma}_{\mathrm{loc}}$ [Hindgren 1997, Mahan 1989, Del Sole 1994] that produce optical spectra very similar to the RPA.

A crucial property of f_{xc} is its long-range behavior,

$$f_{\mathrm{xc}}(\boldsymbol{r}, \boldsymbol{r}', \omega) \sim -\frac{\alpha(\omega)}{|\boldsymbol{r} - \boldsymbol{r}'|}\,. \tag{10.45}$$

The stronger the electron-hole effects are, the larger is the correction embodied in α. In the case of wide-gap insulators like LiF there is a large region of frequencies and transfer momenta $\boldsymbol{q}$ where f_{xc} is stronger than the Hartree term (i.e., $\alpha > 1$). This leads to unphysical linewidths: for a large energy range, $\Im\Sigma$, and hence τ^{-1}, has the wrong sign! This result is visualized by noticing that with respect to a GW calculation a change of sign of τ^{-1} is controlled by $\mathrm{sign}(v_{\mathrm{ee}} + f_{\mathrm{xc}})$, that is proportional to $[1 - \alpha(\omega)]$. A similar result was obtained in [Hindgren 1997] looking at the high $\boldsymbol{q}$ limit of the TDLDA kernel that goes as $f_{\mathrm{xc}} \sim q^2$. The reason for this important failure of the two-point vertex function is connected to the imposed reduction of the non-locality from the original, three-point vertex function. In physical terms, $\tilde{\Gamma}_{\mathrm{loc}}$ overestimates the intensity of the vertex correction because two incoming particles (entering in 1 and 2 in the exact vertex function $\tilde{\Gamma}$) are supposed to coexist at the same space-time point.

To overcome this difficulty, we have to release the constraint on the spatial locality, and define a TDDFT vertex function $\tilde{\Gamma}_{\mathrm{TDDFT}}(12,3)$ such that, for a given approximation of order n in W, $f_{\mathrm{xc}}^{(p)}(1,2)$, $\tilde{\Gamma}_{\mathrm{TDDFT}}$ is consistent with (10.40). By inspecting (10.39) and (10.40) we obtain

$$\tilde{\Gamma}^{(1)}_{\mathrm{TDDFT}}(12,3) \equiv \delta(1,2)\delta(2,3) + \mathrm{i}W(1,2)\int \mathrm{d}4\, G(1,4)G(4,2)\tilde{\Gamma}_{\mathrm{loc}}(4,3)\,, \tag{10.46}$$

It is crucial to observe that $\tilde{\Gamma}^{(1)}_{\mathrm{TDDFT}}$ *is not* a first order vertex, as $\tilde{\Gamma}_{\mathrm{loc}}$, but it sums an infinite number of diagrams. Equation (10.46) can be easily generalized to give higher order approximations for $\tilde{\Gamma}$, consistent with the high order corrections to f_{xc}, $f_{\mathrm{xc}}^{(p)}$.

As it is commonly done, we neglect dynamical effects in the BSE [Marini 2003a], i.e., we assume $W(1,2) \approx W(\boldsymbol{r}_1, \boldsymbol{r}_2, \omega = 0)$ in (10.46). This static

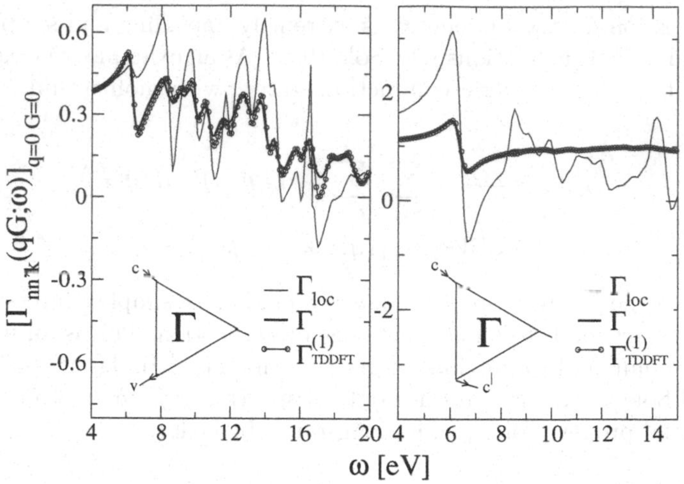

Fig. 10.3. LiF: the exact vertex function $\tilde{\Gamma}$ is compared with the local approximation $\tilde{\Gamma}_{\mathrm{loc}}$ and the first-order TDDFT-based $\tilde{\Gamma}^{(1)}_{\mathrm{TDDFT}}$ in two of the possible scattering configurations. *Left panel*: the vertex describes an electron-hole scattering, involved in the BSE dynamics of an electron-hole pair (polarization function). *Right panel*: the vertex describes an electron-electron scattering, typical of the self-energy process. The dramatically failure of $\tilde{\Gamma}_{\mathrm{loc}}$ in the right panel, that overestimates the true vertex function, shows the crucial importance of the vertex spatial non-locality

approximation has been showed to be well-motivated when the BSE is used to calculate χ [Marini 2003a], while there are no *a priori* reasons to confirm its validity in the vertex case, it remains a reasonable approximation in the present case as the dynamical part of W is not excited when we consider only low-energy excitations in Σ. This analysis is confirmed by an important property of $\tilde{\Gamma}^{(p)}_{\mathrm{TDDFT}}$ that can be devised from (10.46) when a dynamical W is used. From [Marini 2003a] we know that the BSE kernel embodies the correct *unscreening* effect that undresses the QP renormalization factors Z entering the response function. The very same property, extended to $\tilde{\Gamma}$ gives the correct $1/Z$ behavior expected on the basis of the Ward identities [Strinati 1988]. This property suggests and confirms the stringent relation between the dynamical properties of $\tilde{\Gamma}$ and the self-consistent solution of the Dyson equation.

The performance of $\tilde{\Gamma}^{(1)}_{\mathrm{TDDFT}}$ is exemplified in Fig. 10.3, where the quantity

$$\tilde{\Gamma}_{nn'\boldsymbol{k}}(\boldsymbol{q},\boldsymbol{G},\omega) \equiv \int \mathrm{d}t \int \mathrm{d}^3r_1 \int \mathrm{d}^3r_2\, \mathrm{e}^{\mathrm{i}[\omega t+(\boldsymbol{q}+\boldsymbol{G})\cdot \boldsymbol{r}_3]} \varphi^*_{n\boldsymbol{k}}(\boldsymbol{r}_1)\varphi_{n'\boldsymbol{k}-\boldsymbol{q}}(\boldsymbol{r}_2)\tilde{\Gamma}(\boldsymbol{r}_1\boldsymbol{r}_2,\boldsymbol{r}_3,t) \quad (10.47)$$

has been introduced. $\tilde{\Gamma}_{nn'\boldsymbol{k}}$ describes the scattering amplitude of two states $|n\boldsymbol{k}\rangle$, $|n'\boldsymbol{k}-\boldsymbol{q}\rangle$ whose dynamics are governed by the BSE. In the case of

$(n, n') = (c, v)$ ([left panel of Fig. 10.3]) the process is an electron-hole attraction, relevant to the polarization function. From Fig. 10.3 it is clear the $\tilde{\Gamma}_{\rm loc}$ is an excellent approximation to the real vertex $\tilde{\Gamma}$. This confirms and provides a deeper explanation of the cancellations we have seen in Sect. 10.3.2 that appear in the $f_{\rm xc}$ perturbative expansion. However, in the case of $(n, n') = (c, c')$ ([right panel of Fig. 10.3]) $\tilde{\Gamma}_{\rm loc}$ dramatically overestimates the real vertex amplitude (as discussed above) while $\tilde{\Gamma}^{(1)}_{\rm TDDFT}$ is still a very good approximation to $\tilde{\Gamma}$. There is a clear message coming from Fig. 10.3: Even if the vertex function entering the BSE for the response function is analytically identical to the vertex entering the self-energy, the processes described in the two cases are completely different. Consequently, any approximation to $\tilde{\Gamma}$ derived in a TDDFT context, even if corresponding to a correct response function, must be carefully compared with the real vertex, and the approximation involved (especially the imposed spatial nonlocality) must be attentively verified.

The expression for Σ corresponding to $\tilde{\Gamma}^{(1)}_{\rm TDDFT}$ can be calculated explicitly [Marini 2004]. The effect of the vertex corrections on the QP lifetimes of LiF is showed in Fig. 10.4. The overall effect is huge: the linewidths up to 3 eV above the forbidden region display a linear dependence with energy while in the RPA are almost zero because of the slow rise of the RPA loss function. A

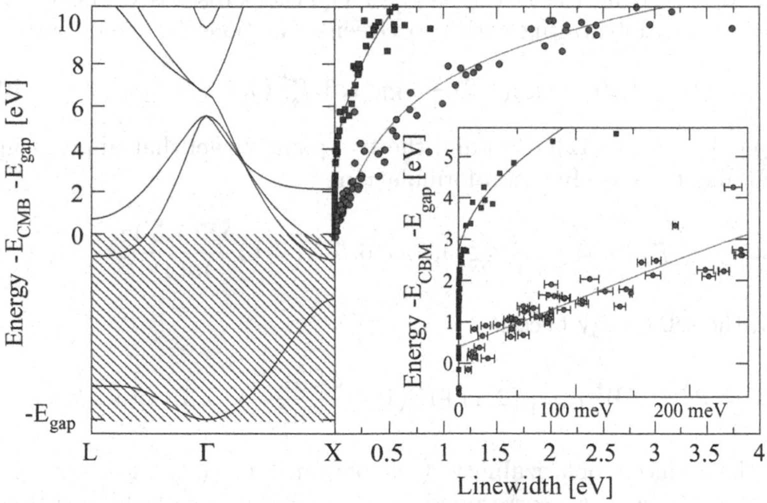

Fig. 10.4. *Left panel*: calculated DFT band-structure of LiF (here $E_{\rm CBM}$ and $E_{\rm gap}$ stand for the DFT conduction band minimum energy and the energy gap). *Right panel*: Electron linewidths calculated "on mass-shell" as function of the single-particle energy. *Boxes*: RPA G_0W_0. *Circles*: TDDFT-based vertex correction to the self-energy, i.e, a $G_0W\tilde{\Gamma}^{(1)}_{\rm TDDFT}$ approach that turn out to be very close to a simpler G_0W calculation (see text). The *dashed* area denotes the forbidden energy region for quasiparticle decay into electron-hole pairs. Error bars represent the theoretical uncertainty. (Adapted from [Marini 2004])

similar energy dependence has been observed in highly correlated materials [Smith 2001]. Instead, the present linear dependence of the linewidths is due to the combination of an almost constant density of states close to the conduction band minimum and to a "step-like" energy dependence of the loss function.

10.4.1 Including Density-Functional Concepts into MBPT

In this section we shortly review a very important step towards the rationalization of previous results, as well as to how to incorporate some concepts from density-functional theory into MBPT, in order to derive simplify expression for the electron self-energy and vertex corrections. The theory has been presented in [Bruneval 2005]. The fundamental idea is that the essential physics to describe excitations is captured in the density variations. Therefore, one can use an alternative chain rule in Hedin's equations to express $\delta\Sigma/\delta V$, namely $(\delta\Sigma/\delta n)(\delta n/\delta V)$. By doing that, the equation for the vertex looks like

$$\tilde{\Gamma}(12,3) = \delta(1,3)\delta(2,3) + \frac{\delta\Sigma(1,2)}{\delta n(4)}\tilde{\chi}(4,3)\,, \tag{10.48}$$

where $\tilde{\chi} = \delta n/\delta V$ is nothing but the irreducible polarizability that is usually calculated by solving the Bethe-Salpeter equations discussed above for the vertex function. Integrating with two Green's functions G, one obtains

$$\tilde{\chi}(1,2) = \chi_{\mathrm{KS}}(1,2) + \chi_{\mathrm{KS}}(1,3)f_{\mathrm{xc}}^{\mathrm{eff}}(3,4)\tilde{\chi}(4,2)\,, \tag{10.49}$$

with $\chi_{\mathrm{KS}}(12) = -\mathrm{i}G(12)G(21)$ and the two-point kernel that appears in the TDDFT-like response function formulation

$$f_{\mathrm{xc}}^{\mathrm{eff}}(3,4) = -\mathrm{i}\chi_{\mathrm{KS}}^{-1}(3,6)G(6,5)G(5',6)\frac{\delta\Sigma(5,5')}{\delta n(4)}\,. \tag{10.50}$$

Finally, the self-energy becomes

$$\Sigma(1,2) = \mathrm{i}G(1,2)W_{\mathrm{TC-TC}}(2,1) + \mathrm{i}G(1,4)\frac{\delta\Sigma(4,2)}{\delta n(5)}\chi(5,3)v_{\mathrm{ee}}(3,1^+)\,. \tag{10.51}$$

where the reduced polarizability χ is obtained by $\chi = \tilde{\chi} + \tilde{\chi}v_{\mathrm{ee}}\chi$, and $W_{\mathrm{TC-TC}} = (1+v_{\mathrm{ee}}\chi)v_{\mathrm{ee}}$ corresponds to the test-charge test-charge screening function [not to be confused with the test-charge test-electron screen function W^{TDDFT} discussed above in the context of a perturbative vertex function $\tilde{\Gamma}$ in (10.46)]. Following [Bruneval 2005], it is convenient to reformulate the vertex equation as

$$\tilde{\Gamma}(12,3) = \delta(1,3)\delta(2,3) + \delta(1,2)f_{\mathrm{xc}}(1,4)\tilde{\chi}(4,3) + \Delta\tilde{\Gamma}(12,3)\,, \tag{10.52}$$

where

$$\Delta\tilde{\Gamma}(12,3) = \left\{\frac{\delta\Sigma(1,2)}{\delta n(4)} - \delta(1,2) f_{\text{xc}}(1,4)\right\}\tilde{\chi}(4,3)\,. \tag{10.53}$$

The dominant terms in the polarization function dynamics are in fact contained in the first two (one- and two-point) contributions to $\tilde{\Gamma}$. On the contrary, as discussed in Sect. 10.4, the highly non–local character of $\tilde{\Gamma}$ cannot be neglected in the self–energy operator. Nevertheless, (10.51) is a promising path to go beyond *GW* for many interesting applications as it represents a consistent approach to derive efficient approximations to the vertex function.

From the previous section, it is now clear that the TDDFT f_{xc} kernel consists of two terms, namely $f_{\text{xc}}^{(1)}$ and $f_{\text{xc}}^{(2)}$: $f_{\text{xc}}^{(1)}$ changes the Kohn-Sham response function into the independent QP one (band gap problem); $f_{\text{xc}}^{(2)}$ accounts for the electron-hole interaction [Tokatly 2002, Stubner 2004]. By using the *GW* approximation for the self-energy, as it is customary done in the MBPT calculations, we can reproduce an approximate xc kernel [Bruneval 2005] that accounts for the electron-hole part, and looks like

$$\begin{aligned} f_{\text{xc}}^{(2)}(3,4) &= f_{\text{xc}}^{\text{eff}}(3,4) \\ &= \chi_{\text{KS}}^{-1}(3,6)G(6,5)G(5',6)W(5,5')G(5,7)G(7,5')\chi_{\text{KS}}^{-1}(7,4)\,. \end{aligned} \tag{10.54}$$

This kernel coincides with the approximated kernels presented in previous works [Marini 2003b, Adragna 2003, Reining 2002, Sottile 2003], derived here in a more elegant, DFT–based approach.

10.5 Conclusions and Perspectives

In this chapter we have reviewed several promising attempts to include Many–Body effects in the TDDFT xc kernel. As it will be shown in 20 already a first order, simplified kernel provides a very good description of excitonic effects in both extended as well as low dimensional systems. As the present theory stands, the main drawback is the use of the quasiparticle wavefunctions and eigenstates for building the kernel. A major step forward would be to device an f_{xc} that describes both effects simultaneously and that can be derived from the functional derivative of a given action functional. Work along those lines is in progress in different groups and we expect some important results to come up in the near future.

In addition to the "standard" use of the TDDFT kernel within the linear response regime we have also discussed how MBPT can benefit of the simple two-points structure of f_{xc} to derive efficient expressions for the self-energy vertex function. These expressions are promising, as they constitute non perturbative schemes to go beyond the usual GW approximation.

In conclusion, the main goal of this chapter is to illustrate that TDDFT and MBPT can be fruitfully combined to provide a better description of correlation effects both in single and in the two-particles dynamics with important application in the predicting power of the *ab-initio* techniques.

11 Exact Conditions

K. Burke

11.1 Introduction

In this chapter, we collect and discuss several of the exact conditions known about time-dependent density functional theory. The subject of TDDFT is still much less developed than its ground-state counterpart, and this is reflected in the number and usefulness of exact conditions known.

11.2 Review of the Ground State

In ground-state DFT, the unknown exchange-correlation energy functional, $E_{\text{xc}}[n]$, plays a crucial role. In fact, it is really this energy that we typically wish to approximate with some given level of accuracy and reliability, and *not* the density itself. Using such an approximation in a modern Kohn-Sham ground-state DFT calculation, we can calculate the total energy of any configuration of the nuclei of the system, and so extract the bond lengths and angles of molecules, and deduce the lowest energy lattice structure of solids. We can also extract forces in simulations, vibrational frequencies, phonons, and bulk moduli. We can also discover response properties to both external electric fields and magnetic fields (using spin DFT). The accuracy of the self-consistent density is irrelevant to most of these uses.

Given the central role of the energy, it makes sense to devote much effort to its study as a density functional. Knowledge of its behavior in various limits can be crucial to restraining and constructing accurate approximations, and to understanding their limitations. This task is greatly simplified by the fact that the total ground-state energy satisfies the variational principle. Many exact conditions use this in their derivation.

Let us quickly review some of the more prominent exact conditions. They almost all concern the energy functional, which, as mentioned above, is crucial for good KS-DFT calculations. We do not give original references here, but refer the interested reader to [Perdew 2003] for a thorough discussion.

11.2.1 Basic Conditions

These conditions are very elementary, and any sensible approximation should satisfy them.

K. Burke: *Exact Conditions*, Lect. Notes Phys. **706**, 181–197 (2006)
DOI 10.1007/3-540-35426-3_11

- **Coordinate scaling:** By defining $n_\gamma(\boldsymbol{r}) = \gamma^3\, n(\gamma \boldsymbol{r})$, one can easily show

$$E_{\rm x}[n_\gamma] = \gamma\, E_{\rm x}[n]\,. \tag{11.1}$$

- **Virial theorem:**

$$E_{\rm xc} + T_{\rm c} = -\int {\rm d}^3r\, n(\boldsymbol{r})\, \boldsymbol{r}\cdot\nabla v_{\rm xc}(\boldsymbol{r})\,, \tag{11.2}$$

where $T_{\rm c}$ is the kinetic contribution to the correlation energy.
- **Coupling constant:** In DFT, one imagines varying the strength of the electron-electron repulsion, while keeping the density fixed, and defines quantities as a function of λ. One finds:

$$E_{\rm xc}^{\lambda}[n] = \lambda^2 E_{\rm xc}[n_{(1/\lambda)}]\,, \tag{11.3}$$

i.e., altering the coupling constant is simply related to scaling the density.
- **Adiabatic connection formula:** By using the Hellmann-Feynman theorem, one can show:

$$E_{\rm c}[n] = \int_0^1 {\rm d}\lambda\, U_{\rm c}^{\lambda}[n]/\lambda\,, \tag{11.4}$$

where $U_{\rm c}^{\lambda}$ is the potential contribution to the correlation energy.

11.2.2 Finite Systems

The next set of conditions are derived for finite systems, just as the Hohenberg-Kohn theorem is.

- **Coordinate scaling:** Coordinate scaling of the correlation is less simple than exchange:

$$\begin{aligned} E_{\rm c}[n_\gamma] &> \gamma\, E_{\rm c}[n] && (\gamma > 1) \\ E_{\rm c}[n_\gamma] &= E_{\rm c}^{(2)}[n] + E_{\rm c}^{(3)}[n]/\gamma + \cdots && (\gamma \to \infty) \\ E_{\rm c}[n_\gamma] &= \gamma B[n] + \gamma^{3/2} C[n] + \cdots && (\gamma \to 0)\,, \end{aligned} \tag{11.5}$$

where $E_{\rm c}^{(2)}[n]$, $E_{\rm c}^{(3)}[n]$, $B[n]$ and $C[n]$ are all scale-invariant functionals. Not all popular approximations satisfy these conditions.
- **Self-interaction:** For any one-electron system,

$$E_{\rm x}[n] = -E_{\rm H}[n], \quad E_{\rm c} = 0 \qquad (N = 1), \tag{11.6}$$

where $E_{\rm H}$ is the Hartree energy.

- **Asymptotic behavior of potential:** Far from a Coulombic system

$$v_{xc}(\mathbf{r}) \to -1/r \quad (r \to \infty), \tag{11.7}$$

and

$$\varepsilon_{HOMO} = -I, \tag{11.8}$$

where ε_{HOMO} is the eigenenergy of the highest occupied KS molecular orbital, and I the ionization potential. These results are intimately related to the self-interaction of one electron.
- **Lieb-Oxford bound:** For any density,

$$E_{xc} \geq 2.273\ E_x^{LDA}. \tag{11.9}$$

11.2.3 Uniform and Nearly Uniform Gas

This last set of conditions involve the properties of the uniform or nearly uniform electron gas.

- **Uniform density:** When the density is uniform, $E_{xc} = e_{xc}^{unif}(n)\ \mathcal{V}$, where $e_{xc}^{unif}(n)$ is the xc energy density of a uniform electron gas of density n, and $\mathcal{V}$ is the volume.
- **Slowly varying density:** For slowly varying densities, E_{xc} should recover the gradient expansion.
- **Linear response of uniform gas:** Another generic limit is when a weak perturbation is applied to a uniform gas, and the resulting change in energy is given by the static response function, $\chi(q, \omega = 0)$. This function is known from accurate quantum Monte Carlo calculations [Moroni 1995], and approximations can be tested against it.

11.2.4 Finite Versus Extended Systems

Note the distinction above between extended systems (like the uniform gas or any bulk system) and finite systems. The basic theorems of DFT are proven for *finite* quantum mechanical systems, with densities that decay at large distances from the center. Their extension to extended systems, even those as simple as the uniform gas, requires careful thought. For ground-state properties, one can usually take results directly to the extended limit without change, but not always. For example, the high-density limit of the correlation energy in (11.5) fails for a uniform gas.

11.2.5 Types of Approximations

Despite a plethora of approximations [Perdew 2005], no present-day approximation satisfies all these conditions, and so one chooses which conditions to impose on a given approximate form. Nonempirical approaches attempt to fix all parameters via exact conditions [Perdew 1996a, Perdew 1996b], while good empirical approaches might include one or two parameters that are fit to some data set [Becke 1988b, Lee 1988, Becke 1993b].

There are two basic flavors of approximations: pure density functionals, which are often designed to meet conditions on the uniform gas, and orbital-dependent functionals [Grabo 1998], which meet the finite-system conditions more naturally. The most sophisticated approximations being developed today use both [Tao 2003].

11.3 Conditions and Approximations

The time-dependent problem is much more diverse than the ground-state problem, making the known exact conditions more difficult to classify. We make the basic distinction between general time-dependent perturbations, of arbitrary strength, and weak fields, where linear response applies. The former give conditions on the xc potential for all time-dependent densities, the latter yield conditions directly on the xc kernel, which is a functional of the ground-state density alone. Of course, all of the former also yield conditions in the special case of weak fields.

11.3.1 Role of the Energy

In the time-dependent problem, we do not have the energy playing a central role. Formally, the action plays an analogous role (see Chap. 2), but in practice, we never evaluate the action in TDDFT calculations (as it is identically zero on the real time evolution). In TDDFT, our focus is truly the time-dependent density itself, and so, by extension, the potential determining that density. Thus many of our conditions are in terms of the potential.

Note also that most pure *density* functionals for the ground-state problem produce poor approximations for the details of the potential. Such approximations work well only for quantities integrated over space, such as the energy. Thus approximations that work well for ground-state energies are sometimes very poor as adiabatic approximations in TDDFT. For example their failure to satisfy (11.7) leads to large errors in the KS energies of higher-lying orbitals.

In place of the energy, there are a variety of physical properties that people wish to calculate. On the one hand, quantum chemists are most often focused on first few low-lying excitations, which might be crucial for determining the photochemistry of some biomolecule. Then the adiabatic generalization of standard ground-state approximations is often sufficient (see Chaps. 17, 22 and 23). At the other extreme, very often people who study matter in strong laser fields are very focused on ionization probabilities, and there the violation of (11.8) makes density approximations too crude, and requires orbital-dependent approximations instead.

11.3.2 Approximations

As we go through the various exact conditions, we will discuss whether the simplest approximations in present use satisfy them. The most important are:

- **ALDA:** Adiabatic local density approximation, the simplest pure density functional, commonly used in many calculations, and described in Chap. 1.
- **AA:** "Exact" adiabatic approximation, in which we imagine using the exact ground-state potential for a given density (see Chap. 1).
- **EXX:** Exact exchange, the orbital-dependent functional, treated as an implicit density functional (see Chap. 9).

A key aim of today's methodological development is to build in correlation memory effects. Exact exchange (for more than two unpolarized electrons) has some memory when considered as a density functional. We will discuss two attempts at memory inclusion, both limited to the linear response regime:

- **GK:** The Gross-Kohn approximation is simply to use the local frequency-dependent kernel of the uniform gas, $f_{\rm xc}^{\rm unif}(q \to 0, \omega)$ instead of its adiabatic limit as used in ALDA.
- **VK:** The Vignale-Kohn approximation is simply the gradient expansion in the current density for a slowly-varying gas (see Chap. 5).

The approximations suggested in the rest of this Part (Chaps. 8 and 10) could be tested for satisfaction of the conditions below, and perhaps improved.

11.4 General Conditions

In this section, we discuss conditions that apply no matter how strong the time-dependent potential is.

11.4.1 Adiabatic Limit

The most essential exact condition in TDDFT is the adiabatic limit. For any finite system, or an extended system with a finite gap, the deviation from the instantaneous ground-state during a perturbation (of arbitrary strength) can be made arbitrarily small, by slowing down the time-evolution, i.e., if the perturbation is $v(t)$, replacing it by $v(t/\tau)$ and making τ sufficiently large. This is the adiabatic theorem of quantum mechanics.

Similarly, as the time-dependence becomes very slow (or equivalently, as the frequency becomes small), for such system the functionals reduce to their ground-state counterparts:

$$v_{\rm xc}(\boldsymbol{r},t) \to v_{\rm xc}[n(t)](\boldsymbol{r}) \qquad (\tau \to \infty) \tag{11.10}$$

where $v_{\text{xc}}[n](\boldsymbol{r})$ is the exact ground-state xc potential of density $n(\boldsymbol{r})$.

Clearly, any adiabatic approximation satisfies this theorem, and so also does EXX, by reducing to their ground-state analogs for slow variations. On the other hand, if an approximation to $v_{\text{xc}}(\boldsymbol{r},t)$ were devised that was not based on ground-state DFT, this theorem could be used in reverse to *define* the corresponding ground-state functional.

11.4.2 Equations of Motion

In this section, we discuss some elementary conditions that any reasonable TDDFT approximation should satisfy. Because almost all approximations do satisfy these conditions, they are best applied as tests of propagation schemes (see Part III). Satisfaction of these conditions by propagation schemes can be used to interpret their quality. For schemes that do not automatically satisfy a given condition, then a numerical check of its validity provides a test on the accuracy of the solution. A simple analog is the check of the virial theorem in ground-state DFT in a finite basis.

These conditions are all found via a very simple procedure. They begin with some operator that depends only on the time-dependent density, such as the total force on the electrons. The equations of motion for the operator in both the interacting and the KS systems are written down, and subtracted. Since the time-dependent density is the same in both systems, the difference vanishes. Usually, the Hartree term also separately satisfies the resulting equation, and so can be subtracted from both sides, yielding a condition on the xc potential alone. This procedure is well-described in Chap. 5 for the zero xc force theorem.

Zero xc Force and Torque: These are very simple conditions saying that interaction among the particles cannot generate a net force [Vignale 1995a, Vignale 1995b]:

$$\int \mathrm{d}^3r\, n(\boldsymbol{r},t)\, \nabla v_{\text{xc}}(\boldsymbol{r},t) = 0 \tag{11.11a}$$

$$\int \mathrm{d}^3r\, n(\boldsymbol{r},t)\, \boldsymbol{r}\times\nabla v_{\text{xc}}(\boldsymbol{r},t) = \int \mathrm{d}^3r\, \boldsymbol{r}\times\frac{\partial \boldsymbol{j}_{\text{xc}}(\boldsymbol{r},t)}{\partial t}\,, \tag{11.11b}$$

where $\boldsymbol{j}_{\text{xc}}(\boldsymbol{r},t)$ is the difference between the interacting current density and the KS current density [van Leeuwen 2001]. The second condition says that there is no net xc torque, *provided* the KS and true current densities are identical. This is not guaranteed in TDDFT (but is in TDCDFT).

xc Power and Virial: By applying the same methodology to the equation of motion for the Hamiltonian, we find [Hessler 1999]:

$$\int \mathrm{d}^3r\, \frac{\mathrm{d}n(\boldsymbol{r},t)}{\mathrm{d}t}\, v_{\text{xc}}(\boldsymbol{r},t) = \frac{\mathrm{d}E_{\text{xc}}}{\mathrm{d}t}\,. \tag{11.12}$$

while another equation of motion yields the virial theorem, which intriguingly has the exact same form as in the ground state, (11.2):

$$E_{\mathrm{xc}}[n](t) + T_{\mathrm{c}}[n](t) = -\int \mathrm{d}^3r\, n(\boldsymbol{r},t)\, \boldsymbol{r}\cdot\nabla v_{\mathrm{xc}}[n](\boldsymbol{r},t)\,. \qquad (11.13)$$

These conditions are so basic that they are trivially satisfied by any reasonable approximation, including the ALDA, AA, and EXX. Thus they are more useful as detailed checks on a propagation scheme. The correlation contribution to the latter is very small, and makes a very demanding test. But because the energy does not play the same central role as in the ground-state problem (and the action is *not* simply the time-integral of the energy – see Chap. 2), that is all they are used for so far.

11.4.3 Self-Interaction

For any One-Electron System,

$$v_{\mathrm{x}}(\boldsymbol{r},t) = -\int \mathrm{d}^3r'\, \frac{n(\boldsymbol{r},t')}{|\boldsymbol{r}-\boldsymbol{r}'|}, \quad v_{\mathrm{c}}(\boldsymbol{r},t) = 0, \qquad (N=1)\,. \qquad (11.14)$$

These conditions are automatically satisfied by EXX, but are violated by ALDA, GK, and VK.

11.4.4 Initial State Dependence

There is a simple condition based on the principle that *any* instant along a given density history can be regarded as the initial moment [Maitra 2002b, Maitra 2005a] (see Chap. 4). This follows very naturally from the fact that the time-dependent Shrödinger equation is first order. When applied to both interacting and non-interacting systems, we find:

$$v_{\mathrm{xc}}[n_{t'}, \Psi(t'), \Phi(t')](\boldsymbol{r},t) = v_{\mathrm{xc}}[n, \Psi(0), \Phi(0)](\boldsymbol{r},t) \qquad \text{for } t > t'\,, \qquad (11.15)$$

This is discussed in much detail in Chap. 4. Here we just mention that any adiabatic approximation, by virtue of its lack of memory and lack of initial-state dependence, automatically satisfies it. Interestingly, although EXX is instantaneous in the orbitals, it will have memory (and so initial-state dependence) as a density functional (when applied to more than two unpolarized electrons).

But this condition provides very difficult tests for any functional with memory. Consider any two known evolutions of an interacting system, which after some time, $\tilde{t}$, become identical. This result states they must have identical xc potentials at that time and forever after, even though they had different histories before then. An approximate functional with memory is unlikely, in general, to produce such identical potentials.

11.4.5 Coupling-Constant Dependence

Because of the lack of a variational principle for the energy, there is no simple analog of the adiabatic connection formula, (11.4), or definite results as limits are approached, (11.5). But there remains a simple connection between scaling and the coupling constant [Hessler 1999]. For exchange, analogous to (11.1), the relation is linear:

$$v_{\mathrm{x}}[n_\gamma, \Phi_\gamma(0)](\boldsymbol{r}, t) = \gamma\, v_{\mathrm{x}}[n, \Phi(0)](\gamma\boldsymbol{r}, \gamma^2 t)\ , \tag{11.16}$$

where $\Phi(0)$ is the initial state of the Kohn-Sham system and, for time-dependent densities,

$$n_\gamma(\boldsymbol{r}, t) = \gamma^3\, n(\gamma\boldsymbol{r}, \gamma^2 t)\ . \tag{11.17}$$

There is no simple correlation scaling, but we can relate the coupling constant to scaling and find, analogous to (11.4):

$$v_{\mathrm{c}}^{\lambda}[n, \Psi(0), \Phi(0)](\boldsymbol{r}, t) = \lambda^2 v_{\mathrm{c}}[n_{(1/\lambda)}, \Psi_{(1/\lambda)}(0), \Phi_{(1/\lambda)}(0)](\lambda\boldsymbol{r}, \lambda^2 t)\ , \tag{11.18}$$

where $\Psi(0)$ is the initial state of the interacting system. It seems likely that, taking the limit $\lambda \to 0$, makes the exchange term dominant for finite systems (just as in the ground-state) [Hessler 2002], but this has yet to be proven.

11.4.6 Translational Invariance

Consider a rigid boost $\boldsymbol{R}(t)$ of a system starting in its ground state at $t = 0$, with $\boldsymbol{R}(0) = \mathrm{d}\boldsymbol{R}/\mathrm{d}t(0) = 0$. Then the exchange-correlation potential of the boosted density will be that of the unboosted density, evaluated at the boosted point, i.e.,

$$v_{\mathrm{xc}}[n'](\boldsymbol{r}, t) = v_{\mathrm{xc}}[n](\boldsymbol{r} - \boldsymbol{R}(t), t)\ , \qquad n'(\boldsymbol{r}, t) = n(\boldsymbol{r} - \boldsymbol{R}(t), t)\ , \tag{11.19}$$

This condition is universally valid [Vignale 1995a], and played a crucial role in the development of the VK approximation.

11.5 Linear Response

In the special case of linear response, all information is contained in the kernel. All the general conditions of Sect. 11.4 also yield results for the xc kernel.

11.5.1 Adiabatic Limit

For any finite system, the exact kernel satisfies:

$$f_{\mathrm{xc}}(\boldsymbol{r}, \boldsymbol{r}', \omega \to 0) \to \frac{\delta^2 E_{\mathrm{xc}}}{\delta n(\boldsymbol{r})\delta n(\boldsymbol{r}')} \tag{11.20}$$

where E_{xc} is the exact xc energy. Obviously, any approximate functional satisfies this, with its corresponding ground-state approximation on the right.

11.5.2 Zero Force and Torque

The exact conditions on the potential of Sect. 11.4.2 also yield conditions on $f_{\rm xc}$, when applied to an infinitesimal perturbation (see Chap. 5). Taking functional derivatives of (11.11a) yields

$$\int d^3r\, n(\boldsymbol{r})\, \nabla f_{\rm xc}(\boldsymbol{r},\boldsymbol{r}',\omega) = -\nabla' v_{\rm xc}(\boldsymbol{r}') \tag{11.21}$$

and

$$\int d^3r\, n(\boldsymbol{r})\, \boldsymbol{r}\times\nabla f_{\rm xc}(\boldsymbol{r},\boldsymbol{r}',\omega) = -\boldsymbol{r}'\times\nabla' v_{\rm xc}(\boldsymbol{r}')\,, \tag{11.22}$$

the latter again assuming no xc transverse currents.

Again, these are satisfied by ground-state DFT with the static xc kernel, so they are automatically satisfied by any adiabatic approximation. Similarly, in the absence of correlation, they hold for EXX.

The equations employing energies do not produce directly useful results in the linear response regime, because the functional derivative of the exact time-dependent xc energy is not the xc potential.

11.5.3 Self-Interaction Error

For one electron, functional differentiation of (11.14) yields:

$$f_{\rm x}(\boldsymbol{r},\boldsymbol{r}',\omega) = -\frac{1}{|\boldsymbol{r}-\boldsymbol{r}'|}\,,\qquad f_{\rm c}(\boldsymbol{r},\boldsymbol{r}',\omega)=0\,,\qquad (N=1) \tag{11.23}$$

These conditions are trivially satisfied by EXX, but violated by the density functionals ALDA, GK, and VK.

11.5.4 Initial-State Dependence

The initial-state condition, (11.15), leads to very interesting restrictions on $f_{\rm xc}$ for arbitrary densities. But the information is given in terms of the initial-state dependence, which is very difficult to find.

11.5.5 Coupling-Constant Dependence

The exchange kernel scales linearly with coordinates, as found by differentiating (11.16):

$$f_{\rm x}[n_\gamma](\boldsymbol{r},\boldsymbol{r}',\omega) = \gamma f_{\rm x}[n](\gamma\boldsymbol{r},\gamma\boldsymbol{r}',\omega/\gamma^2)\,. \tag{11.24}$$

A functional derivative and Fourier-transform of (11.18) yields [Lein 2000b]

$$f_{\rm c}^{\lambda}[n_{\rm GS}](\boldsymbol{r},\boldsymbol{r}',\omega) = \lambda^2 f_{\rm c}[n_{0,(1/\lambda)}](\lambda\boldsymbol{r},\lambda\boldsymbol{r}',\omega/\lambda^2)\,. \tag{11.25}$$

A similar condition has recently been derived for the coupling-constant dependence of the vector potential in TDCDFT [Dion 2005].

These conditions are trivial for EXX. They can be used to test the derivations of correlation approximations in cases where the coupling-constant dependence can be easily deduced. More often, they can be used to *generate* the coupling constant dependence when needed, such as in the adiabatic connection formula of (11.4).

11.5.6 Symmetry

Because the susceptibility is symmetric, so must also be the kernel:

$$f_{\rm xc}(\boldsymbol{r},\boldsymbol{r}',\omega) = f_{\rm xc}(\boldsymbol{r}',\boldsymbol{r},\omega)\,. \tag{11.26}$$

This innocuous looking condition is satisfied by any adiabatic approximation by virtue of the kernel being the second derivative of an energy, and is obviously satisfied by EXX.

11.5.7 Kramers-Kronig

The kernel $f_{\rm xc}(\boldsymbol{r},\boldsymbol{r}',\omega)$ is an analytic function of ω in the upper half of the complex ω-plane and approaches a real function $f_{\rm xc}(\boldsymbol{r},\boldsymbol{r}',\infty)$ for $\omega \to \infty$. Therefore, defining the function

$$\Delta f_{\rm xc}(\boldsymbol{r},\boldsymbol{r}',\omega) = f_{\rm xc}(\boldsymbol{r},\boldsymbol{r}',\omega) - f_{\rm xc}(\boldsymbol{r},\boldsymbol{r}',\infty)\,, \tag{11.27}$$

we find

$$\Re \Delta f_{\rm xc}(\boldsymbol{r},\boldsymbol{r}',\omega) = \mathbb{P}\int \frac{{\rm d}\omega'}{\pi}\frac{\Im f_{\rm xc}(\boldsymbol{r},\boldsymbol{r}',\omega')}{\omega'-\omega} \tag{11.28}$$

$$\Im f_{\rm xc}(\boldsymbol{r},\boldsymbol{r}',\omega) = -\mathbb{P}\int \frac{{\rm d}\omega'}{\pi}\frac{\Re \Delta f_{\rm xc}(\boldsymbol{r},\boldsymbol{r}',\omega')}{\omega'-\omega}\,, \tag{11.29}$$

where $\mathbb{P}$ denotes the principal part of the integral. Also, since $f_{\rm xc}(\boldsymbol{x},\boldsymbol{x}')$ is real-valued,

$$f_{\rm xc}(\boldsymbol{r},\boldsymbol{r}',\omega) = f^*_{\rm xc}(\boldsymbol{r},\boldsymbol{r}',-\omega)\,. \tag{11.30}$$

The simple lesson here is that any adiabatic kernel (no frequency dependence) is purely real, and any kernel with memory has an imaginary part (or else is not sensible). Putting it the other way round, to produce a complex kernel requires memory.

Thus, any adiabatic approximation must have a real kernel, which they do. And EXX, to the extent that it has any frequency-dependence (for more than 2 electrons), must have a complex kernel. Both GK and VK have complex kernels satisfying these conditions.

11.5.8 Adiabatic Connection

A beautiful condition on the exact xc kernel is given simply by the adiabatic connection formula for the ground-state correlation energy:

$$-\frac{1}{2}\int d^3r \int d^3r'\, v_{ee}(\boldsymbol{r}',\boldsymbol{r}) \int_0^1 d\lambda \int_0^\infty \frac{d\omega}{\pi} \left[\chi_\lambda(\boldsymbol{r},\boldsymbol{r}',i\omega) - \chi_{KS}(\boldsymbol{r},\boldsymbol{r}',i\omega)\right] = E_c \tag{11.31}$$

Combined with the Dyson-like equation of Chap. 1 for χ_λ as a function of χ_{KS} and f_{xc}, this is being used to generate new and useful approximations to the ground-state correlation energy (see Chaps. 28, 29 and 30), although at considerable computational cost.

But it provides an obvious exact condition on any approximate xc kernel for *any* system. Thus *every* system for which the correlation energy is known can be used to test approximations for f_{xc}. Note that, e.g., using ALDA for the kernel does *not* yield the corresponding E_{xc}^{LDA}, but rather a much more sophisticated functional. Even insertion of f_x yields correlation contributions to all orders in E_c. And lastly, even the exact adiabatic approximation, $f_{xc}[n_{GS}](\boldsymbol{r},\boldsymbol{r}',\omega = 0)$, does not yield the exact $E_{xc}[n_{GS}]$. In Chap. 28, this condition is used to test several approximations, but only for the uniform gas.

11.6 Finite Versus Extended Systems, and Currents

As mentioned in Sect. 11.2.4, care must be taken when extending exact ground-state DFT results to extended systems. This is even more the case for TDDFT. The first half of the RG theorem (Chap. 1) provides a one-to-one correspondence between potentials and *current* densities, but a surface condition must be invoked to produce the necessary correspondence with densities. With hindsight, this is very suggestive that time-dependent functionals may contain a non-local dependence on the details at a surface. As such, they are more amenable to local approximations in the current rather than the density.

As discussed elsewhere (Chap. 5) and first pointed out by Dobson [Dobson 1994a], the frequency-dependent LDA (GK approximation) violates the translational invariance condition of Sect. 11.4.6. One can trace this failure back to the non-locality of the xc functional in TDDFT. But, by going to a current formulation, everything once again becomes reasonable. The gradient expansion in the current, for a slowly varying gas, was first derived by Vignale and Kohn [Vignale 1996], and later simplified by Vignale, Ullrich, and Conti [Vignale 1997], and is discussed in much detail in Chap. 5.

For our purposes, the most important point is that, by construction, VK satisfies translational invariance. The frequency-dependence shuts off (it reduces to ALDA) when the motion is a rigid translation, but turns on when there is a true (non-translational) motion of the density [Vignale 1996].

11.6.1 Gradient Expansion in the Current

Any functional with memory should recover the VK gradient expansion in this limit, or justify why it does not. However, the VK approximation is *only* the gradient expansion. In the ground-state, the gradient expansion in the density was found to give poor results, and afterwards discovered to violate several important sum rules [Burke 1998a]. Fixing those sum-rules led to the development of generalized gradient approximations.

11.6.2 Response to Homogeneous Field

A decade ago, GGG [Gonze 1995b] caused a stir by pointing out that the periodic density in an insulating solid in an electric field is insufficient to determine the one-body potential, in apparent violation of the Hohenberg-Kohn theorem [Hohenberg 1964]. In fact, this effect appears straightforwardly in TDCDFT, and is even estimated by calculations using the VK approximation [van Faassen 2003a, Maitra 2003a]. When translated back to TDDFT language, one finds a $1/q^2$ dependence in $f_{\rm xc}$, where q is the wavevector corresponding to $\boldsymbol{r} - \boldsymbol{r}'$ (see Chaps. 10 and 20). Thus this effect arises naturally in the static limit of TDCDFT (see Chaps. 5 and 19).

However, it implies a need for a kernel that has the same degree of nonlocality as the Hartree kernel, and this is missed by any local or semilocal approximation, such as ALDA, but *is* built in to EXX [Kim 2002a] or AA and many-body derived kernels (see Chap. 10).

11.7 Odds and Ends

11.7.1 Functional Derivatives

A TDDFT result ought to come from a TDDFT calculation, but this is not always the case. By a TDDFT calculation, we mean the result of an evolution of the TDKS equations with some approximation for the xc potential that is a functional of the density. This implies that the xc kernel should be the functional derivative of some xc potential, which also reduces to the ground-state potential in the adiabatic limit.

All the approximations discussed here satisfy this rule. But calculations that intermix kernels with potentials in the solution of Casida's equations violate this condition, and run the risk of violating underlying sum-rules.

11.7.2 Infinite Lifetimes of Eigenstates

This may seem like an odd requirement. When TDDFT is applied to calculate a transition to an excited state, the frequency should be real. This is obviously true for ALDA and EXX, but not so clear when memory approximations are used. As shown in Sect. 11.5.7, Kramers-Kronig relations mean that memory implies imaginary xc kernels, and these can yield imaginary contributions to the transition frequencies. Exactly these effects were seen in calculations using the VUC for atomic transitions [Ullrich 2004]. Indeed, very long lifetimes were found when VUC was working well, and much shorter ones occurred when VUC was failing badly.

11.7.3 Single-Pole Approximation for Exchange

This is another odd condition, in which two wrongs make something right. Using Görling-Levy perturbation theory [Görling 1993a], one can calculate the exact exchange contributions to excited state energies [Filippi 1997, Zhang 2004a]. To recover these results using TDDFT, one does *not* simply use f_x, and solve the Dyson-like equations. As noted in Sect. 11.5.8, the infinite iteration yields contributions to all orders in the coupling constant.

However, the single-pole approximation (SPA) truncates this series after one iteration, and so drops all other orders. Thus, the correct exact exchange results are recovered in TDDFT from the SPA solution to Casida's equations, and *not* by a full solution [Gonze 1999]. This procedure can be extended to the next order [Appel 2003].

11.8 Memory Correlation Approximations

The first approximation that went beyond adiabatic is the Gross-Kohn approximation, as mentioned above, which was replaced by the VK approximation, when it was realized that it violated translational invariance.

11.8.1 Double Excitations and Branch Cuts

Maitra et al. [Maitra 2004, Cave 2004] argued that a strong ω-dependence in f_{xc} allows double excitation solutions to Casida's equations. Similarly, the second ionization of the He atom implies a branch cut in its f_{xc} at the frequency needed [Burke 2005a]. Under limited circumstances, this frequency dependence can be estimated, but a generalization [Casida 2005] has been proposed. It would be interesting to check its compliance with the conditions listed in this chapter.

11.8.2 Beyond Linear Response

A major point of interest is to go beyond the linear response regime for non-adiabatic approximations. The first such attempt was the bootstrap approach of [Dobson 1997]. More recent attempts are described in Chap. 8 and in [Kurzweil 2004].

11.9 Summary

What lessons can we take away from this brief survey?

1. In the ground-state theory, the xc energy dominates, and many conditions are proven for that functional. This is not so for TDDFT.
2. The adiabatic approximation satisfies many exact conditions by virtue of its lack of memory. Inclusion of memory may lead to violations of conditions that adiabatic approximations satisfy. This is reminiscent of the ground-state problem, where the gradient expansion approximation violates several key sum rules respected by the local approximation. Explicit imposition of those rules led to the development of generalized gradient approximations.
3. Explicit density functionals have poor quality potentials, but it is the time-dependent density that counts. The potential is a far more sensitive functional of the density than vice versa. Thus, while we have enumerated many conditions on the xc potential, it is important to determine which conditions significantly affect the density.

As shown in several chapters in this book, many people are presently testing the limits of our simple approximations, and very likely, these or other exact conditions will provide guidance on how to go beyond them.

Acknowledgments

I gratefully acknowledge support of NSF grant CHE-0355405 and DOE grant DE-FG02-01ER45928, and thank Hardy Gross, Robert van Leeuwen, Giovanni Vignale, and Neepa Maitra for helpful discussions.

Part III

Numerical Aspects

12 Propagators for the Time-Dependent Kohn-Sham Equations

A. Castro and M.A.L. Marques

12.1 Introduction

The main practical result of the Runge-Gross theorem are the time-dependent Kohn-Sham (TDKS) equations: a set of coupled one-particle Schrödinger-like equations with the form (atomic units are used hereafter)

$$\mathrm{i}\frac{\partial \varphi_i(t)}{\partial t} = [\hat{T} + \hat{V}_{\mathrm{KS}}(t)]\varphi_i(t)\,, \tag{12.1a}$$

$$v_{\mathrm{KS}}(\boldsymbol{r},t) = v_{\mathrm{ext}}(\boldsymbol{r},t) + \int \mathrm{d}^3r'\,\frac{n(\boldsymbol{r}',t)}{|\boldsymbol{r}-\boldsymbol{r}'|} + v_{\mathrm{xc}}[n](\boldsymbol{r},t)\,. \tag{12.1b}$$

In these equations, the index i runs through all the occupied Kohn-Sham states φ_i; $\hat{T}$ is the kinetic operator; v_{ext} is any (possibly time-dependent) external potential acting on the electronic system; $v_{\mathrm{xc}}[n]$ is the exchange and correlation potential – which is a functional of the time-dependent density

$$n(\boldsymbol{r},t) = \left\langle \Phi(t) \left| \sum_i \delta(\boldsymbol{r}-\hat{\boldsymbol{r}}_i) \right| \Phi(t) \right\rangle = \sum_i |\varphi_i(\boldsymbol{r},t)|^2\,. \tag{12.2}$$

During the last years, most applications of TDDFT were performed within linear response theory, where the response properties of the system are usually obtained in frequency domain. One may, however, work directly in the time-domain, propagating (12.1a). This has the advantage of allowing the inclusion of intense external perturbations, beyond the linear response regime. Of course, this "real-time" formulation of TDDFT requires the use of an algorithm to propagate Schrödinger-like equations.

Not surprisingly, the study of efficacious algorithms for this purpose has a long history, and multiple answers. We are concerned with a very general problem, yet we must beware of general purpose solutions: one expects that the efficiency depends strongly on the characteristics of the time-independent part of the Hamiltonian, on the time-dependent perturbation, and also on the initial state. From all possible approaches, we focus in this chapter on the ones most relevant to the propagation of the TDKS equations. This case has several important features:

A. Castro and M.A.L. Marques: *Propagators for the Time-Dependent Kohn-Sham Equations*, Lect. Notes Phys. **706**, 197–210 (2006)
DOI 10.1007/3-540-35426-3_12

- The Hamiltonian is intrinsically time-dependent, which is obvious since it depends parametrically on the time-dependent density.
- This time dependence is not known *a priori*, since it is deduced from the solution density itself, $v_{KS} = v_{KS}[n]$. The problem may then be formulated as follows: given $\varphi(\tau)$ and $\hat{H}(\tau)$ for $\tau \le t$, calculate $\varphi(t+\Delta t)$ for some Δt. This unpleasant fact is usually not taken into account in most studies of Schrödinger's equation, and adds extra difficulties, since all approximators will require the knowledge of the Hamiltonian in the interval $[t, t+\Delta t]$.
- Typically, one works with very large basis sets, where the Hamiltonian is represented as a very large, sparse matrix. This happens for instance in the real-space representations we have used.
- The Hamiltonian is usually Hermitian. Also, it is unbounded – and this fact is one of the roots of the numerical difficulties.

In this chapter, we give a pedagogical introduction to the problem of propagating the Kohn-Sham equations, and to some of its solutions. We have been enlightened by several sources, not all of them focused on TDDFT. Most of the literature refers to nuclear wavepacket propagation, either in quantum, semi-classical, or mixed schemes. The equations are, nevertheless, identical, and experience from this field may be translated to others. We learned from Kosloff's review [Kosloff 1988], from the work of Lubich and coworkers [Lubich 2002, Hochbruck 1998, Hochbruck 1999], from the comparisons of Truong and others [Truong 1992], and from other references that will be cited when appropriate. For the particular problem of TDDFT, we would like to refer to the work of Sugino and Miyamoto [Sugino 1999]. It is also important to mention here the advances in the simulation of (adiabatic) molecular dynamics using the Car-Parrinello approach [Car 1985]. The time-integration is effectively performed using modified Verlet and Gaussian dynamics including multiple time-scale methodologies [Tuckerman 1994a, Tuckerman 1994b]. However, those works do not address the real electron dynamics of a system but a fictitious one determined by an effective electron mass, and need to impose the orthogonality constraint for the wavefunctions (which is automatically fulfilled in the unitary propagation schemes to be described below). Finally, we also refer the reader to our previous work on the subject [Castro 2004a], where on top of the algorithmic discussion that we present here, a quantitative analysis of some of the possible solutions is also given.

We have implemented some of the most common approaches to the propagation of a quantum wave-packet in our computer code `octopus`,[1] a general purpose pseudopotential, real-space code.[2] The routines that implement these

[1] The `octopus` project is aimed at describing the electron-ion dynamics in finite and extended systems under the influence of time-dependent electromagnetic fields. The program can be freely downloaded from `http://www.tddft.org/programs/octopus/`. For details see [Marques 2003b].

[2] By real-space, or direct-space, we mean that all functions are discretized on a grid, and that the Laplacian is approximated by finite differences [Beck 2000].

techniques are available from the `octopus` web-site and can be used in more general contexts than the ones discussed here.

12.2 Formulation of the Problem

The Schrödinger equation may be rewritten in terms of its linear propagator $\hat{U}(t,t_0)$, which obeys the equation

$$\mathrm{i}\frac{\mathrm{d}}{\mathrm{d}t}\hat{U}(t,t_0) = \hat{H}(t)\hat{U}(t,t_0)\,. \tag{12.3}$$

The solution of the time-dependent Schrödinger equation, for a given initial state $\varphi(t_0)$, is then written as $\varphi(t) = \hat{U}(t,t_0)\varphi(t_0)$. The differential equation (12.3) may be rewritten as an integral equation

$$\hat{U}(t,t_0) = \hat{1} - \mathrm{i}\int_{t_0}^{t}\mathrm{d}\tau\;\hat{H}(\tau)\hat{U}(\tau,t_0)\,. \tag{12.4}$$

It is then easy to derive a Dyson's series – whose convergence is, unfortunately, not guaranteed – to formally solve the problem:

$$\hat{U}(t,t_0) = \hat{1} + \sum_{n=1}^{\infty}(-\mathrm{i})^n\int_{t_0}^{t}\mathrm{d}t_1\int_{t_0}^{t_1}\mathrm{d}t_2\ldots\int_{t_0}^{t_{n-1}}\mathrm{d}t_n\hat{H}(t_1)\ldots\hat{H}(t_n)\,. \tag{12.5}$$

By making use of the time-ordering operator $\hat{\mathcal{T}}$, this series takes on a form that vaguely reminds us of the definition of the exponential:

$$\hat{U}(t,t_0) = \hat{1} + \sum_{n=1}^{\infty}\frac{(-\mathrm{i})^n}{n!}\int_{t_0}^{t}\mathrm{d}t_1\int_{t_0}^{t}\mathrm{d}t_2\ldots\int_{t_0}^{t}\mathrm{d}t_n\hat{\mathcal{T}}[\hat{H}(t_1)\ldots\hat{H}(t_n)]\,. \tag{12.6}$$

Due to this resemblance, one normally defines the *time-ordered exponential* to encapsulate the expression: $\hat{U}(t,t_0) = \hat{\mathcal{T}}\exp\{-\mathrm{i}\int_{t_0}^{t}\mathrm{d}\tau\hat{H}(\tau)\}$. This, of course, only hides the ugliness of (12.6). The series expansion can be simplified if the Hamiltonian commutes with itself at different times, in which case we are left with a simple exponential: $\hat{U}(t,t_0) = \exp\{-\mathrm{i}\int_{t_0}^{t}\mathrm{d}\tau\hat{H}(\tau)\}$. Moreover, if the Hamiltonian does not depend on time, we can get rid of the integration: $\hat{U}(t,t_0) = \exp\{-\mathrm{i}(t-t_0)\hat{H}\}$. Unfortunately, as mentioned before, none of these simplifications applies in the case of TDDFT.

The evolution operator has some important properties that can be derived directly from its definition, and that any good approximator should preserve:

- For a Hermitian Hamiltonian, the evolution operator is unitary, i.e.

$$\hat{U}^{\dagger}(t+\Delta t,t) = \hat{U}^{-1}(t+\Delta t,t)\,. \tag{12.7}$$

  This mathematical property is linked to the conservation of probability of the wavefunction. Any desirable approximate propagator should be unitary, at least approximately, for Hermitian Hamiltonians.

- Another important property fulfilled by the exact evolution propagator is time-reversal symmetry:

$$\hat{U}(t+\Delta t, t) = \hat{U}^{-1}(t, t+\Delta t) . \tag{12.8}$$

Note that this property does not hold if a magnetic field is present; it must not be enforced if one wants to handle magnetic cases. However, any desirable algorithm should respect this property in the particular case where no magnetic field is applied.

- For any three instants t_1, t_2, t_3, $\hat{U}(t_1, t_2) = \hat{U}(t_1, t_3)\hat{U}(t_3, t_2)$.

This last property permits us to break the simulation into pieces. In practice, it is usually not convenient to obtain $\varphi(t)$ directly from φ_0 for a long interval $[0, t]$. Instead, one breaks $[0, t]$ into smaller time intervals:

$$\hat{U}(t, 0) = \prod_{i=0}^{N-1} \hat{U}(t_i + \Delta t_i, t_i) , \tag{12.9}$$

where $t_0 = 0$, $t_{i+1} = t_i + \Delta t_i$ and $t_N = t$. Typically, the time step is taken to be constant, i.e., $\Delta t_i = \Delta t$. However, it is possible to use variable time-step methods, especially if the implemented algorithm is able to make an optimal choice of the time step, to enhance the efficiency without compromising the accuracy. In any case, we deal with the problem of performing the short-time propagation

$$\varphi(t+\Delta t) = \hat{\mathcal{T}} \exp\left\{ -\mathrm{i} \int_t^{t+\Delta t} \mathrm{d}\tau\, \hat{H}(\tau) \right\} \varphi(t) . \tag{12.10}$$

There are good reasons for dividing $[0, t]$ into smaller intervals: (i) First of all, the time dependence of $\hat{H}$ becomes less critical, and the norm of the exponential argument is reduced (it increases linearly with Δt). This makes it easier to approximate the propagator, as the errors of the approximations depend critically on this norm. (ii) There is a natural limit to the maximum size of Δt: If $\omega_{\max}$ is the maximum frequency that we want to discern, Δt can not be larger than $\approx 1/\omega_{\max} = \Delta t_{\max}$.

Obviously, we seek the most effective algorithm. Below $\Delta t_{\max}$, we are free to choose Δt considering performance reasons. If $p(\Delta t)$ is the cost of propagating Δt for a given method, one should then choose the Δt_{opt} that minimizes $p(\Delta t)/\Delta t$, the cost of propagating the wave function per unit time. The optimal cost number of a given method is $p(\Delta t_{\mathrm{opt}})/\Delta t_{\mathrm{opt}}$, so the method that minimizes this optimal cost number can be viewed as the "best" method.

The value of $\omega_{\max}$ is either determined by the energy spectrum of the ground-state many-body Hamiltonian or by the frequency of the applied electromagnetic field. In the former case, the maximum frequency of the Hamiltonian is typically determined by the kinetic term. If the wave functions are expanded in a plane-wave representation, $\omega_{\max}$ is related to the maximum reciprocal lattice vector $G_{\rm max}$ used in the expansion; On the other hand, if we choose to work with a real-space discretization of the Hamiltonian, $\omega_{\rm max}$ is determined by the mesh spacing h. We have therefore

$$\omega_{\rm max} = \frac{G^2_{\rm max}}{2} = \frac{2\pi^2}{h^2} \,. \tag{12.11}$$

Note that the choice of either h or $G_{\rm max}$ is in turn determined by the *hardness* of the potentials that define the Hamiltonian. In many cases the evolution will not probe the very high-frequencies – the highest-energies states are not significantly populated – and so we can choose Δt to be larger than $1/\omega_{\rm max}$.

From a numerical point of view, the algorithm used to perform the time propagation should be "stable" and "accurate". The term "stable" is frequently used in a rather loose form. It is, however, possible to give it a precise definition: A propagator is stable below $\Delta t_{\rm max}$ if, for any $\Delta t < \Delta t_{\rm max}$ and $n > 0$, $\hat{U}^n(t+\Delta t, t)$ is uniformly bounded. One way to assure that the algorithm is stable is by making it "contractive", which means that $||\hat{U}(t+\Delta t)|| \leq 1$. Of course, if the algorithm is unitary, it is also contractive and hence stable; but if the algorithm is only approximatively unitary, it is better if it is contractive. The reason for this is easy to understand: the error is typically proportional to the norm. A contractive algorithm will reduce the norm, and consequently also reduce the error; on the other hand, a non-contractive scheme will yield larger errors at each time-step. The adverb "unconditionally" is sometimes added to these concepts to refer to algorithms that possess a given property independently of Δt and of the spectral characteristics of $\hat{H}$ (e.g., unconditionally stable, etc.).

As mentioned before, the time dependence of the Hamiltonian does not allow us to write the evolution propagator as a simple exponential: $\hat{U}(t+\Delta t, t) \neq \exp\{-\mathrm{i}\Delta t \hat{H}(t)\}$. Nevertheless, quite a few methods rely on algorithms to approximate the exponential of an operator, $\exp(\hat{A})$ – where for example $\hat{A}$ has the form $-\mathrm{i}\Delta t\hat{H}(\tau)$ for a given time τ. For this reason, our discussion is separated in two parts: First we look at several algorithms to approximate $\exp(\hat{A})$, where $\hat{A}$ is a *time-independent* operator. In particular, polynomial expansions, projection in Krylov subspaces, and split-operator methods. We then discuss different approximations for the full time-evolution operator, like the mid-point and implicit rules, and Magnus expansions. Split-operator techniques can also be modified to approximate the full time-dependent propagator.

12.3 Approximations to the Exponential of an Operator

In principle, the most desirable algorithm to calculate $\exp(\hat{A})v$, where v is an arbitrary vector, would begin with the evaluation of $\exp(\hat{A})$. In this way, we would be able to easily apply the exponential of the matrix $\hat{A}$ to any arbitrary vector. Unfortunately, the methods that exist to calculate the exponential of a matrix are computationally limited to matrices of order less than a few thousand (for a recent review, please consult [Moler 2003]). In a typical plane-wave or real-space calculation the Hamiltonian matrix can be of the order $\approx 10^5 - 10^6$, and therefore way too large to apply any of these methods. In fact, the size of the Hamiltonian may not even permit its full storage in matrix form.[3]

The alternative is to use iterative methods that yield directly $\exp(\hat{A})v$ for a particular choice of the vector v. These methods have a much better scaling with the order of the matrix. We will focus on three different techniques that dominate the literature: polynomial expansion of the exponential – either in the standard base or in the Chebychev base, splitting schemes, and Krylov subspace projection techniques.

12.3.1 Polynomial Expansions

The exponential of a matrix $\hat{A}$ is defined by the Taylor expansion

$$\exp(\hat{A}) = \sum_{n=0}^{\infty} \frac{1}{n!} \hat{A}^n \,, \tag{12.12}$$

a series that is unconditionally convergent for any matrix $\hat{A}$. This suggests an obvious method to approximate the exponential:

$$\text{Taylor}_k\{\hat{A}, v\} = \sum_{n=0}^{k} \frac{1}{n!} \hat{A}^n v \,. \tag{12.13}$$

For a given k, the method is of order k and requires k matrix-vector operations. It amounts to expanding the exponential function in the standard base of polynomials, $\{1, x, x^2, \dots\}$. The truncation of the infinite series at a given k breaks the unitarity of the exponential. It turns out that $k = 4$ is particularly suited for our applications [Giansiracusa 2002]: $k = 2$ is unconditionally unstable; $k = 4$ is conditionally stable; $k = 6$ is also conditionally stable but for smaller values of Δt.

The standard base of polynomials is not the only choice; one can use any given (complete and orthonormal) base $\{P_n(x)\}_{n=0}^{\infty}$. It is well known that Chebychev polynomial approximations are optimal for approximating functions [Smirnov 1968], so we define:

$$\text{Cheb}_k\{\hat{A}, v\} = \sum_{n=0}^{k} c_n \, T_n(\hat{A}) \, v \,, \tag{12.14}$$

---

[3] A similar situation appears when solving the linear system $\hat{A}x = v$: The evaluation of $\hat{A}^{-1}$ would allow the solution of the linear system for any vector v. However, the effort to invert the matrix $\hat{A}$ grows as N^3, where N is the dimension of the matrix.

where c_n are the coefficients of the expansion, and T_n is the Chebychev polynomial of order n. For a skew-Hermitian matrix $\hat{A}$ of the form $-\mathrm{i}\hat{H}\Delta t$ the Chebychev expansion reduces to [Tal-Ezer 1984]:

$$\mathrm{Cheb}_k\{-\mathrm{i}\hat{H}\Delta t, v\} = \sum_{n=0}^{k}(2-\delta_{n0})(-\mathrm{i})^n J_n(\Delta t) T_n(\hat{H})v \ , \qquad (12.15)$$

where J_n are the Bessel functions. The resulting method is also of order k, and, thanks to Clenshaw's algorithm [Clenshaw 1955], requires k matrix-vector operations. As the Chebychev polynomials are only defined in the range $[-1, 1]$, the Hamiltonian has to be scaled so that its spectrum lies within this range before using (12.15). The application of Chebychev polynomials to Chemistry was pioneered by Kosloff [Kosloff 1988]; more recent studies can be found in [Baer 2001, Chen 1999].

12.3.2 Krylov Subspace Projection

The m-th Krylov subspace, $\mathcal{K}_m\{\hat{A}, v\}$, for a given operator $\hat{A}$ and vector v, is defined as:

$$\mathcal{K}_m\{\hat{A}, v\} = \mathrm{span}\{v, \hat{A}v, \hat{A}^2 v, \ldots, \hat{A}^{m-1}v\} \ . \qquad (12.16)$$

Note that $\dim \mathcal{K}_m\{\hat{A}, v\}$ may be smaller than m if v does not have non-null components of at least m distinct eigenvectors of $\hat{A}$. The Lanczos procedure generates recursively an orthonormal base $\{v_i\}_{i=1}^{m}$ such that:

$$\hat{A}\hat{V}_m = \hat{V}_m \hat{H}_m + h_{m+1,m} v_{m+1} e_m^T \ , \qquad (12.17)$$

where $\hat{V}_m = [v_1, \ldots, v_m]$, $\hat{H}_m$ is an $m \times m$ symmetric tridiagonal matrix with components $h_{i,j}$ (upper Heisenberg if $\hat{H}$ is non-Hermitian), and e_i is the i-th unit vector in $\mathbb{C}^m$. $\hat{H}_m$ is the projection of $\hat{A}$ onto $K_m\{\hat{A}, v\}$ and is the upper-left part of $\hat{H}_{m+1}$. By induction, it can be proved [Saad 1992] that for any polynomial p_{m-1} of degree $\leq m-1$

$$p_{m-1}(\hat{A})v = \hat{V}_m p_{m-1}(\hat{H}_m)\hat{V}_m^T v = \hat{V}_m p_{m-1}(\hat{H}_m) e_1 \ . \qquad (12.18)$$

This suggests a method to approximate any function, and specifically the exponential

$$\mathrm{Lanczos}_k\{\hat{A}, v\} = \hat{V}_k \exp(\hat{H}_k) e_1 \ . \qquad (12.19)$$

Very good approximations are often obtained for relatively small k. The calculation of $\exp(\hat{H}_k)$ can be computed by any of the methods described in [Moler 2003]. The Krylov subspace projection is an order k method that requires k matrix-vector operations. To within our knowledge, it was Park and Light [Park 1986] who first applied the Lanczos algorithm to Chemistry; Hochbruck and Lubich [Hochbruck 1997] made a thorough mathematical analysis of the technique.

12.3.3 Splitting Techniques

The split-operator (SO) technique takes advantage of the fact that the Hamiltonian is composed of two terms, one diagonal in Fourier space – the kinetic operator $\hat{T}$, and the other diagonal (or almost diagonal) in real space – the potential operator $\hat{V}$. The idea is to approximate the propagator by the following product of exponentials:

$$\begin{aligned}\text{split}\{-\mathrm{i}\Delta t\hat{H}, v\} &= \mathrm{S}_2(-\mathrm{i}\Delta t\hat{H})v \\ &= \exp\left\{-\mathrm{i}\frac{\Delta t}{2}\hat{T}\right\}\exp\left\{-\mathrm{i}\Delta t\hat{V}\right\}\exp\left\{-\mathrm{i}\frac{\Delta t}{2}\hat{T}\right\}v\,. \end{aligned} \quad (12.20)$$

This decomposition neglects terms involving the commutator $[\hat{T}, \hat{V}]$ and higher order commutators, and is of $\mathcal{O}(\Delta t^2)$. Equation (12.20) is sometimes called "potential referenced split operator", since the potential term appears sandwiched between the two kinetic terms. A "kinetic referenced" scheme is equally legitimate. Since the three exponentials may be computed exactly, it is always unitary and unconditionally stable, providing a very reliable second order method. The split operator was first introduced in Physics and Chemistry by Feit and coworkers [Feit 1982a, Feit 1982b].

Besides the simple SO method, a wide variety of other splitting schemes have been proposed [Suzuki 1993, Suzuki 1992b, Mikhailova 1999]. One of these, the fourth order symmetric decomposition, was studied and applied to TDDFT by Sugino and Miyamoto [Sugino 1999]:

$$\text{Suzuki}\{-\mathrm{i}\Delta t\hat{H}, v\} = \prod_{j=1}^{5}\mathrm{S}_2(-\mathrm{i}p_j\Delta t\hat{H})v\,, \quad (12.21)$$

where the p_j are a properly chosen set of real numbers. Henceforth we will call this scheme "Suzuki-Trotter" (ST), following the nomenclature of [Sugino 1999].

12.4 Analysis of Integrators for the TDSE

Let us now recall the definition of the *full* problem: finding an approximation for $\varphi(t + \Delta t)$ from the knowledge of $\varphi(\tau)$ and $\hat{H}(\tau)$ for $0 \leq \tau \leq t$. In the following, we briefly describe several possible propagators.

However, before moving on, it is worth noting that most methods also require the knowledge of the Hamiltonian at some points in time between $t \leq \tau \leq t + \Delta t$, *and in a TDDFT calculation this is not known a priori, since the Hamiltonian is a functional of the time-dependent density.* To obtain this quantity, one can, e.g., extrapolate the Hamiltonian using a polynomial fit to n previous steps. However, this can reduce the accuracy of the propagator. To be fully consistent the following predictor-corrector method can be employed: (i) obtain $\hat{H}(\tau)$ through extrapolation; (ii) propagate φ to get $\varphi(t+\Delta t)$; (iii) from $\varphi(t+\Delta t)$ calculate $\hat{H}(t+\Delta t)$; (iv) obtain $\hat{H}(\tau)$ by interpolating between $\hat{H}(t)$ and $\hat{H}(t+\Delta t)$; (v) repeat steps (ii)–(iv) until self-consistency is achieved. For small time intervals, step (i) may be sufficient. For the theoretical description of the properties of the propagators (unitarity, time-reversibility), we will assume that $\hat{H}(\tau)$ is properly obtained using the above-mentioned self-consistent procedure, and that all numerical operations (calculation of the exponential of an operator, solution of a linear system, etc.) are performed exactly.

12.4.1 "Classical" Propagators

In the literature of partial differential equations, one can find a family of methods, such as the symmetric second order differencing scheme (SOD), the implicit midpoint rule [also known as Crank-Nicholson (CN) method], implicit or explicit Runge-Kutta, multistep algorithms, etc. This family of approximators are often referred to as "classical propagators". These propagators are of "general purpose", and have well known numerical properties. However, in the Chemistry and Physics communities, the typical form of the Hamiltonian matrix, made up of a kinetic term (diagonal in Fourier space) and a potential term (diagonal or almost diagonal in real space), has traditionally favored the use of splitting techniques or other methods.

We will mention the SOD and CN methods mainly for historical reasons, since they are probably the most popular. They are easy to implement, reliable, and reasonably efficient. However, our experience and that of others [Lubich 2002] suggest the use of different techniques.

The SOD [Kosloff 1988] requires the storage of two time slices:

$$\varphi(t+\Delta t)=\varphi(t-\Delta t)-2\mathrm{i}\Delta t\hat{H}(t)\varphi(t)\,. \tag{12.22}$$

It preserves time-reversal symmetry (in fact, it is designed with this idea in mind), and is conditionally stable.

The implicit midpoint rule is defined by

$$\hat{U}_{\mathrm{CN}}(t+\Delta t,t)=\frac{1-\frac{\mathrm{i}}{2}\Delta t\hat{H}(t+\Delta t/2)}{1+\frac{\mathrm{i}}{2}\Delta t\hat{H}(t+\Delta t/2)}\,. \tag{12.23}$$

The problem of propagating an orbital with this scheme is usually cast in the solution of the linear system:

$$\hat{L}\ \varphi(t+\Delta t)=b\,, \tag{12.24}$$

where $\hat{L}=\hat{I}+\mathrm{i}\frac{\Delta}{2}\hat{H}(t+\Delta t/2)$ and $b=[\hat{I}-\mathrm{i}\frac{\Delta t}{2}\hat{H}(t+\Delta t/2)]\varphi(t)$. The CN scheme is second-order in Δt, and is unitary and preserves time-reversal symmetry.

12.4.2 Exponential Midpoint Rule

The exponential midpoint (EM) rule consists in approximating the propagator by the exponential calculated at time $t + \Delta t/2$

$$\hat{U}_{\mathrm{EM}}(t + \Delta t, t) \equiv \exp\{-\mathrm{i}\Delta t \hat{H}(t + \Delta t/2)\} . \tag{12.25}$$

The actual propagation can then be done by any of the methods described in Sect. 12.3. If we assume that the exponential is calculated exactly, and that $\hat{H}(t + \Delta t/2)$ is obtained self-consistently, then this method is also unitary and time-reversible.

12.4.3 Time-Reversal Symmetry Based Propagator

In a time-reversible method, propagating backwards $\Delta t/2$ starting from $\varphi(t+\Delta t)$ or propagating forwards $\Delta t/2$ starting from $\varphi(t)$ should lead to the same result. By using the simplest approximation to the propagator, this statement leads to the condition

$$\exp\left\{+\mathrm{i}\frac{\Delta t}{2}\hat{H}(t + \Delta t)\right\}\varphi(t + \Delta t) = \exp\left\{-\mathrm{i}\frac{\Delta t}{2}\hat{H}(t)\right\}\varphi(t) . \tag{12.26}$$

Rearranging the terms, we arrive at an approximation to the propagator

$$\hat{U}_{\mathrm{ETRS}}(t + \Delta t, t) = \exp\left\{-\mathrm{i}\frac{\Delta t}{2}\hat{H}(t + \Delta t)\right\}\exp\left\{-\mathrm{i}\frac{\Delta t}{2}\hat{H}(t)\right\} . \tag{12.27}$$

We call this method the *enforced time-reversal symmetry* (ETRS) method.

12.4.4 Splitting Techniques

The splitting techniques have been described in Sect. 12.3.3 as a way to approximate the exponential of a time-independent Hamiltonian. By combining them, e.g., with the above mentioned EM or ETRS methods, one obtains an approximation for the full propagator based on either the split-operator or on the Suzuki-Trotter scheme. There is, however, an alternative way to improve the splitting schemes with small added computational cost.

Watanabe and Tsukada [Watanabe 2002] have recently combined the EM approximation with the split-operator method. In practice, this consists in setting $\hat{V} = \hat{V}(t + \Delta t/2)$ in (12.20). If this potential is obtained accurately we end up with an order two method, otherwise the method is of first order. There is, however, a simpler alternative:

$$\begin{aligned}\hat{U}_{\mathrm{SO}}(t + \Delta t, t) &= \mathrm{S}_2(-\mathrm{i}\Delta t(\hat{T} + \hat{\hat{V}})) \\ &= \exp\left\{-\frac{1}{2}\mathrm{i}\Delta t\hat{T}\right\}\exp\left\{-\mathrm{i}\Delta t\hat{\hat{V}}\right\}\exp\left\{-\frac{1}{2}\mathrm{i}\Delta t\hat{T}\right\} ,\end{aligned} \tag{12.28}$$

where the potential operator $\hat{\tilde{V}}$ is defined by:

$$\widetilde{v} = v_{\rm ext}(\boldsymbol{r}, t + \Delta t/2) + \int {\rm d}^3 r' \, \frac{\widetilde{n}(\boldsymbol{r}')}{|\boldsymbol{r} - \boldsymbol{r}'|} + v_{\rm xc}[\widetilde{n}](\boldsymbol{r}, t) \, . \tag{12.29}$$

In this expression $\widetilde{n}$ is the density built *after* applying the first kinetic exponential in (12.28). In other words, the modified SO method is: (i) apply the first kinetic term; (ii) recalculate the density and obtain the Kohn-Sham potential, and (iii) apply the potential term and the second kinetic term. In this simple way we recover an order two method.

For the higher order Suzuki-Trotter technique, Suzuki provided a time-dependent version:

$$\hat{U}_{\rm ST}(t + \Delta t, t) = \prod_{j=1}^{5} {\rm S}_2(-{\rm i}p_j \Delta t \hat{H}(t_j)) \, , \tag{12.30}$$

where the times t_j are related to the set p_j trough $t_j = t + (p_1 + \cdots + p_j/2)\Delta t$. Once again, the potential between t and $t+\Delta t$ has to be properly extrapolated to obtain a true order four technique (similar for higher order expansions).

12.4.5 Magnus Expansions

As noted previously, $\hat{U}(t+\Delta t, t)$ does not reduce to a simple exponential of the form $\exp\{-{\rm i}\Delta t \hat{H}(t)\}$ unless the Hamiltonian is time-independent. One may ask if there exists an operator $\hat{\Omega}(t+\Delta t, t)$ such that $\hat{U}(t+\Delta t, t) = \exp\{\hat{\Omega}(t+\Delta t, t)\}$. Magnus answered this question affirmatively in 1954 [Magnus 1954]: there exists an infinite series, convergent at least for some local environment of t, that provides us with this operator:

$$\hat{\Omega}(t + \Delta t, t) = \sum_{k=1}^{\infty} \hat{\Omega}_k(t + \Delta t, t) \, . \tag{12.31}$$

There also exists a procedure to generate the exact $\hat{\Omega}_k$ operators [Klarsfeld 1989]:

$$\hat{\Omega}_k(t + \Delta t, t) = \sum_{j=0}^{k-1} \frac{B_j}{j!} \int_t^{t+\Delta t} {\rm d}\tau \hat{S}_k^j(\tau) \, , \tag{12.32}$$

where B_j are Bernoulli numbers, and the operators $\hat{S}$ are generated recursively as

$$\hat{S}_1^0(\tau) = -{\rm i}\hat{H}(\tau) \quad ; \quad \hat{S}_k^0(\tau) = 0 \qquad (k > 1) \tag{12.33a}$$

$$\hat{S}_k^j(\tau) = \sum_{m=1}^{k-j} \left[\hat{\Omega}_m(t + \Delta t, t), \hat{S}_{k-m}^{j-1}(\tau)\right] \; (1 \leq j \leq k-1) \, . \tag{12.33b}$$

For example, the first two terms of the recurrence are:

$$\hat{\Omega}_1(t+\Delta t,t) = \int_t^{t+\Delta t} \mathrm{d}\tau_1 [-\mathrm{i}\hat{H}(\tau_1)] \tag{12.34a}$$

$$\hat{\Omega}_2(t+\Delta t,t) = \int_t^{t+\Delta t} \mathrm{d}\tau_1 \int_t^{\tau_1} \mathrm{d}\tau_2 [-\mathrm{i}\hat{H}(\tau_1), -\mathrm{i}\hat{H}(\tau_2)] \,. \tag{12.34b}$$

In general, the k-th term will be a k-dimensional integral of a sum of commutators of $\hat{H}$ at different times. An approximation of order $2n$ to the full Magnus operator (and hence, to the evolution operator) is achieved by truncating the Magnus series to nth order, and approximating the integrals through a nth order quadrature formula. The exponential midpoint rule can be regarded as the second order Magnus expansion, $\hat{U}_{\mathrm{EM}} = \hat{U}_{\mathrm{M}(2)}$, since

$$\hat{\Omega}_{\mathrm{M}(2)}(t+\Delta t,t) = -\mathrm{i}\hat{H}(t+\Delta t/2)\Delta t \,. \tag{12.35}$$

The fourth order Magnus expression is constructed by taking the first two terms in the Magnus series and using, for example, a two-point Gaussian-Legendre quadrature to approximate the integrals. The result is

$$\hat{\Omega}_{\mathrm{M}(4)}(t+\Delta t,t) = -\mathrm{i}\frac{\Delta t}{2}[\hat{H}(t_1)+\hat{H}(t_2)] - \frac{\sqrt{3}\Delta t^2}{12}[\hat{H}(t_2),\hat{H}(t_1)] \,, \tag{12.36}$$

where $t_{1,2} = t + \left[(1/2) \mp \sqrt{3}/6\right]\Delta t$ are the Gauss quadrature sampling points. For the specific case of the Kohn-Sham Hamiltonian, the fourth order Magnus propagator has the form

$$\hat{U}_{\mathrm{M}(4)}(t+\Delta t,t) = \exp\{-\mathrm{i}\Delta t \hat{H}_{\mathrm{M}(4)}(t,\Delta t)\} \,. \tag{12.37}$$

The modified Hamiltonian operator $\hat{H}_{\mathrm{M}(4)}(t,\Delta t)$ is defined as

$$\hat{H}_{\mathrm{M}(4)}(t,\Delta t) = \overline{H}(t,\Delta t) + \mathrm{i}[\hat{T} + \hat{V}_{\mathrm{ext}}^{\mathrm{nonlocal}}, \overline{\Delta V}(t,\Delta t)] \,, \tag{12.38}$$

where only the non-local components of the Kohn-Sham Hamiltonian contribute to the commutator, and with the definitions

$$\overline{H}(t,\Delta t) = \hat{T} + \frac{1}{2}\{\hat{V}_{\mathrm{KS}}(t_1) + \hat{V}_{\mathrm{KS}}(t_2)\} \,, \tag{12.39a}$$

$$\overline{\Delta V}(t,\Delta t) = \frac{\sqrt{3}}{12}\Delta t \{\hat{V}_{\mathrm{KS}}(t_2) - \hat{V}_{\mathrm{KS}}(t_1)\} \,. \tag{12.39b}$$

Expression (12.38) assumes that the non-local part of the Kohn-Sham Hamiltonian does not vary significantly in the interval of interest $(t, t+\Delta t)$. This non-local component is part of the ionic pseudopotentials used in electronic structure calculations and in consequence its variation is associated with the ionic movement. In principle, this movement should be negligible in the electronic time scale that determines Δt, which justifies the above assumption.

The fourth order Magnus expansion involves the computation of one commutator. The number of such commutators grows rapidly with increasing order, although some work has recently been devoted to significantly reduce this number [Blanes 2000]. The Magnus expansion has received a great deal of attention from the Chemistry and Physics community. A very recent in-depth study of the scheme may be found in [Hochbruck 2003]. The first application to the field of quantum molecular systems was made by Milfeld and Wyatt [Milfeld 1983] in 1983. However, we could not find any application of the Magnus expansions in the field of electronic structure calculations.

12.5 Conclusions

We have dedicated a previous paper [Castro 2004a] to a comparison of the numerical efficiency of some of the algorithms described above, in the context of a TDDFT implementation. Our experience is, however, not conclusive: we did not find an "always optimal" algorithm for the propagation of the time-dependent Kohn-Sham equations. The final choice depends on the internal characteristics of the physical system, and on the frequency and intensity of any existing external fields. Furthermore, the final performance of any method also depends on the specific implementation of the equations, and possibly also on the computer architecture. (For example, we have observed strong variations on the performance of the fast Fourier transforms, which have a definitive influence on the cost of the splitting techniques.) Nevertheless, and keeping in mind all these observations, we believe that the exponential midpoint rule combined with the Lanczos exponential approximator yields a very good algorithm to represent the time propagator for a wide range of systems. Moreover, if the problem involves high frequencies, it is also worth trying the higher order Magnus expansion.

Even if our numerical results were obtained using a TDDFT code based on real-space methods, we expect them to be applicable to other implementations (e.g., plane waves). Furthermore, some of the knowledge may safely be transported to the numerical implementation of other theories where either Schrödinger-like propagators or methods to approximate the action of exponential of operators are sought – we mention the diffusion Monte-Carlo method, or the recently investigated idea of obtaining accurate electronic wave functions through exponentials of two-body operators [Nooijen 2000, Nakatsuji 2000, van Voorhis 2001, Piecuch 2003], as examples where these findings may be useful.

Acknowledgments

The authors were partially supported by the EC 6th framework Network of Excellence NANOQUANTA (NMP4-CT-2004-500198). MALM also acknowledges financial support by the Marie Curie Actions of the EC (Contract No. MEIF-CT-2004-010384).

13 Solution of the Linear-Response Equations in a Basis Set

P.L. de Boeij

13.1 Introduction

The induced density can be obtained within a linear response calculation by solving a coupled set of equations, in which the first order change in the density $\delta n(\boldsymbol{r},\omega)$ follows from the first order change in the self-consistent potential $\delta v_{\mathrm{KS}}(\boldsymbol{r},\omega)$ and *vice versa.* Here the induced density can be given as,

$$\delta n(\boldsymbol{r},\omega) = \int \mathrm{d}^3r'\, \chi_{\mathrm{KS}}(\boldsymbol{r},\boldsymbol{r}',\omega)\delta v_{\mathrm{KS}}(\boldsymbol{r}',\omega)\,, \tag{13.1}$$

in which the Kohn-Sham response kernel can be expressed in terms of the unperturbed orbital functions $\varphi_i(\boldsymbol{r})$, orbital energies ε_i, and occupation numbers n_i, using the Lehmann representation,

$$\chi_{\mathrm{KS}}(\boldsymbol{r},\boldsymbol{r}',\omega) = \sum_{i,j}(n_j - n_i)\frac{\varphi_j^*(\boldsymbol{r})\varphi_i(\boldsymbol{r})\varphi_i^*(\boldsymbol{r}')\varphi_j(\boldsymbol{r}')}{\varepsilon_j - \varepsilon_i + \omega + \mathrm{i}\eta}\,. \tag{13.2}$$

Here each term is a product of factors that depend only on either $\boldsymbol{r}$ or $\boldsymbol{r}'$. The positive infinitesimal η ensures causality of the response function. The first-order change in the self-consistent potential is given in terms of the induced density by,

$$\delta v_{\mathrm{KS}}(\boldsymbol{r},\omega) = \delta v_{\mathrm{ext}}(\boldsymbol{r},\omega) + \int \mathrm{d}^3r' \left\{\frac{1}{|\boldsymbol{r}-\boldsymbol{r}'|} + f_{\mathrm{xc}}(\boldsymbol{r},\boldsymbol{r}',\omega)\right\}\delta n(\boldsymbol{r}',\omega)\,, \tag{13.3}$$

in which we recognize the usual external, Hartree, and exchange-correlation terms. Together with (13.1) and (13.2) this relation completes the self-consistent field scheme for the linear response.

13.2 An Expansion in Orbital Products

By inserting the Lehmann expansion in (13.1), it becomes clear that the factorization of the terms allows to directly integrate the product of the response function and the first order change in the potential. The induced density can then be written as an expansion in terms of orbital products,

P.L. de Boeij: *Solution of the Linear-Response Equations in a Basis Set*, Lect. Notes Phys. **706**, 211–215 (2006)
DOI 10.1007/3-540-35426-3_13

$$\delta n(\boldsymbol{r},\omega) = \sum_{i,j} \delta P_{ij}(\omega)\varphi_j^*(\boldsymbol{r})\varphi_i(\boldsymbol{r}) \,, \tag{13.4}$$

in which the expansion coefficients follow from

$$\delta P_{ij}(\omega) = \frac{n_j - n_i}{\varepsilon_j - \varepsilon_i + \omega + \mathrm{i}\eta} \int \mathrm{d}^3 r\, \varphi_i^*(\boldsymbol{r})\delta v_{\mathrm{KS}}(\boldsymbol{r},\omega)\varphi_j(\boldsymbol{r}) \,. \tag{13.5}$$

These coefficients need to be evaluated only for combinations of occupied and virtual orbitals, i.e., for $n_j \neq n_i$, and only for those combinations for which the integrals do not vanish on the basis of symmetry alone. Moreover, we only need to consider positive ω by using the relation $\delta P_{ij}(\omega) = \delta P_{ji}^*(-\omega)$.

The self-consistent field equations can now in principle be solved directly in terms of these expansion coefficients $\delta P_{ij}(\omega)$. Inserting expansion (13.4) for the induced density in the expression for the induced potential (13.3), and using the result to evaluate the integrals in (13.5), this yields a closed expression, given by

$$\sum_{k,l} \{(\varepsilon_j - \varepsilon_i + \omega)\delta_{ik}\delta_{jl} - (n_j - n_i)K_{ij,kl}^{\mathrm{Hxc}}(\omega)\}\delta P_{kl}(\omega) = (n_j - n_i) \int \mathrm{d}^3 r\, \varphi_i^*(\boldsymbol{r})\delta v_{\mathrm{ext}}(\boldsymbol{r},\omega)\varphi_j(\boldsymbol{r}) \,. \tag{13.6}$$

In this linear set of equations a coupling matrix $K_{ij,kl}^{\mathrm{Hxc}}(\omega)$ enters. It has two contributions being the Hartree term, which is defined as

$$K_{ij,kl}^{\mathrm{H}}(\omega) = \int \mathrm{d}^3 r \int \mathrm{d}^3 r'\, \varphi_i^*(\boldsymbol{r})\varphi_j(\boldsymbol{r}) \frac{1}{|\boldsymbol{r}-\boldsymbol{r}'|} \varphi_k^*(\boldsymbol{r}')\varphi_l(\boldsymbol{r}') \,, \tag{13.7}$$

and the exchange-correlation contribution, which is defined as

$$K_{ij,kl}^{\mathrm{xc}}(\omega) = \int \mathrm{d}^3 r \int \mathrm{d}^3 r'\, \varphi_i^*(\boldsymbol{r})\varphi_j(\boldsymbol{r}) f_{\mathrm{xc}}(\boldsymbol{r},\boldsymbol{r}',\omega)\varphi_k^*(\boldsymbol{r}')\varphi_l(\boldsymbol{r}') \,. \tag{13.8}$$

13.3 An Efficient Solution Scheme

A first (naïve) estimate for the work needed to solve (13.6) is the amount of elementary operations needed to evaluate all coupling matrix elements, which amounts to a number of floating point operations in the order of $N_{\mathrm{atom}}^6 \bar{n}_{\mathrm{occ}}^2 \bar{n}_{\mathrm{unocc}}^2 \bar{n}_{\mathrm{grid}}^2$. Here N_{atom} is the number of atoms in the system, and $\bar{n}_{\mathrm{occ}}$ respectively $\bar{n}_{\mathrm{unocc}}$ are the average number of occupied and unoccupied states per atom. $\bar{n}_{\mathrm{grid}}$ is the average number of grid points per atom needed to do the integrations. This N_{atom}^6-scaling becomes prohibitive for larger systems.

Rather then solving the set of linear equations (13.6) directly, a more efficient scheme is possible by using an iterative algorithm [Olsen 1988] like the conjugate gradient method or the direct inversion of the iterative subspace [Pulay 1980, Pulay 1982] technique. Such methods involve only repeated calculations of the matrix-vector products, which can be performed rather efficiently due to the factorized form of the coupling matrix elements. As initial vector one often chooses to use the uncoupled solution obtained by setting $K^{\mathrm{Hxc}}(\omega) = 0$, i.e.,

$$\delta P_{ij}^{(0)}(\omega) = \frac{n_j - n_i}{\varepsilon_j - \varepsilon_i + \omega + \mathrm{i}\eta} \int \mathrm{d}^3 r \, \varphi_i^*(\boldsymbol{r}) \delta v_{\mathrm{ext}}(\boldsymbol{r}, \omega) \varphi_j(\boldsymbol{r}) \, . \tag{13.9}$$

In the iteration procedure more accurate vectors are constructed in each cycle, until a converged result is obtained. Usually the number of iterations needed is much smaller than the dimension of the linear set of equations that we want to solve.

The matrix-vector multiplications are best performed in three consecutive steps. As first step the induced density is constructed on a grid of points using the expansion in (13.4). Since the molecular orbitals are often expressed as a linear combination of atom-centered basis functions $\phi_\mu(\boldsymbol{r})$,

$$\varphi_i(\boldsymbol{r}) = \sum_\mu c_{\mu i} \phi_\mu(\boldsymbol{r} - \boldsymbol{R}_\mu) \, , \tag{13.10}$$

one can use a similar expansion in terms of the atom-centered functions,

$$\delta n(\boldsymbol{r}, \omega) = \sum_{\nu,\mu} \phi_\nu^*(\boldsymbol{r} - \boldsymbol{R}_\nu) \phi_\mu(\boldsymbol{r} - \boldsymbol{R}_\mu) \delta \bar{P}_{\mu\nu}(\omega) \, . \tag{13.11}$$

The new coefficients follow from $\delta \bar{P}_{\mu\nu}(\omega) = \sum_{ij} c_{\mu i}^* c_{\nu j} \delta P_{ij}(\omega)$. One can choose to work with the total density, or to break it up into atom-pair densities, and treat each of them in the following steps separately. These atom-pair densities are obtained by grouping the terms in (13.11) per atom pair. These densities are well-localized in space and their expansions contain only few numbers of μ, ν-combinations with non-vanishing overlap, as the atomic functions have exponential tails. This will facilitate the use of distance effects needed to get a linear scaling of the computational cost with increasing system size.

The next step is to evaluate the Coulomb potential for this density. This can be done most efficiently [Baerends 1973] by introducing an auxiliary set of (atom-centered) functions, $f_\lambda(\boldsymbol{r})$, for which the Coulomb integrals can be evaluated analytically,

$$f_\lambda^{\mathrm{C}}(\boldsymbol{r}) = \int \mathrm{d}^3 r' \, \frac{f_\lambda(\boldsymbol{r}')}{|\boldsymbol{r} - \boldsymbol{r}'|} \, . \tag{13.12}$$

By fitting the density using these functions, i.e., by expressing the density as $\delta n(\boldsymbol{r}, \omega) \approx \sum_\lambda \bar{c}_\lambda f_\lambda(\boldsymbol{r} - \boldsymbol{R}_\lambda)$ where the coefficients can be obtained by minimizing the total fit error, one gets

$$\delta v_{\mathrm{H}}(\boldsymbol{r},\omega) = \sum_{\lambda} \bar{c}_\lambda f_\lambda^{\mathrm{C}}(\boldsymbol{r}-\boldsymbol{R}_\lambda)\,. \tag{13.13}$$

In order not to introduce spurious long-range terms, it is important to use a constrained fit, in which the charge of the fitted density is exactly reproduced in the fit. If the orbital functions are expressed on a basis of Gaussian-type functions, additional fit functions are not needed, as the atom-pair density can be expressed exactly in terms of Gaussian functions, for which the potentials are known analytically.

The exchange correlation contribution to the potential can now also be evaluated. For the simplest adiabatic local density approximation, for instance, we get,

$$\begin{aligned}\delta v_{\mathrm{xc}}(\boldsymbol{r},\omega) &= \int \mathrm{d}^3r'\, f_{\mathrm{xc}}^{\mathrm{ALDA}}(\boldsymbol{r},\boldsymbol{r}',\omega)\,\delta n(\boldsymbol{r}',\omega)\\ &= \left.\frac{\mathrm{d}^2(n e_{\mathrm{xc}}(n))}{\mathrm{d}n^2}\right|_{n_{\mathrm{GS}}(\boldsymbol{r})} \delta n(\boldsymbol{r},\omega)\,,\end{aligned} \tag{13.14}$$

in which $e_{\mathrm{xc}}(n)$ is the exchange-correlation energy density of the homogeneous electron gas.

The third and final step is the evaluation of the matrix elements of the induced potential,

$$\begin{aligned}\sum_{k,l} K_{ij,kl}^{\mathrm{Hxc}}(\omega)\delta P_{kl}(\omega) &= \int \mathrm{d}^3r\, \varphi_i^*(\boldsymbol{r})\delta v_{\mathrm{Hxc}}(\boldsymbol{r},\omega)\varphi_j(\boldsymbol{r})\\ &= \sum_{\mu,\nu} c_{\mu i}^* c_{\nu j} \cdot \int \mathrm{d}^3r\, \phi_\mu^*(\boldsymbol{r}-\boldsymbol{R}_\mu)\delta v_{\mathrm{Hxc}}(\boldsymbol{r},\omega)\phi_\nu(\boldsymbol{r}-\boldsymbol{R}_\nu)\,.\end{aligned} \tag{13.15}$$

Here too, distance effects can be used by introducing the expansion of the orbitals in terms of atom-centered basis functions.

Each of the steps above involves at maximum in the order of $N_{\mathrm{atom}}^3 \bar{n}_{\mathrm{occ}}$ $\bar{n}_{\mathrm{unocc}} \bar{n}_{\mathrm{grid}}$ floating point operations in case one chooses to work with the total induced density. If instead one uses the atom-pair approach, this factor can be modified to $N_{\mathrm{atom}} \bar{n}_{\mathrm{pairs}} \bar{n}_{\mathrm{bas}}^2 \bar{n}_{\mathrm{grid}}$ for the evaluation of the density and the matrix elements, and to $N_{\mathrm{atom}}^2 \bar{n}_{\mathrm{fit}} \bar{n}_{\mathrm{grid}}$ for the potential. Here the construction of the potential is one order in N_{atom} more expensive as the potential functions $f_\lambda^{\mathrm{C}}(\boldsymbol{r})$ decay only slowly, and need to be evaluated at all grid points in the system. By expressing the long tails in a multipole expansion this unfavorable scaling can be cured [Greengard 1987, White 1994]. Here $\bar{n}_{\mathrm{pairs}}$ is the average number of pairs that have overlapping basis functions, $\bar{n}_{\mathrm{bas}}$ and $\bar{n}_{\mathrm{fit}}$ are the average number of basis and fit functions per atom, and $\bar{n}_{\mathrm{grid}}$ is the average number of grid points needed per atom(-pair). Although the atom-pair approach will become more favorable for larger systems, the total density approach can fully utilize the symmetry. Not only can whole blocks of matrix elements be known in advance to vanish due to symmetry, the grid needed to integrate the remaining terms can also be reduced to cover the irreducible wedge only. In the atom-pair approach symmetry can in general only be used to identify equivalent integrals. A careful analysis of the prefactors involved is needed to determine which approach is more favorable. This will depend on many technical details of the implementation and on the size and symmetry of the system under study.

14 Excited-State Dynamics in Finite Systems and Biomolecules

J. Hutter

14.1 Introduction

Linear response in time-dependent DFT gives access to electronic excitations. By solving Casida's eigenvalue equation [Casida 1996] excitation energies and oscillator strengths of molecular systems can be calculated. Information on vertical excitation spectra is obtained either starting from an optimized ground state geometry or by sampling excitations from configurations gathered during molecular dynamics on the ground state energy surface. We obtain a definition for the total excited state energy by adding the excitation energy to the ground state energy of the system. This energy as a function of the molecular coordinates defines excited state potential energy surfaces

$$E_\omega(\{\boldsymbol{R}\}) = E_{\mathrm{GS}}(\{\boldsymbol{R}\}) + \omega(\{\boldsymbol{R}\}) \,, \qquad (14.1)$$

where $\{\boldsymbol{R}\}$ denotes the coordinates of all atoms. Exploring the potential energy surface is of interest for many fields of application ranging from standard spectroscopy to photochemistry and biochemistry. Understanding photophysical and chemical processes, for example, requires detailed knowledge of at least special points on the potential energy surface. Fluorescence is associated with relaxed geometrical structure in the excited state. The shape of absorption bands may depend on structural changes upon electronic excitation. Normal mode analysis in the excited state can be used to describe the fine structure of absorption bands using the Franck-Condon principle or give direct access to analyze excited state vibrational spectra. The dynamics of excited states is needed for the simulation of photochemical reactions. Newtonian dynamics requires the calculation of first derivatives with respect to atomic positions. These derivatives are also needed in most efficient algorithms to locate special points on potential energy surfaces. Going beyond adiabatic dynamics, for example by using surface hopping techniques requires also the calculation of coupling matrix elements.

In the next sections we will show how derivatives of excitation energies can be calculated efficiently. The derivation is limited to the case of the adiabatic approximation to the xc kernel.

J. Hutter: *Excited-State Dynamics in Finite Systems and Biomolecules*, Lect. Notes Phys. **706**, 217–226 (2006)
DOI 10.1007/3-540-35426-3_14

14.2 Lagrangian of the Excited State Energy

We use the equivalent but more flexible variational formulation [Furche 2001a] of linear response within TDDFT instead of Casida's eigenvalue equation. Excitation energies are defined as the stationary points of the functional

$$G[x,y,w] = \frac{1}{2}\left[x^\dagger(A+B)x + y^\dagger(A-B)y\right] + \frac{w}{2}\left(x^\dagger y + y^\dagger x - 2\right) . \quad (14.2)$$

The Lagrange multiplier w is real valued and is, as we will see below, equal to the excitation energy ω at the stationary points. The indices of vectors x and y are running over all values of q, where q is shorthand for the index triple $(ia\sigma)$. Here σ denotes one of the two spin states (α, β). In this chapter we make use of the convention that indices $i, j, \ldots$ and $a, b, \ldots$ are running over all occupied and unoccupied states of this spin state, respectively, indices $r, s, \ldots$ denote general orbitals and corresponding sums extend over all orbitals, occupied and unoccupied. The first order response of the one-particle density matrix $\gamma^{(1)}(\boldsymbol{r}, \boldsymbol{r}')$ is related to the vectors x and y

$$\gamma^{(1)}(\boldsymbol{r}, \boldsymbol{r}') = \frac{1}{2}\sum_q \left[(x_q + y_q)\Phi_q(\boldsymbol{r}, \boldsymbol{r}') + (x_q - y_q)\Phi_q(\boldsymbol{r}', \boldsymbol{r})\right] . \quad (14.3)$$

The basis functions are defined by $\Phi_q(\boldsymbol{r}, \boldsymbol{r}') = \varphi_{a\sigma}(\boldsymbol{r})\varphi_{i\sigma}(\boldsymbol{r}')$ and the orthogonal KS orbitals $\varphi_{p\sigma}(\boldsymbol{r})$ are solutions to the static KS equations. Canonical KS orbitals are related to eigenvalues $\varepsilon_{p\sigma}$. In the absence of magnetic fields we can assume the KS orbitals to be real. The matrices A and B are defined by

$$A_{qq'} = A_{ia\sigma,jb\sigma'} = \left(F_{ab\sigma}\delta_{ij} - F_{ij\sigma}\delta_{ab}\right)\delta_{\sigma\sigma'} + B_{qq'} , \quad (14.4a)$$

$$B_{qq'} = B_{ia\sigma,jb\sigma'} = \langle q \mid f_{\text{Hxc}}(\boldsymbol{r}, \boldsymbol{r}') \mid q' \rangle . \quad (14.4b)$$

In this equation $F_{pq\sigma}$ denotes the KS matrix in the basis of KS orbitals $\varphi_{p\sigma}(\boldsymbol{r})$.

From the variational principle the following stationarity conditions for $G[x,y,w]$ are obtained

$$\frac{\delta G}{\delta x_q} = \sum_{q'}(A+B)_{qq'}x_{q'} - w x_q = 0 , \quad (14.5a)$$

$$\frac{\delta G}{\delta y_q} = \sum_{q'}(A-B)_{qq'}y_{q'} - w y_q = 0 , \quad (14.5b)$$

$$\frac{\delta G}{\delta w} = \sum_q x_q y_q - 1 = 0 . \quad (14.5c)$$

If canonical KS orbitals are used as a basis set, solving (14.5a) and (14.5b) with the constraint (14.5c) leads to Casida's eigenvalue equation. We can also see from these equations that, within the adiabatic approximation to the xc kernel, the Lagrange multiplier w takes the value of the excitation energy ω. For a more general basis when $A - B$ is not a diagonal matrix, x and y are the right and left eigenvectors of the non-Hermitian matrix $(A - B)(A + B)$.

We assume now that the ground state KS equations and the linear response (14.5a)–(14.5c) have been solved for a specific state ω and its solution vectors x and y are known. In other words, the functional G and its derivatives have to be evaluated at the stationary point (x, y, ω).

Properties are defined as derivatives of ground or excited state energies with respect to an external perturbation. We will denote this perturbation by a superscript η. If η is an atomic coordinate, the derivative leads to nuclear forces needed in molecular dynamics or algorithms to locate special points on potential energy surfaces. However, perturbations like a static electrical field allow the calculation of other properties, dipole moments in this case. Excited state properties are calculated as the sum of derivatives of ground state energy and excitation energy

$$E_{\omega}^{\eta} = E_{\mathrm{GS}}^{\eta} + \omega^{\eta} \, . \tag{14.6}$$

The contribution from the ground state KS energy can be calculated using well established methods. Therefore we will concentrate in the following on the contribution from the excitation energy. In terms of the functional G this derivative is given by

$$\omega^{\eta} = G^{\eta}[x, y, w] = \frac{1}{2} \left\{ x^{\dagger}(A + B)^{\eta} x + y^{\dagger}(A - B)^{\eta} y \right\} \, . \tag{14.7}$$

Due to the variational nature of G we don't have to evaluate the derivatives of x, y, and w with respect to η, only derivatives of the operators A and B appear. However, the straightforward calculation of these derivatives includes derivatives of the KS orbitals. These derivatives are needed as the operators clearly depend on the KS orbitals and contrary to the ground state KS energy are not variational in the KS orbitals. The orbital derivatives pose a problem insofar as they directly depend on the perturbation. For example, the case of nuclear gradients requires the calculation of three times the number of atoms derivatives for each orbital.

The calculation of these derivatives can be circumvented by the introduction of relaxed densities, that can be computed independently of the perturbation. In quantum chemistry this is usually achieved using the Sternheimer interchange method leading to the so-called z-vector equations [van Caillie 1999]. The lengthy derivation of the z-vector equations can be simplified considerably [Furche 2002a] by using auxiliary Lagrangian functions.

We define the excited state Lagrangian L as the auxiliary functional

$$L[x, y, w, \varphi, z, t] = G[x, y, w] + \sum_{q} z_q F_q - \sum_{\substack{rs\sigma \\ r \leq s}} t_{rs\sigma} \left(S_{rs\sigma} - \delta_{rs} \right) \, . \tag{14.8}$$

The last sum runs over all KS orbital index pairs and $S_{rs\sigma} = \langle \varphi_{r\sigma} \mid \varphi_{s\sigma} \rangle$ is the overlap matrix of KS orbitals. This functional is stationary with respect to all its parameters x, y, w, φ, z, and t. The sum of the ground state KS Lagrangian and L is therefore a fully variational expression for the excited state energy functional. Variation of L with respect to x, y, and w leads back to the already considered variation of G. The additional Lagrange multipliers z and t enforce the conditions that the orbitals φ_i satisfy the static KS equations and are orthogonal

$$\frac{\partial L}{\partial z_q} = F_q = 0 \, , \tag{14.9a}$$

$$\frac{\partial L}{\partial t_{rs\sigma}} = S_{rs\sigma} - \delta_{rs} = 0 \, . \tag{14.9b}$$

These conditions fix the KS orbitals for all possible values of the external perturbation. The remaining Lagrange multipliers z and W are then determined from the remaining conditions

$$\frac{\partial L}{\partial \varphi_{p\sigma}(\boldsymbol{r})} = 0 \, . \tag{14.10}$$

Solving this equation replaces the calculation of the KS orbital derivatives. Note that this equation is independent of the perturbation.

Once all the parameters have been determined from the stationary conditions, derivatives of the excitation energy are calculated from

$$\begin{aligned} \omega^{\eta} = L^{\eta}[x, y, w, \varphi, z, t] &= G^{\eta}[x, y, w] \\ &+ \sum_q z_q F_q^{\eta} - \sum_{rs\sigma} t_{rs\sigma} \left(S_{rs\sigma}^{\eta} - \delta_{rs} \right) \, . \end{aligned} \tag{14.11}$$

The advantage of this formulation over the direct calculation of the derivatives from (14.6) is that L is an explicit functional of any external perturbation, whereas G implicitly depends on the perturbation through the KS orbitals. The direct consequence is that for a basis set expansion of the KS orbitals

$$\varphi_i(\boldsymbol{r}) = \sum_{\alpha} c_{\alpha i} \phi_{\alpha}(\boldsymbol{r}) \, , \tag{14.12}$$

only derivatives of the explicit dependence of the basis functions $\phi_{\alpha}(\boldsymbol{r})$ on the perturbations have to be considered.

14.3 Lagrange Multipliers and Relaxed One-Particle Density Matrix

From the stationarity condition of the Lagrangian we can derive a linear equation for z (the z-vector equation)

$$\sum_{q'}(A+B)_{qq'}z_{q'} = -r_q \; . \tag{14.13}$$

The right-hand-side r of this equation is defined by

$$r_{ia\sigma} = \sum_b x_{ib\sigma}K^{(1)}_{ab\sigma}[n_X] - \sum_j x_{ja\sigma}K^{(1)}_{ji\sigma}[n_X] + K^{(1)}_{ia\sigma}[n_U] + 2K^{(2)}_{ia\sigma}[n_X] \; , \tag{14.14}$$

where we have used the following quantities: the unrelaxed difference density matrix U

$$U_{ab\sigma} = \frac{1}{2}\sum_i (x_{ia\sigma}x_{ib\sigma} + y_{ia\sigma}y_{ib\sigma}) \; , \tag{14.15a}$$

$$U_{ij\sigma} = -\frac{1}{2}\sum_a (x_{ia\sigma}x_{ja\sigma} + y_{ia\sigma}y_{ja\sigma}) \; , \tag{14.15b}$$

$$U_{ia\sigma} = U_{ai\sigma} = 0 \; , \tag{14.15c}$$

the kernel functional evaluated for densities $n_X(r)$ and $n_U(r)$

$$K^{(1)}_{rs\sigma}[n] = \int \mathrm{d}^3r \int \mathrm{d}^3r' \, \varphi^*_{r\sigma}(\boldsymbol{r}) \, [f_{\mathrm{Hxc}}(\boldsymbol{r},\boldsymbol{r}')\, n(\boldsymbol{r}')] \, \varphi_{s\sigma}(\boldsymbol{r}) \tag{14.16a}$$

$$n_X(\boldsymbol{r}) = \sum_{rs\sigma} X_{rs\sigma} \, \varphi^*_{r\sigma}(\boldsymbol{r})\varphi_{s\sigma}(\boldsymbol{r}) \tag{14.16b}$$

$$n_U(\boldsymbol{r}) = \sum_{rs\sigma} U_{rs\sigma} \, \varphi^*_{r\sigma}(\boldsymbol{r})\varphi_{s\sigma}(\boldsymbol{r}) \; , \tag{14.16c}$$

and the corresponding matrix elements of the functional derivative of the xc kernel

$$K^{(2)}_{rs\sigma}[n_X] = \int \mathrm{d}^3r \int \mathrm{d}^3r' \int \mathrm{d}^3r'' \varphi^*_{r\sigma}(\boldsymbol{r}) \times \left[\frac{\delta f_{\mathrm{xc}}(\boldsymbol{r},\boldsymbol{r}')}{\delta n(\boldsymbol{r}'')} n_X(\boldsymbol{r}')\, n_X(\boldsymbol{r}'')\right] \varphi_{s\sigma}(\boldsymbol{r}) \; , \tag{14.17}$$

where X is the matrix

$$X_{ia\sigma} = x_{ia\sigma}; \quad X_{ai\sigma} = x_{ia\sigma}; \quad X_{ii\sigma} = X_{aa\sigma} = 0 \, . \tag{14.18}$$

Once the z-vector equation has been solved, the relaxed difference one-particle density matrix P can be calculated from

$$P = U + Z \; , \tag{14.19}$$

and the corresponding relaxed difference density is

$$n_P(\boldsymbol{r}) = \sum_{rs\sigma} P_{rs\sigma} \, \varphi^*_{r\sigma}(\boldsymbol{r})\varphi_{s\sigma}(\boldsymbol{r}) \; , \tag{14.20}$$

where Z is defined by

$$Z_{ia\sigma} = z_{ia\sigma}; \quad Z_{ai\sigma} = z_{ia\sigma}; \quad Z_{ii\sigma} = Z_{aa\sigma} = 0 \,. \tag{14.21}$$

The Lagrange multipliers t can directly be calculated from

$$t_{ij\sigma} = \frac{1}{1+\delta_{ij}} \left\{ \omega \sum_a \left(x_{ia\sigma} y_{ja\sigma} + y_{ia\sigma} x_{ja\sigma} \right) - \sum_a \varepsilon_{a\sigma} \left(x_{ia\sigma} x_{ja\sigma} + y_{ia\sigma} y_{ja\sigma} \right) \right\} + K^{(1)}_{ij\sigma}[n_P] + 2K^{(2)}_{ij\sigma}[n_X] \tag{14.22a}$$

$$t_{ab\sigma} = \frac{1}{1+\delta_{ab}} \left\{ \omega \sum_i \left(x_{ia\sigma} y_{ib\sigma} + y_{ia\sigma} x_{ib\sigma} \right) - \sum_i \varepsilon_{i\sigma} \left(x_{ia\sigma} x_{ib\sigma} + y_{ia\sigma} y_{ib\sigma} \right) \right\} \tag{14.22b}$$

$$t_{ia\sigma} = \sum_j x_{ja\sigma} K^{(1)}_{ji\sigma}[n_X] + \varepsilon_{i\sigma} z_{ia\sigma} \,, \tag{14.22c}$$

where we have given formulas for the special case of a basis from canonical KS orbitals.

14.4 Molecular Dynamics

The Born–Oppenheimer separation between the motions of nuclei and electrons is at the heart of most electronic structure calculations. This separation allows the description of the dynamics of the nuclei with classical mechanics while the electronic structure remains in an adiabatic eigenstate of the system. For systems in the electronic ground state, the coupling of Newtonian dynamics with DFT has become a very successful simulation method. Using a standard integration scheme, Newton's equations of motion

$$M_I \ddot{R}_{I\alpha} = -\frac{\partial E}{\partial R_{I\alpha}} \,, \tag{14.23}$$

are solved starting from an initial configuration. When combined with DFT, this requires the solution of the KS equations and the calculation of the first derivative with respect to atomic positions at each time step. Alternatively, an extended Lagrangian technique introduced by Car and Parrinello [Car 1985] can be used. In this method, the KS orbitals are propagated along with the nuclear positions and a re-optimization of the KS orbitals is avoided. However, smaller time-steps are needed for an accurate integration of orbital degrees of freedom.

The method of adiabatic dynamics can easily be applied to other states than the ground state. Having defined an excited state energy surface as the sum of the ground state energy and an excitation energy from linear response within TDDFT, we only need the nuclear forces to follow the system in time.

Applying (14.23) restricts the dynamics to the initially chosen adiabatic state. This restriction together with the neglect of the quantum mechanical behavior of the nuclei often limits the scope of the simulations. Quantum effects such as tunneling, interference and level quantization may be important for certain systems. The single state approximation will be much more severe for excited states than it is for the ground state, and transitions between electronic states have to be taken into account for most problems of interest. When such transitions occur, the forces experienced by the nuclei may change drastically. This effect has to be properly incorporated into the dynamics and is crucial for describing many dynamical effects like photochemistry, electron transfer in molecules, or radiationless transitions.

In cases where the adiabatic approximation fails, the non-adiabatic coupling vector

$$d_{fi} = \langle \Psi_i \mid \nabla_{\boldsymbol{R}} \mid \Psi_f \rangle \,, \tag{14.24}$$

also called derivative coupling between the adiabatic states Ψ_i and Ψ_f becomes the central quantity linking the change of the electronic state with nuclear motion. Explicit use of the coupling vector is made in molecular dynamics with quantum transitions (surface hopping) method [Tully 1990] for the determination of the switching probabilities and velocity rearrangements. But also in the characterization of the potential energy surface near avoided crossings and conical intersections d_{fi} plays an important role.

The exact or approximate calculation of the non-adiabatic coupling vector in TDDFT will not be discussed here. Recent work on approaches based on DFT can be found in the literature [Baer 2005, Billeter 2005]. We will now return to the derivative calculation of the excited state energy. Equation (14.11) can be turned into the following form

$$\begin{aligned}
\omega^\eta = \sum_{\mu\nu\sigma} h^\eta_{\mu\nu} P_{\mu\nu\sigma} - \sum_{\mu\nu\sigma} S^\eta_{\mu\nu} t_{\mu\nu\sigma} + \sum_{\mu\nu\sigma} V^{\mathrm{xc}\,(\eta)}_{\mu\nu\sigma} P_{\mu\nu\sigma} \\
+ \sum_{\mu\nu\sigma}\sum_{\kappa\lambda\sigma'} (\mu\nu \mid \kappa\lambda)^\eta \left(P_{\mu\nu\sigma} D_{\kappa\lambda\sigma'} + X_{\mu\nu\sigma} X_{\kappa\lambda\sigma'}\right) \\
+ \sum_{\mu\nu\sigma}\sum_{\kappa\lambda\sigma'} + f^{\mathrm{xc}\,(\eta)}_{\mu\nu\sigma\kappa\lambda\sigma'} X_{\mu\nu\sigma} X_{\kappa\lambda\sigma'} \,.
\end{aligned} \tag{14.25}$$

In (14.25) all quantities have been transformed into the basis set representation from (14.12). The quantities $h_{\mu\nu}$ and $S_{\mu\nu\sigma}$ are the core Hamiltonian and overlap matrix, $(\mu\nu \mid \kappa\lambda)$ is a two-electron Coulomb integral in Mulliken notation, and $D_{\mu\nu\sigma}$ is the ground state KS density matrix. Finally, matrix elements of the xc potential and kernel are defined as

$$V^{\mathrm{xc}}_{\mu\nu\sigma} = \int \mathrm{d}^3 r\, \phi^*_\mu(\boldsymbol{r})\, \frac{\delta E_{\mathrm{xc}}}{\delta n_\sigma(\boldsymbol{r})}\, \phi_\nu(\boldsymbol{r}) \,, \tag{14.26a}$$

$$f^{\mathrm{xc}}_{\mu\nu\sigma\kappa\lambda\sigma'} = \int \mathrm{d}^3 r \int \mathrm{d}^3 r'\, \phi^*_\mu(\boldsymbol{r})\, \phi^*_\nu(\boldsymbol{r})\, \frac{\delta^2 E_{\mathrm{xc}}}{\delta n_\sigma(\boldsymbol{r}) \delta n_{\sigma'}(\boldsymbol{r}')}\, \phi^*_\kappa(\boldsymbol{r}')\, \phi^*_\lambda(\boldsymbol{r}') \,. \tag{14.26b}$$

All terms in (14.25) have to be considered for the case of a basis set that depends on atomic positions. This is the case for basis sets in standard quantum chemistry applications, where linear combinations of Gaussian or exponential functions are used. If an othonormal basis set independent of the atomic positions [Hutter 2003] is used, (14.25) reduces to

$$\omega^{\eta} = \sum_{\mu\nu\sigma} h^{\eta}_{\mu\nu} P_{\mu\nu\sigma} = \int \mathrm{d}^3 r \, n_P(\boldsymbol{r}) \frac{\partial v_{\mathrm{ext}}(\boldsymbol{r})}{\partial \eta} \,. \tag{14.27}$$

Plane wave basis sets, as used in computational solid state physics, are a prominent example. However, such basis sets need usually a large amount of functions to accurately describe the KS orbitals. This makes it numerically very inconvenient to calculate all unoccupied KS orbitals as needed to solve the linear response eigenvalue equation and the z-vector equation. The explicit use of the unoccupied KS orbitals can be avoided [Baroni 1987] by employing a projector Q on the space of unoccupied orbitals in the basis function

$$Q_{\mu\nu\sigma} = \delta_{\mu\nu} - D_{\mu\nu\sigma} \,. \tag{14.28}$$

This scheme leads also to algorithms that allow us to make use of locality and therefore reduce the scaling.

14.5 Coupling to Classical Force Fields

Many chemical reactions take place in condensed phase and therefore an accurate theoretical description of solvent effects is of utmost importance in computational quantum chemistry. Enzymatic catalysis is a prominent example of how a solvent environment with complex structure and dynamics can steer the properties and chemical reactions of a solute molecule. Theoretical descriptions have to account for the fact that the solvent structure and dynamics are too complex to be covered by continuum models. However, due to computational limitations, the representation of the environment can often not be based on quantum mechanics. On the other hand, the active center of the system where a chemical reaction or an optical excitation takes place requires an accurate treatment using quantum mechanics. Molecular dynamics simulations have been used and extended for increasingly complicated molecular systems ranging from aqueous solutions of simple ions to those of lipid membranes or proteins. Intramolecular force fields for wide classes of macromolecules have been developed, which decompose these molecules into local chemical bonding motives whose mechanical properties are then described by simple empirical formulas. These bonded forces are combined with non-bonded van der Waals and electrostatic interactions of fixed point charges (see Sect. 22.4.1).

One route to the simulation of large condensed systems is the combination of electronic structure theory methods with empirical force fields (see

Sect. 22.4.2). The idea behind this combined quantum mechanics/molecular mechanics (QM/MM) methods is to describe a part of the molecule/system quantum mechanically and the rest within the empirical force field approach. In this QM/MM approach we can write the total energy of the systems as

$$E_{\mathrm{tot}} = E_{\mathrm{QM}} + E_{\mathrm{MM}} + E_{\mathrm{QM-MM}} \,, \tag{14.29}$$

where E_{QM} is the energy of the QM part of the system, E_{MM} the energy of the MM system, and $E_{\mathrm{QM-MM}}$ describes the coupling of the two parts. We will be concerned here with the case where E_{QM} is given by (14.1) and we don't have to discuss E_{MM} further. However, the coupling term $E_{\mathrm{QM-MM}}$ has to be investigated. If the boundary of the QM and MM regions intersects covalent bonds, special precautions have to be taken. Popular approaches [Sherwood 2000] use either so-called link-atoms or pseudo-potentials to saturate the bonding pattern of the quantum system, and the bonds across the QM/MM boundary are modeled by empirical force fields. The force field parameters, as well as van der Waals parameters were optimized to describe the interaction of a ground state system. It is therefore important to ensure that in the case of the description of excited states with TDDFT, the actual excitation region is well separated from the QM–MM boundary. The important interaction term in our case is the electrostatic interaction of the classical point charges with the electronic charge density of the quantum system

$$E^{\mathrm{el}}_{\mathrm{QM-MM}} = \sum_I \int \mathrm{d}^3 r \, \frac{n_{\mathrm{QM}}(\boldsymbol{r}) q_I(\boldsymbol{R}_I)}{|\boldsymbol{r} - \boldsymbol{R}_I|} \,. \tag{14.30}$$

The sum in (14.30) runs over all point charges q_I at position $\boldsymbol{R}_I$. If we now define the charge density of the QM system as the derivative of the total energy with respect to the external potential

$$n^{\omega}(\boldsymbol{r}) = \frac{\partial E_{\omega}}{\partial v_{\mathrm{ext}}(\boldsymbol{r})} = n(\boldsymbol{r}) + n_P(\boldsymbol{r}) \,, \tag{14.31}$$

we see that it is the relaxed density that acts as the true charge density of the system in the excited state. Replacing $n_{\mathrm{QM}}(\boldsymbol{r})$ with $n^{\omega}(\boldsymbol{r})$ in the definition of $E^{\mathrm{el}}_{\mathrm{QM-MM}}$ also ensures that the calculation of the excitation energy and nuclear forces is still consistent. If we add the interaction term to the Lagrangian (14.8) the additional external potential from the classical point charges appears in the definition of A (14.4a) and the KS matrix F.

The scheme outlined above, based on an interaction energy given by (14.30) applies also to other situations where a subsystem treated by TDDFT is embedded into a charge density. A prominent case of such a method is the KS method with constrained electron density embedding by Wesolowski [Wesolowski 1993]. An extension of this method for TDDFT has been formulated [Casida 2004] and has been used to calculate the solvatochromic shift of excitation energies for molecules in solution [Neugebauer 2005]. The electron density of solvent molecules was frozen, i.e. not allow to relax according to the environment. Different configurations were sampled from a trajectory generated by a standard Car–Parrinello MD simulation. The frozen density approach has also been used by Barker and Sprik [Barker 2003] for molecular dynamics simulations. A combination of these two approaches would allow for a scheme for excited state dynamics based on the methods outlined in this chapter.

15 Time Versus Frequency Space Techniques

M.A.L. Marques and A. Rubio

15.1 Introduction

Let us imagine a young student (or a not so young professor who still has time to do research by him- or herself) who wants to make an ab-initio study of the excitation properties of one of those fashionable nanostructures that fill high-impact journals nowadays. The student has heard of TDDFT, and believes that it's just the right tool for the job. The first thing to do is to make sure that his fancy molecule is not part of the set of the difficult, "pathological" cases – not a bulk semiconductor, check!; the system does not involve charge-transfer excitations, check!; not a "strongly-correlated" system, check! As everything looks fine, the student starts the quest to find an adequate computer program to use in his or her research.

After a couple of hours googling, the student comes up with 15 different programs that seem to be adequate for his problem. All the programs appear to be quite easy to compile/install, and they all have nice, simple interfaces that make working with them a pleasure and not a torture. Digging a bit further, the student finds that these programs use very different techniques to obtain the excitations of the system: some use Green's functions and linear response theory, some use linear response theory but without the Green's functions, others propagate in time the TDDFT equations. What to choose? Reading the documentation of the programs is not much of a help, as they all claim to be the fastest and least memory consuming. So, what is the most efficient method? The answer to this question is quite tricky, not only due to the "political" issues that any answer could provoke, but also because "efficiency" is a very ill-defined concept in the world of numerics. A more pragmatic measurement is computer time, but this, of course, depends on the method used, the implementation, the hardware, the size of the problem, and sometimes even on the phase of the moon!

In this article, we try to give a hand to our student by comparing different methods to calculate excitation energies within TDDFT. Our purpose is to show how these methods scale with the size (i.e. number of atoms) of the system, namely in what concerns CPU time and memory requirements. Clearly, our approach is not exhaustive, and is mostly determined by our own scientific background. Furthermore, we concentrate on finite systems, and leave the exercise of adding the extra factor $n_{\boldsymbol{k}}$ (the number of $\boldsymbol{k}$-points)

M.A.L. Marques and A. Rubio: *Time Versus Frequency Space Techniques*, Lect. Notes Phys. **706**, 227–243 (2006)
DOI 10.1007/3-540-35426-3_15

and use of complex wavefunctions to the Reader. Five different methods are reviewed: (i) real-time propagation; (ii) Sternheimer's approach; (iii) Casida's equation; (iv) solution of the response equation in momentum/frequency space; and (v) solution of the response equation in a mixed space. These schemes can be divided between those that use only the knowledge of the occupied subspace (i and ii) and between those that need both occupied and unoccupied states (iii, iv, and v); and those that work in real time (i and v) or in frequency space (ii, iii, iv, and v). The two latter methods (iv and v) are mostly used for the calculation of the response of solids where they have some advantages. However, as they can and have been used successfully for the study of finite systems using super-cell methods, we have decided to include them in this review.

Finally, a word on basis sets. Different representations of the wave functions are regularly used by the community. Those based on the linear combination of localized orbitals (LO) have been been used in the literature to address the linear and nonlinear response of molecules with quite a good success (see, e.g., [Casida 1998a, van Gisbergen 1997]). As compared to grid or plane-wave based representations, the main advantage of this representation stems from the small number of basis functions needed to expand the wave functions and Hamiltonian matrix elements. On the other hand, the systematic convergence with respect to the size of the localized basis set is rather difficult. Which representation is best depends on the precision required and on the dimensionality and inhomogeneity of the system under study, in particular in which representation the response function is more sparse. In this article, we chose to put more emphasis on methods using the mesh representation (in direct or Fourier space). However, most of our results are completely general and applicable to either case.

While discussing the merits of the different approaches we will not take into account the computational requirements to perform the ground state calculations and to obtain the occupied and unoccupied orbitals. Note, however, that for very large systems the calculation of a large set of unoccupied wave functions can be a major bottleneck, as it scales in some cases with the cube of the number of atoms.

15.1.1 Notation

We shall use the symbol N with subscripts for quantities that scale roughly as the size of the physical system, and M for quantities that may be large but are independent of the system size.[1] Important quantities common to all methods are the number of electrons $N_{\rm e}$ and the number of mesh points, N_R and N_G for real space and reciprocal space, and $N_{\rm LO}$ for the number of

[1] We follow the notation given in [Bertsch 2000], where a preliminary comparison of the different methods discussed here was presented.

Table 15.1. Symbols used for the relevant quantities that determine the computational effort required by the various algorithms for TDDFT calculations

Symbol	Meaning	Symbol	Meaning
N_e	number of occupied states	M_T	number of time steps
N_c	unoccupied states	M_ω	number of frequencies
N_R	real-space points	M_H	nonzero elements in a row of $\hat{H}$
N_G	reciprocal-space points	M_{it}	iterations in the iterative schemes
N_{LO}	size of localized basis set	M_{impl}	implementation dependent number

localized orbitals. On the other hand, some methods require the knowledge of N_c unoccupied states.

Additional quantities that play a role are the number of frequencies to calculate M_ω, and the number of time steps in the real-time method, M_T. Also, in methods that rely on sparse matrix multiplication, we need the number M_H of nonzero entries in a row of the Hamiltonian. Furthermore, for methods that rely on iterative techniques, like conjugated gradients, we need an estimate of the number of iterations M_{it}. Finally, we define a number M_{impl}, independent of the size of the system, but that may vary between different implementations.

All symbols are summarized in Table 15.1.

15.2 Time-Evolution Scheme

This method was introduced in detail in Chaps. 1 and 12 as a direct approach to compute both the linear and non-linear response in physical systems. The idea is to excite the ground-state with a perturbing potential and then follow the evolution of the system by solving the TDKS equations in real time. As a final result, we get the induced density $\delta n(\mathbf{r}, t)$ from which, by Fourier transformation to frequency space, we can compute the different responses of the system. Note that from the Fourier transform we obtain *all* frequencies at once, and therefore this is not the method of choice if only one particular frequency is desired.

The implementation of this idea to obtain absorption spectra requires several steps: (i) First, one needs to get the ground-state occupied wave-functions. (ii) The ground-state is then perturbed by multiplying each of the single-particle Kohn-Sham wave-functions by a phase $e^{i\kappa z}$. This phase-factor shifts the momentum of the electrons, giving then a coherent velocity field that causes the appearance of a polarization as the system evolves in time. Note that to study the linear dipole response, the external wave-number κ, i.e. the strength of the applied homogeneous electric field, has to be much smaller than the inverse radius of the system. (iii) The system is propagated until some finite time T. The dynamics of the system can be analyzed in terms of the time-dependent induced dipole moment of the electron cloud

$$\boldsymbol{D}(t) = \sum_{i=1}^{\text{occ}} \langle \varphi_i(t) | \boldsymbol{r} | \varphi_i(t) \rangle \,, \tag{15.1}$$

where $\varphi_i(t)$ are the TDKS wave-functions. (iv) The linear absorption spectrum can then be computed from the time Fourier transform of $\boldsymbol{D}(t)$.

With this technique, we can extract information on the response functions of the system without as much bookkeeping as in the usual perturbative formalism and using only information concerning the occupied ground-state orbitals. It has been implemented both in real-space [Marques 2003b, Castro 2004b, Yabana 1996, Yabana 1999b, Yabana 1999c, Yabana 1999a] and Fourier-space [Sugino 1999].

Many approximations to the time-evolution operator $\exp(-\mathrm{i}\hat{H}\Delta t)$ require the consecutive multiplication of the Hamiltonian matrix by a (complex) vector representing the wave function [Castro 2004a]. For example, a simple but effective scheme is given by the Taylor series expansion (to fourth order) of the time-evolution operator [Giansiracusa 2002]. A predictor-corrector cycle usually requires two such operations. Therefore, the method requires 8 matrix-vector multiplications. The dimensionality of the vector is given by the number of mesh points N_R times the number of electron orbitals N_e. The operator $\hat{H}$ is a sparse matrix with M_H nonzero elements per row. This sparseness of the Hamiltonian matrix in a real space formulation is determined by the finite difference formula for the kinetic energy (a seven or nine-point formula in most cases [Marques 2003b]); and by the nonlocal parts of the potential. In most cases, however, it is the non-local part of the pseudopotential that determines M_H. Thus, the basic matrix-vector multiplication requires around $M_H N_\mathrm{e} N_R$ complex floating point operations. (In the case of using a Fourier space representation for the wave-functions, the matrix vector multiplication has a slightly different scaling with the number of atoms, namely $N_\mathrm{e} N_R \log(N_R)$, due to the use of fast-Fourier transforms.) To propagate the Kohn-Sham equations for M_T time steps then requires a total of

$$\sim M_H M_T N_\mathrm{e} N_R \tag{15.2}$$

floating point operations in real space.

Another time-consuming part of the time-propagation is the solution of the Poisson equation, which must be done twice at each time step. Several methods exist in the market to solve it, like, e.g., multipole expansions combined with relaxation methods to deal with the higher multipoles [Castro 2003]; multigrid techniques; fast-Fourier transforms, etc. In principle, these algorithms are of order N_R or $N_R \log N_R$, and should not dominate for large systems.

Storage requirements are in principle small: the vector wave functions ($N_\mathrm{e} N_R$ complex numbers) plus the Hartree, external and exchange-correlation potentials (N_R floats), the charge densities (N_R floats) and some other intermediate arrays. The storage size is therefore of the order of $(2N_e + M_{\mathrm{impl}})N_R$.

The time-propagation method has several advantages: The scaling of both CPU time and memory is quadratic with the number of atoms; only occupied states need to be propagated and stored; this method is trivially extended to non-linear response and is ideal to be combined with molecular simulations of the ions [Marques 2003b, Car 1985]. The big downside is the fairly large prefactor, that makes this method quite ineffective for the linear-response calculations of very small systems or if we are interested in a specific frequency value (e.g., static response) where the methods in frequency domain will clearly be more effective.

15.3 Sternheimer's Approach

The calculation of response properties from solely the occupied subspace is not restricted to the time-propagation formalism. In fact, matrix elements of the response function – such as polarizabilities, dielectric constants, etc. – can be computed in terms of only the ground-state occupied wavefunctions. The method is based on the perturbative approach of Dalgarno and Lewis [Dalgarno 1955], and has been used to study static polarizabilities of metallic clusters, static response in solids, and linear and non-linear dynamical responses (for more details and references, please refer to the review [Baroni 2001]).

Denoting by φ_i the ground state one-electron states with eigenvalues ε_i, and $\varphi_i^{\pm}$ the perturbed states projected on the unperturbed unoccupied subspace, we can write the dynamical polarizability

$$\alpha(\omega) = \sum_i^{\mathrm{occ}} \langle \varphi_i | \delta v_{\mathrm{ext}} \left(|\varphi_i^{+}\rangle + |\varphi_i^{-}\rangle \right) \; . \tag{15.3}$$

where the perturbed states $|\varphi_i^{\pm}\rangle$ are the solutions of the linear system

$$\left(\varepsilon_i - \hat{H}_{\mathrm{KS}}[n] \pm \omega\right) \varphi_i^{\pm} = \hat{P}_c \left[\delta v_{\mathrm{ext}}(\boldsymbol{r},\omega) + \int \mathrm{d}^3 r' \, K(\boldsymbol{r},\boldsymbol{r}',\omega) \delta n(\boldsymbol{r}',\omega) \right] \varphi_i \; . \tag{15.4}$$

Here, $\delta n = \Re \sum_i \varphi_i(\varphi_i^{+} + \varphi_i^{-})$, and $\hat{P}_c = \mathbb{1} - \sum_i^{\mathrm{occ}} |\varphi_i\rangle\langle\varphi_i|$ is the projector onto the unoccupied subspace. $\hat{H}_{\mathrm{KS}}[n]$ is the unperturbed Kohn-Sham Hamiltonian for the density $n(\boldsymbol{r}',\omega)$. Note that the r.h.s. of (15.4) corresponds to the response of the Kohn-Sham system to a sinusoidal perturbation potential that combines the external field δv_{ext} and the internal field from the time-varying electronic density. The kernel K introduced in previous chapters is defined by

$$K(\boldsymbol{r},\boldsymbol{r}',\omega) = \frac{1}{|\boldsymbol{r}-\boldsymbol{r}'|} + \frac{\delta v_{\mathrm{xc}}(\boldsymbol{r},\omega)}{\delta n(\boldsymbol{r}',\omega)} \; . \tag{15.5}$$

Equation (15.4) is very similar to the Sternheimer equation that appears in adiabatic perturbation theory, but with an additional piece coming from the $\pm\omega$ term on the left-hand-side. Thus, for each ground-state wave function, we get two first-order wave functions, which are equal in the static case. Note that the solution of (15.4) requires only the unperturbed valence wavefunctions φ_i and their linear variation $\varphi_i^{\pm}$. A similar general expression for the second-order susceptibility in terms of only ground-state and first order perturbed valence orbitals is given in [Dal Corso 1996, Nunes 2001, Gonze 1995a]. This formulation makes the evaluation of linear and non-linear response functions in systems containing up to a hundred atoms feasible.

The solution of (15.4) can be obtained by using an iterative method [Baroni 2001] or by minimizing a suitably defined functional, with a numerical effort similar to a ground-state calculation [Giannozzi 1994, Pasquarello 1993]. Let us discuss in more detail the real-space implementation of these equations as done, for example, in [Iwata 2000]. The algorithm uses a double iteration technique very similar to the one used for the solution of the ground-state Kohn-Sham equations. We start by making a guess for the induced density δn; (15.4) is solved by the conjugated gradients method; δn is then refined from the resulting $\varphi_i^{\pm}$; and so on until convergence. As the conjugated gradients method is again based on Hamiltonian vector multiplications, the total numerical cost depends largely on the cost of this operation ($\approx M_H N_{\text{e}} N_R$ in coordinate space). Remembering that we need M_{it} to reach convergence (preconditioning is a key issue here), and that the calculation has to be repeated for every one of the M_ω frequencies, we arrive at the following computation cost

$$M_{\text{it}} M_\omega M_H N_{\text{e}} N_R \,. \tag{15.6}$$

Unfortunately, it is difficult to give an *a priori* estimate of M_{it} or of its size-scaling properties. (We have nevertheless assumed that it does not grow with the number of atoms in the system.) This scheme is widely used to get many response functions including phonon frequencies, macroscopic dielectric constants, etc. [Baroni 2001].

The operator $\hat{P}_c$ projects on the unoccupied wave-function subspace the right hand side of (15.4). This projector operator ensures that the response is orthogonal to the occupied subspace, and can be implemented using a Gram-Schmidt algorithm. Note that such procedure scales like $N_{\text{e}}^2 N_R$, and will therefore be the dominant step for sufficient large number of atoms.

The storage requirements of this method are: the (real) ground-state wavefunctions φ_i; the (real) linear variations $\varphi_i^{\pm}$, and some working arrays of dimension N_R. This amounts to the total storage of $(3N_{\text{e}} + M_{\text{impl}})N_R$ floats.

In conclusion, the scaling of the Sternheimer method is really competitive, both in terms of computing time and storage. If only a few frequencies are required, it is clearly the method of choice. However, if a whole spectrum is needed, the choice between the Sternheimer and the time-evolution method will basically depend on the respective implementations.

15.4 Casida's Equation

One of the most popular methods for the calculation of excitation energies is due to Marc Casida [Casida 1995a]. For systems with a discrete spectrum of excitation (like finite systems) it is possible to recast the response equation into the pseudo-eigenvalue equation,

$$\hat{R}F_q = \Omega_q^2 F_q \,, \tag{15.7}$$

where the eigenvalues Ω_q^2 are the square of the excitation energies, and the eigenvectors F_n are related to the oscillator strengths. The matrix operator $\hat{R}$ is given in terms of

$$R_{q,q'} = (\varepsilon_{a\sigma} - \varepsilon_{i\sigma})^2 \delta_{qq'} + 2\sqrt{\varepsilon_{a\sigma} - \varepsilon_{i\sigma}} K_{q,q'}(\Omega_q) \sqrt{\varepsilon_{a'\sigma'} - \varepsilon_{i'\sigma'}} \,. \tag{15.8}$$

where the compound index $q = (a, i, \sigma)$ labels the combination of the unoccupied orbital $a\sigma$ and occupied orbital $i\sigma$. The interaction $K_{q,q'}$ is simply the particle-hole matrix element of the induced interaction (15.5)

$$K_{ai\sigma,a'i'\sigma'}(\omega) = \int \mathrm{d}^3r \int \mathrm{d}^3r' \, \varphi_{a\sigma}^*(\boldsymbol{r}) \varphi_{i\sigma}(\boldsymbol{r}) \left[\frac{1}{|\boldsymbol{r} - \boldsymbol{r}'|} + f_{\mathrm{xc}\,\sigma\sigma'}(\boldsymbol{r}, \boldsymbol{r}', \omega) \right] \varphi_{a'\sigma'}(\boldsymbol{r}') \varphi_{i'\sigma'}^*(\boldsymbol{r}') \,. \tag{15.9}$$

There is a substantial computational cost in constructing the interaction matrix K in either Fourier or real-space. From the two terms of K, the second involving f_{xc} is usually quite simple and fast to evaluate, as the exchange-correlation kernel is local in the ALDA. The bottleneck is the calculation of the Coulomb matrix elements. As we have discussed, each one of these matrix elements takes at least $\approx N_R$ operations, and there are $N_\mathrm{e}^2 N_\mathrm{c}^2$ such elements. This means that the construction of the full matrix has a leading order of $N_R N_\mathrm{e}^2 N_\mathrm{c}^2$. Of course, the introduction of a finite basis reduces the problem to a smaller matrix algebra, as $N_{\mathrm{LO}} \ll N_R$). However, this change only affects the prefactor, as the scaling $N_{\mathrm{LO}} N_\mathrm{e}^2 N_\mathrm{c}^2$ still increases with the fifth power of the number of atoms in the system.

Once the matrix is constructed, the diagonalization requires $\approx N_\mathrm{c}^3 N_\mathrm{e}^3$ operations. However, in this context the effort required to diagonalize the matrix is usually small compared to the time necessary to construct it. Furthermore, as we are normally interested in low-lying excitations, we can benefit from the performance of iterative solutions techniques to get the first eigenstates of (15.7).

The requirements for storing the matrix grow as $N_\mathrm{c}^2 N_\mathrm{e}^2$. However, in spite of the fairly bad scaling, the prefactor is quite small. Furthermore, the use of iterative techniques to diagonalize the matrix does not require to explicitly construct or store the matrix. In this case, one can also take maximum advantage of the methodology already implemented for the ground-state (in particular for the expensive calculation of the Coulomb matrix elements, as implemented, e.g., in Turbomole [Furche 2002a]). The storage for the orbitals is $\approx (N_\mathrm{c}+N_\mathrm{e})N_R$ floats, which is comparable to the time-evolution and Sternheimer methods.

15.5 Response Equation in Momentum/Frequency Space

Within linear response theory, the response of an electronic system to an external perturbation δv_ext can be cast in term of the inverse microscopical dielectric function $\varepsilon^{-1}(\omega) = 1 + v_\mathrm{ee}\chi(\omega)$, that relates the total effective potential to the external one. Here the spatial dependence has been omitted for simplicity, v_ee is the bare Coulomb interaction, and χ is the linear density response function that relates the response of the charge δn to the external potential

$$\delta n = \chi \delta v_\mathrm{ext} \,. \tag{15.10}$$

The linear response matrix χ is constructed in momentum space from the following matrix inversion

$$\chi = (1 - \chi_\mathrm{KS} K)^{-1} \chi_\mathrm{KS} \,, \tag{15.11}$$

where the independent particle response χ_KS is defined as

$$\chi_\mathrm{KS}(\boldsymbol{r}, \boldsymbol{r}', \omega) = \lim_{\eta \to 0^+} \sum_{jk}^{\infty} (n_k - n_j) \frac{\varphi_j(\boldsymbol{r})\varphi_j^*(\boldsymbol{r}')\varphi_k(\boldsymbol{r}')\varphi_k^*(\boldsymbol{r})}{\omega - (\varepsilon_j - \varepsilon_k) + \mathrm{i}\eta} \,, \tag{15.12}$$

with elements $\boldsymbol{G}, \boldsymbol{G}'$ given by

$$\chi_\mathrm{KS}(\boldsymbol{G}, \boldsymbol{G}', \omega) = \frac{1}{\mathcal{V}} \sum_{kj} (n_k - n_j) \frac{\langle k|\mathrm{e}^{-\mathrm{i}\boldsymbol{G}\cdot\boldsymbol{r}}|j\rangle\langle j|\mathrm{e}^{\mathrm{i}\boldsymbol{G}'\cdot\boldsymbol{r}}|k\rangle}{\omega - (\varepsilon_j - \varepsilon_i) + \mathrm{i}\eta} \,, \tag{15.13}$$

where $\mathcal{V}$ denotes the unit-cell volume, j and k label Kohn-Sham eigenfunctions and ε_k and n_k are the corresponding eigenenergies and occupation factors. The sum goes over N_e occupied orbitals and N_c empty orbitals. The interaction K is the Fourier transform of (15.5).

We now describe the computation starting from the Kohn-Sham wave functions and energies in a momentum space representation.[2] To evaluate the independent particle response χ_KS using (15.13), one first calculates the particle-hole matrix elements of the momentum operator and stores them in a table (in memory or on disk). The computational effort of this operation

[2] For performing response calculations based on a supercell approach we refer the reader to two available codes SELF (`http://people.roma2.infn.it/~marini/self/`) and DP (`http://theory.lsi.polytechnique.fr/codes/dp/dp.html`).

is of the order of $N_e N_c N_G^2$, and the size of the table to be stored amounts to $N_e N_c N_G$ complex numbers. Then, (15.13) has to be evaluated for N_G^2 pairs of $\boldsymbol{G}$ and $\boldsymbol{G}'$, each requiring a particle-hole summation, which gives $\approx N_e N_c N_G^2$ operations for each frequency. If we had to make a full space calculation, the number of empty orbitals summed in (15.13) would be of the same order as the dimensionality of the space (measured by the number of grid points or basis functions). However, the number of empty orbitals can be severely truncated without affecting the long-wavelength dipole response in the relevant spectral range. This is a reasonable approximation, as we are only interested in getting the optical spectra for excitation energies below 20 eV. An important cost saving in building up the response matrix in frequency domain is achieved in this way.

Another saving comes from the truncation of the response matrix, by assuming that the inhomogeneity of the density variations is smaller than the variations in the density itself. In practice, this implies that the off-diagonal elements of the response function are set to zero for G-vectors outside a sphere of radius much smaller than the plane wave cutoff used for the description of the density.

There are now three steps to evaluate (15.11) either in real or Fourier space representations: two matrix multiplications and a matrix inversion. The matrices are dense, so the matrix multiplications cost N_G^3 arithmetic operations.[3] The matrix inversion is of the same order, requiring N_G^3 operations. The total is $\approx 3N_G^3$. These are the most computationally demanding steps in the method, given the truncation in the N_c.

Then, the total computational effort in the super-cell method in Fourier space is

$$M_\omega (N_c N_e N_G^2 + 3N_G^3) \, , \tag{15.14}$$

with the first term dominant. The storage requirements for all the occupied and unoccupied wave functions plus the whole complex response matrix is of the order of

$$(N_e + N_c) N_G + 2N_G^2 + N_e N_c N_G \, . \tag{15.15}$$

In conclusion, the direct application of this method requires a time effort that scales with the third power of the number of atoms and a memory effort that scales quadratically. However, by making use of the real-space representation we can gain a lot (see below) at the cost of two fast Fourier transforms from G-space to coordinate space (this takes $\approx 2N_G \log N_G$ operations).

[3] A small technical point should be mentioned, regarding the divergence of the Coulomb interaction at $\boldsymbol{G} = \boldsymbol{G}'$. This is dealt with by taking a numerical limit as $|\boldsymbol{G} - \boldsymbol{G}'| \to 0$, the cost of which has to be added to the number of operations for computing the matrix product [Hybertsen 1987].

15.6 Space-Time Method for Response Calculations

Methods based on a space and time representation of the response functions take advantage of the rather sparse Hamiltonian matrix in a coordinate representation [Marques 2003b, Yabana 1996, Yabana 1999b, Yabana 1999c, Yabana 1999a, Sugino 1999, Beck 2000, Rojas 1995, Blase 1995, Chelikowsky 1994a]. For extended systems, the translational periodicity of the lattice can be efficiently taken into account in the description of the dielectric response by means of a mixed space formalism [Blase 1995]. This method has been shown to be advantageous compared to Fourier-space representations for super-cell calculations with large vacuum regions (as in the case of surface calculations) and to the real-space representation for periodic systems with small Wigner-Seitz cells [Blase 1995]. However, for bulk systems with a large unit cell or for nanostructures, it is equivalent to the real-space method.

For large systems, the great advantages of either the real or mixed-space approaches are related to the localization range of the response functions of interest, namely the independent-particle polarizability $\chi_{\rm KS}(\boldsymbol{r},\boldsymbol{r}',\omega)$, and the dielectric matrix $\epsilon(\boldsymbol{r},\boldsymbol{r}',\omega)$. We note that localized objects are easily described in real space and different basis functions can be used to describe different regions of space. In practice, the response functions of nonmetallic systems decay rapidly as $|\boldsymbol{r}-\boldsymbol{r}'| \to \infty$, so that, for each $\boldsymbol{r}$, $\chi_{\rm KS}(\boldsymbol{r},\boldsymbol{r}',\omega)$ needs to be calculated only for $\boldsymbol{r}'$ inside a spherical region of radius $R_{\rm max}$ around $\boldsymbol{r}$ (see Fig. 15.1) [Blase 1995]. This is the origin of the success of recent N-linear methods (where N is the number of atoms in the unit cell) proposed to perform band-structure calculations in solids. However, for metals and small-gap semiconductors, the decay rate may be slow and $R_{\rm max}$ may span many unit cells, so that the computational effort in a pure real-space method would be substantial. This problem is solved in the mixed-space representation $\chi^0_{\boldsymbol{q}}(\boldsymbol{r},\boldsymbol{r}',\omega)$, because $\boldsymbol{r}$ and $\boldsymbol{r}'$ are restricted only to a single Wigner-Seitz cell and $\boldsymbol{q}$ spans the irreducible part of the Brillouin zone. This formalism saves a lot of computing time for large-periodic super-cells and clusters of many-atoms, and entails the calculation of response functions of having a favorable scaling with the number of atoms (see below).

This response method is adequate to be combined with real-space calculations of the ground state by means of the finite-difference pseudopotential method [Chelikowsky 1994a] or with adaptive coordinates [Gygi 1995b, Gygi 1995a]. A real-space discretization of the kinetic energy operator leads to sparse Hamiltonian matrices which do not need to be stored in memory and are easy to handle. The localization of the response functions is used in the implementation of the method to different cases, but we stress that appreciable computational savings are obtained only for systems having a volume larger than the volume of the localization sphere.

In the space-time representation, the non-interacting time-ordered density-density response function is given by

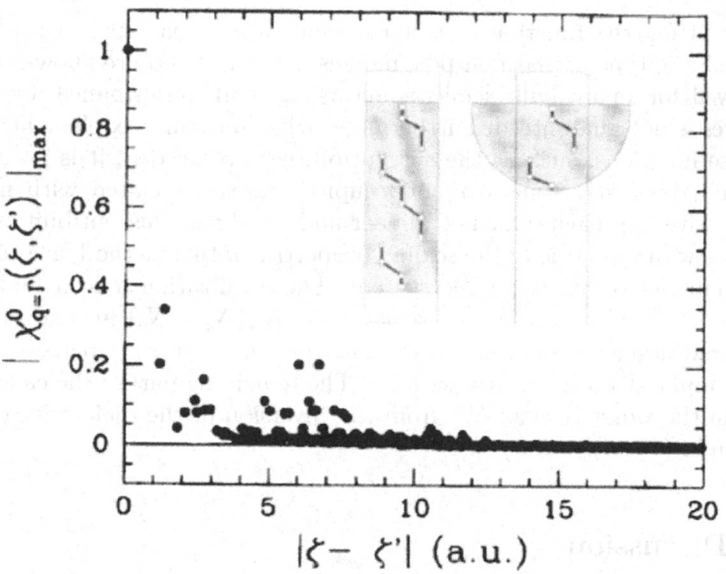

Fig. 15.1. Value of $|\chi^0_{q=\Gamma}(\zeta-\zeta')|$ at the Γ point renormalized to its value at $\zeta-\zeta'$=0 as a function of the distance $|\zeta-\zeta'|$ for the surface of a H/Si(111)-(1x1) 14-layer slab. ζ and ζ' denote points within the Wigner-Seitz cell. A real space cutoff of 18 a.u. can be used keeping the calculated χ^0 in perfect agreement with the exact one. Inset: a symbolic representation of the cross-section in the [110] plane of a Wigner-Seitz cell for the H/Si(111) slab. The *sphere* represents the effective Wigner-Seitz cell spanned by ζ' in the calculation of $\chi^0_{q=\Gamma}(\zeta-\zeta')$ in the real-space approach for ζ on a H atom; however in the mixed-space method only the part of the semi-circle inside the Wigner-Seitz cell needs to be considered. The *filled circles* represent the Si atoms and the *empty circles* the H atoms. Bonds are represented schematically by thick *solid lines*

$$\chi_{\text{KS}}(\boldsymbol{r},\boldsymbol{r}',\tau) = -\text{i}G_{\text{KS}}(\boldsymbol{r},\boldsymbol{r}',\tau)G_{\text{KS}}(\boldsymbol{r}',\boldsymbol{r},-\tau)\,, \tag{15.16}$$

where G_{KS} is the non-interacting time-ordered Green's function defined as

$$G_{\text{KS}}(\boldsymbol{r},\boldsymbol{r}',\tau) = \begin{cases} \text{i}\sum_k^{\text{occ}} \varphi_k(\boldsymbol{r})\varphi_k^*(\boldsymbol{r}')\exp(\text{i}\varepsilon_k\tau)\,, & \tau<0 \\ -\text{i}\sum_k^{\text{unocc}} \varphi_k(\boldsymbol{r})\varphi_k^*(\boldsymbol{r}')\exp(\text{i}\varepsilon_k\tau)\,, & \tau>0 \end{cases} \tag{15.17}$$

We should notice at this stage that the representation in imaginary time makes the evaluation of the summation over unoccupied states in (15.17) converge rapidly for semiconductors because of the decaying exponential factors [Rojas 1995, Blase 1995], as long as small times and large frequencies are not relevant for the physical process under study. However, the price one has to pay for this approach is that imaginary frequencies appear in the response functions and one has to resort to analytical continuation to the real energy axis by fitting the function to a multipolar form [Rojas 1995, Blase 1995]. Although the type of plasmon-pole models normally used are known to work quite well for many bulk semiconductors, a good performance for nanostructures is not guaranteed. Furthermore, when only matrix elements of the response functions, such as the susceptibilities, are needed, it is possible to eliminate the costly sums over unoccupied states associated with normal-perturbative approaches for both linear and non-linear susceptibilities.

We now discuss briefly the scaling properties of this method as a function of the number of atoms in the system. The localization length makes the effort to compute χ_{KS} and G_{KS} scale as $M_H N_R (N_\mathrm{c} + N_\mathrm{e})$ instead of N^3 or N^4 of traditional calculations (note that the number of $\boldsymbol{r}$, $\boldsymbol{r}'$ pairs that have to be computed scales linearly with N). The remaining part of the calculation scales in the worst case as N^3 [from the inversion of the dielectric response function in (15.11)].

15.7 Discussion

In the theory of electronic excitations of finite many-electron systems, the time-dependent Kohn-Sham equations with an ALDA for the exchange-correlation functional offers an attractive compromise towards the goals of accuracy and computational practicality. But even within the TDLDA scheme there are several methods in use, and our objective was to compare them on the same footing. For that purpose, we addressed the problem of the scaling with system size of the different methods.

In comparing the methods, we have deliberately ignored the first step in any approach, the construction of the eigenstates of the static Kohn-Sham operator. In the time-evolution and Sternheimer methods only the occupied orbitals are needed, but for the standard perturbative methods also a large number of unoccupied orbitals. Their calculation scales, in principle, like N^3, but in practice this phase of the computation is short compared to the dynamic calculation and so we can ignore it. Let us now compare the scalings by taking the expressions in Table 15.2 and dropping the subscripts of the N quantities. The time-evolution and Sternheimer methods scale as N^2. Linear response in frequency space has a poorer scaling behavior, namely $N^3 - N^4$ for building χ_{KS} (then we have to add the common matrix multiplication and inversion). The final method we discussed, the matrix method using a mixed space (or real space), seems to have a good N-scaling (N^2 for the independent response and N^3 for inversion and matrix multiplication with a smaller prefactor than the first term), but may be advantageous only in some circumstances and implementations [Furche 2002a] (still the inversion of the response-matrix equation will dominate for very large systems). Besides arithmetic operations, storage can play a role in the practicality of the different algorithms for large systems. Here, we find that the storage requirements favor the time evolution.

Table 15.2. Leading order for the scaling of various algorithms in terms of floating point operations (FPO) and memory requirements. We did not include the prefactors that are important for specific applications

Method	FPO	Memory
Time-evolution	$M_H M_T N_e N_R$	$N_e N_R$
Sternheimer	$M_\omega M_{it} M_H N_e N_R$	$(N_e + N_c) N_R$
Casida	$N_c^2 N_e^2 N_R$	$N_e^2 N_c^2$
Response in Fourier space	$M_\omega (N_c N_e N_G^2 + 3N_G^3)$	N_G^2
Response in mixed/real space	$M_\omega (M_H N_R (N_e + N_c) + 3N_R^3)$	$N_R M_H$

Thus, our results favor the time-evolution method, which offers economy in both storage and arithmetic operations. Closely following comes the Steinheimer approach. However, there are a number of caveats. We have not considered the suitability of the different algorithms for parallel computing. In a parallel computing environment, the frequency-space methods gain favor because the M_ω factor can be trivially absorbed in the parallel processing. In addition, Casida's method can benefit from the parallel computation of different rows of the matrix. Also the sparseness of the Hamiltonian matrix is important for the real space method; this would be lost if for example the energy functional used the full Fock exchange interaction. Finally, we have not discussed the prefactors. These are, of course, dependent on the specific implementation and on the hardware. However, it is frequently true that methods with worst scalings have more favorable prefactors. For example, Casida's method has a much smaller prefactor than the Steinheimer or the time-evolution approaches. Also solving the response equation in Fourier space has a smaller prefactor than solving it in a mixed space.

Let's go back to our student (if he or she hasn't run away by now, scared by this profusion of methods). How to make a judicious choice? All the approaches presented have their advantages and disadvantages, and there is no "optimal" method for all classes of problems. Perhaps the best strategy is to try several different methods and implementations in a test system and benchmark them thoroughly on the hardware available. Besides yielding the "best" choice, this strategy will also educate our student in the ingenuity that led to the variety and diversity of methods that we now have at our disposal.

To conclude, we would like to address a question that goes somewhat beyond the topic of this chapter: Are all these methods really necessary? Will they all survive, or do they follow a Darwinistic pattern of creation and extinction? The answer to this question is far from obvious. If we accept the theory of evolution *à la* Darwin, or its more modern improvements like the theory of punctuated equilibrium, we can expect that over the years one method/program will dominate the market. If, on the other hand, we believe the creationist arguments then "Il responsabile istituzionale della separazioni di ogni genere è infatti il Diavolo, il cui nome deriva appunto dal grego *diabolé*, «divisione» o «disunione»" [Odifreddi 2004].[4] In this latter case, we have to learn to survive in diversity.

Acknowledgments

The authors were partially supported by the EC 6th framework Network of Excellence NANOQUANTA (NMP4-CT-2004-500198). MALM also acknowledges financial support by the Marie Curie Actions of the EC (Contract No. MEIF-CT-2004-010384). The authors wish to thank Silvana Botti for her help in the preparation of this manuscript.

[4] The institutional responsible for every kind of differentiation is, in fact, the Devil, he whose name comes from the Greek *diabolé*, «division» or «separation».

Part IV

Applications: Linear Response

16 Linear-Response Time-Dependent Density Functional Theory for Open-Shell Molecules

M.E. Casida, A. Ipatov, and F. Cordova

16.1 Introduction

While typical stable organic and many stable inorganic molecules have closed-shell ground states, interesting chemistry and molecular physics is by no means limited to these species. For example, O_2 is a common molecule with a triplet ground state and whose spectroscopic importance is dramatically illustrated by its role in the photochemical explanation of the aurora borealis, and the ultraviolet spectra of the high spin d^6 complex that ferrous cation forms in water, $[Fe(H_2O)_6]^{2+}$, is a source of information for fixing ligand field parameters. Excited states of molecules with open-shell ground states also appear as higher energy peaks in photoelectron (ionization) spectra. Moreover nearly all of photochemistry involves some nuclear configurations which may be qualified as having open-shell ground states. It is small wonder that Casida's equations [Casida 1995a] began to be applied to calculate the spectra of open-shell molecules just five years after their introduction [Spielfiedel 1999, Hirata 1999a, Hirata 1999c, Adamo 1999, Guan 2000, Radziszewski 2000, Anduniow 2000]. While it is safe to say that the initial developpers [Casida 1995a, Petersilka 1996a, Jamorski 1996, Bauernschmitt 1996a] of linear-response time-dependent density-functional theory (LR-TDDFT) for the calculation of excitation spectra were thinking about applications to molecules with closed-shell ground states having the same orbitals for different spin (SODS), the original formulation of Casida's equations foresaw their eventual application to molecules with an open-shell ground state by allowing both for different orbitals for different spin (DODS) and for fractional occupation number [Casida 1995a]. Application of the DODS formulation of Casida's equations has led to spectacularly good agreement with experimental spectra in some cases and significant errors in interpretation of calculated results in other cases. This chapter tries to point out where DODS LR-TDDFT is a reasonable approach to the excited states of open-shell molecules and where it is likely to fail. In the cases where it is likely to fail, we give an indication of how the theory may be fixed. Some of the results reported here come from our own unpublished work [Cordova 2006, Ipatov 2006]. Although we do not have the space here to go into the details normally expected for new work, we trust that fuller accounts will eventually be published elsewhere.

M.E. Casida et al.: *Linear-Response Time-Dependent Density Functional Theory for Open-Shell Molecules*, Lect. Notes Phys. **706**, 243–257 (2006)
DOI 10.1007/3-540-35426-3_16

To our knowledge, the first application of Casida's equations to an open-shell molecule was that of Spielfiedel and Handy who used CADPAC to investigate the excited states of PO [Spielfiedel 1999]. At about the same time, both the SODS and DODS version of Casida's equations were programmed in the very popular GAUSSIAN quantum chemistry program [Stratmann 1998]. Users soon began to calculate the spectra of open-shell molecules, though there was still almost no prior experience regarding even the correct interpretation of output in this case. It is to Hirata and Head-Gordon that credit should go for the first systematic study of Casida's equations for open-shell molecules [Hirata 1999a, Hirata 1999c]. This group, who was seeking an efficient but still relatively simple configuration-interaction singles (CIS)-like method for calculating excitation energies, used the QCHEM program to investigate the use of TDDFT for the calculation of the excitation energies of radicals. As Fig. 16.1 shows for two excitation energies of the cyanide radical, DODS CIS (UCIS in the figure) and CIS beginning with a SODS spin-restricted open-shell Hartree-Fock wave function (ROCIS in the figure) show large errors. Inclusion of "extended singles," which are double excitations obtained from single excitations by spin transpositions, helps remarkably (XCIS in the figure). Still, it is the simple TDLDA which is giving the most impressive results for a computational cost similar to the relatively simple CIS. Guan et al. went further and investigated entire spectra for small molecules with open-shell ground states [Guan 2000]. Such spectra are, as a general rule, much more complex than corresponding spectra for closed-shell molecules, showing a plethora of satellite peaks due to oscillator strength fragmentation. The results of that study confirmed that *some* excitations are remarkably well described. Other early applications of Casida's equations to molecules with open-shell ground states include [Radziszewski

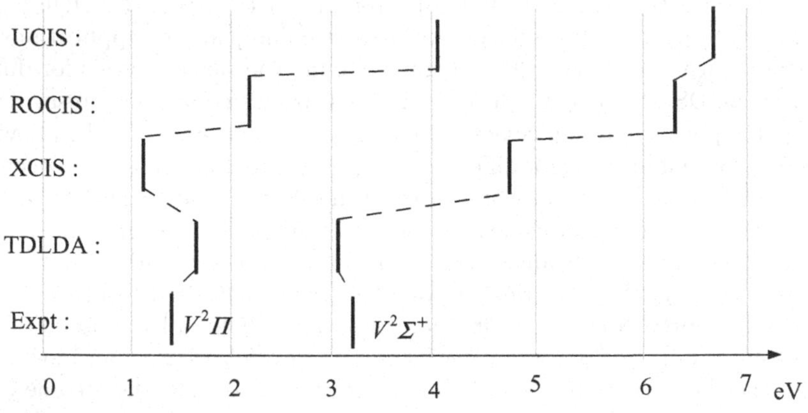

Fig. 16.1. Comparison of CN radical excitation energies calculated with different simple theories. Numerical values taken from Table 1 of [Hirata 1999a]

2000, Anduniow 2000, Brocławik 2001, Andreu 2001, Weisman 2001b, Pou-Amérigo 2002, Rinkevicius 2003].

The remainder of this chapter is organized as follows. Before discussing the excited-states of open-shell molecules, we first consider the performance of DFT for the ground state of open-shell molecules. Perhaps surprisingly one way to judge the quality of a SODS description of the Kohn-Sham ground state is to examine the first excited triplet state from Casida's equations. This also provides a good place to introduce the idea of spin contamination and symmetry-broken DODS calculations. This is followed by a section on TDDFT excitation spectra for open-shell molecules in which it is pointed out that some excited states are simpler than others and that the difficulty that TDDFT has in describing all the excited states of molecules with open-shell ground states is closely related to a failure of the adiabatic approximation. As it turns out, one way to detect and guard against the problem is to calculate spin contamination in the excited states. In the penultimate section, ways to go beyond the adiabatic approximation are very briefly discussed. The final section sums up.

16.2 Open-Shell Ground States

One of the first difficulties one runs into in discussing open-shell molecules is one of definition. While it may seem evident that an open-shell molecule is any molecule which is not closed-shell and that "closed-shell" means that the molecular wave function belongs to the completely symmetric representation of the appropriate symmetry group, the breaking of a bond typically yields a biradical whose wave function belongs to the completely symmetric representation. Nevertheless it has the chemical physics of two open-shell species! We begin first by discussing the problem of biradicals with a singlet ground state and then take a look at molecules with a non-singlet ground state.

One way to generate a biradical ground state is molecular dissociation, the bond breaking of H_2 (H_A-H_B) being the classic textbook case. Our discussion is based upon that of [Casida 2000b]. At the equilibrium geometry the Kohn-Sham wave function has the form of a single determinant, $|\sigma_\uparrow \sigma_\downarrow|$. At large internuclear distance, this wave function takes the form of a linear combination of ionic and covalent parts,

$$
\begin{aligned}
|\sigma_\uparrow \sigma_\downarrow| &= |\frac{1}{\sqrt{2}}(s_{A\uparrow} + s_{B\uparrow}), \frac{1}{\sqrt{2}}(s_{A\downarrow} + s_{B\downarrow})| \\
&= \frac{1}{2}\underbrace{\left(|s_{A\uparrow} s_{A\downarrow}| + |s_{B\uparrow} s_{B\downarrow}|\right)}_{[\mathrm{H}{:}_A^- + \mathrm{H}_B^+ \leftrightarrow \mathrm{H}_A^+ + \mathrm{H}{:}_B^-]} + \frac{1}{2}\underbrace{\left(|s_{A\uparrow} s_{B\downarrow}| + |s_{B\uparrow} s_{A\downarrow}|\right)}_{[\mathrm{H}_A\uparrow + \mathrm{H}_B\downarrow \leftrightarrow \mathrm{H}_A\downarrow + \mathrm{H}_B\uparrow]} \; . \quad (16.1)
\end{aligned}
$$

As H_2 dissociates into neutral H atoms the covalent part should dominate asymptotically, otherwise the energy is too high. This can be simulated by a DODS wave function since it can break symmetry and become $|s_{B\uparrow}s_{A\downarrow}|$. The point where the DODS wave function becomes lower in energy than the SODS wave function is often referred to as the Coulson-Fischer point. Were the exact exchange-correlation functional used in Kohn-Sham theory, the wave function should remain SODS for every bond distance. In practice the exchange-correlation functional is approximate and symmetry breaking does occur.

The Coulson-Fischer point is an example of a triplet instability. Stability conditions for DFT have been presented by Bauernschmitt and Ahlrichs [Bauernschmitt 1996b] but no explicit link was made with TDDFT excitation energies. That link was later put into print by Casida et al. [Casida 2000b]. Our goal here is not to give a complete analysis of the general case but rather to give a simple analysis showing the relation between symmetry breaking in the ground state and imaginary triplet excitation energies for the particular case of the Coulson-Fischer point. To this end, consider the single determinant Kohn-Sham wave function, $|\sqrt{1-\lambda^2}\sigma_\uparrow + \lambda\sigma_\uparrow^*, \sqrt{1-\lambda^2}\sigma_\downarrow - \lambda\sigma_\downarrow^*|$, where λ is a symmetry-breaking parameter. Expanding the energy expression in λ gives, in the notation of [Casida 2000b],

$$E_\lambda = E_0 + 2\lambda^2 \left[\Delta\epsilon + 2\left(K_{\uparrow,\uparrow} - K_{\uparrow,\downarrow}\right)\right] + \mathcal{O}(\lambda^3) \,. \tag{16.2}$$

This DODS energy becomes lower than the SODS energy when the coefficient of λ^2 becomes negative. Since the triplet excitation energy is,

$$\omega_T = \sqrt{\Delta\epsilon\left[\Delta\epsilon + 2\left(K_{\uparrow,\uparrow} - K_{\uparrow,\downarrow}\right)\right]} \,, \tag{16.3}$$

this is exactly the point where the TDDFT triplet excitation energy becomes imaginary, hence nonphysical. Explicit calculations on H_2 [Casida 2000b] confirm that the TDDFT triplet excitation energy becomes grossly underestimated at bond distances shorter than the Coulson-Fischer point and that the degradation of the quality of the excitation energy also shows up in singlet excitation energies. Although an elegant solution in some ways, symmetry breaking is also a problem in other ways. The Coulson-Fischer point is known to occur at larger intermolecular distance in DFT than in Hartree-Fock calculations, and must disappear entirely (i.e., move to infinity) in the limit that the exchange-correlation functional becomes exact. Nevertheless, *the occurace of imaginary excitation energies at some molecular configurations is unphysical and may be considered to be one danger of open-shell TDDFT.*

The ring opening of oxirane (Fig. 16.2) provides a concrete example of how this analysis applies to something less trivial than H_2. The ground and triplet excited states are shown in Fig. 16.3 for conrotatory and disrotatory ring opening [Cordova 2006]. As expected from the famous Woodward-Hoffmann rules, the conrotatory reaction is favored over the disrotatory reaction. More interesting for present purposes are the ridges which correspond to regions of configuration space where the ground state surface approaches

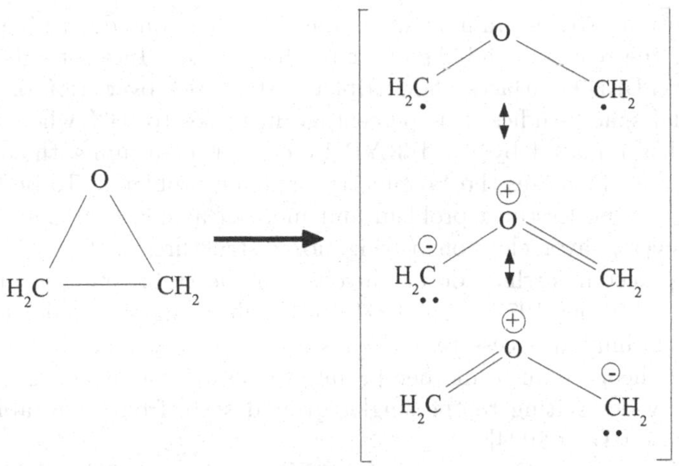

Fig. 16.2. Lewis structures for the CC ring opening of oxirane. If the two methyl groups rotate in the same direction during the ring opening, preserving C_2 symmetry, then the ring opening is said to be conrotatory. If the two methyl groups rotate in the opposite direction during the ring opening, preserving C_s symmetry, then the ring opening is said to be disrotatory

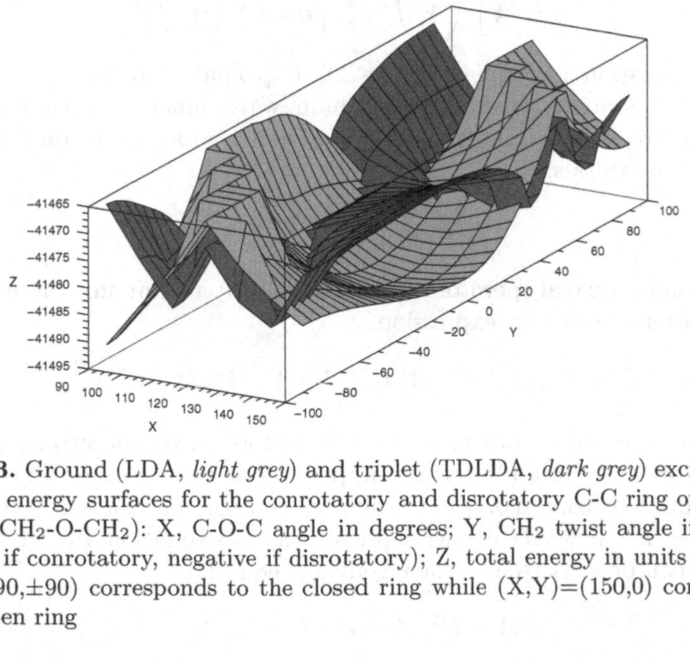

Fig. 16.3. Ground (LDA, *light grey*) and triplet (TDLDA, *dark grey*) excited state potential energy surfaces for the conrotatory and disrotatory C-C ring opening of oxirane (CH_2-O-CH_2): X, C-O-C angle in degrees; Y, CH_2 twist angle in degrees (positive if conrotatory, negative if disrotatory); Z, total energy in units of 10 eV. (X,Y)=(90,±90) corresponds to the closed ring while (X,Y)=(150,0) corresponds to the open ring

an excited state potential energy surface. By convention, a triplet state lying below the ground state ("negative" excitation energy) indicates that the triplet excitation has become imaginary. Imaginary triplet excitation energies are occuring around the ridges. Quantum chemists normally describe these regions with a two-determinant wave function where one determinant corresponds to the reactant and the other to the product. Interestingly enough, while the TDLDA surfaces show triplet instabilities over 51% of the configurational space studies, this percentage increases to 93% when the LDA functional is replaced by the B3LYP functional, confirming that Hartree-Fock exchange increases the "symmetry-breaking problem." To be fair however this may be less of a problem and more of a reflection that the LDA tends to overly favor electron pairing (ionic structures in Fig. 16.2). Typical DFT procedures which do not involve symmetry breaking are multiplet sum theory [Ziegler 1977, Daul 1994] and multiconfigurational DFT. Both also provide limited access to excited states. Spin-flip noncollinear density functional theory is also another promising option for avoiding symmetry breaking by de-exciting to the singlet ground state from a suitable triplet excited state [Wang 2004].

The Tamm-Dancoff approximation (TDA) [Hirata 1999b] decouples the ground state stability problem from the excited-state problem giving qualitatively correct results [Casida 2000b]. The TDA for LR-TDDFT may be understood as an approximation to Casida's equation written as,

$$\begin{bmatrix} \boldsymbol{A} & \boldsymbol{B} \\ \boldsymbol{B} & \boldsymbol{A} \end{bmatrix} \begin{pmatrix} \boldsymbol{X} \\ \boldsymbol{Y} \end{pmatrix} = \omega \begin{bmatrix} \mathbf{1} & \mathbf{0} \\ \mathbf{0} & -\mathbf{1} \end{bmatrix} \begin{pmatrix} \boldsymbol{X} \\ \boldsymbol{Y} \end{pmatrix} . \tag{16.4}$$

Casida's equation is coupled to the DFT ground state stability problem because the stability of the Kohn-Sham wave function with respect to symmetry-breaking can be tested by considering an arbitrary unitary transformation of orbitals,

$$\varphi_r^\lambda(\boldsymbol{r}) = \mathrm{e}^{\mathrm{i}\lambda(\hat{R}+i\hat{I})}\varphi_r(\boldsymbol{r}) , \tag{16.5}$$

where $\hat{R}$ and $\hat{I}$ are real operators [Casida 2002]. After a fair amount of algebra, one arrives at the energy expression,

$$E_\lambda = E_0 + \lambda^2 \left[\boldsymbol{R}^\dagger (\boldsymbol{A}-\boldsymbol{B})\boldsymbol{R} + \boldsymbol{I}^\dagger (\boldsymbol{A}+\boldsymbol{B})\boldsymbol{I}\right] + \mathcal{O}(\lambda^3) , \tag{16.6}$$

where matrix elements of the $\hat{R}$ and $\hat{I}$ operators have been arranged in column vectors and the $\mathcal{O}(\lambda)$ term disappears because the energy has already been minimized before considering symmetry-breaking. The presence of the terms $(\boldsymbol{A} \pm \boldsymbol{B})$ shows the connection with Casida's equation. In fact, Casida's equation can be rewritten as the eigenvalue equation,

$$(\boldsymbol{A}+\boldsymbol{B})(\boldsymbol{A}-\boldsymbol{B})\boldsymbol{Z}_I = \omega_I^2 \boldsymbol{Z}_I . \tag{16.7}$$

For pure DFT, assuming that the *aufbau* principle is obeyed, the matrix $(\boldsymbol{A}-\boldsymbol{B})$ is always positive definite. However $(\boldsymbol{A}+\boldsymbol{B})$ may have negative eigenvalues. In that case, the energy E_λ will fall below E_0 for some value of $\boldsymbol{I}$. At the same time, this will correspond to a negative value of ω_I^2 (i.e., an imaginary value of ω_I.) This curious mathematical relationship is exactly the famous triplet instability. The TDA consists of setting $\boldsymbol{B} = \boldsymbol{0}$. Not only does this decouple the LR-TDDFT excitation energy problem (which no longer involves $\boldsymbol{B}$) from the ground state stability problem (which still involves $\boldsymbol{B}$), but the resulting TDA equation,

$$\boldsymbol{A}\boldsymbol{X} = \omega \boldsymbol{X} \; , \tag{16.8}$$

is the exact TDDFT analogue of the configuration interaction singles (CIS) method (it *is* the CIS method if we accept that Hartree-Fock is a particular case of a hybrid density functional!) Since the CIS method is also a variational method, it is free of the "variational collapse" observed in time-dependent Hartree-Fock when the square of the excitation energy goes first to zero and then becomes imaginary. This is exactly why the Tamm-Dancoff "approximation" to Casida's equation is expected to behave *better* than Casida's equation for calculating excitation energies away from the ground state equilibrium geometry.

Let us now put biradicals aside and focus on open-shell molecules in the sense of spin symmetry. A general many-electron spin eigenfunction must be a simultaneous eigenfunction of the operators,

$$\hat{S}_z = \frac{1}{2}\left(\hat{n}_\uparrow - \hat{n}_\downarrow\right) \tag{16.9a}$$

$$\hat{S}^2 = \sum \hat{P}_{\uparrow,\downarrow} + \hat{n}_\uparrow + \hat{S}_z\left(\hat{S}_z - \hat{1}\right) \; , \tag{16.9b}$$

where the spin number operators may be expressed in second-quantized notation as,

$$\hat{n}_\sigma = \sum r_\sigma^\dagger r_\sigma \; , \tag{16.10}$$

and,

$$\sum \hat{P}_{\uparrow,\downarrow} = \sum r_\downarrow^\dagger s_\uparrow^\dagger s_\downarrow r_\uparrow \; , \tag{16.11}$$

is the spin-transposition operator. All SODS single determinantal wave functions are eigenfunctions of $\hat{S}_z$ but only closed-shell and half-closed-shell determinants are eigenfunctions of $\hat{S}^2$. However DODS determinants often become linear combinations of determinants when expressed in terms of SODS. This means that DODS wave functions may be or nearly may be simultaneous eigenfunctions of $\hat{S}_z$ and $\hat{S}^2$. Since most applications of DFT to open-shell systems are of the spin-unrestricted (DODS) type, it is important to be able to calculate the degree of spin-contamination in the DODS determinant. This is most easily done by realizing that spin-unrestricted calculations generate two molecular orbital basis sets – one for up spin and one for down spin. Denoting the latter by an overbar, the spin-transposition operator becomes,

$$\sum \hat{P}_{\uparrow,\downarrow} = \sum \bar{p}_\downarrow^\dagger s_\uparrow^\dagger \bar{q}_\downarrow r_\uparrow \langle s|\bar{q}\rangle\langle \bar{p}|r\rangle \; . \tag{16.12}$$

Table 16.1. Spin-contamination in LDA calculations of some small molecules

	S^2	Multiplicity $(2S+1)$
BeH	0.7503	2.0003
CN	0.7546	2.0046
CO^+	0.7620	2.0120
N_2^+	0.7514	2.0014
CH_2O^+	0.7512	2.0012

From this it is easy to deduce that, for the ground state, the spin contamination for a DODS determinant is,

$$\langle \hat{S}^2 \rangle = \left(\frac{n_\uparrow - n_\downarrow}{2} \right)^2 + \frac{n_\uparrow + n_\downarrow}{2} - \sum_{i,\bar{j}}^{\text{occ}} |\langle i|\bar{j}\rangle|^2 \, . \tag{16.13}$$

Table 16.1 shows some typical values of the ground state-spin contamination in spin-unrestricted LDA calculations for some small radicals. In this case, the spin contamination is small. Note that, strictly speaking, this gives the $\langle \hat{S}^2 \rangle$ value of the fictious Kohn-Sham system of noninteracting electrons and not necessarily that of the physical system. It is however the best we have for practical purposes and few people, if any, would doubt its diagnostic value.

16.3 Open-Shell Excitation Spectra from TDDFT

We want to understand why LR-TDDFT may fail for the excited states of open-shell molecules. To do so, we begin with a simple three orbital model and solve the spin problem assuming SODS. This provides some guide lines for when open-shell excitation energies should be trusted. Then we return to the DODS problem and show that spin-contamination may be used as a guide to which TDDFT excited states are nonsense.

Figure 16.4 shows which excitations are needed for a minimal description of our three orbital model. A spin-adapted basis set for this model consists of four doublet functions,

$$|D_1\rangle = |i_\uparrow i_\downarrow a_\uparrow\rangle \tag{16.14a}$$
$$|D_2\rangle = |i_\uparrow v_\downarrow v_\uparrow\rangle \tag{16.14b}$$
$$|D_3\rangle = \frac{1}{\sqrt{2}} \left(|i_\downarrow v_\uparrow a_\uparrow\rangle - |i_\uparrow v_\uparrow a_\downarrow\rangle \right) \tag{16.14c}$$
$$|D_4\rangle = \frac{1}{\sqrt{6}} \left(|i_\downarrow v_\uparrow a_\uparrow\rangle + |i_\uparrow v_\uparrow a_\downarrow\rangle - 2|i_\uparrow v_\downarrow a_\uparrow\rangle \right) \, , \tag{16.14d}$$

and one quartet,

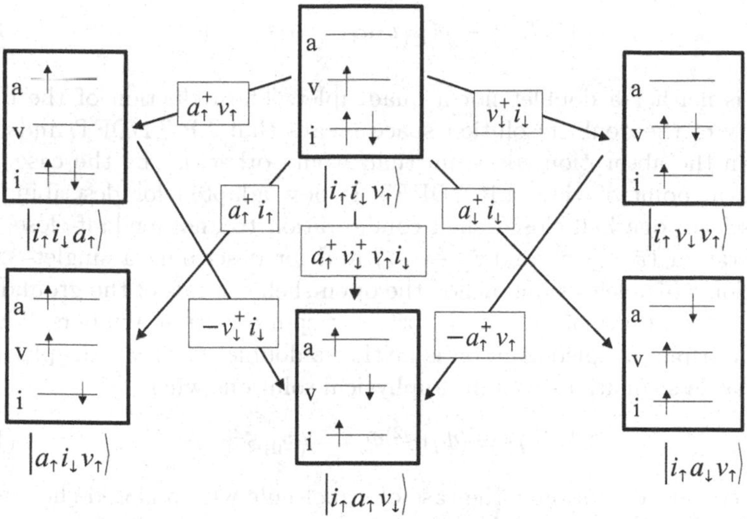

Fig. 16.4. Possible M_S-conserving excitations in a SODS 3-orbital model of a radical

$$|Q\rangle = \frac{1}{\sqrt{3}} \left(|i_\downarrow v_\uparrow a_\uparrow\rangle + |i_\uparrow v_\uparrow a_\downarrow\rangle + |i_\uparrow v_\downarrow a_\uparrow\rangle \right) . \tag{16.15}$$

Note the inclusion of the double (i.e., two-electron excited determinant) $|i_\uparrow v_\downarrow a_\uparrow\rangle$. Such a state has been called an "extended singles" since it differs from a true single (i.e., one-electron excited determinant) only by a permutation of spins among the spatial orbitals. Nevertheless it is a double and not a single and must be there. It is excluded from single excitation methods such as CIS and linear response time-dependent Hartree-Fock. Inspection of Casida's equations [Casida 1995a] shows that explicit double and higher excitations are also excluded from adiabatic LR-TDDFT. That is: Casida's equations are formulated within a finite basis set representation where counting arguments apply. When the adiabatic approximation is made, the number of solutions is exactly equal to the number of single excitations. *This is the second danger of open-shell TDDFT.* Of course adiabatic LR-TDDFT still includes electron correlation effects which Hartree-Fock-based methods describe using multiply-excited determinants. Thus Hartree-Fock-based methods may indicate that an excitation which is well-described by adiabatic LR-TDDFT has substantial double-excitation character even though this double-excitation character is not evident from the TDDFT calculation.

What adiabatic LR-TDDFT actually does is to treat excited states as either singlet-coupled excitations or triplet-coupled excitations. Returning to the three orbital model, one sees that this excludes the D_4 doublet function altogether and produces a triplet coupled excitation,

$$|TC\rangle = \frac{1}{\sqrt{2}} \left(|i_\downarrow v_\uparrow a_\uparrow\rangle + |i_\uparrow v_\uparrow a_\downarrow\rangle \right) , \tag{16.16}$$

which is neither a doublet nor a quadruplet. The reduction of the dimensionality of the doublet solution space means that LR-TDDFT finds fewer peaks in the absorption spectrum than would otherwise be the case. From a physical point of view, LR-TDDFT is best adapted for describing excitations from one half-closed shell configuration to another half-closed shell configuration ($v_\uparrow \to a_\uparrow$ and $i_\downarrow \to v_\downarrow$) and for describing a singlet-coupled excitation which leaves untouched the open-shell orbitals of the ground state ($i \to a$). Both types of excitations preserve spin quantum numbers. The fact that the triplet-coupled solution is neither a doublet nor a quadruplet shows that Casida's equations contain unphysical solutions when,

$$\Delta\langle \hat{S}^2\rangle_I = \langle \Phi_I|\hat{S}^2|\Phi_I\rangle - \langle \Phi_0|\hat{S}^2|\Phi_0\rangle , \tag{16.17}$$

is nonzero. An exception is the case of a molecule with a closed-shell ground state, in which case the triplet-coupled excitations are true triplets.

In practice, DFT calculations for open-shell molecules are of the DODS type. This makes it difficult to select which LR-TDDFT excitations on the basis of physical arguments alone. However we can still try to eliminate unphysical states on the basis of spin contamination. In particular,

$$\Delta\langle \hat{S}^2\rangle_I = \sum \Delta\Gamma^I_{r_\uparrow \bar{q}_\downarrow, \bar{p}_\downarrow s_\uparrow} \langle s|\bar{q}\rangle\langle \bar{p}|r\rangle , \tag{16.18}$$

where,

$$\Gamma_{rq,ps} = \langle p^\dagger s^\dagger q r\rangle , \tag{16.19}$$

is the two-electron reduced density matrix (2-RDM) and $\Delta\Gamma^I$ is the difference between the 2-RDM for the Ith excited state and the 2-RDM for the ground state. We have derived the appropriate expression for linear response time-dependent Hartree-Fock theory using the unrelaxed 2-RDM obtained by taking the derivative,

$$\Delta\Gamma^I_{rs,pq} = \frac{\partial \omega_I}{\partial [pr|qs]/2} . \tag{16.20}$$

The result is,

$$\begin{aligned}
\Delta\langle \hat{S}^2\rangle_I &= \langle \hat{S}^2\rangle_I - \langle \hat{S}^2\rangle_0 \\
&= \sum X^{I*}_{\bar{a}\bar{j}\downarrow} X^I_{\bar{a}\bar{k}\downarrow} \langle \bar{k}|i\rangle\langle i|\bar{j}\rangle + \sum X^{I*}_{ai\uparrow} X^I_{al\uparrow} \langle l|\bar{j}\rangle\langle \bar{j}|i\rangle \\
&\quad + \sum Y^{I*}_{\bar{k}\bar{a}\downarrow} Y^I_{\bar{j}\bar{a}\downarrow} \langle \bar{k}|i\rangle\langle i|\bar{j}\rangle + \sum Y^{I*}_{la\uparrow} Y^I_{ia\uparrow} \langle l|\bar{j}\rangle\langle \bar{j}|i\rangle \\
&\quad - \sum X^{I*}_{bi\uparrow} X^I_{\bar{a}\bar{j}\downarrow} \langle b|\bar{a}\rangle\langle \bar{j}|i\rangle - \sum Y^{I*}_{\bar{j}\bar{a}\downarrow} Y^I_{ib\downarrow} \langle b|\bar{a}\rangle\langle \bar{j}|i\rangle \\
&\quad - \sum X^{I*}_{\bar{b}\bar{k}\downarrow} X^I_{\bar{a}\bar{k}\downarrow} \langle \bar{b}|i\rangle\langle i|\bar{a}\rangle - \sum Y^{I*}_{\bar{k}\bar{a}\downarrow} Y^I_{\bar{k}\bar{b}\downarrow} \langle \bar{b}|i\rangle\langle i|\bar{a}\rangle
\end{aligned}$$

$$
\begin{aligned}
&-\sum X_{bk\uparrow}^{I*}X_{ak\uparrow}^{I}\langle b|\bar{i}\rangle\langle\bar{i}|a\rangle-\sum Y_{ka\uparrow}^{I*}Y_{kb\uparrow}^{I}\langle b|\bar{i}\rangle\langle\bar{i}|a\rangle \\
&-\sum X_{\bar{b}\bar{i}\downarrow}^{I*}X_{aj\uparrow}^{I}\langle j|\bar{i}\rangle\langle\bar{b}|a\rangle-\sum Y_{ja\uparrow}^{I*}Y_{\bar{i}\bar{b}\downarrow}^{I}\langle j|\bar{i}\rangle\langle\bar{b}|a\rangle \\
&+\sum X_{bi\uparrow}^{I*}Y_{\bar{j}\bar{a}\downarrow}^{I}\langle b|\bar{j}\rangle\langle\bar{a}|i\rangle+\sum X_{\bar{a}\bar{j}\downarrow}^{I*}Y_{ib\uparrow}^{I}\langle b|\bar{j}\rangle\langle\bar{a}|i\rangle \\
&+\sum Y_{\bar{i}\bar{b}\downarrow}^{I*}X_{aj\uparrow}^{I}\langle j|\bar{b}\rangle\langle\bar{i}|a\rangle+\sum Y_{ja\uparrow}^{I*}X_{\bar{b}\bar{i}\downarrow}^{I}\langle j|\bar{b}\rangle\langle\bar{i}|a\rangle\,, \qquad (16.21)
\end{aligned}
$$

where we have assumed the normalization,

$$
|\boldsymbol{X}|^2-|\boldsymbol{Y}|^2=1\,, \qquad (16.22)
$$

and i, j, $\bar{i}$, and $\bar{j}$ refer to occupied orbitals while a, b, $\bar{a}$, and $\bar{b}$ refer to unoccupied orbitals. This agrees with the CIS result of Maurice and Head-Gordon [Maurice 1995] when the Y-component is set equal to zero.

The formaldehyde cation, CH_2O^+, is a good example of the type of information provided by calculations of excited-state spin contamination in LR-TDDFT. Experimental data for the excited states of this species are available from ionization spectra of neutral formaldehyde [Bawagan 1988] and high-quality multireference configuration interaction (MRCI) calculated excitation energies are also available [Bawagan 1988, Bruna 1998]. The ground state of the cation and excitations into its singly occupied molecular orbital (SOMO) correspond to principal ionization potentials [one-hole (1h) states], normally treated in DFT using the ΔSCF method or Slater's transition orbital approximation to ΔSCF ionization potentials. However LR-TDDFT also offers the attractive possibility to be able to treat more complex ionization satellites which involve correlation between 1h states and two-hole/one-particle (2h1p) states. Table 16.2 shows the results of our TDLDA calculations on the CH_2O^+ at the equilibrium geometry of neutral formaldehyde using the two programs GAUSSIAN03 [Gaussian 2003] and our own version of DEMON2K [DEMON2K 2005]. Note that *all* of the lowest ten excited states are shown. The main numerical differences between results of the two programs come from the use of an auxiliary function-based method in DEMON2K, not used in our GAUSSIAN03 calculations. More importantly, GAUSSIAN03 automatically assigns the symmetry representation of each molecular orbital which this version of DEMON2K does not do and DEMON2K calculates spin contamination which GAUSSIAN03 does not do. So, between the two programs, we have a powerful set of tools for assigning TDLDA excited states. Spin contamination is small in this example so that interpretation is straightforward. In particular, it is immediately seen from the table that $\Delta\langle\hat{S}^2\rangle$ is close to either zero or two. The former indicates a doublet excited state. The latter indicates an unphysical triplet-coupled (TC) excited state which is neither a doublet nor a quadruplet, but which will ultimately generate a doublet and a quadruplet when coupled with suitable extended singles (i.e., doubly excited determinants). In the case of a doublet excited state, we can go further and distinguish between 1h excited states and 2h1p excited states. Comparison

Table 16.2. Spin contamination in CH_2O^+ excited states. All excited states are well below the TDLDA ionization threshold ($-\epsilon^{\uparrow}_{\text{HOMO}} = 15.1$ eV, $-\epsilon^{\downarrow}_{\text{HOMO}} = 18.2$ eV). SOMO refers to the singly occupied molecular orbital (spin up HOMO). TC refers to a triplet = coupled excitation. See the discussion in the text

	Excitation Energy (eV) ($\Delta\langle\hat{S}^2\rangle$)				
State	TDLDA[a]	TDLDA[b]	TDLDA TDA[b]	Assignment[c]	MRCI[d]
10	9.9685	9.9403	10.0286	$^2B_2$	
		(0.0127)	(0.0067)	$(2b_2^{\text{SOMO}} \to 3b_2)$	
9	9.7157	9.7036	9.7862	$^2B_2$	
		(0.0380)	(0.0367)	$(2b_2^{\text{SOMO}} \to 3b_2,$	
				$5a_1 \to 6a_1)$	
8	8.6708	8.3665	8.5006	$^{TC}B_2$	
		(2.0046)	(1.9834)	$(5a_1 \to 6a_1)$	
7	7.9952	7.8765	7.8980	$^2B_1$	3.86[e], 3.84[f]
		(0.0058)	(0.0047)	$(1b_1 \to 2b_2^{\text{SOMO}})$	
6	7.9510	7.6532	7.7987	$^{TC}B_2$	
		(1.9870)	(1.9619)	$(5a_1 \to 6a_1)$	
5	7.5641	7.5623	7.7839	$^2A_2$	
		(0.0061)	(0.0050)	$(5a_1 \to 2b_1)$	
4	5.4543	5.1002	5.2638	$^{TC}A_2$	
		(0.0064)	(1.9933)	$(5a_1 \to 2b_1)$	
3	5.1707	5.0833	5.1678	$^2A_1$	
		(2.0502)	(0.0045)	$(2b_2^{\text{SOMO}} \to 6a_1)$	
2	4.7445	4.6046	4.6862	$^2A_1$	5.30[e], 5.46[f]
		(0.0019)	(0.0019)	$(5a_1 \to 2b_2^{\text{SOMO}})$	
1	2.7014	2.6439	2.6669	$^2B_1$	5.78[e], 6.46[f]
		(0.0020)	(0.0012)	$(2b_2^{\text{SOMO}} \to 2b_1)$	

[a] GAUSSIAN03. [b] DEMON2K. [c] TDLDA TDA. [d] Multireference configuration interaction. [e] [Bruna 1998]. [f] [Bawagan 1988].

with experimental data and the results of the MRCI calculations is straightforward, at least for some states. States 2 and 7 correspond to principal ionization potentials while state 1 is a shakeup satellite of state 7. There is an obvious disagreement between the TDLDA and MRCI assignments of these excitation energies which comes from an LDA ordering of the cationic molecular orbitals,

$$\underbrace{(1a_1)^2(2a_1)^2}_{\text{core}}\underbrace{(3a_1)^2(1b_2)^2(4a_1)^2(1b_1)^2(5a_1)^2(2b_2)^1}_{\text{valence}}, \tag{16.23}$$

which differs from the expected ordering in the neutral [Bawagan 1988],

$$\underbrace{(1a_1)^2(2a_1)^2}_{\text{core}}\ \underbrace{(3a_1)^2(4a_1)^2(1b_2)^2(5a_1)^2(1b_1)^2(2b_2)^2}_{\text{valence}} .$$

On the other hand, it is an open question whether TDLDA and MRCI assignments *should* agree with each other since they refer to different one-particle reference systems. Perhaps only the total state symmetry should be taken into consideration since this is ultimately related to spectroscopic selection rules. In that case, we might try interchanging the MRCI energies for the TDLDA states 1 and 7 and numerical agreement between the two types of calculations, although still imperfect, looks a lot better.

This example is one of the first where a very detailed comparison has been made between the results of LR-TDDFT and traditional Hartree-Fock-based calculations for open-shell systems. More of this type of work will have to be done before we can be ultimately comfortable with using LR-TDDFT for calculating and assigning the spectra of such systems.

16.4 Beyond the Adiabatic Approximation

It should now be clear that multiple-electron excitations are important for the proper treatment of the excitations of molecules with open-shell ground states. However, as emphasized above, adiabatic LR-TDDFT only includes one-electron excitations (albeit "dressed" to include important electron correlation effects). There is thus a problem. It had been hoped that higher-order response theory might allow the extraction of two-electron excitations within the TDDFT adiabatic approximation [Gross 1996], but it is now clear that this is not the case. In particular, the poles of the dynamic second hyperpolarizability are identical to the poles of the dynamic polarizability, [Tretiak 2003] which is to say the one electron excitations of adiabatic LR-TDDFT. A more successful strategy has been the spin-flip TDDFT developed by Shao, Head-Gordon, and Krylov [Shao 2003, Slipchenko 2003] in which extended singles appear in TDDFT through consideration of perturbations which flip spins. Unfortunately the treatment is restricted to hybrid functionals and depends strongly on the coefficient of Hartree-Fock exchange. Wang and Ziegler have made a potentially major advance by showing that spin-flip TDDFT can be developed for nonhybrid functionals provided one begins with the non-collinear formulation of the exchange-correlation potential which arises naturally in the context of relativistic density-functional theory [Wang 2004]. A more general strategy is to add a non-DFT many-body polarization propagator correction [Casida 2005] and that is briefly described here.

The idea behind the propagator correction approach is the similarity of the relation,

$$\chi^{-1}(\boldsymbol{r}_1,\boldsymbol{r}_2;\omega) = \chi_{\mathrm{KS}}^{-1}(\boldsymbol{r}_1,\boldsymbol{r}_2;\omega) + f_{\mathrm{Hxc}}(\boldsymbol{r}_1,\boldsymbol{r}_2;\omega) , \tag{16.24}$$

which defines the Hartree (Coulomb) and exchange-correlation kernel of LR-TDDFT and the Bethe-Salpeter equation,

$$\Pi^{-1}(\boldsymbol{r}_1,\boldsymbol{r}_2;\boldsymbol{r}_3,\boldsymbol{r}_4;\omega) = \Pi_{\mathrm{KS}}^{-1}(\boldsymbol{r}_1,\boldsymbol{r}_2;\boldsymbol{r}_3,\boldsymbol{r}_4;\omega) + K_{\mathrm{Hxc}}(\boldsymbol{r}_1,\boldsymbol{r}_2;\boldsymbol{r}_3,\boldsymbol{r}_4;\omega)\,, \tag{16.25}$$

which defines the kernel, K_{Hxc}, in terms of the polarization propagator, Π. (See [Onida 2002].) Here we have deliberately chosen the Kohn-Sham orbital hamiltonian as the zero-order hamiltonian. The main difference between the two equations is that the TDDFT equation is a two-point equation (involving only $\boldsymbol{r}_1$ and $\boldsymbol{r}_2$) while the Bethe-Salpeter equation is a four-point equation (involving $\boldsymbol{r}_1$, $\boldsymbol{r}_2$, $\boldsymbol{r}_3$, and $\boldsymbol{r}_4$). However,

$$\Pi(\boldsymbol{r}_1,\boldsymbol{r}_1;\boldsymbol{r}_2,\boldsymbol{r}_2;\omega) = \chi(\boldsymbol{r}_1,\boldsymbol{r}_2;\omega) \tag{16.26a}$$

$$\Pi_{\mathrm{KS}}(\boldsymbol{r}_1,\boldsymbol{r}_1;\boldsymbol{r}_2,\boldsymbol{r}_2;\omega) = \chi_{\mathrm{KS}}(\boldsymbol{r}_1,\boldsymbol{r}_2;\omega)\,, \tag{16.26b}$$

so that we can write that,

$$\begin{aligned} f_{\mathrm{Hxc}}(\boldsymbol{r}_1,\boldsymbol{r}_2;\omega) = \int \mathrm{d}\boldsymbol{r}_3 \cdots \int \mathrm{d}\boldsymbol{r}_9\, \chi_{\mathrm{KS}}^{-1}(\boldsymbol{r}_1,\boldsymbol{r}_3;\omega)\Pi_{\mathrm{KS}}(\boldsymbol{r}_3,\boldsymbol{r}_3;\boldsymbol{r}_4,\boldsymbol{r}_5;\omega) \\ \times K_{\mathrm{Hxc}}(\boldsymbol{r}_4,\boldsymbol{r}_5;\boldsymbol{r}_6,\boldsymbol{r}_7;\omega)\Pi(\boldsymbol{r}_7,\boldsymbol{r}_8;\boldsymbol{r}_9,\boldsymbol{r}_9;\omega)\chi^{-1}(\boldsymbol{r}_9,\boldsymbol{r}_2;\omega)\,. \end{aligned} \tag{16.27}$$

This provides a way to calculate the TDDFT kernel from many-body theory, though it of course does not provide a functional. More importantly it provides a way to determine a non-DFT many-body nonadiabatic correction to the adiabatic TDDFT kernel. Thus only,

$$\Delta f_{\mathrm{Hxc}}(\omega) = f_{\mathrm{Hxc}}(\omega) - f_{\mathrm{Hxc}}(0)\,, \tag{16.28}$$

need be obtained using the polarization propagator formalism, obtaining $f_{\mathrm{Hxc}}(0)$ in the usual way from adiabatic TDDFT. The ultimate usefulness of this method is yet to be determined but will most likely depend upon the complexity of the nonadiabatic correction. Some simplifications already occur because Casida's equation is a four-point formulation of TDDFT. This means that it is reasonable to use,

$$\Delta K_{\mathrm{Hxc}}(\omega) = K_{\mathrm{Hxc}}(\omega) - K_{\mathrm{Hxc}}(0)\,, \tag{16.29}$$

directly in Casida's equation. That is what is proposed in [Casida 2005] where connections are also made with results from Gonze and Scheffler [Gonze 1999], the "dressed TDDFT" of Maitra, Zhang, Cave, and Burke [Maitra 2004, Cave 2004], and spin-flip TDDFT [Shao 2003, Slipchenko 2003]. It is pointed out that the grafting of TDDFT and propagator theory is not smooth in the open-shell case and spin-projectors are recommended to help join the two formalisms.

16.5 Conclusion

With a few notable exceptions (atomic term symbols, dissociation of H_2, ...) quantum chemistry courses place a great deal of emphasis on closed-shell molecules, undoubtedly because the theory is simpler. However molecules with open-shell ground states and their spectra are hardly infrequent in nature. DFT and TDDFT offer a tempting toolbox for their study, provided they are used cautiously by informed users. In practice, this typically means the use of spin-unrestricted DODS formulations of (TD)DFT. We have tried to point out two dangers – namely (i) the existance of triplet instabilities for biradicals and (ii) the lack of explicit multiple excitations in adiabatic TDDFT.

The first danger means that TDDFT excitation energies are often seriously in error in regions of space where ground and excited potential energy surfaces come close together. This is where traditional *ab initio* quantum chemistry prescribes the use of at least a two determinantal wave function. One can argue that the Kohn-Sham determinant should be able to handle even this situation, provided the exchange-correlation functional is exact, but it is not and the lowest energy solution is typically a symmetry-broken DODS solution in practice. On the bright side, the occurance of an imaginary triplet energy in a LR-TDDFT calculation is an indication that there is a problem with the treatment of the ground state. Less optimistically, LR-TDDFT fails for calculating excitation energies in these regions of configurational space. One easy way to reduce this problem is to decouple the excited state problem from the ground state stability problem by invoking the Tamm-Dancoff approximation, though only a few programs seem to have this option.

The second danger means that adiabatic TDDFT produces triplet coupled states which, except in the case of molecules with closed-shell ground states, can be quite unphysical. Moreover too few singlet coupled states are produced. This means that the wise user of LR-TDDFT for the spectra of molecules with open-shell ground states should carefully examine the physical nature of each transition before concluding that it is well-produced by LR-TDDFT. One aid is to calculate ground and excited state spin contamination. While calculating ground state spin contamination is a common option in many programs, we know of no common program which allows spin contamination to be calculated for LR-TDDFT excited states.

In addition to implementing the TDA and calculation of excited state spin contamination, we have pointed out another promising option for developers. This is the inclusion of some type of polarization propagator corrections to account for nonadiabatic effects. The success of this last approach will depend upon how easy it is to develop simple effective corrections. The work has only just begun.

17 Atoms and Clusters

J.R. Chelikowsky, Y. Saad, and I. Vasiliev

17.1 Introduction

Historically, optical properties have played a key role in our understanding of the electronic structure of matter. An obvious example comes from early studies of the optical excitations in the hydrogen atom, which led to the development of the quantum theory of electronic states, first by Bohr and later by Heisenberg and Schrödinger. Indeed, the initial validation of the quantum theory centered on describing the spectral lines of a hydrogen atom and later many-electron atoms. This is reflected by the historical spectroscopic notation of atomic states by s, p, d and f. The letters refer to the hydrogen spectral lines characterized as "sharp," "principal," "diffuse," and "fine," respectively. This mode of discovery has continued over the last century or so. For example, most of our current understanding of the electronic structure of atoms, molecules and solids comes from examining the optical and dielectric properties of these systems. As a more recent example, the energy band structures of semiconductors were first established by using optical properties as the input for electronic structure calculations [Cohen 1989]. Unfortunately, examining optical excitations based on contemporary quantum mechanical methods can be especially challenging because accurate methods for *structural* energies, such as DFT, are often not well suited for *excited* state properties. This requires new methods designed for predicting excited states and new algorithms for implementing them.

In this chapter, we outline some recent advances in computational methods and their implementation for predicting the optical properties of *atoms and clusters*. Our focus is on utilizing the *pseudopotential method* along with DFT implemented within LDA as outlined in this text and in the original literature [Hohenberg 1964, Kohn 1965, Kohn 1983b]. The combination of the LDA and the pseudopotential approach has proved to be very successful for predicting the *structural and cohesive properties* of matter [Chelikowsky 1992, Payne 1992, Pickett 1989, Srivastava 1987]. The pseudopotential approximation removes the chemically inert core electrons from the problem, effectively reducing the number of electronic degrees of freedom in the quantum mechanical equations. Pseudo wave functions are smoothly varying and can be easily represented within any chosen basis (such as a plane wave representation) or grid methods (such as finite element or

J.R. Chelikowsky et al.: *Atoms and Clusters*, Lect. Notes Phys. **706**, 259–269 (2006)
DOI 10.1007/3-540-35426-3_17

finite differences). For localized systems, like molecules or nanoclusters, a direct real-space implementation of this technique is particularly advantageous [Chelikowsky 1994a, Chelikowsky 1994b, Briggs 1995, Gygi 1995b, Zumbach 1996, Fattebert 2000, Pask 2001, Chelikowsky 2003a]. With this approach, the Kohn-Sham equation for the electronic states is solved on a real-space three dimensional grid within a spherical boundary domain. The kinetic energy operator is approximated by a higher-order finite difference expansion on the grid points [Fornberg 1994, Smith 1978]. Unlike "supercell" calculations in momentum space [Andreoni 1990], real-space methods do not produce an artificial periodicity, and do not impose restrictions on the net charge of the system.

The pseudopotential approximation is highly accurate and can be further improved by explicitly including core states. However, implementations of DFT can be more problematic. In principle, DFT is exact; however, in practice approximations must be made. One of the most significant limitations of "conventional" density functional formalism is its inability to deal with electronic excitations. Within time-independent, or static, DFT, a quantum mechanical system is described through the ground state electronic charge density. While this approach can be accurate for the ground state of a many-electron system, the excited electronic states are not adequately represented by the static formalism [Gross 1996, Petersilka 1996a]. The inability to describe excitations severely restricts the range of applications for conventional density functional methods, since many important physical properties such as optical absorption and emission, response to time-dependent fields, the dynamical dielectric function, and the band gap in semiconductors are associated with excited states.

Explicit calculations for excited states can present enormous challenges for theoretical methods. Accurate calculations for excitation energies and absorption spectra typically require computationally intensive techniques, such as the configuration interaction method [Saunders 1983, Buenker 1978], quantum Monte Carlo simulations [Bernu 1990, Williamson 2002] or the Green's function methods [Sham 1966, Hedin 1965, Hybertsen 1986]. While these methods describe electronic excitations properly, they are usually limited to very small systems because of high computational demands.

An alternative approach is to consider methods based on time dependent DFT such as those using TDLDA [Gross 1996, Petersilka 1996a, Casida 1995a, Casida 1996, Vasiliev 1999, Raghavachari 2002, Jaramillo 2000, Hirata 1999b, Yabana 1996, Yabana 1997, Yabana 1999a]. The TDLDA technique can be viewed as a natural extension of the ground state density-functional LDA formalism, designed to include the proper representation of excited states. TDLDA excitation energies of a many-electron system are usually computed from conventional, time independent Kohn-Sham transition energies and wave functions. Compared to other theoretical methods for excited states, the TDLDA technique requires considerably less computational effort. Despite its relative simplicity, the TDLDA method incorporates screening and relevant correlation effects for electronic excitations [Gross 1996, Petersilka 1996a, Casida 1995a, Casida 1996]. In this sense, TDLDA represents a fully *ab initio* formalism for excited states.

We will review this technique by illustrating computations for transition energies and optical absorption spectra for several representative systems such as atoms and atomic clusters. These systems are very useful for testing optical excitation methods. In the case of atoms, highly accurate experimental methods can be used to determine the optical spectra. This is not the case for clusters, which in principle are stable only in isolation. As such, it is very difficult to measure the properties of clusters without special techniques, e.g., examining clusters in atomic or molecular beams, or by embedding them in an inert matrix. However, this novel state of matter allows one to examine changes by adding one atom to the system at a time and assessing how the electronic properties change with size. Our numerical emphasis will be on real space methods for the ground state electronic structure problem and on frequency domain methods for the time dependent response of the system. We note that other methods have been successfully implemented for these systems and are discussed within this volume and elsewhere [Yabana 1996, Yabana 1997, Yabana 1999a, Onida 2002].

17.2 Theoretical Methods

The energies and oscillator strengths of optical transitions in many-electron systems can be obtained by considering the system's response to an external perturbation. In the frequency-based TDDFT, the response of the density matrix to an applied periodic electric field can be used used to derive the density-functional expression for the dynamic polarizability [Casida 1995a]. The excitation energies Ω_q, which correspond to the poles of the dynamic polarizability, are obtained from the solution of an eigenvalue problem generate by the Casida equation [Casida 1995a]:

$$\hat{R}F_q = \Omega_q^2 F_q \,, \tag{17.1}$$

where the matrix $\hat{R}$ is given by

$$R_{ij\sigma,kl\tau} = \delta_{i,k}\delta_{j,l}\delta_{\sigma,\tau}\omega_{kl\tau}^2 + 2\sqrt{\lambda_{ij\sigma}\omega_{ij\sigma}}\,K_{ij\sigma,kl\tau}\sqrt{\lambda_{kl\tau}\omega_{kl\tau}}\,. \tag{17.2}$$

In this equation, the indices i, j, and σ (k, l, and τ) refer to the space and spin components, respectively, of the unperturbed static Kohn-Sham orbitals $\varphi_{i\sigma}(\boldsymbol{r})$, $\omega_{ij\sigma} = (\varepsilon_{j\sigma} - \varepsilon_{i\sigma})$ are the differences between the eigenvalues of the single-particle states, $\lambda_{ij\sigma} = n_{i\sigma} - n_{j\sigma}$ are the difference between their occupation numbers. Atomic units will be used in this chapter, i.e., $e = \hbar = m = 1$. The coupling matrix $\hat{K}$ in the *adiabatic* approximation is

$$K_{ij\sigma,kl\tau} = \int \mathrm{d}^3r \int \mathrm{d}^3r' \, \varphi^*_{i\sigma}(\boldsymbol{r})\varphi_{j\sigma}(\boldsymbol{r}) \left\{ \frac{1}{|\boldsymbol{r}-\boldsymbol{r}'|} + f_{\mathrm{xc}\,\sigma\tau}(\boldsymbol{r},\boldsymbol{r}') \right\} \varphi_{k\tau}(\boldsymbol{r}')\varphi^*_{l\tau}(\boldsymbol{r}') \, , \quad (17.3)$$

where $f_{\mathrm{xc}}[n]$ is the exchange-correlation (xc) kernel of the system and $n_\sigma(\boldsymbol{r})$ is the spin density. The oscillator strengths f_n, which correspond to the residues of the dynamic polarizability, are given by

$$f_n = \frac{2}{3} \sum_{\alpha=\{x,y,z\}} |D_\alpha \, \hat{S}^{1/2} \, F_n|^2 \, , \quad (17.4)$$

where F_n are the eigenvectors of (17.1), $S_{ij\sigma,kl\tau} = \delta_{i,k}\delta_{j,l}\delta_{\sigma,\tau}\lambda_{kl\tau}\omega_{kl\tau}$, and D_α is the dipole matrix element, $D_{\alpha,\,ij\sigma} = \int \mathrm{d}^3r \, \varphi_{i\sigma}(\boldsymbol{r})\alpha\varphi_{j\sigma}(\boldsymbol{r})$, $\alpha = \{x, y, z\}$. The static Kohn-Sham orbitals $\varphi_{i,\sigma}(\boldsymbol{r})$ and their eigenvalues ε_i used in (17.1)–(17.4) are obtained by solving the system of time-independent Kohn-Sham equations [Chelikowsky 1994a, Chelikowsky 1994b, Vasiliev 2002a]:

$$\left\{ -\frac{\nabla^2}{2} + \sum_\alpha v_{\mathrm{ion}}(\boldsymbol{r} - \boldsymbol{R}_\alpha) + v_{\mathrm{H}}[n](\boldsymbol{r}) + v_{\mathrm{xc}}[n](\boldsymbol{r}) \right\} \varphi_{i\sigma}(\boldsymbol{r}) = \varepsilon_i \varphi_{i\sigma}(\boldsymbol{r}). \quad (17.5)$$

The ionic potential of each atom situated at $\boldsymbol{R}_\alpha$ is represented by a pseudopotential $v_{\mathrm{ion}}(\boldsymbol{r} - \boldsymbol{R}_\alpha)$, which accounts for the interaction with core electrons and nuclei. The Hartree potential, $v_{\mathrm{H}}[n](\boldsymbol{r})$, describes the electrostatic interactions among valence electrons. The xc potential, $v_{\mathrm{xc}}[n](\boldsymbol{r})$, represents the non-classical part of the Hamiltonian. The single-electron Kohn-Sham eigenvalues ε_i and eigen wave functions $\varphi_{i\sigma}(\boldsymbol{r})$ in (17.5) pertain to valence electrons only.

The most computationally demanding part of TDDFT calculations is the evaluation of the coupling matrix $\hat{K}$ given by (17.3). However, the computational cost of the TDDFT response formalism can be substantially reduced if the integral of (17.3) is split into two parts. The first part represents a double integral over $1/|\boldsymbol{r} - \boldsymbol{r}'|$. This term can be evaluated by solving a Poisson equation within the boundary domain [Vasiliev 2002a]. The Poisson equation method provides approximately an order of magnitude speed-up compared to the direct summation over grid points. The second part represents a double integral over f_{xc}. The evaluation of this term is linked to the properties of the xc functional used in (17.3). If a local functional is employed, the second term is reduced to a single integral.

Since the exact form of the xc energy functional and its derivatives in (17.3) and (17.5) are not known, this functional has to be approximated. One of the simplest and most commonly used approximations for the xc functional is based on LDA. Within the LDA, the xc energy and potential functionals are replaced by *local functions* of the charge density [Kohn 1965]

$$E_{\mathrm{xc}}[n] = \int \mathrm{d}^3r \, n(\boldsymbol{r}) e_{\mathrm{xc}}(n(\boldsymbol{r})), \qquad v^\sigma_{\mathrm{xc}}[n](\boldsymbol{r}) = \frac{\delta[n(\boldsymbol{r})e_{\mathrm{xc}}(n(\boldsymbol{r}))]}{\delta n_\sigma(\boldsymbol{r})} \, . \quad (17.6)$$

A number of different parametrizations for $e_{\mathrm{xc}}(n)$ are available in literature. In this work we employ the Ceperley-Alder xc functional parametrized by Perdew and Zunger [Ceperley 1980, Perdew 1981]. The Perdew-Zunger parametrization has been slightly adjusted to guarantee a continuous second functional derivative of the correlation energy as required in (17.3). There is a notable issue with the functional form of Ceperley and Alder. Namely, one might argue on classical grounds that the electronic potential should vary as $-1/r$ at large distances. However, potentials constructed within the LDA do not follow such a decay; instead they fall off exponentially. This behavior leads to an electronic potential that compared to the true potential is too weakly bound and for negative ions even unbound [van Leeuwen 1994]. Atomic systems are well suited for testing this issue and several groups have attempted to devise new model functionals.

In the scheme introduced by Leeuwen and Baerends (LB94) [van Leeuwen 1994], the asymptotic tail of the LDA potential is corrected by including an additional term related to the gradient of the charge density

$$v_{\mathrm{xc}}^{\mathrm{LB94}}(\boldsymbol{r}) = v_{\mathrm{xc}}^{\mathrm{LDA}}(\boldsymbol{r}) - \beta n_\sigma^{1/3} \frac{x_\sigma^2}{1 + 3\beta x_\sigma \sinh^{-1} x_\sigma}, \qquad x_\sigma = \frac{|\nabla n_\sigma|}{n_\sigma^{4/3}} . \tag{17.7}$$

Another scheme proposed by Casida and Salahub (ACLDA) [Casida 2000a] further improves the behavior of the Leeuwen and Baerends potential in the core region. It replaces the core of this potential with the LDA potential shifted by the difference between the self-consistent ionization energy calculated as $E_{\mathrm{ion}}^{\mathrm{LDA-SCF}} = E_{\mathrm{total}}^{\mathrm{LDA}(+1)} - E_{\mathrm{total}}^{\mathrm{LDA}(0)}$ and the LDA Kohn-Sham "ionization" threshold, $E_{\mathrm{ion}}^{\mathrm{LDA-KS}}$, defined as the negative value of the energy of the highest occupied single-electron molecular orbital $\varepsilon_{\mathrm{HOMO}}^{\mathrm{LDA}}$:

$$v_{\mathrm{xc}}^{\mathrm{ACLDA}}(\boldsymbol{r}) = \max[v_{\mathrm{xc}}^{\mathrm{LDA}}(\boldsymbol{r}) - \Delta, v_{\mathrm{xc}}^{\mathrm{LB94}}(\boldsymbol{r})]; \qquad \Delta = E_{\mathrm{ion}}^{\mathrm{LDA-SCF}} - E_{\mathrm{ion}}^{\mathrm{LDA-KS}} . \tag{17.8}$$

The linear-response time-dependent formalism and the asymptotically correct LB94 and ACLDA xc functionals can be easily implemented in the framework of a real space higher-order finite-difference pseudopotential code [Vasiliev 2004]. Such calculations can be performed on a real-space Cartesian grid, *without the use of explicit basis functions.* A common choice of pseudopotentials for this implementation is based on the Troullier-Martins recipe [Troullier 1991], which can be modified to include the LB94 and ACLDA functionals.

17.3 Applications to Atoms

A standard method for evaluating the accuracy of excited state methods is to examine atomic transitions in well known systems such as inert gas atoms. Because the outer electronic shells of inert gas elements are completely

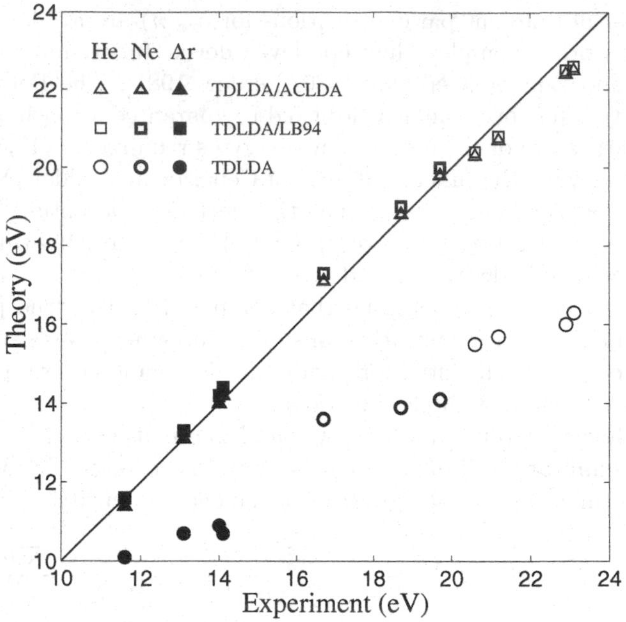

Fig. 17.1. Excitation energies for selected inert gas atoms calculated in the framework of the linear response TDDFT formalism combined with the LDA [Ceperley 1980, Perdew 1981], LB94 [van Leeuwen 1994], and ACLDA [Casida 2000a] xc functionals

filled, these atoms have high excitation and ionization energies. Calculations based on taking eigenvalue differences using the local density approximations routinely underestimate the electronic transition energies of inert gas atoms by ~30–40% or more [van Leeuwen 1994].

Figure 17.1 illustrates the excitation energies of the inert gas atoms He, Ne, and Ar using a variety of energy functionals and a comparison to experiment. It is clear from this figure that the TDLDA energies calculated using the local xc potential without asymptotic corrections substantially underestimate the experimental values. At the same time, the energies computed on the basis of the asymptotically accurate LB94 and ACLDA potentials are in excellent agreement with experiment [Bashkin 1975] and with previous atomic calculations based on quantum chemistry methods [van Gisbergen 1998, Drake 1996].

17.4 Applications to Clusters

The electronic and structural properties of atomic clusters stand as one of the outstanding problems in materials physics. Clusters often possess properties that are characteristic of neither the atomic nor solid state. For example, the energy levels in atoms may be discrete and well-separated in energy relative to $k_B T$. In contrast, solids have a continuum of states (energy bands). Clusters may reside between these limits, i.e., the energy levels may be discrete, but with a separation much smaller than $k_B T$. The most fundamental issue in dealing with clusters is the determination of their structure. Before any accurate theoretical calculations can be performed for a cluster, the atomic geometry of a system must be defined. However, this can be a formidable exercise. Serious problems arise from the existence of multiple local minima in the potential-energy-surface of these systems; many similar structures can exist with very small energy differences, i.e., differences smaller than $k_B T$.

A convenient method to determine the structure of small or moderate sized clusters is *simulated annealing*. Within this technique, atoms are randomly placed within a large cell and allowed to interact at a high (usually fictitious) temperature. The atoms will sample a large number of configurations. As the system is cooled, the number of high energy configurations sampled is restricted. If the annealing is done slowly enough, the procedure should quench out structural candidates for the ground state structures. The interatomic forces used in the simulation can be found from the Hellmann-Feynman theorem [Binggeli 1992].

Silicon clusters and hydrogenated silicon clusters have been extensively examined using this technique [Vasiliev 2001, Vasiliev 2002a, Vasiliev 2002b, Vasiliev 2002c, Proot 1992, Delerue 1993, Reboredo 2000, Wang 1994, Wang 1993, Baierle 1997, Hill 1995, Öğüt 1997, Garoufalis 2001, Williamson 2002, Puzder 2003, Delley 1993, Rohlfing 1998b, Grossman 2001]. In particular, as the silicon clusters become larger they can be compared to quantum dots. For example, SiH_4 (silane), which represents the smallest hydrogenated silicon cluster, has often been used in these studies to compare the accuracy of different computational approaches. The electronic and optical properties of SiH_4 have been calculated by a variety of theoretical methods, including such computationally intensive techniques as the Bethe-Salpeter equation (BSE) for the two-particle Green's function [Benedict 2003, Rohlfing 1998b] and the quantum Monte Carlo method (QMC) [Grossman 2001]. In contrast to larger hydrogenated silicon clusters, SiH_4 is characterized by a large ($\sim$9 eV) absorption gap. In this sense, the electronic structure of SiH_4 resembles that of inert gas atoms. It has been shown that the conventional TDLDA formalism underestimates the excitation energies of SiH_4 [Vasiliev 2002a, Vasiliev 2001]. SiH_4 is also known to have a negative electron affinity, which additionally complicates the problem.

The absorption spectra of the silane molecule calculated with the LDA, ACLDA, and LB94 functionals are shown in Fig. 17.2. The bottom panel of Fig. 17.2, shows the experimental spectrum of SiH_4 [Itoh 1986]. The positions and the relative intensities of the main absorption peaks in the experimental spectra are indicated by vertical lines. The assignment of Rydberg transitions in the experimental spectrum follows the notation of [Itoh 1986].

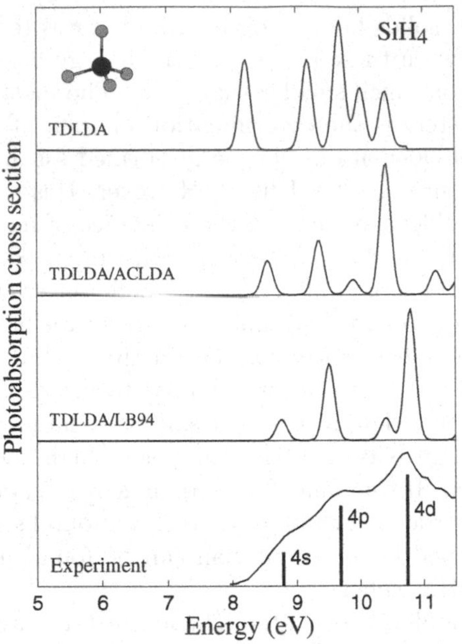

Fig. 17.2. Absorption spectra of the silane molecule SiH_4. The *top* three panels show the theoretical spectra computed in the framework of TDDFT. The xc potential in these calculations is approximated by the LDA, ACLDA, and LB94 functionals, respectively. A Gaussian convolution of 0.1 eV is used to simulate finite broadening of the calculated spectra. The *bottom* panel shows the experimental spectrum of SiH_4 adapted from [Itoh 1986]

A comparison of the calculated and experimental excitation energies indicates that the use of asymptotically accurate xc functionals substantially improves the quality of the theoretical absorption spectrum of SiH_4. While the TDLDA excitation energies are underestimated by (5–10)%, the values calculated with the LB94 and ACLDA functionals agree with experiment to within 2% and 4%, respectively. The lowest TDDFT singlet excitation energy calculated with the LB94 functional is close to the values obtained by the more computationally intensive BSE (9.0 eV) [Rohlfing 1998b] and QMC (9.1 eV) [Grossman 2001] methods. The energies of the second and third absorption peaks in the LB94 absorption spectrum of SiH_4 appear to be in better agreement with experiment than the values obtained in the BSE calculations.

The main difference between the time independent DFT and time dependent DFT calculations for these systems is a strong blue-shift of the oscillator strength. This effect can also be documented for other hydrogenated silicon molecules. In Fig. 17.3, we illustrate the optical spectrum of Si_5H_{12} using the LDA and weighting each transition by the dipole matrix element. The threshold for the LDA transition is approximately 5.8 eV, which is the same as

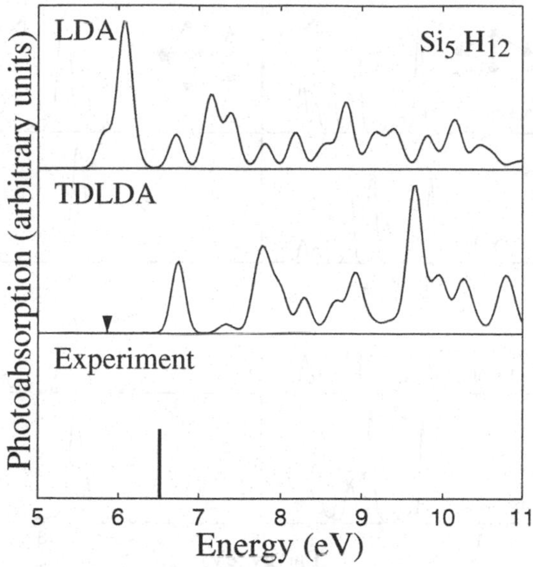

Fig. 17.3. Calculated optical spectra for Si_5H_{12} using LDA and TDLDA. The arrow in the TDLDA panel shows the lowest eigenvalue from 17.1. Experiment shows the measured optical gap as quoted in [Delley 1993]

the TDLDA lowest eigenvalue, Ω_0 as determined from 17.1 to within 0.1 eV. However, the spectrum clearly indicates that significant optical absorption does not occur until nearly 6.6 eV. This value is consistent with experiment. For the majority of the hydrogenated clusters (SiH_4 being an exception), the oscillator strength of the first TDLDA transition vanishes. The difference between the lowest transition and the first allowed optical transition can be significant. This suggests that TDDFT calculations without regard to the oscillator strength will not successfully describe the experimental absorption spectra of these systems.

We can illustrate the utility and generality of this approach by considering clusters composed of a prototypical free electron metal: sodium. Sodium clusters have attracted considerable theoretical attention due to their simple electronic structure [de Heer 1993, Brack 1993, Bonačić-Koutecký 1991] and the availability of a relatively large body of experimental data [de Heer 1993, Wang 1990a, Wang 1990b, Wang 1992]. Because of the presence of delocalized valence electrons, these clusters have low excitation energies. Despite the "incorrect" asymptotic tail of the LDA potential, the TDLDA formalism correctly reproduces the main features of the experimental absorption spectra of small Na clusters [Vasiliev 1999, Vasiliev 2002a]. This suggests that the theoretical TDDFT spectra of sodium clusters are not sensitive to the asymptotic behavior of the xc functional.

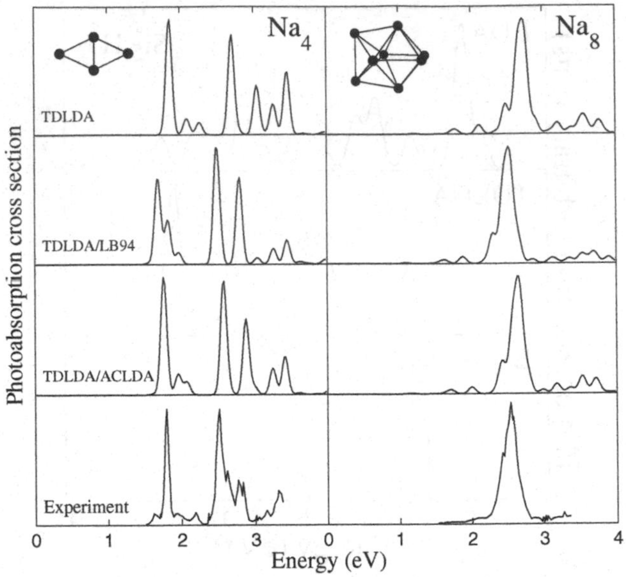

Fig. 17.4. Absorption spectra of small sodium clusters. Experimental spectra are adapted from [Wang 1990a, Wang 1990b]. All theoretical spectra are broadened using a Gaussian convolution of 0.06 eV

Figure 17.4 shows the calculated and experimental spectra of Na_4 and Na_8 clusters. A comparison of the calculated spectra indicates that the addition of the asymptotic correction produces only a slight shift of the low-energy absorption peaks. The differences between the theoretical excitation energies calculated with the LDA, LB94, and ACLDA functionals in this range do not exceed 0.1–0.2 eV. The low energy part of the absorption spectrum of Na_4 calculated with the LB94 functional appears to be somewhat less accurate than the other TDDFT spectra. This can be attributed to a less accurate behavior of the LB94 potential for transition energies below the ionization energy. At the same time, Fig. 17.4 shows that the asymptotically correct ACLDA and LB94 functionals improve the accuracy of the high-energy electronic transitions in the spectrum of Na_4. The spectra computed with the ACLDA potential demonstrate the best agreement with experiment overall. This is not surprising, considering that by construction the ACLDA functional has been designed to combine the best characteristics of the LDA and the LB94 potentials.

The asymptotic tail of the exchange-correlation potential predominantly affects the outer regions of an atomic cluster or a quantum dot. Consequently, the influence of asymptotic corrections on the calculated spectra is expected to disappear in the bulk limit. This can be illustrated by applying the asymptotically corrected TDDFT formalism to hydrogenated silicon quantum dots. A comparison of TDLDA and asymptotically corrected TDDFT spectra for the $Si_{29}H_{36}$ and $Si_{35}H_{36}$ dots reveals that the asymptotic correction has almost no effect on the calculated values of the lower excitation energies [Vasiliev 2004]. The convergence between the spectra of silicon dots calculated with and without asymptotic correction can be explained by a combination of two factors. First, the structures of Si_nH_m clusters become more crystalline in nature as the cluster size increases. The electronic wave functions in bulk silicon are, for the most part, determined by the xc potential in the core region not affected by the asymptotic correction. Second, absorption gaps in silicon quantum dots decrease with increasing dot diameter [Vasiliev 2001]. Owing to the decreasing gap size, the low-energy transitions in the spectra of larger Si_nH_m clusters are shifted below the Kohn-Sham LDA ionization threshold. As a result, the energies of these optical transitions become less sensitive to the asymptotic behavior of the xc potential. This observation can be confirmed by comparing the spectra of the SiH_4 and Si_5H_{12} clusters: The energy of the experimental absorption gap in SiH_4 (8.8 eV) [Itoh 1986] is higher than the Kohn-Sham LDA ionization threshold for this cluster (8.6 eV). The conventional TDLDA method underestimates the energy of the lowest optical transition for SiH_4 (8.2 eV) [Vasiliev 2001]. The TDDFT calculations based on the asymptotically correct ACLDA and LB94 functionals increase the theoretical energy of this absorption peak to 8.5 and 8.8 eV, respectively, and improve agreement with experiment. In contrast, the energy of the experimental absorption gap in Si_5H_{12} (6.5 eV) [Delley 1993] is lower than the Kohn-Sham LDA ionization energy value for this cluster (7.3 eV) [Vasiliev 2001]. Consequently, the energies of the lowest dipole-allowed transitions in the spectra of Si_5H_{12} remain practically unchanged in the TDDFT calculations based on the LDA (6.6 eV), ACLDA (6.6 eV), and the LB94 (6.7 eV) functionals.

Overall, our results for small clusters and molecules are consistent with the TDDFT calculations for single atoms. The asymptotic behavior of the exchange-correlation potential has almost no influence on electronic transitions below the ionization energy. However, the asymptotically correct functionals do improve the quality of the calculated optical spectra above the ionization energy. The linear response TDDFT formalism combined with the LB94 and ACLDA potentials correctly describes the spectra of small localized quantum mechanical systems over a broad range of excitation energies. Moreover, the asymptotically corrected TDDFT formalism appears to work equally well for clusters and molecules with different types of chemical bonding.

18 Semiconductor Nanostructures

C.A. Ullrich

18.1 Introduction

The past decades have witnessed a breathtaking progress in semiconductor device fabrication techniques, with a relentless trend towards miniaturization. Nowadays, semiconductor nanostructures of almost any desired design can be grown with a precision down to a single atomic layer by using epitaxial methods. To confine, manipulate, and control the charge and spin carriers in semiconductor nanostructures, one can vary the material or alloy composition of a sample along the growth direction, which gives rise to quantum wells or superlattices with sharp interfaces. Gradual changes of alloy composition are also possible, for example to grow parabolic quantum wells. Free electrons or holes are supplied by remote or modulation doping (the doping centers are physically separated from the wells), which leads to systems with very high mobilities. Finally, the sample is gated, and static electric fields can be applied. The above methods provide the toolbox used for "band engineering." An easy-to-read introduction to band engineering from a historic perspective is given in the Nobel Lecture by Herbert Kroemer [Kroemer 2001]. Out of the vast number of textbooks and monographs on semiconductor nanostructures, the book by Davies is particularly recommended [Davies 1998].

This chapter will be concerned with the electron dynamics in semiconductor quantum wells and quantum dots. We will assume that these systems are initially filled with a given number of electrons from remote doping. Starting from the ground state, the electrons then carry out highly collective, plasmon-like dynamical processes with energies typically in the infrared, which will be described using TDDFT in the linear-response regime. This is to be distinguished from electronic excitation processes in *undoped* bulk semiconductors and nanostructures. In these systems, free carriers or electron-hole pairs are *created* by interband optical excitations, typically in the visible range, and studied using a variety of experimental approaches such as photoluminescence spectroscopy or ultrafast pump-probe techniques. A full theoretical description of interband excitations is rather involved since it requires detailed input of the electronic band structure, as well as the long-range Coulomb forces that are responsible for exciton formation. This will be the subject of Chaps. 19 and 20.

C.A. Ullrich: *Semiconductor Nanostructures*, Lect. Notes Phys. **706**, 271–285 (2006)
DOI 10.1007/3-540-35426-3_18

18.2 Effective-Mass Approximation for Quantum Wells

Bloch's theorem tells us that electronic states in the periodic potential of a perfect crystal have the form of modulated plane waves, $\psi_{n\boldsymbol{q}}(\boldsymbol{r}) = u_{n\boldsymbol{q}}(\boldsymbol{r})\mathrm{e}^{\mathrm{i}\boldsymbol{q}\boldsymbol{r}}$, where $u_{n\boldsymbol{q}}$ is a lattice-periodic function, with band index n and wavevector $\boldsymbol{q}$ (disregarding spin, for now). The associated single-particle energies, $\varepsilon_{n\boldsymbol{q}}$, can be computed using standard band structure techniques, which rely heavily on the symmetry of the perfect crystal. At first sight it appears that the situation becomes extremely complicated if that symmetry is broken, for example due to an impurity or in a heterostructure. Paradoxically, the opposite happens: by making the so-called effective-mass approximation, we can simplify the problem enormously and arrive at a description of quantum confinement with relatively minor sacrifices in accuracy. This is due to the fact that, for direct semiconductors such as GaAs, the valence and conduction bands are parabolic near the zone center, as shown in Fig. 18.1. The energy dispersion of free electrons and holes can therefore be described in terms of effective electron mass (m^*) and hole masses (m^*_{hh}, m^*_{lh}, m^*_{so}), which are usually much smaller than the free electron mass.

In the following, we will limit our attention to the conduction electrons, and our goal is to describe the electronic intersubband transitions in a single GaAs/AlGaAs quantum well (see Fig. 18.1). We assume that the direction of growth of the quantum well is along the z-axis, and the system is infinitely extended in the xy plane. The effective-mass approximation then consists in making the following ansatz for the single-particle states:

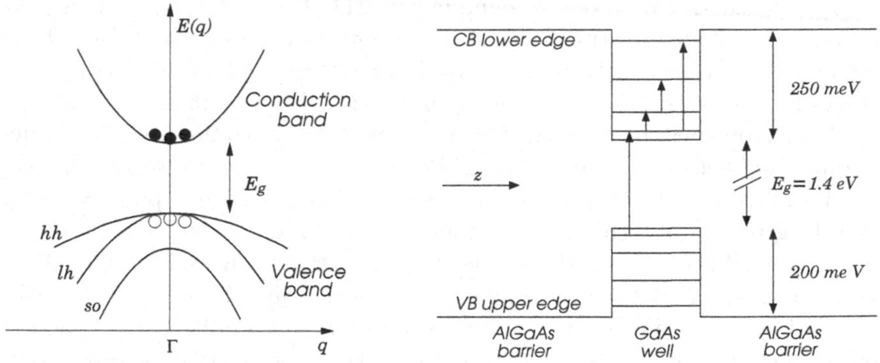

Fig. 18.1. *Left*: Band structure of a direct semiconductor around the zone center. The valence band consists of heavy holes (hh), light holes (lh) and split-off (so) states. Electrons tend to sink to the bottom of the conduction band, holes float on top of the valence band. *Right*: Electronic levels and optical transitions in a GaAs/AlGaAs quantum well. Interband transitions have energies of the order of the band gap E_g. Intersubband transitions have much lower energies, in the range of 10–100 meV

$$\psi_{j\boldsymbol{q}_\parallel}(\boldsymbol{r}) = \frac{1}{\sqrt{A}} \, \mathrm{e}^{\mathrm{i}\boldsymbol{q}_\parallel \boldsymbol{r}_\parallel} \varphi_{jq_\parallel}(z) \,, \tag{18.1}$$

where $\boldsymbol{r}_\parallel = (x, y)$ and $\boldsymbol{q}_\parallel = (q_x, q_y)$ are the in-plane position and wavevector. Notice that we have dropped the band index n, since we consider only conduction band states, and we have introduced the subband index j. The envelope functions $\varphi_{jq_\parallel}(z)$ follow from a one-dimensional (1D) KS equation:

$$\left\{ -\frac{\mathrm{d}}{\mathrm{d}z} \frac{\hbar^2}{2m^*(z)} \frac{\mathrm{d}}{\mathrm{d}z} + \frac{\hbar^2 q_\parallel^2}{2m^*(z)} + v_{\mathrm{KS}}(z) \right\} \varphi_{jq_\parallel}(z) = E_{jq_\parallel} \varphi_{jq_\parallel}(z) \,, \tag{18.2}$$

where $m^*(z)$ accounts for the different effective masses in the well and barrier materials. The peculiar form of the kinetic energy operator ensures conservation of current at the interfaces [Davies 1998]. Often, however, one can ignore the z-dependence of m^*, and simply use the effective mass of the well throughout. This is justified since the subband wavefunctions do not penetrate much into the barrier, and the effective masses of GaAs and $\mathrm{Al_{0.3}Ga_{0.7}As}$ are not that different anyway. In that case, (18.2) becomes

$$\left\{ -\frac{\hbar^2}{2m^*} \frac{\mathrm{d}^2}{\mathrm{d}z^2} + v_{\mathrm{KS}}(z) \right\} \varphi_j(z) = \varepsilon_j \varphi_j(z) \,, \tag{18.3}$$

i.e., the envelope functions are independent of $q_\parallel$, and the subband energy dispersions are parabolic:

$$E_{jq_\parallel} = \varepsilon_j + \frac{\hbar^2 q_\parallel^2}{2m^*} \,. \tag{18.4}$$

As usual, the effective potential is given by $v_{\mathrm{KS}}(z) = v_{\mathrm{conf}}(z) + v_{\mathrm{H}}(z) + v_{\mathrm{xc}}(z)$. Here, v_{conf} is the bare confining potential of the quantum well, typically a square well potential plus a linear potential due to an electric field. The Hartree potential follows from Poisson's equation, $\mathrm{d}^2 v_{\mathrm{H}}/\mathrm{d}z^2 = -4\pi e^{*2} n(z)$, where the effective electronic charge $e^* = e/\sqrt{\epsilon}$ accounts for dielectric screening with static dielectric constant ϵ. Integration of Poisson's equation gives

$$v_{\mathrm{H}}(z) = -2\pi e^{*2} \int \mathrm{d}z' \, |z - z'| n(z') \,. \tag{18.5}$$

For the xc potential, we use the LDA (one wouldn't gain too much from more sophisticated xc functionals, in view of the rather drastic effective-mass approximation):

$$v_{\mathrm{xc}}(z) = \left. \frac{\mathrm{d}\bar{n} e_{\mathrm{xc}}^{\mathrm{hom}}(\bar{n})}{\mathrm{d}\bar{n}} \right|_{\bar{n}=n(z)} \,. \tag{18.6}$$

Finally, the ground-state density $n(z)$ is obtained as

$$n(z) = 2 \sum_{j,q_\parallel} |\varphi_j(z)|^2 \, \theta(E_F - E_{jq_\parallel}) = \frac{m^*}{\pi \hbar^2} \sum_{\substack{j \\ \varepsilon_j < E_F}} |\varphi_j(z)|^2 \, (E_F - \varepsilon_j) \,. \tag{18.7}$$

The conduction band Fermi energy E_F is determined by normalization:

$$N_s = \int dz\, n(z) = \frac{m^*}{\pi\hbar^2}\left[E_F N_{occ} + \sum_{j=1}^{occ} \varepsilon_j\right], \qquad (18.8)$$

where N_{occ} is the number of occupied subband levels, and we assume the φ_j's to be normalized to one. N_s is the sheet density, i.e., the number of conduction electrons per unit area introduced by remote doping (typically, $N_s \sim 10^{10}$–10^{11} cm^{-2}). Figure 18.2 shows a characteristic example of a narrow quantum well containing three bound subband levels.

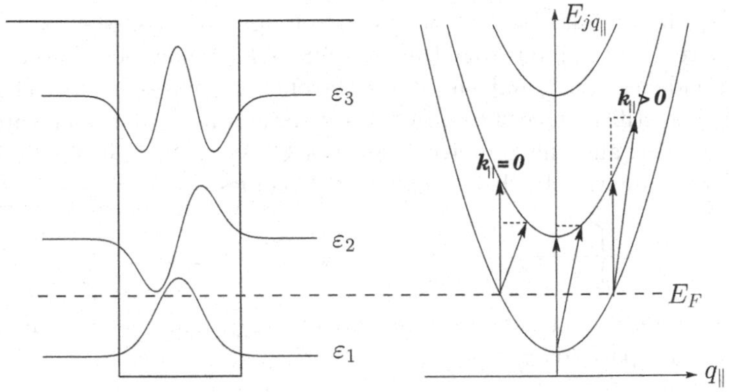

Fig. 18.2. Subband envelope functions and energy dispersions (18.4) in a narrow quantum well. In this example, only the lowest subband is occupied, as indicated by the conduction band Fermi level E_F. Intersubband transitions may occur with zero and finite momentum transfer $\boldsymbol{k}_\parallel$

Equations (18.3)–(18.8) constitute the most elementary DFT approach to describe the self-consistent subband structure in a quantum well, and are in fact quite accurate for the widely studied GaAs/AlGaAs systems. There are, however, situations where one needs to do better than this, for example, in materials with a smaller band gap such as InAs/AlSb. A powerful method to calculate the electronic structure in semiconductors close to the band extrema is the so-called $\boldsymbol{k} \cdot \boldsymbol{p}$ or Kane approach [Kane 1957], where one expands in terms of a basis of valence and conduction band states at the zone center, including some semi-empirical parameters to get the correct band gap. We will not discuss any technical details here (see, e.g., [Davies 1998, Bastard 1988]), but we point out two main consequences emerging from a more rigorous treatment:

1. The subband energy dispersions deviate from parabolicity, due to an energy dependence of the effective mass. The bands are also warped to some extent, although this effect is much stronger for the valence bands.

2. In the presence of spin-orbit coupling, spin is no longer a good quantum number. For systems without inversion symmetry, the quantum well subbands acquire a $\boldsymbol{q}_{\|}$-dependent spin splitting.

18.3 Intersubband Dynamics in Quantum Wells

The THz frequency regime is scientifically rich, but despite recent progress it is technologically still underdeveloped [Liu 2000]. Subband spacings in typical III-V quantum wells are of the order of a few tens of meV. Since a photon energy of 10 meV corresponds to a frequency of 2.4 THz, intersubband (ISB) transitions in quantum wells appear as natural candidates for device applications to fill the "terahertz gap" in the electromagnetic spectrum.

18.3.1 TDDFT Response Theory and Plasmon Dispersions

The fundamental coupling mechanism between electromagnetic waves and carriers in a doped quantum well is through a collective excitation, the so-called *ISB plasmon.* To date, most applications of TDDFT linear response theory in quantum wells have been concerned with calculating ISB plasmon dispersions and linewidths [Bloss 1989, Jogai 1991, Ryan 1991, Dobson 1992, Marmorkos 1993, Ullrich 1998, Ullrich 2001, Ullrich 2002b]. Due to its highly collective nature, and since the dynamics is essentially 1D, ISB plasmons in quantum wells are an ideal testing ground for new TDDFT approaches. We will begin by discussing the TDDFT linear response formalism within the ALDA. Since our quantum wells are translationally invariant in the xy plane, we write the TDDFT linear spin-density response equation as follows:

$$\delta n_\sigma(\boldsymbol{k}_{\|}, z, \omega) = \sum_{\sigma'} \int \mathrm{d}z' \, \chi^{\mathrm{KS}}_{\sigma\sigma'}(\boldsymbol{k}_{\|}, z, z', \omega) \delta v_{\mathrm{KS}\,\sigma'}(\boldsymbol{k}_{\|}, z', \omega) \,. \tag{18.9}$$

The Kohn-Sham response function is diagonal in the spins and given by

$$\chi^{\mathrm{KS}}_{\sigma\sigma'}(\boldsymbol{k}_{\|}, z, z', \omega) = \delta_{\sigma\sigma'} \sum_{\mu=1}^{\mathrm{occ}} \sum_{\nu=1}^{\infty} F_{\mu\nu}(\boldsymbol{k}_{\|}, \omega) \varphi_\mu(z)\varphi_\nu(z)\varphi_\mu(z')\varphi_\nu(z') \,, \tag{18.10}$$

where the envelope functions $\varphi_\mu(z)$ follow from (18.3), and

$$F_{\mu\nu}(\boldsymbol{k}_{\|}, \omega) = \int \frac{\mathrm{d}^2 q_{\|}}{(2\pi)^2} \left[\frac{\theta(E_F - E_{\mu q_{\|}})}{\hbar\omega - \hbar\Omega^{\nu\mu}_{\boldsymbol{k}_{\|}, \boldsymbol{q}_{\|}} + \mathrm{i}\eta} - \frac{\theta(E_F - E_{\mu q_{\|}})}{\hbar\omega + \hbar\Omega^{\nu\mu}_{\boldsymbol{k}_{\|}, \boldsymbol{q}_{\|}} + \mathrm{i}\eta} \right] \tag{18.11}$$

is essentially the 2D Lindhard function, with η a positive infinitesimal and

$$\hbar\Omega^{\nu\mu}_{\boldsymbol{k}_{\|}, \boldsymbol{q}_{\|}} = \varepsilon_\nu - \varepsilon_\mu + \frac{\hbar^2}{m^*} \left(\boldsymbol{k}_{\|} \cdot \boldsymbol{q}_{\|} + k_{\|}^2/2 \right) \,. \tag{18.12}$$

The $\boldsymbol{q}_{\|}$-integral in (18.11) can be carried out analytically for all $\boldsymbol{k}_{\|}$ [Eguiluz 1985]. The linearized effective potential is given by $\delta v_{\mathrm{KS}\,\sigma} = \delta v_{\mathrm{ext}\,\sigma} + \delta v_{\mathrm{H}\,\sigma} + \delta v_{\mathrm{xc}\,\sigma}$, where

$$\delta v_{\mathrm{H}\,\sigma}(\boldsymbol{k}_{\|}, z, \omega) = \frac{2\pi e^{*2}}{k_{\|}} \sum_{\sigma'} \int \mathrm{d}z'\, \mathrm{e}^{-k_{\|}|z-z'|} \delta n_{\sigma'}(\boldsymbol{k}_{\|}, z', \omega) \,, \tag{18.13a}$$

$$\delta v_{\mathrm{xc}\,\sigma}(\boldsymbol{k}_{\|}, z, \omega) = \sum_{\sigma'} \int \mathrm{d}z'\, f^{\mathrm{ALDA}}_{\mathrm{xc},\sigma\sigma'}(z, z', \omega) \delta n_{\sigma'}(\boldsymbol{k}_{\|}, z', \omega) \,, \tag{18.13b}$$

and we choose the following external perturbation:

$$\delta v_{\mathrm{ext}\,\sigma}(\boldsymbol{k}_{\|}, z, \omega) = S^{\pm}_{\sigma} e E_0 \mathrm{e}^{k_{\|} z} \,, \tag{18.14}$$

which can be made to couple to the ISB charge or spin plasmons by using $S^{\pm}_{\sigma} = \delta_{\sigma,\uparrow} \pm \delta_{\sigma,\downarrow}$, respectively. Having solved the response equation (18.9) self-consistently, we calculate the so-called reflection amplitude [Dobson 1992],

$$R(\boldsymbol{k}_{\|}, \omega) = \sum_{\sigma} S^{\pm}_{\sigma} \int \mathrm{d}z\, \mathrm{e}^{k_{\|} z} \delta n_{\sigma}(\boldsymbol{k}_{\|}, z, \omega) \,, \tag{18.15}$$

from which we obtain the absorption cross section

$$\sigma(\boldsymbol{k}_{\|}, \omega) = -\frac{2\omega}{e E_0 k_{\|}^2} \Im R(\boldsymbol{k}_{\|}, \omega) \,. \tag{18.16}$$

Of particular interest is the case of zero-momentum transfer (Fig. 18.2), where

$$F_{\mu\nu}(0, \omega) = \frac{(E_F - \varepsilon_{\mu}) m^*}{2\pi\hbar^2} \left[\frac{1}{\hbar\omega - \varepsilon_{\nu} + \varepsilon_{\mu} + \mathrm{i}\eta} - \frac{1}{\hbar\omega + \varepsilon_{\nu} - \varepsilon_{\mu} + \mathrm{i}\eta} \right] . \tag{18.17}$$

In this case, the external perturbation reduces to the usual dipole approximation for linearly polarized electromagnetic waves, $\delta v_{\mathrm{ext}\,\sigma}(z, \omega) = S^{\pm}_{\sigma} e E_0 z$, and the photoabsorption cross section becomes

$$\sigma(\omega) = -\frac{2\omega}{e E_0} \Im \sum_{\sigma} S^{\pm}_{\sigma} \int \mathrm{d}z\, z\, \delta n_{\sigma}(z, \omega) \,. \tag{18.18}$$

ISB (as well as intrasubband) plasmons correspond to peaks in $\sigma(\boldsymbol{k}_{\|}, \omega)$.

Figure 18.3 shows the spectrum of excitations for a 40 nm GaAs/$\mathrm{Al}_{0.3}\mathrm{Ga}_{0.7}$As square quantum well with $N_{\mathrm{s}} = 10^{11}\,\mathrm{cm}^{-2}$, where only the lowest subband is occupied in the ground state. The shaded areas represent the regions of incoherent particle-hole excitations (Landau damping), which arise from the poles of the KS response function where $\omega = \Omega^{\nu\mu}_{\boldsymbol{k}_{\|}, \boldsymbol{q}_{\|}}$ [see (18.11) and (18.12)]. Comparison with Fig. 18.2 shows that all vertical ($k_{\|} = 0$) single-particle ISB excitations have the same energy, $\omega = \varepsilon_2 - \varepsilon_1 \equiv \omega_{21}$. At finite $k_{\|}$, ISB

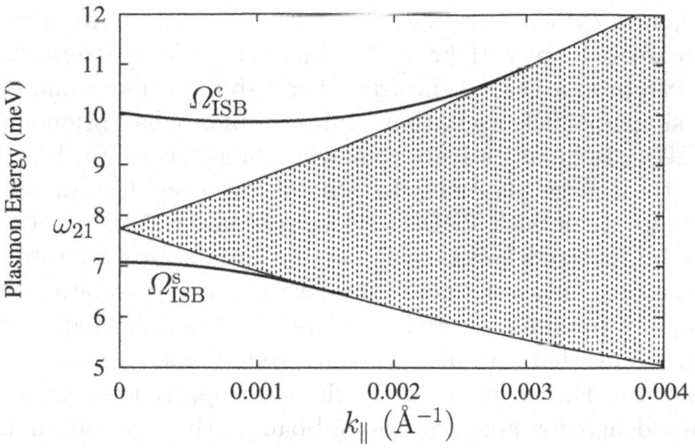

Fig. 18.3. ISB charge- and spin-plasmon dispersions in a 40 nm GaAs/AlGaAs quantum well, Ω^{c}_{ISB} and Ω^{s}_{ISB}. The *shaded* region indicates Landau damping

single-particle excitations with different energy are possible, which accounts for the opening up of the Landau damping region.

The plasmon excitations are separated from the Landau damping region. We can use the so-called small-matrix approximation of TDDFT linear response theory [Appel 2003] to solve the response equation analytically. Keeping only those terms that contain the first and second subband in the KS response function, we have, for $k_\parallel = 0$,

$$\chi^{KS}_{\sigma\sigma'}(z, z', \omega) \approx N_s \delta_{\sigma\sigma'} \left[\frac{1}{\omega - \omega_{21}} - \frac{1}{\omega + \omega_{21}} \right] \varphi_1(z)\varphi_2(z)\varphi_1(z')\varphi_2(z') . \tag{18.19}$$

Defining

$$K_{\sigma\sigma'} = \frac{N_s}{2} \int dz \int dz' \left[-2\pi e^{*2}|z-z'| + f^{ALDA}_{xc,\sigma\sigma'}(z, z') \right] \varphi_1(z)\varphi_2(z)\varphi_1(z')\varphi_2(z') , \tag{18.20}$$

we get the ISB plasmon frequencies

$$\Omega^{c,s}_{ISB} = \sqrt{\omega_{21}^2 + 2\omega_{21}(K_{\uparrow\uparrow} \pm K_{\uparrow\downarrow})} . \tag{18.21}$$

The Hartree contribution in $K_{\sigma\sigma'}$ is known as "depolarization shift" [Allen 1976], and the xc contribution is sometimes (somewhat misleadingly) called "excitonic shift". The Hartree part always induces an up-shift of the plasmon frequency with respect to ω_{21}, and the xc part gives a smaller down-shift. In Ω^{c}_{ISB}, the positive shift dominates, but for the spin plasmon the Hartree parts cancel out in (18.21) and Ω^{s}_{ISB} is redshifted (see Fig. 18.3). This is a remarkable result: the existence of the ISB spin plasmon is purely a consequence of $f_{xc,\sigma\sigma'}$.

Experimental evidence for plasmon excitations in quantum wells can be obtained in several ways [Helm 2000]. Inelastic light scattering techniques [Pinczuk 1989] produce signals for charge and spin plasmons and even for the incoherent single-particle excitations. Infrared photoabsorption spectroscopy [Williams 2001] is quantitatively somewhat more accurate, but allows one only to see the charge plasmons. In general, however, the quantitative and qualitative agreement with TDDFT linear response results is very good, limited mainly by the crudeness of the effective-mass approximation.

Things become obviously more complicated if one goes beyond the simple effective-mass approximation and includes, for example, band nonparabolicity. It turns out that the plasmons are remarkably robust, due to their collective nature. For example, while the single-particle spectrum becomes strongly broadened for nonparabolic subbands, the ISB plasmons are still sharp lines [Warburton 1996]. In quantum wells with inversion asymmetry and spin-orbit coupling, which may exhibit the so-called Rashba or Dresselhaus effects, the spin plasmon dispersion is predicted to have an anisotropic splitting [Ullrich 2002c, Ullrich 2003].

18.3.2 ISB Plasmon Linewidth

ISB transitions are the basis for various THz devices such as detectors and sources of coherent radiation [Liu 2000]. The optical properties of such devices are determined mainly by the frequency of the ISB excitations, but also by their linewidth, e.g., the spectral resolution of a detector. In practice, the linewidth arises from a complex interplay of a variety of scattering mechanisms:

- *intrinsic* damping mechanisms exist even in a perfect device, such as electron-phonon scattering and electronic many-body effects;
- *extrinsic* damping mechanisms are caused by disorder, such as impurity or interface roughness scattering.

A general microscopic theory of dissipation should include all these mechanisms on an equal footing, clearly a very complicated task. TDDFT is beginning to emerge as an interesting alternative to traditional approaches for dissipative quantum dynamics such as non-equilibrium Green's functions [Haug 1996]. In the following, we limit the discussion to wide quantum wells, where the ISB plasmon frequency is below the LO phonon frequency of GaAs (35.6 meV) so that phonon scattering can be ignored. We will also ignore spin.

In Chap. 5, it was discussed how one can go beyond the ALDA and construct a frequency-dependent and local xc vector potential [Vignale 1996, Vignale 1997]:

$$\mathrm{i}\omega\delta a_{\mathrm{xc},j}(\boldsymbol{r},\omega) = \nabla_j\,\delta v_{\mathrm{xc}}^{\mathrm{ALDA}}(\boldsymbol{r},\omega) - \frac{1}{n(\boldsymbol{r})}\sum_k \nabla_k\,\sigma_{\mathrm{xc},\,jk}(\boldsymbol{r},\omega)\ , \qquad (18.22)$$

Table 18.1. ISB plasmon frequency Ω and linewidth Γ (in meV) of single and double GaAs/AlGaAs quantum wells [Ullrich 1998], with $f_{\mathrm{xc}}^{\mathrm{hom}}$ from [Gross 1985][a] and [Nifosì 1998][b]

	ALDA	GK[a]	GK[b]	VUC[a]	VUC[b]	Exp.
Ω (single)	10.25	10.63	10.23	10.31	10.24	10.7
Γ (single)		0.683	0.655	0.128	0.104	0.53
Ω (double)	13.85	14.24	13.88	20.64	12.55	14.6
Γ (double)		1.00	0.403	8.55	4.15	1.17

where the xc viscoelastic stress tensor σ_{xc} depends on the velocity field $\boldsymbol{u} = \delta\boldsymbol{j}/n$. Since quantum wells are quasi-1D systems, we can convert this into a scalar potential (for ISB excitations with $\boldsymbol{k}_{\|} = 0$), by using

$$\delta j(z,\omega) = \mathrm{i}\omega \int_{-\infty}^{z} \mathrm{d}z' \, \delta n(z',\omega) \, , \qquad \delta v_{\mathrm{xc}}(z,\omega) = \mathrm{i}\omega \int_{-\infty}^{z} \mathrm{d}z' \, \delta a_{\mathrm{xc}}(z',\omega) \, . \tag{18.23}$$

We only need the zz component of the viscoelastic stress tensor:

$$\sigma_{\mathrm{xc},\, zz}(z,\omega) = \left(\zeta_{\mathrm{xc}} + \frac{4}{3}\,\eta_{\mathrm{xc}}\right) \frac{\partial u(z,\omega)}{\partial z} \, , \tag{18.24}$$

where

$$\zeta_{\mathrm{xc}} + \frac{4}{3}\,\eta_{\mathrm{xc}} = -\frac{n^2}{\mathrm{i}\omega} \left[f_{\mathrm{xc}}^{\mathrm{hom}}(n,\omega) - \frac{\mathrm{d}^2 n e_{\mathrm{xc}}^{\mathrm{hom}}}{\mathrm{d}n^2} \right]_{n=n(z)} \equiv -\frac{n^2}{\mathrm{i}\omega}\, f_{\mathrm{xc}}^{\mathrm{dyn}}(z,\omega) \, . \tag{18.25}$$

With this, we can derive an explicit expression for the linearized xc potential which directly shows the non-local dependence on the density response δn:

$$\begin{aligned} \delta v_{\mathrm{xc}}(z,\omega) &= f_{\mathrm{xc}}^{\mathrm{hom}}(z,\omega)\, \delta n(z,\omega) - \frac{n'(z)}{n(z)}\, f_{\mathrm{xc}}^{\mathrm{dyn}}(z,\omega) \int_{-\infty}^{z} \mathrm{d}z''\, \delta n(z'',\omega) \\ &- \int_{z}^{\infty} \mathrm{d}z' \, \frac{n'(z')}{n(z')}\, f_{\mathrm{xc}}^{\mathrm{dyn}}(z',\omega) \left\{ \delta n(z',\omega) - \frac{n'(z')}{n(z')} \int_{-\infty}^{z'} \mathrm{d}z''\, \delta n(z'',\omega) \right\} . \end{aligned} \tag{18.26}$$

The first term on the right-hand side of (18.26) is the Gross-Kohn (GK) approximation [Gross 1985]. The other terms are needed to satisfy both the harmonic potential theorem as well as Onsager's reciprocity theorem, i.e., symmetry under interchange of z and z' of the associated xc kernel $f_{\mathrm{xc}}(z, z', \omega)$.

The performance of the non-adiabatic xc potential was analyzed in detail in [Ullrich 1998]. Table 18.1 shows results for a 384 Å single quantum well and an asymmetric double quantum well (85 Å and 73 Å wells, 23 Å barrier), comparing ALDA, GK and (18.26), named VUC after the authors of [Vignale 1997]. For the single well, all functionals give plasmon frequencies close to experiment. The VUC linewidth Γ is *smaller* than the experimental value, as it should be, since disorder is not included. GK, on the other hand, leads to overdamping. For the double quantum well, VUC runs into problems. The analysis of [Ullrich 1998] shows that this is due to a breakdown of the hydrodynamical approach, caused by the presence of the tunnelling barrier.

For sufficiently "hydrodynamic" cases, the non-adiabatic corrections to $\Omega_{\rm ALDA}$ are thus relatively small. Similar conclusions were found in a recent study of atomic excitations [Ullrich 2004]. Generalizing the small-matrix approximation (18.21), one finds a simple expression for the frequency shift and linewidth:

$$\Omega \approx \Omega_{\rm ALDA} + \frac{\Im \dot{E}_{\rm diss}}{2\Omega_{\rm ALDA}} - {\rm i}\,\frac{\Re \dot{E}_{\rm diss}}{2\Omega_{\rm ALDA}}\,, \tag{18.27}$$

where the average rate of energy dissipation is given by

$$\dot{E}_{\rm diss} = -\frac{N_{\rm s}}{2}\int {\rm d}z\; \sigma_{{\rm xc},\,zz}(z,\omega)\frac{\partial u(z,\omega)}{\partial z}\,, \tag{18.28}$$

following a similar expression in classical fluid dynamics [Landau 1987].

To arrive at a more realistic picture, extrinsic damping must be included. A TDDFT linear response approach for weakly disordered systems has recently been put forward [Ullrich 2001, Ullrich 2002b]. The basic idea is to treat extrinsic scattering through the so-called memory function formalism, which is a generalization of Mermin's relaxation time approximation [Mermin 1970]:

$$\chi_0^\tau(\boldsymbol{q},\omega)^{-1} = \frac{\omega}{\omega+{\rm i}/\tau}\,\chi_0(\boldsymbol{q},\omega+{\rm i}/\tau)^{-1} + \frac{{\rm i}/\tau}{\omega+{\rm i}/\tau}\,\chi_0(\boldsymbol{q},0)^{-1}\,. \tag{18.29}$$

Here, χ_0 and χ_0^τ are the Lindhard functions for a homogeneous electron gas with and without disorder. In [Ullrich 2001, Ullrich 2002b], the phenomenological scattering time τ is replaced by the memory function $M(\boldsymbol{q},\omega)$, which accounts for extrinsic damping (interface roughness and charged impurities) via microscopic disorder scattering potentials. By replacing χ_0 with the full TDDFT response function, intrinsic dissipation enters through non-adiabatic xc effects. For the case of ISB plasmons in quantum wells, (18.29) is further modified to account for the inhomogeneity in the z-direction.

Figure 18.4 shows TDDFT ISB plasmon frequencies and linewidths compared with experimental data [Williams 2001]. Under the influence of a static electric field, the ISB plasmon frequency Ω exhibits a quadratic Stark effect, but only TDDFT describes correctly the crossing of the Ω-curves for different values of $N_{\rm s}$. The bottom right panel of Fig. 18.4 shows the intrinsic plasmon linewidth due to the non-adiabatic xc effects discussed above. We see that these effects make a non-negligible contribution (up to 30% of the experimental linewidth, for large $N_{\rm s}$ and small field). Good agreement with

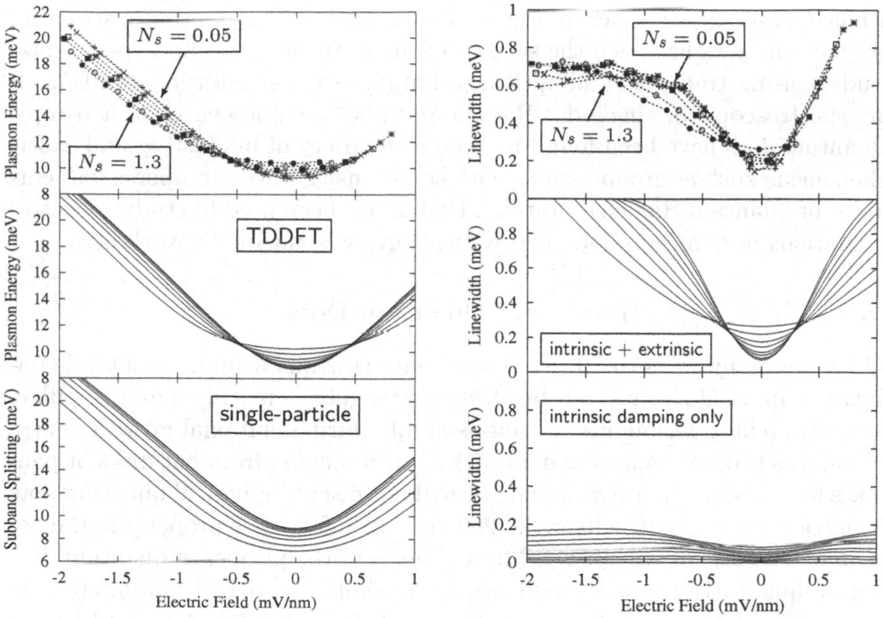

Fig. 18.4. Plasmon frequencies and linewidths in a 40 nm GaAs/AlGaAs quantum well, with N_s between 0.05 and $1.3 \times 10^{11}\,\text{cm}^{-2}$. The electric field "tilts" the quantum well and pushes the electrons more towards an interface

experiment is achieved through the dominant contribution of interface roughness scattering.

We finally mention that first applications of TDDFT to ultrafast, nonlinear ISB dynamics are beginning to emerge. In that case, dissipation can be treated phenomenologically within a density-matrix framework [Wijewardane 2004], or arises from xc memory effects through a nonlinear generalization of (18.22) [Wijewardane 2005].

18.4 Quantum Dots

Over the past 20 years, semiconductor quantum dots or "artificial atoms" have been intensely studied, and have become one of the most fascinating and innovative branches of nanoscale science and technology (for reviews, see e.g., [Ashoori 1996, Jacak 1998, Reimann 2002]). Practical applications of quantum dots span a wide range. Many technologies are already available or currently under development, such as quantum dot lasers or sensors for infrared vision or biological imaging. The use of quantum dots as qubits appears to be a promising avenue toward a scalable architecture for quantum computing. However, quantum dots are also of great fundamental interest, since one can manipulate the number of particles and their interactions to a much larger degree than in atoms and molecules. The electronic structure of quantum dots has been the subject of many theoretical and experimental studies, using transport and optical techniques such as addition and tunneling spectroscopy, or single-dot Raman and photoluminescence spectroscopy. Quantum dots have been found to exhibit a variety of interesting and exotic phenomena such as ground states with spin-density waves and persistent currents, or quantum Hall edge states. TDDFT has been used to study collective excitations in quantum dots, and we will review some of the work here.

18.4.1 Electronic Structure of Quantum Dots

There are many ways to fabricate semiconductor quantum dots: lateral confinement in a 2D electron gas by electrostatic gates, creating nanosize pillars by mesa-etching techniques, or self-assembly during epitaxial growth. These techniques produce quantum dots with shapes ranging from flat disks or pancakes to lens-shaped or pyramidal (we will not discuss spherical quantum dots or nanocrystals here [Chelikowsky 2003b], which fall more properly in the domain of clusters, see Chap. 17). Many aspects of the physics of quantum dots can be qualitatively understood using the simple model of non-interacting conduction electrons confined in a 2D parabolic potential, described by the Hamiltonian

$$H = \frac{\hbar^2}{2m^*}\left(\boldsymbol{p} - \frac{e}{c}\boldsymbol{A}_{\rm ext}\right)^2 + \frac{1}{2}m^*\omega_0^2 r^2 \,. \tag{18.30}$$

In cylindrical coordinates, the external vector potential $\boldsymbol{A}_{\rm ext}(r,\phi) = \frac{1}{2}Br\boldsymbol{e}_\phi$ is associated with a uniform magnetic field B perpendicular to the dot, with cyclotron frequency $\omega_c = eB/m^*c$. The Hamiltonian (18.30) can be exactly diagonalized, resulting in the so-called Fock-Darwin spectrum (see Fig. 18.5):

$$E_{nl} = (2n + |l| + 1)\hbar\sqrt{\omega_0^2 + \frac{\omega_c^2}{4}} - \frac{\hbar\omega_c}{2}\, l \,. \tag{18.31}$$

For large magnetic fields, the E_{nl} energy levels form so-called Landau bands [Reimann 2002]. Here n and l are the principal and azimuthal quantum number, and we have ignored the small Zeeman splitting. The Fock-Darwin spectrum exhibits an atomic-like shell structure, with "magic" electron numbers $2, 6, 12, 20, \ldots$

More realistic models for the electronic structure of quantum dots are obtained along several lines. For example, instead of a circularly symmetric parabolic confining potential, one may consider elliptic deformations [Hirose 1999, Hirose 2004], or electrostatic potentials associated with (possibly non-circular) 2D jellium disks of uniform charge density [Ullrich 2000a]. Much work has been done going beyond the effective-mass approximation, for example including spin-orbit coupling [Voskoboynikov 2001, Governale 2002] or using the $\boldsymbol{k}\cdot\boldsymbol{p}$ approximation, including strain [Pryor 1997, Pryor 1998].

Electron interactions can be treated with exact diagonalization [Maksym 1990, Pfannkuche 1993], but only for $\lesssim$10 electrons. We use current-DFT

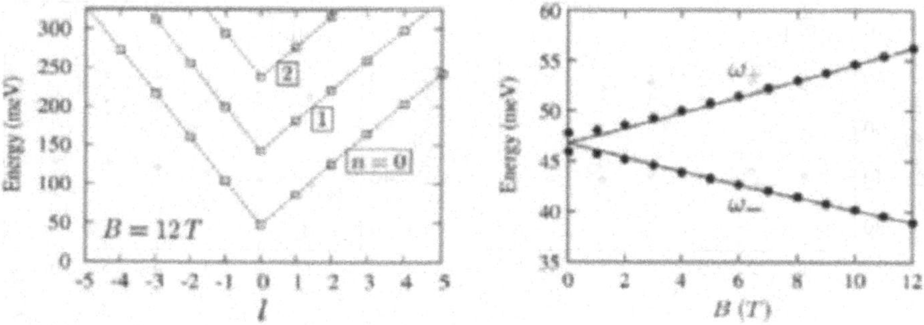

Fig. 18.5. *Left*: Fock-Darwin spectrum (18.31) of a parabolic quantum dot with $\omega_0 = 46.8\,\mathrm{meV}$ and InAs material parameters. *Right*: associated collective excitations (*lines*). The points are for a non-circular, non-parabolic two-electron dot [Ullrich 2000a]

(CDFT) [Vignale 1987, Vignale 1988, Ferconi 1994, Steffens 1998, Pi 1998] to calculate the electronic structure in the presence of magnetic fields. For 2D systems with circular symmetry, we write the KS orbitals as $\varphi_{jl\sigma}(\boldsymbol{r}) = \mathrm{e}^{-\mathrm{i}l\phi}\tilde{\varphi}_{jl\sigma}(r)$, where the radial functions satisfy the following equation:

$$\begin{aligned}\bigg\{ -\frac{\hbar^2}{2m^*}\left(\frac{\mathrm{d}^2}{\mathrm{d}r^2} + \frac{1}{r}\frac{\mathrm{d}}{\mathrm{d}r} - \frac{l^2}{r^2}\right) - \frac{e\hbar l B}{2m^* c} + \frac{e^2 B^2 r^2}{8m^* c^2} + \frac{1}{2}\, g^* \mu_B B \sigma \\ - \frac{e\hbar l}{m^* c}\frac{A_{\mathrm{xc}\,\sigma}(r)}{r} + v_{\mathrm{dot}}(r) + v_{\mathrm{H}}(r) + v_{\mathrm{xc}\,\sigma}(r) \bigg\} \tilde{\varphi}_{jl\sigma}(r) = \varepsilon_{jl\sigma}\, \tilde{\varphi}_{jl\sigma}(r)\end{aligned} \tag{18.32}$$

where $v_{\mathrm{dot}}(r)$ is the bare confining dot potential. It has been shown [Ferconi 1994] that CDFT ground-state energies of two- and three-electron dots agree to within $\lesssim 3\%$ with exact-diagonalization results over a wide range of magnetic fields.

18.4.2 Collective Excitations: Kohn's Theorem and Beyond

Collective excitations in quantum dots have been experimentally observed using FIR absorption [Sikorski 1989, Demel 1990] and inelastic light scattering [Strenz 1994, Schüller 1996]. According to the generalized Kohn's theorem [Kohn 1961, Brey 1989, Yip 1991], in parabolically confined systems the only possible dipole excitation is the center-of-mass mode, independent of electron interactions. In a magnetic field, this results in the two modes $\omega_\pm = \sqrt{\omega_0^2 + \omega_c^2/4} \pm \omega_c/2$ shown in Fig. 18.5. The influence of electron interaction on the charge-density mode spectrum can thus only be observed for anharmonic confinement. Figure 18.6 shows a comparison between experiment [Fricke 1996] and TDDFT linear response theory [Ullrich 2000a] for

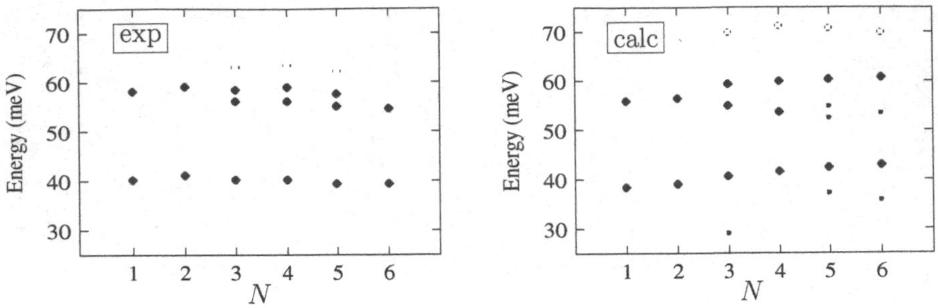

Fig. 18.6. Collective excitations of self-assembled InAs quantum dots with up to $N = 6$ electrons at $B = 12\,\mathrm{T}$. *Left*: FIR data [Fricke 1996]; *Right*: TDDFT calculation [Ullrich 2000a]. The open symbols are excitations which become dipole-allowed for non-circular deformations

InAs self-assembled quantum dots embedded in GaAs. The number of electrons in the dots was varied between $1 \leq N \leq 6$. For one and two electrons, one sees the two modes $\omega_\pm$ as expected, but for $N > 2$, additional peaks show up. The origin of the mode splitting is twofold: (i) due to the finite size of the dots, the confining potential is not parabolic; (ii) due to strain, the dots have a non-circular deformation. Only by including both effects, the threefold splitting of the ω_+ mode is reproduced. The agreement between theory and experiment is not perfect since the 2D calculation ignores the finite height of the dots and the presence of the wetting layer continuum.

TDDFT calculations of the electron dynamics in quantum dots have been performed by several groups. Hirose et al. [Hirose 2004] obtained excitation energies of closed-shell elliptic quantum dots without magnetic fields. A systematic effort to explore the collective electron dynamics in quantum dots and related nanostructures was carried out by Serra, Lipparini and coworkers [Serra 1997, Lipparini 1998, Serra 1999a, Lipparini 1999, Barranco 2000, Serra 1999b, Emperador 1999, Valín-Rodríguez 2000, Puente 2001, Puente 1999, Serra 1999c, Valín-Rodríguez 2001b, Lipparini 2002, Serra 2003a, Valín-Rodríguez 2001a, Valín-Rodríguez 2002, Serra 2003b]. They studied a variety of charge- and spin-density excitations (both longitudinal and transverse) in circular dots [Serra 1997, Lipparini 1998, Serra 1999a, Lipparini 1999, Barranco 2000, Serra 1999b], quantum rings [Emperador 1999, Valín-Rodríguez 2000, Puente 2001], deformed dots of various shapes [Puente 1999, Serra 1999c, Valín-Rodríguez 2001b, Lipparini 2002, Serra 2003a], and quantum-dot molecules [Valín-Rodríguez 2001a]. Collective modes in the presence of spin-orbit coupling were also considered [Valín-Rodríguez 2002, Serra 2003b]. All calculations were carried out using the adiabatic approximation for xc.

To conclude: in addition to their technological significance, semiconductor quantum dots offer many opportunities to study the behavior of confined, interacting electron systems. Just like real atoms, they have discrete shell structures; however, the effects of magnetic fields and spin-orbit coupling are orders of magnitude stronger. Furthermore, since the confining potential is rather smooth, the electron dynamics is highly collective. From the TDDFT viewpoint, the electronic dynamics of semiconductor nanostructures thus provides a wealth of phenomena that still remain to be fully explored.

Acknowledgments

This work was supported in part by the Petroleum Research Fund, administered by the American Chemical Society. The author is a Cottrell Scholar of Research Corporation.

19 Solids from Time-Dependent Current DFT

P.L. de Boeij

19.1 Introduction

The description of the ground state of crystalline systems within density functional theory, and of their response to external fields within the time-dependent version of this theory, relies heavily on the use of periodic boundary conditions. As a model for the bulk part of the system one considers a large region containing N elementary unit cells. Then, while imposing constraints that ensure the single-valuedness and periodicity of the wave function at the boundary, one considers the limit of infinite N to derive properties for the macroscopic samples. In this treatment, one implicitly assumes that the Hohenberg-Kohn theorem [Hohenberg 1964] and the Kohn-Sham approach [Kohn 1965], and their time-dependent equivalents derived by Runge and Gross [Runge 1984], apply separately to the bulk part of the system. This implies that effects caused by density changes at the outer surface, which are artificially removed in this periodic boundary approach, can be neglected. However, this can not be justified as these effects are real. For example, when a real system is perturbed by an external electric field, there will be a macroscopic response: a current density will (momentarily) be induced in the bulk with a nonzero average. By virtue of the continuity relation, this uniform component corresponds to a density change at the outer surface, but not to a density change inside the bulk. The density change at the surface gives rise to a macroscopic screening field, which can not be described as a functional of the bulk density alone [Gonze 1995b, Gonze 1997b]. Implicit in the periodic boundary treatment of the density functional approach is therefore that the system remains macroscopically unpolarized: charges at the surfaces should be compensated and no uniform external field may be present. While these conditions can be met for the ground-state description, similar assumptions may become problematic in the time-dependent case, where charge may be exchanged between surface and bulk regions, and where the bulk may become polarized. For isotropic systems some of these difficulties can be circumvented within the density functional approach by making use of the relation between the density-density response function and the trace of the current-current response function [Onida 2002, Kim 2002b, Kim 2002a, Nozières 1999]. However, for anisotropic materials this relation does not provide enough information to extract all components of the screening field. The induced polarization

P.L. de Boeij: *Solids from Time-Dependent Current DFT*, Lect. Notes Phys. **706**, 287–300 (2006)
DOI 10.1007/3-540-35426-3_19

can of course be described as functional of the current density in the bulk. This is the first reason to consider the use of the periodic boundary approach within the framework of time-dependent current density functional theory (TDCDFT). In this approach, the particle density is replaced by the particle current density as the fundamental quantity [Vignale 1996, Gross 1996] (see Chap. 5), which is allowed based on the observation by Ghosh and Dhara [Dhara 1987, Ghosh 1988] that the Runge-Gross theorems can be extended to systems subjected to general time-dependent electromagnetic fields. An additional bonus is that, within this more general treatment, we can look at the response to transverse fields. In the traditional density formulation, only the response to longitudinal fields can be considered, since only purely longitudinal fields can be described by the scalar potentials entering this theory.

There is a second important reason to consider using the current formulation for extended systems. While TDDFT is mostly used within the adiabatic local density approximation (ALDA), it has become clear that in extended systems nonlocal exchange-correlation effects can be very important [Kim 2002b, Kim 2002a, Sottile 2003, Adragna 2003]. In TDCDFT an approach to go beyond the ALDA is possible if one includes such long range exchange-correlation effects in the effective vector potential $\mathbf{A}_{\text{xc}}(\mathbf{r}, t)$. Vignale and Kohn [Vignale 1996, Vignale 1998] derived such a form by studying the dynamic response of a weakly inhomogeneous electron gas. They showed that in first order a dynamical exchange-correlation functional can be formulated in terms of the current density that is nonlocal in time but still local in space. Van Faassen et al. [van Faassen 2002, van Faassen 2003a] indeed showed that the inclusion of the Vignale-Kohn functional in TDCDFT calculations gives greatly improved polarizabilities for π-conjugated polymers in which similar surface effects occur [van Gisbergen 1999b].

In this chapter we show how intrinsic, i.e., material properties, can be obtained and how extrinsic, i.e., size and shape dependent effects, can effectively be removed from the computational scheme. The result of this analysis is that a macroscopic component of the current-density appears as an extra degree of freedom, which is not uniquely fixed by the lattice periodic density. A natural way to treat the periodic systems is now obtained by changing the basic dynamical variable from the time-dependent density to the induced current-density. Observable quantities, like for instance the induced macroscopic polarization, can then be given in closed form as current functionals. The response to both longitudinal and transverse fields is treated in this unified approach. We will give the linear response formulation for the resulting time-dependent Kohn-Sham system, and show how (non)-adiabatic density and current-density dependent exchange-correlation functionals can be included.

19.2 Surface and Macroscopic Bulk Effects

First we examine the time-dependent Hartree potential, defined for a finite system as

$$v_{\mathrm{H}}(\boldsymbol{r},t) = \int \mathrm{d}^3 r' \, \frac{n(\boldsymbol{r}',t)}{|\boldsymbol{r}-\boldsymbol{r}'|} \,. \tag{19.1}$$

For extended crystalline systems we want to separate the surface and sample-shape dependent contributions to this potential from the bulk intrinsic parts. This separation is however not so trivial. We start by writing the contributions from the surface (S) and the bulk regions ($B = \cup_i \mathcal{V}_i$) separately,

$$v_{\mathrm{H}}(\boldsymbol{r},t) = \int_S \mathrm{d}\sigma' \, \frac{n(\boldsymbol{r}',t)}{|\boldsymbol{r}-\boldsymbol{r}'|} + \sum_i \int_{\mathcal{V}_i} \mathrm{d}^3 r' \, \frac{n(\boldsymbol{r}',t)}{|\boldsymbol{r}-\boldsymbol{r}'|} \,. \tag{19.2}$$

Ideally we would like to consider only the bulk part for an infinite periodic lattice, by extending the sum over unit cells ($\mathcal{V}_i$) to infinity, while effectively removing the surface part. However, the result of this procedure is not uniquely defined. This is easily understood by using a multipole expansion for the contribution of each cell. While the potential of an order-n multipole decays asymptotically at a distance R as $1/R^{n+1}$, the number of such contributions in the lattice sum grows as R^2. As result, the lattice sum diverges for the monopole moment, and a sample-shape dependence arises due to the truncation at the boundary between bulk and surface of the conditionally convergent lattice sums for the dipole and quadrupole moments. Only for higher order moments the lattice sums do converge uniquely. Fortunately, the shape dependent terms can be isolated and removed if we proceed in the following way. First we write the density in the bulk region as a Fourier integral,

$$n(\boldsymbol{r},t) = \int \mathrm{d}^3 q \, n_{\boldsymbol{q}}(\boldsymbol{r},t) \mathrm{e}^{\mathrm{i}\boldsymbol{q}\cdot\boldsymbol{r}} \,, \tag{19.3}$$

where the functions $n_{\boldsymbol{q}}(\boldsymbol{r},t)$ are lattice periodic, and the vector $\boldsymbol{q}$ is restricted to the first Brillouin zone. We will now treat each $\boldsymbol{q}$-component of the density individually, and introduce the following excess quantity,

$$\Delta v(\boldsymbol{r},t) = \sum_i \int \mathrm{d}^3 q \, \Delta v_{\boldsymbol{q},i}(\boldsymbol{r},t) \mathrm{e}^{\mathrm{i}\boldsymbol{q}\cdot\boldsymbol{r}} \,, \tag{19.4}$$

which represents the potential of a background formed by plain monopole, dipole, and quadrupole density waves,

$$\Delta v_{\boldsymbol{q},i}(\boldsymbol{r},t) = \int_{\mathcal{V}_i} \mathrm{d}^3 r' \; \mathrm{e}^{\mathrm{i}\boldsymbol{q}\cdot(\boldsymbol{r}'-\boldsymbol{r})} \times$$

$$\left[\mu_{\boldsymbol{q}}(t) + \sum_\alpha \mu_{\boldsymbol{q},\alpha}(t) \frac{\partial}{\partial r'_\alpha} + \frac{1}{2} \sum_{\alpha\beta} \mu_{\boldsymbol{q},\alpha\beta}(t) \frac{\partial^2}{\partial r'_\alpha \partial r'_\beta} \right] \frac{1}{|\boldsymbol{r}-\boldsymbol{r}'|} \,. \tag{19.5}$$

Here we worked out the full contraction of the rank-n tensors $\mu_{\boldsymbol{q},(n)}(t)$ and the order-n derivative $\partial^n/\partial \boldsymbol{r}'^n$. The uniform densities $\mu_{\boldsymbol{q},(n)}(t)$ have now to give for each cell the same three lowest order terms in the multipole expansion as the corresponding $\boldsymbol{q}$-component of the real density. This ensures that the shape dependence of the potential of the excess densities is identical to that of the real density. Combining the contributions of all bulk cells and integrating by parts using Green's integral theorem, the resulting contribution to the excess potential (19.4) can also be represented using a plain monopole density wave in the bulk in combination with a charge and dipole layer at the boundary between bulk and surface. We can now remove the shape dependence and get a model-independent bulk potential by subtracting the excess potential from the bulk contribution,

$$v_{\mathrm{H,\,mic}}(\boldsymbol{r},t) = \int_B \mathrm{d}^3 r' \, \frac{n(\boldsymbol{r}',t)}{|\boldsymbol{r}-\boldsymbol{r}'|} - \Delta v(\boldsymbol{r},t) \; . \tag{19.6}$$

This potential gives the microscopic variation in the bulk and is completely determined by the bulk density. It is well-defined and model independent: the value obtained does not depend on the particular choice for the unit cell of the periodic system or for the origin. The remaining contribution of the bulk is combined with the surface part of the potential and hence contains all shape-dependent parts,

$$v_{\mathrm{H,\,mac}}(\boldsymbol{r},t) = \int_S \mathrm{d}\sigma' \, \frac{n(\boldsymbol{r}',t)}{|\boldsymbol{r}-\boldsymbol{r}'|} + \Delta v(\boldsymbol{r},t) \; , \tag{19.7}$$

This macroscopic part of the Hartree potential has to be added to the true external potential, and acts as an "externally" determined perturbing potential for the bulk system. The bulk part of the system may now be treated using the periodic boundary approach if only the microscopic part of the Hartree term, as given in (19.6) is retained. In addition, it will be convenient, and for $\boldsymbol{q}=0$ even necessary, to represent the macroscopic potential in the bulk using a macroscopic longitudinal electric field,

$$\boldsymbol{E}_{\mathrm{mac}}(\boldsymbol{r},t) = \nabla v_{\mathrm{ext}}(\boldsymbol{r},t) + \nabla v_{\mathrm{H,\,mac}}(\boldsymbol{r},t) \; . \tag{19.8}$$

This field can be chosen to satisfy the same periodic boundary conditions, even though the corresponding macroscopic potential will then violate these. The construction of the microscopic and macroscopic contributions is depicted in Fig. 19.1.

One of the properties of interest is the induced macroscopic polarization, which is defined as the time-integral of the induced macroscopic current density $\boldsymbol{j}_{\mathrm{mac}}(\boldsymbol{r},t)$,

$$\boldsymbol{P}_{\mathrm{mac}}(\boldsymbol{r},t) = -\int_{t_0}^{t} \mathrm{d}t' \, \boldsymbol{j}_{\mathrm{mac}}(\boldsymbol{r},t') \; . \tag{19.9}$$

For an isotropic system in a longitudinal external field the induced current will be longitudinal as well. For a finite system, the longitudinal part of the

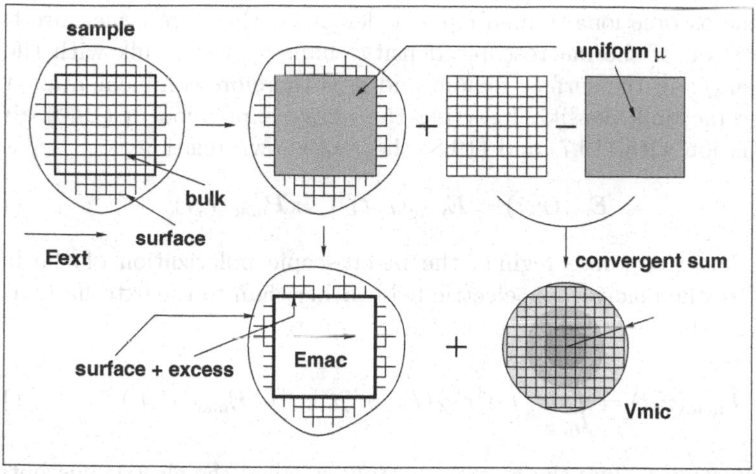

Fig. 19.1. The construction of the microscopic and macroscopic contributions to the Hartree potential

current density can be obtained by integrating the continuity equation,

$$\nabla \cdot \boldsymbol{j}_{\mathrm{L}}(\boldsymbol{r},t) + \frac{\partial}{\partial t} n(\boldsymbol{r},t) = 0, \quad \text{with} \quad \boldsymbol{j}_{\mathrm{L}}(\infty,t) = 0\,. \tag{19.10}$$

This gives the following density functional for the finite systems,

$$\boldsymbol{j}_{\mathrm{L}}(\boldsymbol{r},t) = \frac{1}{4\pi} \nabla \frac{\partial}{\partial t} \int \mathrm{d}^3 r' \, \frac{n(\boldsymbol{r}',t)}{|\boldsymbol{r}-\boldsymbol{r}'|}\,. \tag{19.11}$$

If we want to identify surface and shape-dependent contributions we will encounter the same problems as with the Hartree potential. We again need to introduce an excess contribution,

$$\Delta \boldsymbol{j}_{\mathrm{L}}(\boldsymbol{r},t) = \frac{1}{4\pi} \nabla \frac{\partial}{\partial t} \Delta v(\boldsymbol{r},t)\,, \tag{19.12}$$

with $\Delta v(\boldsymbol{r},t)$ given by (19.4), which allows us to unambiguously define a microscopic longitudinal current density in the bulk,

$$\boldsymbol{j}_{\mathrm{L,\,mic}}(\boldsymbol{r},t) = \frac{1}{4\pi} \nabla \frac{\partial}{\partial t} \left\{ \int_B \mathrm{d}^3 r' \, \frac{n(\boldsymbol{r}',t)}{|\boldsymbol{r}-\boldsymbol{r}'|} - \Delta v(\boldsymbol{r},t) \right\}\,. \tag{19.13}$$

This part contains exactly the same information as the microscopic part of the time-dependent density. The remaining macroscopic part of the current density contains all shape dependent and surface contributions to the longitudinal current in the bulk,

$$\boldsymbol{j}_{\mathrm{L,\,mac}}(\boldsymbol{r},t) = \frac{1}{4\pi} \nabla \frac{\partial}{\partial t} \left\{ \int_S \mathrm{d}^3 r' \, \frac{n(\boldsymbol{r}',t)}{|\boldsymbol{r}-\boldsymbol{r}'|} + \Delta v(\boldsymbol{r},t) \right\}\,. \tag{19.14}$$

The macroscopic longitudinal current density is therefore a measure for the combination of the macroscopic density changes in the bulk with the density changes in the surface region, and it is therefore complementary to the microscopic bulk density. Inserting the above expression in (19.9) gives in combination with (19.7) and (19.8) the well known relation

$$\boldsymbol{E}_{\mathrm{ext}}(\boldsymbol{r},t) = \boldsymbol{E}_{\mathrm{mac}}(\boldsymbol{r},t) + 4\pi \boldsymbol{P}_{\mathrm{mac}}(\boldsymbol{r},t) \,. \tag{19.15}$$

In the linear response regime, the macroscopic polarization of the bulk is related to the macroscopic electric field rather than to the externally applied field, via what is called the constitutive equation,

$$\boldsymbol{P}_{\mathrm{mac}}(\boldsymbol{r},t) = \int_{t_0}^{t} \mathrm{d}t' \int \mathrm{d}^3r' \, \chi(\boldsymbol{r}-\boldsymbol{r}',t-t') \cdot \boldsymbol{E}_{\mathrm{mac}}(\boldsymbol{r}',t') \,. \tag{19.16}$$

This equation defines the material property called the electric susceptibility $\chi(\boldsymbol{r}-\boldsymbol{r}',t-t')$. As the constitutive equation takes the form of a convolution it is more convenient to work with the Fourier transform $\chi(\boldsymbol{q},\omega)$ from which also the macroscopic dielectric function can be derived,

$$\epsilon(\boldsymbol{q},\omega) = 1 + 4\pi\chi(\boldsymbol{q},\omega) \,. \tag{19.17}$$

In general $\chi(\boldsymbol{q},\omega)$ and $\epsilon(\boldsymbol{q},\omega)$ are tensors, which transform as scalars in isotropic systems. One of the aims of the application of TDCDFT to the solids is to calculate these response properties.

19.3 The Time-Dependent Current Density Functional Approach

For finite systems, and for a given initial state, the Runge-Gross theorem [Runge 1984] ensures a one-to-one mapping between the time-dependent density and the time-dependent external potential. As first step in the original proof of this theorem, invertibility is established for the mapping from external potentials to v-representable currents. This is done by using the equation of motion for the current density, without having to refer explicitly to the boundary of the system. To arrive at a one-to-one relation with the time-dependent density, however, one needs to invoke explicitly the finiteness of the system. This second step becomes problematic for the periodic boundary approach [Maitra 2003a].

We have to conclude that, in the periodic-boundary formulation, the microscopic bulk density and microscopic bulk potential are not sufficient, and that the complementary information about the surface region that is contained in $\boldsymbol{j}_{\mathrm{L,mac}}(\boldsymbol{r},t)$ and $\boldsymbol{E}_{\mathrm{L,mac}}(\boldsymbol{r},t)$ has to be included in the description. The justification of the application of the Runge-Gross theorem in the periodic-boundary approach will depend on the existence of a similar exact mapping between on one hand the accessible densities in the bulk region, in this case the microscopic density in combination with the macroscopic longitudinal current density, and on the other hand the external potentials for this region, i.e., the combination of the microscopic external potential with the macroscopic longitudinal electric field,

$$\{n_{\text{mic}}(\boldsymbol{r},t), \boldsymbol{j}_{\text{L, mac}}(\boldsymbol{r},t)\}_{\text{bulk}} \leftrightarrow \{v_{\text{ext, mic}}(\boldsymbol{r},t), \boldsymbol{E}_{\text{L, mac}}(\boldsymbol{r},t)\}_{\text{bulk}} \,. \tag{19.18}$$

Equivalently, one could work with the full longitudinal current density $\boldsymbol{j}_{\text{L}}(\boldsymbol{r},t)$ and the total external longitudinal field $\boldsymbol{E}_{\text{L}}(\boldsymbol{r},t)$. An alternative route is formed by obtaining the current density within the current density functional framework [Dhara 1987, Ghosh 1988, Vignale 1996, Gross 1996]. In this framework, one allows for both longitudinal and transverse external fields, and as basic variable one uses the total current density rather than just the density. We will allow for the more general case of anisotropic systems and/or transverse fields, for which we will assume the existence of an exact mapping between the current density and the total external electric field for the bulk region, similar to the Ghosh-Dhara theorem for finite systems. For the static limit [Gonze 1995b, Gonze 1997b, Martin 1997a], and for a one-dimensional circular geometry, this can indeed be established [Maitra 2003a], but no proof exists for the validity of these theorems in the general periodic boundary case. We will assume these theorems to hold true in the remainder.

In the corresponding time-dependent Kohn-Sham scheme, the true density and current density of the interacting system are reproduced in a non-interacting system with effective scalar and vector potentials, via

$$n(\boldsymbol{r},t) = \sum_{j=1}^{N} |\varphi_j(\boldsymbol{r},t)|^2 \,, \tag{19.19}$$

and,

$$\boldsymbol{j}(\boldsymbol{r},t) = \sum_{j=1}^{N} \Re\{-\mathrm{i}\varphi_j^*(\boldsymbol{r},t)\nabla\varphi_j(\boldsymbol{r},t)\} + \frac{1}{c} n(\boldsymbol{r},t)\boldsymbol{A}_{\text{KS}}(\boldsymbol{r},t) \,. \tag{19.20}$$

The first and second term on the right-hand side are the paramagnetic and diamagnetic currents respectively. Here we merely need to assume that the density and current density are non-interacting $\boldsymbol{A}$-representable, which is a much weaker condition than the non-interacting v-representability, which is indeed problematic for the current density [D'Agosta 2005a]. The orbitals are solutions of the time-dependent Kohn-Sham equations,

$$\mathrm{i}\frac{\partial\varphi_j(\boldsymbol{r},t)}{\partial t} = \left\{\frac{1}{2}\left|-\mathrm{i}\nabla + \frac{1}{c}\boldsymbol{A}_{\text{KS}}(\boldsymbol{r},t)\right|^2 + v_{\text{KS, mic}}(\boldsymbol{r},t)\right\}\varphi_j(\boldsymbol{r},t) \,, \tag{19.21}$$

where periodic boundary constraints are imposed on the orbitals and on the effective potentials, i.e., the gauge is chosen to be compatible with the periodic boundary assumption [Kootstra 2000a, Kootstra 2000b]. At t_0 we assume that $\boldsymbol{j}(\boldsymbol{r}, t_0) = 0$ and $\boldsymbol{A}_{\mathrm{KS}}(\boldsymbol{r}, t_0) = 0$, and that $v_{\mathrm{KS,\,mic}}(\boldsymbol{r}, t_0)$ is the effective scalar potential giving the initial density $n(\boldsymbol{r}, t_0)$, which we choose to be the lattice periodic ground state density $n_{\mathrm{GS}}(\boldsymbol{r})$. The initial potential is then uniquely determined by virtue of the Hohenberg-Kohn. This potential will be lattice periodic, and we can choose the orbitals to be initially of Bloch form. For $t > t_0$ the effective time-dependent potentials are uniquely determined, apart from an arbitrary gauge transform, by the exact time-dependent density and current density as result of the Ghosh-Dhara theorem. To comply with the periodic boundary constraints, the gauge is chosen such that only microscopic Hartree and exhange-correlation contributions are included in the time-dependent scalar potential,

$$v_{\mathrm{KS,\,mic}}(\boldsymbol{r}, t) = v_{\mathrm{ext,\,mic}}(\boldsymbol{r}, t) + v_{\mathrm{Hxc,mic}}[n, \boldsymbol{j}](\boldsymbol{r}, t) \,, \tag{19.22}$$

while all macroscopic terms are included in the effective vector potential,

$$\boldsymbol{A}_{\mathrm{KS}}(\boldsymbol{r}, t) = -c \int_{t_0}^{t} \mathrm{d}t' \boldsymbol{E}_{\mathrm{mac}}(\boldsymbol{r}, t') + \boldsymbol{A}_{\mathrm{xc}}[n, \boldsymbol{j}](\boldsymbol{r}, t) \,. \tag{19.23}$$

The effective vector potential will in general contain exchange-correlation contributions to ensure that, apart from the true time-dependent density, also the true current density is reproduced in the Kohn-Sham system.

In order to obtain the dielectric function of the crystal, we will consider the linear response to a given macroscopic electric field $\boldsymbol{E}_{\mathrm{mac}}(\boldsymbol{r}, t)$, which is $\boldsymbol{q}$- and ω-dependent [Romaniello 2005],

$$\boldsymbol{E}_{\mathrm{mac}}(\boldsymbol{r}, t) = \boldsymbol{E}(\boldsymbol{q}, \omega)\, \mathrm{e}^{\mathrm{i}(\boldsymbol{q}\cdot\boldsymbol{r} - \omega t)} + \mathrm{c.c.} \tag{19.24}$$

The induced density, and similarly the induced current density, can now be given in the following form,

$$\delta n(\boldsymbol{r}, t) = \delta n_{\boldsymbol{q}}(\boldsymbol{r}, \omega) \mathrm{e}^{\mathrm{i}(\boldsymbol{q}\cdot\boldsymbol{r} - \omega t)} + \mathrm{c.c.} \,, \tag{19.25}$$

and

$$\delta \boldsymbol{j}(\boldsymbol{r}, t) = \delta \boldsymbol{j}_{\boldsymbol{q}}(\boldsymbol{r}, \omega) \mathrm{e}^{\mathrm{i}(\boldsymbol{q}\cdot\boldsymbol{r} - \omega t)} + \mathrm{c.c.} \,, \tag{19.26}$$

in which $\delta n_{\boldsymbol{q}}(\boldsymbol{r}, \omega)$ and $\delta \boldsymbol{j}_{\boldsymbol{q}}(\boldsymbol{r}, \omega)$ are lattice periodic. By using first order perturbation theory, the induced density is readily expressed in terms of the unperturbed Bloch orbitals $\varphi_{i\boldsymbol{k}}(\boldsymbol{r})$, orbital energies $\varepsilon_{i\boldsymbol{k}}$, and occupation numbers $n_{i\boldsymbol{k}}$. One arrives at,

$$\delta n_{\boldsymbol{q}}(\boldsymbol{r}, \omega) = \frac{1}{N_k} \sum_{\boldsymbol{k}} \sum_{i,j} \varphi^*_{i\boldsymbol{k}}(\boldsymbol{r}) \hat{n}_{\boldsymbol{q}} \varphi_{j\boldsymbol{k}+\boldsymbol{q}}(\boldsymbol{r}) P_{ij\boldsymbol{k}}(\boldsymbol{q}, \omega) \,, \tag{19.27}$$

where $\hat{n}_{\boldsymbol{q}} = \mathrm{e}^{-\mathrm{i}\boldsymbol{q}\cdot\boldsymbol{r}}$, and

$$P_{ij\boldsymbol{k}}(\boldsymbol{q}, \omega) = \frac{n_{i\boldsymbol{k}} - n_{j\boldsymbol{k}+\boldsymbol{q}}}{\varepsilon_{i\boldsymbol{k}} - \varepsilon_{j\boldsymbol{k}+\boldsymbol{q}} + \omega + \mathrm{i}\eta} \langle \varphi_{j\boldsymbol{k}+\boldsymbol{q}} | \delta \hat{h}(\boldsymbol{q}, \omega) | \varphi_{i\boldsymbol{k}} \rangle \,. \tag{19.28}$$

Here the positive infinitesimal η ensures causality. We have introduced the following short-hand notation for the first-order self-consistent perturbation,

$$\delta\hat{h}(\boldsymbol{q},\omega) = \frac{-\mathrm{i}}{2c}\left[\delta\boldsymbol{A}_{\mathrm{KS}}(\boldsymbol{r},\omega)\cdot\nabla - \nabla^{\dagger}\cdot\delta\boldsymbol{A}_{\mathrm{KS}}(\boldsymbol{r},\omega)\right] + \delta v_{\mathrm{KS,\,mic}}(\boldsymbol{r},\omega)\,, \quad (19.29)$$

where the perturbing effective scalar and vector potentials, $\delta v_{\mathrm{KS,\,mic}}(\boldsymbol{r},\omega)$ and $\delta\boldsymbol{A}_{\mathrm{KS}}(\boldsymbol{r},\omega)$, are linear in the macroscopic field $\boldsymbol{E}_{\mathrm{mac}}(\boldsymbol{q},\omega)$.

For the induced current-density we can derive expressions along the same lines as used for the induced density. We get two contributions to the induced current density, $\delta\boldsymbol{j}_{\boldsymbol{q}}(\boldsymbol{r},\omega) = \delta\boldsymbol{j}^{\mathrm{p}}_{\boldsymbol{q}}(\boldsymbol{r},\omega) + \delta\boldsymbol{j}^{\mathrm{d}}_{\boldsymbol{q}}(\boldsymbol{r},\omega)$. The paramagnetic component can be obtained from,

$$\delta\boldsymbol{j}^{\mathrm{p}}_{\boldsymbol{q}}(\boldsymbol{r},\omega) = \frac{1}{N_k}\sum_{\boldsymbol{k}}\sum_{ij}\varphi^{*}_{i\boldsymbol{k}}(\boldsymbol{r})\hat{\boldsymbol{j}}_{\boldsymbol{q}}\varphi_{j\boldsymbol{k}+\boldsymbol{q}}(\boldsymbol{r})P_{ij\boldsymbol{k}}(\boldsymbol{q},\omega)\,, \quad (19.30)$$

where $\hat{\boldsymbol{j}}_{\boldsymbol{q}} = -\mathrm{i}\left(\mathrm{e}^{-\mathrm{i}\boldsymbol{q}\cdot\boldsymbol{r}}\nabla - \nabla^{\dagger}\mathrm{e}^{-\mathrm{i}\boldsymbol{q}\cdot\boldsymbol{r}}\right)/2$. The diamagnetic contribution to the induced current-density is much simpler and is given by,

$$\delta\boldsymbol{j}^{\mathrm{d}}_{\boldsymbol{q}}(\boldsymbol{r},\omega) = \frac{1}{c}n(\boldsymbol{r})\mathrm{e}^{-\mathrm{i}\boldsymbol{q}\cdot\boldsymbol{r}}\delta\boldsymbol{A}_{\mathrm{KS}}(\boldsymbol{r},\omega)\,. \quad (19.31)$$

In practical calculations it is important to consider the relation between the diamagnetic and paramagnetic contributions, as they tend to cancel one another at small frequency due to the longitudinal conductivity sum rule [Nozières 1999].

Like in the ordinary linear response scheme of TDDFT, we need to obtain the perturbing potentials self-consistently. The contribution of the induced density to the microscopic Hartree potential is evaluated using (19.6), while the macroscopic contribution is by construction already contained in the macroscopic electric field. We can choose to retain only terms linear in the induced density in the microscopic exchange-correlation parts of the first-order scalar potential and gauge transform all other terms to the exchange-correlation vector potential. In this way we keep contact with the traditional TDDFT description. For the first-order exchange-correlation scalar potential we write,

$$\delta v_{\mathrm{xc,\,mic}}(\boldsymbol{r},\omega) = \int \mathrm{d}^3r'\, f_{\mathrm{xc}}(\boldsymbol{r},\boldsymbol{r}',\omega)\,\delta n(\boldsymbol{r}',\omega)\,, \quad (19.32)$$

in which we explicitly assume that the integration over space is converging, i.e., that the kernel is short-range. Here we use the adiabatic local density approximation for this exchange-correlation kernel,

$$f_{\mathrm{xc}}(\boldsymbol{r},\boldsymbol{r}',\omega) = \delta(\boldsymbol{r}-\boldsymbol{r}')\left.\frac{\mathrm{d}v^{\mathrm{LDA}}_{\mathrm{xc}}(n)}{\mathrm{d}n}\right|_{n=n(\boldsymbol{r})}\,. \quad (19.33)$$

All other exchange-correlation effects are included in the exchange-correlation vector potential. As the induced density is a (local) functional of the induced current-density through the continuity equation, we can formally write this vector potential as a pure functional of the induced current-density,

$$\delta A_{\mathrm{xc},\,\alpha}(\boldsymbol{r},\omega) = \int \mathrm{d}^3r' \sum_{\beta} f_{\mathrm{xc},\,\alpha\beta}(\boldsymbol{r},\boldsymbol{r}',\omega) \cdot \delta j_{\beta}(\boldsymbol{r}',\omega) \, . \tag{19.34}$$

It remains to find good approximations for this exchange-correlation contribution. If it is neglected altogether, we retrieve the adiabatic local density approximation. Here we consider a functional proposed by Vignale and Kohn, which takes the form of a viscoelastic field [Vignale 1997] (see Chap. 5),

$$\frac{\mathrm{i}\omega}{c}\delta A_{\mathrm{xc},\,\alpha}(\boldsymbol{r},\omega) = -\frac{1}{n_{\mathrm{GS}}(\boldsymbol{r})}\sum_{\beta}\frac{\partial}{\partial r_{\beta}}\sigma_{\mathrm{xc},\,\alpha\beta}(\boldsymbol{r},\omega) \, , \tag{19.35}$$

where $\sigma_{\mathrm{xc}}(\boldsymbol{r},\omega)$ is a tensor field which has the structure of a symmetric viscoelastic stress tensor,

$$\sigma_{\mathrm{xc},\,\alpha\beta} = \tilde{\eta}_{\mathrm{xc}}\left(\frac{\partial u_{\alpha}}{\partial r_{\beta}} + \frac{\partial u_{\beta}}{\partial r_{\alpha}} - \frac{2}{3}\delta_{\alpha\beta}\nabla\cdot\boldsymbol{u}\right) + \tilde{\zeta}\delta_{\alpha\beta}\nabla\cdot\boldsymbol{u} \, . \tag{19.36}$$

Here the velocity field $\boldsymbol{u}(\boldsymbol{r},\omega)$ is given by

$$\boldsymbol{u}(\boldsymbol{r},\omega) = \frac{\delta\boldsymbol{j}(\boldsymbol{r},\omega)}{n_{\mathrm{GS}}(\boldsymbol{r})} \, . \tag{19.37}$$

The coefficients $\tilde{\eta}_{\mathrm{xc}}(\boldsymbol{r},\omega)$ and $\tilde{\zeta}_{\mathrm{xc}}(\boldsymbol{r},\omega)$ are directly related to the transverse and longitudinal response coefficients $f_{\mathrm{xc},\,\mathrm{T}}(n,\omega)$ and $f_{\mathrm{xc},\,\mathrm{L}}(n,\omega)$ of the homogeneous electron gas, which are evaluated at the local density $n(\boldsymbol{r})$.

19.4 Application to Solids

The lattice-periodicity of $\delta\boldsymbol{j}_{\boldsymbol{q}}(\boldsymbol{r},\omega)$ allows to calculate the macroscopic induced polarization as,

$$\boldsymbol{P}_{\mathrm{mac}}(\boldsymbol{q},\omega) = \chi(\boldsymbol{q},\omega)\cdot\boldsymbol{E}_{\mathrm{mac}}(\boldsymbol{q},\omega) = \frac{-\mathrm{i}}{\mathcal{V}\omega}\int_{\mathcal{V}}\mathrm{d}^3r' \, \delta\boldsymbol{j}_{\boldsymbol{q}}(\boldsymbol{r}',\omega) \, , \tag{19.38}$$

where the average is taken over the unit cell. This result immediately gives the electric susceptibility $\chi(\boldsymbol{q},\omega)$, and hence the dielectric function $\epsilon(\boldsymbol{q},\omega)$.

We have studied the static dielectric constant in the ALDA approximation for several binary compounds [Kootstra 2000a, Kootstra 2000b]. In Figs. 19.2 and 19.3 we have compiled the results for the static dielectric constants ϵ_{∞} for this large set of compounds. The materials have been grouped according to the chemical groups of the constituent elements as well as to their lattice type. For the whole range of materials we see a fairly good agreement of

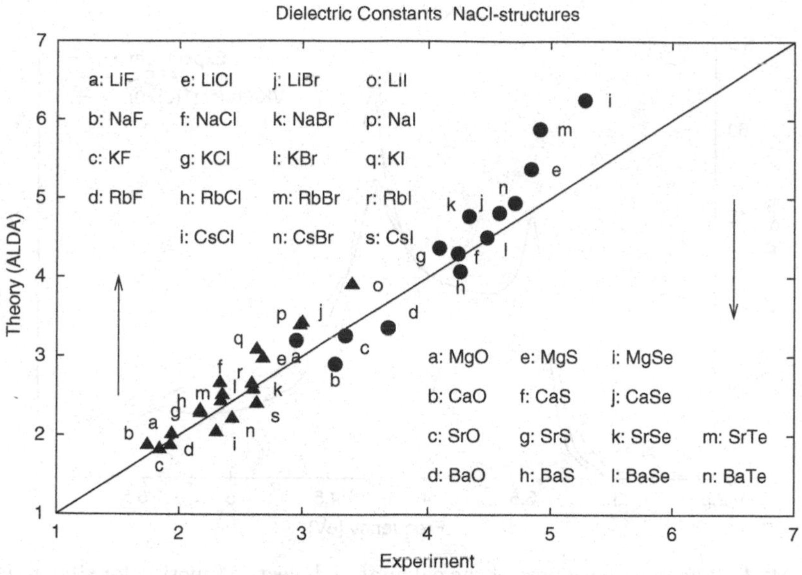

Fig. 19.2. Theoretical (ALDA) versus experimental dielectric constant for binary crystals of the rocksalt lattice type

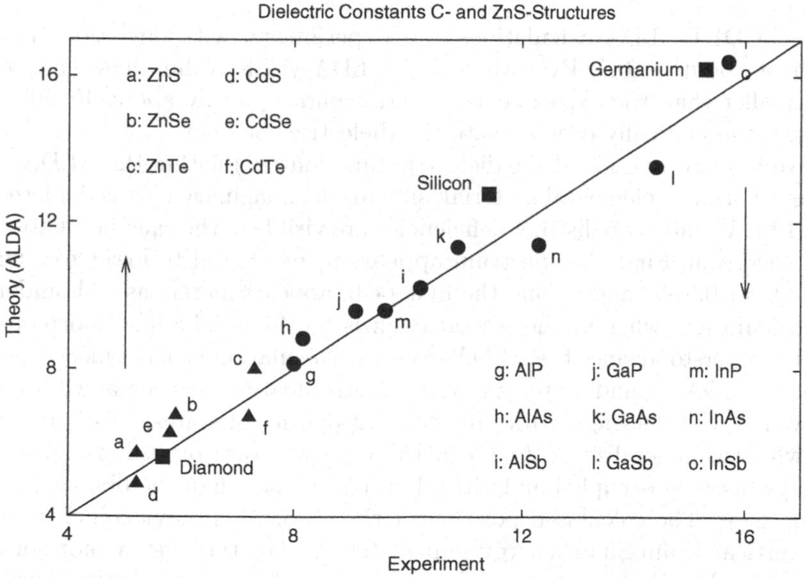

Fig. 19.3. Theoretical (ALDA) versus experimental dielectric constant for the elementary and binary crystals of the diamond and zinc-blend lattice types

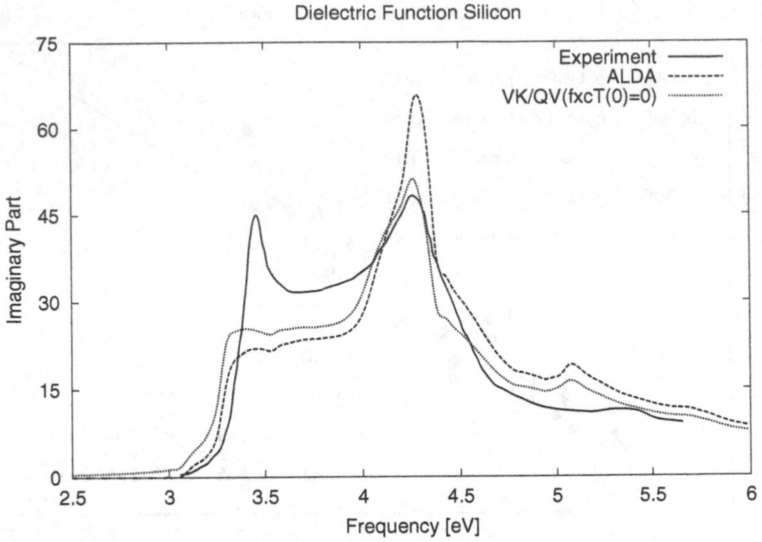

Fig. 19.4. The imaginary part of the calculated dielectric function for silicon using the ALDA and Vignale-Kohn (see text) functionals. The experimental data have been obtained from [Lautenschlager 1987]

the TDCDFT-ALDA calculations and experiments, with deviations in the order of about 5–10%. Even though the LDA yields Kohn-Sham gaps that are smaller than the experimental (fundamental) gap by about 40–50% we do not systematically overestimate the dielectric constant.

We have also obtained the dielectric function $\epsilon(\omega)$ within the ALDA. For the most studied elemental material, silicon, the imaginary part is depicted in Fig. 19.4. Usually two distinct deficiencies are visible in the calculated absorption spectrum. First the spectrum appears to be shifted to lower frequency over about 0.5 eV, and second the first peak appears merely as a shoulder in the calculation, whereas the second peak is too high. The first shortcoming can be understood since the ALDA response calculation is performed starting from the LDA ground state. As most spectral features can be attributed to the van Hove-type singularities in the joint-density of states, this is in keeping with the general trend found in LDA calculations: orbital energies lead to gaps between occupied and virtual states smaller than the observed excitation gaps. The calculated position of the absorption onset coincides with the vertical Kohn-Sham energy gap of 2.6 eV. We thus get a more-or-less uniformly shifted absorption spectrum. This is observed for other materials too. When a uniform shift is applied to the calculated spectra by using a so-called scissors shift, one usually gets a good correspondence between the measured and calculated spectral features [Levine 1989]. By starting from an improved ground state description using an exact-exchange calculation, Kim and Görling [Kim 2002b, Kim 2002a] could indeed correctly describe the absorption edge for silicon. To get the results in Fig. 19.4 we used a scissors shift.

The incorrect description of the first peak within the ALDA, which arises from excitonic effects [Reining 2002, Rohlfing 1998a, Benedict 1998a, Benedict 1998b], is caused by the local nature of the time-dependent density functional, which cannot describe nonlocal effects like the electron-hole attraction. Improved approximate expressions for the exchange-correlation kernel, for instance using the time-dependent exact exchange kernel [Kim 2002b, Kim 2002a], lead to a considerably improved intensity for the exciton peak in silicon. Similar results can be obtained by using a form based on the Kohn-Sham Green's function obtained from a perturbation expansion to first order in the screened interaction, in combination with Kohn-Sham orbital energy corrections [Sottile 2003, Adragna 2003]. Guided by the form of the Vignale-Kohn functional, and by retaining only macroscopic contributions, a simple polarization dependent exchange-correlation functional [Gonze 1995b, Gonze 1997b, Martin 1997a] could be derived. Using an additional empirical prefactor, the spectra could be improved using such a polarization functional [de Boeij 2001]. However, inclusion of the full Vignale-Kohn functional leads to much worse results unless a much reduced transverse kernel $f_{\mathrm{xc,T}}(n, \omega)$ is used to calculate the viscoelastic coefficients [Berger 2006]. In particular, this is true for the static limit of the transverse kernel, which was found to determine to a large extent the strength of the screening field, and hence the absorption spectrum of π-conjugated polyacetylene polymers [Berger 2005]. There a better correspondence with experiment and other calculations were found with reduced values for the static transverse kernel. Figure 19.4 shows the slightly improved result for the case in which the frequency dependence of this kernel is treated [Qian 2002, Qian 2003] but with a vanishing static limit. The static response reduces then again to the ALDA results. We do not observe a correct description of the exciton peak, but the second peak is reduced in strength considerably.

19.5 Conclusion

The description of extended systems using periodic boundary conditions within a density-functional framework is most naturally done using the time-dependent current-density functional approach. In this scheme information about changes of the density in the surface region that may lead to macroscopic screening effects, but that may not show up in the periodic density in the bulk region, is contained in the (macroscopic) current-density in the bulk. Even if the periodic boundary assumption is used within the current functional approach, the exchange-correlation contributions to the potentials may still depend on these surface effects as the potentials are now functionals of the current density. Other than the traditional density formulation, the current functional approach is able to describe the response of not only isotropic but also anisotropic systems to both transverse and longitudinal fields. At the same time, the nonlocal density dependence that is inherent to the nonadiabatic exchange-correlation functionals in the traditional density functional approach can be formulated using a local current-density dependent contribution to the Kohn-Sham vector potential. The common adiabatic local density approximation yields on average reasonable static dielectric constants for nonmetallic compounds, but the frequency dependent dielectric functions exhibit several deficiencies which can be attributed to the incorrect description of the exchange-correlation effects by the presently available density and current functionals.

20 Optical Properties of Solids and Nanostructures from a Many-Body f_{xc} Kernel

A. Marini, R. Del Sole, and A. Rubio

20.1 Introduction

Until the late 1990's, the situation for the ab initio calculation of optical properties of real materials was not nearly as good as that for the quasiparticle properties. As already mentioned in Chap. 10, the description of the optical response of an interacting electron system asks for the inclusion of effects beyond single-particle excitations as electron-hole interactions (excitonic effects). The important consequence of such effects is illustrated below for many different semiconductors and insulators by comparing the computed absorption spectrum neglecting electron-hole interaction with the experimental spectrum. For wide band-gap insulators there is hardly any resemblance between the spectrum from the noninteracting theory to that of experiment.

In contrast to the many-body Bether-Salpeter scheme[1], TDDFT using the standard local and semilocal xc functionals has a number of commonly invoked failures. One example is the difficulty encountered when studying extended systems; another one is the severe underestimation of high-lying excitation energies in molecules. For example, the fact that the strong nonlocality of the exact functional is not captured by the usual approximations for xc leads to a very poor description of the polarisability per unit-length of long-conjugated molecular chains. As we will see below, the main reason of this deficiency is related to the long-range nature of the xc kernel [Onida 2002]. Indeed, the advances in the development of functionals during the last years based on many-body perturbation theory and other schemes (see contributions to Part I), have broaden the field of applicability of TDDFT, as demonstrated by the variety of examples presented in this chapter.

Before discussing the applications of the many-body f_{xc} kernel, it is important to highlight the intrinsic pathologies that should be present in DFT, and that are naturally incorporated into a many-body scheme. We know that the actual functional relation between $n(\boldsymbol{r})$ and $v_{xc}(\boldsymbol{r})$ is highly nonanalytical and highly nonlocal. Some specific problems related to this inherent nonlocalities of the xc functional relevant for the description of optical properties are:

[1] The Bethe-Salpeter scheme is based on approximating the electron self-energy by the GW approximation and solving the two-particle Bethe-Salpeter equation; see Chap. 10 for more details.

A. Marini et al.: *Optical Properties of Solids and Nanostructures from a Many-Body f_{xc} Kernel*, Lect. Notes Phys. **706**, 301–316 (2006)
DOI 10.1007/3-540-35426-3_20

(i) The band-gap in the local density functional theory (Kohn-Sham gap E_g) is typically 30–50% less than the observed band-gap. This so-called band-gap problem is related to the discontinuity of the xc potential with respect to the number of particles [Perdew 1983, Sham 1983]. (ii) The macroscopic xc electric field [Gonze 1995b, Gonze 1997b, Gonze 1997a, Resta 1994, Ortiz 1998], which is related to the spatial nonlocality of the xc-potential. Similarly, the description of how excitonic effects modify the shape and pole-structure of the Kohn-Sham system needs nonlocal and, most likely, frequency dependent xc kernels. A many-body Bethe-Salpeter approach handles properly all those effects and, therefore, provides some guidelines about how to incorporate many-body effects into the xc kernel f_{xc} in order to properly describe the optical spectra of extended and low-dimensional systems (as illustrated, for example, in Chap. 10 of this book).

In this chapter we review the results recently obtained by different groups [Onida 2002, Reining 2002, Sottile 2003, Adragna 2003, Marini 2003b, Marini 2004, Botti 2004, Del Sole 2003, Bruneval 2005] using the many-body derived xc kernel presented in Chap. 10. In particular, we focus the discussion on results within linear response theory for the optical absorption and loss function of extended solids and low-dimensional structures. As a particular case, the new derived kernels reproduce the exact-exchange results of [Kim 2002a, Kim 2002b] obtained by turning-off the screening in the building-up process of the $f_{\rm xc}$ kernel. Following Chap. 10 we focus the discussion here on how can an $f_{\rm xc}$ mimic the electron-hole interaction contribution, absorbing the quasiparticle self-energy shifts in the eigenvalues of our starting independent Kohn-Sham response function $\chi_{\rm KS}$ (see discussion in Chap. 10 for details).

We decided to present the results in terms of the dimensionality of the system under study. The idea behind is to show clearly that the many-body derived $f_{\rm xc}$ is able to reproduce the results obtained from the solution of the many-body Bethe-Salpeter equation, not only for solids but also for molecules, polymers and surfaces. In this way, we provide compelling evidence about the robustness and wide-range of applicability of the new xc kernel and lays down the basic ingredients to build a fully DFT-based approach. Work along those lines is in progress using a variational many-body total energy functionals [Almbladh 1999, Dahlen 2005] in either an optimised effective potential (OEP) or generalized Kohn-Sham scheme[2] [Gruning 2005].

20.2 Applications to Solids and Surfaces

In Chap. 10 a many-body xc kernel $f_{\rm xc}$ was derived by taking as reference the Bethe-Salpeter equation. The kernel is, in principle, nonlocal both in space

[2] The simplest case are functionals derived from the GW approximation to the electron self-energy and that correspond to the so-called random-phase-approximation (RPA)-total energy functionals.

and time. However, by looking at the leading terms of this kernel a simple static long range correction (LRC) model was derived [Reining 2002]:

$$f_{xc}^{LRC}(\boldsymbol{q}, \boldsymbol{G}, \boldsymbol{G}') = -\delta_{\boldsymbol{G},\boldsymbol{G}'} \frac{\alpha}{|\boldsymbol{q}+\boldsymbol{G}|^2} , \tag{20.1}$$

where α is a suitable constant. It is illustrative to see how this simple LRC model suffices to get a proper optical spectra for a wide range of solids. It has been shown [Onida 2002, Reining 2002, Sottile 2003, Botti 2004] that the optical absorption spectrum of solids exhibiting a strong continuum excitonic effect is considerably improved with respect to calculations where the adiabatic local-density approximation is used. However, there are limitations of this simple approach, and in particular the same improvement cannot be found for the whole spectral range including the valence plasmons and bound excitons. For this, a full solution of the many-body kernel including all coupling-terms is needed [Marini 2003b, Reining 2002, Sottile 2003, Olevano 2001]. The frequency dependence of f_{xc} has been clearly shown in [Adragna 2003]. Still, for semiconductors with no strong excitonic effect a simple dynamical extension of the LRC model has been put forward by [Botti 2005].[3]

The advantage of this dynamical model is that both absorption and loss-spectra are covered with the same set of parameters, however for insulators with strong electron-hole effects one has to resort to the solution of the full many-body f_{xc}.

Applications of the LRC model have been performed [Botti 2004] for the real and the imaginary part of the dielectric function of bulk silicon, gallium arsenide, aluminum arsenide, diamond, magnesium oxide and silicon carbide, and for the loss function of silicon. A summary of those results is presented in Figs. 20.1, 20.2, and 20.3. The dot-dashed curves stem from a standard TDLDA calculation. The TDLDA results are close to the RPA one, showing the well-known discrepancies with experiment: peak positions are wrong (the TDDFT spectrum is redshifted), and the intensity of the first main structure (the E_1 peak in Si, GaAs and AlAs) is strongly underestimated. The dashed curve (named GW-RPA) is the result obtained by replacing KS eigenvalues with GW quasiparticle energies in the RPA screening. The calculated spectrum is now blue shifted. Moreover, the lineshape has not been corrected. Finally, the continuous curve is the result of the TDDFT calculation using the LRC xc kernel. An excellent fit to experiment is obtained using $\alpha = 0.22$, 0.2, and 0.35 for Si, GaAs and AlAs, respectively. There is a clear dependence of the parameter α on the material. As one would expect, α is approximatively inversely proportional to the screening in the material. This fact can give a hint of how to estimate the excitonic correction to an absorption spectrum for a material where one has not yet solved the Bethe-Salpeter equation

[3] The form of the dynamical-model for the kernel is: $f_{xc} = -\frac{\alpha+\beta\omega^2}{q^2}$, with α and β related to the macroscopic dielectric constant and bulk plasma frequency.

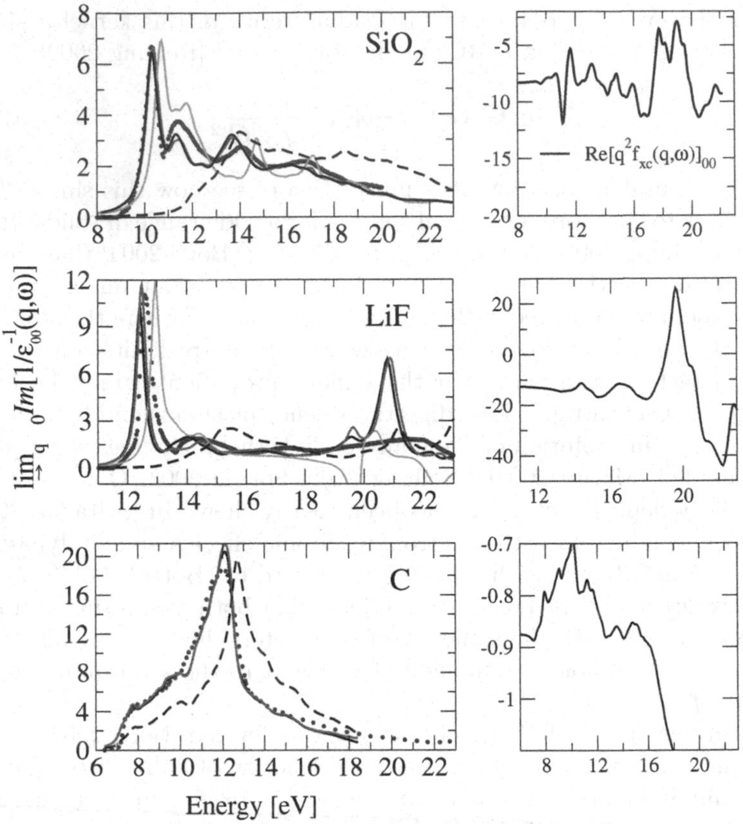

Fig. 20.1. Calculated absorption spectra of different semiconductors and insulators. Panel (**a**) is silicon, (**b**) GaAs, (**c**) AlAs, and (**d**) MgO. *Dots*: experimental data. *Dot-dashed curve*: TDLDA result. *Dashed curve*: GW-RPA (i.e., standard RPA calculation using the GW-quasiparticle energies instead of the Kohn-Sham ones). *Continuous curve*: TDDFT result using the LRC approximation to the kernel (Adapted from [Botti 2004])

(see [Botti 2004] for more details). In particular, we can estimate the spectra of new compounds as semiconducting alloys ("computational alchemy").

Similar agreement is found for the real part of the dielectric function. This fact is illustrated considering the case of GaAs as example (see Figs. 20.1 and 20.2). We see that even for this simple and well-known semiconductor, only with the inclusion of electron-hole interaction we have good agreement between theory and experiment. The influence of the electron-hole interaction extends over an energy range far above the fundamental band gap. As seen from the figure, the optical strength of GaAs is enhanced by nearly a factor of two in the low frequency regime. Also, the electron-hole interaction enhances and shifts the second prominent peak (the so-called E_2 peak) structure at 5 eV

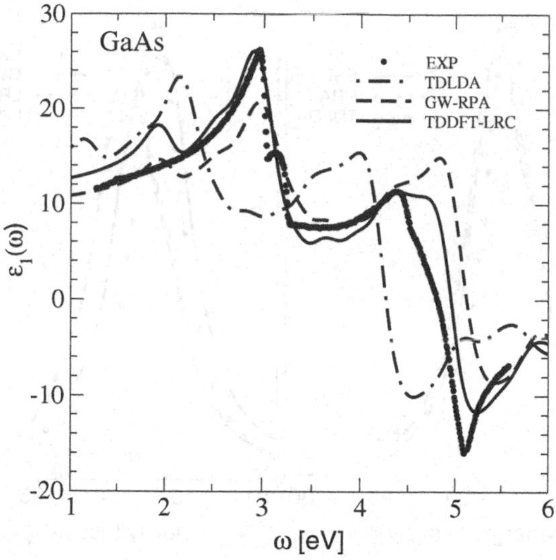

Fig. 20.2. Real part of the macroscopic dielectric response of GaAs computed within the different approximations discussed in the main text and compared to experiments. The labeling is the same as the one in Fig. 20.1

to a value much closer to experiment. This very large shift of about 0.5 eV is not due to a negative shift of the transition energies, as one might naïvely expect from an attractive electron-hole interaction. The changes in the optical spectrum originate mainly from the coupling of different electron-hole configurations in the excited states, which leads to a constructive coherent superposition of the interband transition oscillator strengths for transitions at lower energies and to a destructive superposition at energies above 5 eV [Rohlfing 1998a].

When going to large-gap materials, the screening is smaller and the electron-hole interaction becomes stronger. One can therefore expect that this drastic LRC approximation to the full kernel will break down (indeed this is the case for systems having bound excitons, see below and [Marini 2003b]). This can be clearly seen in Fig. 20.1 for MgO. In this case, the choice of $\alpha = 1.8$ is a compromise which allows to enhance the first excitonic peak to a good fraction of the experimental value, without overestimating too much the strength of the subsequent structures.

From the previous discussion we conclude that low energy part of the macroscopic dielectric function of many semiconductors is extremely well reproduced when just the long-range contribution for the xc kernel is taken into account, whereas the *same* long-range contribution cannot yield good results for the loss function (see Fig. 20.3). The role of long-range interactions

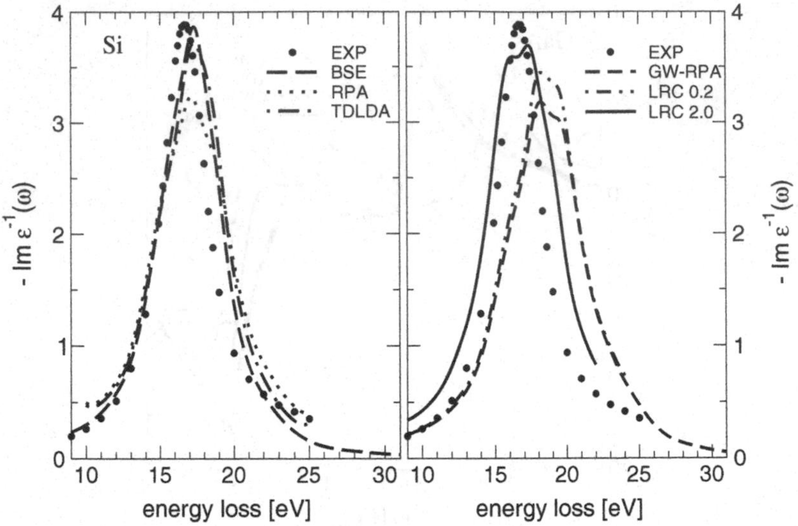

Fig. 20.3. Calculated energy-loss function of Si from different approximation to the xc kernel: RPA, TDLDA, Bethe-Salpeter (BSE) and LRC-model (adapted from [Olevano 2001, Botti 2004]). Two values of the LRC model are used, the one that describes the absorption spectra ($alpha = 0.2$) does not reproduces the loss function that requires a larger value of the constant $\alpha = 2$. The experimental data is given by the *dots*

is fundamentally different in the loss spectra, and one can expect that a small long-range contribution to the kernel will have much less effect than in the case of absorption spectra. Figure 20.3 demonstrates the quite general finding that, in the case of loss spectra, both the RPA (dotted curve) and, even better, the TDLDA (dot-dashed curve), already manage to reproduce reasonably the experimental results (dots) [Olevano 2001] (the Bethe-Salpeter results are also given for completeness; dashed curve). As expected the one-parameter LRC approach breaks down for this application. The dot-dashed curve of the bottom panel of Fig. 20.3 shows the LRC result using the same $\alpha = 0.22$ as for the absorption spectrum. Almost no effect is seen on the loss spectrum, and one is thus left essentially with the rather unsatisfying GW-RPA result. Instead, using the much larger value $\alpha = 2$ (continuous curve), the result is again satisfactory[4]. Therefore, this implies that if one is interested in a large frequency range, the full many-body kernel derived in Chap. 10 should be used. In that case we will recover the same level of accuracy as the Bethe-Salpeter calculations (see below for some results obtained for LiF [Marini 2003b]).

[4] This is the result that one would obtain by using the dynamical extension of the LRC model as done in [Botti 2005] for Si using the same set of parameters for both absorption and electron energy loss spectra.

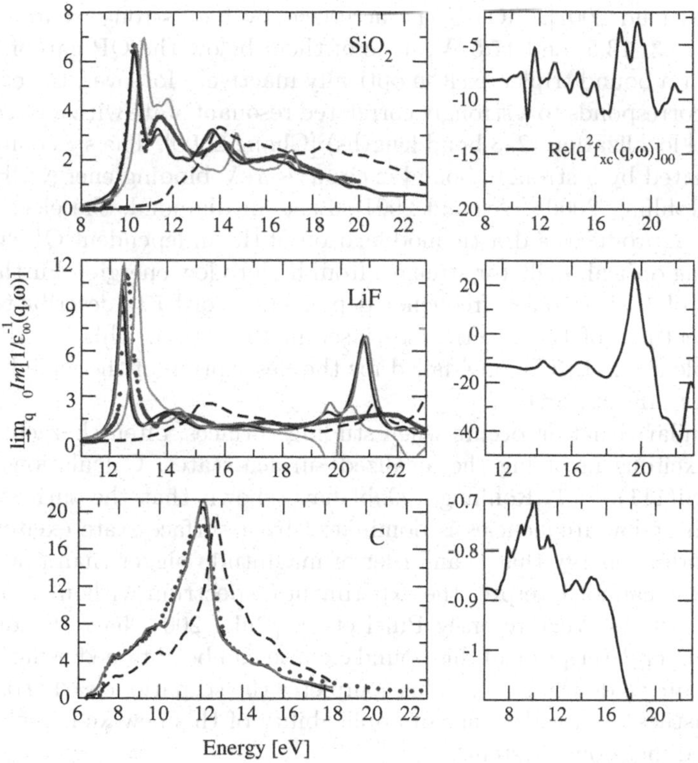

Fig. 20.4. *Left*: Calculated optical absorption spectra within the BSE (*continuous line*) and the new TDDFT xc kernel derived from the BSE (*dashed line*; nearly indistinguishable from the *continuous line*). The comparison between the two theoretical approaches and with experiment (*dots*) is excellent. We also provide the independent-quasiparticle response (GW-RPA; *dashed-dotted line*). *Right*: Frequency dependence of the head of f_{xc} (note the different scale for SiO_2, LiF and C) (Adapted from [Marini 2003b])

Now it is mandatory to address whether or not the description of strongly bound excitons within the TDDFT formalism is possible. This goal has been achieved using a frequency dependent and spatial nonlocal f_{xc} (see [Marini 2003b] and the detailed derivation in Chap. 10 of this book). For all the materials discussed above where the LRC model work quite well, it is easy to show that the full kernel gives even better description of the BSE results than the simple model. Furthermore, it also works for wide band-gap insulators where the LRC model completely fails as showed for the optical absorption and electron-energy loss spectra of LiF, SiO_2 and diamond reproduces quite well the Bethe-Salpeter results and experiments (see Fig. 20.4). In these three systems the role of excitonic effects in the optical spectrum and EELS has been already analyzed within the BSE [Chang 2000, Benedict 1998a, Rohlfing

2000b, Arnaud 2001]. SiO_2 is characterized by four strong excitonic peaks at 10.3, 11.3, 13.5, and 17.5 eV, none of them below the QP gap of 10.1 eV, except for a bound triplet exciton optically inactive. Moreover, the exciton at 10.3 eV corresponds to a strongly correlated resonant state with a large degree of spatial localization (2–3 bond lengths) [Chang 2000]. The spectrum of LiF is dominated by a strongly bound exciton (~ 3 eV binding energy) [Benedict 1998a, Rohlfing 2000b, Arnaud 2001]. Last, in diamond, the electron–hole interaction produces a drastic modification of the independent QP spectrum by shifting optical oscillator strength from high to low energies. Furthermore, the head of f_{xc} is strongly frequency dependent in order to describe the high-energy features of the spectra (see inset in Fig. 20.4). This illustrate why the simple static-LRC model failed for the description of the wide-band gap insulators LiF and SiO_2.

A similar situation occurs when studing surfaces, often characterized by strong excitons involving the localized surface states. Calculations carried out for Si(111)2×1 [Rohlfing 1999b] have shown that the surface optical spectrum at low frequencies is dominated by a surface state exciton which has a binding energy that is an order of magnitude bigger than that of bulk Si, and one cannot interpret the experimental spectrum without considering excitonic effects. Very recently Pulci et al. [Pulci 2005] have demonstrated that a proper description of this bound exciton can be obtained using TDDFT with the above described f_{xc}. This completes the scenario, clearly confirming the robustness and wide-range of applicability of this new xc kernel in three and two-dimensional systems.

To conclude this section we show the calculated EELS of LiF (Fig. 20.5) for a *finite transfer momentum* $\boldsymbol{q}$ along the ΓX direction, where previous BSE calculations and experimental results are available [Caliebe 2001]. This is a stringent test as the description of EELS needs causal response functions, including the anti-resonant part [Marini 2003b]. The results of this causal f_{xc}-calculation are presented in Fig. 20.5 for a 1st order (green line) and a 2nd order (red line) f_{xc}. While a first order, causal f_{xc} gives very good results, almost perfect agreement with BSE results is restored considering the second order correction to f_{xc}. This is not surprising as nothing ensures the same level of cancellation in the higher order expansion of resonant and causal functions. However, the final agreement with experimental data is anyway very good even at the first order. This is important for the predictive power of the many-body f_{xc} kernel for applications in low-dimensional structures.

In all these calculations it turns out that both spatial nonlocality and frequency dependence of the f_{xc} kernel are important in order to properly describe excitonic effects in the optical and electron energy-loss spectra. Still, quasiparticle effects need to be embodied properly within this approximated TDDFT scheme. More work needs to be done along this lines.

As discussed in Chap. 10 of the present book, the perturbative kernel allows to handle the loss function as well as lifetime effects in wide-band

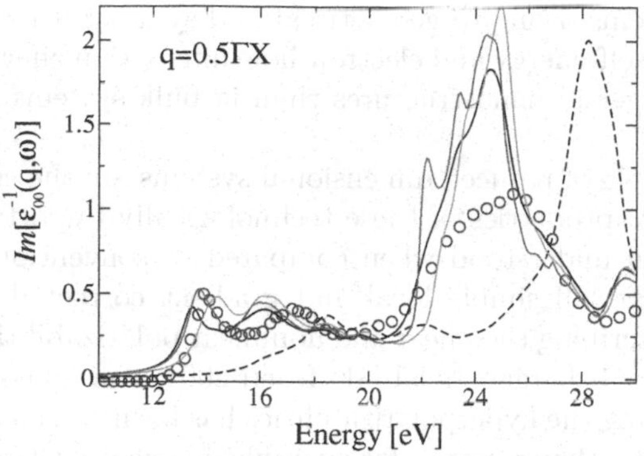

Fig. 20.5. Calculated EELS of LiF for a momentum transfer of $q = 0.5\Gamma X$: BSE (*full line*), 1st order f_{xc} (*dot-dashed line*), 2nd order f_{xc} (*dashed line*), and independent-QP (*dots*). Experiment (*circles*) are taken from [Caliebe 2001] (Adapted from [Marini 2003b])

gap insulators [Marini 2004]. In conclusion, we have provided compiling evidence of the good performance of the many-body derived f_{xc} kernel to describing the optical absorption and electron energy-loss spectra (for arbitrary q-momentum transfer) of bulk systems.

20.3 Applications to One Dimensional Systems and Molecules

In any finite system subject to an electric field, there is accumulation of charge at the surface, which induces a counter-field inside the sample. It is not possible for any local (or semi-local) functional of the density to describe the counter-field produced by the macroscopic polarization of the system [Gonze 1995b, Gonze 1997b, Gonze 1997a, Resta 1994, Ortiz 1998]. To circumvent this problem, it was proposed to use as an extra dynamical variable the surface charge, or equivalently the macroscopic field produced by that charge [Bertsch 2000b]. Another way to take into account the macroscopic polarization is by the current response of the system within a time-dependent current-density functional formalism [de Boeij 2001, van Faassen 2003a]. Those approaches have been discussed at length in other chapters of the present Part IV, therefore we will present here results obtained for one-dimensional systems and small molecules [Varsano 2005] by using the previous many-body xc kernel that has been shown to work for solids. The rational for this choice is that the Bethe-Salpeter approach has been valuable in explaining and predicting the quasiparticle excitations and optical response of reduced dimensional systems and nanostructures. This is because Coulomb interaction effects in general are more dominant in lower dimensional systems owing to geometrical and symmetry restrictions. As illustrated below, self-energy and electron-hole interaction effects can be orders of magnitude larger in nanostructures than in bulk systems made up of the same elements.

A good example of reduced dimensional systems are the conjugated polymers. The optical properties of these technologically important systems are still far from well understood when compared to conventional semiconductors. In particular, all simple local and gradient corrected functionals fail completely in describing the linear and nonlinear polarizabilities of long molecular chains [van Gisbergen 1999b, de Boeij 2001, van Faassen 2003a]. This failure in describing the hyperpolarizabilities has been traced back to the field counteracting term that appears, for example, in exact-exchange calculations or in orbital-dependent functionals, but that is completely absent in LDA or GGA functionals. Current density functional approaches also seem to solve, in part, this problem [de Boeij 2001, van Faassen 2003a]. Indeed, the use of the nonlocal and frequency-dependent f_{xc} derived from the Bethe-Salpeter equation restores the good agreement between the calculated polarizabilities and experimental data. Furthermore, it is able to describe also the H_2-linear chain where the current DFT approach fails.

To illustrate the discussion we show, in Fig. 20.6, the optical absorption spectra of a prototype polymer: polyacethylene. We see that each of the 1D van Hove singularities in the interband absorption spectrum is replaced by a series of sharp peaks due to excitonic states. The lowest optically active exciton is a bound exciton state (singlet-excitation of π-π^* character) but the others are strong resonant exciton states. The peak structure agrees very well with experiment. The resonant part of f_{xc} reproduces very well the BSE results [Rohlfing 1999a] (and experiments). Again this shows the robustness of the f_{xc} for describing low-dimensional structures.

Furthermore, in Fig. 20.7 we show the computed spectra for a finite-system: the molecular unit of polyacethylene. For this finite system we see again that the resonant part of the xc kernel describes very well the BSE calculation (also resonant). However, in contrast to the infinite one-dimensional polymer, for this finite unit the coupling terms of the BSE equation are important accounting for a ~1 eV shift to lower energies of the main singlet-exciton peak (the upper part of the spectra is less sensitive to the coupling terms). In this case, the full causal f_{xc} also gives rise to a red-shift of the spectra but not enough to cope with the full coupling term in the BSE. This is an indication that we need to go beyond a first order xc kernel, an indication that has been also observed in the description of the electron-energy loss function of LiF [Marini 2003b] where the causal response needs a higher order f_{xc} in order to match the equivalent BSE results. Still, we can conclude that the first-order many-body f_{xc} is able to account for excitonic effects in one-dimensional structures, but higher orders are needed in zero-dimensional structures (molecules).

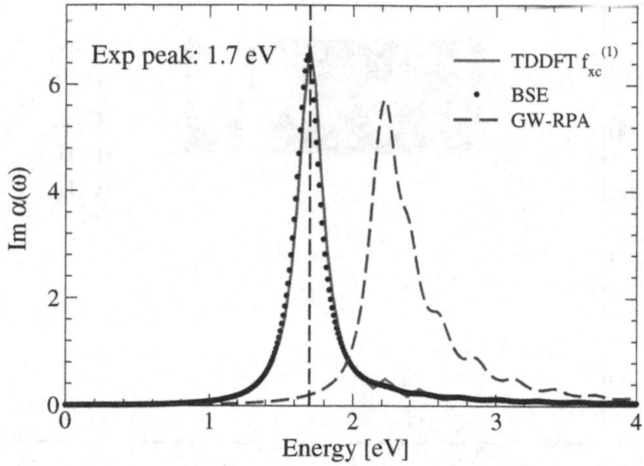

Fig. 20.6. Calculated optical absorption spectra of polyacethylene compared with experiments (*vertical dashed-line*). Clearly, the TDDFT calculation using the many-body f_{xc} is in good agreement with experiments as well as with the BSE calculation (*dots*). We also show the results of an RPA calculation using the *GW* quasiparticles energies (*dashed line*) to illustrate the impact of the electron-hole interaction in this one-dimensional system (Adapted from [Varsano 2005])

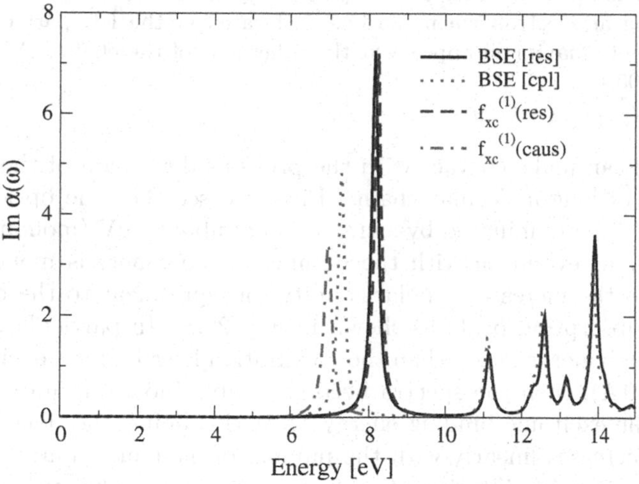

Fig. 20.7. Calculated optical absorption spectra of a single unit of the poly-acethylene polymer. We compare the resonant and full BSE calculation with the TDDFT results using the resonant or the causal f_{xc} kernels (Adapted from [Varsano 2005])

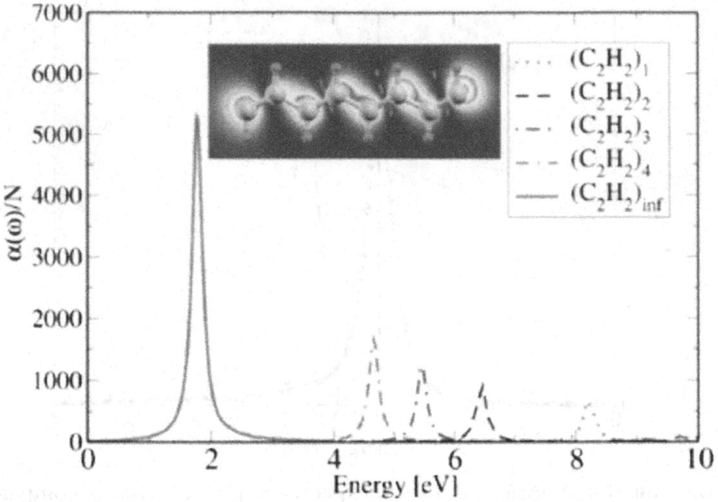

Fig. 20.8. Optical absorption spectra of the infinite chain of polyacethylene compared with the spectra for four different finite chains formed with different number of polyacethylene units. The figure illustrate the evolution of the main excitonic peak from about 7 eV in the monomer to 1.7 eV [Leising 1988] in the polymer chain. The inset shows the calculated excitonic wavefunction corresponding to the main absorption peak of the four-unit polyacethylene chain: we plot the probability of finding an electron when the hole is located at the left part of the chain (the probability maximum appears at the other side of the chain) (Adapted from [Varsano 2005])

Now we can make contact with the previous discussion of the static polarizability of long-molecular chains. First, we see that the optical spectra shifts from being dominated by a transition at about 7 eV (monomer) to one at 1.7 eV. This evolution with the number of monomers is monotonic and gives rise to the increase in polarizability (or equivakent to the decrease of the major absorption peak, as shown in Fig. 20.8). In particular, when the size of the polymer is larger than the localization length of the exciton (of the order of 100 Å), then the spectra becomes stable (no quantum confinement effects in the excitonic binding energy)[5]. At this point, the polarizability of the chain increases linearly with the number of monomer units. For smaller lengths, the polarizability increases faster but with smaller values than the corresponding LDA or GGA, in agreement with experiments and previous

[5] This is clearly illustrated in the inset of Fig. 20.8 where the exciton wavefunction for the four unit polyacethylene chain is plotted. As the size of the molecule is smaller than the localisation lenght of the bulk exciton, then for if the hole is located on the right hand side the maximum probability for the electron is at the other extreme of the chain. This clearly creates a counteracting field that tends to reduce the polarisability.

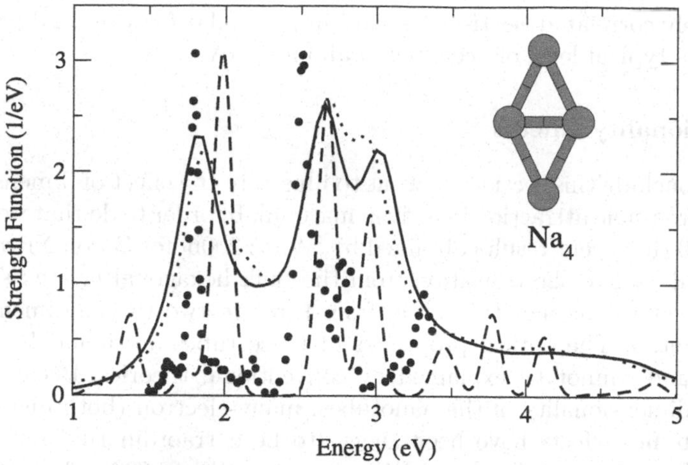

Fig. 20.9. Calculation of the optical absorption (that is proportional to the strength function) of a Na_4 cluster using the BSE scheme (*dashed line*) [Onida 1995] and with TDDFT using different kernels [Marques 2001]: TDLDA (*solid line*), exact-exchange (*dotted line*). *Filled dots* represent the experimental results from [Wang 1990a] (Adapted from [Onida 2002])

calculations [van Gisbergen 1999b, de Boeij 2001, van Faassen 2003a]. A similar trend has been obtained for the H_2-chain [Varsano 2005]; the BSE results for the polarizability per H_2 unit are very close to Hartree-Fock and much lower than the LDA, GGA or current DFT (as it fails for this hydrogen chain [de Boeij 2001, van Faassen 2003a]). In summary, this BSE-derived scheme accounts for the presence of the counteracting field that is responsible for the lower static linear and nonlinear polarizabilities of the molecular chains [van Gisbergen 1999b]. Still work needs to be done to improve the description of higher-order correlation effects by f_{xc}.

Another example of low-dimensional systems are clusters. In Fig. 20.9, we show some results on the optical spectra of the Na_4 cluster calculated using the BSE approach as well as those from TDLDA and experiment. The measured spectrum consists of three peaks in the 1.5–3.5 eV range and a broader feature around 4.5 eV. The agreement between results from TDDFT and BSE calculations is very good. The comparison with the experimental peak positions is also quite good, although the calculated peaks appear slightly shifted to higher energies. The LDA kernel is a good approximation of the xc kernel of small sodium clusters. Good agreement has been obtained for other small semiconductor and metal clusters [Onida 2002] as well as biomolecules [Marques 2003a].

The above are just several selected examples, given to illustrate the current status in ab initio calculations of optical properties of materials. Similar results have been obtained for the spectroscopic properties of many other moderately correlated electron systems, in particular for semiconducting systems, to a typical level of accuracy of about 0.1 eV.

Dimensionality Effects

Just to conclude this section we want to illustrate the effect of dimensionality and electron-hole attraction in a given material. In order to do that we show in Fig. 20.10 the recent results obtained by [Wirtz 2006] for Boron Nitride (BN) compounds where the transition from the bulk hexagonal layered structure to the sheet to the nanotubes, i.e, from three to two to one dimensions, it is highlighted. The optical properties of these tubes are found be to quite unusual and cannot be explained by conventional theories. Because of the reduced dimensionality of the nanotubes, many-electron (both quasiparticle and excitonic) effects have been shown to be extraordinarily important in both carbon [Spataru 2004] and BN nanotubes [Wirtz 2006, Park 2005].

In Fig. 20.10 we compare the results of a simple RPA calculation of the optical absorption spectra with the BSE results (that would be equivalent to the

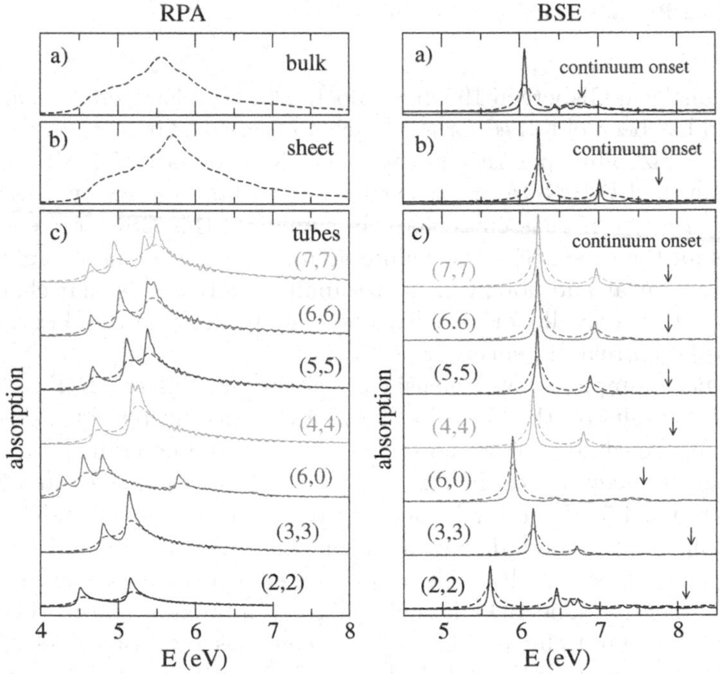

Fig. 20.10. Calculated optical absorption spectra of (**a**) hBN, (**b**) BN-sheet, and (**c**) six different BN tubes with increasing diameter d. The onset of continuum excitations is denoted by the vertical arrow. We compare the results of the BSE approach (*right hand side*) with the RPA (*left hand side*). Light polarization is parallel to the plane/tube axis, respectively (Adapted from [Wirtz 2006])

ones obtained using the many-body f_{xc}). Comparing the two figures a striking effect is observed: the electron-hole attraction modifies strongly the independent particle spectra (RPA) concentrating most of the oscillator strength into one active excitonic peak. The position of this peak seems rather insensitive to the dimensionality of the system in contrast to the RPA calculation where the shape of the spectra depends quite strongly on the tube diameter (the effect becomes smaller for very large tube diameters and it converges to the sheet and bulk values). The main effect of the dimensionality appears in the onset of the continuum excitations and the set of excitonic series above the main active peak. Thus, the binding energy for the first and dominant excitonic peak depends sensitively on the dimensionality of the system varying from 0.7 eV in the bulk BN to 3 eV in the hypothetical (2, 2) tube. However, the position of the first active excitonic peak is almost independent of the tube radius and system dimensionality. The reason for this subtle cancellation of dimensionality effects in the optical absorption stems from the strongly localized nature of the exciton (Frenkel-type) in BN systems [Wirtz 2006]. This band-gap constancy is in agreement with epxeriments[Arenal 2005] and has implications for the application of BN tubes for photoluminescence devices.

We remark that dimensionality effects would be more visible in other spectroscopic measurements as photoemission spectroscopy, where we mainly map the quasiparticle spectra, and this (as the exciton binding itself) is sensitive to the change in screening going from the tube to the sheet to bulk hBN [Wirtz 2006]. In particular the *quasi-particle band-gap* will vary strongly with dimensionality (opening as dimensionality reduces). The situation is different to the case of carbon nanotubes, where also excitonc effects are very important but also they depend on the specific nature of the tube [Spataru 2004].

20.4 Summary

We have discussed applications of an ab initio approach to calculating electron excitation energies, optical spectra, and exciton states in real materials following a many-body derived xc functional for TDDFT. The approach is based on evaluating the one-particle and the two-particle effects in the optical excitations of the interacting electron system, including relevant electron self-energy and electron-hole interaction effects at the GW approximation level (mimicked by an effective two-point xc kernel). It provides a unified approach to the investigation of both extended and confined systems from first principles. Various applications have shown that the method is capable of describing successfully the spectroscopic properties of a range of systems including semiconductors, insulators, surfaces, conjugated polymers, small clusters, and nanostructures. The agreement between theoretical spectra and data from experiments such as photoemission, tunneling, optical and related measurements is in general remarkably good for moderately correlated electron systems.

The main drawback in the many-body-based schemes is that we need to start from a GW-like quasiparticle calculation instead of the Kohn-Sham band-structure. As a matter of principle, both the quasiparticle shift as well as the redistribution of oscillator strength (including creation of new poles) due to excitonic effects (electron-hole attraction) should be accounted for by the nonlocal and frequency dependent f_{xc} kernel. However, until now we have been able to produce relatively good kernels that reproduce the effect of excitons in the response function but fail to include the quasiparticle shift. More developments are needed in the future to have a fully DFT-based theory that can be applied to both ground-state and excite-state properties. One appealing way are the orbital-dependent functionals derived from variational many-body formalism for the total energy (Φ and Ψ derivable functionals) [Almbladh 1999]. Preliminary results using the GW self-energy to build the total-energy functional and ulterior xc potential and kernel have been done for Si and LiF [Gruning 2005]. This approach includes the proper derivative discontinuity of the potential, being the corresponding Kohn-Sham gap very close to the standard local density approximation one. It also provides electron-hole effects and a reasonable description of the structural properties.

More work needs to be done along those lines and towards the inclusion of higher order vertex effects into the many-body description of electron-electron and electron-phonon interactions in order to have a more general ab initio theory of the spectra of weak and strongly correlated electronic systems.

21 Linear Response Calculations for Polymers

P.L. de Boeij

21.1 Introduction

In general, finite field DFT and TDDFT calculations yield accurate values for the response properties of molecular systems when standard approximations for the exchange-correlation functionals are used [Gross 1996]. In combination with their high efficiency, this makes these theoretical approaches ideal candidates for the calculation of physical properties of large molecular systems of technological interest. For an important class of materials, however, this potential is not yet realized. It has been observed that density functional calculations on the static response properties of long π-conjugated molecular chains give large overestimations for the polarizability [Champagne 1998] when local and gradient-corrected exchange-correlation functionals are used. The errors become even worse for their nonlinear response properties [Champagne 1998]. Similarly, the static polarization of π-conjugated push-pull systems is incorrectly described [Champagne 2000], even if it is obtained in the absence of an external field. Closely related to these findings, also large errors have been reported in the calculated excitation energies [Grimme 2003]. The reason for these deviations is by now well understood: both local and gradient-corrected density approximations are unable to correctly describe the induced contribution to the exact exchange-correlation potential (see Chap. 9). It contains a component that increases linearly along the chain counteracting the external field [van Gisbergen 1999b, Gritsenko 2000]. The density, on the other hand, remains more-or-less periodic in the bulk of the chains, and changes only at the chain ends. The same phenomenon that is responsible for the macroscopic exchange-correlation field in insulating solids [Gonze 1995b, Gonze 1997b, Martin 1997a] seems to be at work here. In the infinite systems this additional screening field can be viewed as a polarization dependent exchange-correlation effect. In the finite molecular systems, however, the complete density is known, and no polarization dependence of the exchange-correlation functional needs to be invoked. It becomes clear that the failure of the standard density functionals is related to their local density dependence: the exchange-correlation potential is relatively insensitive to the polarization charge induced by the external electric field at the chain ends. The external field is insufficiently screened, and a too strong response is obtained in these approximations.

P.L. de Boeij: *Linear Response Calculations for Polymers*, Lect. Notes Phys. **706**, 317–322 (2006)
DOI 10.1007/3-540-35426-3_21

21.2 The Counteracting Exchange Potential

The overestimation of the polarizability is readily reproduced in finite field calculations on model chains consisting of a coaxial stack of evenly spaced hydrogen molecules. By using the approximation for the exact exchange potential developed by Krieger, Li, and Iafrate (KLI) [Krieger 1992b], the main cause of the occurrence of the counteracting exchange-correlation field could be traced back to the so-called "response" part of this exchange potential [van Gisbergen 1999b, Gritsenko 2000]. Here the KLI potential has two contributions,

$$v_{\mathrm{x}}^{\mathrm{KLI}}(\boldsymbol{r}) = v_{\mathrm{x}}^{\mathrm{hole}}(\boldsymbol{r}) + \sum_{i=1}^{N-1} f_i \frac{n_i(\boldsymbol{r})}{n(\boldsymbol{r})} \,. \tag{21.1}$$

The first term is the potential of the Fermi hole $v_{\mathrm{x}}^{\mathrm{hole}}(\boldsymbol{r})$, which has the correct asymptotic attractive long-range behavior $-1/r$. The second term is the exchange contribution to the "response part", which depends on the Kohn-Sham orbitals via the orbital densities $n_i(\boldsymbol{r})$ and orbital-energy dependent weights f_i. Here deeper-lying orbitals have bigger weights. The response part is repulsive and short-range as it decays exponentially for large r. Surprisingly, it is not the polarization of the long-range Fermi hole part that causes the global change in the potential as was previously proposed to explain the related effect in solids [Gonze 1995b, Gonze 1997b, Martin 1997a]. Instead, the nonlocal density dependence of the counteracting field arises due to the polarization of the occupied orbitals induced by the external field [van Gisbergen 1999b, Gritsenko 2000]. This effect can be understood in the following way. Suppose that in the zero field case we have two molecular orbitals being the bonding and anti-bonding combinations, $\varphi_{\mathrm{b}}(\boldsymbol{r}) = [\phi_1(\boldsymbol{r}) + \phi_2(\boldsymbol{r})]/\sqrt{2}$ and $\varphi_{\mathrm{a}}(\boldsymbol{r}) = [\phi_1(\boldsymbol{r}) - \phi_2(\boldsymbol{r})]/\sqrt{2}$ respectively of two fragment orbitals $\phi_1(\boldsymbol{r})$ and $\phi_2(\boldsymbol{r})$. In the presence of the perturbing field these orbitals will get mixed, leading to predominantly $\phi_1(\boldsymbol{r})$ and $\phi_2(\boldsymbol{r})$ type molecular solutions. Now, assuming that the fragment orbitals are rigid, the total density will not be changed by this mixing, and it remains $n(\boldsymbol{r}) = |\phi_1(\boldsymbol{r})|^2 + |\phi_2(\boldsymbol{r})|^2$. A local density functional will hence not be sensitive to this orbital polarization. Nevertheless, the response part of the KLI potential will be changed as the weights depend on the orbital energies: the weight f_1 will increase for the stabilized orbital having its density $n_1(\boldsymbol{r}) = |\phi_1(\boldsymbol{r})|^2$ at the lower potential side, and f_2 will decrease for the destabilized orbital with the density $n_2(\boldsymbol{r}) = |\phi_2(\boldsymbol{r})|^2$ at the higher side. The net result takes the form of a counteracting field. Even though the response part is of short range, the dependence on the orbitals makes it sensitive to global changes in the molecule.

Similar results have been obtained using an improved x-only approximation based on a physically motivated common energy denominator approximation (CEDA) for the orbital Green's function [Gritsenko 2001, Grüning 2002]. This approximation differs from the KLI approach in the sense that the occupied-occupied orbital mixing is now completely removed in the corresponding approximation for the density-density response function. The resulting approximation for the exchange potential takes a form similar to the KLI result (21.1), but now it contains also off-diagonal terms,

$$v_{\mathrm{x}}^{\mathrm{CEDA}}(\boldsymbol{r}) = v_{\mathrm{x}}^{\mathrm{hole}}(\boldsymbol{r}) + \sum_{i=1}^{N-1} f_i \frac{n_i(\boldsymbol{r})}{n(\boldsymbol{r})} + \sum_{i,j\neq i}^{N} f_{ij} \frac{\phi_i^*(\boldsymbol{r})\phi_j(\boldsymbol{r})}{n(\boldsymbol{r})} . \tag{21.2}$$

This additional off-diagonal orbital structure of the CEDA potential was shown to be important in particular for the description of the response properties of molecular chains [Grüning 2002].

Though these KLI and CEDA functionals resemble many features of the exact exchange-only, or optimized effective potential [Talman 1976, Ivanov 1999, Görling 1999a], the exact exchange-only potential gives results much closer to Hartree-Fock, with a higher counteracting field strength [Mori-Sanchez 2003, Kümmel 2004]. However, even the Hartree-Fock polarizabilities overestimate the values obtained with correlated methods. It remains to be seen to what extent correlation effects contribute to the global counteracting field, as the optimized potential approaches have until recently [Facco Bonetti 2001] been restricted to the exchange-only approximation.

The orbital dependence of these (approximate) exchange-only potentials allows for the nonlocal density-dependence, which is the basis for their success. In the static linear response approach the change in the exchange potential $\delta v_{\mathrm{x}}(\boldsymbol{r})$ is obtained from the density change $\delta n(\boldsymbol{r})$ via the exchange kernel,

$$\delta v_{\mathrm{x}}(\boldsymbol{r}) = \int \mathrm{d}^3 r' \, f_{\mathrm{x}}(\boldsymbol{r}, \boldsymbol{r}') \delta n(\boldsymbol{r}') , \tag{21.3}$$

with

$$f_{\mathrm{x}}(\boldsymbol{r}, \boldsymbol{r}') = \frac{\partial v_{\mathrm{x}}(\boldsymbol{r})}{\partial n(\boldsymbol{r}')} . \tag{21.4}$$

When the orbital structure of the exchange potential is properly incorporated in the exchange kernel [Gritsenko 2001, Grüning 2002], evidently the same results should be obtained in a response calculation as in the finite field approach. The kernel will have to be nonlocal in order to generate the characteristic contribution counteracting the external electric field. In solids related strategies have been applied. It was shown for several semiconductors that certain excitonic features, which are missing in the calculated absorption spectrum obtained using local functionals, can indeed be described correctly by using the exact-exchange kernel [Kim 2002b, Kim 2002a], or simply by imposing the typical $1/|\boldsymbol{r} - \boldsymbol{r}'|$ long-range dependence of $f_{\mathrm{xc}}(\boldsymbol{r}, \boldsymbol{r}')$ [Reining 2002, Botti 2004].

A very promising new route to improve upon the exchange-only kernels is found in comparing the TDDFT approach and the related Bethe-Salpeter equation from many-body perturbation theory [Onida 2002, Sottile 2003,

Adragna 2003, Marini 2003b, Stubner 2004]. Using this Bethe-Salpeter approach on, e.g., infinite polyacetylene polymers [Rohlfing 1999a, Puschnig 2002], it has been found that the excitonic features found in solids are present also in these one-dimensional π-conjugated systems.

21.3 An Alternative to Orbital-Dependent Potentials

Another route to include the nonlocal density dependence can be followed by changing to current dependent functionals within the framework of time-dependent current density functional theory [Dhara 1987, Ghosh 1988, Vignale 1996, Vignale 1998] (see Chap. 5). The main motivation here is that the current density can be seen as a local indicator of global changes in the system, and that a local exchange-correlation approximation can indeed be used if the basic variable is the induced current density rather than the density. Such a current functional was derived by Vignale and Kohn [Vignale 1996, Vignale 1998]. The current formulation is inherently a dynamical one, in which static response properties can be obtained in the low frequency limit of the time-dependent perturbation approach. Van Faassen et al. [van Faassen 2002, van Faassen 2003a] showed that including the current functional of Vignale and Kohn in the response calculations gives greatly improved static polarizabilities for several π-conjugated polymers, while the results for the nonconjugate polymers, which are already correctly described within the local density approaches, are modified only to a small extent. Unfortunately, however, in this current functional approach the description of the prototype hydrogen chain remained problematic. This is in great contrast with the (approximate) exchange-only orbital potentials and Hartree-Fock results, which do produce a sizable counteracting field in this model system. Mixed quality results were also obtained for the frequency dependent response of small molecules. The excitation energies were improved upon the ALDA results for $\pi^* \leftarrow \pi$ transitions in several π-conjugated systems, but much worse results were sometimes obtained in other systems [van Faassen 2003b].

Figure 21.1 shows the static axial polarizability per oligomer unit for the prototype π-conjugated system, polyacetylene, obtained using the adiabatic local density (ALDA) and Vignale-Kohn (VK) functionals [van Faassen 2002], together with restricted Hartree-Fock (HF) results [Kirtman 1995], and results from second order perturbation calculations (MP2) [Toto 1995]. Exact exchange-only results are expected to be very close to the HF results [Mori-Sanchez 2003, Kümmel 2004]. It becomes clear that the overestimation of the polarizability observed in the ALDA calculations is reduced considerably by using the VK functional, giving values in close agreement with MP2 results that are somewhat smaller than HF results. The same current functional method can also be applied to the absorption spectrum of the infinite chain of the prototype polyacetylene system [Berger 2005]. This gives results that correspond well with the finite oligomer results [van Faassen 2002, van

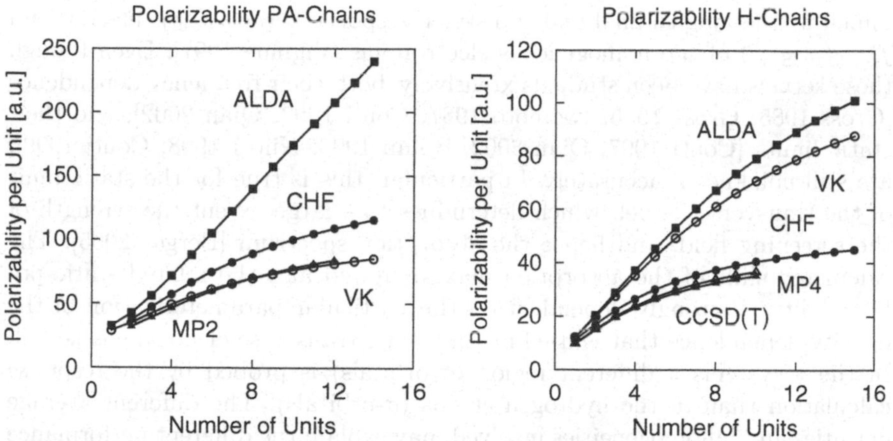

Fig. 21.1. The static axial polarizability per oligomer unit for polyacetylene (*left*) obtained using the adiabatic local density (ALDA) and Vignale-Kohn (VK) functionals [van Faassen 2002], together with spin-restricted coupled Hartree-Fock (CHF) results [Kirtman 1995] and results from second order perturbation calculations (MP2) [Toto 1995]. Similar results for the model hydrogen chains (*right*), now compared with spin-restricted coupled Hartree-Fock (CHF) results, fourth order perturbation calculations (MP4), and coupled cluster calculations with singles, doubles and triples included (CCSD(T)) [Champagne 1995]

Faassen 2003a], but that deviate considerably from Bethe-Salpeter results [Rohlfing 1999a, Puschnig 2002] and experiment. In particular the position of the absorption maximum is shifted upward like in the oligomers [van Faassen 2004] but now too strongly.

One may wonder if the application of the VK functional in molecular systems is justified. After all, the VK functional was derived for the electron gas under the assumption of weak spatial inhomogeneity, which is certainly violated in these systems. However, the VK-functional satisfies a number of important physical constraints which are valid for systems with arbitrary time-dependence and inhomogeneity [Vignale 1996, Vignale 1998]. In fact, the Vignale-Kohn functional can be viewed as the linear regime limit of the exact reformulation of TDDFT in the co-moving Lagrangian frame [Tokatly 2005a, Tokatly 2005b] (see Chap. 8). These exact properties may support the application to molecular systems, but the correctness of the results can not be guaranteed, as the (induced) deformations are certainly not small, and the VK functional is thus applied well beyond the linear regime. The varying quality of the results obtained with this functional for the molecular response properties may also be the result of the following arguments. The functional can be cast as a viscoelastic stress field [Vignale 1997], in which viscoelastic coefficients $\tilde{\eta}_{xc}(n_{GS},\omega)$ and $\tilde{\zeta}_{xc}(n_{GS},\omega)$ enter that are frequency dependent functions of the ground state density, and that are determined by the longitudinal and transverse response kernels $f_{\mathrm{xc,L}}(n_{\mathrm{GS}},\omega)$ and $f_{\mathrm{xc,T}}(n_{\mathrm{GS}},\omega)$ of the homogeneous electron gas [Vignale 1997]. Even though these kernels have been studied extensively, both their frequency dependence [Gross 1985, Gross 1985, Iwamoto 1987, Conti 1997, Qian 2002], and their static limits [Conti 1997, Qian 2002, Böhm 1996, Nifosì 1998, Conti 1999], are still not known accurately. In particular, this is true for the static limit of the transverse kernel, which determines to a large extent the strength of the screening field, and hence the absorption spectrum [Berger 2005]. The calculated shift of the absorption maximum and also the related static polarizability is strongly dependent on the particular parameterization of the density-dependence that is used for the static transverse electron gas kernel. In the π-systems a different region (π-orbitals) is probed by the response calculation than in the hydrogen chains (σ-orbitals). The different average densities and inhomogeneities involved may explain the different performance for these systems. More accurate electron gas kernels will be needed to verify if the application of the Vignale-Kohn functional to molecular response properties is indeed justified.

21.4 Conclusion

In conclusion, we can state that the strong overpolarization observed in DFT calculations that use local and gradient-corrected density functionals for the exchange-correlation potential is the result of the locality of these functionals. The origin of the nonlocal density dependence of the exact exchange-correlation contribution to the induced Kohn-Sham potential has been established as resulting from a polarization of the occupied Kohn-Sham orbitals, which does not modify the density locally. The nonlocal density dependence of the exchange-correlation potential can be introduced via an explicit orbital dependence either via the optimized potential method or by using expansions in terms of the Kohn-Sham Greens function, or by introducing current density functionals. The results obtained with the various methods are not yet in full agreement. Improved approximate exchange-only potentials, beyond KLI and CEDA, should eventually converge to the exact exchange-only results, which in turn are close to the Hartree-Fock results. Further progress can be made if correlation effects are included in the optimized potential. In the current functional approach more accurate electron-gas kernels have to be obtained, and the regime beyond weak inhomogeneity should be described correctly. This may most elegantly be done using the reformulation of TDDFT in the co-moving Lagrangian frame.

22 Biochromophores

X. Lopez and M.A.L. Marques

22.1 Introduction

In this chapter we give a brief introduction to the field of protein chromophores, and how TDDFT can and is being used to study these important systems. As we believe that a large majority of the Readers of this book come from the fields of Physics or quantum Chemistry, with little or no knowledge of Biochemistry, we will try to keep the discussion non-technical and at a fairly basic level.

First of all, how do we define a biochromophore? Biochromophores are molecules, present in many types of cells (plant, animal, bacteria, etc.), that absorb light in the visible or near ultra-violet (UV) part of the spectrum. These molecules are extremely important, as they are responsible for processes fundamental for life as we know it. In fact, the processes of vision, photosynthesis, photoperiodism, bioluminescence, DNA damage, etc., all of them are either governed or triggered by photo absorption. Furthermore, some of these molecules found their way into very important technical applications. An example is the green fluorescent protein (GFP), a protein found in a jellyfish that lives in the cold waters of the north Pacific, that has played a key role as a marker to monitor gene expression and protein localization in living organisms.

22.2 $\pi \to \pi^*$ Transitions and Biochromophores

Most organic molecules are completely transparent to visible light, starting to absorb in the UV regime. However, light emitted by the sun, after passing through the Earth's atmosphere, has its maximum in the visible range. It is quite interesting to see how nature solved this problem.

Upon absorption of a photon, a molecule undergoes a transition between two molecular states. The most common transitions are $\sigma \to \sigma^*$ and $n \to \sigma^*$, that absorb light in the UV. Note that as all stable molecules possess localized σ-bonds, these processes are present in all molecules. Whenever π-symmetry bonds are present, another very important class of transitions appears: $\pi \to \pi^*$. The simplest molecule containing a π-bond, ethylene (C_2H_4), exhibits strong absorption at only 163 nm ($\epsilon_{max} = 1.5 \times 10^4$ L mol^{-1} cm^{-1}),

X. Lopez and M.A.L. Marques: *Biochromophores*, Lect. Notes Phys. **706**, 323–336 (2006)
DOI 10.1007/3-540-35426-3_22

still in the UV region. However, by creating molecular chains with alternate π bonds, absorption can be shifted to longer wavelengths. In these so-called π-conjugated molecules, the electrons are delocalized over the whole π system. Therefore, by increasing the length of the molecule, the π-electrons become more delocalized, and the HOMO-LUMO energy gap decreases. If the π system is long enough, optical absorption may lie in the visible region.

Therefore, biochromophores are usually organic molecules with long conjugated π bonds, resulting in very efficient $\pi \to \pi^*$ transitions at the visible or near-UV range of the spectrum. In general, the prosthetic groups that act as biochromophores must be built specifically for the task, and carefully incorporated into the proteins. Although there is a great variety of prosthetic groups in nature, there are only a few basic designs. These include the pyrroles, the porphyrins (like the hemes present in chlorophyll and hemoglobin), the carotenoids (like retinal, or β-caroten), the flavins, etc. (see Fig. 22.1). As an example, consider β-carotene (see Fig. 22.1d), a molecule common in vegetables and fruits. It consists of a long chain of π-conjugated bonds terminated by two cyclohexane-like rings, that produce the bright orange color of carrots. Another important example is 11-cis-retinal (Fig. 22.1e), a derivative of all-trans-retinol, more commonly known as vitamin A. This molecule is the responsible for vision, being the primary receptor for photons entering the eye.

Another common feature of these biochromophores is that they are not isolated but bound covalently to a protein. For instance, the retinol in our eyes is incorporated in a family of proteins, named opsins. Moreover, the protein environment itself can (and often does) have a decisive influence on the response of the chromophore to light. For instance, a solution of 11-cis-retinal absorbs at about 380 nm; but in combination with the protein, the absorption maximum shifts to about 500 nm. Furthermore, the mechanism of color vision requires three different photoreceptors (the so-called cones) in the eye, each one absorbing at a different wavelength: blue, green and red. However, when analyzing the corresponding opsins, one finds that they all contain the same chromophore: 11-cis-retinal. The absorption at different wavelengths is a consequence of the different interaction of the retinal with the protein in each of the cones.

22.3 Biochromophores in Proteins

As explained in the previous section, the protein structure around the chromophore is often very important to account for the properties of the chromophore itself. Therefore, in this section we give a brief introduction to protein structure and how it can be simulated in the context of biophysical calculations.

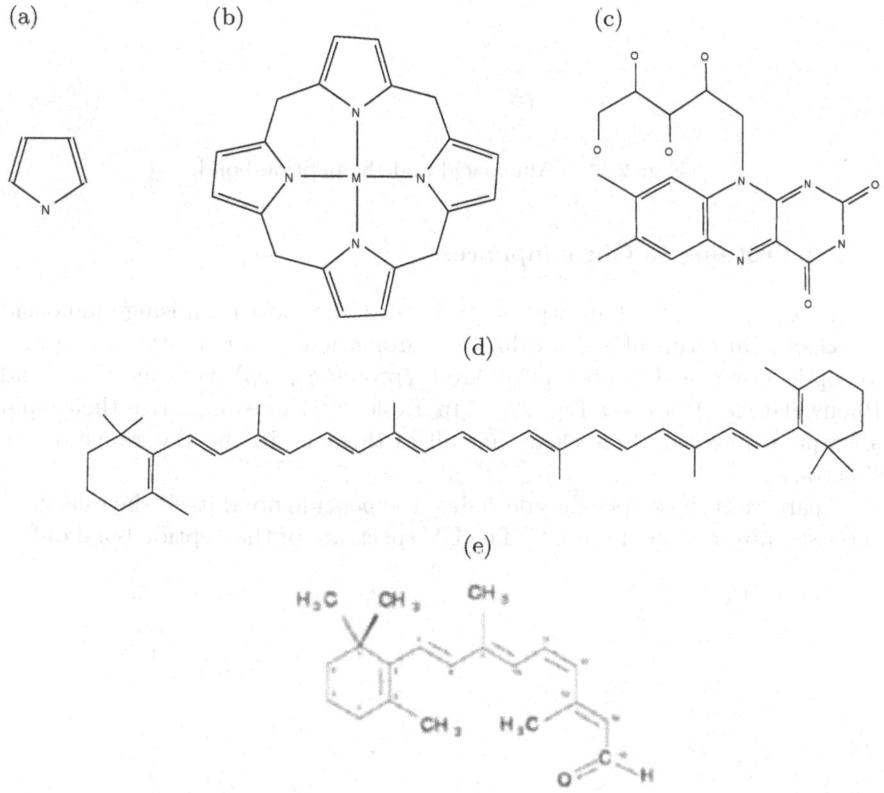

Fig. 22.1. Some of the basic designs of prosthetic groups in nature: (**a**) pyrrole; (**b**) heme, where M denotes a metal atom, like Fe or Mg; (**c**) flavin (riboflavin); (**d**) carotenoid (β-carotene); (**e**) 11-cis retinal

22.3.1 Proteins: Aminoacid Polymers

Proteins can be defined as polymers of aminoacids (see Fig. 22.2), molecules characterized by an amino ($-NH_3^+$), an acid ($-CO_2^-$), and a variable sidechain (R). The amino end of one aminoacid can undergo a condensation reaction with the acid end of the next aminoacid, giving raise to a dipeptide bound by a so-called *peptide bond.* The nature of each aminoacid is ultimately determined by the chemical nature of the sidechain (R in Fig. 22.2). There are only 20 different sidechains found in natural proteins. However, they span a great variety of physical-chemistry properties, and so the combination of these 20 aminoacids suffices to produce the immense diversity of proteins found in living systems.

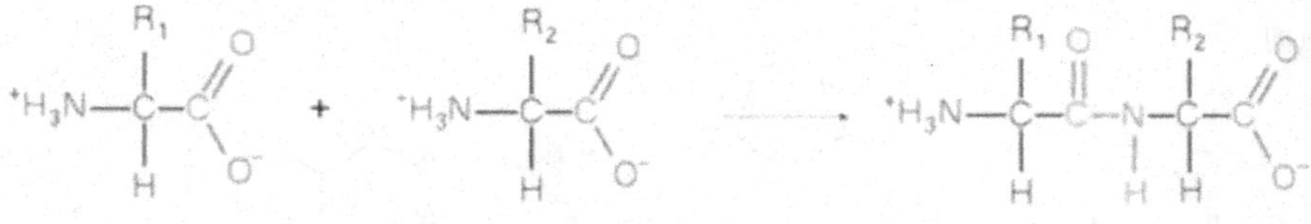

Fig. 22.2. Aminoacid and the peptide bond

22.3.2 Proteins as Chromophores

With respect to optical absorption properties, the most promising aminoacid sidechains (in terms of π delocalization/aromaticity) as potential candidates for optical or near-UV absorption are tryptophan (Trp), tyrosine (Tyr) and Phenylalanine (Phe) (see Fig. 22.3). In Table 22.1, we summarize their main absorption maxima. It is clear that all of them lie in the UV range of the spectra.

Apart from these specific sidechains, the peptide bond itself shows a characteristic absorption in the UV. The UV spectrum of the peptide bond differs

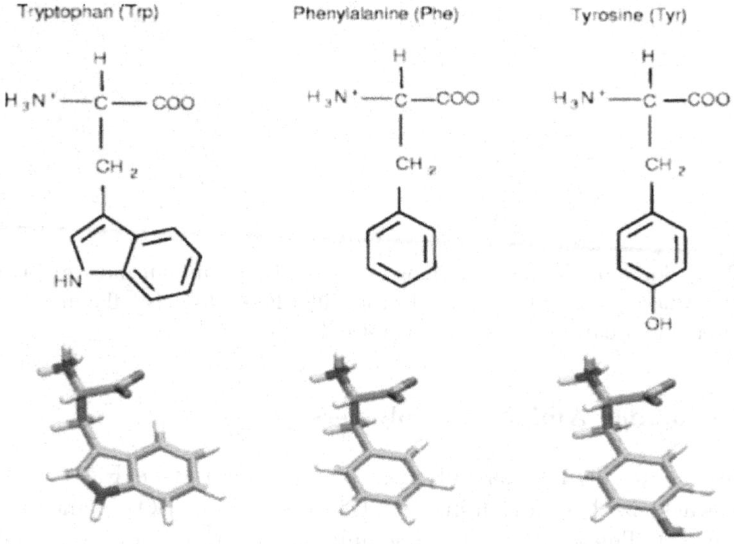

Fig. 22.3. Aromatic side chains

Table 22.1. Absorption of aromatic aminoacid side chains

AA	λ (nm)	Type	Remarks
Trp	240–290	most intense	
Tyr	274	$\pi - \pi^*$ ($\epsilon_{\max} \sim 1400$)	analogous to phenol (271 nm)
Phe	250	weak $\pi - \pi^*$	analogous to benzene (256 nm)

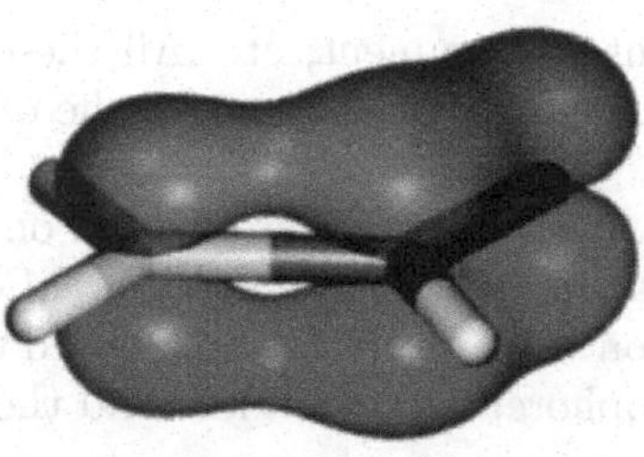

Fig. 22.4. Representation of one of the occupied π orbitals in formamide as a model of the electronic structure in a peptide linkage. Notice the delocalization of the π electronic cloud in formamide among the three atoms: nitrogen, carbon, and oxygen

from the one of a carbonyl group (C=O), and has its origin in the characteristic π electronic delocalization along the N−C=O atoms (see Fig. 22.4). As a result, a strong $\pi - \pi^*$ type transition is observed at 190 nm for each peptide bond, and the transition dipole is not along the C=O direction but along a line between O and N in the plane of the peptide bond. Due to the fact that each linkage between a pair of aminoacids contains one peptide bond, the absorption spectrum of any prosthetic group or aminoacid sidechain at this or higher energies is completely masked by the peptide absorption. On the other hand, each peptide exciton can interact with the exciton of nearby peptide bonds, modifying the overall absorption properties of the protein. This modification is highly sensible to structural changes in the protein backbone, making electronic absorption spectroscopy a useful tool to monitor protein and polypeptide structural changes [van Holde 1998].

22.3.3 Proteins and Phrosthetic Groups

Overall, we see that proteins by themselves are not able to absorb light in the visible range of the spectrum. However, since the main source of light available on Earth's surface is in the visible range, biological systems have to modulate their optical response to be able to absorb in this lower energy regime. To do so, they have two options: (i) incorporate phrosthetic groups with large delocalized π systems inside the protein structure (as those in Fig. 22.1); or (ii) chemically modify some of the aminoacid sidechains to form a chromopeptide structure. In Sect. 22.5, we will analyze one example of each case.

Once the chromophore is installed inside the protein matrix, the protein structure can affect the response of the chromophore through a variety of effects: (i) inducing a structural change on the chromophore that shifts the absorption maximum (see Sect. 22.5.1); (ii) polarizing the electronic cloud differently in the ground and excited state, thereby producing a modification of the energy gap (see Sect. 22.5.2); (iii) effectively shielding the chromophore from the aqueous solvent environment, etc. All these effects act simultaneously and are very often the key to understand the observed behavior of the chromopeptide.

In the next section, we give a brief introduction on how to simulate a protein using molecular dynamics methods and QM/MM Hamiltonians. These techniques, in conjunction with TDDFT, can be used to study the absorption properties of the chromophore, and to understand the effects induced by the protein environment.

22.4 Methods

22.4.1 Molecular Dynamics

One of the most used tools used to study the structure and function of proteins is molecular dynamics [McCammon 1987, Brunger 1988, Karplus 2002]. Briefly, this method is based on the ergodic hypothesis, which states that significant statistical averages can be obtained by averaging over the time evolution of a system.

$$\langle A \rangle_{\text{ensemble}} = \langle A \rangle_{\text{time}} \,. \tag{22.1}$$

To follow the time evolution of a given molecule, one can propagate the Newton's equation of motion for each particle (atom) of the system,

$$\boldsymbol{F}_i = m_i \boldsymbol{a}_i = m_i \frac{\mathrm{d}^2 \boldsymbol{R}_i}{\mathrm{d}t^2} \,. \tag{22.2}$$

The force acting on each particle can be derived from the form of the potential energy (see below)

$$\boldsymbol{F}_i = -\frac{\mathrm{d}V}{\mathrm{d}\boldsymbol{R}_i} \,. \tag{22.3}$$

There are several numerical methods that can be used to propagate these equations. One of the most used is the *Verlet* algorithm, which is based on a Taylor expansion to second order of the trajectories of each atom. Thus,

$$\boldsymbol{R}(t + \Delta t) = \boldsymbol{R}(t) + \left.\frac{\mathrm{d}\boldsymbol{R}}{\mathrm{d}t}\right|_t \Delta t + \frac{1}{2} \left.\frac{\mathrm{d}^2\boldsymbol{R}}{\mathrm{d}t^2}\right|_t (\Delta t)^2 + \cdots \tag{22.4a}$$

$$\boldsymbol{R}(t - \Delta t) = \boldsymbol{R}(t) - \left.\frac{\mathrm{d}\boldsymbol{R}}{\mathrm{d}t}\right|_t \Delta t + \frac{1}{2} \left.\frac{\mathrm{d}^2\boldsymbol{R}}{\mathrm{d}t^2}\right|_t (\Delta t)^2 + \cdots \tag{22.4b}$$

Summing the two equations one has,

$$\boldsymbol{R}(t + \Delta t) = 2\boldsymbol{R}(t) - \boldsymbol{R}(t - \Delta t) + \left.\frac{\mathrm{d}^2\boldsymbol{R}}{\mathrm{d}t^2}\right|_t (\Delta t)^2 + \cdots \tag{22.5}$$

In the context of protein simulations, the starting positions of the atoms are often taken from X-ray structures. The most common source for these structures is the *Protein Data Bank* (`http://www.rcsb.org/pdb`). This is a database of protein structures resolved by NMR, X-ray, or homology modeling. The information is given in a standard "PDB" format containing the Cartesian coordinates for the atoms, important experimental details of the system, B-factors, etc.

22.4.2 Force Fields

Due to the large numbers of atoms in a protein, the majority of molecular dynamics calculations are done using simple analytical forms for the potential energy, in which electrons are not treated explicitly. Two kinds of terms are usually distinguished in these molecular mechanics force fields, the so-called *bonded* and *non-bonded* terms. The former try to reproduce how the energy changes when atoms that are covalently bound are geometrically distorted, and are usually written as harmonic potentials around the equilibrium molecular distances and angles. In addition, periodic potentials are added to account for changes in energy along torsional degrees of freedom. Finally, non-bonded terms simulate van der Waals and electrostatic attraction forces between atoms that are not covalently bound. Thus, a typical force field has the following form,

$$V_{\mathrm{MM}} = \sum_{\text{bonds}} \frac{1}{2} K_b (d-d_0)^2 + \sum_{\text{angles}} \frac{1}{2} K_a (\theta-\theta_0)^2 + \sum_{\text{dihedrals}} K_d [1+\cos(n\phi-\gamma)] + \sum_{i=1}^{N} \sum_{j>i}^{N} 4\epsilon_{ij} \left[\left(\frac{\sigma_{ij}}{R_{ij}} \right)^{12} - \left(\frac{\sigma_{ij}}{R_{ij}} \right)^{6} \right] + \sum_{i=1}^{N} \sum_{j>i}^{N} \frac{q_i q_j}{\epsilon R_{ij}} \qquad (22.6)$$

where d measures the bond distance, θ the angle formed by 3 atoms, ϕ the dihedral angle formed by 4 atoms, and R_{ij} the distance between atoms i and j. (All other quantities are carefully chosen constants.) The first three summations correspond to bonded terms, and the last two to the non-bonded van der Waals and electrostatic interactions. There are various quite comprehensive force fields available, the most common being CHARMM [Brooks 1983], AMBER [Pearlman 1995], OPLS-AA [Jorgensen 1996], etc.

22.4.3 QM/MM Techniques

In certain cases, for special molecules for which there are no available parameterizations, or for studying chemical reactions in which bonds are broken and formed, it is necessary to go beyond the force field approximation. In these cases, a good compromise between accuracy and efficiency can be obtained with a mixed quantum mechanical/molecular mechanical (QM/MM) approach [Field 1990]. In QM/MM the system is partitioned into a small region that is treated quantum mechanically (e.g., the chromophore) and a big region handled by a force field. The resultant Hamiltonian is the sum of the force field, the quantum mechanical energy and a Hamiltonian describing the interaction between the QM and MM regions.

$$\hat{H} = \hat{H}_{\mathrm{QM}} + \hat{H}_{\mathrm{MM}} + \hat{H}_{\mathrm{QM/MM}} \, , \tag{22.7}$$

where

$$\hat{H}_{\mathrm{QM/MM}} = \hat{H}^{\mathrm{elect}}_{\mathrm{QM/MM}} + \hat{H}^{\mathrm{vdW}}_{\mathrm{QM/MM}} + \hat{H}^{\mathrm{bonded}}_{\mathrm{QM/MM}} \, . \tag{22.8}$$

Among the QM/MM coupling terms, we have an electrostatic part that accounts for the electrostatic interaction of the QM region with the set of point charges q_M representing the protein in the MM zone.

$$\hat{H}^{\mathrm{elect}}_{\mathrm{QM/MM}} = -\sum_{i,M} \frac{q_M}{r_{iM}} + \sum_{\alpha,M} \frac{Z_\alpha q_M}{R_{\alpha M}} \tag{22.9}$$

where r_{iM} denotes the distance between the electron i and the MM nucleus M, and $R_{\alpha M}$ is the distance between the QM nucleus α and the MM nucleus M. This is the part that enters the SCF equations, requiring the evaluation of extra one-electron integrals.

Special care must be taken with the QM/MM boundary [Reuter 2000]. One option is to introduce a hydrogen atom whenever the frontier between the QM and MM regions passes through a chemical bond. This so-called H-link atom is forced during the minimization to be aligned with the frontier bond, and does not interact with the MM atoms. Other options include more sophisticated techniques in which the SCF equations are solved in the presence of frozen orbitals located at the QM-MM atom interfaces.

22.5 Practical Cases

In this section, we discuss some practical cases of biochromophores. Due to space limitations, we have only selected two representative cases: (i) the green fluorescent protein and its mutants, where the protein surroundings modulate the response mainly by structural effects; (ii) the astaxanthin (AXT) inside crustacyanin, an example of a polarization mechanism induced by nearby aminoacids with little structural effects. We hope that these two test cases can exemplify the complexity and diversity of ways in which natural proteins can generate and adapt biochromophores to yield the desired sensitivity to visible light.

22.5.1 Green Fluorescent Protein and its Mutants: Structural Effects

The Green Fluorescent Protein

Among photo-active proteins, and due to is unique photophysical properties [Zimmer 2002], the family composed by the green fluorescent protein (GFP) and its mutants has attracted a considerable amount of attention during the last decade. By absorbing UV light, this biomolecule remits it, through fluorescence, as green light. Moreover, the GFP is unique because it does not involve any external prosthetic group. In fact, the chromophore responsible for the photophysics of the GFP is completely generated by the cyclization and oxidation of a sequence of three aminoacids of the protein, Ser65, Tyr66, and Gly67. Furthermore, this transformation is auto-catalyzed, i.e., it does not require any external enzyme. Thus, all the information needed to synthesize the biochromophore is encoded in the corresponding gene. Through genetic engineering, it is possible to literally attach the GFP gene to some other gene. This yields a fused protein in which the GFP is attached to the protein of interest without affecting its function. In this way, it is possible to trace the concentration of the particular protein in an organism merely by exposing it to UV light. The technique is so disseminated, that the fluorescent GFP has already been expressed in bacteria, yeast, slime mold, plants, drosophila, zebrafish, and in mammalian cells.

The protein is composed of 238 amino-acids, and is folded in a β-sheet barrel conformation with the chromophore in its interior (see Fig. 22.5). The

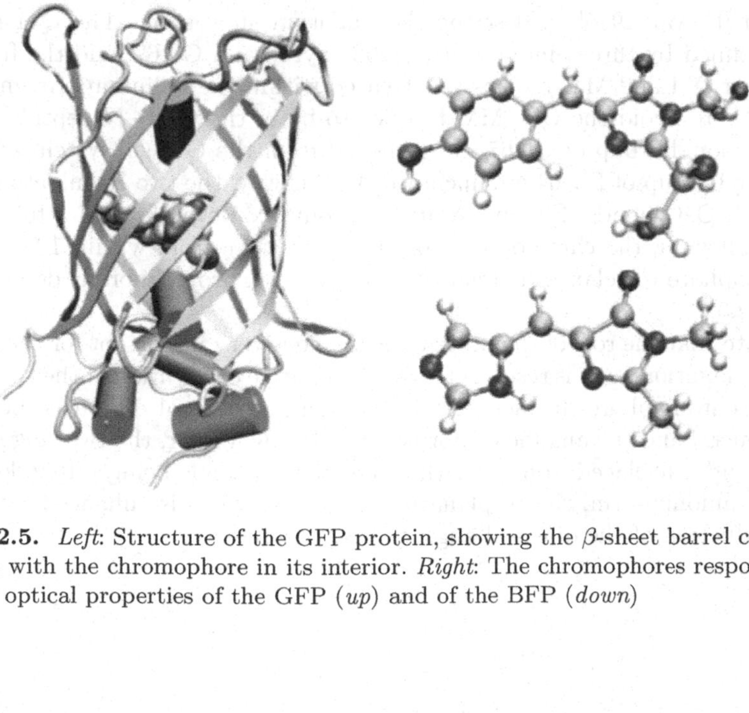

Fig. 22.5. *Left*: Structure of the GFP protein, showing the β-sheet barrel conformation with the chromophore in its interior. *Right*: The chromophores responsible for the optical properties of the GFP (*up*) and of the BFP (*down*)

influence of the barrel is twofold: (i) It isolates and protects the chromophore from the protein environment. In this way, the fluorescence properties of the GFP are largely unaffected by the conditions outside the protein. (ii) It influences the geometry of the chromophore, which in part determines its optical properties.

The optical absorption spectrum of the wild type (wt)-GFP, measured at 1.6 K, shows two main resonances at 2.63 and 3.05 eV [Creemers 2000, Creemers 1999]. These are attributed to the two thermodynamically stable protonation states of the chromophore, namely one negative and one neutral configuration. The equilibrium between those two states can be controlled by external factors such as pH and by mutations that affect the chromophore environment [Creemers 2000, Creemers 1999, Brejc 1997, Haupts 1998]. Excitation at either of these frequencies leads to fluorescent green-light emission, peaked at 2.44 eV [Sinicropi 2005], which is the main mechanism for energy release in wt-GFP. It is clear that the photo-physics of the GFP is governed by a complex equilibrium between the neutral and anionic configurations.

The first step of our study consisted in the preparation of the structures of the GFP. The X-ray structure (1.9 Å resolution) from [Yang 1996] (PDB code: 1GFL) was taken as the reference structure. As the X-ray measurements are not sensitive to the hydrogen atoms, we first had to add hydrogens where needed, using the CHARMM program [Brooks 1983]. In order to relax the geometry of the system we performed 3 successive optimizations. An initial optimization was done with all backbone and chromophore atoms fixed at their crystallographic positions. In a second step we allowed relaxation only of the coordinates of the chromophore. We performed this step with a QM/MM hybrid method [Field 1990], using the AM1 semi-empirical Hamiltonian [Dewar 1985] to describe the quantum subsystem. The QM region was defined by three amino-acids, Ser65, Tyr66 and Gly67, and the frontier between QM and MM regions was treated within the H-link approximation. In order to avoid the QM/MM frontier to be in the C(O)–N peptide bond, the carbonyl group of Gly67 was removed from the QM subsystem and the carbonyl group of Phe64 was included. In this way, the two H-link atoms cut through C–C bonds. Finally, from this geometry we performed a full relaxation allowing the chromophore and every protein atom within 10 Å of the chromophore to relax. The final structure of the chromophore is depicted in Fig. 22.5.

Note that the role of the protein backbone is very important for the structural relaxation, as it is responsible for the relative orientation of the two rings of the chromophore. In fact, this is the main structural difference between the neutral and the anionic conformations. In the former, the two ring planes are slightly displaced from planarity, the dihedral angle being -10.9 degrees. In the anionic form, the co-planarity of the two rings is enhanced, through the reduction of the dihedral angle to -1.8 degrees.

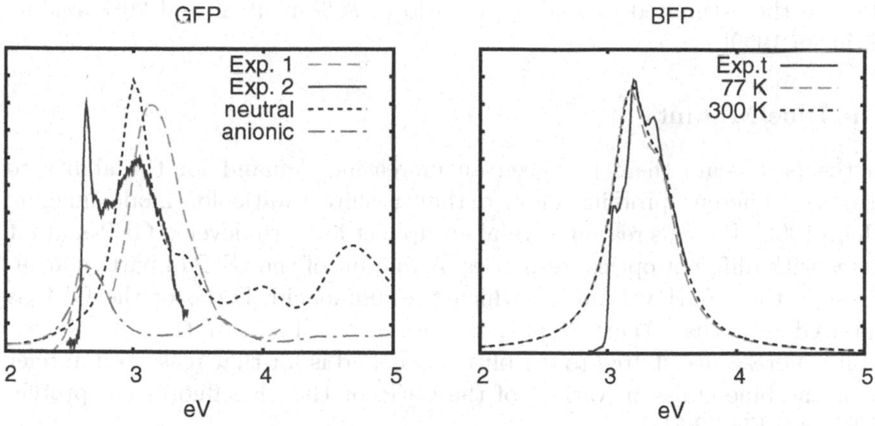

Fig. 22.6. The computed photo-absorption cross-section, σ, compared to the experimental measurement for the GFP (*left*) and the BFP (*right*). The y-axis has arbitrary units. The experimental results are from [Nielsen 2001] (Exp. 1 for the GFP), [Creemers 2000, Creemers 1999] (Exp. 2 for the GFP), and [Bublitz 1998] (BFP)

From these relaxed structures, we extracted the quantum-mechanical subsystem, that we subsequently used in our TDDFT calculations [Marques 2003b, Castro 2004b]. Our results are summarized in Fig. 22.6, where we present the computed spectra of the neutral and anionic conformations of the GFP chromophore, together with the available experimental data. The spectra is averaged over the three spatial directions. However, the GFP is a very anisotropic molecule, and most of the absorption at visible frequencies is due to light polarized in the x-direction, the direction along the line that joins the two rings. The excitations involved in this process correspond to a π-π^* transition in both neutral and anionic forms. The molecule is nearly transparent to visible light polarized along the other two orthogonal directions. The main excitation peaks for the neutral and anionic forms are at 3.01 and 2.67 eV, respectively. These values are in really good agreement with the measured excitation energies, located at 3.05 and 2.63 eV [Creemers 2000, Creemers 1999]. We emphasize that this good agreement is obtained since we take into account the breaking of the planarity of the biochromopeptide caused by the protein surrounding. Note that the measured peaks can be clearly assigned to either the neutral or anionic forms of the GFP.

The calculated oscillator strength of the π-π^* transition is larger in the anionic than in the neutral GFP. However, in the experimental spectrum the peak corresponding to the neutral form has a larger intensity. This is explained by the different concentration of the two species in vivo. To reproduce the experimental ratio between the two peaks, we have to assume a $\sim$4:1 ratio for the concentration of the neutral/anionic forms, which is very close to the estimated experimental ratio of 80% neutral and 20% anionic [Sullivan 1999].

The Blue Mutant

In the past years there has been an increasing demand for the ability to visualize different proteins in vivo that require multicolor mode imaging [Heim 1996]. For this reason, several groups set forth to develop GFP-mutant forms with different optical responses. A mutant of the GFP of particular interest is the Y66H variant, in which the aminoacid Tyr66 of the GFP is mutated to a His [Wachter 1997, Bublitz 1998]. The resultant protein exhibits fluorescence shifted to the blue range, and is for that reason often refer to as the blue emission variant of the GFP, or the blue fluorescent protein (BFP; see Fig 22.5).

The BFP chromophore is considerably more complicated than the GFP. It has four possible protonation states (one anionic, two neutral, and one cationic), each one of them with two possible stable conformations, one cis and one trans. The main candidate to explain the experimental spectrum of the BFP turns out to be a neutral-cis configuration. However, in contrast to the wt-GFP where the response of the anionic and of neutral states occurs at distinct frequency ranges [Zimmer 2002, Chattoraj 1996, Marques 2003a], in the BFP both anionic and neutral configurations have very similar spectra, with only minor differences in the fine structure close to the main peak. On the other hand, even if the absorption spectrum is not conclusive, pKa analysis seems to rule out the existence of the anionic state in vivo. Furthermore, other protonation states (such as the cationic) can not present in the BFP protein as their spectral features are outside the measured absorption spectra.

Like in the GFP, it is quite important to take into account the protein-induced structural changes of the chromophore. In fact, the protein induces the breaking of planarity of the otherwise planar gas-phase structures. Our calculations suggest that this breakdown of planarity could be responsible for the 0.13 eV red-shift observed between folded and unfolded conformations of the BFP protein [Wachter 1997]. Furthermore, there is a subtle cancellation between the shielding of the electromagnetic field acting on the chromophore due to its closest residues. This cancellation effect makes that the calculated spectra for the isolated (twisted) chromophore and for the chromophore in the protein are nearly identical. Therefore we can conclude that in the BFP the polarization mechanism is of reduced influence.

Finally, we studied the effect of temperature in the BFP, by performing constant T molecular-dynamics simulations at 77 and 300 K. (Note that the experiment was performed at 77 K [Bublitz 1998]). We observed that the chromophore fluctuates quite a lot, mainly in what concerns the relative orientation between the two rings. Using this information, it was simple to incorporate the effects of the temperature in the spectrum of the BFP. We

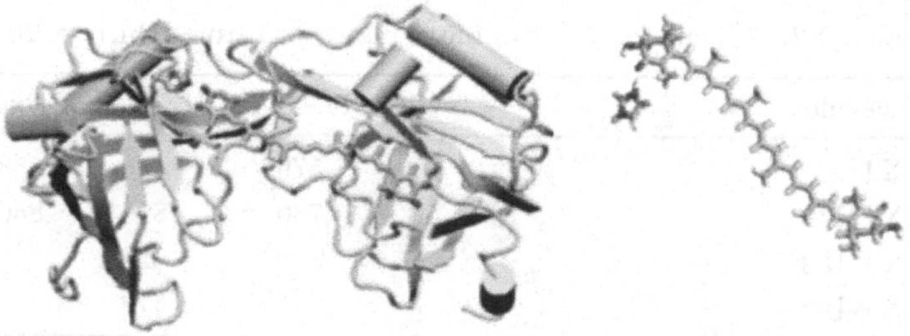

Fig. 22.7. *Left*: X-ray structure of AXT in β-crustacyanin [Cianci 2002]. *Right*: AXT hydrogen bonded to protonated imidazole, as a model for the effect of protonated histidines in the absorption spectra of AXT. Geometry taken from [Durbeej 2003]

obtained a ~0.1 eV shift of the main absorption peak. Our results are shown in Fig. 22.6, and are, like previously for the GFP, in excellent agreement with experimental data.

22.5.2 Astaxanthin and the Colour of the Lobster's Shell

The blue colouration of the shell of the lobster *Homarus gammarus* is due to the binding of the carotenoid astaxanthin (AXT) to the protein complex crustacyanin (See Fig. 22.7). The blue color is produced by the α-crustacyanin (consisting of eight β-crustacyanin dimers as in Fig. 22.7), that exhibits its absorption maximum at 632 nm. However, the absorption spectrum of free dilute AXT peaks between 472 and 506 nm, depending on the type of solvent. Thus, the (non-covalent) binding of AXT to the apoprotein subunits of β-crustacyanin gives rise to a significant bathochromic shift, in fact, among the largest shifts recorded in nature. In principle, various mechanisms could be responsible for the shift: (i) a twist around double bonds induced by the protein environment and (ii) a polarization mechanism in which proximal charged groups (aminoacid sidechains) and hydrogen bonding to AXT induced a significant charge rearrangement of the π electronic cloud. There are several nearby aminoacids and water molecules that could contribute to this polarization mechanism, including histidine, threonine, serine, and aspargine, all of them hydrogen-bonded to AXT according to the X-ray structure [Cianci 2002]. In Fig. 22.7, we show the structure of AXT embedded in the protein matrix. It is evident from the figure that AXT is quite distorted inside the protein. In addition, there is a variety of aminoacid sidechains in direct hydrogen-bond contact with the protein (only histidine is indicated).

Durbeej and Eriksson [Durbeej 2003] have analyzed the bathochromic shift of astaxanthin in crustacycanin by means of CIS (single excited configuration interaction), TDDFT [Stratmann 1998], and the semiempirical ZINDO/S [Ridley 1973] method. Some of their results are shown in Table 22.2. The role that nearby residues could have in the absorption spectrum

Table 22.2. Calculated λ_{max} in nm. Data taken from [Durbeej 2003]

Molecule	CIS	TDDFT	ZINDO/S	Exp
AXT	394	579	468	488
$AXTH^+$	582	780	816	840
$AXT\text{-}His^+$	–	–	623	
AXT-His	–	–	473	

AXT in α-crustacyanin: 632 nm

of AXT was considered by calculating its spectrum in the presence of reduced sidechain models representing the whole aminoacid.

It was found that only when His was included (see Fig. 22.7) there was a significant shift in the spectrum, from 468 nm to 623 nm. Moreover, the actual protonation state of the histidine (which at physiological pH can vary between protonated and neutral) had a profound effect: histidine needed to be protonated in order to trigger the shift. Other aminoacids that are hydrogen bonded to AXT, such as threonine, serine, and aspargine were found to have only a minor influence in the spectra. On the other hand, geometrical distortions of the AXT carotene by the protein matrix, leading to a co-planarity of the β-rings, had also a small effect compared to the polarization caused by the protonated histidine. These results support the view of a polarization mechanism as the origin of the bathochromic shift of AXT in crustacyanin, largely due to one of the astaxanthin keto groups being hydrogen bonded to a histidine residue.

22.6 Conclusions

With these examples, we demonstrated the precision of TDDFT for the calculation of absorption spectra of biochromophores, and its usefulness in extracting chemical information on the nature and state of these molecules. However, we still lack a complete understanding of the excited state dynamics of protein chromophores. To have a proper description of those systems would require an extension of the present QM/MM techniques to allow for a better description of the environment excitations and the structural transformations in the excited state. Work along those lines is already in progress.

Acknowledgments

MALM acknowledges financial support by the Marie Curie Actions of the European Commission, Contract No. MEIF-CT-2004-010384, and partial support by the EC Network of Excellence NANOQUANTA (NMP4-CT-2004-500198). We also thank Prof. Leif A. Eriksson for providing us with material regarding the astaxanthin.

23 Excited States and Photochemistry

D. Rappoport and F. Furche

23.1 Introduction

The treatment of electronically excited states of molecules and clusters is by far the most common application of time-dependent density functional theory (TDDFT) in chemistry. TDDFT calculations are increasingly used by non-experts to support and interpret experimental results. Important reasons for the success of TDDFT in photochemistry are its cost/performance ratio which is unmatched by traditional methods and the relatively wide applicability range. In this chapter we briefly survey the technology which underlies these applications. We discuss what accuracy can be expected for excited state properties with contemporary functionals, and where they fail. Finally, we present an up-to-date survey of TDDFT applications in photochemistry. Since the literature is growing rapidly, we limit ourselves to exemplary work and "hot" topics. Strong fields, applications to molecular dynamics, and the optical response of clusters, extended systems, and biochromophores are treated in Chaps. 22–26, 27, 17–18, 20–21, and 22, respectively. The present work complements and extends earlier reviews [Furche 2005a, Furche 2005c].

23.2 Excited State Properties from TDDFT

23.2.1 Lagrangian Approach

The Lagrangian approach [Furche 2001a, Furche 2002a] is probably the most convenient route to excited state properties in the TDDFT framework. Since the formalism is covered in Chap. 14, we can focus on its implementation and applications to molecular systems here.

We recall from Chap. 14 that the Lagrangian of the excitation energy takes the form

$$\begin{aligned} L[x,y,w,\varphi,z,W] = \frac{1}{2}\left[x^\dagger(A+B)x + y^\dagger(A-B)y\right] + \frac{w}{2}\left(x^\dagger y + y^\dagger x - 2\right) \\ + \sum_{ia\sigma} z_{ia\sigma}F_{ia\sigma} - \sum_{\substack{rs\sigma \\ r\leq s}} W_{rs\sigma}(S_{rs\sigma} - \delta_{rs}) \,. \end{aligned} \tag{23.1}$$

The so-called orbital rotation Hessians $(A+B)$ and $(A-B)$ read

D. Rappoport and F. Furche: *Excited States and Photochemistry*, Lect. Notes Phys. **706**, 337–357 (2006)
DOI 10.1007/3-540-35426-3_23

$$\begin{aligned}(A+B)_{ia\sigma jb\sigma'} &= (\varepsilon_{a\sigma}-\varepsilon_{i\sigma})\delta_{ij}\delta_{ab}\delta_{\sigma\sigma'} + 2(ia\sigma|jb\sigma') + 2f^{\mathrm{xc}}_{ia\sigma,jb\sigma'} \\ &\quad - c_{\mathrm{x}}\delta_{\sigma\sigma'}[(ja\sigma|ib\sigma) + (ab\sigma|ij\sigma)] \qquad (23.2a)\\ (A-B)_{ia\sigma jb\sigma'} &= (\varepsilon_{a\sigma}-\varepsilon_{i\sigma})\delta_{ij}\delta_{ab}\delta_{\sigma\sigma'} \\ &\quad + c_{\mathrm{x}}\delta_{\sigma\sigma'}[(ja\sigma|ib\sigma) - (ab\sigma|ij\sigma)]\,. \qquad (23.2b)\end{aligned}$$

$(pq\sigma|rs\sigma')$ is a two-electron repulsion integral in Mulliken notation, and $f^{\mathrm{xc}}_{ia\sigma,jb\sigma'}$ represents a matrix element of the exchange-correlation kernel in the AA. The hybrid mixing parameter c_{x} [Becke 1993a, Becke 1993b] is used to interpolate between the "pure" or non-hybrid density functionals ($c_{\mathrm{x}} = 0$) and TDHF theory ($c_{\mathrm{x}} = 1$, $f^{\mathrm{xc}}_{ia\sigma,jb\sigma'} = 0$). This extends the formalism of Chap. 14 to hybrid functionals, e.g. B3LYP [Becke 1993b] or PBE0 [Perdew 1996c], for which $c_{\mathrm{x}} = 0.2$ or $c_{\mathrm{x}} = \frac{1}{4}$, respectively.

23.2.2 Implementation

The LCAO-MO expansion, see (14.12), reduces the computation of excited state energies and properties to a finite-dimensional optimization problem for L, which can be handled algebraically. The stationarity conditions for L lead to the following problems which have to be solved subsequently in an excited state property calculation.

1. *The ground-state KS equations* (in unitary invariant form),

$$\frac{\delta L}{\delta Z_{ia\sigma}} = F_{ia\sigma} = 0\,, \qquad (23.3a)$$

$$\frac{\delta L}{\delta W_{pq\sigma}} = S_{pq\sigma} - \delta_{pq} = 0\,. \qquad (23.3b)$$

Results are the ground-state KS MO coefficients $c_{\alpha i}$ and their eigenvalues, as well as the ground state energy. Computational strategies to solve this problem have been developed over decades. Efficient excited state methods take advantage of this technology as much as possible.

2. *The TDKS eigenvalue problem* (Casida's equations [Casida 1995a])

$$\frac{\delta L}{\delta x_{ia\sigma}} = \sum_{jb\sigma'}(A+B)_{ia\sigma jb\sigma'}x_{jb\sigma'} - w x_{ia\sigma} = 0\,, \qquad (23.4a)$$

$$\frac{\delta L}{\delta y_{ia\sigma}} = \sum_{jb\sigma'}(A-B)_{ia\sigma jb\sigma'}y_{jb\sigma'} - w y_{ia\sigma} = 0\,, \qquad (23.4b)$$

$$\frac{\delta L}{\delta w} = \sum_{ia\sigma} x_{ia\sigma}y_{ia\sigma} - 1 = 0\,, \qquad (23.4c)$$

together with the non-standard normalization condition for the transition vectors which is enforced by w. The results are the excitation energies and transition vectors. They are used (i) to compute transition moments and (ii) to analyze the character of a transition (e.g., $\pi \to \pi^*$) in terms of occupied and virtual MOs. The complete solution of the TDKS EVP for all excited states leads to a prohibitive $O(N^6)$ scaling of CPU time and to $O(N^5)$ I/O (N is the dimension of the one-particle basis). In most applications only the lowest states are of interest; iterative diagonalization methods such as the Davidson method are therefore the first choice [Olsen 1988, Stratmann 1998, Chernyak 2000]. In these iterative procedures, the time-determining step are two matrix-vector operations per excited state and iteration, $(A+B)x$ and $(A-B)y$, which can be cast into a form closely resembling a ground state Fock matrix construction [Weiss 1993]. In this way, a single-point excitation energy can be computed with similar effort as a single-point ground state energy. Block algorithms lead to additional savings if several states are computed at the same time [Furche 2000a, Furche 2005c].

It is instructive to compare the iterative solution of Casida's equations in frequency space to the real-time propagation methods discussed in Chap. 12 of this book (see also Sect. 15.1 for an extended discussion). If the above techniques are used, a Davidson-type iteration in frequency space is about as expensive as a single time-step, because both are computationally equivalent to a Fock matrix construction. Since the iterative diagonalization requires typically less than 10 iterations to converge, it is more efficient than real-time methods which require hundreds of time-steps to compute an excitation energy. The domain of real-time methods are strong fields and highly excited states which are difficult or impossible to treat in the linear response framework.

3. The z-vector equation and the determining equations for the energy-weighted density matrix W. They follow from the stationarity condition

$$\frac{\delta L}{\delta c_{\alpha p \sigma}} = 0 \,, \tag{23.5}$$

where L depends on the MO coefficients $C_{ap\sigma}$ through the molecular orbitals φ. The z-vector equation is a static perturbed KS equation of the form

$$\sum_{jb\sigma'} (A+B)_{ia\sigma jb\sigma'} z_{jb\sigma'} = -r_{ia\sigma} \,. \tag{23.6}$$

The expressions for r and W involve third order functional derivatives and are explicity given in (14.13) and in [Furche 2002a]. The difference between the excited and ground state density matrices is given by

$$P = T + Z \,, \tag{23.7}$$

where the "unrelaxed" part T contains products of the excitation vectors only. Z accounts for relaxation of the ground state orbitals; it can be of the same order of magnitude as T. P is the functional derivative of the excitation energy with respect to the external potential; therefore, the difference between the *interacting* excited and ground state densities is accessible from P (since the density is the functional derivative of the energy as a functional of the external potential [Lieb 1983, Lieb 1985]). For example, if μ denotes the dipole moment operator, $\mathrm{Tr}\{P\mu\}$ is the change of the dipole moment upon excitation from the ground state; by adding the ground state density matrix to P, excited state properties can be computed in this way. Population analysis or graphical representation of P can give insight in the re-distribution of the electronic charge due to the excitation process.

Once all the parameters of L have been determined, first-order properties are straightforward, because the Hellmann-Feynman theorem implies that, to first order, L depends on a perturbation η only through operator matrix elements in the atomic orbital basis (see Chap. 14 and [Furche 2002a] for details),

$$\omega^{\eta} = \sum_{\mu\nu\sigma} h^{\eta}_{\mu\nu} P_{\mu\nu\sigma} - \sum_{\mu\nu\sigma} S^{\eta}_{\mu\nu} W_{\mu\nu\sigma} + \sum_{\mu\nu\sigma} V^{\mathrm{xc}\,(\eta)}_{\mu\nu\sigma} P_{\mu\nu\sigma}$$
$$+ \sum_{\mu\nu\sigma} \sum_{\kappa\lambda\sigma'} (\mu\nu|\kappa\lambda)^{\eta} \Gamma_{\mu\nu\sigma\kappa\lambda\sigma'} + \sum_{\mu\nu\kappa\lambda\sigma\sigma'} f^{\mathrm{xc}\,(\eta)}_{\mu\nu\sigma\kappa\lambda\sigma'} X_{\mu\nu\sigma} X_{\kappa\lambda\sigma'} \,. \quad (23.8)$$

η may represent, e.g., a component of an external electric field, in which case all terms except the first are zero; or it may represent a nuclear coordinate. An important consequence of the Lagrangian approach is that in the gradient expression the derivatives need to be taken *only* with respect to basis functions as indicated by parentheses, (η), for quantities depending on the density; MO coefficient derivatives do not occur. ω^{η} has nearly the same form as the ground state energy gradient [Johnson 1993] and thus should be treated on the same footing [Furche 2002a]. Total excited state properties are obtained by simply adding the ground state contributions to P, W, and two-particle difference density matrix Γ. The gradient of the excited state energy is thus not substantially more expensive to compute than the gradient of the ground state energy.

23.2.3 Efficiency

TDDFT implementations benefit from the fact that they can exploit nearly the entire machinery of existing ground state DFT algorithms. Efficient techniques may be transferred to the TDKS EVP and the z-vector equation, e.g., integral-direct methods [Almlöf 1982], pre-screening [Häser 1989], efficient quadrature [Treutler 1995] and exploitation of point group symmetry [Taylor 1985, Weiss 1993, Furche 2005c]. Similarly, the expression for the excited state gradient takes the same form as the ground state gradient (apart from the additional last term in 23.8). The excited state gradient can thus be processed along the same lines as the ground state gradient [Furche 2002a].

When non-hybrid functionals are used, the resolution of the identity (RI-J) approximation [Eichkorn 1995, Bauernschmitt 1997] leads to a significant speed-up at virtually no loss of accuracy. The RI-J approximation corresponds to an expansion of the density in a basis of auxiliary, atom-centered Gaussians (denoted by indices P, Q) which is used to evaluate the Hartree or Coulomb energy. The error in the Coulomb energy is minimized by choosing the expansion coefficients with respect to the so called Coulomb metric [Dunlap 1979, Eichkorn 1995]. The variational stability may be used to optimize auxiliary basis sets [Eichkorn 1995, Eichkorn 1997].

The RI-J approximation provides an efficient algorithm for the Coulomb contribution to the matrix-vector products $(A+B)x$, which is the bottleneck of TDDFT calculations in the absence of Hartree-Fock exchange: Four-center two-electron integrals are completely avoided, and the calculation of the Hartree response is reduced to a multiplication of transition vectors in the atomic orbital (AO) basis $x_{\mu\nu\sigma}$ with three- and the inverse of two-center electron repulsion integrals (in Mulliken notation),

$$K^{(1)\,\mathrm{H}}_{\kappa\lambda\sigma'} = \sum_{Q} (\kappa\lambda|Q) \sum_{P} (Q|P)^{-1} \sum_{\mu\nu\sigma} (P|\mu\nu)\, x_{\mu\nu\sigma}\,. \tag{23.9}$$

Thus, the calculation of the Hartree response, which formally scales as $O(N^4)$, is replaced by two steps with formal $O(N^3)$ scaling. Integral screening leads to an asymptotic $O(N^2)$ scaling with and without RI-J; nevertheless, the prefactor is much smaller within the RI-J approximation. In addition, a large number of three-center integrals $(\mu\nu|P)$ can be pre-computed and stored in memory for later use in the iterative solution algorithm.

An often used approximation to TDDFT excitation energies is the Tamm-Dancoff approximation (TDA) [Grimme 1996, Hirata 1999b, Hutter 2003], which amounts to restricting $y = x$ in the variation of the Lagrange functional L. As a result, the TDKS EVP reduces to the symmetric EVP

$$Ax^{\mathrm{TDA}} = \omega^{\mathrm{TDA}} x^{\mathrm{TDA}}. \tag{23.10}$$

At first sight, the TDA offers considerable computational advantage due to the reduction of dimensionality by a factor of 2. This argument overlooks that the rate-determining step in efficient integral-direct algorithms is the multiplication of AO-transformed vectors $x_{\mu\nu\sigma}$ by the four-center integrals contained in A. In an integral-direct algorithm, the cost for computing a single matrix-vector-product Ax is approximately the same as the cost for computing *two* matrix-vector-products in the full approach: Since the AO-transformed vector $x_{\mu\nu\sigma}$ is neither symmetric nor skew-symmetric, computing Ax effectively amounts to computing [Weiss 1993]

$$Ax = \frac{1}{2}(A+B)\,x + \frac{1}{2}(A-B)\,x\;. \tag{23.11}$$

While the TDA excitation energies are generally quite close to the full TDDFT excitation energies, transition moments do not satisfy the common sum rules [Furche 2001a] and are generally less accurate than the full TDDFT transition moments. However, even though the TDA is hardly less expensive than the full TDDFT treatment, it can be more robust with respect to triplet instabilities [Čížek 1967, Bauernschmitt 1996b].

With the present technology, the size limit for applications (using a single-processor workstation) is ca. 500–1000 atoms for few- or single-state calculations, and 200–500 atoms for computing spectra over a range of several eV. Systems of this size have been the domain of semi-empirical methods in the past.

23.2.4 Basis Set Requirements

Flexible Gaussian basis sets developed for ground states are usually suited for excited state calculations. The smallest recommendable basis sets are of split valence quality and have polarization functions on all atoms except H, e.g., SV(P) [Schäfer 1992, Weigend 2005] or 6-31G* [Hariharan 1973, Francl 1982, Rassolov 1998]. Especially in larger systems, these basis sets can give useful accuracy, e.g., for simulating UV spectra (see below). However, excitation energies may be overestimated by 0.2–0.5 eV, and individual oscillator strengths may be qualitatively correct only. A useful (but not sufficient) indicator of the basis set quality is the deviation between the oscillator strengths computed in the length and in the velocity gauge, which approaches zero in the basis set limit [Furche 2001a]. Triple-zeta valence basis sets with two sets of polarization functions, e.g., cc-pVTZ [Dunning 1989, Woon 1993, Wilson 1999] or TZVPP [Schäfer 1994, Weigend 2005], usually lead to basis set errors well below the functional error; larger basis sets [Weigend 2003, Dunning 1989, Woon 1993, Wilson 1999] are used to benchmark. Higher excitations and Rydberg states may require additional diffuse functions.

Standard auxiliary basis sets for ground state RI-J calculations [Eichkorn 1995, Eichkorn 1997] are well suited for excited states, although diffuse augmentation may be necessary. With such optimized basis sets, the errors introduced by the RI-J approximation are generally less than 10% of the basis set errors for excitation energies [Bauernschmitt 1997, Rappoport 2005] and even less for excited state structures and first-order properties [Rappoport 2005].

23.3 Performance

23.3.1 Vertical Excitation and CD Spectra

Most benchmark studies agree that vertical excitation energies of low-lying valence states are predicted with maximum errors of ~0.4 eV by local density approximation (LDA) and generalized gradient approximation (GGA) functionals [Bauernschmitt 1996a, Hirata 1999b, Parac 2002, Grimme 2004].

Table 23.1. Mean absolute deviations from experimental excited state properties. CIS denotes configuration interaction singles, LDA is the local density approximation in the Perdew-Wang parameterization [Perdew 1992a], BP86 is the Becke-Perdew 1986 GGA [Becke 1988b, Perdew 1986], PBE is the Perdew-Burke-Ernzerhof GGA [Perdew 1996b], and B3LYP is Becke's three-parameter hybrid [Becke 1993b]. The data are taken from [Furche 2002a], to where the reader is referred for further details

Property	Exp. Mean	CIS	LDA	BP86	PBE	B3LYP
34 ad. exc. energies [eV]	4.5	0.6[a]	0.3[b]	0.3	0.3	0.3
40 bond distances [pm]	142.2	3.5[a]	1.5[b]	1.3	1.3	1.3
10 dipole moments [D]	1.3	0.4[a]	0.1[a]	0.1	0.1	0.2
80 vib. frequencies [cm^{-1}]	1258	169[c]	62[b]	49	49	61

[a] Excludes the 1 $^2\Sigma^+$ state of NO (instability).
[b] Excludes the 1 $^1B_1$ state of CCl_2 and the 1 $^2\Sigma^+$ state of NO (instabilities).
[c] Excludes the 1 $^2\Sigma^+$ state of NO (instability) and the $\nu_{13}(1a_2)$ frequency of the 1 $^1B_2$ state of pyridine (saddle point).

Hybrid functionals are sometimes more accurate, but may display a less systematic error pattern. A limitation of these benchmarks is that vertical excitation energies are not well-defined experimentally; however, adiabatic excitation energies show the same trends, as displayed in Table 23.1.

Traditional methods such as TDHF or configuration interaction singles (CIS) produce substantially larger errors at comparable or higher computational cost. Bearing in mind that UV-VIS spectra of larger molecules are mostly low-resolution spectra recorded in solution, and in view of the relatively low cost of a TDDFT calculation, errors in the range of 0.4 eV are acceptable for many purposes.

Calculated oscillator strengths may be severely in error for individual states, but the global shape of the calculated spectra is often accurate. Because semi-local functionals often predict the onset of the continuum to be 1–2 eV too low (due to the lack of derivative discontinuity), this is especially true for excitations in the continuum (excitation energy > |HOMO energy|) [Casida 1998a], unless special techniques are used [Wasserman 2003]. Rotatory strengths which determine electronic circular dichroism (CD) spectra can be computed from magnetic transition moments in the density matrix based approach to TDDFT response theory [Furche 2000b, Furche 2001a]. The simulated CD spectra predict the absolute configuration of chiral compounds in a simple and mostly reliable way. TDDFT also works well for inherently chiral chromophores and transition metal compounds (see below) where semi-empirical methods tend to fail.

The formalism for computing two-photon absorption cross sections in a TDDFT framework has first been outlined in 2001 [Furche 2001a]. So far, implementations based on the coupled-perturbed KS approach (CPKS) approach [Sałek 2003] (including solvent effects [Frediani 2005]) and the sum-over-states approach [Masunov 2004] have been reported. The latter has been applied to the two-photon properties of push-pull oligo-(phenylenevinylene) chromophores [Kobko 2004].

23.3.2 Excited State Structure and Dynamics

An adequate description of most photophysical and photochemical properties requires information on excited potential energy surfaces beyond vertical excitation energies. Early benchmark studies indicated at least qualitative agreement of excited potential surfaces calculated using TDDFT and correlated wavefunction methods [Casida 1998b, Sobolewski 1999, Aquino 2005]. An increasing number of excited state reaction path calculations using TDDFT have been reported. However, the proposed reaction paths often do not correspond to minimum energy paths (MEPs), i.e., the internal degrees of freedom other than the reaction coordinate are not optimized.

Analytical gradients of the excited state energy with respect to the nuclear positions are a basic prerequisite for systematic studies of excited state potential energy surfaces even in small systems. Several implementations are available now [van Caillie 2000, Furche 2002a, Hutter 2003]. While errors in adiabatic excitation energies are similar to errors in vertical excitation energies (see above), the calculated excited state structures, dipole moments, and vibrational frequencies are more accurate on a relative scale, with errors in the range of those observed in ground state calculations (see Table 23.1). The traditional CIS method, which has almost exclusively been used for excited state optimizations in larger systems, is comparable in cost, but significantly less accurate. Moreover, the KS reference is much less sensitive to stability problems than the HF reference, which is an important advantage especially if the ground and excited state structures differ strongly.

Individual excited states of larger molecules can be selectively investigated by pump-probe experiments. The resulting time-dependent absorption, fluorescence, IR, and resonance Raman spectra can be assigned by TDDFT excited state calculations. Applications have demonstrated that calculated vibrational frequencies are often accurate enough to determine the excited state structure by comparison with experiment [Rappoport 2004]. The combination of TDDFT and transient spectroscopy methods thus offers a promising strategy for excited state structure elucidation in larger systems. Computed normal modes of excited states can be used to study the vibronic structure of UV spectra within the Franck-Condon and Herzberg-Teller approximation [Dierksen 2004, Dierksen 2005]. For a detailed understanding of photochemical reactions beyond MEPs, excited state nuclear dynamics simulations (including non-adiabatic effects) will be necessary.

23.3.3 Shortcomings of Present TDDFT

In certain situations contemporary functionals produce much larger errors in excitation energies, or entirely miss important aspects of the spectrum.

Inaccurate Ground-State KS Potentials

It had been well-known for many years that the xc potentials of LDA and GGA are inaccurate. At large distances, they decay exponentially rather than as the correct $-1/r$. This can be a severe problem for TDDFT, since the orbital energies can be very sensitive to the details of the potential. This is not a problem if only low-lying valence excitations of large molecules are required, but the energy of low-lying diffuse states is often considerably underestimated, while higher Rydberg states are completely missing in the bound spectrum [Casida 1998a].

There now exist several schemes for imposing the correct asymptotic decay of the xc potential [Casida 2000a]. But such potentials are not the functional derivative of any xc energy. While this has no direct effect on vertical excitation energies, other excited state properties are not well defined. Exact exchange DFT methodology is developing rapidly [Della Sala 2003a], which does not suffer from this problem. Furthermore, when correctly interpreted, even the time-dependent LDA recovers the correct oscillator strength despite these difficulties [Wasserman 2003].

Multiple Excitations

In principle, the exact electronic response functions contain all levels of excitation. But the TDKS response spans the space of KS single-particle excitations only, and this is unchanged by a frequency-independent xc kernel, i.e., within the AA. First attempts have been made to incorporate the effect of doubles excitations perturbatively [Maitra 2004, Casida 2005].

Charge Transfer Problems

Charge transfer (CT) excitations are notoriously predicted too low in energy, sometimes by more than 1 eV [Dreuw 2003b]. In chain-like systems such as polyenes, polyacenes, or other conjugated polymers, the error in CT excitation energies increases with the chain length [Pogantsch 2002, Grimme 2003, Tretiak 2005]. In the limit of complete charge separation, this can be related to the lack of derivative discontinuities in semi-local functionals [Tozer 2003]. Various correction schemes for CT excitation energies have been suggested [Tozer 2003, Dreuw 2004]. For short-range CT between overlapping subsystems, hybrid functionals are a potential remedy to the problem. In a genuine TDKS treatment, strongly frequency-dependent xc kernels appear to be necessary for a proper description of CT [Maitra 2005b].

23.4 Applications

23.4.1 Aromatic Compounds and Fullerenes

Aromatic hydrocarbons have been the favorite class of medium-size molecules in pioneering TDDFT studies [Heinze 2000, Niehaus 2001, Nguyen 2002a, Halasinski 2003, Parac 2003, Beenken 2005] which generally show good accuracy compared with more expensive correlated methods. However, as recently pointed out by Parac and Grimme [Parac 2003, Grimme 2003], the energies of predominantly ionic L_a states [Platt 1949] of aromatic hydrocarbons are considerably underestimated by TDDFT. The revived interest in photophysics of polycyclic aromatic hydrocarbons is due to their proposed occurrence in dark interstellar matter [Weisman 2001a, Hirata 1999c, Hirata 2003, Weisman 2003, Malloci 2004, Bauschlicher 2005].

TDDFT methods have proven increasingly useful for many kinds of investigations lying beyond the scope of accurate correlated methods. Studies of the photophysical behavior of dye molecules [Cave 2002a, Cave 2002b, Prieto 2004, Abbott 2004, Wang 2005a] may serve as an example. Effects of solvation have been examined both by explicit taking into account the solvent molecules [Cai 2002b, Cai 2005, Sulpizi 2003, Odelius 2004, Bartkowiak 2005] and by continuous solvation models [Adamo 2000, Cossi 2001, Mennucci 2001, Gutierrez 2005, Improta 2005]. Mechanistic studies of excited state reactions are still at an the early stage but gain strong support from recent methodical developments (see Chap. 14). Several TDDFT studies have been performed on photoinduced cis–trans isomerization of azobenzene [Schultz 2003, Tiago 2005] and stilbene [Dietl 2005, Improta 2004] which are considered model compounds for molecular switches. Spiropyrans are another class of light-switchable compounds which are reversibly converted to merocyanines.by UV irradiation. The mechanism of this ring opening reaction has been the subject of a combined TDDFT/CIS study [Sheng 2004].

Determination of absolute configuration of chiral molecules relies increasingly on the comparison of measured chiroptical properties with theoretical predictions [Diedrich 2003]. Calculations of electronic CD spectra with TDDFT and DFT/SCI [Grimme 1996] methods have been performed on molecules with different kinds of chirality: central [Macleod 2004], helical [Furche 2000b, Lebon 2004], planar [Grimme 1998, Stephens 2004], and axial chirality [Toyota 2003, Toyota 2004]. The TDDFT results have been used to assign the absolute configuration of helicenes [Furche 2000b], rotationally hindered bianthryls [Toyota 2003], and cyclophanes [Grimme 1998].

Optical spectroscopy studies of fullerenes aided by calculations have substantially added to our understanding of their structures and properties [Bauernschmitt 1998, Rubio 1993, Castro 2002, Castro 2004b]. The optical properties can be largely varied by functionalization and substitution [Li 2003, Xie 2004a, Xie 2004b]. Fullerenes [Li 2004] and carbon nanotubes [Guo 2004] hold promise as non-linear optical materials for optical devices.

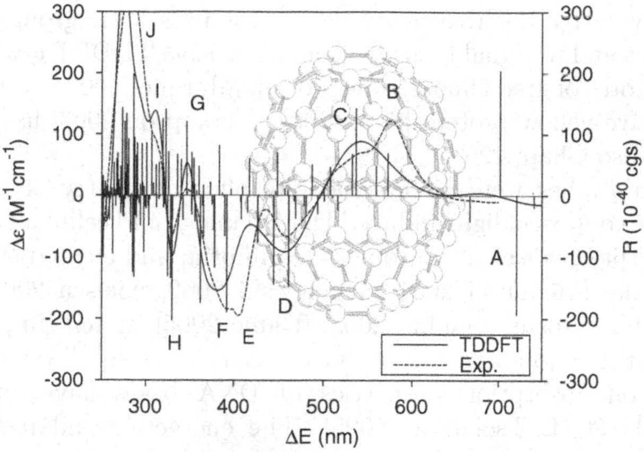

Fig. 23.1. Calculated and experimental CD spectrum of (^{f}A)-C_{76}. TDDFT calculations were performed with the BP86 functional and an augmented SVP basis set [Furche 2000a] using the TURBOMOLE program package [Ahlrichs 1989]. The RI-J approximation together with TZVP auxiliary basis sets [Eichkorn 1997] was used. Experimental data (in CH_2Cl_2) are from [Goto 1998]

Small-gap fullerenes are highly reactive and can at present only be studied theoretically. For example, of the seven isomers of C_{80} obeying the isolated pentagon rule (IPR), only three have a large gap, and two of those have been observed so far [Furche 2001b]. Absolute configurations of chiral fullerenes can be assigned by TDDFT, as has been shown for D_2-C_{84} [Furche 2002b] as well as for C_{76} and C_{78} isomers [Furche 2000a]. In Fig. 23.1, the calculated circular dichroism (CD) spectrum of the smallest chiral IPR fullerene D_2-C_{76} is compared to the experimental spectrum in CH_2Cl_2 solution [Goto 1998]. Very good agreement is found between the experimental spectrum and the TDDFT results, and the absolute configuration of the fullerene can be unequivocally assigned, in contrast to previous semiempirical calculations. TDDFT calculations on C_{76}^-, C_{78}^-, and C_{84}^- have been used to interpret the photoelectron spectra of the corresponding fullerene dianions [Ehrler 2003, Ehrler 2005].

23.4.2 Biological Systems

Biological systems have been one of the major goals in recent TDDFT investigations. Vertical excitations has been calculated for a series of indole derivatives occurring as intermediates of tryptophane metabolism and melanin production [Crespo 2002, Il'iechev 2003]. Further investigations addressed the photochemical behavior of urocanic acid [Danielsson 2001, Dmitrenko 2004] which is the major UV absorber in human epidermis and plays a role in photoimmunosuppression and skin cancer. Electronic absorption spectra of flavins and related alloxazins have been simulated by several groups [Neiss 2003, Sikorska 2004a, Sikorska 2004b, Sikorska 2005]. Flavins are an important class of biologically active heterocycles occurring as prosthetic groups of various redox enzymes and in signal transduction. Extensive TDDFT calculations on the chromophores of green fluorescent protein [Marques 2003a, Vendrell 2004] and photoactive yellow protein [Sergi 2001, Thompson 2003] have appeared recently, see also Chap. 22.

Numerous studies were directed at the photochemistry of nucleic acid bases in in the context of light-induced DNA damage and cellular repair mechanisms. Absorption spectra, tautomeric equilibria, and excited state geometries of adenine [Mennucci 2001, Sobolewski 2002, Nielsen 2005], guanine [Shukla 2005b], cytosine [Shukla 2002, Tomic 2005], uracil [Improta 2005], and hypoxanthine [Shukla 2005a] have been reported. Two comprehensive studies on absorption properties of DNA bases have appeared recently [Shukla 2004, Tsolakidis 2005]. The energetics and mechanism of thymine dimer formation and photoinduced cycloreversion reactions occurring in DNA repair mechanisms have been analyzed in a model system [Durbeej 2000, Durbeej 2002]. The photosensitizing effect of indanone derivatives on the thymine dimerization has been studied [Gutierrez 2005]. TDDFT calculations on complexes of thymine with psoralens have been performed to explain the phototoxic action of psoralens [Llano 2003] which find application in photochemotherapy [Nakata 2003]. Conjugated cytotoxic dyes showing strong two-photon absorption have potential use in photodynamic therapy [Badaeva 2005]. The fluorescence of DNA base analogues pyrrolocytosine [Thompson 2005], 6,8-dimethylisoxanthopterin [Seibert 2002], ethenoadenine [Major 2003], and wybutine [Lahiri 2004] has been explored with regard to applications as fluorescent nucleic acid markers.

23.4.3 Porphyrins and Related Compounds

Porphyrins, phthalocyanines, porphyrazines, and similar heterocyclic systems show a variety of optical and photochemical properties that are of interest from a biochemical as well as a technological point of view. The first rationale of the characteristic features observed in the absorption spectra of porphyrins was given by Gouterman [Gouterman 1961, Gouterman 1978]. It is based on a simple perimeter model for [18]-annulene, the basic building unit of porphyrins. In Gouterman's scheme two energetically close pairs of orbitals, the two highest occupied molecular orbitals (HOMO and HOMO-1) and the two lowest virtual MOs (LUMO and LUMO+1), are involved in the lowest singlet transitions and are responsible for the so-called Q- and B-bands of porphyrins.

The first TDDFT results on free base porphin were reported by Bauernschmitt and Ahlrichs [Bauernschmitt 1996a] and later confirmed by Scuseria and co-workers [Stratmann 1998]. Subsequent studies by Baerends and Sundholm groups addressed the validation of the four-orbital model of Gouterman for the free base porphin and the assignment of its UV/VIS spectrum [van Gisbergen 1999c, Sundholm 2000c, Baerends 2002]. Investigations by Parusel and co-workers employed the DFT/SCI [Parusel 2000a] and DFT/MRCI methods [Parusel 2001a] for the same purpose. While a correspondence to the Gouterman model could be established for the Q bands, the origin of the intense B band is still under discussion. It appears that lower occupied orbitals are significantly involved in these transitions [Sundholm 2000c, Baerends 2002], and a non-negligible contribution from double excitations is suggested from DFT/MRCI results [Parusel 2001a]; the simple four-orbital model does not hold. A similar picture emerges for carbaporphyrins [Liu 2003], porphyrazine [Baerends 2002, Infante 2003], corrphycene [Gorski 2002], hemiporphyrazine [Persico 2004], and corrin [Jaworska 2003a] where Gouterman's model provides a rough description of low-lying electronic transitions.

Electronic excitation energies of porphyrins are strongly affected by conformational flexibility of the macrocycle, deviations from planarity leading to red shifts of Q- and B-bands. The suggestion that non-planarity of hemes in hemoproteins and photosynthetic proteins may influence their biological activity [Shelnutt 1998] stimulated extended research on conformationally distorted porphyrins [Chen 2003, Rosa 2003, Parusel 2000b, Wertsching 2001, Ryeng 2002, Cramariuc 2004]. Spectral properties of porphyrin heteroanalogues [Delaere 2003, Wan 2004] and conjugated oligomers of porphyrins [Yamaguchi 2002b, Yamaguchi 2004] have been successfully treated with TDDFT methods.

Porphyrinoid systems have a tendency to form chelate complexes with various metal cations. Two large groups of complexes can be distinguished by their spectral behavior, described as regular and irregular porphyrins by Gouterman [Gouterman 1978]. Main group and closed-shell transition metal cations form regular complexes that largely resemble the parent macrocycles because the contribution of the metal to the frontier orbitals is small. This was shown by Nguyen, Baerends, and co-workers for Zn^{II} [Ricciardi 2001, Nguyen 2001, Nguyen 2002b, Nguyen 2003] and by Sundholm for Mg^{II} complexes [Sundholm 2000a]. Interesting non-linear optical properties are found for a number of phthalocyanine complexes of group 13 and 14 metals [Wu 2004, Ricciardi 2000, Ricciardi 2002]. Irregular metal complexes contain transition metal cations with incomplete *d*-shells which iteract strongly with the π-system of the ligand leading to significantly modified optical properties [Rosa 2001, Wasbotten 2002, Rogers 2003, Jaworska 2003b, Gunarante 2005, Petit 2005]. The most important representatives of this class are iron and cobalt complexes which are closely related to heme [Dreuw 2002, Dunietz 2003] and vitamin B_{12} [Andruniow 2001, Stich 2003, Stich 2004].

Calculations of optical properties of chlorophylls and bacteriochlorophylls by Sundholm [Sundholm 1999, Sundholm 2000a, Sundholm 2000b, Sundholm 2003] and Yamaguchi [Yamaguchi 2002a, Yamaguchi 2002c] showed that good accuracy can be achieved with the BP86 and B3LYP functionals. The behavior of bacteriochlorophyll in an electric field was the subject of a study by Pullerits and co-workers [Kjellberg 2003]. The interaction between chlorophyll molecules and carotenoids and dynamics of the photosynthetic apparatus have been extensively studied by Head-Gordon and coworkers [Hsu 2001, Dreuw 2003a, Dreuw 2003c, Dreuw 2004, Vaswani 2003]. The mechanism of the adenosylcobalamin-dependent methylmalonyl-CoA mutase has been investigated by a joint spectroscopic and TDDFT study [Brooks 2004].

23.4.4 Transition Metal Compounds

The number of TDDFT investigations on transition metal complexes is rapidly increasing in the last few years, and the first comprehensive reviews have appeared recently [Rosa 2004, Daniel 2004]. The accuracy of TDDFT is sufficient for an assignment of electronic excitations in many closed-shell metal complexes like metal carbonyls and cyclopentadienyls [van Gisbergen 1999a, Rosa 1999, Boulet 2001, Full 2001, Jaworska 2004, Hummel 2005]. The self-interaction error of the present functionals is the major limitation to the accuracy of excitation energies leading to underestimation of charge-transfer and overestimation of ligand field $d \to d$ excitation energies [Autschbach 2003].

The diversity of photophysical and photochemical properties of transition metal complexes is reflected in the breadth of TDDFT studies on this class of compounds. The most common application of TDDFT so far is the simulation of optical absorption spectra which are often used for characterization of complex properties. Interesting perspectives of transition metal complexes with aromatic ligands in photovoltaics or as photocatalysts stimulated a large number of studies on ruthenium pyridyl complexes [Monat 2002, Guillemoles 2002, De Angelis 2004a, Zhou 2005, Ciofini 2004, Fantacci 2004a, Zalis 2004, Batista 2005], Cu(I) phenanthroline complexes [Zgierski 2003, Siddique 2003, Wang 2005b], as well as on complexes of platinum [Stoyanov 2003] and iron [De Angelis 2004b]. Complexes of ruthenium [Fantacci 2004b], rhodium [Ghizdavu 2003] and rhenium [Dyer 2003] have been proposed as DNA probes. Ir metallacycles are promising electron donors in organic light-emitting devices [Polson 2005].

A number of metal-containing catalysts derived from titanocene [Wang 2003], zirconocene [Cavillot 2002, Belelli 2005, Wang 2003], and nickel diimine complexes [Cavillot 2005] have been the subject of TDDFT calculations. TDDFT calculations for neutral dithiolene complexes of nickel, palladium, and platinum were used to explain the unusual strong absorption in the near IR region [Romaniello 2003a, Romaniello 2003b]. Mechanistic studies have been performed on photoinduced ligand dissociation reactions in carbonyls of chromium and iron [Goumans 2003], tungsten [Zakrzewski 2004, Zalis 2003], and ruthenium [Gabrielsson 2004]. The mechanism of hydrogen elimination from $[Ru(PH_3)_3(CO)(H_2)]$ in the first excited singlet state has been investigated by dynamics calculations [Torres 2003].

Electronic circular dichroism calculations on chiral transition metal complexes are still limited in number but show encouraging results [Autschbach 2003, Jorge 2003, Jorge 2005, Bark 2004, Le Guennic 2005]. By means of sum-over-states methods, first calculations on non-linear optical properties have been performed on complexes of Cr [Liyanage 2003], Ru [Coe 2004a, Coe 2004b], and Zn [Baev 2004], as well as on ferrocene derivatives [Liao 2005].

23.4.5 Organic Polymers

Two different theoretical approaches have been used for polymers: solid state methods employing periodical boundary conditions and oligomer methods considering discrete fragments of increasing size. For calculations of excitation energies of organic polymers, the latter seem to be more widespread, although a LCAO-crystalline orbital implementation of excitation energies of extended systems has been reported [Hirata 1999d, Tobita 2001] (see also Chap. 21).

For oligomer methods, the convergence of the calculated properties to the bulk limit and the quality of the extrapolated properties have been of primary interest. Several papers by Ratner, Zojer, and co-workers summarize computational results on different classes of polymers [Hutchison 2002, Pogantsch 2002], e.g., polyenes, polythiophenes, and polyphenylenes. Smooth convergence is found for band gaps and bond length alternation parameters for oligomers of increasing size [Hutchison 2003], however, full band gap structures of polymers cannot be obtained by extrapolation from oligomers. Often used empirical extrapolation procedures with respect to $1/n$ [Ma 2002], where n is the number of monomer units, were shown to have significant errors. Reimers and co-workers [Cai 2002a] note a tendency to spurious metallic behavior and wrong ground state multiplicities in large conjugated π-systems. The relative stability of 1^1B_u and 2^1A_g states (in C_{2h} symmetry) in polyenes and carotenoids plays an important role in the photochemistry of the light harvesting complex but remains under discussion [Hsu 2001, Wanko 2004, Catalan 2003, Catalan 2004].

Driven by the continuing demand for new polymeric materials, the number of theoretical investigations on optical properties of polymers is constantly growing. Polythiophenes [Kwon 2000, Della Sala 2001b, Elmaci 2002, Casado 2002, Della Sala 2003c, Fabiano 2005], polypyrroles [Yurtsever 2003, Zhu 2003], polyynes [Zahradník 2005], and poly(phenylenevinylene)s [Fratiloiu 2004, Han 2004] are important industrial materials for optoelectronic devices such as light emitting diodes. Recent studies by Della Sala and co-workers addressed the effects of oxidation [Raganato 2004, Pisignano 2004, Anni 2005] and functionalization [Vitale 2004] of oligothiophenes on their photophysics. Design of molecular logical devices based on oligothiophenes has been discussed [Tamulis 2003, Tamuliene 2004, Tamulis 2004]. Carbon chains and related conjugated polymers have received much interest as the prime example of one-dimensional wires [Zhang 2004b, Weimer 2005, Magyar 2005].

Non-linear optical properties have been studied for various donor-acceptor substituted polymers [Bartholomew 2004, Sun 2004, Katan 2005].

23.4.6 Charge and Proton Transfer

Electronic excitation often leads to changes in proton or electron affinity of molecules. Excited state charge and proton transfer are thus among the simplest and the most common photochemical reactions. In the work of Parusel, Grimme, and others, photoinduced intramolecular charge transfer (ICT) in donor-acceptor substituted aromatic systems was investigated by TDDFT [Parusel 2002], DFT/SCI [Parusel 1998, Bulliard 1999], and DFT/MRCI [Parusel 2000c, Parusel 2001b, Parusel 2001c] methods (see [Parusel 2002] for an overview). Most of the studies addressed 4-(N,N)-dimethylaminobenzonitrile (DMABN), a prototypical dual fluorescent compound showing a strong emission from the ICT state in polar solutions. In extensive studies by Jamorski and co-workers [Jamorski 2002a, Jamorski 2002b, Jamorski 2002c, Jamorski 2003a, Jamorski 2004], the accuracy of TDDFT for exploration of intramolecular charge transfer phenomena has been assessed, and a classification for the emission properties of these compounds was presented [Jamorski 2003b, Jamorski 2003c]. A definite assignment of the structure of the two lowest singlet states has recently been given by means of TDDFT [Rappoport 2004] and confirmed by coupled cluster calculations [Köhn 2004]. In several TDDFT studies, the ICT mechanism in donor-substituted pyridines [Szydlowska 2003a, Szydlowska 2003b], benzophenone [Duan 2005], and in a substituted proton sponge 1,8-bis(dimethylamino)-4-cyano-naphthalene [Szemik-Hojniak 2005] have been considered. The distinct non-linear optical properties of donor-acceptor substituted chromophores have received much attention in recent TDDFT studies [Ray 2004, Reis 2004, Datta 2005, Day 2005, Jensen 2005].

Excited state proton transfer phenomena are very fast reactions and play an important role in biological processes. Tautomer equilibria and excited state proton transfer in salicylic acid [Sobolewski 1999, Sobolewski 2004], hydroxy substituted flavonols [Falkovskaia 2002, Falkovskaia 2003], 2-(2′-pyridyl)pyrrole [Kijak 2004], and [2,2′-bipyridine]-3,3′-diol [Barone 2003] have been explored with TDDFT. Excited state potential surfaces of a series of hydrogen-bonded systems have been calculated with TDDFT [Aquino 2005] and compared with results from approximate coupled cluster method RI-CC2 [Christiansen 1995, Hättig 2000]. As an example, the minimum energy profiles for intramolecular proton transfer in $n \to \pi^*$ and $\pi \to \pi^*$ excited states of ortho-hydroxybenzaldehyde (o-HBA) are displayed in Fig. 23.2. The $n \to \pi^*$ excited state exhibits two minima corresponding to the enol and the keto form of o-HBA, respectively. The keto tautomer is by 0.13 eV more stable than the enol with B3LYP/TZVP which is in excellent agreement with the RI-CC2/TZVP result of 0.12 eV. In the $\pi \to \pi^*$ excited states, the potential surfaces are rather flat with only one minimum. Very good agreement

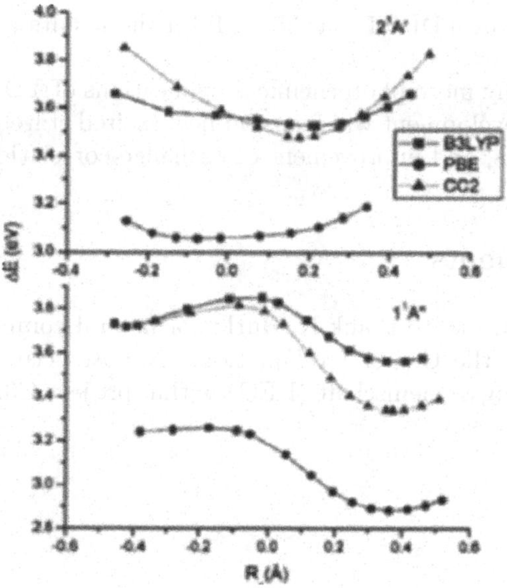

Fig. 23.2. Excited state minimum energy profiles for $1^1A''$ ($n \rightarrow \pi^*$) and $2^1A'$ ($\pi \rightarrow \pi^*$) excited states of ortho-hydroxybenzaldehyde along the proton transfer coordinate $R\_$. TDDFT calculations were performed with B3LYP and PBE functionals and SVP basis sets. RI-CC2 calculations used SVP basis sets and the corresponding auxiliary basis sets [Weigend 1998]. Reprinted with permission from [Aquino 2005]. Copyright 2005 American Chemical Society

is found between B3-LYP and RI-CC2 minimum energy profiles, the profiles from PBE calculations agree in the shape of the potential but are shifted to lower energies by ca. 0.4 eV.

7-Azaindole is an often used model compound for DNA base pairs; its tautomerization is considered a model for point mutations via mispairing. Investigations of the excited state behavior of the 7-azaindole dimer [Catalan 2004b, Catalan 2005] and 7-azaindole–water complexes [Casadesús 2003] are of great interest both from theoretical point of view and for its implication on biology.

23.5 Conclusions and Outlook

The number of TDDFT applications in photochemistry is growing at an unprecedented rate. We would like to stress here that most chemists use TDDFT not because they like the theory, but because TDDFT outperforms other methods for medium-size and large molecules, and because efficient codes are available. Present TDDFT is *not* a black box method, and its known shortcomings should always be kept in mind; nevertheless, even in critical cases present TDDFT can be useful if the results are validated and interpreted with care.

We expect many more photochemical applications of TDDFT in the near future. Method development will focus on new excited state properties, more efficient algorithms, and improvement of exchange-correlation functionals.

Acknowledgments

The authors would like to thank K. Burke for helpful comments. This work was supported by the Center for Functional Nanostructures (CFN) of the Deutsche Forschungsgemeinschaft (DFG) within project C3.9.

Part V

Applications: Beyond Linear Response

24 Atoms and Molecules in Strong Laser Fields

C.A. Ullrich and A.D. Bandrauk

24.1 Introduction

The field of research dealing with the interactions of superstrong and ultrashort laser pulses with atoms and molecules has reached a considerable degree of maturity, as summarized in a number of books [Gavrila 1992, Piraux 1993, Delone 2000] and review articles [Mainfray 1991, Freeman 1991, Burnett 1993, Protopapas 1997, Salières 1999, Joachain 2000, Brabec 2000, Dörner 2002, Becker 2002]. As shown in Fig. 24.1, the attainable laser intensity has increased by 12 orders of magnitude since the invention of the laser in 1960, as a result of a series of technological advances. Today, focused intensities in excess of 10^{21} W/cm^2 can be produced from both large-scale and lab-scale lasers. By comparison, the atomic unit of intensity is $I_0 = 3.52 \times 10^{16}$ W/cm^2 which, by the relation $I = c\mathcal{E}^2/8\pi$, corresponds to the atomic unit of the electric field strength $\mathcal{E}_0 = e/a_0^2 = 5.14 \times 10^{11}$ V/m, i.e., the electric field which an electron experiences in the 1s orbital of a hydrogen atom. Thus, the forces produced by the laser field match and even exceed the Coulomb forces that attract the electrons to the nucleus, or that bind the atoms together in a molecule.

On the other hand, the minimum attainable pulse lengths have dramatically decreased. Whereas at the end of the 20th century the focus was on femtosecond (fs) photochemistry and photophysics, culminating with the 1999 Nobel Prize to A. H. Zewail for "femtochemistry," a major effort is now underway to develop attosecond optical pulses [Drescher 2001, Paul 2001, Hentschel 2001]. Passing the attosecond frontier makes it possible to image and to manipulate the dynamics of electrons on their natural inneratomic time scales [Kienberger 2002, Drescher 2002, Kienberger 2004, Wickenhauser 2005], defined by the period of the 1s hydrogen orbit, 24 attoseconds.

Section 24.2 presents an overview of the physics of atoms in strong laser fields, giving a flavor of the interesting and surprising phenomena that take place in this highly nonlinear regime (see Fig. 24.2). In Sect. 24.3 we review the contributions that TDDFT has made to this exciting field of research, pointing out the successes as well as the difficulties that still need to be overcome. Section 24.4 discusses molecules in strong fields. We conclude in Sect. 24.5 by summarizing the challenges and opportunities for TDDFT in

C.A. Ullrich and A.D. Bandrauk: *Atoms and Molecules in Strong Laser Fields*, Lect. Notes Phys. **706**, 357–375 (2006)
DOI 10.1007/3-540-35426-3_24

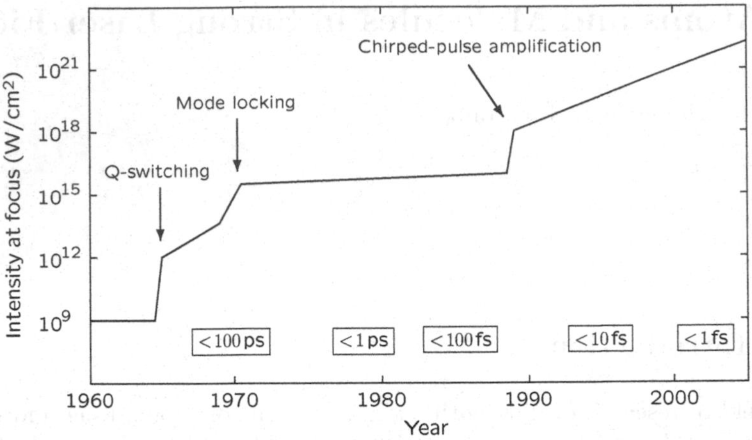

Fig. 24.1. The attainable pulsed-laser peak intensity has increased by 12 orders of magnitude over the last 40 years. Ultrashort pulses can be generated with durations of a few femtoseconds and, of lately, even down to the attosecond regime [Drescher 2001, Paul 2001, Hentschel 2001]

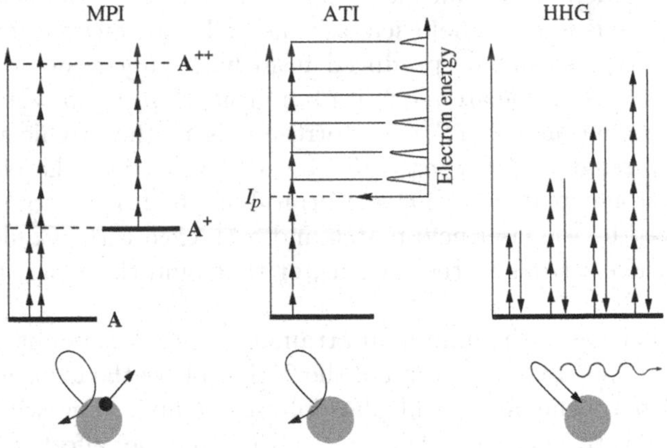

Fig. 24.2. Schematic representation of three nonlinear phenomena of atoms in intense laser fields [L'Huillier 2002]: sequential and non-sequential multiphoton ionization (MPI), above-threshold ionization (ATI), and high-harmonic generation (HHG). A simple interpretation is based on a semiclassical recollision model [Corkum 1993, Kulander 1993, Lewenstein 1994, Yudin 2001]

the study of strong-field atomic and molecular processes. Unless indicated otherwise, atomic units ($e = \hbar = m = 1$) are used in this chapter.

24.2 Atoms in Strong Laser Fields: An Overview

24.2.1 Multiphoton Ionization

An intense laser pulse can ionize an atom even when the photon energy is much smaller than the ionization potentials [L'Huillier 1983, Perry 1988]. Figure 24.3 shows experimental data for the number of xenon ions as a function of laser intensity, at a wavelength of 585 nm and a pulse duration of 1 ps [Perry 1988]. At this wavelength, 6 photons are necessary to ionize the Xe atom, whereas the ionization process leading from, e.g., Xe^{+5} to Xe^{+6} requires already 34 photons. Notice that once the ion has reached the charge

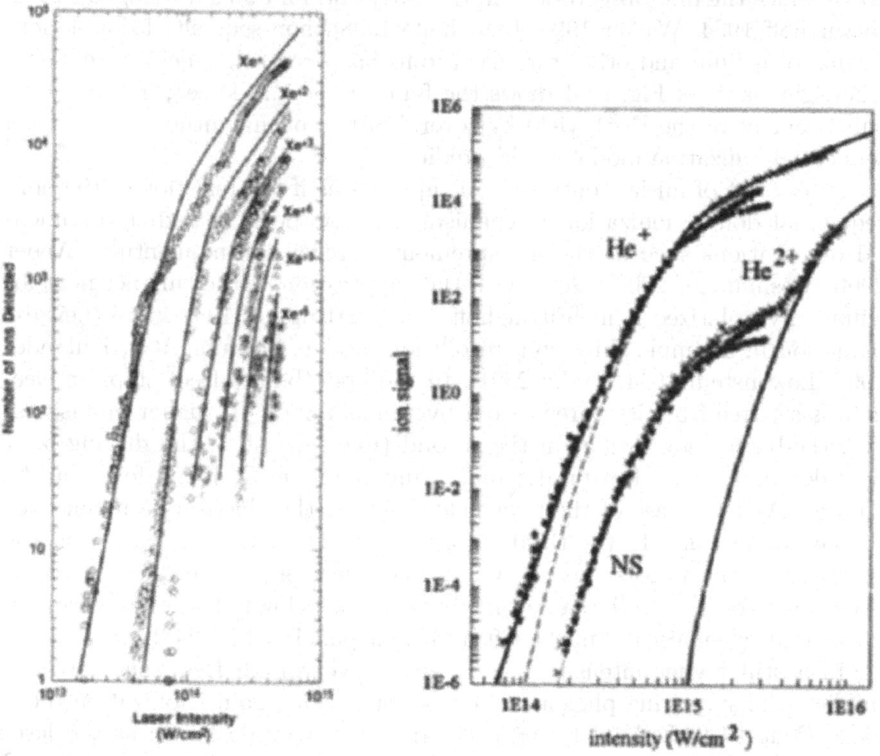

Fig. 24.3. *Left*: Sequential multiphoton ionization of Xe by 1 ps laser pulses of wavelength 585 nm (from [Perry 1988]). *Right*: Non-sequential double ionization of He by 160 fs laser pulses of wavelength 780 nm (from [Walker 1994]). In both panels, the full lines follow from rate equation models including only sequential processes

state $+q$, increasing the laser intensity by about 50% is enough to remove a further electron, while the number of photons which should be absorbed at each sequential ionization step rapidly grows with the degree of ionization. Recently, multiphoton ionization up to Xe^{+20} was observed [Yamakawa 2004].

Under the experimental conditions of [L'Huillier 1983, Perry 1988], ionization proceeds via stepwise removal of the electrons. Ionization yields such as shown in Fig. 24.3 can be theoretically explained by kinetic rate-equation models employing ionization rates obtained from lowest order perturbation theory. For each ion species, the ion yield rises very steeply with intensity ($\sim I^N$, where N is the minimum number of photons required). At the saturation intensity, a marked change appears in the slope of the curves, associated with a depletion of an ion species when the ionization probability reaches unity.

For highly intense (around 10^{15} W/cm^2) and ultrashort (around 100 fs) laser pulses, this simple picture of multiphoton ionization becomes more complicated, and highly correlated ionization mechanisms come into play [Dörner 2002]. Since the first observations in the early and mid-90's [Fittinghoff 1992, Fittinghoff 1994, Walker 1994, Larochelle 1998], non-sequential double ionization of helium and other rare gas atoms has been a hot field of research. The right panel of Fig. 24.3 shows the famous "helium-knee", indicating an enhancement of the He^{2+} yield by several orders of magnitude over what a sequential ionization model would predict.

After a lot of initial controversy, it appears as if the question of the non-sequential double ionization mechanism has now been settled. Experimental observations such as the measurement of recoil ion-momentum [Weber 2000, Moshammer 2000], along with the suppression of the enhancement for elliptically polarized light [Fittinghoff 1992, Fittinghoff 1994], find their explanation in a simple three-step recollision model [Corkum 1993, Kulander 1993, Lewenstein 1994, Yudin 2001]. In the first (bound-free) step, an electron is set free from its parent atom by tunnelling or (at higher intensities) over-the-barrier ionization. In the second (free-free) step, the driving laser field dominates the electron dynamics, and the ionic Coulomb force can be ignored. As the phase of the laser field reverses, the electron is driven back to the atomic core. In the third step, the electron can then scatter off the core and in the process knock out another electron, or produce harmonic generation through radiative recombination (see below). These processes are illustrated schematically in the left and right panel of Fig. 24.2.

In a still higher intensity regime and at very high frequencies, theory predicts the surprising phenomenon of stabilization against ionization [Pont 1988, Pont 1990, Kulander 1991a, Eberly 1993, Gavrila 2002]: as the laser intensity is increased, atomic ionization rates pass through a maximum and then *decrease*. Experimental evidence of stabilization has been reported for atomic Rydberg states [de Boer 1993, de Boer 1994, van Druten 1997].

24.2.2 Above-Threshold Ionization

The so-called above-threshold ionization (ATI) [Becker 2002] is another highly nonlinear phenomenon which occurs when electrons absorb a large number of extra photons in addition to those needed to overcome the ionization barrier. As illustrated in the center panel of Fig. 24.2, one detects a whole sequence of equally spaced peaks in the kinetic-energy distribution of the photoelectrons [Agostini 1979, Kruit 1983], with energies $n\hbar\omega - I_p$. For short and intense pulses, the atomic ionization potential I_p gets an additional shift by the ponderomotive potential $U_p = e^2\mathcal{E}^2/4m\omega^2$ associated with the wiggle motion of a free electron in a laser field. Most of the early work in the 1980's concentrated on the low-energy part of the ATI spectrum, investigating the role of the ponderomotive potential, the AC-Stark shifted resonant excited states, and the transition from multiphoton to tunnelling regime.

In the mid-90's the experimental precision in recording photoelectron spectra increased significantly, and it was found that ATI spectra extend over many tens of eV, with a decrease for the first orders up to $\sim 2\,U_p$, followed by a large plateau extending up to $\sim 10\,U_p$ [Paulus 1994, Paulus 2001, Grasbon 2003] (see Fig. 24.4). This can again be explained in the semiclassical recollision model [Corkum 1993, Kulander 1993, Lewenstein 1994, Yudin 2001], where electrons are lifted into the continuum at some phase of the laser's electric field and start from the atom with zero velocity. $2\,U_p$ is the resulting classical maximum kinetic energy for electrons leaving the atom without rescattering. Ionized electrons that do re-encounter the ion and elastically rescatter may acquire a maximal classical energy of $10\,U_p$. With linear polarization, electrons are generated along the polarization direction. However, it was found that angular distributions exhibit a much more complex off-axis structure at the edge of the plateau ("scattering rings"), which are another consequence of the rescattering of the electron wavepacket on the parent ion.

24.2.3 Harmonic Generation

When an atom interacts with a laser field, a dipole moment is induced which in turn acts as a source of radiation. At high laser intensities, the atomic response becomes extremely nonlinear. As a result, pronounced signals at multiples of the driving frequency appear in the photoemission spectrum (high-harmonic generation or HHG [Salières 1999]). Because atoms have inversion symmetry, only odd multiples of the driving frequency are emitted. In contrast to the perturbative picture prevailing at weak laser fields, strong laser pulses can yield a very large number of harmonics [McPherson 1987, Ferray 1988, Li 1989, L'Huillier 1992, Sarukura 1991, Miyazaki 1992, Wahlström 1993, Kondo 1994, L'Huillier 1993, Macklin 1993]. Figure 24.4 shows experimental data on harmonic generation of rare-gas atoms [L'Huillier 1993], exhibiting the typical rapid decrease over the first

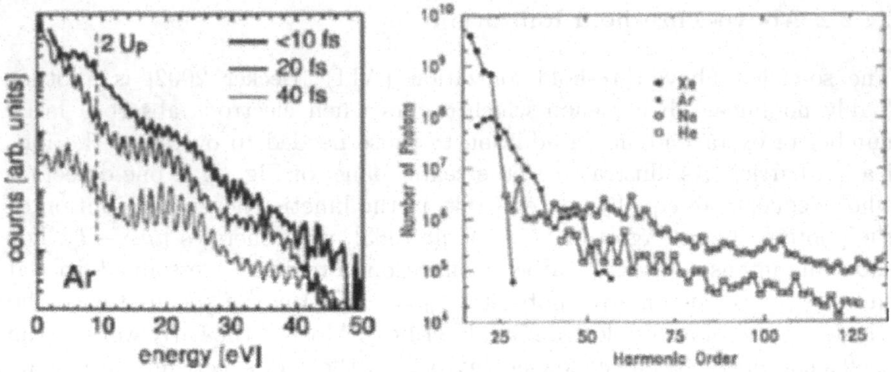

Fig. 24.4. *Left*: ATI spectrum for argon, for various pulse lengths, at 800 nm and 0.8×10^{14} W/cm^2 (from [Grasbon 2003]). *Right*: HHG in rare-gas atoms driven by 1 ps laser pulses of wavelength 1053 nm and intensity 1×10^{15} W/cm^2 (from [L'Huillier 1993])

few harmonics followed by an extended plateau. The highest harmonic observed here is the 135th harmonic of 1053 nm (which corresponds to 7.8 nm) with He as target atom. We find that the width of the plateau decreases going from He to Xe, while at the same time the absolute intensity of the observed harmonics becomes larger. This behavior is linked to differences in the static polarizabilities of the target atoms [Liang 1994, Chin 1995]. Harmonic orders of around 300 have been observed using ultrashort, high-intensity laser pulses, where the atoms experience only a few laser cycles [Zhou 1996, Spielmann 1997, Schnürer 1998]. Under these conditions, harmonic frequencies extend beyond 500 eV, reaching into the water window with wavelengths around 2.7 Å.

Theoretically, the cutoff of the HHG plateaus is predicted to occur at $\hbar\omega_c = I_p + 3.2\,U_p$, following the simple three-step recollision model discussed above [Corkum 1993, Kulander 1993, Lewenstein 1994, Yudin 2001].

24.2.4 Theoretical Methods

There exists a large body of theoretical work devoted to the various aspects of the nonlinear physics of atoms in strong fields (see the recent review by Joachain and coworkers [Joachain 2000] and references therein). From the experimental phenomenology it is clear that perturbative and semi-perturbative methods fail to capture the extremely high degree of nonlinearity, and one must resort to non-perturbative theories. Many approaches, most prominently Floquet [Chu 2004a] and R-matrix Floquet [Burke 1990, Burke 1991] theories, are based on the assumption that the Hamiltonian of the atom-laser field system is periodic in time. Although this is not true for a realistic laser pulse, one can nevertheless incorporate some pulse shape effect into

(R-matrix) Floquet calculations, provided that the atom remains in a Floquet eigenstate that is adiabatically connected to the initial state. Clearly, such an assumption can be expected to break down for ultrashort, femtosecond pulses. The same is true for the so-called many-body S-matrix theory [Becker 2005], which can be viewed as a low-frequency approach.

In general, therefore, one needs to resort to a direct integration of the time-dependent Schrödinger equation (TDSE). Since the pioneering work by Kulander and coworkers [Kulander 1987, Kulander 1992a, Kulander 1993], most activity in this area has focused on the hydrogen atom, where the TDSE can be numerically solved without restrictions on large grids [Cormier 1997, Tong 1997]. This strategy becomes of course tremendously involved as soon as one deals with atoms having more than one electron; a propagation of the full two-electron wave function of helium in all three spatial dimensions was carried out by Parker et al. [Parker 1996], using a massively parallel supercomputer. In general, however, the TDSE for two-electron systems can be solved only on a restricted basis (see the review by Lambropoulos et al. [Lambropoulos 1998]). As an alternative one can treat one- and two-electron atoms and molecules as one-dimensional (1-D) model systems (see Sect. 24.4.2 below), which has been particularly useful to elucidate the mechanism of nonsequential MPI [Eberly 1992, Bauer 1997, Lappas 1998, Lein 2000a, Dahlen 2001, Dahlen 2002].

All studies dealing with many-electron atoms in strong laser fields have made use of more or less severe approximations to reduce the problem to a tractable size. Most conspicuously, in the single-active electron (SAE) model [Kulander 1988, Kulander 1991b, Tang 1991, Xu 1992, Xu 1993] the TDSE is solved for only one "active" electron while the remaining electrons are frozen in their initial configuration, their influence on the active electron being simulated by a *static* model potential. This strategy successfully models the screening of the nuclear charge by the inner electrons, but cannot describe collective effects arising from electronic correlation.

24.3 TDDFT for Atoms in Strong Laser Fields

The first TDDFT studies of atoms in strong laser fields were carried out in the mid-90's [Ullrich 1995b, Ullrich 1996, Ullrich 1997, Erhard 1997, Tong 1998, Tong 2001], solving the TDKS equations for He, Be, and Ne atoms in the presence of time-dependent potentials of the form

$$v_{\text{laser}}(\boldsymbol{r},t) = \mathcal{E} z f(t) \sin(\omega t)\ , \tag{24.1}$$

describing a laser field in the length gauge [Joachain 2000]. Here, $\mathcal{E}$ is the electric-field amplitude, ω is the laser frequency, and $f(t)$ is a function between 0 and 1 which describes the switching-on or pulse envelope of the laser.

The example of a Ne atom in intense laser fields of wavelength $\lambda = 248\,\text{nm}$ was discussed in detail in [Ullrich 1997]. Due to the linear polarization of the laser field, the rotational symmetry of the atom around the z-axis is preserved at all times. It is thus appropriate to solve the TDKS equations using cylindrical coordinates. For this example, the complete outer shell is propagated, consisting of 2s, $2p_0$ and $2p_1$ orbitals, each doubly occupied. Here, the $2p_0$ orbital is oriented along the laser polarization axis, whereas the two $2p_1$ orbitals are perpendicular, and have thus an identical time evolution. The inner 1s orbital is kept frozen in its initial state, i.e., its time evolution is given by $\varphi_{1s}(\boldsymbol{r},t) = \varphi_{1s}(\boldsymbol{r},t_0)\exp[-\mathrm{i}\varepsilon_{1s}(t-t_0)]$. The TDKS equations are solved on a grid with a finite-difference representation, and using the Crank-Nicholson algorithm for the time propagation (see the discussion in Chap. 12). The results in this section were obtained ignoring correlation effects, using the x-only ALDA and TDKLI methods. The TDKLI approach [Ullrich 1995a] arises from an adiabatic approximation to the x-only TDOEP equation (see Chap. 3), or, equivalently, the time-dependent EXX equation (see Chap. 9). One obtains a time-dependent version of the x-only KLI potential [Krieger 1992b], given by

$$v_{x\sigma}^{\mathrm{KLI}}(\mathbf{r},t) = w_{x\sigma}(\mathbf{r},t) + \frac{1}{n_\sigma(\mathbf{r},t)}\sum_j^{N_\sigma} n_{j\sigma}(\mathbf{r},t)\int d^3r'\, n_{j\sigma}(\mathbf{r}',t)v_{x\sigma}^{\mathrm{KLI}}(\mathbf{r}',t) \tag{24.2}$$

with

$$\begin{aligned} w_{x\sigma}(\mathbf{r},t) = -\frac{1}{n_\sigma(\mathbf{r},t)}\sum_{j,k}^{N_\sigma} \varphi_{j\sigma}(\mathbf{r},t)\varphi_{k\sigma}^*(\mathbf{r},t)\int \mathrm{d}^3r'\, \frac{\varphi_{k\sigma}(\mathbf{r}',t)\varphi_{j\sigma}^*(\mathbf{r}',t)}{|\mathbf{r}-\mathbf{r}'|} \\ - n_{j\sigma}(\mathbf{r},t)\int \mathrm{d}^3r''\int \mathrm{d}^3r'\, \frac{\varphi_{j\sigma}(\mathbf{r}'',t)\varphi_{k\sigma}^*(\mathbf{r}'',t)\varphi_{k\sigma}(\mathbf{r}',t)\varphi_{j\sigma}^*(\mathbf{r}',t)}{|\mathbf{r}''-\mathbf{r}'|} \end{aligned}, \tag{24.3}$$

where $n_{j\sigma}(\mathbf{r},t) = |\varphi_{j\sigma}(\mathbf{r},t)|^2$. Equation (24.2) can be solved for $v_{x\sigma}^{\mathrm{KLI}}(\mathbf{r},t)$ in analogy to the static case [Krieger 1992b], the solution only involving the inversion of an $(N_\sigma - 1)\times(N_\sigma - 1)$ matrix.

Under the influence of the intense driving field, the entire valence shell gets strongly excited, and the KS orbitals acquire substantial continuum contributions. To prevent electronic flux from being reflected back from the edges of the numerical grid, absorbing boundary conditions are introduced in the form of a so-called mask function (a complex potential would be an alternative). Over the course of time, the norm of the KS orbitals $N_{j\sigma}(t)$ thus *decreases*, even though the time propagation algorithm is unitary. This allows us to describe ionization in a straightforward manner, by calculating the number of electrons remaining in a finite volume:

$$N_{j\sigma}(t) = \int_{\mathcal{V}} \mathrm{d}^3r\, |\varphi_{j\sigma}(\boldsymbol{r},t)|^2\,, \qquad N(t) = \sum_{j\sigma} N_{j\sigma}(t) = \int_{\mathcal{V}} \mathrm{d}^3r\, n(\boldsymbol{r},t)\,. \tag{24.4}$$

Here, $\mathcal{V}$ refers to a volume centered about the nucleus which contains essentially the entire wavefunction at $t = t_0$. The total number of escaped elec-

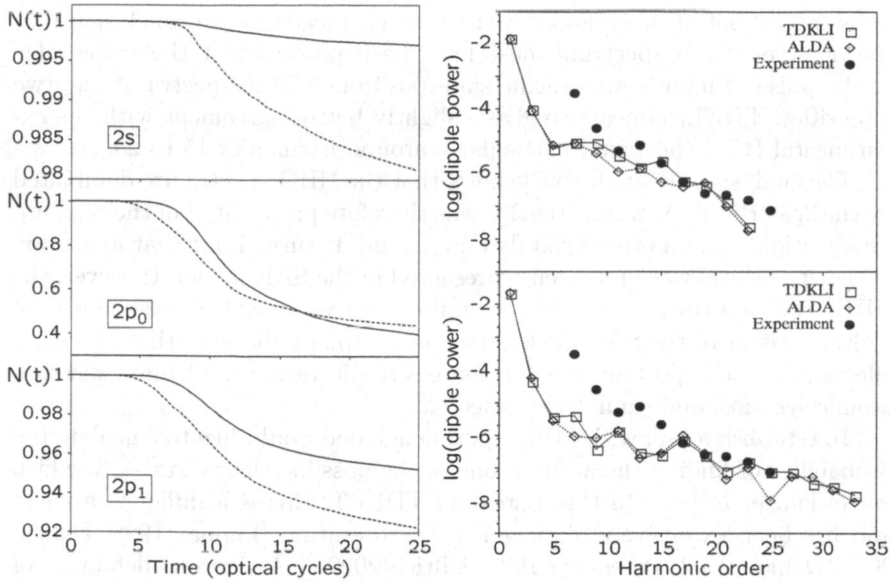

Fig. 24.5. *Left*: time-dependent norm of the Ne 2s, $2p_0$ and $2p_1$ orbitals, calculated with TDKLI (*full lines*) and ALDA (*dashed lines*). The laser parameters are $\lambda = 248\,\text{nm}$, $I = 3 \times 10^{15}\,\text{W/cm}^2$, and a 10-cycle linear ramp switching-on. *Right*: Harmonic distributions for Ne at $I = 3 \times 10^{15}\,\text{W/cm}^2$ (*top*) and $I = 5 \times 10^{15}\,\text{W/cm}^2$ (*bottom*). Experimental data from [Sarukura 1991]

trons, $N_{\text{esc}}(t) = N_0 - N(t)$, is thus a simple functional of the time-dependent density. The left panel of Fig. 24.5 shows $N_{2s}(t)$, $N_{2p_0}(t)$, and $N_{2p_1}(t)$ for the Ne atom. As expected, the 2s orbital is the least ionized of the three orbitals. The $2p_0$ and $2p_1$ orbitals differ by about an order of magnitude in their degree of ionization, which is a typical observation and due to the fact that the $2p_0$ orbital is oriented along the polarization axis which makes it easier for the electrons to escape. The difference between ALDA and TDKLI can be understood from the differences of the KS orbital eigenvalues (the electrons are more weakly bound in LDA than in KLI [Ullrich 1997]).

Another observable which is a straightforward density functional is the dipole moment $D(t) = \int d^3r\, z n(\boldsymbol{r}, t)$. The associated power spectrum $|D(\omega)|^2$ yields the contribution to the HHG spectrum of a single atom, which is proportional to the experimentally observed HHG spectra to within a reasonable approximation (in general, however, one needs to include propagation effects within the interaction volume [Salières 1999]). The right panel of Fig. 24.5 shows HHG spectra for Ne calculated at two intensities, $I = 3$ and $5 \times 10^{15}\,\text{W/cm}^2$, together with experimental data [Sarukura 1991] taken with a pulsed laser of the same wavelength ($\lambda = 248$ nm) but with a much higher peak intensity, $4 \times 10^{17}\,\text{W/cm}^2$. One finds that the essential features of the experimental HHG plateau are reproduced fairly well by the simulations carried out at much lower intensities. This leads to the conclusion that the observed HHG spectrum must have been produced at the rising edge of the pulse. Taking equal-weight superpositions of the spectra at the two intensities, TDKLI appears to give a slightly better agreement with the experimental HHG spectrum, particularly around harmonics 15 to 25.

The analysis of [Ullrich 1997] shows that the HHG spectra are dominated by the $2p_0$ orbital. As a test, the $2p_0$ was therefore propagated in the SAE approximation, i.e., all other orbitals were frozen. It turned out that ionization of the $2p_0$ orbital was quite well represented in the SAE model. However, the HHG spectrum turned out to have unphysical resonance features, which are a clear artifact of the SAE model. This is a strong indication that collective effects play an important role for the electronic response of many-electron atomic systems, and cannot be neglected.

To establish a link with MPI experiments, one would like to calculate the probability of finding the atom in one of the possible charge states to which it can ionize, $P^{+n}(t)$. In the context of TDDFT, this is a difficult problem and has been extensively discussed in the literature [Lappas 1998, Dahlen 2001, Dahlen 2002, Petersilka 1999, Ullrich 2000b]. A rigorous definition of the $P^{+n}(t)$ involves the time-dependent many-body wave function [Ullrich 2000b]:

$$P^0(t) = \sum_{\sigma_1\cdots\sigma_N} \int_{\mathcal{V}} d^3r_1 \ldots \int_{\mathcal{V}} d^3r_N \, |\Psi(\boldsymbol{r}_1\sigma_1, \ldots, \boldsymbol{r}_N\sigma_N, t)|^2 \tag{24.5a}$$

$$P^{+1}(t) = \binom{N}{1} \sum_{\sigma_1\cdots\sigma_N} \int_{\overline{\mathcal{V}}} d^3r_1 \int_{\mathcal{V}} d^3r_2 \ldots \int_{\mathcal{V}} d^3r_N \, |\Psi(\boldsymbol{r}_1\sigma_1, \ldots, \boldsymbol{r}_N\sigma_N, t)|^2 \tag{24.5b}$$

and similarly for all other P^{+n}, where $\overline{\mathcal{V}}$ refers to the region outside the integration volume $\mathcal{V}$. In the case of a two-electron system, one can rewrite these formulas introducing the pair correlation function

$$g(\boldsymbol{r}_1, \boldsymbol{r}_2, t) = 2\frac{|\Psi(\boldsymbol{r}_1, \boldsymbol{r}_2, t)|^2}{n(\boldsymbol{r}_1, t)n(\boldsymbol{r}_2, t)} \tag{24.6}$$

as follows:

$$P^0(t) = \frac{1}{2}\int_{\mathcal{V}} d^3r_1 \int_{\mathcal{V}} d^3r_2 \, n(\boldsymbol{r}_1, t)n(\boldsymbol{r}_2, t)g(\boldsymbol{r}_1, \boldsymbol{r}_2, t) \tag{24.7a}$$

$$P^{+1}(t) = \int_{\mathcal{V}} d^3r \, n(\boldsymbol{r}, t) - \int_{\mathcal{V}} d^3r_1 \int_{\mathcal{V}} d^3r_2 \, n(\boldsymbol{r}_1, t)n(\boldsymbol{r}_2, t)g(\boldsymbol{r}_1, \boldsymbol{r}_2, t) \tag{24.7b}$$

$$P^{+2}(t) = 1 - P^{+1}(t) - P^0(t) \, . \tag{24.7c}$$

The task is thus a twofold one: one needs to find an accurate xc functional for the TDKS equation, as well as an accurate expression for the pair correlation

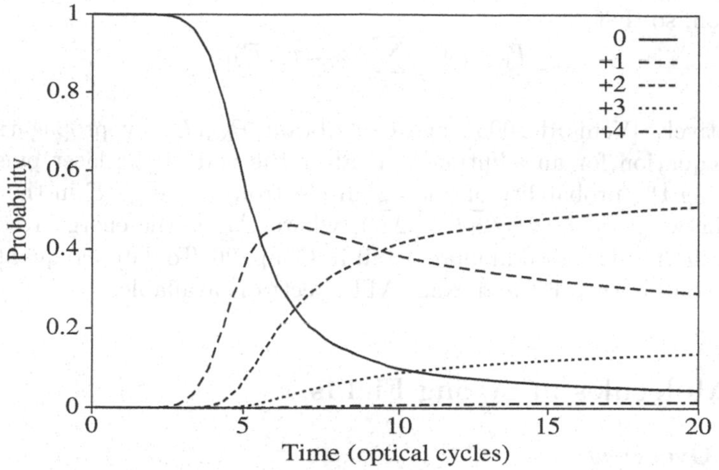

Fig. 24.6. Population of the ionized states of Ne (TDKLI), using the independent-particle approximation for $P^{+n}(t)$. Laser: $\lambda = 248\,\mathrm{nm}$, $I = 5 \times 10^{15}\,\mathrm{W/cm^2}$

function. A straightforward approximation is to set the latter equal to its x-only limit of 1/2, which amounts to an independent-particle approximation. For a two-electron system with a doubly occupied TDKS orbital, this gives

$$P^0(t) = N(t)^2\,, \quad P^{+1}(t) = 2N(t)[1-N(t)]\,, \quad P^{+2}(t) = [1-N(t)]^2\,, \tag{24.8}$$

where $N(t) = \frac{1}{2}\int_V \mathrm{d}^3r\, n(\boldsymbol{r},t)$. Figure 24.6 shows results for the Ne atom, where up to Ne^{+4} ionic states are found. However, it turns out that the independent-particle approximation for $P^{+n}(t)$ leads to significant errors for the non-sequential double ionization probabilities [Lappas 1998, Dahlen 2001, Dahlen 2002], even if one uses the *exact* TDDFT orbital densities, as we will demonstrate below for the case of a 1-D H_2 molecule. Petersilka and Gross [Petersilka 1999] have tried to remedy this situation by employing a static density-functional expression for the pair correlation function $g_c[n](\boldsymbol{r}_1,\boldsymbol{r}_2)$, but without much quantitative improvement. This is indicative of a sizable degree of "correlation" of the two-electron dynamics, in particular for the case of longer wavelengths. At 248 nm, on the other hand, the independent-particle approximation works much better [Dahlen 2002]. Similar findings have been reported by Bauer et al. [Bauer 1997, Bauer 2001, Ceccherini 2001].

Similar issues arise for the calculation of ATI spectra, which are rigorously defined in terms of the many-body wavefunction, and can therefore in principle be expressed as functionals of $n(\boldsymbol{r},t)$. In practice, however, all TDDFT approaches calculate the kinetic-energy spectra associated with the KS single-particle orbitals [Pohl 2000, Pohl 2001, Nguyen 2004], despite the fact that the KS wave function has no strict physical meaning. One obtains the kinetic energy distribution $P_{\mathrm{KS}}(E)$ by recording the KS orbitals $\varphi_{j\sigma}(\boldsymbol{r},t)$ over time at a point $\boldsymbol{r}_b$ near the grid boundary and subsequent Fourier transformation, so that

$$P_{\mathrm{KS}}(E) = \sum_j |\varphi_{j\sigma}(\boldsymbol{r}_b, E)|^2 \,. \tag{24.9}$$

Alternatively [Véniard 2003], one can obtain $P_{\mathrm{KS}}(E)$ by propagating the TDKS equation for an additional Δt after the end of the laser pulse, and calculating the probability of finding an electron of energy E in the spatial region between $r_{\pm} = \Delta t\sqrt{2(E \pm \Delta E)}$, where ΔE is the energy resolution. This method is discussed in more detail in Chap. 26. To date, no quantitative comparison between KS and exact ATI spectra is available.

24.4 Molecules in Strong Fields

24.4.1 Overview

The interaction of molecules with intense laser pulses introduces new challenges due to the presence of the extra degrees of freedom of the nuclear motion, and the associated additional time scales. The shortest nuclear motion period is that of the proton, 10–15 fs, which is comparable to the ionization times at intensities of the order of $10^{14}\,\mathrm{W/cm^2}$ at wavelengths of 800–1064 nm. Thus, the nuclear motion needs to be treated on an equal footing with the radiative processes induced by intense laser fields, in order to study the photochemical dynamics in the nonperturbative multiphoton regime. This section summarizes a few highlights of the physics of molecules in strong fields; more details can be found in recent review articles [Bandrauk 2003a, Marangos 2004]. The topic of molecular alignment in strong fields has been reviewed in [Corkum 1999, Stapelfeldt 2003].

At intensities below $10^{13}\,\mathrm{W/cm^2}$ where ionization is negligible, the dressed-molecule picture [Bandrauk 1994a, Bandrauk 1994b] has been a successful approach to visualize the change of nuclear dynamics due to a laser field. In this representation, the semiclassical radiation field is explicitly included in the molecular Hamiltonian, and resonant processes are pictured as crossing of "dressed" molecular potential curves. At higher intensities, the dynamics of photophysical processes can be conveniently described using laser-induced molecular potentials (LIMPs) [Wunderlich 1997]. In the intensity range 10^{14}–$10^{15}\,\mathrm{W/cm^2}$, rapid ionization starts to set in, accompanied by considerable distortions of intermolecular potentials, creating LIMPs that can lead to bond softening via laser-induced avoided crossings. The molecular ions can also undergo above-threshold dissociation (ATD), the equivalent of ATI in atoms.

Compared to atoms, molecules offer new perspectives of laser-induced electron recollision (LIERC) with parent molecular ions, such as "diffraction" from more than one nuclear center. This leads to a new molecular phenomenon, laser-induced electron diffraction (LIED) [Zuo 1996a], a new tool for probing molecular geometry changes on ultrashort time scales [Bandrauk

2001, Itatani 2004]. Much of the theoretical understanding of LIERC and LIED in molecules is based on exact solutions of the one-electron TDSE for the H_2^+ and H_3^{++} molecules for static (Born-Oppenheimer) [Bandrauk 2001, Itatani 2004] and moving nuclei (non-Born-Oppenheimer) [Bandrauk 2003b].

For multielectron atoms, double ionization in intense laser fields is a highly correlated process (see Sect. 24.2.1), with LIERC a dominant mechanism for the nonsequential steps [Yudin 2001], transforming to sequential double ionization at higher intensities [Walker 1994, Liu 2004]. Molecules differ from atoms by the multicenter Coulomb nature of the electron recollision process, leading to diffraction [Zuo 1996a, Itatani 2004] and even collision with neighboring ions [Bandrauk 1999, Bandrauk 1998]. In particular, at large internuclear distances, tunnelling ionization, which is the first step in atomic LIERC, becomes more complicated as a consequence of charge transfer and charge resonance effects, first predicted by Mulliken [Mulliken 1939] due to the existence of excited ion-pair states [Martin 1997b] which cross the ground state at high intensities in both 1-D [Kawata 2000] and recent 3-D [Harumiya 2002] simulations of H_2. Similar effects have been found in the 1-D models of H_3^+ at high intensities [Kawata 2001].

The H_2 molecule is the prototype model of the two-electron chemical bond with bonding and anti-bonding molecular orbitals. Earlier 3-D calculations of the nonlinear response of H_2 in an intense laser field were performed using a frozen core approximation [Krause 1991], and TDHF using finite-element basis sets [Yu 1995]. Exact 1-D [Kawata 2001] and 3-D [Harumiya 2002] TDSE numerical solutions of H_2 were obtained on large finite grids at equilibrium and large intermolecular distances in order to confirm the universal molecular phenomena of charge resonance enhanced ionization (CREI) first discovered in 3-D [Zuo 1995, Zuo 1996b] and 1-D [Zuo 1995, Zuo 1996b, Chelkowski 1995, Seideman 1995] one-electron molecular model simulations. As illustrated in Fig. 24.7, an electron in a diatomic molecule experiences essentially a double potential well, which becomes distorted in the presence of a laser field by the gradient of the optical potential across the molecule. There exists a range of critical separations between the two nuclei, R_c, at which the bound state in the upper well is raised to the point that it barely sees any barrier to tunnel through into the second well and from then on to ionization. As a result, for certain nuclear distances the ionization rate of the molecule can be an order of magnitude higher than for the individual atoms. At even larger nuclear separations ($R > 10$ a.u. for H_2^+), the LUMO again becomes trapped in the ascending well, resulting in the disappearance of over-barrier ionization.

A number of recent experiments have addressed the issue of sequential and non-sequential ionization in diatomic molecules such as H_2 [Alnaser 2004b], D_2 [Sakai 2003, Légaré 2003], O_2, N_2, NO, and I_2 [Guo 2001, Eremina 2004, Alnaser 2004a, Suzuki 2004]. These studies have confirmed the validity of the CREI and LIERC mechanisms for multielectron systems, and have explored the phenomenon of Coulomb explosion following rapid ionization with few-

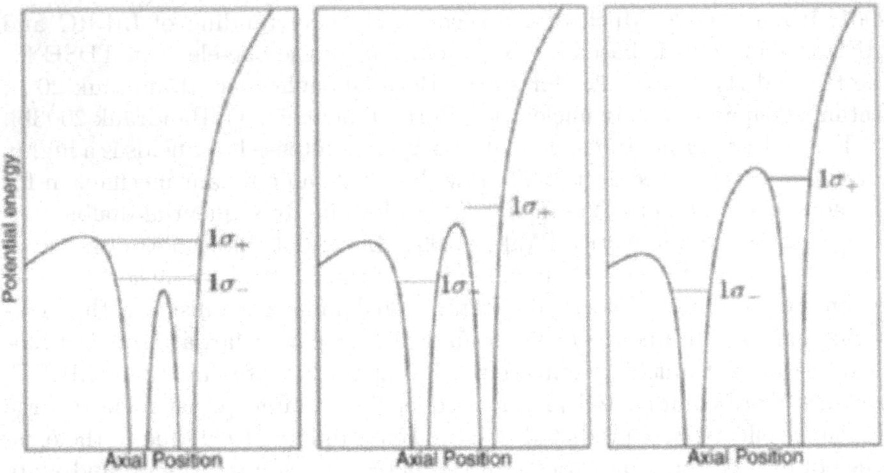

Fig. 24.7. Schematic illustration of charge-resonance enhanced ionization (CREI) in a diatomic molecule. For a range of critical internuclear separations R_c, the LUMO is raised above the inner barrier (*middle panel*). Ionization proceeds via radiative coupling between the HOMO $1\sigma_-$ and LUMO $1\sigma_+$ ("essential" or "doorway" states) [Bandrauk 2003a]

cycle pulses, combined with a possible imaging of the fragments. Other recent studies have focused on MPI of larger, polyatomic molecules [Markevitch 2003, Markevitch 2004].

24.4.2 A 1-D Example: H_2 with Fixed Nuclei

1-D models have proven to be useful to recover the essential physics of laser-molecular interactions at high intensities, i.e., tunnelling ionization and laser control of the ionized electron trajectory. For linear molecules such as H_2 and H_3^+, the linear arrangement of nuclear charges enhances refocusing of the ionized electron along the internuclear axis [Bandrauk 1999, Kawata 2001, Villeneuve 1996].

For H_2 with static nuclei, the 1-D Born-Oppenheimer Hamiltonian for two electrons with coordinates x_1, x_2 with respect to the center of mass of two protons situated at positions $\pm R/2$ is written as [Bandrauk 2005a]

$$\hat{H}_0 = \sum_{i=1}^{2} \left[-\frac{1}{2}\frac{\partial^2}{\partial x_i^2} - \frac{1}{\sqrt{(x_i \pm R/2)^2 + c}} \right] + \frac{1}{\sqrt{(x_1 - x_2)^2 + d}} . \quad (24.10)$$

The softening parameters c and d remove Coulomb singularities and allow for the use of high-order split-operator methods for solving the TDSE [Bandrauk 1993, Bandrauk 1994c]. Propagating the TDSE with $\hat{H}_0$ (24.10) in imaginary time yields the ground-state electronic energies and corresponding molecular potentials. It was found that, putting $c = 0.7$ and

$d = 1.2375$, (24.10) reproduces accurately the first three electronic states of H_2 [Bandrauk 2005b]. The TDSE for H_2 in an intense laser field then becomes

$$\mathrm{i}\frac{\partial \Psi(x_1, x_2, t)}{\partial t} = \left[\hat{H}_0 + v(x_1, x_2, t)\right] \Psi(x_1, x_2, t) , \quad (24.11)$$

where $v(x_1, x_2, t) = -(x_1 + x_2)\mathcal{E}(t)\cos(\omega t)$ is the dipolar (long wavelength) form of the electron-laser interaction for a laser pulse of frequency ω and electric-field envelope $\mathcal{E}(t)$. A third-order time propagation scheme is used:

$$\Psi(t + \Delta t) = \exp\left[-\frac{\mathrm{i}\Delta t w}{2}\right] \exp\left[\frac{\mathrm{i}\Delta t}{2}\left(\frac{\partial^2}{\partial x_1^2} + \frac{\partial^2}{\partial x_2^2}\right)\right] \exp\left[-\frac{\mathrm{i}\Delta t w}{2}\right] \Psi(t) , \quad (24.12)$$

$w = v + v_c$, where v_c is the electron-nuclear attraction in (24.10).

In order to properly separate single and double ionization, a large grid with $|x_i| \leq 1000$ a.u. is used. This allows us to capture 100% of the first ionized electrons and identify double ionization due to recollision processes. We now discuss an example for ionization of the ground $\mathrm{X}^1\Sigma_g^+$ state of H_2 at laser wavelength $\lambda = 800$ nm and intensities $10^{13} < I < 10^{15}$ W/cm^2. Single and double ionization probabilities, P^{+1} and P^{+2}, are obtained by numerical integration of the two-electron probabilities $|\Psi(x_1, x_2, t)|^2$, see (24.5a), (24.5b), and (24.7c), where the integration volume $\mathcal{V}$ is the region $|x_i| \leq 6$ a.u. The numerical integration procedure is verified by the procedure of Pindzola et al. [Pindzola 1997], which consists in projecting the total final wavefunction $\Psi(x_1, x_2, T)$ on complete sets of field free bound and continuum states of H_2^+.

Figure 24.8 shows a comparison of P^{+1} and P^{+2} obtained from the exact one-electron density, but evaluated using the independent-particle expressions (24.8), versus exact results. The probabilities for one-electron ionization, $H_2 \rightarrow H_2^+ + e^-$, and sequential double ionization, $H_2^+ \rightarrow H_2^{++} + e^-$, agree very well in the two different methods. Both P^{+2} by density and exact integration show a knee (plateau) at the saturation of the first ionization, $I \sim 10^{15}$ W/cm^2. The knee coincides with 100% ($P^{+1} = 1$) of the first ionization. The exact double ionization P^{+2} is parallel to P^{+1} at low intensities, confirming its source from recollision of the first electron with the ion core. The efficiency of double ionization through recollision is $P^{+2}/P^{+1} \sim 10^{-2}$. Notice that the ratio is underestimated at low intensities and overestimated in the knee region using the independent-particle formulas, thus confirming the necessity of introducing the correlation factor in (24.7a) and (24.7b).

Figure 24.9 compares exact probabilities versus TDEHF and ALDA (using a 3-D xc functional, since the system is intended to model a 3-D H_2 molecule). In the TDEHF method, one uses two different, non-orthogonal initial orbitals, $1\sigma_g$ and $1\sigma_g'$, which represent the propensity for electrons to be inequivalent in the ground state, i.e., inner and outer electrons. This distinction becomes amplified upon ionization and reflects better the physics than TDHF-like theories where both electrons are initially equivalent. Figure 24.9 shows that

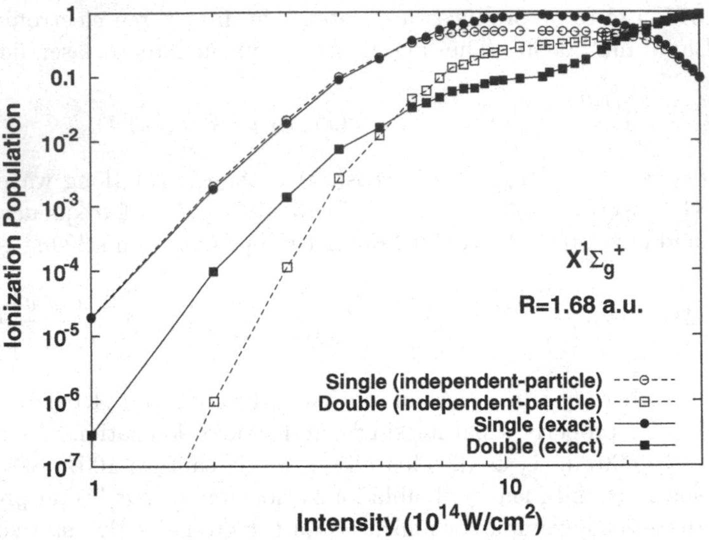

Fig. 24.8. Population of singly and doubly ionized states of 1-D H_2, $\lambda = 800\,\text{nm}$, comparing exact results and independent-particle results based on (24.8), but evaluated with the exact density [Bandrauk 2005b]

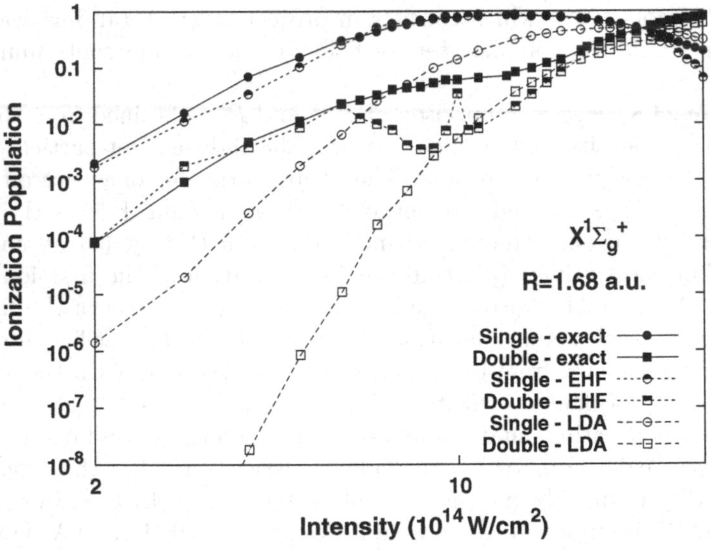

Fig. 24.9. Population of singly and doubly ionized states of 1-D H_2, $\lambda = 800\,\text{nm}$: exact results, extended Hartree-Fock (TDEHF) and ALDA

TDEHF agrees fairly well with the exact P^{+1} and P^{+2} below the knee. Anomalous behavior of the TDEHF probabilities appears in the knee region where the first electron has already ionized and cannot recollide. A similar behavior has been observed in 1-D models of He [Dahlen 2001, Dahlen 2002]. The ALDA single and double ionization are grossly underestimated, reflecting the improper asymptotic behavior of the electronic wavefunction and the ionization potential, as well as an insufficient screening. These results for H_2, as well as our discussion of MPI for atoms in Sect. 24.3, confirm once again the need for developing high-quality time-dependent xc functionals beyond the ALDA.

24.4.3 TDDFT for Molecules in Strong Fields

Even in 1-D, exact numerical calculations of strong-field molecular processes are presently limited to at most two electrons plus moving nuclei. There exist a number of recent studies using TDDFT and related theories to describe the dynamics of multielectron molecular systems.

Chu and Chu [Chu 2001a, Chu 2001b, Chu 2004b] studied HHG and MPI for various dimers with fixed nuclei in strong laser fields, using xc potentials with the correct asymptotic behavior. The role of the binding energy and orientation of individual molecular orbitals was explored, and a variety of correlation and interference effects between these orbitals in HHG spectra was discussed, as well as the role of inner valence electrons in determining the total ionization. Along similar lines, Dundas and Rost [Dundas 2005] used an all-electron, x-only ALDA approach to investigate the suppression of single ionization in N_2, O_2, and F_2 due to destructive interference of outgoing electron waves from the ionized electron orbitals. Overall, these studies again point to the insufficiency of the SAE in describing multielectron systems.

Baer et al. [Baer 2003] considered ionization and HHG in benzene by short circularly-polarized pulses, propagating 15 TDKS orbitals within ALDA, using a 3-D grid with pseudopotentials. An interesting interplay between bound-bound and bound-continuum transitions as well as multielectron dynamics was found to cause some unique features in the HHG spectra, specific to the ring geometry of the molecule and the circularly polarized light.

Several TDDFT studies have explored the route leading from molecules to clusters, e.g. using models based on chains of rare-gas atoms [Véniard 2002], or small silver molecules with fixed nuclear positions [Nobusada 2004]. A massively parallel TDKS calculations combining the dynamics of the highly excited electron cloud with the (classical) motion of the nuclei was carried out by Calvayrac et al. [Calvayrac 1998], simulating the Coulomb explosion of a Na_{12} cluster. A recent study of two-color photoionization in Na_4 and Na_4^+ clusters [Nguyen 2004] has found counterintuitive asymmetries and unexpectedly large plateaus in the ATI spectra. The subject of TDDFT for metal clusters is discussed in more detail in Chap. 26.

The dynamics of the nuclei has traditionally been treated by expanding the total molecular wavefunction in terms of Born-Oppenheimer states, and considering the lowest few potential energy surfaces. Fields of higher intensity require the inclusion of more and more electronic states, and calculations become increasingly time-consuming. Moveover, ionization processes are difficult to describe. Highly excited and ionized electrons have been treated under the condition of fixed nuclear positions, but this obviously cannot describe dissociation. Numerical studies of the strong-field ionization and dissociation of H_2^+ [Bandrauk 2003b] have shown the need for a unified treatment of the interplay between electronic excitations and nuclear vibrations. TDDFT faces severe challenges when it comes to a fully quantum mechanical, non-Born-Oppenheimer description of correlated electron and nuclear dynamics. First steps in this direction have recently been taken by Kreibich et al. [Kreibich 2001a, Kreibich 2001b, Kreibich 2003, Kreibich 2004]. A comparison [Kreibich 2004] of different forms of the electron-nuclear wavefunctions of H_2^+ shows that a simple Hartree mean-field approximation for the nuclear wavefunction is unable to describe nuclear dissociation processes, and much better results are achieved with a variational calculation based on a correlated ansatz for the electron-nuclear wavefunction. This is in line with other recent calculations for the ionization dynamics of multielectron systems based on TDEHF approaches [Dahlen 2001, Dahlen 2002, Kitzler 2004, Zanghellini 2004, Zanghellini 2005].

24.5 Conclusion and Perspectives

In this chapter, we have given a (by no means exhaustive) review of the rich phenomenology and the often counterintuitive effects of intense laser-matter interaction. Experimental and theoretical research continues to move at a rapid pace towards ever increasing intensities and decreasing pulse lengths, giving access to new regimes of electronic and nuclear dynamics. For example, Coulomb explosions of molecular clusters, induced by table-top lasers, have been observed to trigger nuclear fusion reactions [Ditmire 1999]. Another example: when an electron is suddenly removed, the remaining xc hole is filled on a 50-attosecond time scale, which should be a universal phenomenon [Breidbach 2005]. Thus, the nonlinear dynamics of atoms and molecules in short, intense laser pulses provides many fascinating and challenging applications for TDDFT.

Simple semiclassical models give a good intuitive, and often quantitative, understanding of many important strong-field phenomena, such as double photoionization via laser-induced recollision. Unfortunately, these simple pictures appear to be extremely hard to capture with traditional TDKS approaches such as ALDA, where a closed-shell atom or molecule is described by doubly occupied KS orbitals [Maitra 2002b]. Correlated-wavefunction methods such as TDEHF (i.e., using more than one determinant) allow here for more flexibility, as several benchmark calculations for small systems have shown.

The true power of TDDFT emerges when dealing with multielectron systems, and the available results for atoms and molecules in strong laser pulses are definitely encouraging. However, it is clear that we need xc functionals that do a better job in capturing highly correlated dynamical processes. Progress in this direction may come through xc functionals that are orbital-based (see Chap. 9), or functionals that depend on the current and include non-adiabatic memory effects (see Chaps. 5 and 8). Recent studies [Lein 2005, Mundt 2005] suggest that a TDDFT description of ionization requires a more careful treatment of the variations in the total number of particles, resulting in a time-dependent xc potential which changes discontinuously when the particle number passes through an integer value.

A particularly tough problem for TDDFT is the proper description of molecular dissociation and of CREI processes that occur when molecules are stretched. The problem of stretched H_2 has recently been addressed with an xc orbital functional based on the RPA [Fuchs 2005] (see Chap. 29).

Finally, even if one solves the TDKS equations with an extremely good xc functional, the problem remains that there are observables that are very hard to extract from the density, such as ionization probabilities or ATI photoelectron spectra. Lacking alternatives at present, the best we can do is to evaluate these observables using the TDKS orbitals, thus committing one of the four "deadly sins" of TDDFT [Burke 2005a]. We hope that future work in this direction will eventually lead to redemption, and thus to a successful TDDFT description of the many interesting processes of atoms and molecules in strong fields.

25 Highlights and Challenges in Strong-Field Atomic and Molecular Processes

V. Véniard

25.1 Introduction

The interaction of short and intense laser pulses with atoms has revealed a variety of interesting effects (see Chap. 24). Among them, accurate measurements of the multiple ionization yield of atoms have been shown to be orders of magnitude higher than predicted by the single active electron approximation [Fittinghoff 1992, Walker 1994]. The "knee" structure in the ion yield curve is interpreted as a nonsequential process, where the energy absorbed from the field is shared by the two electrons and considered as the most distinctive manifestation of electron correlations. Theoretical models based on simplified two-electron interaction [Watson 1994] or perturbative S-matrix approaches [Becker 1996] have approximately reproduced the experimental data. However, the importance of the time-dependent electron-electron correlation was emphasized recently through the numerical solution of the time-dependent Schrödinger equation of a one-dimensional two-electron model atom [Lappas 1998]. In principle, a direct comparison with the experimental data requires the solution of the three-dimensional time-dependent Schrödinger equation for the atom, but these calculations are numerically too demanding and results have been obtained only for a limited range of parameters in the case of helium [Parker 1998, Parker 2001]. Several approaches using time-dependent extended Hartree-Fock wave functions have been proposed in this framework [Pindzola 1995, Dahlen 2001]. However, TDDFT, which in principle allows one to incorporate correlation effects due to electron-electron interaction, appears to be the most valuable tool to study ionization of atoms or molecules in strong laser pulses. But, for the time being, it was shown that these calculations fail to reproduce the correct ionization dynamics in the low frequency regime, where the experiment was performed [Petersilka 1999].

In the framework of TDDFT, several types of approximations are made in the description of the interaction between atoms (or molecules) and an intense laser field. The most drastic is probably the adiabatic approximation. It amounts to neglecting any explicit time dependence of the xc potential. Then, $v_{\rm xc}$ depends on time only through the instantaneous density $n(\boldsymbol{r},t)$. This approximation can be critical for ultra-short laser pulses or for ultrafast processes. It also leads us (see Chap. 1) to use functional approximations

V. Véniard: *Highlights and Challenges in Strong-Field Atomic and Molecular Processes*, Lect. Notes Phys. **706**, 377–389 (2006)
DOI 10.1007/3-540-35426-3_25

that have been defined and optimized for the ground state of the system. Such functionals reproduce the properties of the ground state correctly but their validity has not been assessed in time-dependent cases, and perhaps other choices would be more appropriate. This point represents the second approximation. Finally, while in principle all the observables can be deduced from the density, their evaluation can be difficult in practice. For example, for the emission of radiation, such as high-order harmonic generation, the dipole can be easily expressed in terms of the density. However, the determination of the ionization rates is more complex. Even if good approximations are known to describe simple ionization, a correct evaluation of the double (or multiple) ionization rates remains problematic. Moreover, no functional exists for the determination of the kinetic energy spectra.

The description of the response of a helium atom subject to a strong infrared laser pulse is a remarkable example of the problems raised by these three approximations. Furthermore, this case is particularly instructive, as an exact calculation, involving two active electrons, can be performed and used as a test case to asses the validity of the TDDFT approach. In this framework, Petersilka and Gross [Petersilka 1999] have shown that the nonsequential double ionization was not accurately described. By comparing the predictions of TDDFT with those deduced from a fully-correlated two-electron calculation, they showed that the approximation for $v_{\mathrm{xc}}[n]$ and for the ionization yields were both crucial.

For several years, the only quantity accessible in experiments has been the ionization yield. Likewise, theoretical calculations have concentrated on this yield. However, this does not allow very accurate comparisons between theoretical calculations and experimental results or among different theories. Kinetic-energy photoelectron spectra can give more insight on the dynamics of the processes [Lafon 2001] or more accurate information into the approximations involved in calculations. However, while single active electron simulations can provide the electron spectra, it is not straightforward to do so with other approaches. The main reason is that, in principle, one needs the true wave function of the system at the end of the pulse to extract the energy of the photoelectrons. DFT, as well as extended Hartree Fock approximations [Dahlen 2001], involve orbitals having no physical meaning, which cannot be used, as they stand, to calculate transition probabilities.

The purpose of this chapter is twofold. First, we briefly describe a method to calculate electron energy spectra, based on the electron density only. This will allow us to analyze in detail the many-electron dynamics as described through TDDFT. We will use a simplified one-dimensional model atom which has proven to be able to reproduce most of the features of the real 3-D atom and provides a good understanding of the many-particle system. Furthermore, for the two-electron case, the fully correlated wave function can be propagated in time, allowing for comparisons with simplified approaches.

The organization of this chapter is as follows: in Sect. 2 we will briefly describe the method used to calculate electron energy spectra. In Sect. 3, we will use these spectra to compare the results obtained in the framework of TDDFT to those obtained with an exact 2-electron simulation for the one-photon double ionization of helium. A model for a lithium atom will be proposed in Sect. 4 and single ionization will be analyzed. Finally, we will show in Sect. 5 that molecular dynamics can be used as a valuable tool to study the ionization process. A brief conclusion will be given in Sect. 6.

25.2 Determination of the Spectra

According to classical mechanics, an electron at time $t = 0$ at the position $x = 0$, with an energy E, will be found at $x = \Delta T\sqrt{2E}$ after propagating freely during ΔT. The quantum analog will be described through a wave packet, initially localized in $x = 0$ at $t = 0$ with a mean momentum k and $E = k^2/2$:

$$\Psi(x, t = 0) = A\mathrm{e}^{\mathrm{i}kx}\mathrm{e}^{-x^2/a^2} , \tag{25.1}$$

where a is the spatial width of the wave packet. Thus, we define the quantity

$$P(E, \Delta E, \Delta T) = \int_{x_<}^{x_>} \mathrm{d}x \; n(x, \Delta T) + \int_{-x_>}^{-x_<} \mathrm{d}x \; n(x, \Delta T) , \tag{25.2}$$

where $n(x, t)$ is the electronic density $n(x, t) = |\psi(x, t)|^2$ and $x_<$ and $x_>$ are defined as

$$x_< = \Delta T\sqrt{2(E - \Delta E)} \tag{25.3a}$$

$$x_> = \Delta T\sqrt{2(E + \Delta E)} . \tag{25.3b}$$

$P(E, \Delta E, \Delta T)$ can be identified as the probability that the energy of the electron, described by (25.1), lies in the interval $[E - \Delta E, E + \Delta E]$. In the limit of large values for ΔT, $P(E, \Delta E, \Delta T)$ depends only slightly on ΔT.

If different momenta, k_1 and k_2, are present in the wave function, as in

$$\Psi(x, t = 0) = A_1\mathrm{e}^{\mathrm{i}k_1x}\mathrm{e}^{-x^2/a_1^2} + A_2\mathrm{e}^{\mathrm{i}k_2x}\mathrm{e}^{-x^2/a_2^2} , \tag{25.4}$$

the wave function has to be propagated in time to let the different components separate spatially, and the following prescription can be given for the time interval ΔT:

$$\Delta T \ll \frac{a_1 + a_2}{|k_1 - k_2|} . \tag{25.5}$$

To validate this method in more detail, we have compared the density-based spectra, obtained from the photoionization of atoms in the presence of a laser field, with those derived from a standard spectral analysis [Véniard

2003]. This comparison shows that an excellent agreement between the two methods can be expected and that the sensitivity is even improved for the density-based spectra. Note that, as the photoelectrons are emitted during the whole duration of the laser pulse T_{pulse}, it can be shown that a lower bound for the energy resolution is given by

$$\frac{\Delta E_{\text{min}}}{E} \approx \frac{T_{\text{pulse}}}{\Delta T} \,. \tag{25.6}$$

More details about the method can be found in [Véniard 2003].

25.3 One-Photon Double Ionization of Helium

In this section, we compare results obtained by solving the TDKS equations in the adiabatic approximation with the fully correlated Schrödinger equation for a model 1-D 2-electron atom submitted to a laser pulse. This one-dimensional model atom has proven to be able to reproduce most of the features of the real 3-D atom and provides a good understanding of the many-particle system.

In the high-frequency range, and for moderate laser intensity, the simple analysis of the ionization probability as a function of the intensity (not shown here) does not lead to significant differences for the two simulations. However, when the laser frequency is above the two-electron threshold, more insight can be gained when looking at the double ionization process.

The TDKS equation for the orbital $\varphi(x,t)$ has been solved in the adiabatic approximation. The external potential,

$$v_{\text{ext}}(x,t) = x\mathcal{E}(t)\sin(\omega t) - \frac{Z}{\sqrt{a^2+x^2}} \,, \tag{25.7}$$

contains two parts: one is the the static Coulomb interaction with the nucleus, while the other is the time-dependent coupling with the electric field $\mathcal{E}(t)\sin(\omega t)$ of the laser, where $\mathcal{E}(t)$ denotes the temporal laser-pulse shape and ω is the laser frequency. The electron-electron interaction $v_{\text{ee}}(x)$ reads, for a one-dimensional system,

$$v_{\text{ee}}(x) = \frac{1}{\sqrt{b^2+x^2}} \,. \tag{25.8}$$

The values of the adjustable parameters a in (25.7) and b in (25.8) are determined numerically in order to reproduce the relevant physical properties for the "atom" considered as closely as possible [Javanainen 1988, Su 1991].

The xc potential $v_{\text{xc}}(x,t)$ has been approximated by

$$v_{\text{xc}}(x,t) = -\frac{1}{2}\int \mathrm{d}x'\, n(x',t)v_{\text{ee}}(|x-x'|) + v_{\text{c}}(x,t) \,, \tag{25.9}$$

where the exchange part is evaluated exactly as it corresponds to the restricted Hartree-Fock exchange potential, and we have used the correlation potential $v_c(x,t)$ originally proposed by Becke [Becke 1988a].

The orbital φ is represented on a grid in coordinate space. The initial state is obtained by solving iteratively the TDKS equation in imaginary time. The propagation in time is done through a unitary Peaceman-Rachford propagator [Kulander 1992a, Kulander 1992b] (see Chap. 12). The photoelectron spectrum is then extracted from the time-dependent density,

$$n(x,t) = 2|\varphi(x,t)|^2 . \tag{25.10}$$

For comparison, we have also solved numerically the TDSE for a one-dimensional 2-electron system [Pindzola 1991, Grobe 1992]

$$\mathrm{i}\frac{\partial}{\partial t}\psi(x_1,x_2,t) = [H_0 + (x_1+x_2)\mathcal{E}(t)\sin(\omega t)]\psi(x_1,x_2,t) , \tag{25.11}$$

where

$$H_0 = \frac{1}{2}p_1^2 + \frac{1}{2}p_2^2 + v_{\mathrm{en}}(x_1) + v_{\mathrm{en}}(x_2) + v_{\mathrm{ee}}(|x_1 - x_2|) , \tag{25.12}$$

and v_{en} denotes the electron-nuclear interaction.

In this numerical simulation, we have not used the electronic density to calculate the photoelectron energy spectrum: it was not possible, for numerical reasons, to propagate the fully correlated wave function after the end of the pulse for a sufficiently long time. As a consequence, the energy resolution was not very good. Instead, to determine the electron energy, the final wave function was divided into three parts, see Fig. 25.1. The ground-state contribution corresponding to $|x_1|, |x_2| < X_c$ was first removed. The two other parts, ψ_1, corresponding to $|x_1| < X_c, |x_2| > X_c$ or $|x_1| > X_c, |x_2| < X_c$ and ψ_2, corresponding to $|x_1| > X_c$ and $|x_2| > X_c$ were Fourier transformed:

$$\bar{\psi}_i(k_1,k_2) = \int \mathrm{d}x_1 \int \mathrm{d}x_2 \, \mathrm{e}^{\mathrm{i}(k_1x_1+k_2x_2)} \, \psi_i(x_1,x_2,T_{\mathrm{pulse}}) . \tag{25.13}$$

Note that ψ_1 is identified as that part of the wave function corresponding to single ionization (one electron is far from the nucleus while the other one remains close), whereas ψ_2 corresponds to double ionization (the two electrons have moved far from the nucleus). The probability that one of the two electrons has the energy E_1 is then defined as [Grobe 1992]

$$P_i(E_1) = \frac{1}{\sqrt{2E_1}} \int \mathrm{d}k_2 \, |\bar{\psi}_i(k_1,k_2)|^2 . \tag{25.14}$$

P_1 (P_2) corresponds to a single (double) ionization process. We have checked that the results depend only slightly on the numerical value chosen for X_c.

The photoelectron spectra are displayed in Fig. 25.2, corresponding to $a = 1$, $Z = 2$ in (25.7) and $b = 1$ in (25.8). The ground-state energy is

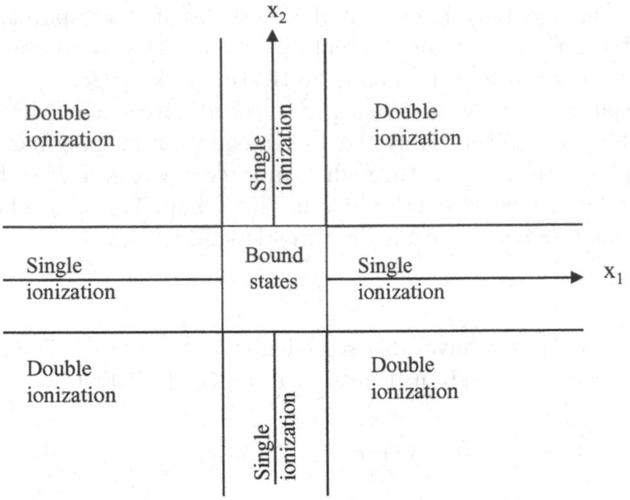

Fig. 25.1. Spatial partition of the final wave function at the end of the pulse. The central part of the figure $|x_1|, |x_2| < X_c$ corresponds to bound states. If $|x_1|$ and $|x_2| > X_c$, then the two electrons are ionized. The remaining part corresponds to single ionization

$E_i = -2.25$ a.u. and the first ionization threshold is $E_{\text{ion}} = 0.75$ a.u. The photon energy is $\omega = 3$ a.u., which exceeds the two-electron threshold, and the laser intensity is $I = 10^{14}\,\text{W/cm}^2$. The pulse shape has a gaussian envelope $\exp[-(t/\tau)^2]$ with $\tau = 100$ optical cycles. The results corresponding to the 2-electron simulation are shown in Fig. 25.2a, where the upper curve represents the single ionization $P_1(E)$. The main peak, located at $E = 2.25$ a.u., corresponds to direct single ionization, the ion being left in its ground state. The peaks seen at lower energies can be interpreted as the ionization of one electron and excitation of the other one, as already found in [Grobe 1992]. The lower curve shows the double ionization of the model atom, as derived from the evaluation of P_2. Here the structureless feature visible in the low energy range ($E < 0.75$ a.u.) results from the one-photon double ionization. By varying the laser field intensity, we have checked that all these features scale linearly with the intensity, as expected from a one-photon process in the perturbative regime. The magnitudes of some of the peaks at higher energy in P_2 depend on the value chosen for X_c. In fact, these peaks are reminiscent of the single ionization with excitation, which is not completely removed from ψ_2. Moreover, the magnitude of the peak located at $E = 2.25$ a.u. scales with I^2 and thus can be interpreted as resulting from a two-photon process [Grobe 1992].

In Fig. 25.2b, we show the results coming from TDDFT for the same set of parameters and corresponding to the total ionization $P_1 + P_2$. Only two peaks are visible in this spectrum. The first one, located at $E = 2.25$ a.u.,

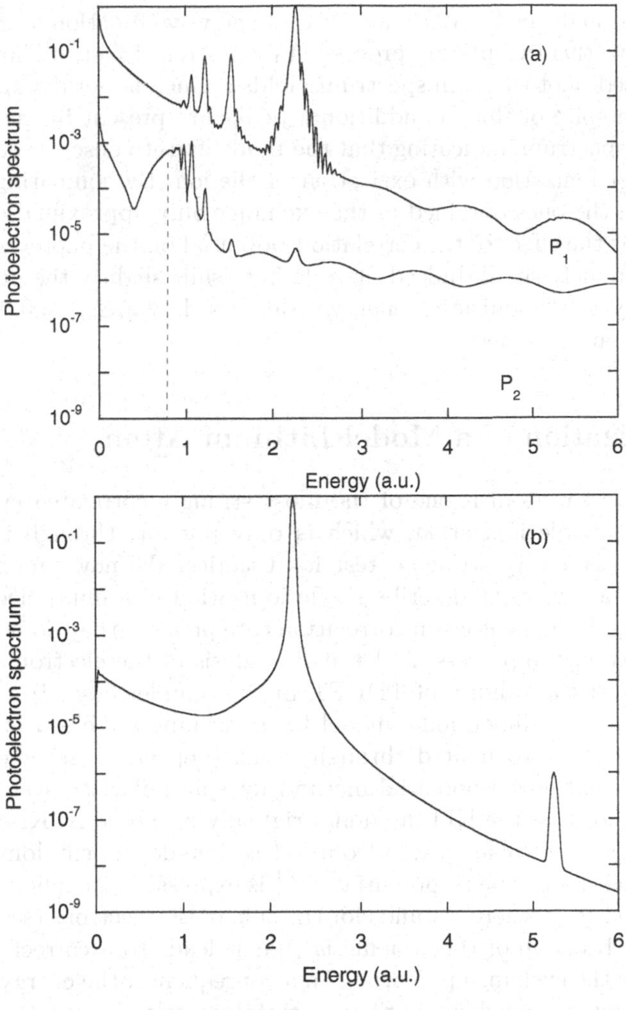

Fig. 25.2. Photoelectron spectrum for a one-dimensional 2-electron atom (ground-state energy $E_i = -2.25$ a.u., first ionization threshold $E_{\mathrm{ion}} = 0.75$ a.u.). Panel (**a**) shows results of a fully correlated 2-electron simulation followed by Fourier analysis of the wave function at the end of the pulse. P_1 and P_2 correspond to single and double ionization. The *dashed* line indicates the maximum energy an electron can have when ejected through a one-photon double ionization process. Panel (**b**) shows the total ($P_1 + P_2$) ionization resulting from a TDDFT simulation. The photon frequency is $\omega = 3$ a.u. and the laser intensity is $I_L = 10^{14}$ W/cm^2

corresponds to the one-photon single ionization, with the ion left in its ground state. An interesting feature is the fact that a second peak is present, visible in the high-energy part at $E = 5.25$ a.u., corresponding to the two-photon single ionization. The slow decrease in the high-frequency range of the spectrum, obtained through the Fourier transform of the wave function, does not allow us to observe this two-photon process. As a matter of fact, it shows that the density-based photoelectron spectrum yields a much larger dynamical range. However, in spite of this, no additional peaks are present in the low-energy part of the spectrum, indicating that the model fails to describe either double ionization or ionization with excitation of the ion. By comparing the above results with the ones obtained in the exchange-only approximation, we have checked that the effect of the correlation potential on the photoelectron spectrum is extremely small. Indeed, its role is to shift slightly the peak towards lower energy, as the ionization energy is decreased by $\Delta E_{\rm ion} \approx 0.04$ a.u. when correlations are included.

25.4 Ionization of a Model Lithium Atom

The two-electron atom is one of the most strongly correlated systems, and one-photon double ionization, which is only possible through the electron correlations, is a very stringent test for theories. We now present a model atom which allows us to describe single ionization in a multi-electron atom. In principle, although electron correlations are present, they do not play such a crucial role in the process. A detailed analysis of the electron spectra can help us to test the validity of TDDFT in this simpler case. To this end, we now turn to a one-dimensional model for describing a 3-electron atom. The xc potential is approximated through a time-dependent scheme, based on the so-called optimized potential method for spin-polarized systems. In this approach, known as the KLI method, originally proposed by Krieger, Li and Iafrate [Krieger 1992b] and extended into the time-dependent domain [Ullrich 1995a, Ullrich 1996], the xc potential $v_{\rm xc}^{\rm KLI}$ is expressed as explicit functionals of the orbitals $\varphi_{j\sigma}$, where σ stands for the spin of the electrons (see Chap. 24). The main advantage of this scheme is that it leads to a correct asymptotic behavior for the exchange potential. As a consequence, the energy eigenvalue of the highest occupied Kohn-Sham orbital corresponds to the ionization energy of the system, satisfying Koopmans' theorem [Krieger 1992b]. As in the preceding section, the adiabatic approximation is used to incorporate the time dependence in the exchange potential.

In the model developed here, we have two electrons in a spin-up state and one electron in a spin-down state. Note that when only one electron is present in a spin state, $v_{\rm xc}^{\rm KLI}$ reduces to the restricted Hartree-Fock exchange potential.

We have solved numerically the three TDKS equations on a grid and analyzed the electronic density. In the following, we will be concerned only with single ionization processes. To this end, the laser frequency will be chosen between the one-electron and two-electron threshold energies and we will restrict ourselves to an intensity regime where sequential and non-sequential multiple ionization rates are expected to be small. A photoelec-

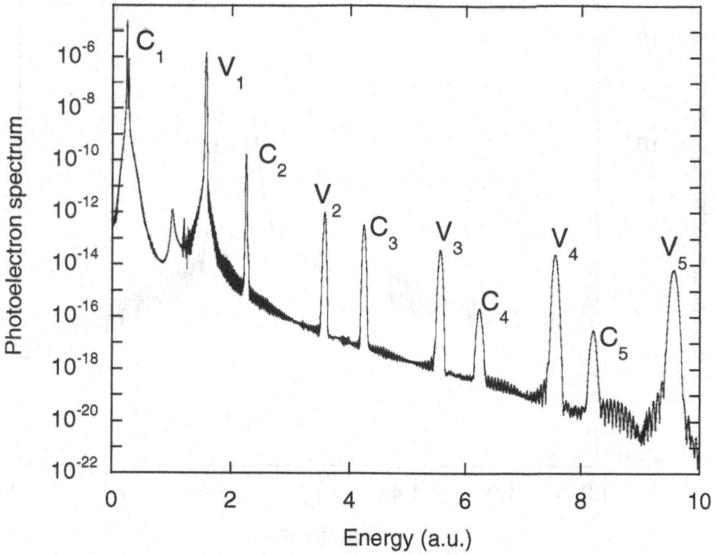

Fig. 25.3. Photoelectron spectrum for a one-dimensional 3-electron atom. The photon frequency is $\omega = 2$ a.u. and the laser intensity is $I_L = 10^{11}$ W/cm^2. The peaks labelled (**C**) correspond to the ionization of the core electrons, while the peaks labelled (**V**) come from the valence electron

tron spectrum is displayed in Fig. 25.3 for the set of parameters: $a = 0.6$ a.u., $b = 0.6$ a.u., $\omega = 2$ a.u., and $I = 10^{11}$ W/cm^2. The pulse has a gaussian envelope $\exp[-(t/\tau)^2]$ with $\tau = 100$ optical cycles. Two series of peaks can be identified in the spectrum, each of them consisting of peaks separated by the photon energy ω. The ones labelled C_N correspond to the ionization of the core electrons through the absorption of N photons, while the peaks labelled V_N correspond to the same process for the valence electron. One can check, by comparing the magnitude of the peaks C_1 and V_1, that in this high-frequency regime, as expected, it is easier to ionize the core electrons than the valence electron.

We have varied the laser field intensity, and the behavior of the peak V_1 is shown in Fig. 25.4. As the magnitude of this peak, corresponding to a one-photon absorption process, scales linearly with intensity, we present only normalized spectra (divided by the intensity). On can see that the shape of the peak changes with intensity: while the high-energy part is not sensitive to the laser intensity, the peaks broaden towards low energies. Note that the ponderomotive energy $U_p = I/4\omega^2$ is very low ($<10^{-4}$ a.u.), so no significant ponderomotive shift should be expected in these spectra [Bucksbaum 1987]. This broadening can be interpreted as an artefact due to the poor approximation of the xc potential we have chosen here. In the ground state, the electronic density n is strongly localized around $x = 0$. Therefore the electron-

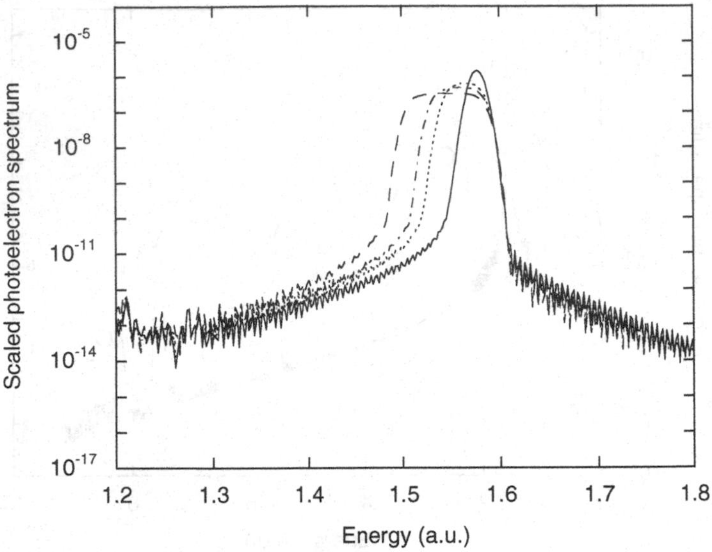

Fig. 25.4. One-dimensional 3-electron atom: behavior of the peak labelled V_1 in the photoelectron spectrum for various laser intensities: $I_L = 10^{11}\,\mathrm{W/cm^2}$ (*thin line*), $I_L = 2 \times 10^{13}\,\mathrm{W/cm^2}$ (*dotted line*), $I_L = 3 \times 10^{13}\,\mathrm{W/cm^2}$ (*dash-dotted line*), $I_L = 5 \times 10^{13}\,\mathrm{W/cm^2}$ (*dashed line*). The photon frequency is $\omega = 2$ a.u

electron repulsion $\int \mathrm{d}x'\, n(x') v_{\mathrm{ee}}(|x - x'|)$ partially compensates (screens) the electron-nuclear interaction v_{en} in the Kohn-Sham equations. As ionization takes place, the orbitals φ_i $(i = 1 - 3)$ start to delocalize, the screening decreases during the laser pulse and the energy of the photoelectrons is shifted toward lower energy as a function of time. We have verified that, if the density used to determine the electron energy is recorded only during the last part of the pulse, only the low energy part of the peak remains, as presented in Fig. 25.5 for $I = 5 \times 10^{13}\,\mathrm{W/cm^2}$. This shows that, even in the perturbative regime, the atom is more difficult to ionize at the end than at the beginning of the pulse. This feature conflicts with the Fermi golden rule, which states that for a given intensity, the ionization probability is constant in time. Note that an exact calculation, as shown in the preceding section, would give sharp peaks.

25.5 Dynamics of an H_2 Molecule in a Strong Laser Field

Finally, we discuss the interaction between molecules and intense laser fields, which leads to a wide range of phenomena: as compared to atoms, molecules present additional degrees of freedom through the nuclear motion and specific features, such as dissociation or Coulomb explosion, can

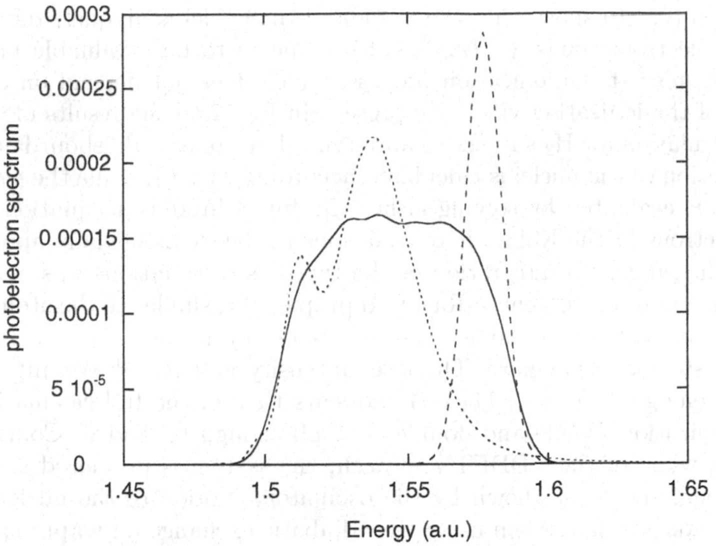

Fig. 25.5. Peak labelled V_1 in the photoelectron spectrum when the electron density is recorded during the whole duration (*thin line*) or during the last half (*dotted line*) of the pulse. The photon frequency is $\omega = 2\,\text{a.u.}$ and the laser intensity is $I_L = 5 \times 10^{13}\,\text{W/cm}^2$. The results obtained for $I_L = 10^{11}\,\text{W/cm}^2$ (*dashed line*) are also shown for comparison

be observed. The difficulty inherent to the theoretical description of these processes is increased, and simplified approaches have been proposed [Giusti-Suzor 1995, Chelkowski 1992, Zuo 1996c]. However, we mention that recently a semi-classical approach, reminiscent of Molecular Dynamics treatments of complex systems, has been proposed to address several questions related to the dynamical response of H_2^+ and H_2 molecules to a strong and short laser pulse, [Rotenberg 2002, Ruiz 2005]. One of the main results of this study has been to show that it is essential to include the motion of the nuclei to account for the details of the dynamics and to discuss the relative probabilities of excitation, dissociation and Coulomb explosion, see also [Feuerstein 2003].

Within the semi-classical approach, the motion of the nuclei is governed by the classical equations:

$$M\frac{\mathrm{d}^2\boldsymbol{R}_\alpha}{\mathrm{d}t^2} = \boldsymbol{F}_{\beta\alpha}(t) + \boldsymbol{F}_{\mathrm{e},\,\alpha}(t) + \boldsymbol{F}_{\mathrm{laser}}(t)\,, \tag{25.15}$$

where $M = 1836\,\text{a.u.}$ is the proton mass. The force $\boldsymbol{F}_{\beta\alpha}(t)$ stands for the interaction between the nuclei α and β, $\boldsymbol{F}_{\mathrm{e},\,\alpha}(t)$ for the interaction electron-nucleus α and $F_{\mathrm{laser}}(t)$ for the interaction nucleus-laser. One has:

$$\boldsymbol{F}_{\mathrm{e},\,\alpha}(t) = \int \mathrm{d}^3r\,\frac{[\boldsymbol{R}_\alpha(t) - \boldsymbol{r}]n(\boldsymbol{r},t)}{|\boldsymbol{R}_\alpha(t) - \boldsymbol{r})|^3}\,. \tag{25.16}$$

Equation (25.16) shows that the motion of the nuclei is uniquely determined by the electronic density $n(\boldsymbol{r}, t)$ and it appears to be a valuable probe for the dynamics of the ionization process, which does not depend on the definition of the ionization yield. We present in Fig. 25.6 the results of a model 1-D calculation for H_2 subject to an intense laser pulse, [Boisbourdain 2005]. The motion of the nuclei is calculated according (25.16), while the electronic density is evaluated by solving either the full Schrödinger equation for the two electrons or the Kohn-Sham equations in the exchange-only approximation. The numerical integration of the two sets of equations was performed through a unitary Peaceman-Rachford propagator, similar to the atomic case. We have used a trapezoidal laser pulse with one cycle linear turn-on and turn-off and six cycles flat part. The laser intensity is $2 \times 10^{14}\text{W/cm}^2$ and the photon energy is $\hbar\omega = 1.17\,\text{eV}$. It appears that in the full calculation, the total ionization (single and double) is high enough to lead to Coulomb explosion, while in the TDDFT approach, the system is promoted to excited vibrational states, as shown by the oscillatory motion of the nuclei. These results indicate that, when using the adiabatic exchange-only approximation for the Kohn-Sham potential, ionization (single and/or double) is strongly underestimated. These results have been obtained in the framework of the semi-classical approximation, but it was shown that this approximation is valid in a wide range of laser frequency and intensity [Ruiz 2005].

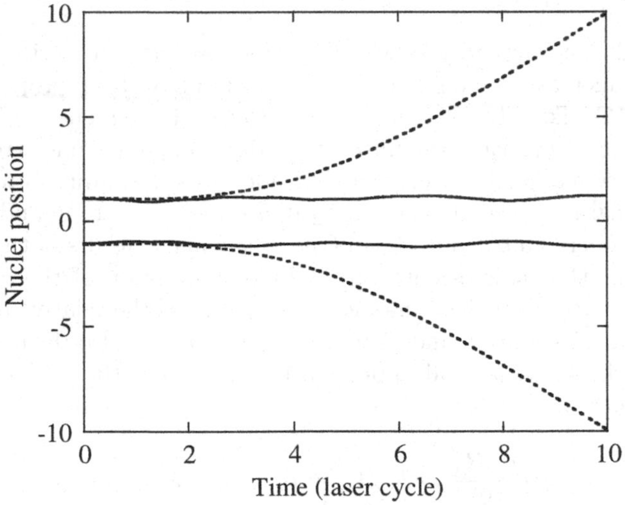

Fig. 25.6. Time-dependent nuclear positions of a 1-D H_2 molecule subject to a laser pulse of intensity 2×10^{14} W/cm^2 and photon energy $\hbar\omega = 1.17\,\text{eV}$. *Broken line*: full two-electron calculation. *Full line*: x-only TDDFT

25.6 Conclusion

The results presented here show that several problems arise when describing the interaction of a multi-electron system with a strong laser field. We have shown that among the approximations involved in the TDDFT calculations, the determination of a correct time-dependent xc potential is crucial. Up to now, it appears that in the high-frequency range, one-photon single ionization, which is the dominant process, is correctly described in the perturbative regime. However, TDDFT still fails to describe more complex processes, such as single-photon double ionization or excitation-ionization. A careful analysis of the photoelectron spectra has shown that a broadening of the peaks appears when increasing the laser intensity. The broadening has no physical meaning, and this artefact demonstrates that the dynamics of the ionization process is not properly described. In the low-frequency range, it is difficult to disentangle single and multiple ionization, without any further approximations on the observables. Furthermore, it has been shown in molecular calculations, where the motion of the nuclei can be used as a probe of the ionization process, that the description of the dynamics of the processes is not correctly reproduced. Most of the results presented here have been obtained in the exchange-only limit, but we have checked that including correlation, see for instance [Perdew 1992a], does not improve significantly the results. We believe that the most critical point is the adiabatic approximation and that substantial improvements are required in the construction of new functionals for time-dependent problems [Lein 2005].

26 Cluster Dynamics in Strong Laser Fields

P.-G. Reinhard and E. Suraud

26.1 Introduction

This article deals with applications of time-dependent density functional theory (TDDFT) to metal clusters. The field of cluster physics is only about three decades old and TDDFT was a key tool in the theoretical description from the beginning. We think, e.g., of the pioneering calculations of [Beck 1984, Ekardt 1984] which provided a fully microscopic description of the dominant Mie plasmon resonance in metal clusters on the grounds of the time-dependent local density approximation (TDLDA). Since then, TDDFT has been widely used as a major tool for computing both the structure and dynamics of clusters, side by side with more macroscopic approaches [Kreibig 1993] and with fully fledged ab initio quantum chemical methods [Bonačić-Koutecký 1991]. The early TDDFT treatments dealt with optical response and used, in fact, a linearized TDLDA. A next step in development went to fully fledged TDLDA propagated in the time domain [Calvayrac 1995, Saalmann 1996, Yabana 1996], first, as an efficient alternative to compute optical response in complex geometries, and second, as a necessary basis for non-linear dynamics. In a final step, the simultaneous propagation of ionic motion at the level of molecular dynamics (MD) was included for the description of strongly excited (non-adiabatic) cluster dynamics yielding together TDLDA-MD [Calvayrac 1998, Kunert 2001]. Highly excited clusters allow a semi-classical approximation in terms of the Vlasov equation yielding Vlasov-LDA where LDA in the naming indicates that the Vlasov mean field is computed in LDA [Gross 1995b, Feret 1996]. This, in turn, allows the description of dynamical electron correlations through a collision term (Vlasov-Ühling-Uhlenbeck method, VUU) [Domps 1998]. The majority of applications, however, deals with standard quantum-mechanical TDLDA.

It becomes obvious when reading this volume that recent developments of TDDFT are focussed on going beyond TDLDA, e.g., by providing a better description of exchange (see Chap. 9). However, cluster physicists are reluctant to adopt more involved schemes. Clusters are already rather complex objects and call for simple methods. In that spirit, one crucial and still simple extension has found its way into many applications. It is the self-interaction correction (SIC) which tries to cure the worst deficiencies of LDA

P.-G. Reinhard and E. Suraud: *Cluster Dynamics in Strong Laser Fields*, Lect. Notes Phys. **706**, 391–406 (2006)
DOI 10.1007/3-540-35426-3_26

for exchange [Perdew 1981]. An early adaption for linear response is found in [Pacheco 1992a]. It becomes unavoidable for a proper dynamical computation of electron emission. The then emerging problems with non-hermiticity can be overcome by combination with the optimized-potential method (OPM) [Ullrich 1995a] or proper averaging [Ullrich 2000c, Legrand 2002]. These methods, TDLDA or TDLDA-MD, possibly augmented with SIC, provide a reliable, robust and efficient tool for large simulations of various dynamical scenarios in cluster physics. For a review on non-linear dynamics see [Calvayrac 2000]. A very broad discussion of cluster dynamics including due reference to experiments and competing methods can be found in our book [Reinhard 2003].

This contribution aims to demonstrate the capabilities of fully fledged TDLDA(-MD) with several examples of non-linear dynamics. We consider in Sect. 26.3 the evaluation of observables from electron emission, namely the photo-electron spectra (PES) and angular distributions. For the PES we have chosen a test case where the collectivity of the Mie plasmon mode which is typical for metal clusters plays a role. For the angular distributions, we demonstrate the complexity of clusters, which is enhanced by thermal shape fluctuations. Section 26.4 discusses one scenario for pump and probe analysis of ionic motion in metal clusters which takes advantage of the dominance of the Mie plasmon. It also demonstrates the capability of TDLDA-MD to perform large scale calculations over long time intervals. A summary of the necessary formal framework is given before all that in Sect. 26.2. This is kept very short because formal and numerical aspects are discussed at several other chapters. More details specific to our treatment can be found in [Calvayrac 2000, Reinhard 2003].

26.2 Formalities

26.2.1 Coupled Ionic and Electronic Dynamics

We describe the cluster dynamics at the level of the TDLDA coupled to ionic molecular dynamics. As this is a much discussed topic in this book, we can keep the formal presentation short. Figure 26.1 summarizes the input and emerging equations of motion. Everything is specified with defining the total energy of the system. The electrons are described quantum mechanically by single-electron wavefunctions $\varphi_i(\boldsymbol{r}, t)$. Their energy is composed by the kinetic energy $\propto |\nabla\varphi_i|^2$, the direct part of the Coulomb energy $E_{\rm H}[n]$ and exchange as well as correlations in LDA. Actually, we are using the exchange-correlation functional of [Perdew 1992a]. The description of electron emission requires taking into account a correction for the self-energy error in the LDA. This will be discussed in more detail in Sect. 26.2.2. The ions are described as classical particles with their kinetic energy given through the velocities $\dot{\boldsymbol{R}}_I$. The ionic potential energy is summarized in $E_{\rm ion}$. The ion-ion interaction is taken as the

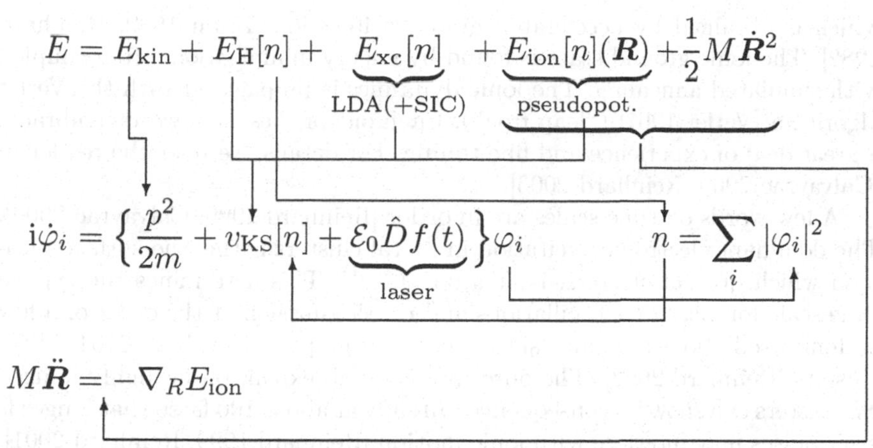

Fig. 26.1. Short summary of the energy functional of TDLDA-MD and the emerging coupled equations of motion for the occupied single-electron wave function $\varphi_i(\boldsymbol{r}, t)$ and the ionic coordinates $\boldsymbol{R}_\alpha$

Coulomb interaction between point particles. On the electronic side, only the valence electrons in each atom are considered. The coupling between ions and these valence electrons is mediated through pseudopotentials which effectively account for the inert core electrons [Reinhard 2003]. Simple metals are here particularly convenient, as they allow the use of purely local pseudopotentials for which we use a particularly soft form, appropriate for efficient numerical handling on coordinate space grids [Kümmel 1999].

Once the total energy is specified, the coupled equations of motion follow variationally in a straightforward manner, as indicated in Fig. 26.1. The dynamical excitation through a laser is mediated through an external time-dependent dipole field added to the Kohn-Sham equations. Its temporal profile is taken as $f(t) = \cos(\omega_{\rm phot}t)\sin^2(t\pi/T_{\rm pulse})$ in the interval $0 \le t \le T_{\rm pulse}$. The $\sin^2$ envelope is a smooth way to switch the laser in a finite time interval. The full width at half maximum (FWHM) for the field strength is $T_{\rm pulse}/2$ and for the field intensity about $T_{\rm pulse}/3$. We will refer in the following to $T_{\rm pulse}/2$. The field strength $\mathcal{E}_0$ determines the amplitude of the external field. It is related to the laser intensity as $I \propto \mathcal{E}_0^2$. We consider here medium strong fields in the range around $I = 10^{10}$ W/cm^2 equivalent to electric fields of $\mathcal{E}_0 = 0.0006$ a.u., as they are typically realized by fs lasers. These fields excite metal clusters far above the linear regime but safely below immediate destruction [Calvayrac 2000, Reinhard 1998].

The numerical solution employs a representation of the electronic wavefunctions on a coordinate space grid. We consider three dimensional Cartesian grids as well as an axially averaged approximation to the electronic mean fields [Montag 1995b]. The dynamical propagation is done with the time splitting technique [Feit 1982b]. The initial condition is the ground state which is obtained by accelerated gradient iteration [Blum 1992, Reinhard 1982]. The ionic ground state is found by energy minmization, often coupled with simulated annealing. The ionic dynamics is propagated with the Verlet algorithm [Verlet 1967] ("leap frog"). Each one of these key words embraces a great deal of experience and fine tuning. For details, we refer the reader to [Calvayrac 2000, Reinhard 2003].

A few words on time scales are in order [Reinhard 1999, Calvayrac 2000]. The dominant electronic excitation in metal clusters is the Mie surface plasmon which, for Na clusters, is at around 3 eV. This determines the typical time scale for electronic oscillations and direct emission at the order of a few fs. Ionic oscillations are much slower, e.g., with a period of about 250 fs in Na clusters [Reinhard 2002]. The onset of Coulomb expansion in highly ionized Na clusters can show its consequences already at about 100 fs, so that longer fs laser pulses may interfere with ionic motion [Reinhard 1999, Reinhard 2001]. In order to keep processes separated, we confine the studies on electron emission in Sect. 26.3 to short laser pulses with FWHM $\approx$ 50 fs which allows us to ignore the ionic motion. The full ion-electron coupling is taken up in Sect. 26.4, where we discuss pump and probe analysis of cluster dynamics.

26.2.2 Self Interaction Correction

The dynamical modeling of electron emission requires that the single-electron levels in the Kohn-Sham ground state have the correct energies relative to the continuum threshold. This is violated by the LDA due to what is called the self-interaction error. Consider, e.g., a neutral cluster. The density entering the Kohn-Sham equations is the total density summed over all electrons, which neutralizes all positive ionic charges. Thus the asymptotic density falls off exponentially. But one electron when departing should see asymptotically the Coulomb field of the remaining positive charge. The contribution of the "observer" electron has to be discounted from the density before computing the Kohn-Sham field for that electron. This is automatically done in the exact exchange functional and that is one of the reasons for the great interest in reintroducing exact exchange into DFT, as documented in several contributions to this volume.

Large scale calculations call for simple solutions, and one of them is provided by the self-interaction correction (SIC) [Perdew 1981]. The idea is simply to subtract all self-interactions in the electronic Hartree, exchange, and correlation energy $E_{\rm Hxc} = E_{\rm H} + E_{\rm xc}$ by modifying it to

$$E^{\rm SIC}_{\rm Hxc}[n] = E_{\rm Hxc}[n] - \sum_i E_{\rm Hxc}[n_i]\,, \quad n = \sum_i n_i\,, \quad n_i = |\varphi_i|^2\;. \tag{26.1}$$

Although conceptually simple, the SIC leads to state dependent Kohn-Sham potentials which cause technical complications, particularly in fully dynamical calculations. With the help of the optimized potential method (OPM) (see Chap. 9 and reference [Ullrich 1995a]), one has developed implementations of SIC in terms of state-independent local potentials which lead to the KLI method [Krieger 1992b] and when skipping the optimization feed-back to the Slater approximation for SIC [Sharp 1953] (for a detailed discussion in connection with clusters see [Legrand 2002]).

Now, the valence electrons in a simple metal cluster have all very similar spatial extension and are also comparatively close in energy. We therefore replace the detailed single-electron densities by one averaged, $\bar{n}_{1e} = n/N_{\rm el}$, which then defines the energy functional for average-density SIC (ADSIC) as

$$E_{\rm Hxc}^{\rm ADSIC}[n] = E_{\rm Hxc}[n] - N_{\rm el} E_{\rm Hxc}[n/N_{\rm el}] \ . \qquad (26.2)$$

This provides the correct asymptotics of the Kohn-Sham potential, while it is formally as simple to handle as the LDA, i.e., there is no orbital dependence. The concept of single-electron densities is not needed anymore, so that ADSIC is also comparable in spirit to semi-classical schemes [Fermi 1934] such as the Thomas-Fermi method or Vlasov dynamics.

Figure 26.2 demonstrates the performance of the various SIC approaches for the anionic cluster Na_{19}^-. The Kohn-Sham potentials in the left and right panels show the correction of the asymptotic behavior by virtue of SIC. Pure LDA produces an unphysical Coulomb barrier for that anion while the ADSIC asymptotic potential performs as it should. The single electron energies

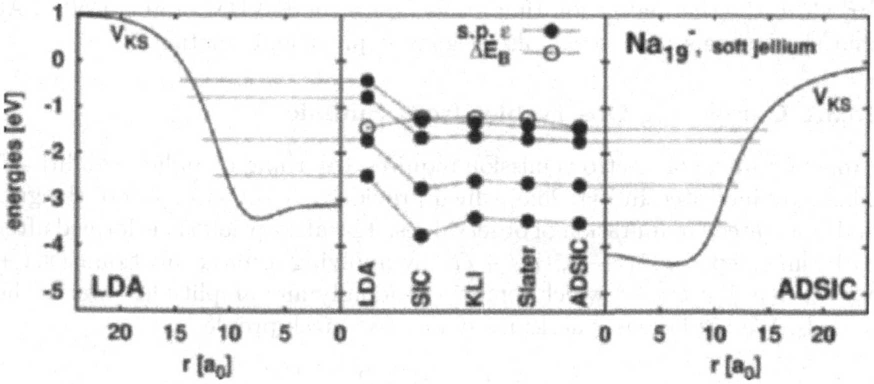

Fig. 26.2. The single electron spectrum in the cluster Na_{19}^- computed at different levels of approximation. The *left* panel shows the Kohn-Sham potentials and the levels for a pure LDA calculation. The *right* panel show the same quantities for ADSIC. The *middle* panel shows the single electron levels for LDA and various stages of the self-interaction correction. SIC means the full recipe according to [Perdew 1981], KLI is an optimized potential SIC following the recipe of [Krieger 1992b], "Slater" stands for the Slater approximation to SIC [Sharp 1953], and ADSIC is the simple average-density recipe [Fermi 1934, Legrand 2002]. The ionic background was modeled with soft jellium using a Wigner-Seitz radius of $4\,a_0$ and a surface parameter of $0.8\,a_0$ (Woods-Saxon profile) [Reinhard 2003]

are compared with one-electron separation energies directly computed from total binding energies $\Delta E(N_{\rm el}) = E(N_{\rm el}) - E(N_{\rm el}-1)$. The definition via ΔE is reliable within the LDA because it involves total energies, rather than orbital energies. The comparison in Fig. 26.2 shows that the LDA levels are not correctly placed, while all SIC schemes provide a very nice agreement of single-electron energies and separation energies. Thus ADSIC allows the computation of, e.g., photo-electron spectra, directly from TDLDA. Furthermore, it is comforting that the simplest ADSIC performs comparably to the more elaborate SIC methods. This holds, however, only for compact simple metal clusters. ADSIC is not applicable to covalent materials with their different length and energy scales, and it is also inappropriate for cluster fission where a spatial separation has to be accounted for.

26.3 Distributions of Emitted Electrons

In this section, we are going to discuss detailed observables from direct electron emission. With direct emission we mean those processes that are caused without delay by the electronic excitation process. A competing mechanism is thermal electron evaporation which requires first a thermal relaxation and then time for subsequent stochastic emission. The thermal times scale with temperature as $\propto T^{-2}$ and thus strongly depend on the excitation energy of the process. We consider here moderate excitations and short laser pulses for which the thermalization time stays above the FWHM of about 50 fs. At that short time scale, we can also neglect explicit ionic motion.

26.3.1 Computing Observables from Emission

Proper handling of electron emission requires absorbing boundary conditions. These are indicated in Fig. 26.3, which provides a schematic view of the grid and associated computation of observables. The absorption is performed after each time step $\varphi(\boldsymbol{r},t) \longrightarrow \tilde{\varphi}(\boldsymbol{r},t+\delta t)$ by applying a mask function $\varphi(\boldsymbol{r},t+\delta t) = \mathcal{M}(\boldsymbol{r})\tilde{\varphi}(\boldsymbol{r},t+\delta t)$ which removes gradually any amplitude towards the bounds. We use here a spherically symmetric mask profile

$$\mathcal{M} = \cos\left(\frac{\pi}{2}\frac{|\boldsymbol{r}|-R_{\rm in}}{R_{\rm out}-R_{\rm in}}\right)^{1/2} \qquad (26.3)$$

which is active in an absorbing margin $R_{\rm in} < |\boldsymbol{r}| < R_{\rm out}$. The spherical profile is needed to minimize grid artifacts when computing angular distributions [Pohl 2004].

The absorbing boundaries reduce gradually the norm of the wavefunctions. This is, however, a desirable physical effect. It mimics the dynamical ionization of the system. The net ionization, i.e., the number of escaping electrons, can be computed simply from the single-particle norms as $N_{\rm esc} = N_{\rm el}(t=0) - \sum_i \langle\varphi_i|\varphi_i\rangle$. This quantity will play a role in Sect. 26.4.

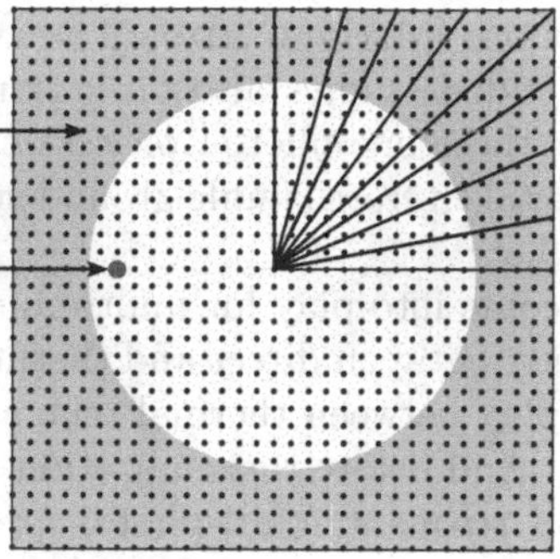

Fig. 26.3. Schematic view of the numerical calculation of electronic dynamics on a grid and subsequent observables. The area where the absorbing boundary conditions are active is indicated by grey shading. The angular distribution of emitted electrons is collected in angular bins as indicated. The photo-electron spectra (PES) are deduced from collecting phase information at an analyzing point shortly before the absorbing boundaries

The PES are evaluated for each state separately from the phase oscillations of the single-electron wavefunction at an analyzing point that is placed near the absorbing boundaries (see Fig. 26.3). The function $\varphi_i(\boldsymbol{r}_{\mathrm{anl}}, t)$ is Fourier transformed to $\tilde{\varphi}_i(\boldsymbol{r}_{\mathrm{anl}}, \omega)$ and the spectrum $\mathcal{P}(\omega) = |\tilde{\varphi}_i(\boldsymbol{r}_{\mathrm{anl}}, \omega)|^2$ is translated to a kinetic-energy spectrum of emitted electrons by identifying $\hbar\omega = \varepsilon_{\mathrm{kin}}$. This simple identification is possible by virtue of the absorbing boundaries which leave only outgoing waves (thus only one sign of electron momentum) in its vicinity.

The angular distributions are evaluated in angular bins as indicated in Fig. 26.3. We collect all probability which was removed by the absorption step (26.3) and accumulate it in the bin to which $\boldsymbol{r}$ belongs. That is done for each wavefunction separately. At the end, we dispose of the angular distribution for emission from each state and, of course, of the total angular distribution as well. Experimentally, the state selective distributions can be deduced from the angle and energy dependent cross section $\mathrm{d}^2\sigma/\mathrm{d}\Omega\mathrm{d}E$ when considering a peak in energy which corresponds to emission from one specific state. The total cross section sums over all states and corresponds to the energy integrated cross section $\mathrm{d}\sigma/\mathrm{d}\Omega$.

26.3.2 Multi-Plasmon Features in Photo-Electron Spectra

Photo-electron spectra carry a lot of useful information about cluster structure and dynamics. The structural aspects prevail in one-photon processes at rather low laser intensity. The one photon maps the bound state energy to a free kinetic energy as $\varepsilon_{\mathrm{kin}} = \varepsilon_\alpha + \hbar\omega_{\mathrm{phot}}$. Measuring the photo-electron spectra $\varepsilon_{\mathrm{kin}}$ together with the well-known photon energy $\hbar\omega_{\mathrm{phot}}$ allows one to deduce the single-electron levels in the cluster. This technique has long been used for cluster anions [McHugh 1989]. The availability of efficient UV sources now allows extended studies also on cations [Wrigge 2002]. As the reaction mechanism for these one-photon processes is extremely simple, the theoretical input can be taken from simple ground state calculations, see e.g., [Bonačić-Koutecký 1989].

Dynamical aspects come into play for multi-photon processes. For example, there is a competition between direct and thermal multi-photon emission depending on the interplay between thermal relaxation time and laser pulse length [Campbell 2000]. And before thermalization sets in, one has a distinction between direct and resonant multi-photon processes, where the latter means that the laser frequency matches a bound excited state which, in turn, has measurable consequences on the outgoing electron wave [Kornberg 1999]. Metal clusters with their pronounced Mie plasmon mode provide another very interesting mechanism. The plasmon is a collective mode that allows multiple excitations of itself with the same frequency, i.e., multi-plasmon excitations [Calvayrac 1995, Calvayrac 1997]. We can thus dream of a "super resonant" process where multi-photon and multi-plasmon channels compete. Such a scenario is demonstrated in Fig. 26.4. We consider laser frequencies

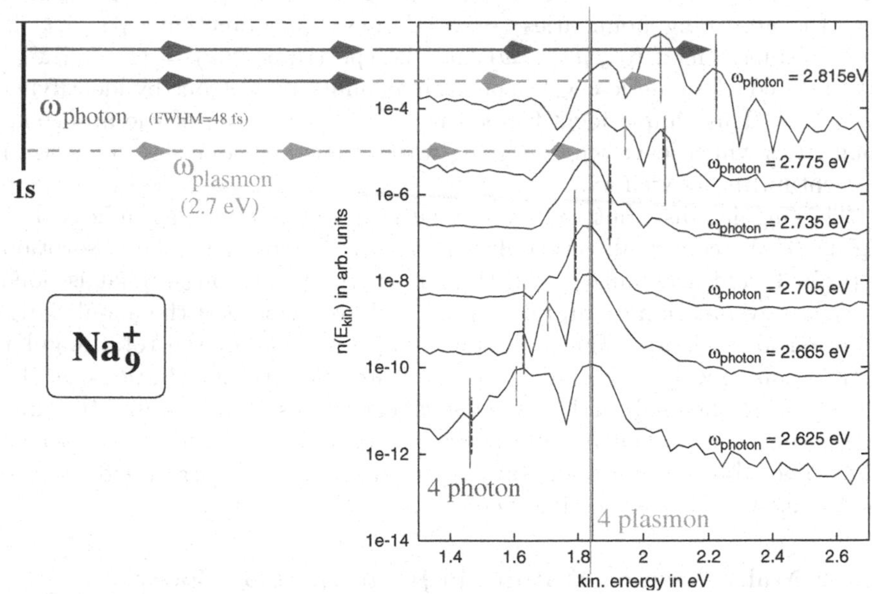

Fig. 26.4. PES from Na_9^+ for various laser frequencies in the vicinity of the Mie plasmon frequency. The PES are shown in a narrow energy range covering four-photon excitation out of the 1s state. The figure is augmented by a schematic indication of the competing mechanisms, four-photon versus two-photon-two-plasmon and four-plasmon processes where photons are indicated by *full lines* with arrow and plasmons by *dashed lines* with arrow

in the vicinity of the Mie plasmon frequency of about 2.7 eV. At least four photons/plasmons are required to lift a bound $1s$ electron above its ionization threshold. The figure shows the PES in a narrow window around that four photon peak. The four plasmon energy is indicated by a vertical line. The PES taken at various different laser frequencies show nicely the coexistence between n-photon plus m-plasmon excitation with $n + m = 4$ and how one could possibly discriminate them experimentally. The four plasmon peak remains invariant while the four photon peak moves when scanning laser frequencies.

The example given here leaves a few questions open. On the theoretical side, one wonders why processes with odd photon/plasmon numbers are suppressed. On the experimental side, such measurements rely on a very high resolution of the PES. This can only be achieved for small clusters at very low temperatures to provide a well defined, narrow plasmon peak.

26.3.3 Angular Distributions – Low Intensity Domain

The analysis of systems with simple geometry in the perturbative regime (i.e., low laser intensities) is conceptually simple. The ground state wavefunctions in a spherical system have well-defined angular profiles $Y_{lm}(\theta, \phi)$. Each photon adds a well-defined angular momentum of one with profile Y_{10}. The angular mix of the outgoing wave can then be computed by exploiting the standard rules of angular momentum coupling. Simple global deformations add a well defined portion of angular profile, e.g., a quadrupole shape mixes a certain amount of Y_{2M} into the amplitudes. This can still be disentangled in a straightforward manner, particularly for one photon processes. Real clusters, however, are very complex molecules with a rich mix of spatial profiles. Metal clusters, being close to a "metal drop", are still dominated by a few global multipole moments, and those with magic electron numbers are quite spherical. Nonetheless, the ionic structure adds all sorts of perturbations, and the question remains what structures finally emerge in the angular distribution of emitted electrons.

A further complication is that the ionic structure in metals is known to exhibit sizeable thermal fluctuations. Measurements are done with ensembles of size-selected clusters. Although there is practically no ionic motion during laser impact and direct electron emission at any given instant, thermally induced motion produces a broad selection of thermally shaken ionic configurations. To include that effect, we produce an ensemble of Na_9^+ clusters at $T = 400$ K by starting from the fully relaxed ground state, initializing the ionic motion in several samples with a Maxwell distribution of ionic velocities, running TDLDA-MD over 10 ps, and sampling the configurations at stochastically chosen times as probes for subsequent computation of laser excitation. The impact of thermal fluctuations on angular distributions is demonstrated in Fig. 26.5. A case with high laser frequency is chosen, as it is more sensitive to details of the cluster shape [Pohl 2004]. It is obvious that the various

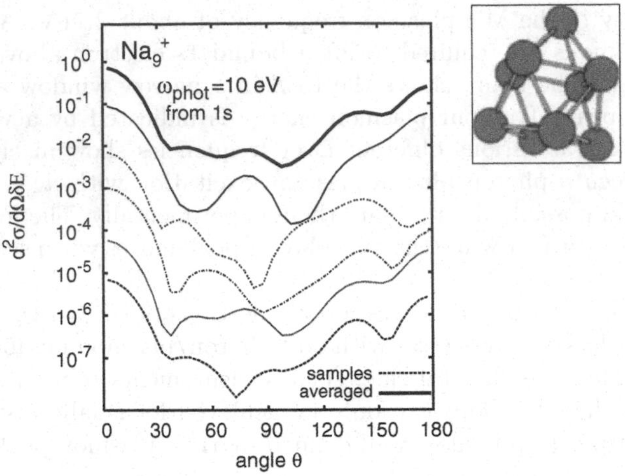

Fig. 26.5. Angular distributions from a thermal ensemble of Na_9^+ clusters at $T = 400$ K. Electrons are emitted from of the 1*s* state by a laser with $\omega_{\text{phot}} = 10\,\text{eV}$. The *dashed* and *dotted lines* show results from a few selected members of the thermal ensemble. The heavy line is the thermal average. The inset shows the ionic configuration of Na_9^+ at $T = 0$

samples from the ensemble can yield rather different angular distributions. A thermal average wipes out all details leaving only gross structures. Figure 26.5 suggests that this could cover up all interesting structures from emission for high laser frequency. Similar plots for low frequency lasers show much less effect from thermal fluctuations. A thermal average may be appropriate here. In that context, we recall that the jellium model is a widely used approximation for the ionic background of metal clusters above melting temperature. We check its performance by building an equivalent jellium model which uses a soft Woods-Saxon profile [Reinhard 2003] with surface deformation chosen as to reproduce the leading multipole moments of the ionic configuration (up to $L = 4$, which is the recommended level of expansion [Montag 1995a]). The question is: To what extent can the jellium results reproduce the full ionic ensemble calculation?

Figure 26.6 shows thermally averaged distributions for a low and a high frequency case (for a more detailed view, see [Pohl 2004]). The distributions are shown for each occupied single electron state separately. The results are compared with those from the equivalent jellium model. The low energy emission produces smooth patterns which agree very well with the jellium predictions and which have few thermal fluctuations (not shown here). The very different profiles of the three different states are clearly seen and can be understood from the dominant spherical quantum numbers plus some correction from global background deformation. Electrons emitted from a high

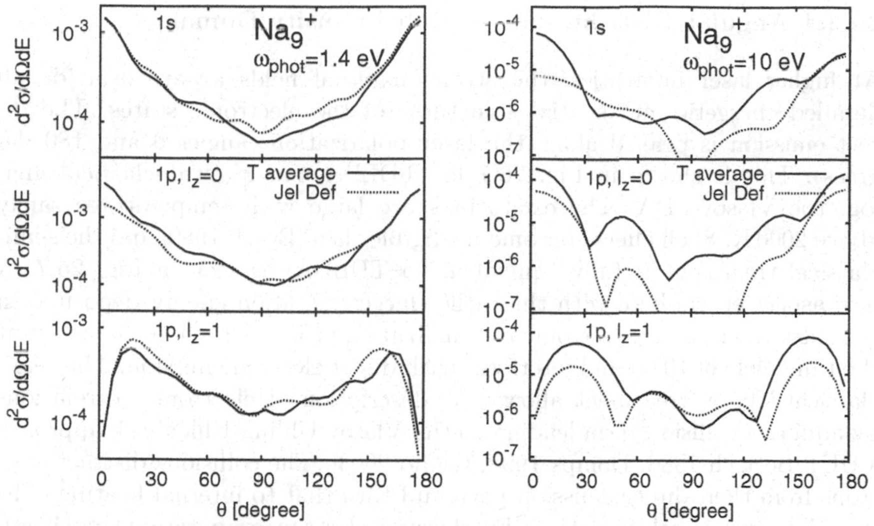

Fig. 26.6. Thermally averaged angular distributions for emission from the three different occupied electron state of Na_9^+ for two laser frequencies as indicated. The cluster was considered as an ensemble of ionic configurations at a temperature of $T = 400$ K. The results from the equivalent jellium model are shown for comparison. The intensities were $I = 10^{11}$ W/cm^2 (*left*) or $I = 10^{12}$ W/cm^2 (*right*)

frequency laser have larger kinetic energies and thus resolve much more spatial detail. This produces more detailed patterns in the angular distributions and significant differences from the jellium model. It is obvious that the measurement in that regime becomes sensitive to the ionic structure. However, as we have learned from Fig. 26.5, the full resolution will be realized only if we suppress thermal fluctuations by choosing sufficiently low temperatures. Furthermore, the outgoing electron wavefunctions at a given energy have a fixed structure of nodes inside the cluster. They are thus not equally sensitive to all regions. One needs to scan a range of frequencies to collect all necessary information. After all, we can state that the gross structure of the cluster can be deduced already from angular distributions produced with low frequency lasers. More detailed information on ionic structure is contained in measurements with high frequencies, particularly when measuring clusters at very low temperatures, scanning a band of laser frequencies, and tagging on the specific single electron states. The tagging can be achieved experimentally by measuring energy selective angular distributions, as done, e.g., in [Baguenard 2001], and setting windows at the different single-electron peaks. The information is then contained in a rather involved manner and requires theoretically supported scattering analysis, similar to what is done, e.g., in low-energy electronic diffraction (LEED) [Heinz 1995].

26.3.4 Angular Distributions – High Intensity Domain

At higher laser intensities, the strong external fields always override all detailed energetic and spatial structures of the electronic states. The direct emission is peaked along the laser polarization (angles 0 and 180 degrees). This is shown in Fig. 26.7 for TDLDA and its semi-classical analog, the Vlasov-LDA. The excitations are large with temperatures safely above 2000 K. Shell effects become negligible then [Brack 1989] and the semi-classical treatment is fully equivalent to TDLDA, as seen in Fig. 26.7. A new aspect enters here with the rather large excitation energy deposited in the cluster. The inter-electronic thermalization proceeds then much faster (at the time scale of 10 fs) and interferes with direct electron emission. The semi-classical Vlasov treatment allows the description of electronic correlations by adding a collision term leading to the Vlasov-Ühling-Uhlenbeck approach VUU) [Bertsch 1988, Domps 1998, Giglio 2003]. The collisions distract electrons from their direct emission path and turn that to internal heating. The stored energy is released later by thermal electron evaporation, stochastically in time and orientation. Such thermal electron emission is, of course, isotropic. Therefore we expect a competition between aligned and isotropic emission depending on the detailed conditions such as laser pulse length, frequency, and intensity. The VUU results in Fig. 26.7 show clearly a strong isotropic component. depending somewhat on the laser frequency. A more detailed analysis shows that the isotropy develops with time at the typical

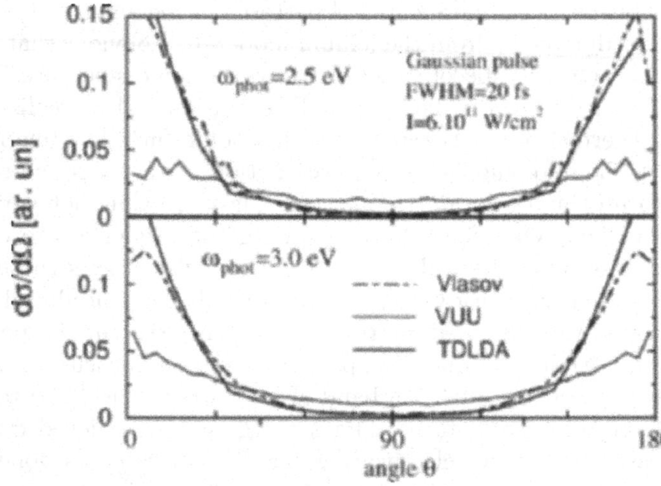

Fig. 26.7. Total angular distributions after excitation of Na_9^+ with an intense laser pulse. Two frequencies are considered as indicated. Results from TDLDA are compared with those from the semi-classical Vlasov-LDA approximation and from the Vlasov-Ühling-Uhlenbeck treatment which includes dynamical electron-electron correlations

scale of electronic relaxation [Giglio 2003]. Thus one can gain information on relaxation times by measurement of angular distributions in the regime of stronger excitations (average ionization of a few electrons) combined with varying time structures, e.g., by using varying pulse length [Campbell 2000].

26.4 Pump-Probe Analysis of Ionic Dynamics

Pump-probe experiments have become a powerful tool for a detailed analysis of time evolution at a microscopic level for a huge variety of systems, from molecules and chemical reactions [Zewail 1994] up to bulk materials [Garraway 1995]. There are many conceivable scenarios for time resolved experiments with clusters due to their complexity (for a summary and more detailed discussion see [Reinhard 2003]). For metal clusters, the strong dominance of the Mie plasmon resonance offers its service as a handle for pump-probe analysis. The position of the plasmon peaks provides immediate information on the global extension and quadrupole deformation of the cluster [Andrae 2002, Andrae 2004]. Fuzzy fragmentation patterns give clues to complicated non-compact geometries [Dinh 2005].

There are basically two different ways to probe the time evolution of the Mie plasmon resonance as sketched in Fig. 26.8. Just as is usually done

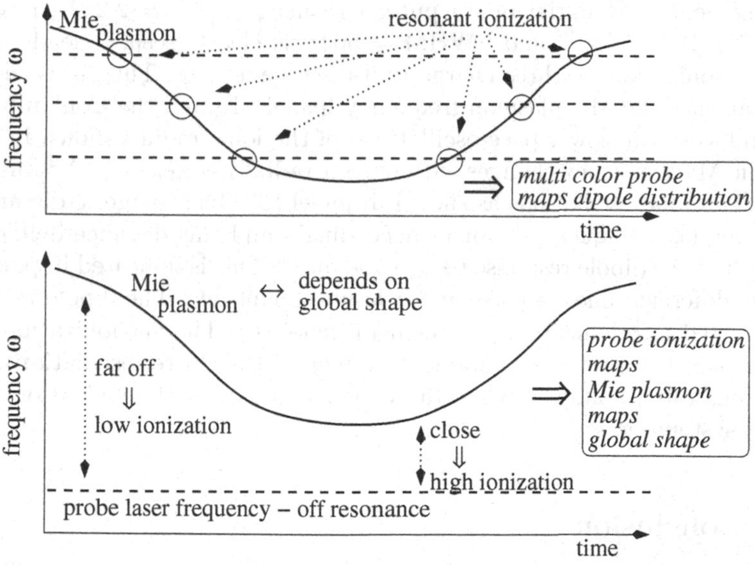

Fig. 26.8. Schematic view of two different strategies for pump-probe setups to analyze ionic dynamics of metal clusters via the strong electronic dipole response (Mie plasmon). The *heavy* line represents the average Mie plasmon frequency which depends on time due to ionic motion, typically at a ps time scale. The *dashed lines* represent the (constant) frequencies of the probe laser

in molecular physics, one can probe the point where the laser comes into resonance with the (time dependent) excitation energy. This is indicated in the upper panel of the figure. One probe frequency thus resonates at most twice during one oscillation and just once in a monotonic motion, such as Coulomb explosion. A more complete mapping can be achieved with the resonant mechanism by using various probe frequencies. An alternative strategy is sketched in the lower panel of Fig. 26.8. One places the probe laser frequency safely below the Mie plasmon resonance and "maps" the changes of the plasmon energy by the variations of probe response with delay time. This maps the whole evolution with one laser frequency. This strategy is applicable in simple spectral situations, i.e., if the Mie plasmon remains well concentrated over the whole time evolution, as is typical e.g. for vibrations [Andrae 2002, Andrae 2004]. The resonant strategy (upper panel) with multicolor probes is more involved, but delivers also much more information. It allows the study of the time evolution of spectral fragmentation when it is important, e.g., in fission [Dinh 2005].

We illustrate the "mapping scenario" for the case of monopole oscillations in Na_{41}^+. This cluster has a nearly perfect spherical shape, i.e., the other low multipole moments $L = 1, 2, 3, 4$ are extremely small. The Coulomb induced motion is then dominated by simple breathing oscillations of the cluster radius. Figure 26.9 demonstrates the process in terms of several observables. Panel (c) shows the electronic response in terms of the envelope of the dipole moment. The initial pump pulse (frequency $\omega_{\rm phot} = 2.2\,\text{eV}$, intensity $I = 1.1 \times 10^{12}\,\text{W/cm}^2$, and FWHM $= 50\,\text{fs}$) is clearly seen. It leads to immediate ionization by three charge units [see panel (b)]. This, in turn, blue shifts at once the Mie plasmon frequency [panel (d)] and the Coulomb pressure induces at a slower pace oscillations of the ionic radius, shown in panel (e). The Mie frequency changes with cluster radius as $\omega_{\rm Mie} \propto R^{-3/2}$ and thus maps the radial oscillations as shown in panel (d). The energetic distance to the probe laser frequency changes accordingly and this distance defines the strength of the dipole response to a probe pulse. This is indicated in panel (c) for two different times, a more and a less resonant one. The dipole response is converted into ionization as shown in panel (b). The net ionization yield after probe then provides an immediate map of the cluster oscillations. This is demonstrated finally in panel (a) which summarizes the ionization yields from a systematic series of probe pulses.

26.5 Conclusion

From a practitioner's perspective, we have discussed three examples of applications of time-dependent density functional theory (TDDFT) to cluster dynamics: photo-electron spectroscopy in a regime of competition between multi-photon and multi-plasmon excitations, angular distributions after laser excitation of a thermal ensemble of metal clusters, and pump-and-probe

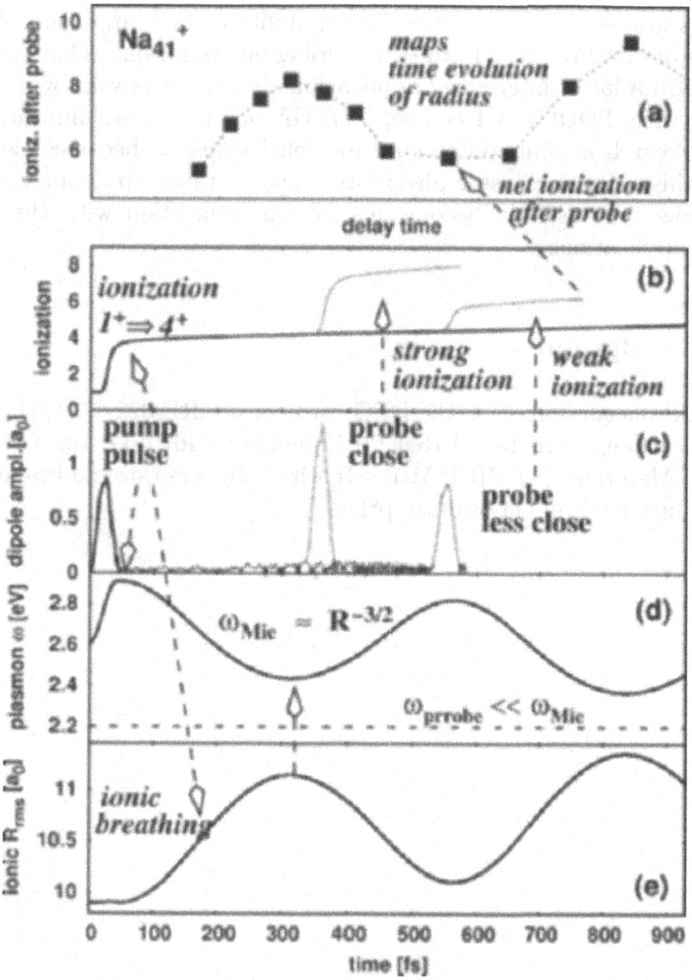

Fig. 26.9. Pump and probe analysis of ionic breathing in Na_{41}^{+}. The lower two panels show the time evolution of the ionic radius [panel (e)] and the Mie plasmon frequency [panel (d)] after initial strong ionization by a pump pulse. The ionization is shown as *solid line* in panel (b). The two other lines in panel (b) show the extra ionization as caused by two different probe pulses. Panel (c) shows the electronic response to the laser pulses in terms of the dipole amplitude. Finally, panel (a) shows the net ionization after probe pulse with a given time delay

analysis of ionic oscillations in highly ionized clusters. The examples were chosen to demonstrate some features that are special to metal clusters, e.g., the dominance of the strongly collective Mie plasmon or the large shape fluctuations and softness of ionic motion. All examples employ the time-dependent local density approximation (TDLDA) augmented with a self-interaction correction (SIC) and with simultaneous molecular dynamics (TDLDA-MD). The test cases also serve to illustrate the capability of that approach. All three examples are freshly taken from still ongoing investigations. That shows that there is still a lot of interesting applications in cluster physics which can be attacked with TDDFT. Of course, TDLDA has its known limitations and intense research is going on to cure its deficiencies, as becomes clear when reading this volume. Cluster physicists must be ready to adopt new developments as soon as they become feasible in connection with these rather complex applications.

Acknowledgments

This work was supported by the DFG, project no. RE 322/10-1, the French-German exchange program PROCOPE number 04670PG, the CNRS Programme "Matériaux" (CPR-ISMIR), Institut Universitaire de France, Humboldt foundation and Gay-Lussac prize.

27 Excited-State Dynamics in Extended Systems

O. Sugino and Y. Miyamoto

27.1 Introduction

There are many subjects in condensed matter physics and chemistry related to excited-state dynamics. For example, photo-induced electronic excitations in the bulk trigger the formation of self-trapped excitons [Song 1996]. These, in the case of $SrTiO_3$, are thought to induce a ferroelectric phase transition [Takesada 2003, Hasegawa 2003]. Photo-excitation in surfaces induces chemical reactions rclated to photography [Jacobson 1976], solar energy conversion [Hangfeldt 1995], and so on. Electron transfer in surfaces induces electrochemical reactions where the excited-state potential surfaces are concerned [Miller 1995]. Owing to recent experimental progresses, such reaction processes have been measured with increasing resolution in time [Asbury 2001].

Excited-state dynamics in extended systems is different from that in finite systems as it involves energy bands, which have a continuous energy spectrum. For example, an initial photo-excitation localized near an adatom on the surface or a defect in the bulk can be transferred, not only to nearby localized states, but also to delocalized bulk states. This means that the system can decay into different final states. Changing the electrode potential, or the bulk chemical potential, can induce donation (acceptance) of an electron to (from) a localized state, triggering the excited-state dynamics.

Owing to the differences between finite and extended systems, different computational approaches have been taken. For finite systems, approaches suitable for handling a limited number of excited-states were developed. Casida [Casida 1996] derived a TDDFT linear response scheme to compute the excitation energies $\Omega_i(\boldsymbol{R})$, for a given ionic configuration $\boldsymbol{R}$, by solving a generalized eigenvalue problem of super-operators. The algorithm was further simplified, in the framework of plane-wave basis sets, using the Tamm-Dancoff approximation [Hutter 2003], and was recently used to conduct a simulation on the excited-state hyper-surfaces [Rohrig 2003], see Sect. 23.

For extended systems, on the other hand, real-time schemes were developed. Yabana and Bertsch [Yabana 1996] simulated the time-evolution of the electron density $n(\boldsymbol{r}, t)$ of molecules such as C_{60} after suffering a pulsed electronic perturbation. This method proved to be efficient in getting the overall shape of the excitation energy spectrum of large molecules. However,

O. Sugino and Y. Miyamoto: *Excited-State Dynamics in Extended Systems*, Lect. Notes Phys. **706**, 407–423 (2006)
DOI 10.1007/3-540-35426-3_27

its suitability for getting very accurate spectra is questionable, due to the long simulation times it requires.

To simulate excited-state dynamics in extended systems, the real-time scheme was coupled with molecular dynamics [Sugino 1999, Miyamoto 1999, Miyamoto 2000, Yokozawa 2000, Miyamoto 2001, Miyamoto 2002, Miyamoto 2004a, Miyamoto 2004b, Miyamoto 2004c]. Sugino and Miyamoto [Sugino 1999] formulated a plane-wave-based scheme in which the time-dependent Kohn-Sham equation

$$\mathrm{i}\frac{\mathrm{d}}{\mathrm{d}t}\varphi_i(\boldsymbol{r},\boldsymbol{R},t) = \hat{H}(\boldsymbol{r},\boldsymbol{R},t)\varphi_i(\boldsymbol{r},\boldsymbol{R},t) \tag{27.1}$$

is used for the electrons and Newton's equation

$$M_\alpha \frac{\mathrm{d}^2}{\mathrm{d}t^2}\boldsymbol{R}_\alpha = -\sum_{i=1}^{N}\langle\varphi_i|\,\frac{\partial \hat{H}}{\partial \boldsymbol{R}_\alpha}\,|\varphi_i\rangle \tag{27.2}$$

is used for the ions. This Ehrenfest-type dynamics [Ehrenfest 1927], which will be called TDDFT-MD from hereon, is based on a mean-field approximation of the electron-nuclear interaction

$$E_{\mathrm{en}} = \sum_\alpha \int \mathrm{d}^3r\, \frac{n(\boldsymbol{r})}{|\boldsymbol{r}-\boldsymbol{R}_\alpha|}\,. \tag{27.3}$$

The TDDFT-MD dynamics follows an adiabatic potential surface, $\varepsilon_i(\boldsymbol{R})$, when the coupling with other potential surfaces $\boldsymbol{d}_{ij}(\boldsymbol{R}) = \langle\varphi_i|\,\nabla_{\boldsymbol{R}}\,|\varphi_j\rangle$ is negligibly small. However, in the presence of the coupling, the Kohn-Sham orbitals φ_i deviate from the adiabatic ones by

$$\propto \int_0^t \mathrm{d}t' \sum_{\alpha j}\frac{\mathrm{d}}{\mathrm{d}t}\boldsymbol{R}_\alpha(t')\cdot\boldsymbol{d}_{ij}(\boldsymbol{R}_\alpha)\,\mathrm{e}^{\mathrm{i}(\varepsilon_i-\varepsilon_j)t'} \overset{t\to\infty}{\longrightarrow} \sum_{\alpha j}\frac{\frac{\mathrm{d}}{\mathrm{d}t}\boldsymbol{R}_\alpha(t)\cdot\boldsymbol{d}_{ij}}{\varepsilon_i-\varepsilon_j}\,. \tag{27.4}$$

This non-adiabatic effect stems from the fact that the electrons cannot completely follow the ionic motion. The change in the electronic structure affects the ionic motion through (27.2).

Note that this formulation is based on constrained DFT, where the potential surface for the excited states is obtained solving self-consistently the KS Hamiltonian in a promoted electron occupancy. One can alternatively use TDDFT linear response theory for the excited states, as is done in Sect. 23. The latter, albeit formally correct, is significantly affected by the approximation to the xc interaction, e.g., ALDA, that substantially underestimates the Rydberg and the charge-transfer excitations. The former, although justified only for the lowest energy state of a given symmetry, describes the localized and extended states in a relatively unbiased manner and often provides us with reasonable potential surfaces [Casida 2000b]. When an external time-dependent perturbation is applied, the electrons deviate from the eigenstate. Corresponding non-adiabatic couplings may be obtained within each scheme, constrained DFT and TDDFT linear response, on the basis of response theory [Billeter 2005]. Note that a rigorous DFT-based formulation is still lacking, although this naive scheme looks like a plausible first step. Full description of the non-adiabatic transition is, on the contrary, more difficult due to finite density changes associated with the transition and corresponding non-orthogonality of the ground and excited states. These make the treatment quite complicated [Niehaus 2005]. In view of these open problems, we will restrict our formulation to constrained DFT and will follow the dynamics until the onset of a non-adiabatic transition to get information on the couplings.

The approximation used in TDDFT-MD can be viewed as a semi-classical approximation [Santer 2001, Ando 2003] to the full electron-ion dynamics, where the equation of motion (EOM) for the electronic density matrix $\rho_{ij}(\{\boldsymbol{R}\},\{\boldsymbol{P}\},t)$ in the phase space of ions with position $\{\boldsymbol{R}\}$ and momentum $\{\boldsymbol{P}\}$ is expanded to lowest order in $\hbar$ giving

$$\begin{aligned}\frac{\mathrm{d}}{\mathrm{d}t}\rho_{ij}(\boldsymbol{R},\boldsymbol{P},t) + \mathrm{i}\left[\varepsilon_i(\boldsymbol{R}) - \varepsilon_j(\boldsymbol{R})\right]\rho_{ij}(\boldsymbol{R},\boldsymbol{P},t) = \\ -\frac{\boldsymbol{P}}{M}\sum_k \left[\boldsymbol{d}_{ik}(\boldsymbol{R})\rho_{kj}(\boldsymbol{R},\boldsymbol{P},t) - \rho_{ik}(\boldsymbol{R},\boldsymbol{P},t)\boldsymbol{d}_{kj}(\boldsymbol{R})\right] \\ -\frac{\boldsymbol{P}}{M}\nabla_{\boldsymbol{R}}\rho_{ij}(\boldsymbol{R},\boldsymbol{P},t) + \frac{\nabla_{\boldsymbol{R}}\varepsilon_i(\boldsymbol{R}) + \nabla_{\boldsymbol{R}}\varepsilon_j(\boldsymbol{R})}{2}\nabla_{\boldsymbol{P}}\rho_{ij}(\boldsymbol{R},\boldsymbol{P},t)\,. \quad (27.5)\end{aligned}$$

For simplicity, the EOM for the one-nucleus system is shown here. This equation indicates that a wave-packet centered at $(\boldsymbol{R},\boldsymbol{P})$ on a potential surface moves classically on that surface when the coupling $\boldsymbol{d}_{ij}/(\varepsilon_i-\varepsilon_j)$ is negligibly small. In the presence of coupling, on the other hand, the wave-packet splits into wave-packets evolving in different potential surfaces and interfering with each other. Therefore, when the splitting becomes significant, the classical approximation of (27.2) cannot be used. In that case, the ions need multiple states to be properly described. We point out, however, that TDDFT-MD yields nevertheless important information on "when" and "to which state" a significant nonadiabatic transition will take place.

The above problem is common to the mean-field-type theories like TDDFT-MD. Equation (27.5) and more practical algorithms [Tully 1971] have been proposed to overcome the problem. In these schemes, the non-adiabatic dynamics is conducted on potential energy surfaces (PES) prepared in advance. However, as in TDDFT the PES depends on the electron density, these algorithms cannot be used in a simple manner. Moreover, due to the memory effect, in TDDFT the PES also depends on the trajectory. A detailed discussion is beyond the scope of this chapter, but we point out that a way to go beyond the mean-field may be to use a path-integral formalism or the Monte Carlo wave-function method [May 2004].

27.2 Real-Time Evolution of the Kohn-Sham Orbitals

In a numerical solution of the time-dependent Kohn-Sham equations, (27.1), the Kohn-Sham orbitals are updated at each time-step using

$$\varphi_i(t + \Delta t) = \hat{\mathcal{T}} \exp\left[-\mathrm{i}\int_t^{t+\Delta t} \mathrm{d}t'\, \hat{H}(t')\right] \varphi_i(t)\,, \tag{27.6}$$

where $\hat{\mathcal{T}}$ is the time-ordering operator. Explicit algorithms, such as the second-order differencing method, (12.22) or higher order schemes, are numerically less stable than implicit algorithms, such as the Crank-Nicholson method, (12.23), where "implicit" means that the result of the application of $\hat{H}$ to the orbitals at $t + \Delta t$, or its polynomial form, appears in the finite difference equation. Semi-implicit algorithms, such as the second order split-operator method [Feit 1982b, Feit 1982a] or higher order schemes, e.g. (12.30), known as Suzuki-Trotter schemes [Suzuki 1992b, Suzuki 1993], are as stable as the implicit ones, but are more efficient because they do not involve matrix inversions [cf. (12.23)]. Furthermore, in the plane-wave scheme, the operators $\hat{T}$ and $\hat{V}$ are diagonal in reciprocal and real space, respectively. Note that the orbitals evolve unitarily in (12.23) and (12.30), allowing us to skip the time-consuming ortho-normalization step that tends to hamper parallel computing calculations.

The Suzuki-Trotter scheme consists in decomposing general non-commuting operators into

$$\exp\left[-\mathrm{i}\sum_{i=1}^{N} \hat{A}_i \Delta t\right] = \mathrm{e}^{-\mathrm{i}\frac{\Delta t}{2}\hat{A}_1}\mathrm{e}^{-\mathrm{i}\frac{\Delta t}{2}\hat{A}_2}\ldots\mathrm{e}^{-\mathrm{i}\frac{\Delta t}{2}\hat{A}_{N-1}}\mathrm{e}^{-\mathrm{i}\Delta t\hat{A}_N}$$
$$\times\, \mathrm{e}^{-\mathrm{i}\frac{\Delta t}{2}\hat{A}_{N-1}}\ldots\mathrm{e}^{-\mathrm{i}\frac{\Delta t}{2}\hat{A}_2}\mathrm{e}^{-\mathrm{i}\frac{\Delta t}{2}\hat{A}_1} + O(\Delta t^3)\,. \tag{27.7}$$

This allows us to use the pseudopotential projection operator

$$\hat{V}_{\text{ps}} = \sum_{\tau l m} |\phi_{\tau l m}\rangle\, D_{\tau l m}\, \langle\phi_{\tau l m}|\;, \tag{27.8}$$

whose components do not always commute with each other. Note that the exponential in each term can be simply written as

$$\mathrm{e}^{|\phi_{\tau l m}\rangle D_{\tau l m}\langle\phi_{\tau l m}|} = 1 + |\phi_{\tau l m}\rangle\, \mathrm{e}^{D_{\tau l m}+1}\, \langle\phi_{\tau l m}| \tag{27.9}$$

for normalized ϕ's.

The Suzuki-Trotter scheme has another advantage: higher order schemes are available even when the potential $\hat{V}(t)$ is time-dependent. This is particularly important because the self-consistent potential of TDDFT can exhibit a significant time-dependence. This can be understood by expressing the time-dependent Kohn-Sham orbitals $\varphi_i(\boldsymbol{r},t)$ as a linear combination of time-independent (adiabatic) orbitals $\bar{\varphi}_i(\boldsymbol{r})$

$$\varphi_i(\boldsymbol{r},t) = \sum_j c_{ij}(t)\bar{\varphi}_j(\boldsymbol{r})\mathrm{e}^{-\mathrm{i}\varepsilon_j t} . \tag{27.10}$$

Since the coefficients c and the adiabatic orbitals are usually weakly time-dependent, the electron density,

$$n(\boldsymbol{r},t) = \sum_{i=1}^{N} |\varphi_i(\boldsymbol{r},t)|^2 = \left| \sum_{jk} \left(\sum_{i=1}^{N} c_{ij}^* c_{ik} \right) \bar{\varphi}_j^*(\boldsymbol{r})\bar{\varphi}_k(\boldsymbol{r})\mathrm{e}^{-\mathrm{i}(\varepsilon_j-\varepsilon_k)t} \right|^2 \tag{27.11}$$

has oscillatory components. These components appear unless the electrons move on a perfectly adiabatic potential surface. This is reflected in the highly time-dependent Hartree-exchange-correlation potential.

The highly time-dependent terms need to be carefully handled even if their magnitude is small, as inappropriate numerical algorithms tend to amplify the errors, destabilizing the calculation. To effectively avoid this problem, a robust predictor-corrector algorithm for obtaining the self-consistent potential and a scheme to cut off unimportant higher Fourier components have been proposed [Sugino 1999]. First, the proposed predictor-corrector algorithm adopts a cubic and time-reversal interpolation scheme, called "railway interpolation," to obtain the potential in the interval between t and $t + \Delta t$. This is then used to evolve the Kohn-Sham orbitals according to (27.6). Although this requires a self-consistent determination of the potential, it greatly stabilizes the simulation and prevents the total energy (which is a constant of motion) from drifting. In practice, thanks to this robust interpolation scheme, the corrections in the predictor-corrector loop are usually small, allowing us, in most cases, to skip the loop.

The second prescription is to discard the physically unimportant higher Fourier components. Namely, for the kinetic energy operator, the electron density, and the Hartree-exchange-correlation potential, higher G components are reduced in magnitude at each iteration step. Note that the large G components are associated with high frequency oscillations in these quantities, and thus would require excessively small Δt. This cut-off scheme is similar to the standard cut-off scheme employed in plane-wave calculations, where the only FFT-grid points used in reciprocal space are those whose corresponding kinetic energy is less than a prescribed value, the cut-off energy (E_{cut}). On the contrary, the present cut-off scheme uses all grid points, but the contribution from the points exceeding the cut-off energy is gradually smoothed away. The full grid calculation is more time-consuming and requires large memory (by a factor of 18 for each wave function), but is necessary for stability. The cut-off scheme is also necessary for other basis sets, as was discussed in [Baer 2001].

27.3 Computational Procedures

TDDFT-MD simulations start with the preparation of an initial electronic excited-state for a chosen ionic coordinate and by assigning initial velocities to the ions. The initial excited-state is obtained through a constrained self-consistent field (SCF) DFT calculation with the time-*independent* Kohn-Sham equation. When considering photo-excited dynamics, occupied and unoccupied Kohn-Sham orbitals, say φ_i and φ_a, respectively, having non-vanishing dipole matrix element, $\langle\varphi_i|\, \boldsymbol{r}\, |\varphi_a\rangle$, are chosen for the electron-hole pair. The SCF calculation is the performed while keeping them half-occupied. The calculation requires continuous attention to the orbital character of the pair, because this character can be exchanged with that of the other orbital when a level crossing occurs in the course of the SCF calculation. When that happens, we need to repopulate the electronic levels in order to keep the electron and hole states half-occupied. Figure 27.1 shows an example for the single vacancy in a carbon nanotube (CNT), where significant level exchange occurs during the SCF calculation [Miyamoto 2002]. Then, the SCF convergence becomes very slow near the level exchange point requiring us to use case-by-case techniques. For example, we found it was effective, although not perfect, to allow the orbitals concerned with the exchange to be equally occupied with fractional number of electrons only in the middle of the exchange.

Note that such level alternation occurs not only in the initial time-independent calculation but also in the TDDFT-MD simulation. Figure 27.2 is an example for the single vacancy in a CNT [Miyamoto 2004a]. As was mentioned above, and contrary to the static case, the convergence during

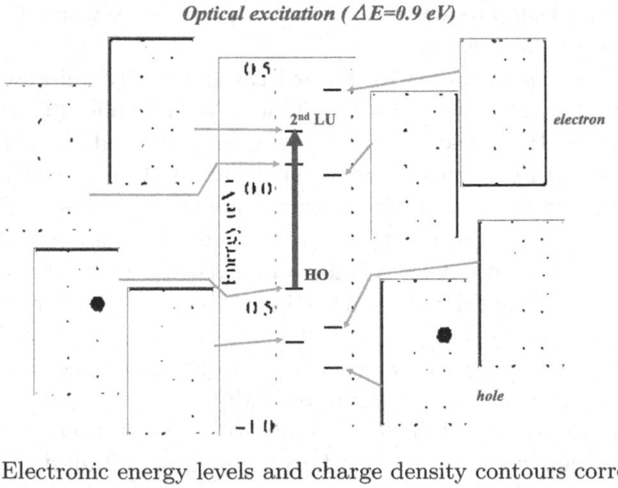

Fig. 27.1. Electronic energy levels and charge density contours corresponding to the ground (*left panels*) and excited states (*right panels*). Note the change in the ordering of the energy levels. See for details [Miyamoto 2002]

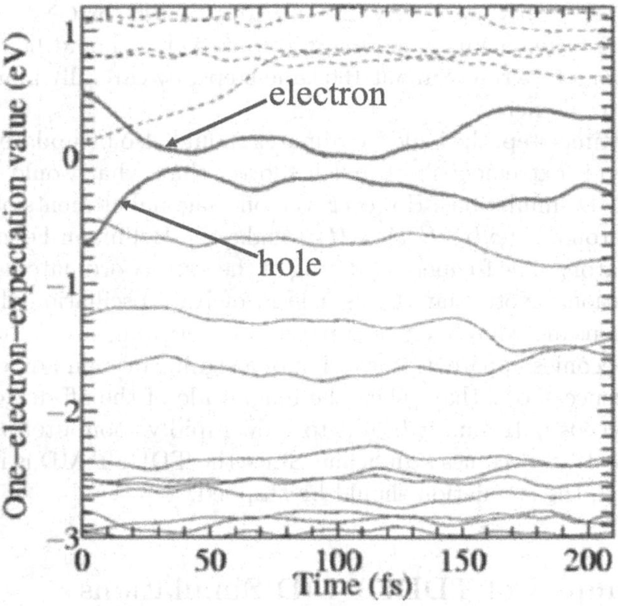

Fig. 27.2. Example of level alternation among differently occupied levels throughout the MD simulation. Displayed is the time-evolution of the expectation values of electronic levels of defected nanotubes (see for details [Miyamoto 2004a]). The *solid* and *dotted lines* are valence and conduction bands, while solid lines denoted by arrows are the excited electron and hole. Level alternation between hole-state and valence band occurs around 10 fs, while level alternation between the excited-electron and conduction band occurs around 20 fs

the dynamical time-steps is very fast. This is an important advantage of the real-time scheme. Static approaches for the simulation, where the constrained DFT calculation is performed at every MD step, can be performed with larger time-steps Δt, but suffer from the very slow SCF convergence problem.

It is important to stress here that we need to start from a well-converged SCF calculation since the long-term stability of the TDDFT-MD simulation is quite sensitive to the degree of SCF convergence. The use of the cut-off scheme proper to TDDFT-MD requires the initial SCF calculation to be performed adopting the same smoothing parameters and the same (full) grid.

In the TDDFT-MD simulation, the Suzuki-Trotter scheme is used for the electrons, while the ionic equations of motion are propagated using the Verlet algorithm. To follow the time-dependent Kohn-Sham orbitals, $\varphi_i e^{-i\omega_i t}$, the time-step ($\Delta t$) needs to be small. Since the orbital energy, ω_i, is bound by the largest kinetic energy, or the cut-off energy, as discussed in Sect. 12.2 (12.11), the proper value in atomic units of time (1 a.u. = 2.42×10^{-2} fs) is roughly estimated as $(2\pi/6)E_{\text{cut}}$. Note that this value is typically a hundred times shorter than that of conventional *ab-initio* MD simulations. For example, for silicon surfaces with $E_{\rm cut} = 8\,$Ry, Δt is 0.40 a.u., and for CNT with $E_{\rm cut} = 40\,$Ry, it is $\Delta t = 0.08$ a.u. A more appropriate value should be determined, however, from a test-run of about 100 time-steps, by carefully monitoring the drift in the total-energy.

At every time-step, the ionic coordinates should also be updated, although the time-step is extremely short, much shorter than what would be required in classical MD simulations. However, the off-diagonal elements of the Kohn-Sham Hamiltonian, $\langle\varphi_i(t)|\,\hat{H}(t)\,|\varphi_j(t)\rangle$, make the Hellmann-Feynman forces highly oscillatory. The frequent updating of the ionic coordinates averages out these oscillations. Note that this is a kind of Rabi oscillation that appears unless electrons are completely on a potential energy surface. When the Rabi oscillation becomes apparent, it is a sign of a significant non-adiabatic transition. Experience shows that, when the magnitude of the off-diagonal matrix elements exceeds 0.01 a.u., it begins to grow rapidly soon after and a deviation from the PES becomes significant. Since the TDDFT-MD is invalid from that point on, the simulation should be stopped.

27.4 Examples of TDDFT-MD Simulations

In the following, we show several applications of TDDFT-MD simulations induced by electronic excitations. The simulation time required for the investigation is at least sub-picosecond, which is a typical time-scale for the fastest ionic motion. Note that this is also the typical time-scale for photo-induced chemical reactions. This time-scale is substantially shorter than that of thermally activated reactions like atomic diffusion, but it is the upper limit accessible with the recent large-scale supercomputers.

27.4.1 Semiconductor Bulk and Surfaces

First, we focus on defects and surfaces of semiconductors. Defects in semiconductors can generate deep levels in the band gap. When the electron localized near the defect is excited, it can be delocalized soon after or it can remain localized for a while, inducing atomic diffusion.

Figure 27.3 shows a gallium arsenide (GaAs) crystal in which silicon (Si) and hydrogen (H) impurity atoms form a complex. Without the H atom, the Si atom generates a shallow donor level. However, when the complex is formed, the donor level goes down in the band gap. The shallow donor carrier density is recovered upon illumination of the crystal by a laser [Loridant-Bernard 1998]. This can be attributed to laser-induced Si-H dissociation. It is likely that the H atom can dissociate efficiently on one of the excited-state potential surfaces. On the ground state, however, the large activation barrier of 2 eV that has to be overcome inhibits this process. This technology is expected to be an alternative to conventional thermal annealing [Pearton 1986].

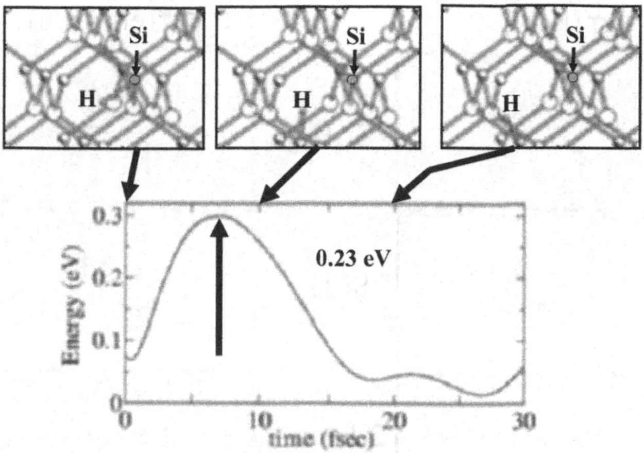

Fig. 27.3. Si–H complex in GaAs. Three snapshots for the simulation where an initial kinetic energy of 0.75 eV was assigned to the H atom in the direction opposite to the Si–H bond. For more details, see [Miyamoto 1999]

We performed TDDFT-MD simulations [Miyamoto 2000] for this system. From the constrained LDA total-energy calculation, the energy required to promote an electron from a Si-H bonding state to a nearby Si-H anti-bonding state was found to be 4 eV, in agreement with the experimental value of 3.5 eV.

When the simulation started without giving an initial velocity to the ions, the ions were found to oscillate around their initial positions without showing any nonadiabatic transition over 30 fs. From an analysis of the trajectory, the Si-H bond was found weakened by 20%. To obtain the activation barrier for the dissociation, another simulation was performed, this time giving the ions an initial kinetic energy of 0.45 eV. The activation barrier height was found to be 0.23 eV from an analysis of the potential energy shown in Fig. 27.3. For the ground state, the corresponding value is 1.79 eV. Such significant reduction of the activation barrier can explain the efficient recovery of the carrier density by laser illumination.

Another example is the H desorption from a Si(111) surface. The simulation was done in order to understand the mechanism for a scanning tunneling microscopy (STM) experiment, in which the desorption was controlled by injecting carriers [Shen 1995, Foley 1998]. The mechanism had been interpreted as a direct dissociation of H along the potential energy surface corresponding to Si–H $\sigma \rightarrow \sigma^*$ excitation. The extremely low yield for the H-desorption found experimentally was attributed to spontaneous recombination of the σ^* electron and σ hole [Shen 1995, Foley 1998].

The TDDFT-MD simulation showed that, contrary to the direct recombination, the excited-state decayed with the σ-hole state being mixed with the valence band [Miyamoto 2000]. In contrast, the σ^*-electron state was not

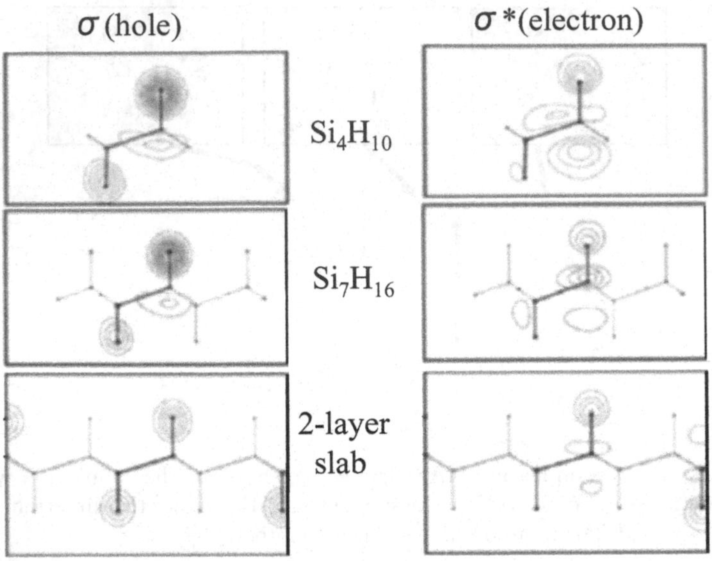

Fig. 27.4. Charge contour maps of an excited electron and holes of Si_4H_{10} and Si_7H_{16} clusters and a 2-layer slab for the Si(111) surface. The *left panels* are hole state (σ) and the *right panels* are excited electron (σ^*)

mixed with other conduction bands. The calculated lifetime of the $\sigma \rightarrow \sigma^*$ excitation is dependent on the size of the system and becomes shorter when the system becomes larger.

The models used for the simulation were clusters, i.e., Si_4H_{10} and Si_7H_{16}, periodically located in a large enough unit cell, and repeated slabs consisting of 2 and 8 layers. Figure 27.4 shows charge contour maps for the σ and σ^* levels for the three clusters and for the 2-layered slab. Note that, as the system size is increased, the degree of localization of the σ and σ^* states is weaker. This makes the force for the Si–H dissociation weaker and the lifetime of the excitation shorter. The lifetime for the 8-layer slab was found to be shorter than 10 fs. Because of this, the Si–H bond length started oscillating soon after the beginning of the simulation.

Note that the results depend on the functional form for the exchange-correlation potential (as shown in Fig. 27.5), although the influence on the qualitative aspects is not so obvious: the amplitude for the Si–H bond length oscillation was found to be larger for the GGA, as the excited-state PES of the GGA is steeper than that of the LDA.

27.4.2 Carbon Nanotubes

After intense research on carbon fibers, it was found that a graphene sheet could be rolled into a tubular form [Iijima 1991]. These particularly thin fibers were called "carbon nanotubes" (CNTs). Theoretical works

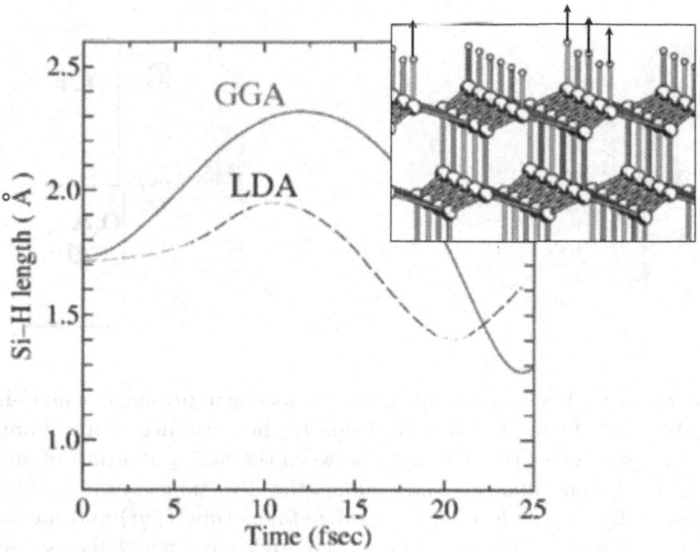

Fig. 27.5. Time evolution of a bond length of one of the surface Si–H bonds, denoted by arrows shown in the inset. The LDA and GGA results are shown. For more details, see [Miyamoto 2000]

[Hamada 1992, Saito 1992, Mintmire 1992, Tanaka 1992] predicted that the electronic properties of CNTs depended on the chirality of the winding. This finding opened a door for electronic applications of CNTs. Real CNTs have imperfections (vacancies, structural defects, and impurities) which affect the conductivity [Igami 1999, Choi 2000]. Since the experimental study of defects and impurities in CNTs has just begun, *ab initio* simulations play an important role guiding the experiments. To address the issues of defects and impurities, we propose (theoretically) to eliminate impurity atoms with the aid of electronic excitations.

Efficient Elimination of Oxygen Impurities from CNTs

Recent growth technology using chemical vapor deposition (CVD) enabled us to fabricate CNTs attached to metal electrodes. This is expected to be an indispensable technique for building devices with CNTs. In order to eliminate the carbon soot among the produced CNTs, CVD needs either alcohol [Maruyama 2002] or water [Hata 2004] in addition to hydrocarbons, but there is a risk of O-contamination. Indeed, an impurity O atom can form a C–O–C complex as a part of the honeycomb network, especially in defected nanotubes [Mazzoni 1999].

Figure 27.6 (a) shows a theoretically determined structure of an O-impurity in a CNT, and the potential modulation of the valence electrons

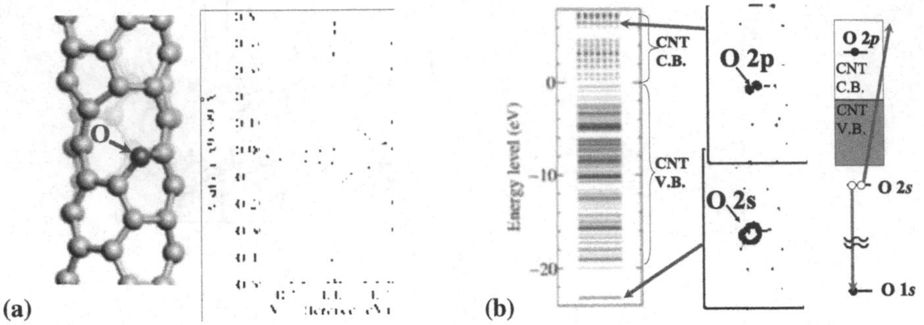

Fig. 27.6. (**a**) *Left*: Atomic structure of the nanotube in presence of an O-impurity. *Right*: Modulation of the SCF potential due to the existence of an O-impurity in the CNT. The plot shows the difference between the SCF potentials of an oxidized and non-oxidized nanotube, averaged along the directions perpendicular to the nanotube axis. (**b**) Energy level of the oxidized nanotube (*left*) and partial charge density of the O 2*s* and 2*p* levels (*center*). The rightmost panel shows a schematic picture of the Auger process initiated by the O 1*s* to 2*p* core excitation

due to the existence of the O-impurity. This modulation is obtained by taking the difference between the SCF potentials of the CNTs with and without the oxygen impurity. This modulation indicates a strong perturbation of the electronic structure, which causes a reduction in the conductivity of the CNTs. Since the C–O chemical bond is much stronger than the C–C bond, oxygen extraction using thermal energy has the risk of breaking the C–C bond network. We proposed the use of electronic excitations to eliminate this O-impurity without damaging the rest of the C–C bond network [Miyamoto 2004c].

Figure 27.6 (b) shows the electronic energy levels of the oxidized CNT shown in Fig. 27.6 (a), labeled as O 2*s* (below the bottom of CNT valence bands) and O 2*p* (in the resonance of the CNT conduction bands). These are highly localized states at an O-impurity site in the CNT. Although notations of the O atomic orbitals are used in this figure, these levels (2*s* and 2*p*) indeed contain 2*p* admixture of the neighboring C atoms, respectively in bonding and anti-bonding phases. One can thus expect that electronic excitation from the 2*s* level to the 2*p* level can be used as a tool to break the C–O–C complex. However, according to our TDDFT-MD simulations [Miyamoto 2004c], this excitation only provokes large oscillations of the position of the O-impurity atom and do not break the complex.

However, we think that a resonant Auger process can induce the breaking of the C–O–C complex. The Auger process can be initiated by the O 1*s* core excitation into the O 2*p* level. Here, the O 2*p* level is the one displayed in Fig. 27.6 (b) instead of the atomic O 2*p* level. This excitation should be

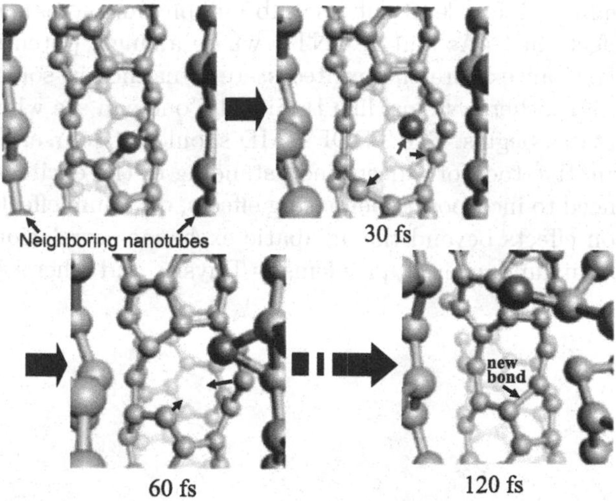

Fig. 27.7. Snapshots of O-emission from the nanotube wall with the Auger final state displayed in the right of (b). In the last snapshot, a new C–C bond, formed upon O emission, is indicated. For more details, see [Miyamoto 2004c]

achieved by either X-ray or electron beam irradiation of the nanotube with excitation energy of 530 eV, and is very likely to cause Auger decay of the remaining two holes in the O 2*s* level as shown in the right of Fig. 27.6 (b).

Since present TDDFT simulations use a single-particle representation for the electronic wave functions, they are not able to treat directly many-body Auger processes. However, it is reasonable to assume that this Auger process is spontaneous. Therefore we start the TDDFT-MD simulation by setting the Auger final state as the initial condition. Figure 27.7 shows ultra-fast dynamics of O-emission from the CNT wall, starting from the Auger final state. This calculation also shows formation of a new C–C bond which heals the large vacancy formed just after oxygen emission. We further tested introduction of an H_2 molecule which can capture the emitted O atom to prevent re-oxidation of the CNT by the emitted O atom.

We believe that this simulation shows a feasible technology to clean up CNTs fabricated by recent CVD technology as a part of electronic devices, namely as a "post-fabrication process".

27.5 Concluding Remarks

We presented the TDDFT-MD scheme, an approach to simulate excited-state dynamics in extended systems. This scheme adopts a classical approximation for the ions and a mean-field approximation for the electron-ion interaction. When using robust numerical algorithms, it is possible to follow, in a stable way, the dynamics for at least sub-ps (sub-million time-steps). For such systems like defects in GaAs and in CNTs, where a single potential surface is concerned, one can explore the excited-state dynamics in some detail with this scheme. For other systems, like H/Si(111), one can see when a decay of the excited-states begins. The TDDFT-MD should, of course, be recognized as only a primitive tool for a deep understanding of the excited-state. To go beyond, we need to incorporate dephasing effects, quantum effects of the ions, many-electron effects beyond the adiabatic exchange-correlation, etc. These are difficult, but fundamental, problems in Physics and Chemistry.

Part VI

New Frontiers

28 Back to the Ground-State: Electron Gas

M. Lein and E.K.U. Gross

28.1 Introduction

In this chapter, we explore how concepts of time-dependent density functional theory can be useful in the search for more accurate approximations of the *ground-state* exchange-correlation (xc) energy functional.

Within stationary density functional theory, all observables related to the ground-state of an interacting many-electron system can be written as functionals of the ground-state density $n(\boldsymbol{r})$ by virtue of the Hohenberg-Kohn theorem [Hohenberg 1964]. Using the Kohn-Sham approach [Kohn 1965], one considers a non-interacting many-electron system with the same ground-state density as the interacting system. The total energy E of the interacting system is then split into

$$E = T_{\mathrm{KS}} + V_{\mathrm{ext}} + E_{\mathrm{Hxc}} = T_{\mathrm{KS}} + V_{\mathrm{ext}} + E_{\mathrm{H}} + E_{\mathrm{xc}} \,. \tag{28.1}$$

Here T_{KS} is the kinetic energy of the Kohn-Sham system, the second contribution,

$$V_{\mathrm{ext}} = \int \mathrm{d}^3r \; n(\boldsymbol{r}) v_{\mathrm{ext}}(\boldsymbol{r}) \,, \tag{28.2}$$

is the potential energy due to the external potential $v_{\mathrm{ext}}(\boldsymbol{r})$, and

$$E_{\mathrm{H}} = \frac{1}{2} \int \mathrm{d}^3r \int \mathrm{d}^3r' \, \frac{n(\boldsymbol{r})n(\boldsymbol{r}')}{|\boldsymbol{r} - \boldsymbol{r}'|} \tag{28.3}$$

is the Hartree energy. The xc energy E_{xc} as well as the Hartree-xc energy E_{Hxc} are defined by (28.1). In the stationary theory, one usually tries to find approximate expressions for the xc energy in terms of the density or the Kohn-Sham orbitals. A different approach is presented in the following. We outline the derivation of the adiabatic-connection fluctuation-dissipation formula which links the ground-state energy to the dynamical response function (cf. [Langreth 1975, Gunnarsson 1976]). We then use TDDFT to relate the correlation energy to the exchange-correlation kernel f_{xc}, and we test the resulting formula by applying it to the uniform electron gas using various approximate exchange-correlation kernels (cf. [Lein 2000b]).

M. Lein and E.K.U. Gross: *Back to the Ground-State: Electron Gas*, Lect. Notes Phys. **706**, 423–434 (2006)
DOI 10.1007/3-540-35426-3_28

28.2 Adiabatic Connection

The principle of adiabatic connection refers to a smooth turning-on of the electron-electron coupling constant λ from zero to unity while keeping the ground-state density fixed, i.e., for each value of λ, the external potential $v_\lambda(\boldsymbol{r})$ is chosen such that the ground-state density of the system with electron-electron interaction $\lambda v_{\mathrm{ee}}(\boldsymbol{r},\boldsymbol{r}') = \lambda/|\boldsymbol{r}-\boldsymbol{r}'|$ equals the ground-state density of the fully interacting system. The Hamiltonian

$$\hat{H}(\lambda) = \hat{T} + \hat{V}(\lambda) + \lambda \hat{V}_{\mathrm{ee}}\,, \tag{28.4}$$

with $\hat{V}(\lambda) = \int \mathrm{d}^3r\, \hat{n}(\boldsymbol{r}) v_\lambda(\boldsymbol{r})$, thus interpolates between the Kohn-Sham Hamiltonian $\hat{H}_{\mathrm{KS}} = \hat{H}(0)$ and the fully interacting Hamiltonian $\hat{H} = \hat{H}(1)$. We first rewrite the Hartree-xc energy as

$$\begin{aligned} E_{\mathrm{Hxc}} &= E - V_{\mathrm{ext}} - T_{\mathrm{KS}} = E - V_{\mathrm{ext}} - E_{\mathrm{KS}} + V_{\mathrm{KS}} \\ &= \int_0^1 \mathrm{d}\lambda\, \frac{\mathrm{d}}{\mathrm{d}\lambda}[E(\lambda) - V(\lambda)]\,. \end{aligned} \tag{28.5}$$

According to the Raleigh-Ritz principle, the ground state $\Psi(\lambda)$ minimizes the expectation value of $\hat{H}(\lambda)$ so that we have

$$\frac{\mathrm{d}E(\lambda)}{\mathrm{d}\lambda} = \frac{\mathrm{d}}{\mathrm{d}\lambda}\langle\Psi(\lambda)|\hat{H}(\lambda)|\Psi(\lambda)\rangle = \langle\Psi(\lambda)|\frac{\mathrm{d}\hat{H}(\lambda)}{\mathrm{d}\lambda}|\Psi(\lambda)\rangle\,. \tag{28.6}$$

Similarly, the fact that the density is independent of λ leads to

$$\frac{\mathrm{d}V(\lambda)}{\mathrm{d}\lambda} = \int \mathrm{d}^3r\; n(\boldsymbol{r})\frac{\mathrm{d}v_\lambda(\boldsymbol{r})}{\mathrm{d}\lambda} = \langle\Psi(\lambda)|\frac{\mathrm{d}\hat{V}(\lambda)}{\mathrm{d}\lambda}|\Psi(\lambda)\rangle\,. \tag{28.7}$$

After inserting $\mathrm{d}\hat{H}(\lambda)/\mathrm{d}\lambda = \mathrm{d}\hat{V}(\lambda)/\mathrm{d}\lambda + \hat{V}_{\mathrm{ee}}$ into (28.6) and substituting (28.6) and (28.7) into (28.5), we arrive at the adiabatic-connection formula

$$E_{\mathrm{Hxc}} = \int_0^1 \mathrm{d}\lambda\; \langle\Psi(\lambda)|\hat{V}_{\mathrm{ee}}|\Psi(\lambda)\rangle = \int_0^1 \mathrm{d}\lambda\; V_{\mathrm{ee}}(\lambda)\,. \tag{28.8}$$

To relate this energy to the response function (cf. Sect. 1.4), we first write the electron-electron interaction

$$\hat{V}_{\mathrm{ee}} = \frac{1}{2}\sum_{j\neq k}^{N} \frac{1}{|\boldsymbol{r}_j - \boldsymbol{r}_k|} = \frac{1}{2}\sum_{j\neq k}^{N} \int \mathrm{d}^3r \int \mathrm{d}^3r'\, \frac{\delta(\boldsymbol{r}-\boldsymbol{r}_j)\,\delta(\boldsymbol{r}'-\boldsymbol{r}_k)}{|\boldsymbol{r}-\boldsymbol{r}'|} \tag{28.9}$$

in terms of the density operator $\hat{n}(\boldsymbol{r}) = \sum_j \delta(\boldsymbol{r}-\boldsymbol{r}_j)$,

$$\hat{V}_{\mathrm{ee}} = \frac{1}{2}\int \mathrm{d}^3r \int \mathrm{d}^3r'\, \frac{1}{|\boldsymbol{r}-\boldsymbol{r}'|}\{\hat{n}(\boldsymbol{r})\hat{n}(\boldsymbol{r}') - \hat{n}(\boldsymbol{r})\delta(\boldsymbol{r}-\boldsymbol{r}')\}\,. \tag{28.10}$$

We then find

$$V_{\mathrm{ee}}(\lambda) = \frac{1}{2}\int \mathrm{d}^3r \int \mathrm{d}^3r' \, \frac{1}{|\boldsymbol{r}-\boldsymbol{r}'|} \times \left\{ \langle \Psi(\lambda)|\hat{n}(\boldsymbol{r})\hat{n}(\boldsymbol{r}')|\Psi(\lambda)\rangle - n(\boldsymbol{r})\delta(\boldsymbol{r}-\boldsymbol{r}') \right\} . \quad (28.11)$$

The expectation value of the product of density operators can be written as

$$\langle \Psi(\lambda)|\hat{n}(\boldsymbol{r})\hat{n}(\boldsymbol{r}')|\Psi(\lambda)\rangle = n(\boldsymbol{r})n(\boldsymbol{r}') + S^{\lambda}(\boldsymbol{r}t,\boldsymbol{r}'t')\big|_{t=t'} , \quad (28.12)$$

where the direct correlation function

$$S^{\lambda}(\boldsymbol{r}t,\boldsymbol{r}'t') = \langle \Psi(\lambda)|\hat{\tilde{n}}(\boldsymbol{r},t)_{\mathrm{H}}\hat{\tilde{n}}(\boldsymbol{r}',t')_{\mathrm{H}}|\Psi(\lambda)\rangle \quad (28.13)$$

characterizes the density fluctuations in the system. Here, $\hat{\tilde{n}}(\boldsymbol{r},t)_{\mathrm{H}} = \hat{n}(\boldsymbol{r},t)_{\mathrm{H}} - n(\boldsymbol{r})$ is the density deviation operator in the Heisenberg picture. The direct correlation function can be expressed in terms of its temporal Fourier transform as

$$S^{\lambda}(\boldsymbol{r}t,\boldsymbol{r}'t') = \int_0^{\infty} \frac{\mathrm{d}\omega}{2\pi} S^{\lambda}(\boldsymbol{r},\boldsymbol{r}',\omega) e^{-i\omega(t-t')} . \quad (28.14)$$

The lower boundary of the integration in (28.14) has been set to zero because $S^{\lambda}(\boldsymbol{r},\boldsymbol{r}',\omega)$ vanishes for $\omega < 0$. The direct correlation function is related to the response function $\chi^{\lambda}(\boldsymbol{r},\boldsymbol{r}',\omega)$ by the zero-temperature fluctuation-dissipation theorem [Pines 1966],

$$-2\Im\{\chi^{\lambda}(\boldsymbol{r},\boldsymbol{r}',\omega)\} = S^{\lambda}(\boldsymbol{r},\boldsymbol{r}',\omega), \quad \omega > 0 . \quad (28.15)$$

Equation (28.12) can therefore be transformed into

$$\langle \Psi(\lambda)|\hat{n}(\boldsymbol{r})\hat{n}(\boldsymbol{r}')|\Psi(\lambda)\rangle = n(\boldsymbol{r})n(\boldsymbol{r}') - \frac{1}{\pi}\Im \int_0^{\infty} \mathrm{i}\,\mathrm{d}u \, \chi^{\lambda}(\boldsymbol{r},\boldsymbol{r}',\mathrm{i}u) , \quad (28.16)$$

where we have moved the integration path onto the imaginary axis in the complex-frequency plane. For numerical evaluations, the integration over imaginary frequencies is more suitable than the real-frequency integration because it avoids the poles in the response function related to the excitation energies of the system. By combining (28.8), (28.11), and (28.16) and exploiting that $\chi^{\lambda}(\boldsymbol{r},\boldsymbol{r}',\mathrm{i}u)$ is real-valued, we obtain the xc energy

$$E_{\mathrm{xc}} = -\frac{1}{2}\int_0^1 \mathrm{d}\lambda \int \mathrm{d}^3r \int \mathrm{d}^3r' \, \frac{1}{|\boldsymbol{r}-\boldsymbol{r}'|} \times \left\{ n(\boldsymbol{r})\delta(\boldsymbol{r}-\boldsymbol{r}') + \frac{1}{\pi}\int_0^{\infty} \mathrm{d}u \, \chi^{\lambda}(\boldsymbol{r},\boldsymbol{r}',\mathrm{i}u) \right\} . \quad (28.17)$$

One can verify by explicit evaluation that the exchange energy is recovered by inserting the response function $\chi_{\mathrm{KS}}(\boldsymbol{r},\boldsymbol{r}',\mathrm{i}u)$ of the non-interacting KS system into (28.17),

$$E_{\mathrm{x}} = -\frac{1}{2}\int_0^1 \mathrm{d}\lambda \int \mathrm{d}^3r \int \mathrm{d}^3r' \, \frac{1}{|\boldsymbol{r}-\boldsymbol{r}'|} \times \left\{ n(\boldsymbol{r})\delta(\boldsymbol{r}-\boldsymbol{r}') + \frac{1}{\pi}\int_0^\infty \mathrm{d}u \; \chi_{\mathrm{KS}}(\boldsymbol{r},\boldsymbol{r}',iu) \right\} . \quad (28.18)$$

Comparing (28.17) and (28.18), we obtain an expression for the correlation energy,

$$E_{\mathrm{c}} = -\frac{1}{2\pi}\int_0^1 \mathrm{d}\lambda \int \mathrm{d}^3r \int \mathrm{d}^3r' \, \frac{1}{|\boldsymbol{r}-\boldsymbol{r}'|} \times \int_0^\infty \mathrm{d}u \; \left\{ \chi^\lambda(\boldsymbol{r},\boldsymbol{r}',iu) - \chi_{\mathrm{KS}}(\boldsymbol{r},\boldsymbol{r}',iu) \right\} . \quad (28.19)$$

In order to use the last equation for practical calculations, we have to approximate the response function $\chi^\lambda(\boldsymbol{r},\boldsymbol{r}',iu)$. A possible route to such approximations is provided by the Dyson-type equation (cf. Sect. 1.4)

$$\chi^\lambda(\boldsymbol{r},\boldsymbol{r}',\omega) - \chi_{\mathrm{KS}}(\boldsymbol{r},\boldsymbol{r}',\omega) = \int \mathrm{d}^3r_1 \int \mathrm{d}^3r_2 \, \chi_{\mathrm{KS}}(\boldsymbol{r},\boldsymbol{r}_1,\omega) \, f^\lambda_{\mathrm{Hxc}}(\boldsymbol{r}_1,\boldsymbol{r}_2,\omega) \, \chi^\lambda(\boldsymbol{r}_2,\boldsymbol{r}',\omega) , \quad (28.20)$$

where

$$f^\lambda_{\mathrm{Hxc}}(\boldsymbol{r}_1,\boldsymbol{r}_2,\omega) = \frac{\lambda}{|\boldsymbol{r}_1-\boldsymbol{r}_2|} + f^\lambda_{\mathrm{xc}}(\boldsymbol{r}_1,\boldsymbol{r}_2,\omega) . \quad (28.21)$$

One way of calculating E_{c} is to approximate χ^λ and f^λ_{xc} independently of each other on the right-hand side of (28.20) and then substitute into (28.19). In another approach, one chooses a given approximation for f^λ_{xc} and solves the integral equation (28.20) for χ^λ. The solution of the Dyson-type equation is demanding in general. In the uniform electron gas, however, the translational invariance dictates that the response functions and the xc kernel do not depend independently on two positions but only on the difference between the two coordinates. These quantities can then be expressed in terms of their Fourier transforms:

$$\chi^\lambda(\boldsymbol{r},\boldsymbol{r}',\omega) = \int \frac{\mathrm{d}^3q}{(2\pi)^3} \, \chi^\lambda(q,\omega) e^{i(\boldsymbol{r}-\boldsymbol{r}')\boldsymbol{q}} , \quad (28.22a)$$

$$f^\lambda_{\mathrm{xc}}(\boldsymbol{r},\boldsymbol{r}',\omega) = \int \frac{\mathrm{d}^3q}{(2\pi)^3} \, f^\lambda_{\mathrm{xc}}(q,\omega) e^{i(\boldsymbol{r}-\boldsymbol{r}')\boldsymbol{q}} . \quad (28.22b)$$

The integral in the Dyson equation is then transformed into a simple product, and the solution is found to be

$$\chi^\lambda(q,\omega) = \frac{\chi_{\mathrm{KS}}(q,\omega)}{1 - \chi_{\mathrm{KS}}(q,\omega) f^\lambda_{\mathrm{Hxc}}(q,\omega)} . \quad (28.23)$$

The response function $\chi_{\mathrm{KS}}(q,\omega)$ of the non-interacting electron gas is the well-known Lindhard function. At imaginary frequency iu, it is given by [von Barth 1972]

$$\chi_{\mathrm{KS}}(q,iu) = \frac{k_{\mathrm{F}}}{2\pi^2}\left\{\frac{Q^2-\tilde{u}^2-1}{4Q}\ln\frac{\tilde{u}^2+(Q+1)^2}{\tilde{u}^2+(Q-1)^2}\right.$$
$$\left. -1+\tilde{u}\arctan\frac{1+Q}{\tilde{u}}+\tilde{u}\arctan\frac{1-Q}{\tilde{u}}\right\}, \tag{28.24}$$

with

$$Q=\frac{q}{2k_{\mathrm{F}}},\quad \tilde{u}=\frac{u}{qk_{\mathrm{F}}},\quad k_{\mathrm{F}}^3=3\pi^2 n\,. \tag{28.25}$$

The correlation energy per electron e_{c} follows from (28.19) and (28.23):

$$e_{\mathrm{c}} = -\frac{1}{\pi^2 n}\int_0^\infty \mathrm{d}q\int_0^1 \mathrm{d}\lambda\int_0^\infty \mathrm{d}u\,\frac{[\chi_{\mathrm{KS}}(q,iu)]^2 f_{\mathrm{Hxc}}^{\lambda}(q,iu)}{1-\chi_{\mathrm{KS}}(q,iu)f_{\mathrm{Hxc}}^{\lambda}(q,iu)}\,. \tag{28.26}$$

Since the Lindhard function is known, only the xc kernel has to be approximated in (28.26).

28.3 Scaling Properties

In the following, we show that the evaluation of the correlation energy is simplified by a scaling property of the xc kernel: given an approximation for the xc kernel in the fully interacting non-uniform system, the xc kernel for any value of the coupling constant follows immediately.

We are interested in the xc kernel of an interacting system in its ground state. This quantity describes the infinitesimal change of the xc potential due to the influence of a small perturbation. Provided that the time evolution of the slightly perturbed system starts from the ground state, and that we also choose the initial Kohn-Sham state to be the ground state of the Kohn-Sham system, the xc potential can be written as a functional of the time-dependent density only. The xc potential then obeys a scaling relation in the form [Hessler 1999]

$$v_{\mathrm{xc}}^{\lambda}[n](\boldsymbol{r},t) = \lambda^2\, v_{\mathrm{xc}}[n'](\lambda\boldsymbol{r},\lambda^2 t)\,, \tag{28.27}$$

with

$$n'(\boldsymbol{r},t) = \lambda^{-3}\, n(\boldsymbol{r}/\lambda, t/\lambda^2)\,. \tag{28.28}$$

A similar relation for the xc kernel follows by taking the functional derivative of (28.27) with respect to the density:

$$f_{\mathrm{xc}}^{\lambda}[n]\,(\boldsymbol{r}t,\boldsymbol{r}'t') = \lambda^4 f_{\mathrm{xc}}[n']\,(\lambda\boldsymbol{r}\;\lambda^2 t,\lambda\boldsymbol{r}'\;\lambda^2 t')\,. \tag{28.29}$$

As we are dealing with the linear-response regime, the xc kernel depends on the difference $(t-t')$ only. Hence we can evaluate the Fourier transform of (28.29) with respect to $(t-t')$:

$$f_{\mathrm{xc}}^{\lambda}[n]\,(\boldsymbol{r},\boldsymbol{r}',\omega) = \lambda^2\, f_{\mathrm{xc}}[n']\,(\lambda\boldsymbol{r},\lambda\boldsymbol{r}',\omega/\lambda^2)\,. \tag{28.30}$$

In the uniform electron gas, the density is constant in space and the xc kernel depends only on the difference $(\boldsymbol{r}-\boldsymbol{r}')$. Then Fourier transformation of (28.30) with respect to $(\boldsymbol{r}-\boldsymbol{r}')$ yields

$$f_{\mathrm{xc}}^{\lambda}[n]\,(q,\omega) = \lambda^{-1}\, f_{\mathrm{xc}}[n/\lambda^3]\,(q/\lambda,\,\omega/\lambda^2)\,. \tag{28.31}$$

The electron-gas literature often uses the local-field factor $G(q,\omega)$ instead of the xc kernel. At coupling constant λ the two quantities are related by

$$G^{\lambda}(q,\omega) = -\frac{q^2}{4\pi\lambda}\, f_{\mathrm{xc}}^{\lambda}(q,\omega)\,. \tag{28.32}$$

The scaling law for the local-field factor reads

$$G^{\lambda}[n]\,(q,\omega) = G[n/\lambda^3]\,(q/\lambda,\omega/\lambda^2)\,. \tag{28.33}$$

Equation (28.33) shows that the limit $\lambda \to 0$ is closely connected to the high-density limit of $G(q,\omega)$. This becomes even more apparent if we write the local-field factor as a function of the Wigner-Seitz radius r_{s}, the reduced wave vector q/k_{F}, and the reduced frequency $\omega/\omega_{\mathrm{F}}$, with

$$\frac{4\pi}{3}{r_{\mathrm{s}}}^3 = \frac{1}{n}, \qquad \omega_{\mathrm{F}} = \frac{k_{\mathrm{F}}^2}{2}\,. \tag{28.34}$$

We then obtain

$$G^{\lambda}(r_{\mathrm{s}},q/k_{\mathrm{F}},\omega/\omega_{\mathrm{F}}) = G\,(\lambda r_{\mathrm{s}},q/k_{\mathrm{F}},\omega/\omega_{\mathrm{F}})\,. \tag{28.35}$$

28.4 Approximations for the xc Kernel

In the uniform electron gas, a number of approximations are available for the xc kernel. Denoting the exact wave-vector dependent and frequency dependent xc kernel of the uniform gas as $f_{\mathrm{xc}}^{\mathrm{hom}}(q,\omega)$, we consider the following approximations:

Random Phase Approximation (RPA) : $f_{\mathrm{xc}}(q,\omega) \equiv 0$.

Adiabatic Local Density Approximation (ALDA) : This is the long-wavelength limit of the static xc kernel:

$$f_{\mathrm{xc}}^{\mathrm{ALDA}}(q,\omega) = \lim_{q\to 0} f_{\mathrm{xc}}^{\mathrm{hom}}(q,0)\,. \tag{28.36}$$

It can readily be expressed in terms of the xc energy per electron e_{xc}:

$$f_{\mathrm{xc}}^{\mathrm{ALDA}}(q,\omega) = \frac{\mathrm{d}^2}{\mathrm{d}n^2}\,[n e_{\mathrm{xc}}(n)]\,. \tag{28.37}$$

Parametrization by Corradini et al. (see [Corradini 1998]): This approximation for the static xc kernel of the uniform electron gas is a fit to the quantum Monte Carlo data published by Moroni, Ceperley, and Senatore [Moroni 1995]. It satisfies the known asymptotic small-q and large-q limits. Since it interpolates between different values of r_s, it can be evaluated for arbitrary values of the density, in contrast to the original parametrization given in [Moroni 1995].

Parametrization by Richardson and Ashcroft (RA) (see [Richardson 1994]; see also [Lein 2000b] for corrections of typographical errors in the original parametrization): This approximation for the xc kernel at imaginary frequencies is based not upon Monte Carlo data but upon results of numerical calculations by Richardson and Ashcroft. It is constructed to satisfy many known exact conditions. The xc kernel is constructed from Richardson and Ashcroft's local-field factor contributions G_n and G_s via

$$f_{\rm xc}^{\rm RA}(q, {\rm i}u) = -\frac{4\pi}{q^2}\Big[G_s(Q, {\rm i}U) + G_n(Q, {\rm i}U)\Big], \tag{28.38}$$

with

$$Q = \frac{q}{2k_{\rm F}}, \quad U = \frac{u}{4\omega_{\rm F}}. \tag{28.39}$$

For the present application, we are fortunate that the RA kernel was derived for imaginary frequencies. Due to its complicated structure near the real axis, the analytic continuation of the xc kernel between imaginary and real axis is not straightforward [Sturm 2000], although it was demonstrated that the continuation of the RA kernel into the complex plane yields good results for the plasmon excitation of the homogeneous electron gas [Tatarczyk 2001]. We also test the static limit of the RA kernel,

$$f_{\rm xc}^{\rm static\,RA}(q, {\rm i}u) = f_{\rm xc}^{\rm RA}(q, 0), \tag{28.40}$$

in order to compare with the static Corradini approximation. (For a comparison of the RA and Monte Carlo $f_{\rm xc}$ in the static limit, see Fig. 3 of [Moroni 1995].) As a dynamic but spatially local approximation we may use the long-wavelength limit of $f_{\rm xc}^{\rm RA}(q, {\rm i}u)$,

$$f_{\rm xc}^{\rm local\,RA}(q, {\rm i}u) = f_{\rm xc}^{\rm RA}(0, {\rm i}u), \tag{28.41}$$

which we refer to as "local RA".

An approximate xc kernel that is readily applicable to inhomogeneous systems is given by the Petersilka-Gossmann-Gross (PGG) kernel [Petersilka 1996a]. This frequency-independent exchange-only approximation was derived in the context of the time-dependent optimized effective potential method [Ullrich 1995a]. Its real-space version reads

$$f_{\rm x}^{\rm PGG}(\boldsymbol{r}, \boldsymbol{r}', \omega) = -\frac{2}{|\boldsymbol{r}-\boldsymbol{r}'|}\frac{|\sum_i n_i \varphi_i(\boldsymbol{r})\varphi_i^*(\boldsymbol{r}')|^2}{n(\boldsymbol{r})n(\boldsymbol{r}')}, \tag{28.42}$$

where $\varphi_i(\boldsymbol{r})$ and n_i are the ground-state KS orbitals and their occupation numbers (0 or 1). In the uniform gas, transformation to momentum space yields:

$$f_{\mathrm{x}}^{\mathrm{PGG}}(q,\omega) = -\frac{3\pi}{10k_{\mathrm{F}}^2}\left\{\left(\frac{2}{Q}-10Q\right)\ln\frac{1+Q}{|1-Q|}\right.$$
$$\left. + (2Q^4-10Q^2)\ln\left[(1+\frac{1}{Q})\left|1-\frac{1}{Q}\right|\right] + 11 + 2Q^2\right\}, \quad (28.43)$$

where $Q = q/(2k_{\mathrm{F}})$. Due to its exchange-only nature, the PGG kernel, taken at coupling constant λ, is simply proportional to λ.

In the following, we evaluate the correlation energy of the uniform electron-gas, (28.26), for the different xc kernels. We expect that RA's parametrization is close to the exact uniform-gas xc kernel and that the Corradini parametrization is close to the exact static limit. The ALDA is the exact long-wavelength limit of the static xc kernel. Hence, a comparison between these three cases will clarify the importance of both wave-vector and frequency dependence of the xc kernel (for the correlation energy).

Accurate correlation energies are given for example by the parametrization of Perdew and Wang in [Perdew 1992a]. We refer to these values as the "exact" correlation energy $e_{\mathrm{c}}^{\mathrm{exact}}$. For each choice of xc kernel, the difference between the correlation energy e_{c} and the exact value $e_{\mathrm{c}}^{\mathrm{exact}}$ is shown in Fig. 28.1 as a function of the density parameter r_{s} in the range $r_{\mathrm{s}} = 0\ldots15$.

The RA results differ by less than 0.02 eV from the exact values, i.e., the RA kernel reproduces the exact correlation energy nearly perfectly. With a deviation of less than 0.1 eV, the Corradini approximation gives a good estimate as well. We note that the result of the *static* version of the RA

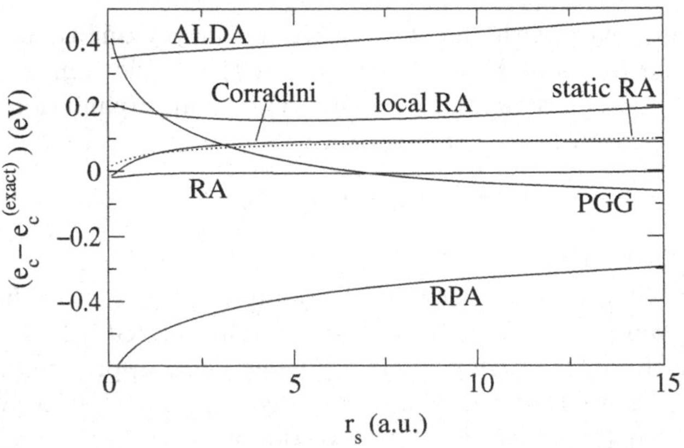

Fig. 28.1. Difference between approximate correlation energies and the exact correlation energy per electron in the uniform electron gas

formula lies almost on top of the Corradini curve. From this we infer that the small error produced by the Corradini parametrization is in fact due to its static nature. We conclude that neglecting the frequency dependence causes an error typically smaller than 0.1 eV.

It is clearly seen in Fig. 28.1 that the RPA approximation, which neglects the xc kernel completely, makes the correlation energy too negative. The inclusion of the simplest possible choice of kernel, the ALDA kernel, severely over-corrects e_c, so that the absolute deviation from the exact correlation energy remains about the same as in RPA. Furthermore, Fig. 28.1 shows that the local RA (dynamic approximation) performs better than ALDA, but worse than Corradini or static RA. Therefore, it seems that the wave-vector dependence of the xc kernel should be taken into account in order to obtain accurate correlation energies. In other words, the xc kernel is very non-local.

The PGG approximation behaves somewhat differently in that it yields an underestimate of the absolute value of e_c for small r_s and an overestimate for large r_s. It is a very good approximation in the range $r_s = 5 \ldots 10$. The behavior near $r_s = 0$ indicates that the PGG kernel differs from the exact exchange-only kernel, since exchange effects should dominate over correlation effects in the high-density limit.

To gain further insight into the effects of the q-dependence and the u-dependence of $f_{xc}(q, iu)$ in (28.26), we analyze the correlation energy into contributions from density fluctuations of different wave vectors q and imaginary frequencies iu. Equation (28.26) naturally defines a wave-vector analysis $e_c(q)$ if only the q-integration is written explicitly while the other integrations are incorporated in $e_c(q)$:

$$e_c = \int_0^\infty \mathrm{d}\left(\frac{q}{2k_F}\right) e_c(q) . \tag{28.44}$$

The exact wave-vector analysis is essentially given by the Fourier transform of the exact coupling-constant averaged correlation-hole density $n\bar{g}_c(r)$:

$$e_c^{\text{exact}}(q) = \frac{2k_F}{\pi} n\bar{g}_c(q) , \tag{28.45}$$

where

$$\bar{g}_c(q) = \int \mathrm{d}^3r \, \bar{g}_c(r) \exp(-i\boldsymbol{qr}) . \tag{28.46}$$

A parametrization of $\bar{g}_c(r)$ has been given by Perdew and Wang [Perdew 1992b]. (Although this parametrization misses the non-analytic behavior of $\bar{g}_c(q)$ at $q = 2k_F$, it is otherwise almost "exact".)

In Fig. 28.2 we compare approximate and "exact" wave-vector analyses for $r_s = 4$. While the RPA curve is too negative for all q, we note that ALDA is rather accurate for small q. The ALDA over-correction to e_c comes from

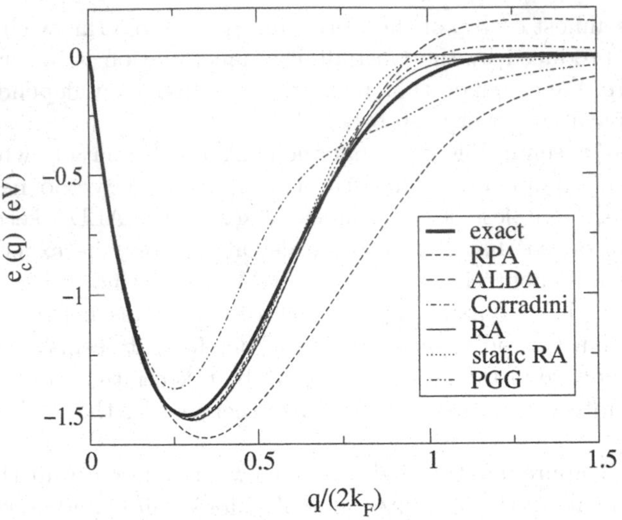

Fig. 28.2. Wave-vector analysis (28.44) of the correlation energy per electron of the uniform gas at $r_s = 4$. Approximations are compared to the "exact" wave-vector analysis of [Perdew 1992b]

positive contributions at large q. To a much smaller extent, the Corradini curve also exhibits this behavior. In general, however, it is close to the exact wave-vector analysis, as are RA and static RA. In the case of PGG, we note a substantial error cancellation between small and large q-values.

From the last three equations, it is apparent that the large-q behavior of $e_c(q)$, and therefore $f_{xc}(q, iu)$, is intimately related to the limit of $\bar{g}_c(r)$ at small inter-electron distances r ("on-top" limit). An unphysical divergence for $r \to 0$ is implied by the ALDA kernel and other semilocal approximations as was emphasized by Furche and Van Voorhis [Furche 2005b].

As a complement to the wave-vector analysis, we define the imaginary-frequency analysis $e_c(u)$ of the correlation energy by writing (28.26) as an integral over u:

$$e_c = \int_0^\infty d\left(\frac{u}{\omega_p}\right) e_c(u) \,, \tag{28.47}$$

with the plasma frequency ω_p given by

$$\omega_p^2 = 4\pi n \,. \tag{28.48}$$

Since $e_c(u)$ is not known exactly, we must restrict ourselves to a comparison among different approximations, as displayed in Fig. 28.3 (low imaginary frequencies) and Fig. 28.4 (high imaginary frequencies) for $r_s = 4$. In all cases, $e_c(u)$ starts with a finite negative value at $u = 0$ and then smoothly approaches zero. Assuming that the RA result is the most accurate one, we

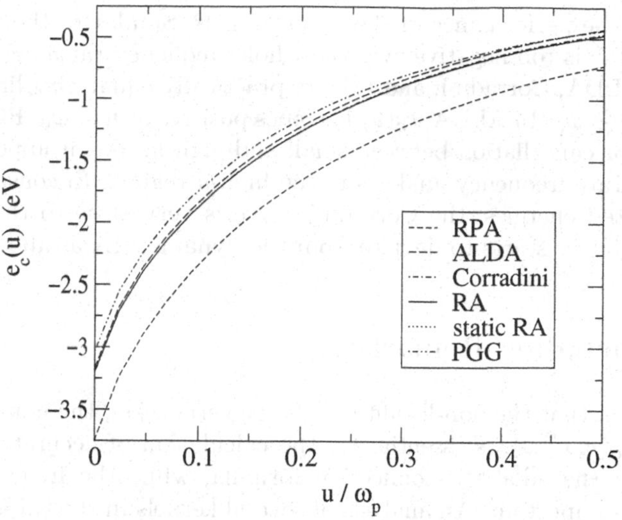

Fig. 28.3. Imaginary-frequency analysis (28.47) of the correlation energy per electron of the uniform gas at $r_s = 4$ in various approximations (low-frequency regime)

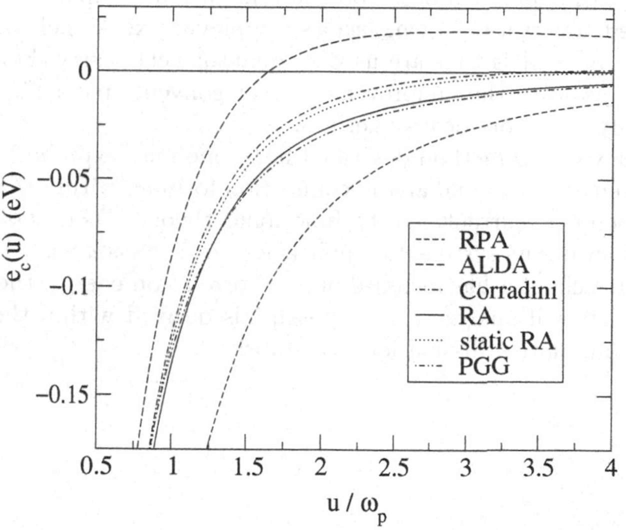

Fig. 28.4. Imaginary-frequency analysis (28.47) of the correlation energy per electron of the uniform gas at $r_s = 4$ in various approximations (high-frequency regime)

may assess the performance of the other kernels. Similar to the wave-vector analysis, RPA is too negative over the whole frequency range. For small frequencies, ALDA, Corradini, and RA are practically equal; the differences are located at $u \gtrsim \omega_\text{p}$. In ALDA, $e_\text{c}(u)$ becomes positive at $u \approx \omega_\text{p}$. PGG exhibits a slight error cancellation between small and high u. Yet, it appears to have a very accurate frequency analysis in the high-u regime. In consistency with the integrated energies, the Corradini curve is very close to the static RA curve. The latter starts to deviate from the dynamic RA at about $u \sim \omega_\text{p}$.

28.5 Concluding Remarks

We have seen that the non-locality of the xc kernel, i.e., the non-zero spatial range of $f_\text{xc}(\boldsymbol{r}, \boldsymbol{r}', \omega)$, is essential for the calculation of accurate correlation energies via the adiabatic-connection formula, while the frequency dependence is less important. An analysis of several kernels in the two-dimensional electron gas has been carried out by Asgari et al. [Asgari 2003]. Similar to the three-dimensional case, the frequency dependence was found to be of minor relevance. Contrary to the 3D gas, the over-correction by the ALDA kernel turned out to be less severe.

Recently, the approach based on the fluctuation-dissipation theorem has been applied to a variety of molecules, employing xc kernels derived from standard xc potentials that are used in quantum chemistry [Furche 2005b]. The results indicate that improvement over conventional DFT methods is achieved only with non-local xc kernels.

In analogy to the method described here, one may express the exchange-correlation energy by an adiabatic-connection formula within the framework of time-dependent current-density functional theory [Dion 2005]. This approach still awaits its systematic application to real systems.

While this chapter has focussed on the correlation energy, the next chapter (Chap. 29) will analyze the xc potentials derived within the adiabatic-connection fluctuation-dissipation framework.

29 The Exchange-Correlation Potential in the Adiabatic-Connection Fluctuation-Dissipation Framework

Y.M. Niquet and M. Fuchs

29.1 Introduction

As shown in the previous chapter, time-dependent DFT – which is basically an excited-state theory – also provides useful insight into the ground-state properties. Indeed, non-local approximations for the correlation energy $E_c[n]$ can be built upon TDDFT using the adiabatic connection and fluctuation dissipation (ACFD) theorems [Langreth 1975, Langreth 1977]. The random-phase approximation (RPA) is the prototype of these ACFD functionals. It has been applied to the homogeneous electron gas more than three decades ago [Nozières 1958, von Barth 1972, Vosko 1980], then to jellium slabs and surfaces [Pitarke 1998, Pitarke 2001, Dobson 1999, Kurth 1999]; the calculation of the ground-state energy surface of simple diatomic molecules [Furche 2001c, Fuchs 2002, Aryasetiawan 2002, Fuchs 2003, Fuchs 2005b] (H_2, N_2 ...) and solids [Miyake 2002, Marini 2006] (Si, Na, NaCl, *h*-BN ...) has been achieved much more recently. Though demanding, the RPA improves over the LDA and GGA in many respects: for example, the RPA accounts for long-range effects such as van der Waals interactions [Dobson 1996, Kohn 1998, Lein 1999] and properly dissociates molecules with electron pair bonds such as H_2 [Fuchs 2005b]. The RPA however misses important short-range correlations [Singwi 1968], the total energy being usually too low [Fuchs 2002] (though isoelectronic total energy differences are believed to be quite accurate [Yan 2000]). This deficiency can however be cured with LDA-like corrections [Kurth 1999, Yan 2000] or using refined time-dependent DFT kernels [Fuchs 2002, Lein 2000b]. As a matter of fact, the RPA is just a particular realization of a large class of functionals, leaving plenty of opportunities for improvement.

The correlation potential $v_c(\boldsymbol{r})$ that is derived from a given approximation for $E_c[n]$ is a key ingredient of density functional theory. ACFD potentials would open the way for self-consistent calculations and could provide valuable information about the underlying functionals. In this chapter, we derive the expression for the RPA exchange-correlation potential $v_{xc}^{RPA}(\boldsymbol{r})$. We show that $v_{xc}^{RPA}(\boldsymbol{r})$ satisfies the so-called linear-response Sham-Schlüter equation [Sham 1983, Sham 1985, Casida 1995b]. We also provide an approximate solution of this equation, that is much simpler to compute while being likely of reasonable accuracy. We then discuss the asymptotic behavior of $v_{xc}^{RPA}(\boldsymbol{r})$

Y.M. Niquet and M. Fuchs: *The Exchange-Correlation Potential in the Adiabatic-Connection Fluctuation-Dissipation Framework*, Lect. Notes Phys. **706**, 435–442 (2006)
DOI 10.1007/3-540-35426-3_29

in finite systems such as atoms [Niquet 2003a, Niquet 2003b, Niquet 2003c, Niquet 2005], which reveals much physics as well as the merits and deficiencies of the RPA. A more extensive discussion of the ACFD potentials (including the PGG and ALDA kernels) can be found in [Niquet 2003c].

Another v_{xc}-related issue is the well-known "bandgap problem". Indeed, the Kohn-Sham (KS) bandgap energy $\varepsilon_{\mathrm{g}} = \varepsilon_{\mathrm{c}} - \varepsilon_{\mathrm{v}}$ is usually found much lower than the experimental bandgap (ε_{v} and ε_{c} being respectively the highest occupied and lowest unoccupied KS energies). As a matter of fact, ε_{g} might differ from the interacting bandgap energy $E_{\mathrm{g}} = E(N+1) + E(N-1) - 2E(N)$, where $E(N)$ is the total energy of the $N-$electron system, due to the existence of a derivative discontinuity in the exchange-correlation functional [Sham 1983, Perdew 1982, Perdew 1983, Sham 1985b]. In practice, the fundamental bandgap energy and quasiparticle band structure of a solid are thus computed with many-body Green function techniques such as the GW method [Hedin 1969, Mattuck 1967]. In principle, the Green function G should be updated through Dyson equation until many-body self-consistency is achieved [Baym 1961, Baym 1962]. In most cases, however, the GW self-energy is calculated using KS orbitals and energies as input, thus leaving out self-consistency. This "$G_{\mathrm{KS}}W_{\mathrm{KS}}$" (also known as "$G_0W_0$") approach has been successfully applied to a wide variety of materials [Hybertsen 1985, Hybertsen 1986, Godby 1988, Aulbur 2000]. There has been, however, recent controversy about the effects of many-body self-consistency on the quasiparticle band structure and about the rationale behind the $G_{\mathrm{KS}}W_{\mathrm{KS}}$ approach [Schöne 1998, Ku 2002]. In this chapter, we show how the RPA functional provides such a rationale within a consistent DFT framework [Niquet 2004].

The chapter is organized as follows: the expression of the RPA potential is first derived in Sect. 29.2; its asymptotic properties are then discussed in Sect. 29.3; the RPA bandgap problem is investigated in Sect. 29.4.

29.2 The RPA Exchange-Correlation Potential

In this section, we show that $v_{\mathrm{xc}}^{\mathrm{RPA}}(\boldsymbol{r})$ satisfies the so-called linear-response Sham-Schlüter equation [Sham 1983, Sham 1985, Casida 1995b]. We also discuss an approximate (but reasonably accurate) solution of this equation. For the sake of simplicity, we restrict ourselves to spin-compensated N-electron systems.

We first focus on the RPA correlation potential $v_{\mathrm{c}}^{\mathrm{RPA}}(\boldsymbol{r}) = \delta E_{\mathrm{c}}^{\mathrm{RPA}}/\delta n(\boldsymbol{r})$. The RPA correlation energy reads, after integration over the coupling constant[1] [Niquet 2003c]:

[1] Please note that RPA-like approximations can also be derived from many-body perturbation theory starting from variational functionals of the Green function [Almbladh 1999]. Different flavours of these functionals (Luttinger-Ward

$$E_c^{\mathrm{RPA}}[n] = \frac{1}{2}\int \frac{\mathrm{d}u}{2\pi}\,\mathrm{Tr}\left\{\ln[1 - v_{\mathrm{ee}}\chi_{\mathrm{KS}}(\mathrm{i}u)] + v_{\mathrm{ee}}\chi_{\mathrm{KS}}(\mathrm{i}u)\right\} . \tag{29.1}$$

At variance with the LDA or GGA, (29.1) is an *explicit* functional the KS orbitals and energies, hence an *implicit* functional of the density. The derivation of $v_c^{\mathrm{RPA}}(\boldsymbol{r})$ is thus somewhat involved. Using the relation:

$$\frac{\delta\,\mathrm{Tr}\{\ln A(\mathrm{i}u)\}}{\delta n(\boldsymbol{r})} = \mathrm{Tr}\left\{A^{-1}(\mathrm{i}u)\frac{\delta A(\mathrm{i}u)}{\delta n(\boldsymbol{r})}\right\} \tag{29.2}$$

we indeed get at once:

$$v_c^{\mathrm{RPA}}(\boldsymbol{r}) = -\frac{1}{2}\int \frac{\mathrm{d}u}{2\pi}\,\mathrm{Tr}\left\{[W_{\mathrm{KS}}(\mathrm{i}u) - v_{\mathrm{ee}}]\frac{\delta\chi_{\mathrm{KS}}(\mathrm{i}u)}{\delta n(\boldsymbol{r})}\right\} , \tag{29.3}$$

where $W_{\mathrm{KS}}(\mathrm{i}u) = [1 - v_{\mathrm{ee}}\chi_{\mathrm{KS}}(\mathrm{i}u)]^{-1}v_{\mathrm{ee}}$ is the RPA screened Coulomb interaction. We are now left with $\delta\chi_{\mathrm{KS}}(\mathrm{i}u)/\delta n(\boldsymbol{r})$. We proceed taking advantage of the much simpler dependence of $\chi_{\mathrm{KS}}(\mathrm{i}u)$ on the KS Green function $G_{\mathrm{KS}}(\mathrm{i}u)$. Indeed (as can be verified by straightforward integration),

$$\chi_{\mathrm{KS}}(\boldsymbol{r},\boldsymbol{r}',\mathrm{i}u) = 2\int \frac{\mathrm{d}v}{2\pi} G_{\mathrm{KS}}(\boldsymbol{r},\boldsymbol{r}',\mathrm{i}u + \mathrm{i}v)G_{\mathrm{KS}}(\boldsymbol{r}',\boldsymbol{r},\mathrm{i}v) , \tag{29.4}$$

where:

$$G_{\mathrm{KS}}(\boldsymbol{r},\boldsymbol{r}',\mathrm{i}u) = \frac{1}{\mathrm{i}u - \hat{H}_{\mathrm{KS}}} = \sum_j \frac{\varphi_j(\boldsymbol{r})\varphi_j^*(\boldsymbol{r}')}{\mathrm{i}u - \varepsilon_j} . \tag{29.5}$$

Here $\hat{H}_{\mathrm{KS}} = \hat{T} + v_{\mathrm{KS}}$ is the KS Hamiltonian, $\hat{T}$ being the kinetic energy operator and v_{KS} the Kohn-Sham potential. The derivative chain rule then yields:

$$\frac{\delta\chi_{\mathrm{KS}}(\boldsymbol{r},\boldsymbol{r}',\mathrm{i}u)}{\delta n(\boldsymbol{r})} = \int \mathrm{d}w \int \mathrm{d}^3r_1 \int \mathrm{d}^3r_2 \int \mathrm{d}^3r_3 \\ \frac{\delta\chi_{\mathrm{KS}}(\boldsymbol{r},\boldsymbol{r}',\mathrm{i}u)}{\delta G_{\mathrm{KS}}(\boldsymbol{r}_1,\boldsymbol{r}_2,\mathrm{i}w)}\frac{\delta G_{\mathrm{KS}}(\boldsymbol{r}_1,\boldsymbol{r}_2,\mathrm{i}w)}{\delta v_{\mathrm{KS}}(\boldsymbol{r}_3)}\frac{\delta v_{\mathrm{KS}}(\boldsymbol{r}_3)}{\delta n(\boldsymbol{r})} . \tag{29.6}$$

The last derivative on the right-hand side, $\delta v_{\mathrm{KS}}(\boldsymbol{r}_3)/\delta n(\boldsymbol{r})$, is just the inverse of the static KS density-density response function, $\chi_{\mathrm{KS}}^{-1}(\boldsymbol{r}_3,\boldsymbol{r},\mathrm{i}u = 0)$ [since by definition $\chi_{\mathrm{KS}}(\boldsymbol{r},\boldsymbol{r}_3,\mathrm{i}u = 0) = \delta n(\boldsymbol{r})/\delta v_{\mathrm{KS}}(\boldsymbol{r}_3)$]. As for the second derivative, we have (in a compact, matrix-like notation):

$$\frac{\delta[G_{\mathrm{KS}}(\mathrm{i}w)G_{\mathrm{KS}}^{-1}(\mathrm{i}w)]}{\delta v_{\mathrm{KS}}(\boldsymbol{r}_3)} = 0 \Rightarrow \frac{\delta G_{\mathrm{KS}}(\mathrm{i}w)}{\delta v_{\mathrm{KS}}(\boldsymbol{r}_3)} = -G_{\mathrm{KS}}(\mathrm{i}w)\frac{\delta G_{\mathrm{KS}}^{-1}(\mathrm{i}w)}{\delta v_{\mathrm{KS}}(\boldsymbol{r}_3)}G_{\mathrm{KS}}(\mathrm{i}w) . \tag{29.7}$$

[Luttinger 1960], Nozières [Nozières 1964]...) however yield different approximations for the correlation energy, and thus different potentials. The original RPA as given by (29.1) can actually be derived from Nozières' functional [Nozières 1964].

Moreover, the functional derivative of the inverse KS Green function $G_{\mathrm{KS}}^{-1}(\mathrm{i}w) = \mathrm{i}w - \hat{T} - v_{\mathrm{KS}}$ is just:

$$\frac{\delta G_{\mathrm{KS}}^{-1}(\boldsymbol{r}_4, \boldsymbol{r}_5, \mathrm{i}w)}{\delta v_{\mathrm{KS}}(\boldsymbol{r}_3)} = -\delta(\boldsymbol{r}_3 - \boldsymbol{r}_4)\delta(\boldsymbol{r}_3 - \boldsymbol{r}_5) . \tag{29.8}$$

Last, the derivative of $\chi_{\mathrm{KS}}(\mathrm{i}u)$ with respect to $G_{\mathrm{KS}}(\mathrm{i}w)$ directly follows from (29.4):

$$\begin{aligned}\frac{\delta \chi_{\mathrm{KS}}(\boldsymbol{r}, \boldsymbol{r}', \mathrm{i}u)}{\delta G_{\mathrm{KS}}(\boldsymbol{r}_1, \boldsymbol{r}_2, \mathrm{i}w)} &= \frac{2}{2\pi}\delta(\boldsymbol{r} - \boldsymbol{r}_1)\delta(\boldsymbol{r}' - \boldsymbol{r}_2) G_{\mathrm{KS}}(\boldsymbol{r}_2, \boldsymbol{r}_1, \mathrm{i}w - \mathrm{i}u) \\ &+ \frac{2}{2\pi}\delta(\boldsymbol{r}' - \boldsymbol{r}_1)\delta(\boldsymbol{r} - \boldsymbol{r}_2) G_{\mathrm{KS}}(\boldsymbol{r}_2, \boldsymbol{r}_1, \mathrm{i}w + \mathrm{i}u) . \end{aligned} \tag{29.9}$$

Backward substitution of (29.9), (29.8), (29.7) and (29.6) into (29.3) finally yields the following expression for the RPA correlation potential:

$$v_{\mathrm{c}}^{\mathrm{RPA}}(\boldsymbol{r}) = \int \mathrm{d}^3 r' \, \chi_{\mathrm{KS}}^{-1}(\boldsymbol{r}, \boldsymbol{r}', \mathrm{i}u = 0) \rho_{\mathrm{c}}(\boldsymbol{r}') , \tag{29.10}$$

where:

$$\rho_{\mathrm{c}}(\boldsymbol{r}) = 2 \int \frac{\mathrm{d}u}{2\pi} \int \mathrm{d}^3 r_1 \int \mathrm{d}^3 r_2 \, G_{\mathrm{KS}}(\boldsymbol{r}, \boldsymbol{r}_1, \mathrm{i}u) \Sigma_{\mathrm{c}}(\boldsymbol{r}_1, \boldsymbol{r}_2, \mathrm{i}u) G_{\mathrm{KS}}(\boldsymbol{r}_2, \boldsymbol{r}, \mathrm{i}u) . \tag{29.11}$$

and $\Sigma_{\mathrm{c}}(\boldsymbol{r}_1, \boldsymbol{r}_2, \mathrm{i}u)$ is the correlation part of the so-called imaginary-frequency $G_{\mathrm{KS}} W_{\mathrm{KS}}$ self-energy [Hedin 1969] (also see Sect. 29.4):

$$\Sigma_{\mathrm{c}}(\boldsymbol{r}, \boldsymbol{r}', \varepsilon) = -\int \frac{\mathrm{d}v}{2\pi} G_{\mathrm{KS}}(\boldsymbol{r}, \boldsymbol{r}', \varepsilon + \mathrm{i}v) [W_{\mathrm{KS}}(\boldsymbol{r}, \boldsymbol{r}', \mathrm{i}v) - v_{\mathrm{ee}}(\boldsymbol{r}, \boldsymbol{r}')] . \tag{29.12}$$

Equations (29.10) and (29.11) have a clear OEP-like structure (see Chap. 9) [Krieger 1992b, Grabo 1998]. As a matter of fact, the exact-exchange potential $v_{\mathrm{x}}(\boldsymbol{r}) = \delta E_{\mathrm{x}}[n]/\delta n(\boldsymbol{r})$ also satisfies (29.10) and (29.11) with $\Sigma_{\mathrm{c}}(\mathrm{i}u)$ replaced by the exchange-only self-energy:

$$\begin{aligned}\Sigma_{\mathrm{x}}(\boldsymbol{r}, \boldsymbol{r}') &= -\int \frac{\mathrm{d}v}{2\pi} \, \mathrm{e}^{\mathrm{i}v\eta} G_{\mathrm{KS}}(\boldsymbol{r}, \boldsymbol{r}', \mathrm{i}v) v_{\mathrm{ee}}(\boldsymbol{r}, \boldsymbol{r}') \\ &= -\sum_{j=1}^{N/2} \varphi_j(\boldsymbol{r}) \varphi_j^*(\boldsymbol{r}') v_{\mathrm{ee}}(\boldsymbol{r}, \boldsymbol{r}') , \end{aligned} \tag{29.13}$$

with $\eta \to 0^+$. Moreover, (29.10) for $v_{\mathrm{xc}}^{\mathrm{RPA}}(\boldsymbol{r}) = v_{\mathrm{x}}(\boldsymbol{r}) + v_{\mathrm{c}}^{\mathrm{RPA}}(\boldsymbol{r})$ can be cast in a more familiar form applying $\chi_{\mathrm{KS}}(\mathrm{i}u = 0)$ on both sides then using (29.4):

$$\begin{aligned} 0 = \int \frac{\mathrm{d}u}{2\pi} \int \mathrm{d}^3 r_1 \int \mathrm{d}^3 r_2 \, G_{\mathrm{KS}}(\boldsymbol{r}, \boldsymbol{r}_1, \mathrm{i}u) \Big\{ &\Sigma_{\mathrm{xc}}(\boldsymbol{r}_1, \boldsymbol{r}_2, \mathrm{i}u) \\ &- v_{\mathrm{xc}}^{\mathrm{RPA}}(\boldsymbol{r}_1) \delta(\boldsymbol{r}_1 - \boldsymbol{r}_2) \Big\} G_{\mathrm{KS}}(\boldsymbol{r}_2, \boldsymbol{r}, \mathrm{i}u) , \end{aligned} \tag{29.14}$$

where

$$\Sigma_{\mathrm{xc}}(\boldsymbol{r},\boldsymbol{r}',\varepsilon) = -\int \frac{\mathrm{d}v}{2\pi}\, \mathrm{e}^{\mathrm{i}v\eta} G_{\mathrm{KS}}(\boldsymbol{r},\boldsymbol{r}',\varepsilon+\mathrm{i}v) W_{\mathrm{KS}}(\boldsymbol{r},\boldsymbol{r}',\mathrm{i}v)\,. \tag{29.15}$$

is the full imaginary-frequency $G_{\mathrm{KS}}W_{\mathrm{KS}}$ self-energy [Hedin 1969]. This equation is known as the linear-response Sham-Schlüter equation [Sham 1983, Sham 1985, Casida 1995b] in the GW approximation for the self-energy.[2]

Solving the linear-response Sham-Schlüter equation is a formidable task. This has nonetheless been done in bulk silicon by Godby et al. [Godby 1986, Godby 1988] and for a jellium surface by Eguiluz et al. [Eguiluz 1992]. An approximate but hopefully reasonnable solution of the linear-response Sham-Schlüter equation [Niquet 2003b, Niquet 2003c, Niquet 2005, Kotani 1998] can be drawn from the static COHSEX approximation to the $G_{\mathrm{KS}}W_{\mathrm{KS}}$ self-energy, that neglects some dynamical contributions to (29.15) [Hedin 1969]. The static COHSEX self-energy splits into a statically screened exchange part Σ_{sex} and a static Coulomb hole term Σ_{coh}:

$$\Sigma_{\mathrm{sex}}(\boldsymbol{r},\boldsymbol{r}') = -\sum_{j=1}^{N/2} \varphi_j(\boldsymbol{r})\varphi_j^*(\boldsymbol{r}') W_{\mathrm{KS}}(\boldsymbol{r},\boldsymbol{r}',iu=0) \tag{29.16a}$$

$$\Sigma_{\mathrm{coh}}(\boldsymbol{r},\boldsymbol{r}') = \frac{1}{2}\delta(\boldsymbol{r}-\boldsymbol{r}') W^s_{\mathrm{KS}}(\boldsymbol{r},\boldsymbol{r},iu=0)\,, \tag{29.16b}$$

where $W^s_{\mathrm{KS}}(\boldsymbol{r},\boldsymbol{r}',iu) = W_{\mathrm{KS}}(\boldsymbol{r},\boldsymbol{r}',iu) - v_{\mathrm{ee}}(\boldsymbol{r},\boldsymbol{r}')$ is the response part of the RPA screened Coulomb interaction. $\Sigma_{\mathrm{sex}}(\boldsymbol{r},\boldsymbol{r}')$ has the same functional form as the exchange-only self-energy [(29.13)], but with $v_{\mathrm{ee}}(\boldsymbol{r},\boldsymbol{r}')$ replaced by $W_{\mathrm{KS}}(\boldsymbol{r},\boldsymbol{r}',iu=0)$. $\Sigma_{\mathrm{coh}}(\boldsymbol{r},\boldsymbol{r}')$ on the other hand is just half the potential felt by an electron at $\boldsymbol{r}$ and due to the "Coulomb hole" that forms around it (the factor 1/2 comes from the adiabatic building of this Coulomb hole [Hedin 1969]). Accordingly, the approximate RPA potential that solves (29.14) in the static COHSEX approximation for the self-energy also splits in two parts $\tilde{v}^{\mathrm{RPA}}_{\mathrm{xc}}(\boldsymbol{r}) = \tilde{v}_{\mathrm{sex}}(\boldsymbol{r}) + \tilde{v}_{\mathrm{coh}}(\boldsymbol{r})$, where $\tilde{v}_{\mathrm{coh}}(\boldsymbol{r}) = \frac{1}{2}W^s_{\mathrm{KS}}(\boldsymbol{r},\boldsymbol{r},iu=0)$ and $\tilde{v}_{\mathrm{sex}}(\boldsymbol{r})$ satisfies the same equation as the exchange-only potential with $v_{\mathrm{ee}}(\boldsymbol{r},\boldsymbol{r}')$ replaced by $W_{\mathrm{KS}}(\boldsymbol{r},\boldsymbol{r}',iu=0)$. Therefore $\tilde{v}^{\mathrm{RPA}}_{\mathrm{xc}}(\boldsymbol{r})$ can be calculated with any existing x-OEP code once $W_{\mathrm{KS}}(\boldsymbol{r},\boldsymbol{r}',iu=0)$ is known. This has been done by Kotani [Kotani 1998] for example in some bulk metals and silicon.

29.3 Asymptotic Behavior of the RPA Potential in Finite Systems

In this section, we discuss the asymptotic behavior of the RPA potential in finite systems such as atoms. This indeed provides useful insight into the physics of the RPA functional. We focus on the helium atom as a test case.

[2] Note that the potential is defined by (29.10) and (29.14) up to an additive constant, because $\int \mathrm{d}^3r\, \chi_{\mathrm{KS}}(\boldsymbol{r},\boldsymbol{r}',\mathrm{i}u) = 0$.

Let us first recall the asymptotic behavior ($r \to \infty$) of the *exact* exchange-correlation potential of the He atom [Almbladh 1985]:

$$v_{\mathrm{xc}}(\boldsymbol{r}) = -\frac{1}{r} - \frac{\alpha_{\mathrm{He}^+}}{2r^4} + \mathcal{O}\left(\frac{1}{r^5}\right), \tag{29.17}$$

where $\alpha_{\mathrm{He}^+} = 9/32$ a.u. is the static polarizability of the He^+ ion. The $-1/r$ term [that comes from $v_{\mathrm{x}}(\boldsymbol{r})$] is the bare potential felt by an electron dragged away from the He atom, while the $\propto 1/r^4$ term [that comes from $v_{\mathrm{c}}(\boldsymbol{r})$] arises from the polarization of the resulting He^+ ion.

We have carefully investigated the asymptotic behavior of the RPA potential in [Niquet 2003b, Niquet 2003c, Niquet 2005]. We find that $v_{\mathrm{xc}}^{\mathrm{RPA}}(\boldsymbol{r})$ indeed behaves as (29.17), but with α_{He^+} replaced by the RPA polarizability of the He atom:

$$\alpha_{\mathrm{He}}^{\mathrm{RPA}} = -\int \mathrm{d}^3r \int \mathrm{d}^3r'\, z\chi_{\mathrm{RPA}}(\boldsymbol{r}, \boldsymbol{r}', iu = 0)z', \tag{29.18}$$

$\chi_{\mathrm{RPA}}(iu) = [1 - \chi_{\mathrm{KS}}(iu)v_{\mathrm{ee}}]^{-1}\chi_{\mathrm{KS}}(iu)$ being the atomic RPA density-density response function. Note that this result can be (most) easily inferred from a multipole expansion of $W_{\mathrm{KS}}(\boldsymbol{r}, \boldsymbol{r}', iu)$ in (29.16), $\tilde{v}_{\mathrm{xc}}^{\mathrm{RPA}}(\boldsymbol{r})$ having the same asymptotic behavior as $v_{\mathrm{xc}}^{\mathrm{RPA}}(\boldsymbol{r})$.

The asymptotic behavior of $v_{\mathrm{x}}(\boldsymbol{r})$ in the exact exchange approximation would be $v_{\mathrm{x}}(\boldsymbol{r}) = -1/r + \mathcal{O}(\exp.)$ [Krieger 1992b, Grabo 1998]. The RPA thus improves over x-OEP in giving the next, $1/r^4$ term of the asymptotic expansion, that accounts for polarization effects. Of course this $1/r^4$ term will not change the long-range behavior of the KS orbitals (as compared to x-OEP), but the underlying physics might still impact the vicinity of the atom. Nonetheless, the $1/r^4$ prefactor, $\alpha_{\mathrm{He}}^{\mathrm{RPA}} = 1.22$ a.u. [van Gisbergen 1998], while consistent with the RPA approximation, is quite far from the expected $\alpha_{\mathrm{He}^+} = 0.28$ a.u. This is reminiscent of the self-correlation problem in the RPA density-density response function and correlation energy: the system still polarizes as a two-electron system upon ionization because (i) the RPA fails to suppress self-correlation and (ii) the RPA fails to account for orbital relaxation. The ACFD-PGG functional, that include exact-exchange effects in two-electron systems, corrects for most of this self-correlation error, yielding $v_{\mathrm{xc}}(\boldsymbol{r}) = -1/r - 0.68/(2r^4) + \mathcal{O}(1/r^5)$ [Niquet 2005]. It still misses orbital relaxation however.

29.4 The Bandgap Energy of Solids

As mentioned in the introduction, DFT, which is a ground-state theory, is not meant for the calculation of quasiparticle band structures. Even the bandgap energy, a linear combination of ground-state total energies, may not be correctly given by the KS band structure[3] [Sham 1983, Perdew 1982, Perdew 1983, Sham 1985b]. The KS RPA bandgap energies themselves, calculated in $v_{\mathrm{xc}}^{\mathrm{RPA}}(\boldsymbol{r})$ or $\tilde{v}_{\mathrm{xc}}^{\mathrm{RPA}}(\boldsymbol{r})$, are too low [Godby 1986, Godby 1988, Kotani 1998]. The quasiparticle band structure of a solid is thus usually computed with the GW method [Hedin 1969, Mattuck 1967]. In principle, the Dyson equation should be solved iteratively to find the self-consistent Green function G and the self-consistent GW self-energy Σ_{xc}. This would notably make the quasiparticle band structure independent of the initial guess for the Green function [Baym 1961, Baym 1962]. Such a self-consistent GW calculation has however long been untractable for real materials. The GW self-energy is thus usually computed using KS orbitals and energies as input. This "$G_{\mathrm{KS}}W_{\mathrm{KS}}$" approach has been successfully applied to a wide variety of materials [Hybertsen 1985, Hybertsen 1986, Godby 1988, Aulbur 2000]. In most cases it shifts the KS conduction bands with respect to the KS valence bands in a nearly rigid way. It has long been thought that many-body self-consistency would degrade the quality of the GW band structure. The first self-consistent GW calculations in real materials [Schöne 1998, Ku 2002] have however revived this debate and questioned the rationale behind $G_{\mathrm{KS}}W_{\mathrm{KS}}$ calculations. In this paragraph we show that the RPA functional provides such a rationale within a consistent DFT framework.

The interacting fundamental bandgap energy of a N-electron system is the difference $E_{\mathrm{g}} = I - A$ between its first ionization potential I and its electron affinity A. These are defined as total energy differences between the N- and $(N \pm 1)$-electron systems:

$$I = E(N-1) - E(N) \tag{29.19a}$$

$$A = E(N) - E(N+1)\,. \tag{29.19b}$$

The calculation of I and A as RPA total energy differences yields, after tedious manipulations [Niquet 2004]:

$$-I = \varepsilon_{\mathrm{v}} + \langle \varphi_{\mathrm{v}} | \Sigma_{\mathrm{xc}}(\varepsilon_{\mathrm{v}}) - v_{\mathrm{xc}}^{\mathrm{RPA}} | \varphi_{\mathrm{v}} \rangle + \mathcal{O}(\mathcal{V}^{-1/4}) \tag{29.20a}$$

$$-A = \varepsilon_{\mathrm{c}} + \langle \varphi_{\mathrm{c}} | \Sigma_{\mathrm{xc}}(\varepsilon_{\mathrm{c}}) - v_{\mathrm{xc}}^{\mathrm{RPA}} | \varphi_{\mathrm{c}} \rangle + \mathcal{O}(\mathcal{V}^{-1/4})\,. \tag{29.20b}$$

$\varphi_{\mathrm{v}}(\boldsymbol{r})$ and $\varphi_{\mathrm{c}}(\boldsymbol{r})$ are the highest occupied and lowest unoccupied orbital KS orbitals, Σ_{xc} is the $G_{\mathrm{KS}}W_{\mathrm{KS}}$ self-energy[4] [(29.15)] and $\mathcal{V}$ is the volume of the system. Also, one may show [Niquet 2003b] that $\langle \varphi_{\mathrm{v}} | \Sigma_{\mathrm{xc}}(\varepsilon_{\mathrm{v}}) - v_{\mathrm{xc}}^{\mathrm{RPA}} | \varphi_{\mathrm{v}} \rangle = 0$ if the KS potential $v_{\mathrm{KS}}(\boldsymbol{r})$ has been shifted so that $\lim_{r\to\infty} v_{\mathrm{KS}}(\boldsymbol{r}) = 0$.

[3] The KS x-OEP bandgap energies of many semiconductors are found to be in good agreement with the experiment [Städele 1997, Städele 1999]. However, the interacting x-OEP bandgap energies of these materials are close to the Hatree-Fock bandgaps and thus too large.

[4] Equation (29.15) is equal to the usual (real-frequency) expression for the $G_{\mathrm{KS}}W_{\mathrm{KS}}$ self-energy [Hedin 1969] as long as ε is in the KS bandgap (including $\varepsilon = \varepsilon_{\mathrm{c}}$ and $\varepsilon = \varepsilon_{\mathrm{v}}$).

This yields $I = -\varepsilon_\mathrm{v} + \mathcal{O}(\mathcal{V}^{-1/4})$: the highest occupied KS energy tends to the negative of the DFT-RPA first ionization potential in bulk systems[5] [Almbladh 1985, Levy 1984]. This result only holds if ε_v has been calculated self-consistently [i.e., using $v_\mathrm{xc}^\mathrm{RPA}(\boldsymbol{r})$]; Equations (29.20) remain valid though if LDA/GGA orbitals, energies and potential are used as input for the RPA total energy. The interacting RPA bandgap energy of a solid ($\mathcal{V} \to \infty$) thus differs from the KS RPA bandgap energy $\varepsilon_\mathrm{g} = \varepsilon_\mathrm{c} - \varepsilon_\mathrm{v}$ by a $G_\mathrm{KS}W_\mathrm{KS}$-like self-energy correction. This shows that the usual $G_\mathrm{KS}W_\mathrm{KS}$ approach is not a mere practical recipe and that it has a well-defined and meaningful interpretation within DFT. This does not, however, settle the debate about the need and effects of many-body self-consistency.[6] The interacting RPA bandgap energies ($E_\mathrm{g} = 1.35$ eV for Si and $E_\mathrm{g} = 5.88$ eV for diamond [Niquet 2004]) are usually in reasonnable agreement with the experiment (1.17 eV and 5.48 eV respectively), while the KS RPA bandgap energy of silicon for example is close to the LDA [Godby 1986, Godby 1988, Kotani 1998], showing that the RPA potential has a sizeable derivative discontinuity.

29.5 Conclusion

In this chapter, we have derived the expression of the RPA exchange-correlation potential $v_\mathrm{xc}^\mathrm{RPA}(\boldsymbol{r})$. We have shown that the latter satisfies the so-called linear-response Sham-Schlüter equation in the GW approximation for the self-energy. We have provided an approximate but hopefully reasonnable solution of this equation for practical use. We have then discussed the asymptotic behavior of the RPA potential in finite systems. We have shown that $v_\mathrm{xc}^\mathrm{RPA}(\boldsymbol{r})$ has the expected $-1/r - \alpha/(2r^4)$ tail in the He atom, though α is the RPA polarizability of He instead of the polarizability of the He^+ ion. This is reminiscent of the self-correlation problem in the RPA density-density response function, which can be (partly) corrected in more refined ACFD approximations such as the exact-exchange TD kernel. Last, we have calculated the RPA bandgap energy of a solid. We have shown that the latter differs from the Kohn-Sham bandgap energy by a $G_\mathrm{KS}W_\mathrm{KS}$-like self-energy correction, computed using Kohn-Sham orbitals and energies as input. This provides a clear rationale behind $G_\mathrm{KS}W_\mathrm{KS}$ quasiparticle bandgap calculations, that can be consistently interpreted and analyzed within density functional theory.

---

[5] Let us recall that $I = -\varepsilon_\mathrm{v}$ for the exact exchange-correlation functional provided the external potential $v_\mathrm{ext}(\boldsymbol{r}) \to 0$ when $r \to \infty$ [Almbladh 1985, Levy 1984].

[6] There is, moreover, a subtle but important difference between (29.20) and the usual expression for the $G_\mathrm{KS}W_\mathrm{KS}$ bandgap energy [Hybertsen 1985, Hybertsen 1986, Aulbur 2000]: the former lack the so-called renormalization factors that appear in the latter. The interacting RPA bandgap energies are thus a bit larger than the usual $G_\mathrm{KS}W_\mathrm{KS}$ figures. A detailed discussion of this point can be found in [Niquet 2004].

30 Dispersion (Van Der Waals) Forces and TDDFT

J.F. Dobson

30.1 Introduction

By "dispersion forces" [Mahanty 1976] we mean the part of the attractive van der Waals (vdW) interaction that cannot be attributed to any permanent electric multipoles. These ubiquitous forces are typically weaker than ionic and covalent bonding forces, but are of longer range than the latter, typically decaying algebraically rather than exponentially with separation. They are important in soft condensed matter and in rare-gas chemistry, for example. We will work in the electromagnetically non-retarded (non-Casimir [Milton 2001]) limit, which means in practice that we can treat interacting systems at separations from about a micron down to full overlap of electronic clouds. We do not aim for a complete review of vdW phenomena and theories, but will rather concentrate on adiabatic connection / fluctuation-dissipation (ACFD) approaches. These work with dynamic electron density-density response functions, which in turn can be calculated by TDDFT methods. An interesting alternative is suggested in Chap. 3.

30.2 Simple Models of the vdW Interaction between Small Systems

30.2.1 Coupled-Fluctuation Model

It is worthwhile to consider first a very simple picture of the vdW interaction between neutral spherical atoms at separation $R \gg b$ where b is an atomic size. (For more detail see, e.g., [Dobson 2001] or [Langbein 1974].) The Hartree field of a neutral spherical atom decays exponentially with distance, and so the Hartree energy cannot explain the algebraic decay of the vdW interaction. However, the zero-point motions of the electrons (or thermal motions where significant) can cause a temporary fluctuating dipole moment d_2 to arise on atom #2. The nonretarded Coulomb interaction energy between this dipole, and another dipole of order $\alpha_1 d_2 R^{-3}$ that it induces on atom #1, has a nonzero average value that can be estimated [Dobson 2001] as

$$E = -C_6 R^{-6}\,, \qquad C_6 = K\hbar\omega_0\alpha_1\alpha_2\,. \tag{30.1}$$

J.F. Dobson: *Dispersion (Van Der Waals) Forces*, Lect. Notes Phys. **706**, 443–462 (2006)
DOI 10.1007/3-540-35426-3_30

Here α_1 and α_2 are the dipolar polarizabilities of the atoms. The "Hamaker constant" C_6 for this geometry contains a dimensionless constant K, not specifiable from the above qualitative argument. The factor R^{-6} can be understood as arising from *two* actions of the dipolar field, each proportional to R^{-3}, showing that this approach relates to *second*-order perturbation theory.

30.2.2 Model Based on the Static Correlation Hole: Failure of LDA/GGA at Large Separations

The spontaneous dipole d_2 invoked above would be implied *if* we had found an electron at a position $\boldsymbol{r}$ on one side of atom #2. The induced dipolar distortion on atom #1 then represents a very distant part of the correlation hole density $n_c(\boldsymbol{r}'|\boldsymbol{r})$ [Gunnarsson 1976] due to discovery of the electron at $\boldsymbol{r}$. The shape of this hole is entirely determined by the shape of atom #1, and is thus quite unlike the long-ranged part of the xc hole present in a uniform electron gas of density $n(\boldsymbol{r})$. It is therefore unsurprising that the local density approximation (LDA) misses the long-ranged tail of the vdW interaction. In fact, the LDA and the GGAs can only obtain the vdW tail via the distortion of the density of each atom. This distortion is predicted by these theories to decay exponentially with separation of the two atoms, thus ruling out the correct algebraic decay of the energy. The situation with GGA is less clear when the densities of the interacting fragments overlap. If the principal attractive correlation energy contribution comes from electrons near the overlap region, then treating this region as part of a weakly nonuniform gas might be reasonable. In keeping with this, various different GGAs can give qualitatively reasonable results for vdW systems such as rare-gas dimers. The results are not consistent or reliable, however [Perez-Jorda 1995, Zhang 1997, Patton 1997a, Patton 1997b], though surprisingly good results *near the energy minimum* are obtained [Perez-Jorda 1999, Walsh 2005] with Hartree-Fock exchange plus the Wilson-Levy functional [Wilson 1998, Wilson 1990]. Some discussion is given in a recent review [Dobson 2001].

30.2.3 Model Based on Small Distortions of the Ground State Density

Instead of considering the energy directly, Feynman [Feynman 1939] and Allen and Tozer [Allen 2002] considered the small separation-dependent changes $\delta n(\boldsymbol{r}, \boldsymbol{R})$ in the groundstate density $n(\boldsymbol{r})$ of each fragment, caused by the Coulomb interaction v_{ee}^{12} between the fragments. The Coulomb field acting at the nucleus of each fragment, created by $\delta n(\boldsymbol{r}, \boldsymbol{R})$ as source, leads to a force which was identified as the vdW force, in the distant limit. One can then obtain the correct result $\boldsymbol{F} = -\nabla_R(-C_6 R^{-6})$ in the widely-separated limit, in agreement with (30.1). Such a result emerges, for example, if $\delta n(\boldsymbol{r}, \boldsymbol{R})$ is calculated from a many-electron wavefunction correct to *second* order in

v_{ee}^{12}, involving a double summation with two energy denominators. [The *first-order* wavefunction perturbation makes zero contribution to $\delta n(\boldsymbol{r}, \boldsymbol{R})$.] By contrast, looking at the total *energy* to second order in v_{ee}^{12} one already obtains the dispersion interaction with only a single summation and one energy denominator, a substantially easier task of the same order as obtaining the *first-order* perturbed wavefunction. From here on we restrict attention to approaches based directly on the energy.

30.2.4 Coupled-Plasmon Model

Another simple way to obtain the R^{-6} interaction is to regard the coupled fluctuating dipoles invoked above as forming a coupled plasmon mode of the two systems [Langbein 1974]. One solves coupled equations for the time-dependent density distortions on the two systems, leading to two normal modes (in- and out-of-phase plasmons) of free vibration of the electrons. The R dependence of the sum of the zero-point plasmon energies $\sum_i \hbar\omega_i/2$ gives an energy of form $-C_6R^{-6}$, in qualitative agreement with the coupled-fluctuation approach described above, for the case of two small separated systems (see, e.g., [Dobson 2005a, Mahanty 1976, Langbein 1974]). A strength of the coupled-plasmon approach is that it is not perturbative, and is equally valid for large or small systems, even for metallic cases where zero energy denominators could render perturbation theory suspect. The coupled-plasmon theory is linked to the correlation-hole approach by the fluctuation-dissipation theorem to be discussed starting from Sect. 30.6.2 below.

30.3 The Simplest Models for vdW Energetics of Larger Systems

There is a large early literature (see, e.g., [Mahanty 1976]) calculating forces between macroscopic bodies by adding R^{-6} energy contributions between pairs of atoms, or pairs of volume elements. To describe close contacts or chemical bonds, the small-R divergence has to be "saturated" or cut off, and substituted by another form for small R. The well-known "6–12" or Lennard-Jones potential $\phi_{6-12}(R) = -C_6R^{-6} + C_{12}R^{-12}$ is an example used both in chemical and biological situations, and also quite recently for graphitic structures. Typically, the coefficients C_6 and C_{12} are fitted to experimental data, and may differ even between different structures of the same (e.g., graphitic) type [Girifalco 2000]. Other saturation procedures have been considered [Wu 2002]. Recent work has provided a non-empirical way to achieve short-distance saturation [Dion 2004] that appears to be valid at least for small finite systems (see Sect. 30.9.2). We show in Sect. 30.5 below, however, that the $\sum R^{-6}$ tail of fitted Lennard-Jones potentials is not in principle correct for the asymptotic vdW interaction of an important class of condensed matter systems. It is also difficult to see how a more recent fitting scheme [von Lilienfeld 2004] can deal with the severe non-additivity required. These considerations reinforce the need to develop the seamless ACFD approach to be described below in Sect. 30.6.

30.4 Formal Perturbation Theory Approaches

30.4.1 Second Order Perturbation Theory for Two Finite Nonoverlapping Systems

A more precise result than (30.1) can be obtained by treating the inter-fragment Coulomb interaction $v_{\mathrm{ee}}^{12} = e^2/r_{12}$ in second-order perturbation theory [Longuet-Higgins 1965, Zaremba 1976]:

$$E_{12}^{(2)} = -\frac{\hbar}{2\pi}\int \mathrm{d}^3 r_1 \int \mathrm{d}^3 r_1' \int \mathrm{d}^3 r_2 \int \mathrm{d}^3 r_2' \frac{e^2}{r_{12}}\frac{e^2}{r_{12}'} \times \int_0^\infty \mathrm{d}u\, \chi_1(\boldsymbol{r}_1, \boldsymbol{r}_1', \mathrm{i}u)\chi_2(\boldsymbol{r}_2, \boldsymbol{r}_2', \mathrm{i}u)\,. \quad (30.2)$$

Here χ_1 and χ_2 are the density-density response functions (see Chap. 1) of the two fragments separately (each treated in the complete absence of the other, but including all interactions inside each fragment, and evaluated at imaginary frequency $\omega = \mathrm{i}u$). The quantities $\boldsymbol{r}_1$ and $\boldsymbol{r}_1'$ are positions inside the first fragment, while $\boldsymbol{r}_2$ and $\boldsymbol{r}_2'$ are inside the second fragment. There are many equivalent forms [McWeeny 1989, Zaremba 1976] of the second-order perturbation energy. The form (30.2) has been chosen for display here because it is useful in establishing connections between different approaches to the dispersion interaction to be discussed below.

Equation (30.2) is not justified when the electrons on the two systems overlap, so that electrons in system 1 and system 2 cannot be treated as distinguishable. A more complex v_{ee}^{12} perturbation theory, the symmetry adapted perturbation theory (SAPT [Jeziorski 1994]) is available in cases of overlap. Some doubts about (30.2) also arise for spatially large systems (see Sect. 30.5).

The charge-neutrality and constant-potential conditions $\int \mathrm{d}r'\, \chi(\boldsymbol{r}, \boldsymbol{r}', \mathrm{i}u) = \int \mathrm{d}r\, \chi(\boldsymbol{r}, \boldsymbol{r}', \mathrm{i}u) = 0$ are automatically satisfied [Dobson 1998, Dobson 1996] if one writes the density-density responses as the gradient of nonlocal dynamic polarizability tensors α [Hunt 1983], $\chi(\boldsymbol{r}, \boldsymbol{r}', \mathrm{i}u) = -e^{-2}\partial_{r_i}\partial_{r_j'}\alpha_{ij}(\boldsymbol{r}, \boldsymbol{r}', \mathrm{i}u)$. Using integration by parts and defining $t_{ij}(r) = r^{-2}(3r_i r_j - \delta_{ij} r^2)$ one then obtains from (30.2)

$$E^{(2)} = -\frac{\hbar}{2\pi}\int d^3r_1\int d^3r_2\int d^3r_1'\int d^3r_2' \sum_{ijkl=1}^{3} t_{ik}(\boldsymbol{r}_{12})t_{jl}(\boldsymbol{r}_{12}')$$
$$\times\, r_{12}^{-3}r_{12}'^{-3}\int_0^\infty du\, \alpha_{ij}^{(1)}(\boldsymbol{r}_1,\boldsymbol{r}_1',iu)\alpha_{kl}^{(2)}(\boldsymbol{r}_2,\boldsymbol{r}_2',iu)\,. \quad (30.3)$$

From (30.3) it is clear that the leading dependence of the dispersion energy $E^{(2)}$ is of $O(R^{-6})$ at large separations (as indicated by the simple arguments in Sect. 30.2 above), but that $E^{(2)}$ depends in general on the orientation of the two systems as embodied in the angular dependence of the polarizability tensors $\alpha^{(1)}$, $\alpha^{(2)}$. By writing $r_{12} = R + x_1 - x_2$, expanding t_{ij} in powers of x_i and x_j', and assuming isotropic dipole polarizabilities $A_{ij} = \delta_{ij}A(iu)$ for each system, one obtains from (30.3) in lowest order the following more familiar formula generalizing (30.1):

$$E^{(2)} = -C_6/R^6\,, \qquad C_6 = \frac{3\hbar}{\pi}\int du\, A^{(1)}(iu)A^{(2)}(iu)\,. \quad (30.4)$$

Higher terms in the multipolar expansion in powers of x_1 and x_2 give corrections to (30.3) of order R^{-8}, R^{-10}, etc. [Jeziorski 1994, Osinga 1997].

For well-separated finite systems, (30.2) reduces the calculation of accurate vdW interaction parameters to the calculation of sufficiently accurate dynamical density response functions of the isolated components. The simplest case is that of two atoms, but atomic polarizabilities are in fact notoriously difficult to treat via density functional methods, because they involve extreme inhomogenity of the density. Straightforward ALDA response calculation for atoms, based on LDA KS potentials, lead to dimer C_6 coefficients that are up to 20% too large for rare gases [van Gisbergen 1995] and 50% for the Be dimer [Dobson 2005b]. RPA response tends to underestimate C_6. Some of the problems with ALDA are due to the need for self-interaction correction [Mahan 1990] perhaps both in the groundstate calculation and in the response [Pacheco 1992a, Pacheco 1992b, Pacheco 1997]. Use of the LB94 groundstate KS potential, which has the correct asymptotic behavior like SIC theories, leads to improved C_6 and C_8 vdW coefficients [Osinga 1997].

Equation (30.2) should not be confused with Moeller-Plesset (MP) perturbation theory [Szabo 1989]. The latter goes systematically beyond the Hartree-Fock theory by treating the *entire* bare electron-electron interaction v_{ee} as a perturbation. This is in contrast to (30.2), in which the exact susceptibilities χ_1 and χ_2 contain the *intra*-subsystem interactions v_{ee}^{11}, v_{ee}^{22} in principle to *all* orders: only the *inter*-subsystem interaction v_{ee}^{12} is treated perturbatively. Correspondingly, the separation-dependent part of the second-order MP (MP2) energy is equivalent to (30.2) but with the *bare* (Kohn-Sham or HF) responses χ_{01}, χ_{02} on the right-hand side. On the other hand, in MP2 and related approaches the overlapped case is not excluded. There have been a number of approaches related to this idea

[Engel 2000, Engel 2001, Angyan 2005]. Lein et al. [Lein 1999] used perturbation of (30.14) within the ACFD and so could include f_{xc} directly, obtaining reasonable results for noble-gas and metal dimers.

30.4.2 vdW and Higher-Order Perturbation Theory

For non-overlapping electronic systems one can go further within perturbation theory with respect to the inter-system Coulomb interactions v_{ee}^{ij}. In third order one finds an interaction between three separated systems, which cannot be expressed as the pairwise sum of R^{-6} Hamaker terms such as (30.4). At large separations and spherical systems the leading (dipolar) contribution to this third-order term has the Axilrod-Teller form (see e.g. [Rapcewicz 1991]) $E^{\mathrm{vdW},(3)} \approx C_9 R_{12}^{-3} R_{23}^{-3} R_{13}^{-3}$, where C_9 contains some angular dependence. There are also corrections to the pair interaction (30.3) from perturbation orders beyond 2 [Jeziorski 1994].

30.5 Nonuniversality of vdW Asymptotics in Layered and Striated Systems

Unfortunately, the finite-order perturbation approaches discussed in the previous section, as well as $\sum R^{-6}$ formulas partly justified by this perturbative approach, are questionable for many (infinite) solid-state systems of current technological interest [Dobson 2001, White 2003, Dobson 2005a, Dobson 2006]. The most severe cases are those where there is a zero electronic energy gap, leading to zero energy denominators in perturbation theory. One example is the case of parallel nanoscopically thin metallic sheets separated by distance D: here, by summing the zero point energies of coupled 2D plasmons, one obtains [Bostrom 2000, Dobson 2001, Dobson 2005a] an interaction $E^{\mathrm{vdW}} \sim -C_{5/2} D^{-5/2}$. By contrast, the $\sum R_{ij}^{-6}$ approach gives $E^{\mathrm{vdW}} \sim -C_4 D^{-4}$, which is necessarily much smaller at large separations. For two parallel metallic nanotubes or nanowires of radius b, separated by distance $D \gg 2b$, the coupled plasmon approach was recently found [White 2003, Dobson 2005a, Dobson 2006] to give $E^{\mathrm{vdW}} \sim -C_a D^{-2} \left[\ln(D/b)\right]^{-3/2}$, whereas the $\sum R_{ij}^{-6}$ method gives $E^{\mathrm{vdW}} \sim -C_5 D^{-5}$, different by almost three powers of D. The important case of two graphene planes (zero-gap semiconductors) was shown recently [Dobson 2005c, Dobson 2006] to give $E^{\mathrm{vdW}} \sim -C_3 D^{-3}$, whereas the $\sum R_{ij}^{-6}$ approach gives $E^{\mathrm{vdW}} \sim -C_4 D^{-4}$, the same result as for regular 2D insulators [Rydberg 2000]. A finite number of higher perturbation terms (see Sect. 30.4.2), added to $\sum R_{ij}^{-6}$, cannot reproduce these unconventional power laws.

In the above cases where $\sum R_{ij}^{-6}$ fails qualitatively, in addition to the zero energy gap there is an incomplete metallic screening because the systems are nanoscopically small in at least one space dimension. Three-dimensional metals (e.g., thick metallic plates) do not exhibit vdW decay laws of unusual form [Dobson 2001], and this seems to be associated with complete metallic screening, leading to a finite polarizability at small frequency and wavenumber [Dobson 2006], and a finite plasmon energy gap.

The above considerations apply to widely separated sub-systems. Recently, evidence has also been given that standard theories (LDA/GGA/fitted Lennard-Jones potentials [Charlier 1994, Girifalco 2002]) do not give reliable answers for the energetics of layered metallic or semi-metallic systems near their equilibrium spacing, either. This is despite the fact that the LDA predicts good equilibrium geometries. For theoretical evidence, see [Dobson 2004], Fig. 4 of [Jung 2004], and [Rydberg 2003a, Hasegawa 2004, Tournus 2005]. There is experimental evidence also [Benedict 1998c, Zacharia 2004]. These are technologically important systems (graphite and derivatives, fullerenes, conducting nanotubes), so these discrepancies are significant.

In principle, therefore, and especially for large solid-state systems, one would like a theory of vdW forces that is not local nor perturbative and is "seamless" – that is, it gives good results at all separations. Below we make the case that appropriate theories can be constructed from the fluctuation-dissipation theorem along with a density-density response function $\chi(\boldsymbol{r}, \boldsymbol{r}', \omega)$ from TDDFT: indeed even the random phase approximation (RPA) version of χ (the case $f_{\text{xc}} = 0$) can sometimes give good results. We first present the necessary theory in some detail.

30.6 Correlation Energies from Response Functions: The Fluctuation-Dissipation Theorem

The approach to be described here has antecedents in theories of the Lifshitz type [Dzyaloshinskii 1961] that included electromagnetic retardation but ultimately approximated the response function of electrons in a local macroscopic fashion. That theory was mainly applied to bulk systems at macroscopic separation, and represented the response of the electrons through a spatially local but frequency-dependent dielectric function obtained principally from experiment. Here we consider only the electromagnetically non-retarded case but retain the full nonlocal microscopic response functions of the electrons, permitting a "seamless" treatment at any separation where electromagnetic retardation is unimportant, right down to bonding with overlap of the electronic clouds.

30.6.1 Basic Adiabatic Connection Fluctuation-Dissipation Theory

As explained in Chap. 1, the exact adiabatic connection formula (ACF) for the combined Hartree, exchange, and correlation energy in the ground state of an inhomogeneous many-electron system is [Langreth 1975, Gunnarsson 1976, Harris 1975]

$$E_{\text{Hxc}} = \frac{1}{2}\int_0^1 d\lambda \int d^3r \int d^3r' \, \frac{e^2}{|\boldsymbol{r}-\boldsymbol{r}'|} n_{2\lambda}(\boldsymbol{r},\boldsymbol{r}') \,. \tag{30.5}$$

Here $n_{2\lambda}(\boldsymbol{r},\boldsymbol{r}')$ is the pair density in the reduced-interaction many-electron ground state $|\lambda\rangle$ with electron-electron Coulomb potential $\lambda e^2/r_{12}$. The probability of finding an electron in a small volume d^3r near $\boldsymbol{r}$, and simultaneously another electron in d^3r' near $\boldsymbol{r}'$, is $n_{2\lambda}(\boldsymbol{r},\boldsymbol{r}')d^3r\, d^3r'$. The λ integration restores the necessary kinetic correlation energy arising from quantal zero-point motions, and is performed at constant ground-state density $n(\boldsymbol{r})$.

Introducing an operator $\hat{\boldsymbol{r}}_a$ for the position of the ath electron, and remembering that the second electron found at $\boldsymbol{r}'$ cannot be the same as the first one ($a \neq b$) we have

$$\begin{aligned} n_{2\lambda}(\boldsymbol{r},\boldsymbol{r}') &= \langle\lambda| \sum_{a\neq b} \delta^3(\boldsymbol{r}-\hat{\boldsymbol{r}}_a)\delta^3(\boldsymbol{r}'-\hat{\boldsymbol{r}}_b)\,|\lambda\rangle \\ &= \langle\lambda| \sum_{a,b} \delta^3(\boldsymbol{r}-\hat{\boldsymbol{r}}_a)\delta^3(\boldsymbol{r}'-\hat{\boldsymbol{r}}_b)\,|\lambda\rangle - \delta^3(\boldsymbol{r}-\boldsymbol{r}')\,\langle\lambda|\sum_a \delta^3(\boldsymbol{r}-\hat{\boldsymbol{r}}_a)\,|\lambda\rangle \\ &= \langle\lambda|\,\hat{n}(\boldsymbol{r})\hat{n}(\boldsymbol{r}')\,|\lambda\rangle - \delta^3(\boldsymbol{r}-\boldsymbol{r}')n(\boldsymbol{r}) \,. \end{aligned} \tag{30.6}$$

Here we used the electron density operator $\hat{n}(\boldsymbol{r}) = \sum_a \delta^3(\boldsymbol{r}-\hat{\boldsymbol{r}}_a)$, whose expectation is the electron density, $\langle\hat{n}(\boldsymbol{r})\rangle = n(\boldsymbol{r})$. We now also introduce the density fluctuation operator

$$\delta\hat{n}(\boldsymbol{r}) = \hat{n}(\boldsymbol{r}) - n(\boldsymbol{r}) \,. \tag{30.7}$$

(This operator corresponds to spontaneous density fluctuation processes such as those invoked in Sect. 30.2.1, in which a density fluctuation away from the expectation value $n(\boldsymbol{r})$ produces dipole fields that initiate the van der Waals energy.) Then, putting $\hat{n} = n + \delta\hat{n}$ into (30.6) and noting that $\langle\lambda|\,\delta\hat{n}(\boldsymbol{r})\,|\lambda\rangle = 0$, we have

$$n_{2\lambda}(\boldsymbol{r},\boldsymbol{r}') = \langle\lambda|\,\delta\hat{n}(\boldsymbol{r})\delta\hat{n}(\boldsymbol{r}')\,|\lambda\rangle + n(\boldsymbol{r})n(\boldsymbol{r}') - \delta^3(\boldsymbol{r}-\boldsymbol{r}')n(\boldsymbol{r}) \,. \tag{30.8}$$

When the second term on the right side of (30.8) is put into the ACF (30.5) it simply yields the Hartree electron-electron energy. The other two terms of (30.8) thus yield the exact xc energy:

$$E_{\text{xc}} = \frac{1}{2}\int_0^1 d\lambda \int d^3r \int d^3r' \, \frac{e^2}{|\boldsymbol{r}-\boldsymbol{r}'|} \left[\langle\lambda|\,\delta\hat{n}(\boldsymbol{r})\delta\hat{n}(\boldsymbol{r}')\,|\lambda\rangle - \delta^3(\boldsymbol{r}-\boldsymbol{r}')n(\boldsymbol{r})\right] \,. \tag{30.9}$$

[The correlation term $\langle\lambda|\,\delta\hat{n}(\mathbf{r})\delta\hat{n}(\mathbf{r}')\,|\lambda\rangle$ represents the fact that density fluctuations at $\boldsymbol{r}'$ can be tied to (correlated with) fluctuations at $\boldsymbol{r}$, just the kind of process described in Sect. 30.2.1].

The idea now is that the correlations represented by the fluctuation term in (30.9) are *caused* by interactions between the electrons, and the description of this physics is rather subtle: If a fluctuation occurs, it will cause some interaction energy beyond the Hartree description. It is easier conceptually, and helpful in the construction of approximation schemes, to relate this process to the interaction between the *non*-random density changes caused *when* a small externally-controlled field is applied. Thus we are led to introduce time-dependent density response theory, and we have converted a tricky "if" scenario into a conceptually simpler "when" scenario. The mathematical tool that justifies this shift in philosophy is the fluctuation-dissipation theorem [Callen 1951], derived in sufficient generality (i.e., for two unequal operators) in the book by Landau and Lifshitz [Landau 1969]. For completeness we now give a simple direct derivation of the frequency integrated, zero-temperature form of the theorem needed here.

Suppose that $\hat{A}$ and $\hat{B}$ [e.g., $\delta\hat{n}(\boldsymbol{r})$ and $\delta\hat{n}(\boldsymbol{r}')$] are Hermitian operators with zero groundstate average, $\langle 0|\hat{A}|0\rangle = \langle 0|\hat{B}|0\rangle = 0$, where $|0\rangle$ is the groundstate of an interacting system with hamiltonian $\hat{H}$. In the presence of an additional externally applied time-dependent "potential" $\delta v \exp(ut)$ coupling to $\hat{A}$, the hamiltionian is $\hat{H} - \delta v \exp(ut)\hat{A}$. Calculating the perturbed state $|\Psi(t)\rangle$ by standard first-order time-dependent perturbation methods, one finds that the expectation of property $\hat{B}$ at time t is of form $\langle\Psi(t)|\hat{B}|\Psi(t)\rangle = \chi_{BA}(\mathrm{i}u)\delta v \exp(ut)$, where

$$\chi_{AB}(\mathrm{i}u) + \chi_{BA}(\mathrm{i}u) = -2\sum_J E_{IJ}\frac{\langle 0|\hat{A}|J\rangle\langle J|\hat{B}|0\rangle + \langle 0|\hat{B}|J\rangle\langle J|\hat{A}|0\rangle}{E_{IJ}^2 + \hbar^2u^2}. \tag{30.10}$$

Here $|J\rangle$ is an eigenstate, and $E_{IJ} = E_I - E_J$ is an eigenenergy difference, of the interacting hamiltonian $\hat{H}$. The quantity χ_{AB} is the "AB response function", e.g., the electronic density-density response in the case at hand. Note that the use of a real external potential $\exp(ut)$ corresponds to the more usual choice $\exp(-\mathrm{i}\omega t)$ but with a positive imaginary frequency $\omega = \mathrm{i}u$. Using the arctan integral $\int_0^\infty \mathrm{d}u\,(E_{IJ}^2 + \hbar^2u^2)^{-1} = \pi/(2\hbar E_{IJ})$ we have

$$\begin{aligned}\int_0^\infty \mathrm{d}u\,(\chi_{AB} + \chi_{BA}) &= -\frac{\pi}{\hbar}\sum_J\left\{\langle 0|\hat{A}|J\rangle\langle J|\hat{B}|0\rangle + \langle 0|\hat{B}|J\rangle\langle J|\hat{A}|0\rangle\right\}\\ &= -\frac{\pi}{\hbar}\langle 0|\hat{A}\hat{B} + \hat{B}\hat{A}|0\rangle\,. \end{aligned}\tag{30.11}$$

This is a frequency-integrated, $T = 0\,\mathrm{K}$ form of the very general *fluctuation-dissipation theorem* [Callen 1951, Landau 1969] (FDT). The FDT is more usually quoted as a result at finite temperature and for a single frequency lying just above the real axis, where $\Im\chi_{AA}$ is known to represent energy absorption ("dissipation"). Applying (30.11) to the density fluctuation operators $\hat{A}, \hat{B} = \delta\hat{n}(\boldsymbol{r}), \delta\hat{n}(\boldsymbol{r}')$ from (30.7), we have

$$\langle\lambda|\,\delta\hat{n}(\mathbf{r})\delta\hat{n}(\mathbf{r}')+\delta\hat{n}(\mathbf{r}')\delta\hat{n}(\mathbf{r})\,|\lambda\rangle = -\frac{\hbar}{\pi}\int_0^\infty (\chi_\lambda(\mathbf{r},\mathbf{r}',iu)+\chi_\lambda(\mathbf{r}',\mathbf{r},iu))\,du \tag{30.12}$$

where the density-density response χ_λ is just that introduced in Chap. 1, applied to the system with reduced Coulomb interaction $\lambda e^2/r_{12}$.

Combining (30.12) with (30.9) and noting that the $\boldsymbol{r}$ and $\boldsymbol{r}'$ integrations can be interchanged, we obtain

$$E_{\mathrm{xc}} = \frac{1}{2}\int_0^1 \mathrm{d}\lambda \int \mathrm{d}^3r \int \mathrm{d}^3r' \,\frac{e^2}{|\boldsymbol{r}-\boldsymbol{r}'|} \left\{-\frac{\hbar}{\pi}\int_0^\infty \mathrm{d}u\,\chi_\lambda(\boldsymbol{r},\boldsymbol{r}',iu) - \delta^3(\boldsymbol{r}-\boldsymbol{r}')n(\boldsymbol{r})\right\}. \tag{30.13}$$

We will refer to the important exact formula (30.13) as the "ACFD" xc energy. Its generalization to finite temperature is straightforward and can be expressed using discrete Matsubara frequencies instead of a continuous frequency integral.

In TDDFT the linearized interacting density-density response $\chi_\lambda \equiv \chi_\lambda(\boldsymbol{r},\boldsymbol{r}',\omega)$ and the KS response $\chi_{\mathrm{KS}} \equiv \chi_{\lambda=0}$ are related by the Dyson-like screening equation [Gross 1985]

$$\chi_\lambda = \chi_{\mathrm{KS}} + \chi_{\mathrm{KS}} * (\lambda v_{\mathrm{ee}} + f_{\mathrm{xc},\lambda}) * \chi_\lambda\,, \tag{30.14}$$

where stars represent convolution in $(\boldsymbol{r},\boldsymbol{r}')$ space and v_{ee} is the bare electron-electron Coulomb potential (see Chap. 1). In practice, the inputs to (30.14) (namely χ_{KS} and – in almost all existing approximations – f_{xc}) can be computed from the groundstate KS orbitals $\{\varphi_i\}$. In turn, the $\{\varphi_i\}$ are directly computable from the groundstate KS potential $v_{\mathrm{KS}}(\boldsymbol{r})$. *Thus the ACFD energy (30.13) is often best regarded as a functional* $E^{\mathrm{ACFD}}[v_{\mathrm{KS}}(r)]$ *of the groundstate KS potential.* In principle, one should introduce a high level of self-consistency (the OEP level) by choosing v_{KS} to minimize the nonlocal functional (30.13), with the external potential fixed. In this sense, most detailed calculations to date have been "post-functionals": that is, they have computed $E[v_{\mathrm{KS}}^{\mathrm{LDA}}]$ rather than $E[v_{\mathrm{KS}}^{\mathrm{OEP}}]$. In Sect. 30.7.1 below we will argue that this does not significantly affect the predictions of (30.13) for the distant vdW correlation energy, at least in the case $f_{\mathrm{xc}} = 0$ (corresponding to the RPA). Thus the energy functionals are not overly sensitive to the v_{KS} used as input. On the other hand, if one uses the Feynman approach described in Sect. 30.2.3 above, in which the vdW force is calculated directly as a force on the nuclei, then the small selfconsistent changes in v_{KS} as a function of distance, calculated beyond the LDA, are crucial [Allen 2002].

30.6.2 Exact Exchange: A Strength of the ACFD Approach

In order for a theory to work well for interacting systems with overlap of electronic clouds, it is essential that it accounts adequately for the Pauli exchange energy: indeed, the Hartree-Fock theory already describes many covalent bonds quite well. In fact, when the independent-electron (Kohn-Sham) response $\chi_{\mathrm{KS}} \equiv \chi_{\lambda=0}$ is substituted for χ_λ, (30.13) yields the exact nonlocal exchange energy $E_{\mathrm{x}}^{\mathrm{Exact,\,DFT}}$, defined in the DFT sense. The simplest proof assumes real KS orbitals φ_i, so that

$$\chi_{\mathrm{KS}}(\boldsymbol{r},\boldsymbol{r}',iu) = 2\sum_{i\neq j} w_{ij} n_i \frac{\varphi_i(\boldsymbol{r})\varphi_j(\boldsymbol{r}')\varphi_j(\boldsymbol{r})\varphi_i(\boldsymbol{r}')}{w_{ij}^2+\hbar^2u^2}\,, \tag{30.15}$$

independent of λ, where $w_{ij} = \varepsilon_i - \varepsilon_j$ is a KS eigenvalue difference and n_i is a Fermi occupation factor. Putting this into (30.13) one obtains

$$E_{\mathrm{xc}}[\chi_{\mathrm{KS}}] = -\frac{1}{2}\int \mathrm{d}^3r \int \mathrm{d}^3r' \,\frac{e^2}{|\boldsymbol{r}-\boldsymbol{r}'|}\sum_i n_i\varphi_i(\boldsymbol{r})\varphi_i(\boldsymbol{r}')\varphi_i(\boldsymbol{r})\varphi_i(\boldsymbol{r}')\,. \tag{30.16}$$

Here the u integration in (30.13) gave $\pi/(2\hbar)$, and the completeness relation $\delta^3(\boldsymbol{r}-\boldsymbol{r}') = \sum_j \varphi_j(\boldsymbol{r})\varphi_j(\boldsymbol{r}')$ was used to show that the delta function term in (30.13) simply restores the $j = i$ term excluded from the sum (30.15) for χ_{KS}. The orthonormality of the $\{\varphi_i(\boldsymbol{r})\}$ was also used. Equation (30.16) is just the exact DFT exchange energy $E_{\mathrm{x}}^{\mathrm{Exact,\,DFT}}$, showing that any ACFD energy formula based on a numerically accurate χ_{KS} for a pair of interacting systems is already a candidate to be a useful, seamless vdW theory as far as the overlapped regime is concerned. This is true even if the screening of χ_{KS} to create χ_λ is done crudely, as for example in the RPA (time-dependent Hartree, $f_{\mathrm{xc}} = 0$) approximation.

30.7 The xc Energy in the Random Phase Approximation

The energy in the RPA is defined by (30.13) and (30.14) with f_{xc} set to zero. Numerical evaluation of the RPA energy is non-trivial for inhomogeneous systems. It has been evaluated numerically for small molecules where it gave a reasonable account of "static correlation" [Furche 2001c]. It performed better than the GGA for the binding of Be_2 [Fuchs 2002]. For Na and Si solids it gave a reasonable geometry [Miyake 2002]. Numerical RPA calculations for layered jellium systems [Dobson 1999] exhibited a vdW energy tail (unlike LDA calculations) and also exposed significant flaws in the LDA predictions even in the overlapped regime [Dobson 2004, Jung 2004] (see also Sects. 30.9 and 30.8.3). Here we first explore some formal vdW properties of the RPA in the well-separated "dispersion force" limit.

30.7.1 Testing the RPA Correlation Energy For vdW in the Well-Separated Limit: The Second-Order Perturbation Regime

The general physical arguments above show that *basic* physics of the vdW interaction is present within the ACFD energy with mean-field-like approximations for the response function. However it is not clear a priori that such an approach can produce *accurate* predictions of vdW energies. It was therefore reassuring to find [Dobson 1994b] that the formula (30.2) is exactly true within the ACFD, at least with the RPA approximation for the response. The proof [Dobson 1994b] involved direct perturbation of the RPA version of the screening equation (30.14) with respect to v_{ee}^{12}. It was necessary to include: (i) terms involving first-order changes to the cross response $\chi_{\lambda,\,12}$, but also (ii) terms involving second order changes in the intra-system responses χ_{11} and χ_{22} (recently called "spectator" terms [Langreth 2005]).

After an integration with the Coulomb potential, both of these terms yielded contributions of second order in v_{ee}^{12}. The outcome was that the vdW interaction from the RPA correlation energy is of form (30.2) but with the individual responses of the isolated systems, naturally, replaced by their RPA versions:

$$E_{12}^{\mathrm{RPA},\,2} \sim -\frac{\hbar}{2\pi}\int \mathrm{d}^3 r_1 \ldots \int \mathrm{d}^3 r_2' \,\frac{e^2}{r_{12}}\frac{e^2}{r_{12}'}\int_0^\infty \mathrm{d}u\; \chi_1^{\mathrm{RPA}}(\boldsymbol{r}_1,\boldsymbol{r}_1',\mathrm{i}u)\chi_2^{\mathrm{RPA}}(\boldsymbol{r}_2,\boldsymbol{r}_2',\mathrm{i}u)\,. \tag{30.17}$$

That is, the formula (30.2) is true within the seamless ACFD/RPA formalism for a pair of widely separated systems.

It is also worth noting that the proof of (30.17) in [Dobson 1994b] assumed that the KS (independent-electron) responses $\chi_{11}^{(0)}$, $\chi_{22}^{(0)}$ of the subsystems did not vary with separation D, as $D \to \infty$. This corresponds to the most common way of implementing RPA correlation energy as a "post-functional," that is, using χ_{KS} deduced from the KS orbitals obtained from a groundstate LDA or GGA calculation. Of couse, in a fully selfconsistent (OEP) RPA correlation energy calculation, one would vary the groundstate KS potential v_{KS} (and correspondingly χ_{KS}), to minimize the total nonlocal RPA energy. An estimate of the effect of the change in the KS potential in the subsystems can be made as follows. Suppose that we start with the selfconsistent OEP-RPA solution $v_{\mathrm{KS}}^{\mathrm{OEP}}$ as described above and make a small change δv_{KS} in the KS potential (e.g., toward the LDA post-functional value described above). The change in RPA energy due to this change is of *second* order in δv_{KS} because the energy is minimized by $v_{\mathrm{KS}}^{\mathrm{OEP}}$. If we assume that $\delta v_{\mathrm{KS}} = O(D^{-p})$, as required to produce the nuclear D^{-7} vdW force via the Feynman-Tozer approach described in Sect. 30.2.3 above, then the (second-order) correction energy is of $O(D^{-2p})$ and so is negligible compared with the post-functional energy (30.17).

30.7.2 Coupled Plasmons and the ACFD Approach

Another strength of the ACFD approach is that it contains the sum of plasmon zero point energies (see Sect. 30.2.4 above), in the sense to be described below. In large systems, where the single-particle excitations form a continuum, any isolated poles of χ_λ as a function of frequency, lying near to the real frequency axis, are generally considered to correspond to plasmons – collective electron density oscillations. Using the analyticity of causal response functions and the f-sum rule (which ensures that χ falls off as ω^{-2} at large frequency), one can convert the imaginary frequency integral in the ACFD formula (30.13) to an integral along the positive real frequency axis. Hence it is apparent that any plasmon poles lying close below the real axis will have a large influence on the xc energy. At least for the RPA [the case $f_{\rm xc} = 0$ in the screening equation (30.14)], one can show that the contributions of these poles (via their *semi*-residues) is just the sum of the zero-point energies $\sum_q \hbar\omega_q/2$ of the plasmons. For more detail, see [Dobson 2005a] or [Mahanty 1976]. Thus the ACFD(RPA) correlation energy *contains* a contribution from the plasmon zero point energies, and hence contains the basic physics discussed in Sect. 30.2.4. It is not in general true, however, that the ACFD(RPA) correlation energy *equals* the sum of plasmon zero-point energies: the integration contour is not closed, and Cauchy's theorem does not give a definitive answer for the integral. Portions of the integration contour away from the poles can be important as well. Indeed there are cases where there are no real plasmons, but the RPA correlation energy still yields an interesting vdW interaction [Dobson 2005c, Dobson 2006].

So far we have motivated the idea that the ACFD is at least qualitatively correct in both the overlapped and the distant regimes, even when used with quite simple models for the response function χ_λ.

30.8 Beyond the RPA: The ACFD with a Nonzero xc Kernel

30.8.1 The Case of Two Small Distant Systems in the ACFD with a Nonzero xc Kernel

The asymptotic situation is not entirely simple when one uses a TDDFT response with a nonzero kernel $f_{\rm xc}$ within the ACFD energy (30.13), rather than the RPA response. Suppose that one assumes (i) that the dynamic kernel $f_{\rm xc}(\mathbf{r}_1, \mathbf{r}_2, \omega)$ is zero when $\mathbf{r}_1$ and $\mathbf{r}_2$ lie in different fragments, to second order in $v_{\rm ee}^{12}$, and (ii) that the presence of the other fragment does not affect the value of $f_{\rm xc}(\mathbf{r}_1, \mathbf{r}_1', \omega)$ inside a given fragment, again to second order in $v_{\rm ee}^{12}$ – i.e., we neglect "spectator" effects in $f_{\rm xc}$. (These assumptions are satisfied within the ALDA provided one uses a "post-functional" approach in which the changes in density and $\chi_{\rm KS}$ in each fragment, beyond the isolated case, are ignored.) Then in fact one can show by generalizing the methods described in [Dobson 1994b] that the seamless ACFD approach does *not* yield the perturbative result (30.2) in the well-separated limit, there being an additional energy term involving $\partial f_{\mathrm{xc}}/\partial\lambda$. When used with simple local approximations for f_{xc}, the additional term is non-vanishing, and it is not yet known whether it improves or worsens the agreement of the seamless ACFD TDDFT approach with the exact C_6 asymptotics. For the exact nonlocal f_{xc}, the $\partial f_{\mathrm{xc}}/\partial\lambda$ term must cancel to second order with the nonlocal and spectator terms neglected in the above analysis. It is even possible that this requirement could lead to a useful new constraint on model f_{xc} kernels. More work is required to clarify these points.

30.8.2 Beyond the RPA in the ACFD: Energy-Optimized $\mathbf{f_{xc}}$ Kernels

The standard non-memory approximation for f_{xc} has been the adiabatic local density approximation [Zangwill 1980a] (ALDA) given by $f_{\mathrm{xc}}^{\mathrm{ALDA}}(\boldsymbol{r},\boldsymbol{r}',\omega) = \delta^3(\boldsymbol{r}-\boldsymbol{r}')\mathrm{d}^2[ne_{\mathrm{xc}}^{\mathrm{hom}}(n)]/\mathrm{d}n^2$. It is "optimized" for describing low-frequency, long-wavelength excitations in near-homogeneous systems, and is therefore quite unsuitable for the calculation of ground-state xc energies from (30.13) and (30.14), because these formulae effectively sample all the frequency and wavenumber space. For example, in a uniform electron gas, $f_{\mathrm{xc}}^{\mathrm{ALDA}}(\boldsymbol{r},\boldsymbol{r}',\omega)$ leads to a very poor evaluation of the correlation energy when substituted into (30.13) and (30.14), (see, e.g., Fig. 1 of [Dobson 2000a]).

The simplest way to remedy this is to find a local frequency-independent kernel $f_{\mathrm{xc}}^{\mathrm{opt}}(\lambda,n)$ that *does* give the correct uniform-gas E_{xc} when substituted into (30.13) and (30.14). It turns out [Dobson 1994b, Dobson 2000a] that this requirement, applied for every homogeneous density n, together with the scaling rule $f_{\mathrm{xc}}(\lambda,n) = \lambda^{-1}\hbar^4 e^{-2} m^{-2} F(\lambda^{-1} r_{\mathrm{s}})$, $4\pi r_{\mathrm{s}}^3 a_B^3/3 \equiv n^{-1}$, uniquely determines the dimensionless energy-optimized kernel $f_{\mathrm{xc}}^{\mathrm{opt}}(r)$. The optimized kernel has to produce a magnitude of corrrelation energy lying between the (too large) RPA value corresponding to $f_{\mathrm{xc}}=0$ and the (too small) value corresponding to the ALDA kernel. Unsurprisingly, then, $f_{\mathrm{xc}}(r_{\mathrm{s}})$ lies between zero and $f_{\mathrm{xc}}^{\mathrm{ALDA}}$. A parametrized form accurate for metallic densities is [Dobson 2000a] $f_{\mathrm{xc}}^{\mathrm{opt}}(r_{\mathrm{s}}) = -0.5004_4 r_{\mathrm{s}}^2 + 4.5365_3\times10^{-3} r_{\mathrm{s}}^3 - 3.366_0\times10^{-5} r_{\mathrm{s}}^4$.

The theory is of course constrained to give accurate correlation energy for the uniform gas. Its performance for a *nonuniform* gas has been tried in the context of layered jellium [Dobson 2000a] where it had little effect on the energy *differences* required for a vdW energy calculation, compared with a pure RPA calculation. Other than this, there has been little direct testing of this kernel in other systems.

The purely local (q-independent) character of the $f_{\mathrm{xc}}^{\mathrm{opt}}$ kernel can cause numerical difficulties in the solution of the screening equation (30.14), when a q-space algorithm is used. For this reason, and to include some of the physical aspects of nonlocality, a static but spatially nonlocal energy-optimized kernel was tried by Jung et al. [Jung 2004]. They used the λ-scaling law

$$f_{\mathrm{xc},\lambda}(\boldsymbol{r}_1,\boldsymbol{r}_2,\omega) = \lambda^{-1} f_{\mathrm{xc}}^{(0)}(\lambda^{-3}\tilde{n}, \lambda r_{12}), \tag{30.18}$$

where $\tilde{n}(\boldsymbol{r}_1,\boldsymbol{r}_2)$ is an effective density taken in their applications to be $[n(\boldsymbol{r}_1)+n(\boldsymbol{r}_2)]/2$ or $\sqrt{n(\boldsymbol{r}_1)n(\boldsymbol{r}_2)}$, both leading to similar results. The forms (30.18) (and indeed the scaling in the Dobson-Wang scheme [Dobson 2000a]) is consistent with the frequency-independent limit of the general scaling law derived by Lein, Gross, and Perdew [Lein 2000b]. The spatial nonlocality was assumed by Jung et al. to take a Hubbard-like form

$$f_{\mathrm{xc}}^{(0)} = f_{\mathrm{xc}}^{(0)}(r_{\mathrm{s}}, q) = f_{\mathrm{xc}}^{\mathrm{ALDA}}(r_{\mathrm{s}}) \left\{ 1 + \alpha(r_{\mathrm{s}}) \left(\frac{q}{k_F} \right)^2 \right\}^{-1}, \tag{30.19}$$

where an excellent fit to the numerically exact correlation energy of the uniform gas with $0 < r_{\mathrm{s}} < 50$ was obtained by choosing $\alpha(r_{\mathrm{s}}) = (8.26 + r_{\mathrm{s}})/(100+5r_{\mathrm{s}})$ (see Fig. 2 of [Jung 2004]). With the kernel (30.19), the minimum region of the energy-versus separation curve of two high-density jellium slabs was found to be very similar to the pure RPA result, but the minimum region was lower than the LDA curve, representing an additional binding energy of about 50% compared to LDA [Jung 2004]. The only source of error (other than numerics) in the ACFD calculations is the use of a simple f_{xc} in place of the exact one. Since the results of the various nonlocal theories with and without the xc kernel were very similar, it is presumably the LDA result that is in error.

30.8.3 Beyond the RPA in the ACFD: More Realistic Uniform-Gas Based $\mathbf{f_{xc}}$ Kernels

Recently more realistic q- and/or ω-dependent kernels have been obtained for the uniform electron gas [Richardson 1994, Corradini 1998] that rather accurately [Lein 2000b] give e_{c} for the 3DEG, when substituted into the ACFD formulae (30.13) and (30.14). The static kernel of Corradini et al. [Corradini 1998] has been used with the ACFD under the appropriate λ scaling formula (30.18) for layered nonuniform jellium systems [Pitarke 2003, Jung 2004]. The surface energy results [Pitarke 2003] based on a single jellium slab were not very different from the pure RPA nor from the LDA results. The vdW results for two jellium layers [Jung 2004] were not very different from the energy-optimized schemes described above, nor from the pure RPA calculation. There were some differences from the LDA, and in one case [Jung 2004] all the microscopic schemes including the Corradini version did give about a 50% increase in binding energy of the slabs, compared with an LDA energy calculation. The equilibrium separation D_0 of the slabs was quite similar in all of the f_{xc}-corrected schemes and the LDA, but was slightly smaller in the pure RPA energy calculation.

It should also be possible to graft the more sophisticated frequency-dependent Richardson-Ashcroft uniform-gas kernel [Richardson 1994] onto an ACFD energy calculation for nonuniform systems, but that has not yet been done to the author's knowledge.

Another consideration not yet explored in the vdW context is the use of tensor memory kernels such as that used in the Vignale-Kohn-Ullrich current-current response formula [Vignale 1996, Vignale 1997, Conti 1999]. These kernels are based on the near-homogeneous gas with sinusoidal density variation, and have been proposed [van Faassen 2002] for description of the polarization response in 1D systems. This method obtains rather good polymer properties (see Chap. 21), and so should be a good candidate for seamless van der Waals energetics of these systems.

30.8.4 xc Kernels not Based on the Uniform Electron Gas

All of the xc kernels discussed above have been based on the uniform or near-uniform electron gas, and grafted onto an ACFD energy calculation for a nonuniform system, in a modified local-density manner. Kernels have also been developed that are based specifically on the KS orbitals of the nonuniform system. Perhaps the simplest of these is the kernel of Petersilka, Gossman and Gross [Petersilka 1996a, Petersilka 1998] which is a static, spatially nonlocal exchange-only kernel. It has been tested [Fuchs 2002] on the energetics of the Be-Be dimer, where it performed somewhat similarly to the RPA for the binding energy, and better than the RPA for the bond length. The PGG kernel is self-interaction free, which suggests that it might out-perform the RPA more clearly when used with an appropriate SIC groundstate calculation.

The nonlocal kernel of Reining and coworkers [Reining 2002] is designed for finite-frequency response of semiconductors, and may be relevant to vdW interactions between these systems.

A further possibility, similar in spirit to a system-specific f_{xc}, is the use of the inhomogeneous Singwi-Tosi-Land-Sjolander (ISTLS) formalism [Dobson 2002] to generate χ_λ. To date no vdW calculations have been performed in this way, however.

30.8.5 Is the ACFD Energy Insensitive to $\mathbf{f_{xc}}$ in Layered and Striated Systems having Zero Bandgap?

Perhaps the most surprising vdW results discussed above were the non-standard power laws described in Sect. 30.5 for the interaction energy of layered and striated systems that have no energy gap between occupied and unoccupied KS orbitals. The asymptotic ($D \to \infty$) vdW attraction in these cases is dominated by excitations with low-wavenumber ($q \sim D^{-1} \to 0$). At these wavenumbers the Coulomb potential is $v_{\text{ee}} = 2\pi e^2 \exp(-qD)/q$ (for sheet geometry) or $v_{ee} \approx -2e^2 \ln q$ (for tube or wire geometry), i.e., it is divergent as $q \to 0$. Provided that the kernel f_{xc} is spatially short-ranged, $f_{xc}(q)$ does not diverge as $q \to 0$ and so may be neglected in these non-overlapping geometries, compared with the bare Coulomb term λv_{ee} in (30.14). Limited testing in the layered jellium case [Dobson 2000a, Jung 2004] found that f_{xc} has little effect on the energy-versus separation curves, even near the overlapped equilibrium spacing, suggesting that perhaps low-q excitations are still important even at small spatial separations. It should be stressed that in other cases such as prediction of the vdW attraction between distant atoms from (30.2), it is essential to include f_{xc}: atomic respose functions can be significantly under-estimated in the RPA. Also, in finite-gapped systems (semiconductors and insulators), highly nonlocal kernels may be needed [Reining 2002] to reproduce dielectric behavior correctly: this is not relevant to the zero-gapped systems of Sect. 30.5, however.

30.9 Density-based Approximations for the Response Functions in ACFD vdW Theory

Recently it has become possible to evaluate nonlocal TDDFT response functions and the corresponding ACFD energy for inhomogeneous systems, both in the asymptotic and the overlapped vdW regimes. In simple layered (jellium slab) geometry, this type of calculation is now numerically tractable for vdW energetics, without essential approximations except for the choice of f_{xc} [Dobson 1999, Dobson 2004, Jung 2004]. This has led to the provision of benchmarks quantifying successes and failures of the LDA, and of other approximate theories, for these systems, at all separations of the slabs. ACFD calculations for large systems with full *three*-dimensional inhomogeneity remain near the limit of computational feasibility, but some results are starting to appear [Fuchs 2002, Miyake 2002, Furche 2001c, Marini 2006]. If this approach is to be widely used, it will be necessary to find some suitable approximations that retain the seamless vdW character of the energy functional. These are discussed in Sect. 30.9.2 below.

30.9.1 Density-Based Approximations for the Non-Overlapping Regime

Consider first the result (30.3) relating to non-ovelapping systems, in which the long-ranged effects of v_{ee}^{12} in (30.14) have already been taken into account via perturbation theory. The remaining response functions $\alpha_1(\boldsymbol{r}_1, \boldsymbol{r}_1{}', iu)$ and $\alpha_2(\boldsymbol{r}_2, \boldsymbol{r}_2{}', iu)$ can be approximated in a local fashion based on the local fragment density, without losing the vdW tail. This leads to an approximation to the second-order vdW energy of two small systems by a functional of their individual groundstate electron densities $n_1(\boldsymbol{r}_1)$, $n_2(\boldsymbol{r}_2)$:

$$E^{(2)} \approx -\frac{3\hbar e}{2(4\pi)^{3/2} m^{1/2}} \int_1 \mathrm{d}^3 r_1 \int_2 \mathrm{d}^3 r_2 \, \frac{1}{r_{12}^6} \frac{\sqrt{n_1 n_2}}{\sqrt{n_1} + \sqrt{n_2}} \, . \tag{30.20}$$

Equation (30.20) is a highly nonlocal functional of the groundstate electron density n. It was first obtained [Andersson 1996], by considering some asymptotics, starting from a semi-empirical plasmon-motivated form due to Ashcroft and Rapcewicz [Rapcewicz 1991]. Independently, (30.20) was derived straightforwardly [Dobson 1996] by a constrained local approximation for the response functions in (30.3). In practice, (30.20) gives answers sometimes greatly too large because of the contributions from the tails of the atomic densities, and it requires a cutoff in the spatial integrations, based on gradients of n and described in detail in Andersson et al. [Andersson 1996]. The answers are sensitive to this cutoff, but do provide results for C_6 that are mostly quite good for distant atom-atom interactions, and err in the worst cases by about a factor of 2. Given the relatively poor performance of the RPA and ALDA for atomic polarizabilities via (30.4) based on an LDA KS potential, this kind of error does not look so bad. However, one must also note the very simple and rather successful formulas in the chemistry literature [Atkins 1997], involving the electron affinity and based loosely on the arguments given in Sect. 30.2.1 above.

Another approach [Dobson 2000b, Dinte 2004] allows one to avoid a cutoff by using an exact "force theorem" to constrain model response functions in (30.3). With the simplest (fully local) model polarizability, this approach obtained correct trends but was inferior to the cutoff approach described above. Better results may result from a less local ansatz for the polarizabilities.

30.9.2 "Seamless" Density-Based vdW Approximations Valid into the Overlapped Regime

Within the ACFD, the essential nonlocality of the vdW energy arises from the long range of the electron-electron interaction v_{ee} in the Dyson-like screening equation (30.14). This fact is the key to obtaining approximations that do not suppress the tail of the vdW interaction.The approach of Dobson and Wang [Dobson 1994b, Dobson 1999] to the general case, including overlap, is to evaluate the ACFD energy by approximating the "bare" (Kohn-Sham) response χ_{KS} in the screening equation (30.14), using the groundstate density as input. The screening equation (30.14) is still solved numerically with retention of the long-ranged character of the electron-electron interaction v_{ee}. Care has to be taken to conserve charge, and to ensure that the approximate χ_{KS} does not introduce unphysical flow of electrons from one subsystem to the other, in the non-overlapping limit. This approach was tested successfully [Dobson 1999, Dobson 2000a] for the vdW interaction of a pair of jellium slabs. Here comparison could be made with an accurate numerical solution of the full ACFD equations at all separations from the overlapping contact situation out to the asymptotic non-overlapping vdW regime. The correct vdW-RPA interaction was obtained at large separations where the usual LDA/GGA produces no interaction at all, and the RPA energies near the equilibrium separation were reproduced somewhat better than in the LDA. The tests were subsequently extended [Dobson 2000a] to the inclusion of a local f_{xc} which turned out to make a negligible difference in the cases studied. This type of approach has not yet been tried in other geometries.

A promising path to seamless theories is the separation of the bare electron-electron interaction into short-ranged and long-ranged parts, $v_{\text{ee}} = v_{\text{ee}}^{\text{sr}} + v_{\text{ee}}^{\text{lr}}$, with different approximations made in treating $v_{\text{ee}}^{\text{sr}}$ and $v_{\text{ee}}^{\text{lr}}$ [Stoll 1985, Savin 1991, Savin 1988, Pollet 2002, Angyan 2005]. Kohn, Meir and Makarov [Kohn 1998] gave a systematic way to do this within the ACDF approach, but it seems that so far no corresponding seamless energy calculations have appeared.

A series of approximations to the ACFD have been proposed by a group based at Chalmers and Rutgers Universities. First, for layered jellium systems (with only one-dimensional inhomogeneity), aproximations were developed [Rydberg 2000] to simplify the numerical solution of the screening equation (30.14). Then, in order to deal with layered systems beyond the jellium model, approximations were developed for the susceptibility that took into account the non-jellium nature of the layers. This was achieved by constraining an approximate density response function to reproduce the static polarizability of each layer, to an electric field *perpendicular* to the layer, obtained from a groundstate LDA/GGA calculation [Rydberg 2000, Hult 1999, Rydberg 2003b]. Recently this type of approach has been extended to more general geometries [Dion 2004, Langreth 2005, Schroder 2003]. A good feature of this approach is that it treats overlapped cases while also providing a vdW energy "tail" of classic R^{-6} form

$$E \approx -\int \mathrm{d}^3 r_A \int \mathrm{d}^3 r_B \, f(n_A(\boldsymbol{r}_A), \nabla n_A(\boldsymbol{r}_A); n_B(\boldsymbol{r}_B), \nabla n_B(\boldsymbol{r}_B)) \, |\boldsymbol{r}_A - \boldsymbol{r}_B|^{-6} \tag{30.21}$$

for separated subsystems A and B. These seamless theories are promising for weakly bonded finite molecules and small clusters, giving a natural "saturation" of the R^{-6} van der Waals energy contributions at shorter distances, a long-anticipated goal [Wu 2002]. Low-order perturbation theory also supports (30.21).

On the other hand, one version of the Chalmers/Rutgers approach gives results close to the LDA for the near-equilibrium energetics of graphite [Rydberg 2003b], and the LDA result is now widely believed to be questionable (see Sect. 30.5 above). A later version [Dion 2004] gives a larger graphite binding, intermediate between the two rather divergent experimental numbers [Zacharia 2004, Benedict 1998c]. Furthermore, the asymptotic form (30.21) is known to be incorrect for layered and striated metallic and semimetallic systems at large separations (see Sect. 30.5 above). It remains to be seen how important this discrepancy is in practice: the question can be answered (for example) by numerically exact ACDF calculations currently underway for "stretched" layered semi-metallic (graphitic) systems. If this consideration does turn out to be important, then it seems unlikely that a single theory dependent only on the local density and its gradient can be adequate. For example, the Dobson-Wang approach [Dobson 1999, Dobson 2000a] obtains correct asymptotics for thin metals, but its behaviour has not yet been explored for insulators. The Chalmers/Rutgers approach is appropriate for small systems, semiconductors and insulators, but does not obtain the correct asymptotics for layered and striated metals or semi-metals. It seems that a general theory may have to take into account at least some qualitative data on the states near the (possibly zero) bandgap that determine the metallic or semi-metallic properties of the system. These details of the electronic energy bandstructure are presumably reflected in the groundstate density, but only in rather subtle ways. This area of research is still very much open. One possible approach [Dobson 2005a] is to use an existing vdW-corrected LDA energy formalism, with a further ACFD/TDDFT correction only for the long-wavelength components of the response.

30.10 Summary

The prediction of vdW energetics is still a controversial area, especially for large highly anisotropic "soft" systems with small or zero energy gaps. Even for such "difficult" systems the adiabatic connection fluctuation dissipation (ACFD) approach, based on various TDDFT approximations for the density-density response χ, appears to be very promising. Unfortunately even the RPA ($f_{\mathrm{xc}} = 0$) theory, applied without further approximations, leads to energy calculations at the edge of current computational capabilities. Approximation schemes are becoming available, but so far none is confirmed to be valid in all the known limits. ACFD calculations via inclusion of exchange-correlation kernels f_{xc} have so far received only very limited testing. This has been done for small dimers near their equilibrium spacing, for more general small systems at large separations, and for large layered systems in planar (jellium) geometry. There are difficult issues in all these cases. Ultimately, when these issues have been resolved, TDDFT can be expected to be an important aspect of accurate calculations of the delicate energetics of vdW systems.

31 Kohn-Sham Master Equation Approach to Transport Through Single Molecules

R. Gebauer, K. Burke, and R. Car

31.1 Introduction

In recent years it has become possible to study experimentally charge transport through single molecules [Nitzan 2003]. Typically, the devices in which such experiments can be realized consist of metal-molecule-metal junctions, where two metallic leads are connected by some molecule. Such junctions are expected to be at the basis of future molecular-based electronic devices. But apart from technological applications, these experiments can also be taken as a basis for understanding electron tunneling in general.

The theoretical modeling of molecular transport, however, is a very challenging task: On one hand, it is intuitively clear that the conduction properties of a molecular junction will crucially depend on details of the chemical bonding, particularly at the interface between the metal electrodes and the molecule. Such properties are routinely studied using methods based on density-functional theory (DFT) [Hohenberg 1964]. On the other hand, ground-state theories like DFT cannot be directly applied to systems with a finite current, because such devices are out of equilibrium.

A standard methodology that allows to calculate transport properties from first principles is based on non-equilibrium Green's functions (NEGF) [Keldysh 1965]. In this framework, which is today adopted for almost all calculations in this field, the current is obtained by solving an elastic scattering problem with open boundaries through which electrons are injected from the leads. The NEGF formalism is often used together with DFT in order to achieve chemical accuracy also in systems containing several tens to hundreds of atoms. The NEGF method is explained in detail in Chap. 32 of this book.

Here we present an alternative formulation that can be seen as a generalization of the Boltzmann transport equation (BTE) [Ashcroft 1976] to the fully quantum mechanical case. The BTE is a very successful approach to study transport in non-equilibrium systems. In contrast to NEGF formalisms, where energy-dependent scattering problems are solved, the BTE is formulated in the time-domain: Electrons are accelerated by an external driving force (an electric field, for instance), and the energy which is injected in this way into the system is dissipated by inelastic scattering events. The interplay between acceleration and dissipation leads to a *steady state* in which

R. Gebauer et al.: *Kohn-Sham Master Equation Approach to Transport Through Single Molecules*, Lect. Notes Phys. **706**, 463–477 (2006)
DOI 10.1007/3-540-35426-3_31

a finite, time-independent current flows through the system. The BTE is a semi-classical theory that is formulated in terms of *wave packets* whose dimensions in space are typically of the order of several tens of nanometers. It is therefore clear that the BTE cannot be used to study molecular junctions where chemical details on the atomistic scale need to be taken correctly into account.

In this chapter we show how it is possible to generalize the Boltzmann approach to quantum systems at the nanoscale, and how transport can be treated as a time-dependent problem in this framework.

31.2 Modeling a Molecular Junction

The first step towards a computational treatment of electron transport through single molecules is the choice of a suitable model geometry in which the calculations are performed. Experimentally the nanojunctions are placed between two metallic contacts. Such contacts can be for example the tip of a scanning tunneling microscope (STM), or a metallic surface on which a molecular layer is assembled. Those metallic contacts are then connected via the leads to an external power source, e.g., to a battery. In standard NEGF calculations, this setup is modeled using *open boundary conditions* in the contacts, and the leads outside the computational box on both sides of the molecule are considered semi-infinite, as shown in the upper panel of Fig. 31.1.

31.2.1 Periodic Boundary Conditions

In contrast to the usual approach, we model the molecular junction using a so-called *ring geometry*: the molecule together with a finite piece of the metallic contacts is repeated periodically. This setup, which represents a closed circuit in which no electrons can be exchanged with the exterior, is shown in the middle and lower panel of Fig. 31.1.

In order to induce an electrical current in this ring, one needs to apply an external electromotive force. In our calculations we use a spatially constant electric field $\boldsymbol{\mathcal{E}}$ to accelerate the electrons. Often such a field is represented using a scalar potential $v_{\boldsymbol{\mathcal{E}}}(\boldsymbol{r}) = -\boldsymbol{\mathcal{E}} \cdot \boldsymbol{r}$. This choice, called *position gauge*, is however impossible in a ring geometry with periodic boundary conditions, because the $\boldsymbol{r}$-operator is not periodic and not bounded from below. Instead, we represent $\boldsymbol{\mathcal{E}}$ using a time-dependent vector potential $\boldsymbol{A}(t) = -c\boldsymbol{\mathcal{E}}t$ that is, like the electric field, uniform in space and therefore compatible with periodic boundary consitions. The price to pay for the use of this so-called *velocity gauge* is that the Hamiltonian now becomes explicitly time-dependent. In our framework this is not a problem, because we aim at a kinetic approach in the spirit of the BTE, in which the quantum system is propagated in time until it reaches a steady state.

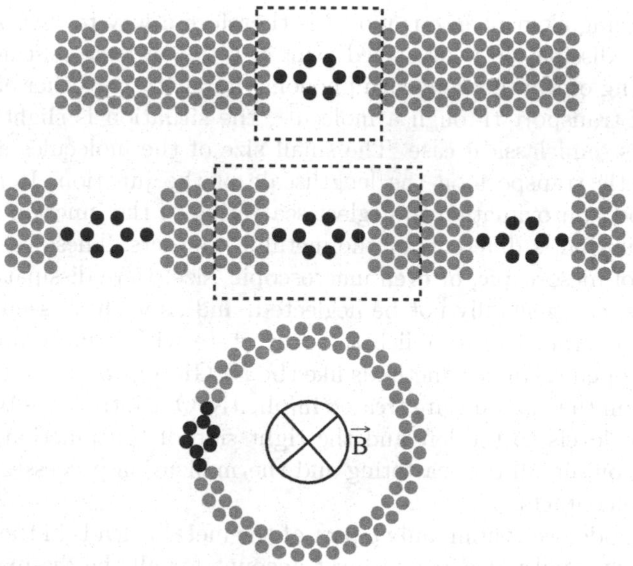

Fig. 31.1. *Upper panel*: Molecular junction with semi-infinite leads and open boundary conditions. *Middle panel*: Periodic geometry where the molecule and a piece of metal is repeated periodically. This can also be imagined as a ring geometry (*lower panel*). The electric field in the ring is induced by a magnetic flux inside the ring. The dashed boxes represent the computational cells with open and closed boundary conditions, respectively

One way to visualize the origin of the vector potential $\boldsymbol{A}(t)$ is by imagining a magnetic field in the center of the ring (but outside the material itself), as schematically shown in Fig. 31.1. If that field increases with time, then an electric field is induced that accelerates the electrons in the ring.

One obvious problem arising from this choice of gauge is that $\boldsymbol{A}(t)$ is not bounded, and as the vector potential increases indefinitely, the Hamiltonian becomes singular. We avoid this by performing gauge transformations in regular intervals, as explained in Sect. 31.4.

31.2.2 The Role of Dissipation

The external electric field that is applied to the ring accelerates the electrons. In a system through which a finite current is flowing, this means that energy is injected and that the electrons are driven away from equilibrium. They will however not settle in a steady state with constant current flowing through the junction, but continue to be accelerated and gain more and more energy. This situation does not correspond to the physics we want to model: In the experimental setup a finite voltage applied to the molecule generally results in a constant current flowing through the junction.

The missing element in our model is therefore a way to *dissipate energy.* In the BTE dissipation is described using a *collision term* that accounts for the scattering of wavepackets with phonons, impurities or other electrons. In the case of transport through a molecule, the situation is slightly different than in this semiclassical case. The small size of the molecules allows for a quasi ballistic transport on the lengthscale of the junction. It is therefore often a good approximation to neglect scattering in the junction itself.

The situation is different for the metallic contacts. These regions of the device are of mesoscopic, or even macroscopic, size. Here dissipation, mainly by phonons, can generally not be neglected, and as we have seen above this dissipation is crucial to establish a steady state with a finite and constant current. In open boundary methods like the NEGF approach, all these effects far away from the molecule are treated implicitly. One introduces two different quasi-Fermi levels to the left and the right side of the junction, and these effectively contain all the scattering and thermalization processes happening in the large contacts.

In our model we contain only a part of the metallic leads in the simulation cell, and this small metal region must account for all the dissipative effects needed to reach steady state. Since practical computations limit the size of the simulation cell, one can treat in practice only metal regions much smaller than the electronic mean free paths that would be required for thermalization. Computationally, one applies therefore a much stronger dissipation inside small metal contacts than what would be given by a realistic electron-phonon coupling strength. In this way, one can achieve thermalization and steady state with relatively small computational cells.

31.3 Master Equations

As discussed above, dissipation is a crucial ingredient for our quantum kinetic approach. One can consider that the system (the electrons in our case) is in contact with a *heat bath* (the phonons), that allows energy transfer between the system and its environment and tends to bring the system towards thermal equilibrium at a given bath temperature T. From that point of view it is clear that the electrons cannot be described simply by wavefunctions: At finite T the equilibrium is given by a statistical average of electronic eigenstates with different energies. Such statistical averages are commonly described using *density matrices.* Let $\hat{\rho}$ be the density matrix of our electronic system, then we have in equilibrium

$$\hat{\rho}^{\mathrm{eq}} = \sum_i n_i^0 \, |i\rangle\langle i| \; , \tag{31.1}$$

where $|i\rangle$ are the eigenstates of the system and n_i^0 are thermal occupation numbers.

31.3.1 Master Equation

If an external bias is applied to the electrons, then $\hat{\rho}$ is driven away from $\hat{\rho}^{\mathrm{eq}}$. The dynamics followed by the density matrix is given by the *master equation*, which in our case has the form

$$\frac{\mathrm{d}}{\mathrm{d}t}\hat{\bar{\rho}}(t) = -\mathrm{i}\left[\hat{H}(t), \hat{\bar{\rho}}(t)\right] + \breve{\mathcal{C}}\left[\hat{\bar{\rho}}(t)\right] . \tag{31.2}$$

Here, $\hat{H}(t)$ contains the external electric field, and the collision term $\breve{\mathcal{C}}[\cdots]$ describes the inelastic scattering with the heat bath. Equation (31.2) with only the first term on the r.h.s. corresponds simply to the time-dependent Schrödinger equation, written for density matrices rather than wavefunctions.

The collision term, which tends to bring the system to its thermal equilibrium, induces transitions between eigenstates of the system. An electron that is placed initially in an excited eigenstate remains there forever if no interaction with the bath is present. With the term $\breve{\mathcal{C}}[\hat{\bar{\rho}}]$ however, there will be a finite probability for the electron to lose energy and go towards a lower unoccupied eigenstate. These transitions between eigenstates are incorporated using Fermi's golden rule that determines which transitions are energetically allowed in presence of a phonon with a given frequency. Because Fermi's golden rule relies implicitly on an averaging procedure in time (all energetically forbidden transitions are oscillatory and average therefore to zero), also the action of $\breve{\mathcal{C}}[\cdots]$ implies such an averaging. We indicate this fact using a *bar* for the averaged $\hat{\bar{\rho}}(t)$ in (31.2).

The derivation of an explicit form for the collision term relies on two major approximations. The first is to assume that the interaction between the electrons and the bath is *weak* and that one can therefore apply perturbation theory. This approximation is usually a very good one, the electron-phonon interaction, for example, can indeed be very well described using a perturbative approach. The second approximation, which is often much more delicate, is the so-called Markov approximation. One assumes that the heat bath is always in thermal equilibrium and that all interactions with the bath are instantaneous processes. In this way the dynamics of the system at time t is only given by its state $\hat{\rho}(t)$ at the same time, and not by its history. This approximation is of course closely related to the use of time-averaging in $\breve{\mathcal{C}}[\hat{\bar{\rho}}(t)]$ and the application of Fermi's golden rule, as discussed above.

Using these approximations, an explicit form of the collision term is derived in many textbooks [Louisell 1973, Cohen-Tannoudji 1992]. Here we only state the result, which can be cast in the form

$$\breve{\mathcal{C}}\left[\hat{\bar{\rho}}(t)\right] = -\sum_{l,m} \gamma_{lm} \left\{ \hat{L}_{ml}\hat{L}_{lm}\hat{\bar{\rho}}(t) + \hat{\bar{\rho}}(t)\hat{L}_{ml}\hat{L}_{lm} - 2\hat{L}_{lm}\hat{\bar{\rho}}(t)\hat{L}_{ml} \right\} . \tag{31.3}$$

Here the $\hat{L}_{ij}$ are operators representing an electronic transition from eigenstate $|j\rangle$ to $|i\rangle$: $\hat{L}_{ij} = |i\rangle\langle j|$. The numbers γ_{ij} define the strength of the coupling to the bath for each possible electronic transition. They are given by

$$\gamma_{ij} = \begin{cases} \left|\langle i|\hat{V}_{\mathrm{e-ph}}|j\rangle\right|^2 [\bar{n}(\varepsilon_j - \varepsilon_i) + 1] & \varepsilon_i < \varepsilon_j \\ \left|\langle i|\hat{V}_{\mathrm{e-ph}}|j\rangle\right|^2 [\bar{n}(\varepsilon_i - \varepsilon_j)] & \varepsilon_i > \varepsilon_j \end{cases} . \tag{31.4}$$

$V_{\mathrm{e-ph}}$ is the electron-phonon interaction potential, ε_i, ε_j are the energies of the electronic eigenstates $|i\rangle$ and $|j\rangle$, respectively, and $\bar{n}(\omega)$ is the thermal mean occupation number of phonons with energy ω: $\bar{n}(\omega) = 1/(\mathrm{e}^{\omega/kT} - 1)$. It is important to note that in (31.4) the probability of an electronic transition downwards in energy is higher than the probability for a jump upwards. This symmetry breaking is a true quantum effect: Induced emission and absorption of phonons have the same probability, but an electron can also jump to a lower energy level by spontaneous emission of a phonon. These spontaneous emission processes are the origin of the breaking of symmetry, and allow the system to reach thermal equilibrium.

As already mentioned in Sect. 31.2.2, we limit our $\hat{V}_{\mathrm{e-ph}}$ to the metal region of the ring. Moreover, we increase the coupling constants γ_{ij} by an overall factor γ_0 that can vary from roughly 100 to 1000. This allows us to reach a steady state using a relatively small metallic region.

31.3.2 TDDFT and a KS Master Equation

In order to carry out realistic calculations, the scheme set up in the previous sections needs to be made computationally tractable. The computational cell typically contains hundreds to thousands of electrons, and it is of course impossible to work in terms of the exact many-particle wavefunctions or density matrices.

It is possible to describe the dissipative many-electron system, evolving under the master equation (31.2) using a generalization of DFT to dissipative systems [Burke 2005b]. In the same way as in standard DFT one can prove that the potential is uniquely determined (up to an arbitrary additional constant) by the charge density $n(\boldsymbol{r})$, it is possible to show that for a dissipative quantum system no two different one-body potentials can give rise to the same time-dependent density $n(\boldsymbol{r}, t)$, given the superoperator $\breve{\mathcal{C}}$ in (31.2) and an initial density matrix $\hat{\rho}(0)$. This theorem establishes a DFT for quantum systems evolving under a master equation. For the use in practical calculations one constructs a corresponding KS system. In that system of non-interacting electrons, the one-body potential $v_{\mathrm{KS}}[n, \hat{\rho}_{\mathrm{KS}}(0), \breve{\mathcal{C}}]$ is defined such that it yields the exact density $n(\boldsymbol{r}, t)$ of the interacting system.

Defined in this way, the KS system has certain pathologies. The superoperator in the many-body master equation is guaranteed to vanish only on the many-body equilibrium density matrix, but not necessarily on the equlibrium KS density matrix. To avoid this problem, we modify the master equation by introducing a KS form of the superoperator $\breve{\mathcal{C}}$ in terms of the *single particle reduced density matrix* $\hat{\rho}_{\rm KS}(t)$. To this end, we define the single-particle potential $v_{\rm KS}(T)(\boldsymbol{r})$ as the KS potential in the Mermin functional at temperature T, i.e., the potential that, when thermally occupied with non-interacting electrons, reproduces the exact density at thermal equilibrium. The KS states $|\alpha\rangle$ defined in this way (which we designate with Greek letters to distinguish them from the many particle eigenstates $|i\rangle$ above) are used as a basis for the density matrix:

$$\hat{\rho}_{\rm KS}(t) = \sum_{\alpha,\beta} f_{\alpha\beta}(t)|\alpha\rangle\langle\beta| \,. \tag{31.5}$$

To find the KS master equation itself, we reduce the many-body (31.2) to a single-particle form by tracing out all other degrees of freedom, and use a Hartree-style approximation for the two-particle correlation functions appearing in the collision term. In this way we find for the coefficients $f_{\alpha\beta}$ in (31.5):

$$\begin{aligned}\frac{\mathrm{d}}{\mathrm{d}t} f_{\alpha\beta} = -\mathrm{i}\sum_{\lambda}\left(H_{\alpha\lambda} f_{\lambda\beta} - f_{\alpha\lambda} H_{\lambda\beta}\right) + \left(\delta_{\alpha\beta} - f_{\alpha\beta}\right)\sum_{\lambda}\frac{1}{2}\left(\gamma_{\alpha\lambda} + \gamma_{\beta\lambda}\right) f_{\lambda\lambda} \\ - f_{\alpha\beta}\sum_{\lambda}\frac{1}{2}\left(\gamma_{\lambda\alpha} + \gamma_{\lambda\beta}\right)\left(1 - f_{\lambda\lambda}\right) \,. \end{aligned} \tag{31.6}$$

Equation (31.6) is the final form for the master equation used in our calculations. It can be viewed as a generalization of the BTE to the full quantum case, as it is formulated in terms of the density matrix rather than in terms of wavepackets. The first term in the r.h.s. represents the Hamiltonian propagation. The $H_{\alpha\beta}$ are the matrix elements of the time-dependent KS Hamiltonian in the basis of the unperturbed orbitals. The second and third terms in the r.h.s. represent the collision terms. One can verify that (31.6) satisfies a series of important properties for a master equation: The trace of $\hat{\rho}$ is invariant, guaranteeing a constant number of electrons in the system. Furthermore, $\hat{\rho}$ stays Hermitian during the time propagation and, in the absence of external perturbations, tends to the Fermi-Dirac distribution as its thermal equilibrium state. Finally, one can show that the collision terms act to reduce the off-diagonal elements $f_{\alpha\beta}$ while pushing the system towards equilibrium.

31.4 Practical Aspects

31.4.1 Time Propagation

The complete single particle density matrix $\hat{\rho}$ has the dimension $(N_{\rm occ} + N_{\rm unoc}) \times (N_{\rm occ} + N_{\rm unocc}) = M \times M$, where M is the size of the basis set and $N_{\rm occ}$, $N_{\rm unocc}$ are the number of occupied and unoccupied KS states, respectively. In practice, it is often impossible to treat the full $M \times M$ matrix in the master equation (31.6). In such cases, the computational load can be significantly reduced by explicitly including only the basis states $|\alpha\rangle$ within a given energy window around the Fermi energy. All states with an energy below this range are considered occupied, all higher lying states are neglected. In this way the time propagation can be carried out even in systems where a large basis set (like, e.g., plane waves) is used to describe the ground state electronic structure.

In practice, the time integration in (31.6) is carried out in the following way: From a given time t the density matrix is propagated for a finite period $\tau_{\mathcal{E}}$ using only the Hamiltonian propagator with the time-dependent vector potential, corresponding to the first term in the r.h.s. of (31.6). Let us choose the x-direction along the circumference of the ring, and assume that the unit cell is of size L in this direction. The electric field $\boldsymbol{\mathcal{E}} = \mathcal{E}\mathbf{e}_x$ is then also directed along $\mathbf{e}_x$, and we can choose $\tau_{\mathcal{E}} = 2\pi/(\mathcal{E}L)$. In this case, the vector potential changes in that period from $\boldsymbol{A}(t)$ to $\boldsymbol{A}(t) - 2\pi c\mathbf{e}_x/L$. At this point, is is possible to perform a *gauge transformation* by adding to the vector potential $2\pi c\mathbf{e}_x/L$ and at the same time multiplying all electronic wavefunctions with a phase factor $\exp(-\mathrm{i}2\pi x/L)$. These phase factors are compatible with periodic boundary conditions. In this way one avoids a vector potential that increases indefinitely. After the Hamiltonian time propagation plus gauge transformation, the density matrix is propagated for the same time interval $\tau_{\mathcal{E}}$ using the collision term only, corresponding to the second and third terms in the r.h.s. of (31.6). After this procedure one obtains the density matrix $\bar{\rho}(t + \tau_{\mathcal{E}})$. In the limit $L \to \infty$, the time $\tau_{\mathcal{E}}$ tends to zero, and this integration procedure converges to the exact solution of the master equation.

31.4.2 Calculating Currents

One important quantity to calculate in this kinetic scheme is of course the current, as it is the central quantity characterizing transport through the molecule. The current $\boldsymbol{j}(\boldsymbol{r},t)$ must satisfy the current continuity equation

$$\frac{\mathrm{d}}{\mathrm{d}t} n(\boldsymbol{r},t) + \nabla \cdot \boldsymbol{j}(\boldsymbol{r},t) = 0\,, \tag{31.7}$$

which translates the fact that charges are locally conserved $[n(\boldsymbol{r},t)$ is the charge density$]$. It is well known that the Hamiltonian propagation does satisfy (31.7), if the current is defined as

$$\boldsymbol{j}_H(\boldsymbol{r},t) = \mathrm{Tr}\left[\hat{\bar{\rho}}(t)\hat{\boldsymbol{J}}(\boldsymbol{r})\right], \tag{31.8a}$$

$$\hat{\boldsymbol{J}}(\boldsymbol{r}) = \frac{1}{2}\left\{[\hat{\boldsymbol{p}} + \boldsymbol{A}(t)/c]\,\delta(\boldsymbol{r} - \hat{\boldsymbol{r}}) + \delta(\boldsymbol{r} - \hat{\boldsymbol{r}})\,[\hat{\boldsymbol{p}} + \boldsymbol{A}(t)/c]\right\}. \tag{31.8b}$$

In the case of our master dynamics however, the current $\boldsymbol{j}_H$ defined in this way does not satisfy the continuity equation. The reason for this is again the time-averaging procedure that underlies the collision term $\breve{\mathcal{C}}[\hat{\rho}]$ that we mentioned above in Sect. 31.3. As explained there, the use of Fermi's golden rule in our scheme implies performing time integrals over many oscillations of electronic transitions. Mapping this time-dependent process on an instantaneous scattering event amounts to neglecting all electronic currents that flow during that time, and, as a consequence, the continuity equation is not satisfied.

It is possible to recover the missing term of the current that originates from the interaction with the bath. This term, which we call *collision* or *dissipative* current $\boldsymbol{j}_C(\boldsymbol{r},t)$ restores the local charge conservation, and the total *physical* current (which is measuread in experiments) is the sum $\boldsymbol{j}(\boldsymbol{r},t) = \boldsymbol{j}_H(\boldsymbol{r},t) + \boldsymbol{j}_C(\boldsymbol{r},t)$. An explicit form for the dissipative current is given elsewhere [Gebauer 2004a].

31.5 Results

This quantum kinetic scheme has been used for both realistic and model calculations, which we will briefly outline in the following.

31.5.1 Model Calculation

To illustrate our methodology, we consider in the following a one-dimensional test system consisting of a double-barrier resonant tunneling structure (DBRTS) [Gebauer 2004b]. We treat electron-electron interactions on the Hartree level, allowing therefore to study the key elements of this scheme only, without discussing more sophisticated exchange and correlation effects.

The external $v_{\rm ext}(x)$ in absence of an applied electric field is shown in the upper panel on the left of Fig. 31.2. The double barrier constitutes the junction and is set into a periodically repeated unit cell as shown in the figure. Outside the barrier region, we assume a carrier density of $4.3\times10^{18}\,\mathrm{cm}^{-3}$, and choose an effective mass of $0.1\,\mathrm{m_e}$ and a dielectric constant of 10. The phonon density of states $\mathcal{D}(\omega) \propto \omega^2$, and the electron-phonon couplings $\gamma_{\alpha\beta} = \gamma_0$ for all states α, β. We choose values of 0.136 meV and of 0.218 meV, respectively, for γ_0, and a temperature of 25.3 K, so that k_BT is comparable to the electronic level spacing at the Fermi energy in our finite system ($L = 244\,\mathrm{nm}$). We solve (31.6) for steady state behavior, i.e. $\mathrm{d}f_{\alpha\beta}/\mathrm{d}t = 0$, and determine the Hartree potential self-consistently, for different values of the applied external field $\mathcal{E}$. The total potential acting on the electrons when $\mathcal{E}$ corresponds to peak current is shown in the middle panel on the left of Fig. 31.2. The total potential includes $v_{\rm ext}(x)$, the externally applied bias [represented by a linearly varying potential $v_{\mathcal{E}}(x) = -\mathcal{E}\cdot x$], and the Hartree potential. We have transformed to the position gauge here only for illustration.

The total potential has the same qualitative behavior found in self-consistent calculations using open boundary conditions for similar model

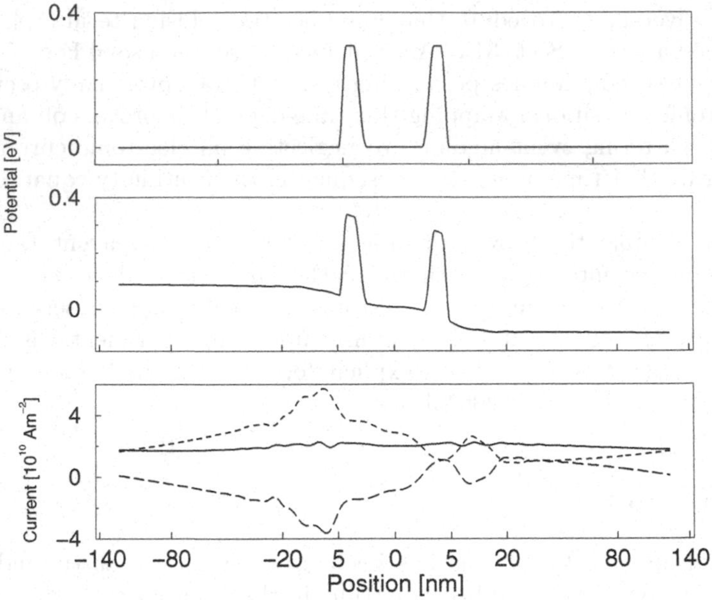

Fig. 31.2. *Upper panel*: the double-barrier resonant tunneling structure in the absence of an electric field. *Middle panel*: total potential at steady state, in presence of an external electric field. *Lower panel*: The currents at steady state. Solid line: total current $j(x)$, long dashed line: collision current $j_C(x)$, short dashed line: Hamiltonian current $j_H(x)$. Please note the non-linear scale of the horizontal axis

systems. In our approach, however, the voltage drop across the barrier is an output of the calculation.

In the lower panel on the left of Fig. 31.2 we plot the expectation values of current densities $j(x)$, $j_C(x)$, and $j_H(x)$ corresponding to the electric field and electron-phonon coupling of the middle panel. In this one-dimensional system at steady state the continuity equation amounts to $\mathrm{d}j(x)/\mathrm{d}x = 0$, the current must therefore be a constant in space. It can be clearly seen in Fig. 31.2 that this is not the case for j_C or j_H separatly, but to a good approximation for the total physical current $j(x)$.

Let us now discuss the current-voltage (I-V) characteristics of this system. From a naïve model in which the bands to the left and right of the barrier are simply shifted rigidly, one obtains the following picture (see also Fig. 31.3): At low bias no current can flow through the system because the electrons see a high barrier blocking any transport (left panel in Fig. 31.3). Only once the bias has reached a threshold value, a localized level between the two barriers enters in resonance with the incoming electrons and allows for a finite current flow (right panel). At very large bias, the localized level falls below lower limit of the occupied bandwidth for the incoming electrons, and current is again suppressed (lower panel).

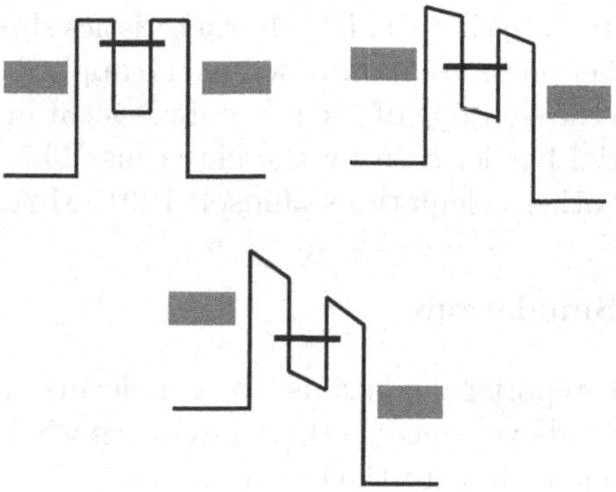

Fig. 31.3. Schematic representation of the transmission behavior of a double-barrier resonant tunneling structure (see text)

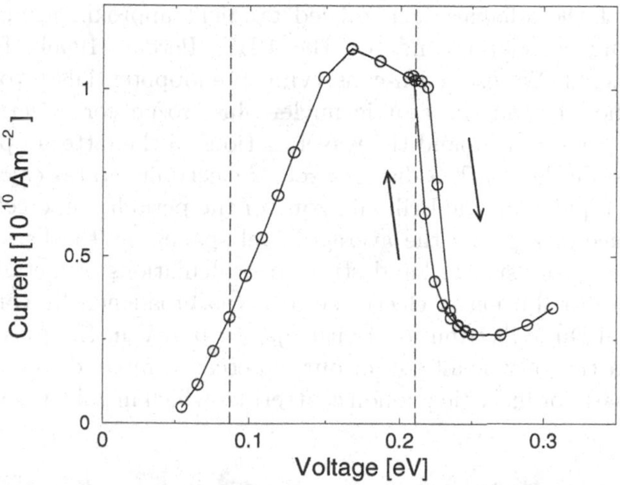

Fig. 31.4. Calculated I-V characteristics of the double-barrier resonant tunneling structure

As shown in Fig. 31.4, our calculated I-V characteristics follow the expected trend. Note that the voltage shown on the horizontal axis is the drop across the barrier structure, rather than the total voltage drop $\mathcal{E} \cdot L$ in the simulation cell. That voltage is an outcome of the self-consistent calculation, rather than an input quantity. The dashed vertical lines indicate the voltages where one would expect current to start respectively stop flowing according to the above considerations.

From this figure we can clearly see that our kinetic model represents correctly the non-linear behavior of the current and intriguing quantum effects like negative differential resistance (NDR) when the current *decreases* with *increasing* applied voltage.

It is interesting to note that the I-V characteristics show a hysteresis loop, depending on whether one goes up or down in the applied electromotive force. This effect is due to the charging of the resonant level at high bias which alters the effective potential barrier seen by the electrons. This hysteresis loop has also been found by other calculations [Jensen 1991, Mizuta 1991].

31.5.2 Realistic Simulations

In the following we report calculations on a molecular electronic device to illustrate further the above concepts. Our circuit consists of a self-assembled monolayer of benzene dithiolate (BDT) molecules in contact with two gold electrodes according to the geometrical setup in Fig. 31.5. Here we briefly mention only a few key computational aspects of these calculations, which are reported elsewhere [Piccinin 2006].

We adopt the adiabatic generalized gradient approximation (GGA) for exchange and correlation and use the PBE [Perdew 1996b] form of the GGA functional. We use norm-conserving pseudopotentials [Troullier 1991] to model the effect of the atomic nuclei plus frozen core electrons on the valence electrons and expand the wavefunctions of the latter in plane waves. Numerically we deal with a discrete set of electronic states calculated at a finite set of k-points in the Brillouin zone of the periodic supercell. With the present choice of k-points the average level spacing at the Fermi energy is around 0.1 eV. As usual in band structure calculations of metallic systems, the discrete distribution of electronic levels is broadened by convoluting it with a Fermi-Dirac distribution (with $k_B T = 0.6\,\text{eV}$ in the present calculation). Given the very small size of our supercell compared to the electronic mean free-path for inelastic phonon scattering, which in gold at room temper-

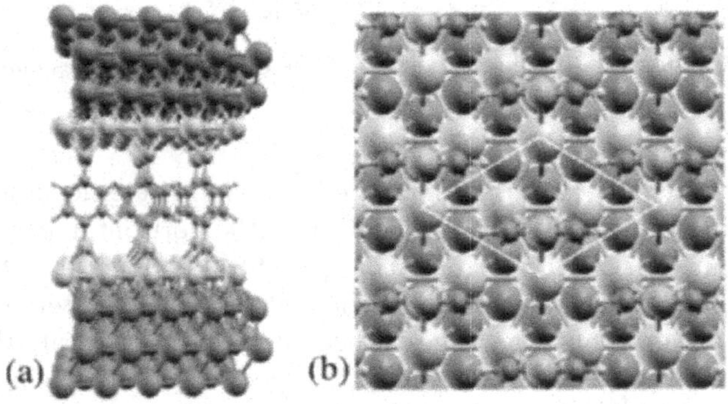

Fig. 31.5. (**a**) Lateral view of the benzene dithiolate (BDT) monolayer between two Au(111) surfaces. A total of 8 layers are included in the unit cell of the simulation. The dark atoms in the slabs indicate the region where dissipation is applied. (**b**) Top view of the BDT monolayer on Au(111): the box indicates the unit cell used in the simulation

ature is of a few hundred Ångström, we use a very crude model for the bath. In particular we neglect inelastic scattering processes in the molecular junction inside the non-shaded region in Fig. 31.5, since electrons traverse this region ballistically to a very good approximation. Inelastic processes needed to thermalize the electrons are confined to the shaded region in Fig. 31.5. To achieve thermalization in such a small space, we adopt an artificially large coupling between the electrons and the bath, similar to what is usually done in non-equilibrium molecular dynamics simulations of classical systems. We also hold the bath at an unphysically large temperature corresponding to the adopted broadening of the energy levels. In spite of these very crude approximations dictated by numerical limitations, the calculation reproduces well a number of the physical features of a real molecular device.

The calculated steady state potential and currents corresponding to an applied electromotive force of 1 eV are reported in Fig. 31.6 [Gebauer 2005]. The potential shows a small linear drop inside the electrodes, as expected

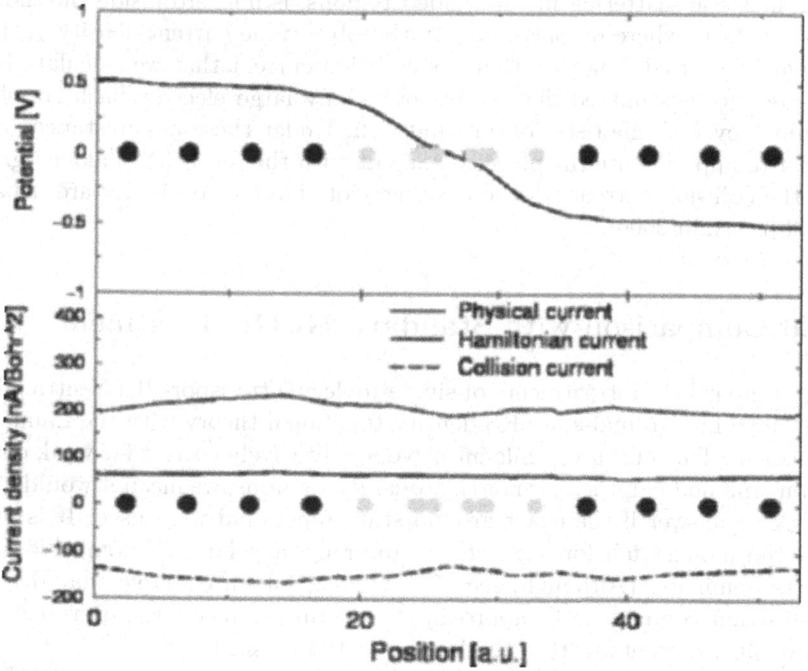

Fig. 31.6. *Top panel*: sum of the externally applied potential and the induced potential at steady state. The externally applied potential is visualized in the position gauge (see text). *Bottom panel*: Hamiltonian current $j_H(r)$, quantum collision current $j_C(r)$ and physical current $j(r) = j_H(r) + j_C(r)$. The electromotive force is applied along the (111) direction, which is shown in the plot. All quantities are averaged over planes perpendicular to the (111) direction. The black dots indicate gold atomic planes. The gray dots indicate BDT atoms

from Ohm's law for a wire of finite resistance when a current circulates. However, most of the potential drop is due to the contact resistance and correctly occurs across the molecular junction. This drop is not purely linear but shows a well defined shoulder within the aromatic ring of the BDT molecules. This reflects the polarization of the ring under the applied bias. The calculated dc current I is about 3.1 μA per DBT molecule, a value that compares well with some recent experiments [Xiao 2004]. Our artificial bath model keeps the distribution of the electrons close to a thermal equilibrium distribution, mimicking real experimental conditions, where the distribution of the electrons injected in a molecular device is close to a thermal equilibrium distribution. Having fixed the strength of the inelastic coupling, the main factor controlling the I-V characteristics of a molecular device like the one in Fig. 31.5 is electron transmission through the molecular junction. Some calculations using an open circuit geometry, which ignore explicit dissipative effects, give a current I which is rather close to the one that we calculate here [Ke 2004]. Notice that the quantum collision current, which originates from inelastic scattering in the shaded regions, is nonzero inside the molecular junction, where it contributes to the observable current density $j(r)$. It should be noticed, however, that the collision current that we calculate here is grossly overestimated due to the artificially large electron-bath coupling required by the small size of our unit cell. Under these circumstances it is better to approximate the physical current with the term (31.8) alone, ignoring the collision current, as the ensuing violations of continuity are usually small [Piccinin 2006].

31.6 Comparison with Standard NEGF Treatment

The standard DFT treatments of single-molecule transport [Di Ventra 2000, Xue 2002] mix ground-state KS density functional theory with the Landauer formalism. This mixture, while intuitive, qualitatively correct for weak correlation, and non-empirical, is not rigorous. By rigorous, we mean it would yield the exact answer if the exact ground-state functional were used. It is clear that the it must fail for, e.g., strong interaction, where Coulomb blockade effects dominate. (Although, see [Toher 2005] for an example that demonstrates that common DFT approximations can produce large overestimates in tunelling currents, within the Landauer formalism.)

Our new formalism *is* rigorous, in the sense that it is based on a DFT theorem that does apply to the present situation. If we knew the exact XC functional for the TDKS Master equation, it would produce the exact answer. The same is true for the work of Chap. 32. Thus, in the limit as $\gamma \to 0$, but keeping γ always finite and the ring large enough to ensure enough scattering occurs in the metal region such that the electrons do not remember their previous encounter with the constriction, the two should yield identical results.

Given the difference in formalisms, it is a major task to compare them to each other, or either to the standard treatment when XC effects are present.

It is much simpler to make the comparison for weak electric fields, where Kubo response theory applies, and dissipative effects can be ignored. Recent work, in the linear response regime, has shown that certain XC electric field effects are missing from the standard treatment [Burke 2005b]. Our dissipative approach recovers the Kubo response in the limit of weak fields [Burke 2005b], and so takes these effects into account.

31.7 Conclusions

In conclusion, we have shown how the semiclassical Boltzmann transport equation can be generalized to treat fully quantum mechanical systems. The resulting master equation allows one to propagate an electronic system in time, under the combined influence of an external driving force and dissipation due to inelastic scattering. We have shown how this general scheme can be applied to the calculation of transport properties, both in model systems and in realistic molecular devices. Further applications of this method to other molecular devices, as well as to electronic conduction through carbon nanotubes, suspended between gold electrodes, are currently in progress.

32 Time-Dependent Transport Through Single Molecules: Nonequilibrium Green's Functions

G. Stefanucci, C.-O. Almbladh, S. Kurth, E.K.U. Gross, A. Rubio,
R. van Leeuwen, N.E. Dahlen, and U. von Barth

32.1 Introduction

The nomenclature *quantum transport* has been coined for the phenomenon of electron motion through constrictions of transverse dimensions smaller than the electron wavelength, e.g., quantum-point contacts, quantum wires, molecules, etc. To describe transport properties on such a small scale, a quantum theory of transport is required. In this Chapter we focus on quantum transport problems whose experimental setup is schematically displayed in Fig. 32.1a. A central region of meso- or nanoscopic size is coupled to two metallic electrodes which play the role of charge reservoirs. The whole system is initially in a well defined equilibrium configuration, described by a *unique* temperature and chemical potential (thermodynamic consistency). No current flows through the junction, the charge density of the electrodes being perfectly balanced. In the previous Chapter, Gebauer et al. proposed to join the left and right remote parts of the system so to obtain a ring geometry, see Fig. 30.1. In their approach the electromotive force is generated by piercing the ring with a magnetic field that increases linearly in time. Here, we consider the longitudinal geometry of Fig. 32.1a and describe an alternative approach. As originally proposed by Cini [Cini 1980], we may drive the system out of equilibrium by exposing the electrons to an external time-dependent potential which is local in time and space. For instance, we may switch on an electric field by putting the system between two capacitor plates far away from the system boundaries, see Fig. 32.1b. The dynamical formation of dipole layers screens the potential drop along the electrodes and the total potential turns out to be uniform in the left and right bulks. Accordingly, the potential drop is entirely limited to the central region. As the system size increases, the remote parts are less disturbed by the junction, and the density inside the electrodes approaches the equilibrium bulk density.

There has been considerable activity to describe transport through these systems on an *ab initio* level. Most approaches are based on a self-consistent procedure first proposed by Lang [Lang 1995]. In this steady-state approach, based on density functional theory (DFT), exchange and correlation is approximated by the static local density potential, and the charge density is obtained self-consistently in the presence of the steady current. However, the original justification involved subtle points such as differ-

G. Stefanucci et al.: *Time-Dependent Transport Through Single Molecules: Nonequilibrium Green's Functions*, Lect. Notes Phys. **706**, 479–492 (2006)
DOI 10.1007/3-540-35426-3_32

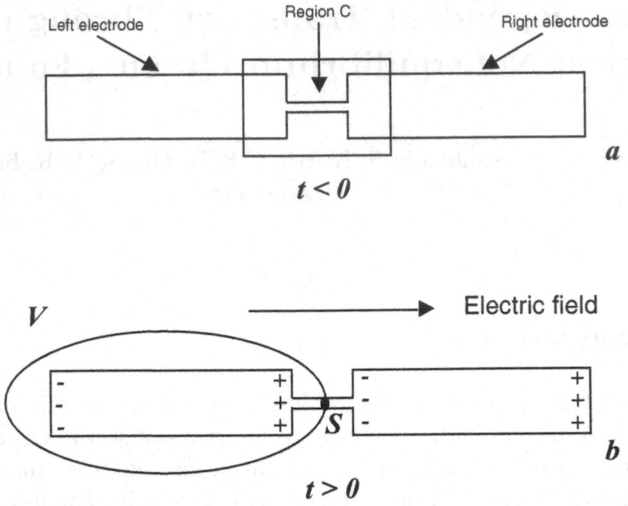

Fig. 32.1. Schematic sketch of the experimental setup described in the main text. A central region which also includes few layers of the left and right electrodes is coupled to macroscopically large metallic reservoirs. (**a**) The system is in equilibrium for negative times. (**b**) At positive times the electrons experience an electric field generated by two capacitor plates far away from the system boundaries. Discarding retardation effects, the screening of the potential drop in the electrodes is instantaneous and the total potential turns out to be uniform in the left and right electrodes separately

ent Fermi levels deep inside the left and right electrodes and the implicit reference of nonlocal perturbations such as tunneling Hamiltonians within a DFT framework. (For a detailed discussion we refer to [Stefanucci 2004b].) The steady-state DFT approach has been further developed [Derosa 2001, Brandbyge 2002, Xue 2002, Calzolari 2004] and the results have been most useful for understanding the qualitative behavior of measured current-voltage characteristics. Quantitatively, however, the theoretical I-V curves typically differ from the experimental ones by several orders of magnitude [Di Ventra 2000]. Several explanations are possible for such a mismatch: models are not sufficiently refined, parasitic effects in measurements have been underestimated, the characteristics of the molecule-contact interfaces are not well understood and difficult to address given their atomistic complexity. Another theoretical reason for this discrepancy might be the fact that the transmission functions computed from static DFT have resonances at the non-interacting Kohn-Sham excitation energies, which in general do not coincide with the true excitation energies. Furthermore, different exchange-correlation functionals lead to DFT-currents that vary by more than an order of magnitude [Krstić 2003].

On the other hand, excitation energies of interacting systems are accessible via time-dependent TDDFT [Runge 1984, Petersilka 1996a]. In this theory, the time-dependent density of an interacting system moving in an external, time-dependent local potential can be calculated via a fictitious system of non-interacting electrons moving in a local, effective time-dependent potential. Therefore, this theory is in principle well suited for the treatment of nonequilibrium transport problems [Stefanucci 2004a, Stefanucci 2004b]. Below, we combine the Cini scheme with TDDFT and we describe in detail how TDDFT can be used to calculate the time-dependent current in systems like the one of Fig. 32.1. The theoretical formulation of an exact theory based on TDDFT and nonequilibrium Green functions (NEGF) has been developed in [Stefanucci 2004b] and shortly after used for conductance calculations of molecular wires [Evers 2004]. A practical scheme to go beyond static calculations and perform the full time evolution has been recently proposed by Kurth et al. [Kurth 2005]. The theory was originally developed for systems initially described by a thermal density matrix. An extension to unbalanced (out of equilibrium) initial states can be found in [Di Ventra 2004].

Here we also mention that another thermodynamically consistent scheme has been proposed by Kamenev and Kohn [Kamenev 2001]. They consider a closed system (ring) and drive it out of equilibrium by switching an external vector potential. As the Cini scheme, this approach also overcomes the problem of having two or more chemical potentials. Since the Kamenev-Kohn approach uses a vector potential rather than a scalar potential, TD current DFT (TDCDFT) would be the natural density-functional extension to use.

32.2 An Exact Formulation Based on TDDFT

In quantum transport problems like the one discussed in the previous section, we are mainly interested in calculating the total current through the junction rather than the current density in some point of the system. Assuming that the electrons can leave the region of volume $\mathcal{V}$ in Fig. 32.1b only through the surface S, then the total time-dependent current $I_S(t)$ is given by the time derivative of the total number of particles in volume $\mathcal{V}$. Denoting by $n(\boldsymbol{r},t)$ the particle density we have

$$I_S(t) = -e \int_{\mathcal{V}} \mathrm{d}^3 r \, \frac{\mathrm{d}}{\mathrm{d}t} n(\boldsymbol{r},t) \,. \tag{32.1}$$

Runge and Gross have shown that $n(\boldsymbol{r},t)$ can be computed in a one-particle manner provided that it falls off rapidly enough for $\boldsymbol{r} \to \infty$ (this theory applies only to those cases where the external disturbance is local in space). Therefore, we may calculate $n(\boldsymbol{r},t)$, and in turn $I_S(t)$, by solving a fictitious non-interacting problem described by an effective Hamiltonian $\hat{H}_{\mathrm{KS}}(t)$. The potential $v_{\mathrm{KS}}(\boldsymbol{r},t)$ experienced by the electrons in $\hat{H}_{\mathrm{KS}}(t)$ is called the

Kohn-Sham (KS) potential and it is given by the sum of the external potential, the Coulomb potential of the nuclei, the Hartree potential and the exchange-correlation potential $v_{\rm xc}$. The latter accounts for the complicated many-body effects and is obtained from an exchange-correlation action functional, $v_{\rm xc}(\boldsymbol{r},t) = \delta A_{\rm xc}[n]/\delta n(\boldsymbol{r},t)$ (as pointed out in [van Leeuwen 1998], the causality and symmetry properties require that the action functional $A_{\rm xc}[n]$ is defined on the Keldysh contour – see Chap. 3). $A_{\rm xc}$ is a functional of the density and of the initial density matrix. In our case, the initial density matrix is the thermal density matrix which, due to the extension of the Hohenberg-Kohn theorem [Hohenberg 1964] to finite temperatures [Mermin 1965], is also a functional of the density.

Without loss of generality we will assume that the external potential vanishes for times $t \leq 0$. The initial equilibrium density is then given by $\sum_i f(\varepsilon_i)|\langle \boldsymbol{r}|\varphi_i(0)\rangle|^2$, where f is the Fermi function. The KS states $|\varphi_i(0)\rangle$ are eigenstates of $\hat{H}_{\rm KS}(0)$ with KS energies ε_i. For positive times, the time-dependent density can be calculated by evolving the KS states according to the Schrödinger equation

$$\mathrm{i}\frac{\mathrm{d}}{\mathrm{d}t}|\varphi_i(t)\rangle = \hat{H}_{\rm KS}(t)|\varphi_i(t)\rangle\,. \tag{32.2}$$

Thus,

$$n(\boldsymbol{r},t) = \sum_i f(\varepsilon_i)\,|\langle \boldsymbol{r}|\varphi_i(t)\rangle|^2\,, \tag{32.3}$$

and the continuity equation, $\dot{n}(\boldsymbol{r},t) = -\nabla\cdot\boldsymbol{j}_{\rm KS}(\boldsymbol{r},t)$, can be written in terms of the KS current density

$$\boldsymbol{j}_{\rm KS}(\boldsymbol{r},t) = -\sum_i f(\varepsilon_i)\,\Im[\varphi_i^*(\boldsymbol{r},t)\nabla\varphi_i(\boldsymbol{r},t)]\,, \tag{32.4}$$

where $\varphi_i(\boldsymbol{r},t) = \langle \boldsymbol{r}|\varphi_i(t)\rangle$ are the time-dependent KS orbitals. Using Gauss theorem and the continuity equation it is straightforward to obtain

$$I_S(t) = e\sum_i f(\varepsilon_i)\int_S \mathrm{d}\sigma\;\hat{\boldsymbol{n}}\cdot\Im[\varphi_i^*(\boldsymbol{r},t)\nabla\varphi_i(\boldsymbol{r},t)]\,, \tag{32.5}$$

where $\hat{\boldsymbol{n}}$ is the unit vector perpendicular to the surface element $\mathrm{d}\sigma$.

The switching on of an electric field excites plasmon oscillations which dynamically screen the external disturbance. Such metallic screening prevents any rearrangements of the initial equilibrium bulk density, provided that the time-dependent perturbation is slowly varying during a typical plasmon time-scale (which is usually less than a fs). Thus, the KS potential $v_{\rm KS}$ undergoes a uniform time-dependent shift deep inside the left and right electrodes and the KS potential drop is entirely limited to the central region.

Let us now consider an electric field constant in time. After the transient phase, the current will slowly decrease. We expect a very long plateau with superimposed oscillations, whose amplitude is inversely proportional to the system size. As the size of the electrodes increases the amplitude of the oscillations decreases and the plateau phase becomes successively longer. The steady-state current is defined as the current at the plateau for infinitely large electrodes.

What is the physical mechanism leading to a steady-state current? In the real system, dissipative effects like electron-electron or electron-phonon scatterings provide a natural explanation for the damping of the transient oscillations and the onset of a steady state. However, in the fictitious KS system the electrons are noninteracting and the damping mechanisms of the real problem are described by the local potential v_{xc}. We conclude that, for any non-interacting system having the geometry of Fig. 32.1, there must be a class of time-dependent local potentials leading to a steady current. Below, we use the NEGF techniques to study under what circumstances a steady-state current develops and what is the underlying physical mechanism. We also show that the steady-current can be expressed in a Landauer-like formula in terms of fictitious transmission coefficients and one-particle energy eigenvalues.

32.3 Non-Equilibrium Green Functions

The one-particle scheme of TDDFT corresponds to a fictitious Green function $\hat{\mathcal{G}}(z,z')$ that satisfies a one-particle equation of motion on the Keldysh contour of Fig. 32.2,

$$\left\{ \mathrm{i}\frac{\mathrm{d}}{\mathrm{d}z} - \hat{H}_{\mathrm{KS}}(z) \right\} \hat{\mathcal{G}}(z,z') = \delta(z,z') . \tag{32.6}$$

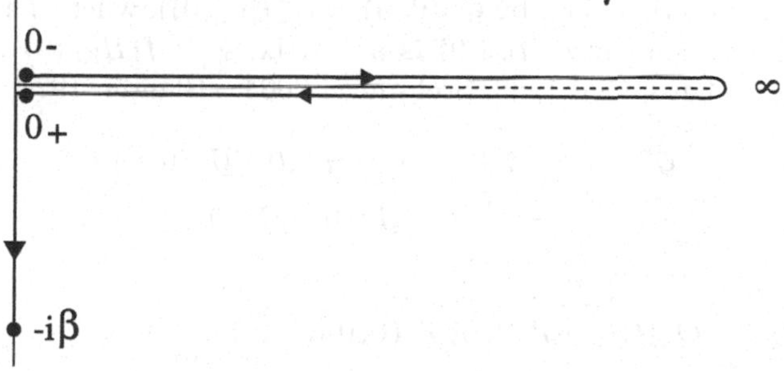

Fig. 32.2. The Keldysh contour γ is an oriented contour with endpoints in 0_- and $-\mathrm{i}\beta$, β being the inverse temperature. It constitutes of a forward branch going from 0_- to ∞, a backward branch coming back from ∞ to 0_+ and a vertical (thermic) track on the imaginary time axis between 0_+ and $-\mathrm{i}\beta$. The variables z and z' run on γ

It is convenient to define the projectors $P_\alpha = \int d^3r_\alpha |\boldsymbol{r}\rangle\langle\boldsymbol{r}|$ onto the left or right electrodes ($\alpha = \mathrm{L,R}$) or the central region ($\alpha = \mathrm{C}$). Although the $\boldsymbol{r}$ basis is not differentiable, the diagonal and off-diagonal matrix elements of the kinetic energy remain well defined in a distribution sense. We introduce the notation

$$\mathcal{O}_{\alpha\beta} \equiv P_\alpha \hat{\mathcal{O}} P_\beta \,, \tag{32.7}$$

where $\hat{\mathcal{O}}$ is an arbitrary operator in one-body space. The uncontacted KS Hamiltonian is $\hat{\mathcal{E}} \equiv \hat{H}_{\mathrm{KS,\,LL}} + \hat{H}_{\mathrm{KS,\,CC}} + \hat{H}_{\mathrm{KS,\,RR}}$ while $\hat{V} \equiv \hat{H}_{\mathrm{KS}} - \hat{\mathcal{E}}$ accounts for the contacting part. Since $\hat{V}_{\mathrm{LR}} = \hat{V}_{\mathrm{RL}} = 0$, from (32.1)–(32.6) the current from the $\alpha = \mathrm{L,R}$ electrode to the central region is

$$\begin{aligned} I_\alpha(t) &= e \int d^3r \; i\frac{d}{dt}\langle\boldsymbol{r}|\hat{\mathcal{G}}^<_{\alpha\alpha}(t,t)|\boldsymbol{r}\rangle \\ &= e \int d^3r \,\langle\boldsymbol{r}|\hat{V}_{\alpha C}\hat{\mathcal{G}}^<_{C\alpha}(t,t) - \hat{\mathcal{G}}^<_{\alpha C}(t,t)\hat{V}_{C\alpha}|\boldsymbol{r}\rangle \,. \end{aligned} \tag{32.8}$$

We define the one-particle operator $\hat{Q}_\alpha(t)$ in the central subregion C as

$$\hat{Q}_\alpha(t) = \hat{\mathcal{G}}^<_{\mathrm{C}\alpha}(t,t)\hat{V}_{\alpha\mathrm{C}} \tag{32.9}$$

and write the total current in (32.8) as

$$I_\alpha(t) = 2e\,\Re\,\mathrm{Tr}\left\{\hat{Q}_\alpha(t)\right\}, \quad \alpha = \mathrm{L,R}\,, \tag{32.10}$$

where the symbol "Tr" denotes the trace over a complete set of one-particle states of C.

For the noninteracting system of TDDFT everything is known once we know how to propagate the one-electron orbitals in time and how they are populated before the system is perturbed. The time evolution is fully described by the retarded or advanced Green functions $\hat{\mathcal{G}}^{\mathrm{R,A}}$, and by the initial population at zero time, i.e., by $\hat{\mathcal{G}}^<(0,0) = \mathrm{i}f(\hat{H}_{\mathrm{KS}}(0))$, where f is the Fermi distribution function [since $\hat{H}_{\mathrm{KS}}(0)$ is a matrix, so is $f(\hat{H}_{\mathrm{KS}}(0))$]. Then, for any $t, t' > 0$ we have [Cini 1980, Stefanucci 2004a, Blandin 1976]

$$\begin{aligned} \hat{\mathcal{G}}^<(t,t') &= \mathrm{i}\,\hat{\mathcal{G}}^{\mathrm{R}}(t,0)f(\hat{H}_{\mathrm{KS}}(0))\hat{\mathcal{G}}^{\mathrm{A}}(0,t') \\ &= \hat{\mathcal{G}}^{\mathrm{R}}(t,0)\hat{\mathcal{G}}^<(0,0)\hat{\mathcal{G}}^{\mathrm{A}}(0,t')\,, \end{aligned} \tag{32.11}$$

and hence

$$\hat{Q}_\alpha(t) = \left[\hat{\mathcal{G}}^{\mathrm{R}}(t,0)\hat{\mathcal{G}}^<(0,0)\hat{\mathcal{G}}^{\mathrm{A}}(0,t)\right]_{\mathrm{C}\alpha}\hat{V}_{\alpha\mathrm{C}}\,. \tag{32.12}$$

The above equation is an *exact* result. For noninteracting electrons, (32.12) agrees with the formula obtained by Cini [Cini 1980]. Indeed, the derivation by Cini does not depend on the details of the noninteracting system and therefore it is also correct for the Kohn-Sham system, which however has the extra merit of reproducing the exact density. The advantage of this approach is that the interaction in the leads and in the conductor are treated on the same footing via self-consistent calculations on the current-carrying system. It also allows for detailed studies of how the contacts influence the conductance properties. We note in passing that (32.12) is also gauge invariant since it does not change under an overall time-dependent shift of the external potential which is constant in space. It is also not modified by a simultaneous shift of the classical electrostatic potential and the chemical potential for $t < 0$.

Let us now focus on the long-time behavior and work out a simplified expression. We introduce the uncontacted Green function $\hat{g}$ which obeys (32.6) with $\hat{V} = 0$,

$$\left\{ \mathrm{i}\frac{\mathrm{d}}{\mathrm{d}z} - \hat{\mathcal{E}}(z) \right\} \hat{g}(z, z') = \delta(z, z') . \tag{32.13}$$

The $\hat{g}$ can be expressed in terms of the one-body evolution operator $\hat{\mathcal{U}}(t)$ which fullfils

$$\mathrm{i}\frac{\mathrm{d}}{\mathrm{d}t}\hat{\mathcal{U}}(t) = \hat{\mathcal{E}}(t)\hat{\mathcal{U}}(t), \qquad \text{with} \quad \hat{\mathcal{U}}(0) = \hat{1} . \tag{32.14}$$

The retarded and advanced components are

$$\hat{g}^{\mathrm{R}}(t, t') = -\Theta(t - t')\hat{\mathcal{U}}(t)\hat{\mathcal{U}}^{\dagger}(t') \tag{32.15a}$$

$$\hat{g}^{\mathrm{A}}(t, t') = \quad \Theta(-t + t')\hat{\mathcal{U}}(t)\hat{\mathcal{U}}^{\dagger}(t') , \tag{32.15b}$$

while the lesser component $\hat{g}^{<}(t, t') = \mathrm{i}\hat{g}^{\mathrm{R}}(t, 0) f(\hat{\mathcal{E}}(0))\hat{g}^{\mathrm{A}}(0, t)$, since also the uncontacted system is initially in equilibrium [cf. (32.11)].

We convert the equation of motion for $\hat{\mathcal{G}}$ into an integral equation

$$\hat{\mathcal{G}}(z, z') = \hat{g}(z, z') + \int_{\gamma} \mathrm{d}\bar{z}\, \hat{g}(z, \bar{z})\hat{V}\hat{\mathcal{G}}(\bar{z}, z') , \tag{32.16}$$

γ being the Keldysh contour of Fig. 32.2. The TDDFT Green function $\hat{\mathcal{G}}$ projected in a subregion $\alpha = \mathrm{L, R}$ or C can be described in terms of self-energies which account for the hopping in and out of the subregion in question. Considering the central region, the self-energy can be written as

$$\hat{\Sigma}(z, z') = \sum_{\alpha=\mathrm{L,R}} \hat{\Sigma}_{\alpha}, \quad \hat{\Sigma}_{\alpha}(z, z') = \hat{V}_{\mathrm{C}\alpha}\, \hat{g}(z, z')\hat{V}_{\alpha\mathrm{C}} . \tag{32.17}$$

Equations (32.16)–(32.17) allow to express $\hat{Q}_{\alpha}$ in terms of the projected Green function onto the central region, $\hat{G} \equiv \hat{\mathcal{G}}_{\mathrm{CC}}$, and $\hat{\Sigma}$. Below we shall make an extensive use of the Keldysh book-keeping of Chap. 3. After some tedious algebra one finds

$$\begin{aligned}\hat{Q}_\alpha(t) = & \sum_{\beta=L,R} \left[\hat{G}^{\mathrm{R}} \cdot \hat{\Sigma}^<_\beta \cdot \left(\delta_{\beta\alpha} + \hat{G}^{\mathrm{A}} \cdot \hat{\Sigma}^{\mathrm{A}}_\alpha\right)\right](t,t) \\ & + \sum_{\beta=L,R} \left[\hat{G}^{\mathrm{R}} \cdot \hat{\Sigma}^\rceil \star \hat{G}^{\mathrm{M}} \star \hat{\Sigma}^\lceil_\beta \cdot \left(\delta_{\beta\alpha} + \hat{G}^{\mathrm{A}} \cdot \hat{\Sigma}^{\mathrm{A}}_\alpha\right)\right](t,t) \\ & + \mathrm{i} \sum_{\beta=L,R} \hat{G}^{\mathrm{R}}(t,0) \left[\hat{G}^{\mathrm{M}} \star \hat{\Sigma}^\lceil_\beta \cdot \left(\delta_{\beta\alpha} + \hat{G}^{\mathrm{A}} \cdot \hat{\Sigma}^{\mathrm{A}}_\alpha\right)\right](0,t) \\ & + \left\{\hat{G}^{\mathrm{R}}(t,0)\hat{G}^{\mathrm{M}}(0,0) - \mathrm{i}\left[\hat{G}^{\mathrm{R}} \cdot \hat{\Sigma}^\rceil \star \hat{G}^{\mathrm{M}}\right](t,0)\right\} \left[\hat{G}^{\mathrm{A}} \cdot \hat{\Sigma}^{\mathrm{A}}_\alpha\right](0,t) \,.\end{aligned} \tag{32.18}$$

Here we briefly explain the notation used. The symbol "$\cdot$" is used to write $\int_0^\infty \mathrm{d}\bar{t}\, f(\bar{t})g(\bar{t})$ as $f \cdot g$, while the symbol "$\star$" is used to write $\int_0^{-\mathrm{i}\beta} \mathrm{d}\bar{\tau}\, f(\bar{\tau})g(\bar{\tau})$ as $f \star g$. The superscripts "M", "$\rceil$", "$\lceil$" in Green functions or self-energies denote the Matsubara component (both arguments on the thermal imaginary track), the Keldysh component with a real first argument and an imaginary second argument and the Keldysh component with an imaginary first argument and a real second argument, respectively.

Let us now take both the left and right electrodes infinitely large and thereafter consider the limit of $t \to \infty$. Then, only the first term on the r.h.s. of (32.18) does not vanish as both $\hat{G}$ and $\hat{\Sigma}$ tend to zero when the separation between their time argument increases. Thus, the long-time limit washes out the initial effect induced by the conducting term $\hat{V}$. Moreover, the asymptotic current is independent of the initial equilibrium distribution of the central device. We expect that for small bias the electrons at the bottom of the left and right conducting bands are not disturbed and the transient process is exponentially short. On the other hand, for strong bias the transient phase might decay as a power law, due to possible band-edge singularities.

Using the asymptotic ($t, t' \to \infty$) relation [Stefanucci 2004a]

$$\hat{G}^<(t,t') = \left[\hat{G}^{\mathrm{R}} \cdot \hat{\Sigma}^< \cdot \hat{G}^{\mathrm{A}}\right](t,t') \,, \tag{32.19}$$

we may write the asymptotic time-dependent current as

$$I_\alpha(t) = 2e\, \Re\, \mathrm{Tr} \left\{ \left[\hat{G}^{\mathrm{R}} \cdot \hat{\Sigma}^<_\alpha\right](t,t) + \left[\hat{G}^< \cdot \hat{\Sigma}^{\mathrm{A}}_\alpha\right](t,t) \right\} . \tag{32.20}$$

Equation (32.20) is valid for interacting devices connected to interacting electrodes, since the noninteracting TDDFT Green function gives the exact density. It also provides a useful framework for studying the transport in interacting systems from first principles. It can be applied both to the case of a constant (dc) bias as well as to the case of a *time-dependent* (e.g., ac) one. For noninteracting electrons, the Green function $\hat{\mathcal{G}}$ of TDDFT coincides with the Green function of the real system and (32.20) agrees with the formula by Wingreen et al. [Wingreen 1993, Jauho 1994].

32.4 Steady State

Let us now consider an external potential having a well defined limit when $t \to \infty$. Taking first the thermodynamic limit of the two electrodes and afterward the limit $t \to \infty$, we expect that the KS Hamiltonian $\hat{H}_{\mathrm{KS}}(t)$ will *globally* converge to an asymptotic KS Hamiltonian $\hat{H}^{\infty}_{\mathrm{KS}}$, meaning that $\lim_{t\to\infty} \hat{\mathcal{E}}(t) = \hat{\mathcal{E}}^{\infty} = \mathrm{const}$. In this case it must exist a unitary operator $\hat{\bar{\mathcal{U}}}$ such that

$$\lim_{t\to\infty} \hat{\mathcal{U}}(t) = \exp[-\mathrm{i}\hat{\mathcal{E}}^{\infty} t]\, \hat{\bar{\mathcal{U}}} \,. \qquad (32.21)$$

Then, in terms of diagonalizing one-body states $|\varphi^{\infty}_{m\alpha}\rangle$ of $\hat{\mathcal{E}}^{\infty}_{\alpha\alpha}$ with eigenvalues $\varepsilon^{\infty}_{m\alpha}$ we have

$$\hat{\Sigma}^{<}_{\alpha}(t,t') = \mathrm{i}\sum_{m,m'} \mathrm{e}^{-\mathrm{i}[\varepsilon^{\infty}_{m\alpha}t - \varepsilon^{\infty}_{m'\alpha}t']} \hat{V}_{\mathrm{C}\alpha}|\varphi^{\infty}_{m\alpha}\rangle\langle\varphi^{\infty}_{m\alpha}|f(\hat{\bar{\mathcal{E}}})|\varphi^{\infty}_{m'\alpha}\rangle\langle\varphi^{\infty}_{m'\alpha}|\hat{V}_{\alpha\mathrm{C}} \,, \qquad (32.22)$$

where $\hat{\bar{\mathcal{E}}} = \hat{\bar{\mathcal{U}}}\ \hat{\mathcal{E}}^{0}\ \hat{\bar{\mathcal{U}}}^{\dagger}$ and $\hat{\mathcal{E}}^{0} \equiv \hat{\mathcal{E}}(t=0)$. For $t, t' \to \infty$, the left and right contraction with a nonsingular $\hat{V}$ causes a perfect destructive interference for states with $|\varepsilon^{\infty}_{m\alpha} - \varepsilon^{\infty}_{m'\alpha}| \gtrsim 1/(t+t')$ and hence the restoration of translational invariance in time

$$\hat{\Sigma}^{<}_{\alpha}(t,t') = \mathrm{i}\sum_{m} f_{m\alpha}\hat{\Gamma}_{m\alpha}\mathrm{e}^{-\mathrm{i}\varepsilon^{\infty}_{m\alpha}(t-t')} \,, \qquad (32.23)$$

where $f_{m\alpha} = \langle\varphi^{\infty}_{m\alpha}|f(\hat{\bar{\mathcal{E}}})|\varphi^{\infty}_{m\alpha}\rangle$ while $\hat{\Gamma}_{m\alpha} = \hat{V}_{\mathrm{C}\alpha}|\varphi^{\infty}_{m\alpha}\rangle\langle\varphi^{\infty}_{m\alpha}|\hat{V}_{\alpha\mathrm{C}}$.[1] The above *dephasing mechanism* is the key ingredient for the appearance of a steady state. Substituting (32.23) into (32.20) we get the steady state current

$$I_{\alpha} = -2e\sum_{m\beta} f_{m\beta}\Big[\mathrm{Tr}\Big\{\hat{G}^{\mathrm{R}}(\varepsilon^{\infty}_{m\beta})\hat{\Gamma}_{m\beta}\hat{G}^{\mathrm{A}}(\varepsilon^{\infty}_{m\beta})\Im[\hat{\Sigma}^{\mathrm{A}}_{\alpha}(\varepsilon^{\infty}_{m\beta})]\Big\} + \delta_{\beta\alpha}\,\mathrm{Tr}\Big\{\hat{\Gamma}_{m\alpha}\Im[\hat{G}^{\mathrm{R}}(\varepsilon^{\infty}_{m\alpha})]\Big\}\Big] \,, \qquad (32.24)$$

with $\hat{G}^{\mathrm{R,A}}(\varepsilon) = [\varepsilon - \hat{\mathcal{E}}^{\infty}_{CC} - \hat{\Sigma}^{\mathrm{R,A}}(\varepsilon)]^{-1}$. Using the equalities

$$\Im[\hat{G}^{\mathrm{R}}] = \frac{1}{2\mathrm{i}}[\hat{G}^{\mathrm{R}} - \hat{G}^{\mathrm{A}}], \qquad [\hat{G}^{\mathrm{R}} - \hat{G}^{\mathrm{A}}] = [\hat{G}^{>} - \hat{G}^{<}] \,, \qquad (32.25)$$

together with

$$[\hat{G}^{>}(\varepsilon) - \hat{G}^{<}(\varepsilon)] = -2\pi\mathrm{i}\sum_{m\alpha}\delta(\varepsilon - \varepsilon^{\infty}_{m\alpha})\hat{G}^{\mathrm{R}}(\varepsilon^{\infty}_{m\alpha})\hat{\Gamma}_{m\alpha}\hat{G}^{\mathrm{A}}(\varepsilon^{\infty}_{m\alpha}) \qquad (32.26)$$

and

[1] In principle, there may be degeneracies which require a diagonalization to be performed for states on the energy shell.

$$\Im[\hat{\Sigma}^{\mathrm{A}}_{\alpha}(\varepsilon)] = \pi \sum_m \delta(\varepsilon - \varepsilon^{\infty}_{m\alpha})\hat{\Gamma}_{m\alpha}\,, \tag{32.27}$$

the steady-state current in (32.24) can be rewritten in a Landauer-like [Imry 2002] form

$$J_{\mathrm{R}} = -e \sum_m [f_{m\mathrm{L}}\mathcal{T}_{m\mathrm{L}} - f_{m\mathrm{R}}\mathcal{T}_{m\mathrm{R}}] = -J_{\mathrm{L}}\,. \tag{32.28}$$

In the above formula $\mathcal{T}_{m\mathrm{R}} = \sum_n \mathcal{T}^{n\mathrm{L}}_{m\mathrm{R}}$ and $\mathcal{T}_{m\mathrm{L}} = \sum_n \mathcal{T}^{n\mathrm{R}}_{m\mathrm{L}}$ are the TDDFT transmission coefficients expressed in terms of the quantities

$$\mathcal{T}^{n\beta}_{m\alpha} = 2\pi\delta(\varepsilon^{\infty}_{m\alpha} - \varepsilon^{\infty}_{n\beta})\mathrm{Tr}\left\{\hat{G}^{\mathrm{R}}(\varepsilon^{\infty}_{m\alpha})\hat{\Gamma}_{m\alpha}\hat{G}^{\mathrm{A}}(\varepsilon^{\infty}_{n\beta})\hat{\Gamma}_{n\beta}\right\} = \mathcal{T}^{m\alpha}_{n\beta}\,. \tag{32.29}$$

Despite the formal analogy with the Landauer formula, (32.28) contains an important conceptual difference, since $f_{m\alpha}$ is not simply given by the Fermi distribution function. For example, if the induced change in effective potential varies widely in space deep inside the electrodes, the band structure $\bar{\hat{\mathcal{E}}}_{\alpha\alpha}$ may be completely different from that of $\hat{\mathcal{E}}^{\infty}_{\alpha\alpha}$. However, if we asymptotically have equilibrium far away from the central region, as we would expect for electrodes with a macroscopic cross section, the change in effective potential must be uniform. To leading order in $1/N$ we then have

$$\hat{\mathcal{E}}_{\alpha\alpha}(t) = \hat{\mathcal{E}}^{0}_{\alpha\alpha} + \delta v_{\alpha}(t)\,, \tag{32.30}$$

and $\hat{\mathcal{E}}^{\infty}_{\alpha\alpha} = \hat{\mathcal{E}}^{0}_{\alpha\alpha} + \delta v_{\alpha,\infty}$. Hence, except for corrections which are of lower order with respect to the system size, $\bar{\hat{\mathcal{E}}}_{\alpha\alpha} = \hat{\mathcal{E}}^{0}_{\alpha\alpha}$ and

$$f_{m\alpha} = f(\varepsilon^{\infty}_{m\alpha} - \delta v_{\alpha,\infty})\,. \tag{32.31}$$

We emphasize that the steady-state current in (32.28) results from a pure dephasing mechanism in the fictitious noninteracting problem. The damping effects of scattering are described by A_{xc} and v_{xc}. Furthermore, the current depends only on the asymptotic value of the KS potential, $v_{\mathrm{KS}}(\boldsymbol{r}, t \to \infty)$, provided that (32.30) holds. However, $v_{\mathrm{KS}}(\boldsymbol{r}, t \to \infty)$ might depend on the history of the external applied potential and the resulting steady-state current might be history dependent. In these cases the full time evolution can not be avoided. In the case of the time-dependent local density approximation (TDLDA), the exchange-correlation potential v_{xc} depends only locally on the instantaneous density and has no memory at all. If the density tends to a constant, so does the KS potential v_{KS}, which again implies that the density tends to a constant. Owing to the non-linearity of the problem there might still be more than one steady-state solution or none at all.

32.5 A Practical Implementation Scheme

The total time-dependent current $I_S(t)$ can be calculated from the KS orbitals according to (32.5). However, before a TDDFT calculation of transport can be tackled, a number of technical problems have to be addressed. In particular, one needs a practical scheme for extracting the set of initial states of the infinitely large system and for propagating them. Of course, since one can in practice only deal with finite systems this can only be achieved by applying the correct boundary conditions. The problem of so-called "transparent boundary conditions" for the time-dependent Schrödinger equation has been attacked by many authors. For a recent overview, the reader is referred to [Moyer 2004] and references therein. Below, we sketch how to compute the initial extended states and how to propagate them (we refer to [Kurth 2005] for the explicit implementation of the algorithm).

The KS eigenstate φ_i of the Hamiltonian $\hat{H}_{\mathrm{KS}}(0)$ is uniquely specified by its eigenenergy ε_i and a label i for the degenerate orbitals of this energy. It is possible to show that the eigenfunctions of $\Im\hat{\mathcal{G}}^{\mathrm{R}}_{\mathrm{CC}}(E)$ can be expressed as a linear combination of the φ_i projected onto the central region. If we use N_{g} grid points to describe the central region, the diagonalization in principle gives N_{g} eigenvectors, but only a few have the physical meaning of extended eigenstates at this energy. It is, however, very easy to identify the physical states by looking at the eigenvalues: only few eigenvalues are nonvanishing. The corresponding states are the physical ones. All the other eigenvalues are zero (or numerically close to zero) and the corresponding states have no physical meaning. This procedure gives the correct extended eigenstates in the central region only up to a normalization factor. When diagonalizing $\Im\hat{\mathcal{G}}^{\mathrm{R}}_{\mathrm{CC}}(E)$ with typical library routines, one obtains eigenvectors that are normalized in the central region. Physically this might be incorrect. Therefore, the normalization has to be fixed separately. This can be done by matching the wavefunction for the central region to the known form (and normalization) of the wavefunction in the macroscopic leads.

Once the initial states have been calculated, we need a suitable algorithm for propagating them. The explicitly treated region C includes the first few atomic layers of the left and right electrodes. The boundaries of this region are chosen in such a way that the density outside C is accurately described by an equilibrium bulk density. It is convenient to write $\hat{\mathcal{E}}_{\alpha\alpha}(t)$, with $\alpha = \mathrm{L}, \mathrm{R}$, as the sum of a term $\hat{\mathcal{E}}_\alpha$ which is constant in time and another term $\hat{\mathcal{V}}_\alpha(t)$ which is explicitly time-dependent, $\hat{\mathcal{E}}_{\alpha\alpha}(t) = \hat{\mathcal{E}}_\alpha + \hat{\mathcal{V}}_\alpha(t)$. In configuration space $\hat{\mathcal{V}}_\alpha(t)$ is diagonal at any time t since the KS potential is local in space. Furthermore, the diagonal elements $\mathcal{V}_\alpha(\boldsymbol{r}, t)$ are spatially constant for metallic electrodes. Thus, $\hat{\mathcal{V}}_\alpha(t) = \mathcal{V}_\alpha(t)\hat{1}_\alpha$ and $\mathcal{V}_{\mathrm{L}}(t) - \mathcal{V}_{\mathrm{R}}(t)$ is the total potential drop across the central region. Here $\hat{1}_\alpha$ is the unit operator for region α. We write $\hat{H}_{\mathrm{KS}}(t) = \hat{\mathcal{E}}(t) + \hat{V} = \hat{\hat{H}}(t) + \hat{\mathcal{V}}(t)$, with $\hat{\mathcal{V}}(t) = \hat{\mathcal{V}}_{\mathrm{L}}(t) + \hat{\mathcal{V}}_{\mathrm{R}}(t)$. For any given initial state $\varphi(0) = \varphi^{(0)}$ we calculate $\varphi(t_m = m\Delta t) = \varphi^{(m)}$ by using a generalized form of the Cayley method

$$(\hat{1} + \mathrm{i}\delta\hat{\hat{H}}^{(m)})\frac{\hat{1} + \mathrm{i}\frac{\delta}{2}\hat{\mathcal{V}}^{(m)}}{\hat{1} - \mathrm{i}\frac{\delta}{2}\hat{\mathcal{V}}^{(m)}}\varphi^{(m+1)} = (\hat{1} - \mathrm{i}\delta\hat{\hat{H}}^{(m)})\frac{\hat{1} - \mathrm{i}\frac{\delta}{2}\hat{\mathcal{V}}^{(m)}}{\hat{1} + \mathrm{i}\frac{\delta}{2}\hat{\mathcal{V}}^{(m)}}\varphi^{(m)}, \quad (32.32)$$

with $\hat{\hat{H}}^{(m)} = \frac{1}{2}[\hat{\hat{H}}(t_{m+1}) + \hat{\hat{H}}(t_m)]$, $\hat{\mathcal{V}}^{(m)} = \frac{1}{2}[\hat{\mathcal{V}}(t_{m+1}) + \hat{\mathcal{V}}(t_m)]$ and $\delta = \Delta t/2$. It should be noted that our propagator is norm conserving (unitary) and accurate to second-order in δ, as is the Cayley propagator. Denoting by φ_α the projected wave function onto the region $\alpha = \mathrm{R, L, C}$, we find from (32.32)

$$\varphi_{\mathrm{C}}^{(m+1)} = \frac{\hat{1} - \mathrm{i}\delta \hat{H}_{\mathrm{eff}}^{(m)}}{\hat{1} + \mathrm{i}\delta \hat{H}_{\mathrm{eff}}^{(m)}} \varphi_{\mathrm{C}}^{(m)} + S^{(m)} - M^{(m)} \,. \tag{32.33}$$

Here, $\hat{H}_{\mathrm{eff}}^{(m)}$ is the effective Hamiltonian of the central region:

$$\hat{H}_{\mathrm{eff}}^{(m)} = \hat{\mathcal{E}}_{\mathrm{CC}}^{(m)} - \hat{V}_{\mathrm{CL}} \frac{\mathrm{i}\delta}{\hat{1} + \mathrm{i}\delta \hat{\mathcal{E}}_{\mathrm{L}}} \hat{V}_{\mathrm{LC}} - \hat{V}_{\mathrm{CR}} \frac{\mathrm{i}\delta}{\hat{1} + \mathrm{i}\delta \hat{\mathcal{E}}_{\mathrm{R}}} \hat{V}_{\mathrm{RC}} \,, \tag{32.34}$$

with $\hat{\mathcal{E}}_{\mathrm{CC}}^{(m)} = \frac{1}{2}[\hat{\mathcal{E}}_{\mathrm{CC}}(t_{m+1}) + \hat{\mathcal{E}}_{\mathrm{CC}}(t_m)]$. The source term $S^{(m)}$ describes the injection of density into the region C. For a wave packet initially localized in C the projection onto the left and right electrode $\varphi_\alpha^{(0)}$ vanishes and $S^{(m)} = 0$ for any m. The memory term $M^{(m)}$ is responsible for the hopping in and out of the region C. Equation (32.33) is the central result of our algorithm for solving the time-dependent Schrödinger equation in extended systems.

As an example, we consider a one-dimensional system of noninteracting electrons at zero temperature where the electrostatic potential vanishes both in the left and right leads. The electrostatic potential in the central region is modeled by a double square potential barrier. Initially, all single particle levels are occupied up to the Fermi energy ε_F. At $t = 0$, a bias is switched on in the leads and the time-evolution of the system is calculated. The numerical parameters are as follows: the Fermi energy is $\varepsilon_F = 0.3$ a.u., the bias is $\mathcal{V}_L = 0.15, 0.25$ a.u. and $\mathcal{V}_R = 0$, the central region extends from $x = -6$ to $x = +6$ a.u. with equidistant grid points with spacing $\Delta x = 0.03$ a.u. The electrostatic potential $v_{\mathrm{ext}}(x) = 0.5$ a.u. for $5 \le |x| \le 6$ and zero otherwise. For the second derivative of the wavefunction (kinetic term) we have used a simple three-point discretization. The energy integral in (32.5) is discretized with 100 points which amounts to a propagation of 200 states. The time step for the propagation is $\Delta t = 10^{-2}$ a.u.

In Fig. 32.3 we have plotted the total current at $x = 0$ as a function of time for two different ways of applying the bias in the left lead: in one case the constant bias $\mathcal{V}_L = \mathcal{V}_0$ is switched on suddenly at $t = 0$, in the other case the constant $\mathcal{V}_0$ is achieved with a smooth switching $\mathcal{V}_L(t) = \mathcal{V}_0 \sin^2(\omega t)$ for $0 < t < \pi/(2\omega)$. As a first feature we notice that a steady state is achieved and that the steady-state current does not depend on the history of the applied bias, in agreement with the results obtained in Sect. 32.4. Second, we notice that the onset of the current is delayed in relation to the switching time $t = 0$. This is easily explained by the fact that the perturbation at $t = 0$ happens in the leads only, e.g., for $|x| > 6$ a.u., while we plot the current at $x = 0$. In other words, we see the delay time needed for the perturbation to propagate

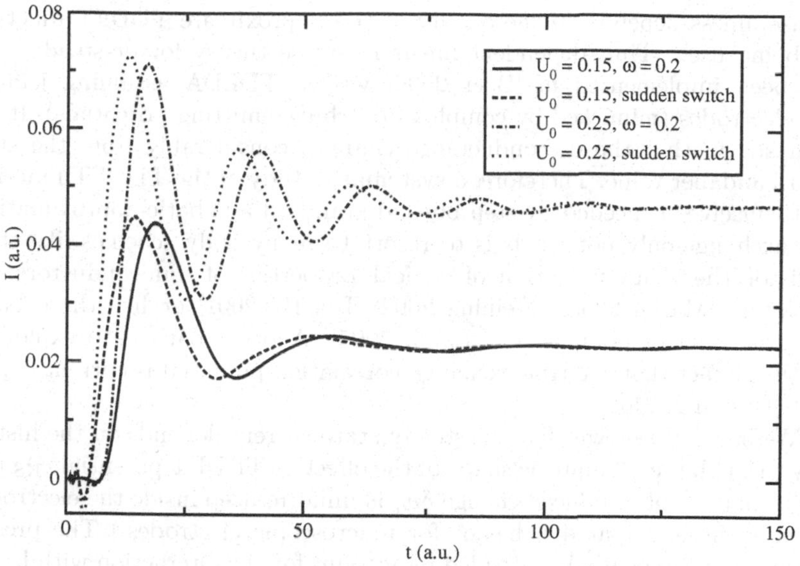

Fig. 32.3. Time evolution of the current for a double square potential barrier when the bias is switched on in two different manners: in one case, the bias $\mathcal{V}_{\rm L} = \mathcal{V}_0$ is suddenly switched on at $t = 0$ while in the other case the same bias is achieved with a smooth switching $\mathcal{V}_{\rm L}(t) = \mathcal{V}_0 \sin^2(\omega t)$ for $0 < t < \pi/(2\omega)$. The parameters for the double barrier and the other numerical parameters are described in the main text

from the leads to the center of our device region. We also note that the higher the bias the more the current exceeds its steady-state value for small times after switching on the bias.

32.6 Conclusions

In conclusion, we have described a formally exact, thermodynamically consistent scheme based on TDDFT and NEGF in order to treat the time-dependent current response of electrode-junction-electrode systems. Among the advantages, we stress the possibility of including the electron-electron interaction not only in the central region but also in the electrodes. We have shown that the steady state develops due to a *dephasing mechanism* without any reference to many-body damping and interactions. The damping mechanism (due to the electron-electron scatterings) of the real problem is described by $v_{\rm xc}$. The nonlinear steady-state current can be expressed in a Landauer-like formula in terms of fictitious transmission coefficients and one-particle energy eigenvalues. Our scheme is equally applicable to time-dependent responses and also allows the calculation of the (transient) current shortly after switching on a driving external field. Clearly, its usefulness depends on the quality of the approximate TDDFT functionals being used. Time-dependent linear response theory for dc-steady state has been implemented in [Baer 2004] within TDLDA assuming jellium-like electrodes (mimicked by complex absorbing/emitting potentials). It has been shown that the dc-conductance changes considerably from the standard Landauer value. Therefore, a systematic study of the TDDFT functionals themselves is needed. A step beyond standard adiabatic-approximations and exchange-only potentials is to resort to many-body schemes like those used for the characterization of optical properties of semiconductors and insulators [Marini 2003b, Reining 2002, Tokatly 2001] or like those based on variational functionals [von Barth 2005]. Another path is to explore in depth the fact that the true exchange-correlation potential is current dependent [Ullrich 2002b].

We have also shown that the steady-state current depends on the history only through the asymptotic shape of the effective TDDFT potential v_{KS} provided that the bias-induced change δv_α is uniform deep inside the electrodes. (This is the anticipated behavior for macroscopic electrodes.) The present formulation can be easily extended to account for the interaction with lattice vibrations at a semiclassical level. The inclusion of phonons might give rise to hysteresis loops due to different transient electronic/geometrical device configurations (e.g., isomerisation or structural modification). This effect will be more dramatic in the case of ac-driving fields of high frequencies where the system might not have enough time to respond to the perturbation.

Acknowledgments

This work was supported by the European Community 6th framework Network of Excellence NANOQUANTA (NMP4-CT-2004-500198) and by the Research and Training Network EXCiT!NG. AR acknowledges support from the EC project M-DNA (IST-2001-38051), Spanish MCyT and the University of the Basque Country. We have benefited from enlightening discussions with L. Wirtz, A. Castro, H. Appel, M. A. L. Marques, and C. Verdozzi.

33 Scattering Amplitudes

A. Wasserman and K. Burke

33.1 Introduction

Electrons are constantly colliding with atoms and molecules: in chemical reactions, in our atmosphere, in stars, plasmas, in a molecular wire carrying a current, or when the tip of a scanning tunneling microscope injects electrons to probe a surface. When the collision occurs at low energies, the calculations become especially difficult due to correlation effects between the projectile electron and those of the target. These *bound-free* correlations are very important. For example, it is due to bound-free correlations that ultra-slow electrons can break up RNA molecules [Hanel 2003] causing serious genotoxic damage. The accurate description of correlation effects when the targets are so complex is a major challenge. Existing approaches based on wavefunction methods, developed from the birth of quantum mechanics and perfected since then to reach great sophistication [Morrison 1983, Burke 1994, Winstead 1996], cannot overcome the exponential barrier resulting from the many-body Schrödinger equation when the number of electrons in the target is large. Wavefunction-based methods can still provide invaluable insights in such complex cases, provided powerful computers and smart tricks are employed (see, e.g., [Grandi 2004] for low-energy electron scattering from uracil), but a truly ab-initio approach circumventing the exponential barrier would be most welcome. The purpose of this chapter is to describe several results relevant to this goal.

Imagine a slow electron approaching an atom or molecule that has N electrons, and is assumed to be in its ground state, with energy E^N_{GS}. The asymptotic kinetic energy of the incoming electron is ε, so the whole system of target plus electron has a total energy of $E^N_{\mathrm{GS}} + \varepsilon$. This is an excited state of the $(N+1)$-electron system, and, as such, it can be described by the linear response formalism of TDDFT starting from the *ground state* of the $(N+1)$-electron system. We will explain how.

The targets we will consider must be able to bind an extra electron. For example, take the target to be a positive ion, so that the $(N+1)$-electron system, with ground-state energy E^{N+1}_{GS}, is neutral. Previous chapters in this book have described how to employ TDDFT to calculate, e.g., excitation energies corresponding to bound $\rightarrow$ bound transitions from the ground state. However, in the scattering situation considered here, the excitation energy is

A. Wasserman and K. Burke: *Scattering Amplitudes*, Lect. Notes Phys. **706**, 493–505 (2006)
DOI 10.1007/3-540-35426-3_33

known in advance: it is $I + \varepsilon$, where I is the first ionization energy of the $(N+1)$-system, $I = E_{\rm GS}^{N} - E_{\rm GS}^{N+1}$. It is the scattering *phase shifts*, rather than the energies, which are of interest in the scattering regime.

The TDDFT approach to scattering that we are about to discuss [Wasserman 2005b] is very different from wavefunction-based methods, yet *exact* in the sense that if the ground-state exchange-correlation potential ($v_{\rm XC}$) and time-dependent exchange-correlation kernel ($f_{\rm XC}$) were known exactly, we could then (in principle) calculate the *exact* scattering phase shifts for the system of $N+1$ interacting electrons. Any given approximation to $v_{\rm XC}$ and $f_{\rm XC}$ leads in turn to definite predictions for the phase shifts. The method involves the following three steps: (i) Finding the ground-state Kohn-Sham potential of the $(N+1)$-electron system, $v_{\rm KS}^{N+1}(\boldsymbol{r})$; (ii) Solving a *potential scattering* problem, namely, scattering from $v_{\rm KS}^{N+1}(\boldsymbol{r})$; and (iii) Correcting the Kohn-Sham scattering phase shifts towards the true ones, via linear response TDDFT.

We start by reviewing those aspects of the linear response formalism of TDDFT that were introduced in Chap. 1 and will be used in the following sections. We then derive TDDFT equations for one-dimensional scattering, and work out in detail two simple examples to show how to calculate transmission and reflection amplitudes in TDDFT. The discussion is then generalized to three dimensions, where we explain how the familiar single pole approximation for bound → bound transitions can be continued to describe bound → continuum transitions to get information about scattering states. We end with a brief summary and outlook.

33.2 Linear Response for the $(N+1)$-Electron System

For a thorough treatment, see Chap. 1. Here we only review what will be needed for the following sections. The central equation of the linear response formalism of TDDFT is the Dyson-like response equation relating the susceptibility $\chi^{N+1}(\boldsymbol{r},\boldsymbol{r}',\omega)$ of a system of interacting electrons with that of its ground-state Kohn-Sham analog, $\chi_{\rm KS}^{N+1}(\boldsymbol{r},\boldsymbol{r}',\omega)$ [Petersilka 1996a], see (1.23). The $N+1$ superscript was added in order to emphasize that we are going to perturb the ground-state of the $(N+1)$-electron system, where N is the number of electrons of the target. In what follows, however, for notational simplicity, the $(N+1)$ superscript will be dropped from all quantities. We write the spin-decomposed susceptibility in the Lehman representation:

$$\chi_{\sigma\sigma'}(\boldsymbol{r},\boldsymbol{r}',\omega) = \lim_{\eta\to 0^+}\left[\sum_i \frac{F_{i\sigma}(\boldsymbol{r})F^*_{i\sigma'}(\boldsymbol{r}')}{\omega - \Omega_i + {\rm i}\eta^+} + {\rm c.c.}(\omega\to-\omega)\right] , \qquad (33.1)$$

with

$$F_{i\sigma}(\boldsymbol{r}) = \langle \Psi_{\rm GS}|\hat{n}_\sigma(\boldsymbol{r})|\Psi_i\rangle \quad ; \quad \hat{n}_\sigma(\boldsymbol{r}) = \sum_{l=1}^{N+1}\delta(\boldsymbol{r}-\hat{\boldsymbol{r}}_l)\delta_{\sigma\hat{\sigma}_l} \qquad (33.2)$$

where $\Psi_{\rm GS}$ is the ground state of the $(N+1)$-electron system, Ψ_i its $i^{\rm th}$ excited state, and $\hat{n}_\sigma(\boldsymbol{r})$ is the σ-spin density operator. In (33.1), Ω_i is the $\Psi_{\rm GS} \to \Psi_i$ transition frequency. The term "c.c.$(\omega \to -\omega)$" stands for the complex conjugate of the first term with ω substituted by $-\omega$. The sum in (33.1) should be understood as a sum over the discrete spectrum and an integral over the continuum. *All* excited states (labelled by i) with non-zero $F_{i\sigma}(\boldsymbol{r})$ contribute to the susceptibility. In particular, the scattering state discussed in the introduction consisting on a free electron of energy ε and an N-electron target, contributes too. How to extract from the susceptibility the scattering information about this single state? The question will be answered in the following sections, starting in one dimension.

33.3 One Dimension

33.3.1 Transmission Amplitudes from the Susceptibility

Consider large distances, where the $(N+1)$-electron ground-state density is dominated by the decay of the highest occupied Kohn-Sham orbital [Katriel 1980]; the ground-state wavefunction behaves as [Ernzerhof 1996]:

$$\Psi_{\rm GS} \underset{x\to\infty}{\longrightarrow} \psi^N_{\rm GS}(x_2,\ldots,x_{N+1})\sqrt{\frac{n(x)}{N+1}}S_{\rm GS}(\sigma,\sigma_2,\ldots,\sigma_{N+1}) \tag{33.3}$$

where $\psi^N_{\rm GS}$ is the ground-state wavefunction of the target, $S_{\rm GS}$ the spin function of the $(N+1)$-electron ground-state and $n(x)$ the $(N+1)$-electron ground-state density. Similarly, the asymptotic behavior of the $i^{\rm th}$ excited state is:

$$\Psi_i \underset{x\to\infty}{\longrightarrow} \psi^N_{i_t}(x_2,\ldots,x_{N+1})\frac{\phi_{k_i}(x)}{\sqrt{N+1}}S_i(\sigma,\sigma_2,\ldots,\sigma_{N+1})\,, \tag{33.4}$$

where $\psi^N_{i_t}$ is an eigenstate of the target (labeled by i_t), S_i is the spin function of the $i^{\rm th}$ excited state of the $(N+1)-$system, and $\phi_{k_i}(x)$ a one-electron orbital (not to be confused with φ, notation reserved for Kohn-Sham orbitals).

The contribution to $F_{i\sigma}(x)$ (33.2) from channels where the target is excited vanishes as $x \to \infty$ due to orthogonality. We therefore focus on *elastic* scattering only. Inserting (33.3) and (33.4) into the 1D-version of (33.2), and taking into account the antisymmetry of both $\Psi_{\rm GS}$ and Ψ_i,

$$F_{i\sigma}(x) \underset{x\to\infty}{\longrightarrow} \sqrt{n(x)}\phi_{k_i}(x)\delta_{0,i_t} \sum_{\sigma_2\ldots\sigma_{N+1}} S^*_{\rm GS}(\sigma\ldots\sigma_{N+1})S_i(\sigma\ldots\sigma_{N+1}) \tag{33.5}$$

The susceptibility at large distances is then obtained by inserting (33.5) into the 1D-version of (33.1):

$$\chi(x,x',\omega) = \sum_{\sigma\sigma'} \chi_{\sigma\sigma'}(x,x',\omega) \underset{x,x'\to\pm\infty}{\longrightarrow} \sqrt{n(x)n(x')}$$
$$\times \sum_i \frac{\phi_{k_i}(x)\phi^*_{k_i}(x')}{\omega - \Omega_i + \mathrm{i}\eta} \delta_{0,i_t} \delta_{S_{\mathrm{GS}},S_i} + \mathrm{c.c.}(\omega \to -\omega) \quad (33.6)$$

Since only scattering states of the $(N+1)$-electron optical potential contribute to the sum in (33.6) at large distances, it becomes an integral over wavenumbers $k = \sqrt{2\varepsilon}$, where ε is the energy of the projectile electron:

$$\sum_i \frac{\phi_{k_i}(x)\phi^*_{k_i}(x')}{\omega - \Omega_i + \mathrm{i}\eta} \underset{x,x'\to\pm\infty}{\longrightarrow} \frac{1}{2\pi} \int_{0[\mathrm{R}],[\mathrm{L}]}^{\infty} \mathrm{d}k \, \frac{\phi_k(x)\phi^*_k(x')}{\omega - \Omega_k + \mathrm{i}\eta} \quad (33.7)$$

In this notation, the functions ϕ_{k_i} are box-normalized, and $\phi_{k_i}(x) = \phi_k(x)/\sqrt{L}$, where $L \to \infty$ is the length of the box. The transition frequency $\Omega_i = E_i^{N+1} - E_{\mathrm{GS}}^{N+1}$ is now simply $\Omega_k = E^N_{\mathrm{GS}} + k^2/2 - E^{N+1}_{\mathrm{GS}} = k^2/2 + I$, where I is the first ionization potential of the $(N+1)$-electron system, and E^M_{GS} and E^M_i are the ground and i^{th} excited state energies of the M-electron system. The subscript "[R],[L]" indicates that the integral is over both orbitals satisfying *right* and *left* boundary conditions:

$$\phi_k^{\substack{[\mathrm{R}]\\[\mathrm{L}]}}(x) \to \begin{cases} \mathrm{e}^{\pm ikx} + r(k)\mathrm{e}^{\mp ikx} \,, & x \to \mp\infty \\ t(k)\mathrm{e}^{\pm ikx} \,, & x \to \pm\infty \end{cases} . \quad (33.8)$$

When $x \to -\infty$ and $x' = -x$ the integral of (33.7) is dominated by a term that oscillates in space with wavenumber $2\sqrt{k^2 - 2I}$ and amplitude given by the transmission amplitude for spin-conserving collisions $t(k)$ at that wavenumber. Denoting this by χ^{osc}, and setting $\varepsilon = \frac{1}{2}k^2$ we obtain:

$$t(\varepsilon) = \lim_{x\to-\infty} \left[\frac{\mathrm{i}\sqrt{2\varepsilon}}{\sqrt{n(x)n(-x)}} \chi^{\mathrm{osc}}(x,-x,\varepsilon+I) \right] . \quad (33.9)$$

Therefore, in order to extract the transmission amplitude $t(\varepsilon)$ from the susceptibility when an electron of energy ε collides with an N-electron target in one dimension, one should first construct the ground-state density of the $(N+1)$-electron system, perturb it in the far left with frequency $I + \varepsilon$, and then look at the oscillations of the density change in the far right: the amplitude of these oscillations ["amplified" by $\mathrm{i}\sqrt{2\varepsilon}n(x)^{-1}$] is the transmission amplitude $t(\varepsilon)$ (see Fig. 33.1).

The derivation of (33.9) does not depend on the interaction between the electrons. Therefore, the same formula applies to the Kohn-Sham system:

$$t_{\mathrm{KS}}(\varepsilon) = \lim_{x\to-\infty} \left[\frac{\mathrm{i}\sqrt{2\varepsilon}}{\sqrt{n(x)n(-x)}} \chi^{\mathrm{osc}}_{\mathrm{KS}}(x,-x,\varepsilon+I) \right] . \quad (33.10)$$

In practice, the Kohn-Sham transmission amplitudes $t_{\mathrm{KS}}(\varepsilon)$ are obtained by solving a *potential scattering* problem, i.e., scattering from the $(N+1)$-electron ground-state KS potential.

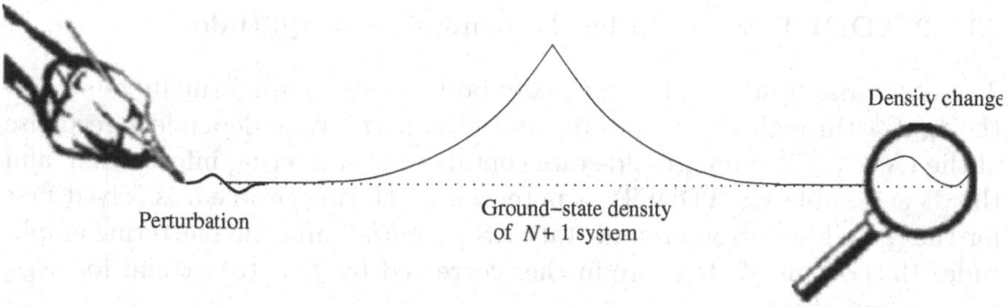

Fig. 33.1. Cartoon of (33.9). To extract the transmission amplitude for an electron of energy ε scattering from an N-electron target: apply a perturbation of frequency $\varepsilon + I$ on the far left of the $(N+1)$-ground-state system (I is its first ionization energy), and look at how the density changes oscillate on the far right. Once amplified, the amplitude of these oscillations correspond to $t(\varepsilon)$. Reproduction from the roof of the Sistine Chapel, with permission from artist

Illustration of (33.10)

For one electron, the susceptibility is given by [Maitra 2003b]:

$$\chi_{\mathrm{KS}}(x, x', \varepsilon + I) = \sqrt{n(x)n(x')}\left[g_{\mathrm{KS}}(x, x', \varepsilon) + g^*_{\mathrm{KS}}(x, x', -\varepsilon - 2I)\right] , \quad (33.11)$$

where the Green's function $g_{\mathrm{KS}}(x, x', \varepsilon)$ has a Fourier transform satisfying

$$\left\{ -\mathrm{i}\frac{\partial}{\partial t} - \frac{1}{2}\frac{\partial^2}{\partial x^2} + v_{\mathrm{KS}}(x) \right\} g_{\mathrm{KS}}(x, x', t - t') = -\mathrm{i}\delta(x - x')\delta(t - t') . \quad (33.12)$$

Let's find the transmission amplitude for an electron scattering from a double delta-function well, $v_{\mathrm{ext}}(x) = -Z_1\delta(x) - Z_2\delta(x - a)$. The Green's function can be readily obtained in this case as

$$g_{\mathrm{KS}}(x, x') = g_1(x, x') - \frac{Z_2 g_1(x, a) g_1(a, x')}{1 + Z_2 g_1(a, a)} , \quad (33.13)$$

where g_1 is the Green's function for a single delta-function of strength Z_1 at the origin. It is given by [Szabo 1989]:

$$g_1(x, x') = \frac{1}{ik}\left\{ \mathrm{e}^{ik|x - x'|} - \frac{Z_1 \mathrm{e}^{ik(|x| + |x'|)}}{ik + Z_1} \right\} , \quad (33.14)$$

with $k = \sqrt{2\varepsilon}$. Having constructed χ_{KS} explicitly, application of (33.10) yields the correct answer

$$t_{\mathrm{KS}} = \frac{ik/(Z_1 + ik)}{1 + Z_2 g_1(a, a)} . \quad (33.15)$$

33.3.2 TDDFT Equation for Transmission Amplitudes

The exact amplitudes $t(\varepsilon)$ of the many-body problem are formally related to the $t_{\mathrm{KS}}(\varepsilon)$ through (33.9), (33.10) and (1.23): the time-dependent response of the $(N+1)$-electron ground-state contains the scattering information, and this is accessible via TDDFT. A potential scattering problem is solved first for the $(N+1)$-electron ground-state KS potential, and the scattering amplitudes thus obtained (t_{KS}) are further corrected by f_{Hxc} to account for, e.g., polarization effects.

Even though (33.9) is impractical as a basis for computations (one can rarely obtain the susceptibility with the desired accuracy in the asymptotic regions, as we did in the previous example) it leads to practical approximations. The simplest of such approximations is obtained by iterating (1.23) once, substituting χ by χ_{KS} in the right-hand side of (1.23). This leads through (33.9) and (33.10) to the following useful distorted-wave-Born-type approximation for the transmission amplitude:

$$t(\varepsilon) = t_{\mathrm{KS}}(\varepsilon) + \frac{1}{\mathrm{i}\sqrt{2\varepsilon}}\langle\langle \mathrm{HOMO}, \varepsilon|\hat{f}_{\mathrm{Hxc}}(\varepsilon + I)|\mathrm{HOMO}, \varepsilon\rangle\rangle\,. \tag{33.16}$$

In (33.16), and from now on, the double-bracket notation stands for:

$$\langle\langle \mathrm{HOMO}, \varepsilon|\hat{f}_{\mathrm{Hxc}}(\varepsilon + I)|\mathrm{HOMO}, \varepsilon\rangle\rangle = \int \mathrm{d}x \int \mathrm{d}x'\, \varphi^*_{\mathrm{HOMO}}(x)\varphi^{[\mathrm{L}]*}_{\varepsilon}(x) f_{\mathrm{Hxc}}(x, x', \varepsilon + I)\varphi_{\mathrm{HOMO}}(x')\varphi^{[\mathrm{R}]}_{\varepsilon}(x')\,, \tag{33.17}$$

where φ_{HOMO} is the highest-occupied molecular orbital of the $(N+1)$-electron system, and $\varphi^{[\mathrm{R}]}_{\varepsilon}(x)$ is the energy-normalized scattering orbital of energy ε satisfying [R]-boundary conditions (see (33.8)). This is reminiscent of the single-pole approximation for excitation energies of bound $\rightarrow$ bound transitions, (1.31). Many other possibilities spring to mind for approximate solutions to (33.9).

33.3.3 A Trivial Example, $N = 0$

The method outlined above is valid for any number of particles. In particular, for the trivial case of $N = 0$ corresponding to *potential scattering*. Consider an electron scattering from a negative delta-function of strength Z in one dimension [Fig. 33.2]. The transmission amplitude as a function of ε is given by (see Sect. 2.5 of [Griffiths 1995]):

$$t(\varepsilon) = \frac{\mathrm{i}k}{Z + \mathrm{i}k} \quad ; \quad k = \sqrt{2\varepsilon}\,. \tag{33.18}$$

How would TDDFT get this answer?: (i) Find the ground-state KS potential of the $(N+1) = 1-$electron system. The external potential admits one bound

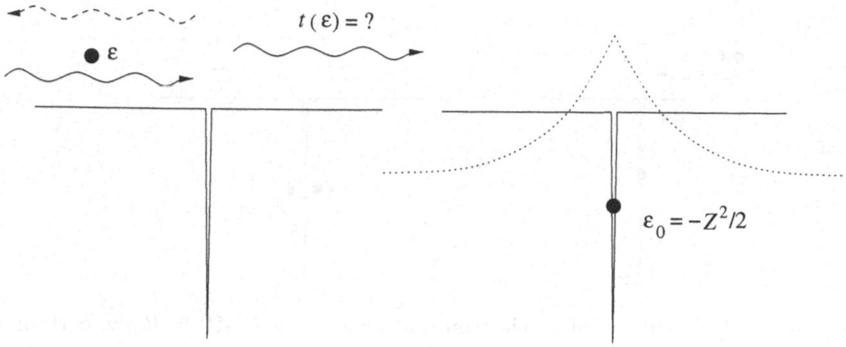

Fig. 33.2. *Left*: cartoon of an electron scattering from a negative delta-function potential. *Right*: cartoon of an electron bound to the same potential; the ground-state density decays exponentially, just as in hydrogenic ions in 3D

state of energy $-Z^2/2$. The ground-state KS potential is given by $v_{\rm KS}(x) = v_{\rm ext}(x) + v_{\rm Hxc}(x)$, but $v_{\rm Hxc} = 0$ for one electron, so $v_{\rm KS}(x) = v_{\rm ext}(x) = -Z\delta(x)$; (ii) Solve the ground-state KS equations for positive energies, to find $t_{\rm KS}(\varepsilon) = ik/(Z + ik)$. (iii) In this case, $f_{\rm Hxc} = 0$, so $\chi = \chi_{\rm KS}$, and $t = t_{\rm KS}$. Notice that approximations that are not self-interaction corrected (to guarantee $v_{\rm Hxc} = 0$) would give sizable errors in this simple case.

33.3.4 A Non-Trivial Example, $N = 1$

Now consider a simple 1D model of an electron scattering from a one-electron atom of nuclear charge Z (Fig. 33.3) [Rosenthal 1971]:

$$\hat{H} = -\frac{1}{2}\frac{\mathrm{d}^2}{\mathrm{d}x_1^2} - \frac{1}{2}\frac{\mathrm{d}^2}{\mathrm{d}x_2^2} - Z\delta(x_1) - Z\delta(x_2) + \lambda\delta(x_1 - x_2)\,, \tag{33.19}$$

The two electrons interact via a delta-function repulsion, scaled by λ. With $\lambda = 0$ the ground state density is a simple exponential, analogous to hydrogenic atoms in 3D.

(i) *Exact solution in the weak interaction limit:* First, we solve for the exact transmission amplitudes to first order in λ using the static exchange method [Bransden 1983]. The total energy must be stationary with respect to variations of both the bound (ϕ_b) and scattering (ϕ_s) orbitals that form the spatial part of the Slater determinant: $\left[\phi_b(x_1)\phi_s(x_2) \pm \phi_b(x_2)\phi_s(x_1)\right]/\sqrt{2}$, where the upper sign corresponds to the singlet, and the lower sign to the triplet case. The static-exchange equations are:

$$\left[-\frac{1}{2}\frac{\mathrm{d}^2}{\mathrm{d}x^2} + \gamma|\phi_{s,b}(x)|^2 - Z\delta(x)\right]\phi_{b,s}(x) = \mu_{b,s}\phi_{b,s}(x)\,, \tag{33.20}$$

where $\gamma = 2\lambda$ for the singlet, and 0 for the triplet. Thus the triplet transmission amplitude is that of a simple δ-function, (33.18). This can be understood

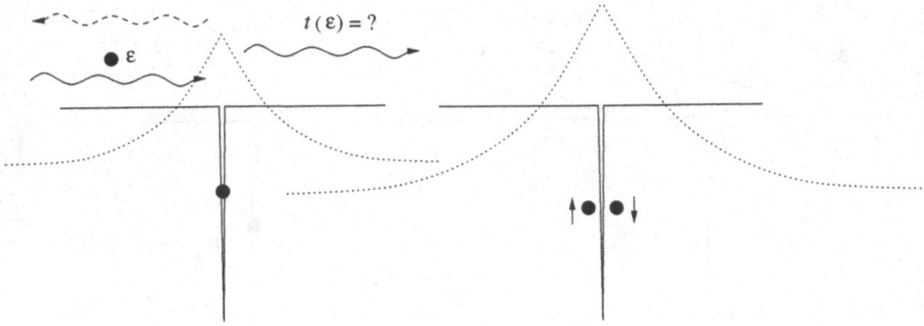

Fig. 33.3. *Left*: cartoon of an electron scattering from 1D-He$^+$. *Right*: cartoon of two electrons bound to the delta function in a singlet state

by noting that in the triplet state, the Hartree term exactly cancels the exchange (the two electrons only interact when they are at the same place, but they cannot be at the same place when they have the same spin, from Pauli's principle). The results for triplet (t_{triplet}) and singlet (t_{singet}) scattering are therefore:

$$t_{\text{triplet}} = t_0 \quad , \quad t_0 \equiv \frac{ik}{Z + ik} \tag{33.21a}$$

$$t_{\text{singlet}} = t_0 + 2\lambda t_1 \quad , \quad t_1 \equiv \frac{-ik^2}{(k - iZ)^2(k + iZ)} \tag{33.21b}$$

(ii) *TDDFT solution:* We now show, step by step, the TDDFT procedure yielding the same result, (33.21). The first step is finding the ground-state KS potential for two electrons *bound* by the δ-function. The ground-state of the $(N+1)$-electron system ($N = 1$) is given to $\mathcal{O}(\lambda)$ by:

$$\Psi_{\text{GS}}(x_1\sigma_1, x_2\sigma_2) = \frac{1}{\sqrt{2}}\varphi_{\text{GS}}(x_1)\varphi_{\text{GS}}(x_2)\left[\delta_{\sigma_1\uparrow}\delta_{\sigma_2\downarrow} - \delta_{\sigma_1\downarrow}\delta_{\sigma_2\uparrow}\right] , \tag{33.22}$$

where the orbital $\varphi_{\text{GS}}(x)$ satisfies [Lieb 1992, Magyar 2004b]:

$$\left[-\frac{1}{2}\frac{\mathrm{d}^2}{\mathrm{d}x^2} - Z\delta(x) + \lambda|\varphi_{\text{GS}}(x)|^2\right]\varphi_{\text{GS}}(x) = \mu\varphi_{\text{GS}}(x) \tag{33.23}$$

To first order in λ,

$$\varphi_{\text{GS}}(x) = \sqrt{Z}\mathrm{e}^{-Z|x|} + \frac{\lambda}{8\sqrt{Z}}\left\{2\mathrm{e}^{-3Z|x|} + \mathrm{e}^{-Z|x|}(4Z|x| - 3)\right\} . \tag{33.24}$$

The bare KS transmission amplitudes $t_{\text{KS}}(\varepsilon)$ characterize the asymptotic behavior of the continuum states of $v_{\text{KS}}(x) = -Z\delta(x) + \lambda|\varphi_{\text{GS}}(x)|^2$, and can be obtained to $\mathcal{O}(\lambda)$ by a distorted-wave Born approximation (see, e.g., Sect. 4.1.4 of [Friedrich 1991]):

$$t_{\text{KS}} = t_0 + \lambda t_1 . \tag{33.25}$$

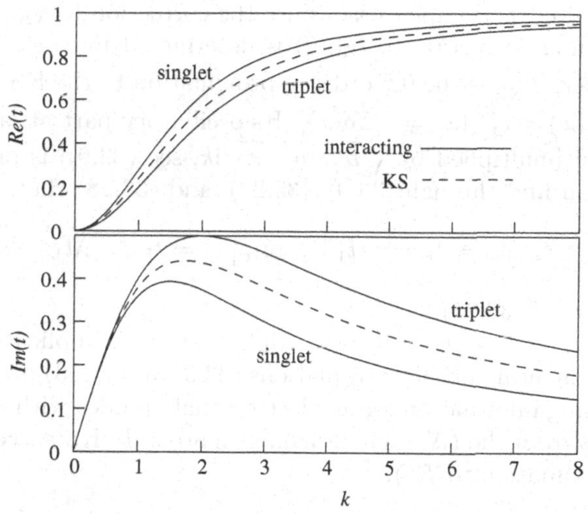

Fig. 33.4. Real and imaginary parts of the KS transmission amplitude t_{KS}, and of the interacting singlet and triplet amplitudes, for the model system of (33.19). $Z = 2$ and $\lambda = 0.5$ in this plot. Reprinted with permission from [Wasserman 2005b]. Copyright 2005, American Institute of Physics

The result is plotted in Fig. 33.4, along with the interacting singlet and triplet transmission amplitudes, (33.21). The quantity λt_1 is the *error* of the ground-state calculation. The interacting problem cannot be reduced to scattering from the $(N+1)$-KS potential, but this is certainly a good starting point; in this case, the KS transmission amplitudes are the exact average of the true singlet and triplet amplitudes [compare (33.25) with (33.21)].

We now apply (33.9) to show that the f_{Hxc}-term of (1.23) corrects the t_{KS} values to their exact singlet and triplet amplitudes. The kernel f_{Hxc} is only needed to $\mathcal{O}(\lambda)$:

$$f_{\mathrm{Hx},\,\sigma\sigma'}(x,x',\omega) = \lambda\delta(x-x')(1-\delta_{\sigma\sigma'})\,, \tag{33.26}$$

where the f_{Hxc} of (1.23) is given to $\mathcal{O}(\lambda)$ by $f_{\mathrm{Hx}} = f_{\mathrm{H}} + f_{\mathrm{x}} = \frac{1}{4}\sum_{\sigma\sigma'} f_{\mathrm{Hx},\,\sigma\sigma'}$ $(= \frac{1}{2} f_{\mathrm{H}}$ here). Equation (33.26) yields:

$$\chi(x,x',\omega) = \chi_{\mathrm{KS}}(x,x',\omega) + \frac{\lambda}{2}\int \mathrm{d}x''\, \chi_{\mathrm{KS}}(x,x'',\omega)\chi(x'',x',\omega)\,. \tag{33.27}$$

Since the ground state of the 2-electron system is a spin-singlet, the Kronecker delta $\delta_{S_{\mathrm{GS}},S_i}$ in (33.6) implies that only *singlet* scattering information may be extracted from χ, whereas information about triplet scattering requires the magnetic susceptibility $\mathcal{M} = \sum_{\sigma\sigma'}(\sigma\sigma')\chi_{\sigma\sigma'}$, related to the KS susceptibility by spin-TDDFT [Petersilka 1996b]:

$$\mathcal{M}(x,x',\omega) = \chi_{\mathrm{KS}}(x,x',\omega) - \frac{\lambda}{2}\int \mathrm{d}x''\, \chi_{\mathrm{KS}}(x,x'',\omega)\mathcal{M}(x'',x',\omega)\,. \tag{33.28}$$

For either singlet or triplet case, since the correction to $\chi_{\rm KS}$ is multiplied by λ, the leading correction to $t_{\rm KS}(\varepsilon)$ is determined by the same quantity, $\hat{\chi}^{(0)}_{\rm KS} * \hat{\chi}^{(0)}_{\rm KS}$, where $\hat{\chi}^{(0)}_{\rm KS}$ is the $0^{\rm th}$ order approximation to the KS susceptibility [i.e., with $v_{\rm KS}(x) = v^{(0)}_{\rm KS}(x) = -Z\delta(x)$]. Its oscillatory part at large distances [Maitra 2003b] [multiplied by $\sqrt{n(x)n(-x)}/ik$, see (33.9)] is precisely equal to λt_1. We then find through (33.9), (33.27), and (33.28) that

$$t_{\rm singlet} = t_{\rm KS} + \lambda t_1 \quad , \quad t_{\rm triplet} = t_{\rm KS} - \lambda t_1 \, , \tag{33.29}$$

in agreement with (33.21).

The method illustrated in the preceeding example is applicable to any one-dimensional scattering problem. Equations (33.9) and (1.23) provide a way to obtain scattering information for an electron that collides with an N-electron target *entirely* from the $(N+1)$-electron ground-state KS susceptibilty (and a given approximation to $f_{\rm xc}$).

33.4 Three Dimensions

33.4.1 Single-Pole Approximation in the Continuum

We have yet to prove an analog of (33.9) for Coulomb repulsion in three dimensions. But we can use quantum-defect theory [Seaton 1958] to deduce the result at zero energy. Consider the $l = 0$ Rydberg series of bound states converging to the first ionization threshold I of the $(N+1)$-electron system:

$$E_i - E_{\rm GS} = I - 1/\left[2(i-\mu_i)^2\right] \, , \tag{33.30}$$

where μ_i is the quantum defect of the ith excited state. Let

$$\varepsilon_i = -1/\left[2(i-\mu_{{\rm KS},\, i})^2\right] \tag{33.31}$$

be the KS orbital energies of that series. The true transition frequencies $\omega_i = E_i - E_{\rm GS}$, are related through TDDFT to the KS frequencies $\omega_{{\rm KS},\, i} = \varepsilon_i - \varepsilon_{\rm HOMO}$, where $\varepsilon_{\rm HOMO}$ is the HOMO energy. Within the single-pole approximation (SPA) [Petersilka 1996a], applicable to Rydberg excitations according to the criteria of applicability discussed in [Appel 2003]:

$$\omega_i = \omega_{{\rm KS},\, i} + 2\langle\langle {\rm HOMO}, i | \hat{f}_{\rm Hxc}(\omega_i) | {\rm HOMO}, i \rangle\rangle \tag{33.32}$$

Numerical studies [Al-Sharif 1998] suggest that $\Delta\mu_i = \mu_i - \mu_{{\rm KS},\, i}$ is a small number when $i \to \infty$. Expanding ω_i around $\Delta\mu_i = 0$, and using $I = -\varepsilon_{\rm HOMO}$, we find:

$$\omega_i = \omega_{{\rm KS},\, i} - \Delta\mu_i/(i-\mu_{{\rm KS},\, i})^3 \, . \tag{33.33}$$

We conclude that, within the SPA,

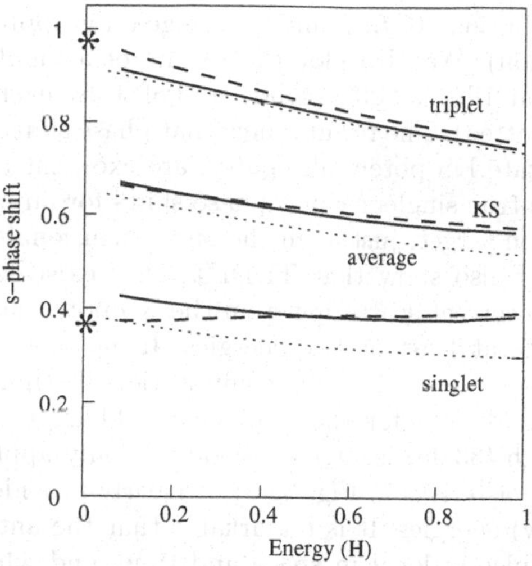

Fig. 33.5. *s*-phase shifts as a function of energy for electron scattering from He^+. *Dashed lines*: the line labeled KS corresponds to the phase shifts from the *exact* KS potential of the He atom; the other dashed lines correspond to the TDDFT singlet and triplet phase shifts calculated in the present work according to (33.35). *Solid lines*: accurate wavefunction calculations of electron-He^+ scattering from [Bhatia 2002]. The solid line in the center is the average of singlet and triplet phase shifts. *Dotted lines*: Static exchange calculations, from [Lucchese 1980]. The asterisks at zero energy correspond to extrapolating the bound → bound results of [Burke 2002]. Reprinted with permission from ref.[Wasserman 2005b]. Copyright 2005, American Institute of Physics

$$\Delta\mu_i = -2(i - \mu_{\mathrm{KS},\,i})^3 \langle\langle \mathrm{HOMO}, i|\hat{f}_{\mathrm{Hxc}}(\omega_i)|\mathrm{HOMO}, i\rangle\rangle\,. \tag{33.34}$$

Letting $i \to \infty$, Seaton's theorem $[\pi \lim_{i\to\infty} \mu_i = \delta(\varepsilon \to 0^+)]$ [Seaton 1958] implies:

$$\delta(\varepsilon) = \delta_{\mathrm{KS}}(\varepsilon) - 2\pi \langle\langle \mathrm{HOMO}, \varepsilon|\hat{f}_{\mathrm{Hxc}}(\varepsilon + I)|\mathrm{HOMO}, \varepsilon\rangle\rangle \tag{33.35}$$

a relation for the phase-shifts δ in terms of the KS phase-shifts δ_{KS} applicable when $\varepsilon \to 0^+$. The factor $(i - \mu_{\mathrm{KS},\,i})^3$ of (33.34) gets absorbed into the energy-normalization factor of the KS continuum states.

We illustrate in Fig. 33.5 the remarkable accuracy of (33.35) when applied to the case of electron scattering from He^+. For this system, an essentially exact ground-state potential for the $N = 2$ electron system is known. This was found by inverting the KS equation using the ground-state density of an extremely accurate wavefunction calculation of the He atom [Umrigar 1994]. We calculated the low-energy KS *s*-phase shifts from this potential, $\delta_{\mathrm{KS}}(\varepsilon)$ (dashed line in the center, Fig. 33.5), and then corrected these phase shifts according to (33.35) employing the BPG approximation to f_{Hxc} [Burke 2002] (which amounts to using the adiabatic local density approximation for the antiparallel contribution to $f_{\rm Hxc}$ and exchange-only approximation for the parallel contribution). We also plot the results of a highly accurate wave-function calculation [Bhatia 2002] (solid), and of static-exchange calculations [Lucchese 1980] (dotted). The results show that phase shifts from the $(N+1)$-electron ground-state KS potential, $\delta_{\rm KS}(\varepsilon)$, are excellent approximations to the average of the true singlet/triplet phase shifts for an electron scattering from the N-electron target, just as in the one-dimensional model of the previous section; they also show that TDDFT, with existing approximations, works very well to correct scattering from the KS potential to the true scattering phase shifts, at least at low energies. In fact, for the singlet phase shifts, TDDFT does better than the computationally more demanding static exchange method, and for the triplet case TDDFT does only slightly worse. Even though (33.35) is, strictly speaking, only applicable at zero energy (marked with asterisks in Fig. 33.5), it clearly provides a good description for finite (low) energies. It is remarkable that the antiparallel spin kernel, which is completely local in space and time, and whose value at each point is given by the exchange-correlation energy density of a uniform electron gas (evaluated at the ground-state density at that point), yields phase shifts for e-He^+ scattering with less than 20% error. Since a signature of density-functional methods is that with the same functional approximations, exchange-correlation effects are often better accounted for in larger systems, the present approach holds promise as a practical method for studying large targets.

33.4.2 Partial-Wave Analysis

For the case of spherically symmetric $(N+1)$-electron ground states, useful expressions can be derived for the transition matrix elements (t-matrix) in the angular momentum representation. For example, the matrix elements in the usual definition [Gonis 1992] $t_l \equiv -k^{-1}\exp[-ik\delta_l]\sin\delta_l$ are given by:

$$t_l = t_l^{\rm KS} + 4\langle\langle f_{\rm Hxc}\rangle\rangle_l \ , \tag{33.36}$$

where the $t_l^{\rm KS}$ are the Kohn-Sham t-matrix elements, and

$$\begin{aligned}\langle\langle f_{\rm Hxc}\rangle\rangle_l = \int d\boldsymbol{r}_1 \int d\boldsymbol{r}_2 \frac{\varphi_{\rm HOMO}(r_1)\phi(r_2)}{(r_1 r_2)^2} f_{\rm Hxc}(\boldsymbol{r}_1,\boldsymbol{r}_2;\varepsilon+I)\times \\ \times\, \varphi_{kl}(r_1)\phi_{kl}(r_2)\mathcal{Y}^{\hat{\boldsymbol{r}}_1\hat{\boldsymbol{r}}_2}_{l_{\rm HOMO}m_{\rm HOMO}}\mathcal{Y}^{*\hat{\boldsymbol{r}}_1\hat{\boldsymbol{r}}_2}_{l0} \ ,\end{aligned} \tag{33.37}$$

with $\mathcal{Y}^{\hat{\boldsymbol{r}}\hat{\boldsymbol{r}}'}_{lm} \equiv Y_l^m(\hat{\boldsymbol{r}})Y_l^{m*}(\hat{\boldsymbol{r}}')$. In (33.37), the φ's are *radial* Kohn-Sham orbitals regular at the origin, and the ϕ's are quasiparticle amplitudes determined by the asymptotic behavior of the interacting radial Green's function (see Sect. 2.3.2 of [Wasserman 2005a]). These are generally difficult to obtain in practice, but approximating them by the corresponding Kohn-Sham orbitals yields a simple prediction for the t-matrix elements. Furthermore, the single-pole approximation of (33.35) is obtained from (33.36) after expanding it to first order in $\delta_l - \delta_l^{\mathrm{KS}}$.

33.5 Summary and Outlook

Based on the linear response formalism of TDDFT we have discussed a new way of calculating elastic scattering amplitudes for electrons scattering from targets that can bind an extra electron. In one dimension, transmission amplitudes can be extracted from the (N+1)-electron ground-state susceptibility, as indicated by (33.9). Since the susceptibility of the interacting system is determined by the Kohn-Sham susceptibility within a given approximation to the exchange-correlation kernel, the transmission amplitudes of the interacting system can be obtained by appropriately correcting the bare Kohn-Sham scattering amplitudes. Equation (33.16), reminiscent of the single-pole approximation for bound $\rightarrow$ bound transitions, provides the simplest approximation to such a correction. A similar formula for scattering phase shifts near zero energy, (33.35), was obtained in three dimensions by applying concepts of quantum defect theory.

These constitute first steps towards the ultimate goal, which is to accurately treat bound-free correlation for low-energy electron scattering from polyatomic molecules. An obvious limitation of the present approach is that it can only be applied to targets that bind an extra electron because the starting point is always the $(N+1)$-ground-state Kohn-Sham system, which may not exist if the N-electron target is neutral, and certainly does not exist if the target is a negative ion. In addition to extending the formalism to treat such cases, there is much work yet to be done: a general proof of principle in three dimensions, testing of the accuracy of approximate ground-state KS potentials, developing and testing approximate solutions to the TDDFT Dyson-like equation, extending the formalism to inelastic scattering, etc. Thus, there is a long and winding road connecting the first steps presented here with the calculations of accurate cross sections for electron scattering from large targets when bound-free correlations are important. The present results show that this road is promising. Of course,"the road goes ever on and on..." [Baggins 1973] but this section looks worthwhile.

Acknowledgements

The origin of this book is two summer schools on TDDFT that were independently organized, one in the USA, another in Europe.

The USA summer school took place in Santa Fe, New Mexico, June 5–10, 2004. It was organized by Carsten A. Ullrich, Kieron Burke, and Giovanni Vignale, supported by a generous grant from the Petroleum Research Fund of the American Chemical Society, which covered all the operating expenses. The school was organized as a 6-day intensive course, and we were very fortunate to enjoy the hospitality of St John's college, whose conference center staff were extremely helpful and professional. The setting of the school was beautiful, in the mountains overlooking Santa Fe. As a reaction to the extremely positive feedback from students as well as lecturers to this summer school, it was decided to initiate a Gordon research conference on TDDFT. The first Gordon conference is planned for the Summer of 2007 in New England, organized by Carsten A. Ullrich and Kieron Burke (for further information, consult the Gordon conferences web site, http://www.grc.uri.edu/).

The European school, which lasted from August 28 to September 8, was organized in the beautiful city of Benasque, Spain, in the middle of the highest peaks of the Pyrenees. The school lasted 10 days, and was followed by a 4 day International Workshop on TDDFT that gathered most of the world experts in this field. These two events were organized by Miguel A. L. Marques, Fernando Nogueira, Angel Rubio, and Eberhard K. U. Gross, with the invaluable help of the Benasque Center for Science, and were sponsored by CECAM, the NANOQUANTA network of excellence, the Ψ_k program of the European Science Foundation, the Spanish Ministery of Science, the National Science Foundation, and the University of the Basque Country. Due to the success of this event and the extremely positive input from the students, it was decided to repeat it on a regular basis (every second year), keeping the same format. Therefore, the next Benasque Summer Schools and International Workshops will take place in 2006 and 2008. Generous support from different national and European sources have been already awarded. Further information and updates can be found in the web page of the Benasque Center for Science (http://benasque.ecm.ub.es/).

Finally, we would like to warmly thank all contributors to this volume, and of course, the participants and teachers of the two previous schools on TDDFT.

References

Abbott, L., S. N. Batchelor, J. Oakes, J. R. L. Smith, and J. N. Moore, 2004, "Semi-empirical and Ab Initio Studies of the Structure and Spectroscopy of the Azo Dye Direct Blue 1: Comparison with Experiment", *J. Phys. Chem. A*, vol. 108, pp. 10208–10218.

Abrikosov, A. A., L. P. Gorkov, and E. Dzyaloshinskii, 1975, *Methods of Quantum Field Theory in Statistical Physics* (Dover, New York).

Adamo, C., and V. Barone, 1999, "Accurate excitation energies from time-dependent density functional theory: assessing the PBE0 model for organic free radicals", *Chem. Phys. Lett.*, vol. 314, pp. 152–157.

Adamo, C., and V. Barone, 2000, "A TDDFT study of the electronic spectrum of s-tetrazine in the gas-phase and in aqueous solution", *Chem. Phys. Lett.*, vol. 330, pp. 152–160.

Adragna, G., R. del Sole, and A. Marini, 2003, "Ab initio calculation of the exchange-correlation kernel in extended systems", *Phys. Rev. B*, vol. 68, pp. 165108-1–5.

Agostini, P., F. Fabre, G. Mainfray, G. Petite, and N. K. Rahman, 1979 "Free-Free Transitions Following Six-Photon Ionization of Xenon Atoms", *Phys. Rev. Lett.*, vol. 42, pp. 1127–1130.

Ahlrichs, R., M. Bär, M. Häser, H. Horn, and C. Kölmel, 1989, "Electronic structure calculations on workstation computers: The program system Turbomole", Chem. Phys. Lett., vol. 162, pp. 165–169.

Al-Sharif, A. I., R. Resta, and C. J. Umrigar, 1998, "Evidence of physical reality in the Kohn-Sham potential: The case of atomic Ne", *Phys. Rev. A*, vol. 57, pp. 2466–2469.

Albrecht, S., L. Reining, R. D. Sole, and G. Onida, 1998, "Ab Initio Calculation of Excitonic Effects in the Optical Spectra of Semiconductors", *Phys. Rev. Lett.*, vol. 80, pp. 4510–4513.

Allen, S. J., D. C. Tsui, and B. Vinter, 1976, "On the absorption of infrared radiation by electrons in semiconductor inversion layers", *Solid State Commun.*, vol. 20, pp. 425–428.

Allen, M. J., and D. J. Tozer, 2002, "Helium dimer dispersion forces and correlation potentials in density functional theory", *J. Chem. Phys.*, vol. 117, pp. 11113–11120.

Almbladh, C.-O., and U. von Barth, 1985, "Exact results for the charge and spin densities, exchange-correlation potentials, and density-functional eigenvalues", *Phys. Rev. B*, vol. 31, pp. 3231–3244.

Almbladh, C. O., U. von Barth, and R. van Leeuwen, 1999, "Variational Total Energies from ϕ and ψ- Derivable Theories", *Int. J. Mod. Phys. B*, vol. 13, pp. 535–542.

Almlöf, J., K. Faegri Jr., and K. Korsell, 1982, "Principles for a direct SCF approach to LCAO-MO *ab initio* calculations", *J. Comput. Chem.*, vol. 3, pp. 385–399.

Alnaser, A. S., S. Voss, X.-M. Tong, C. M. Maharjan, P. Ranitovic, B. Ulrich, T. Osipov, B. Shan, Z. Chang, and C. L. Cocke, 2004a, "Effects Of Molecular Structure on Ion Disintegration Patterns In Ionization of O_2 and N_2 by Short Laser Pulses", *Phys. Rev. Lett.*, vol. 93, pp. 113003-1–4.

Alnaser, A. S., X. M. Tong, T. Osipov, S. Voss, C. M. Maharjan, P. Ranitovic, B. Ulrich, B. Shan, Z. Chang, C. D. Lin, and C. L. Cocke, 2004b, "Routes to Control of H_2 Coulomb Explosion in Few-Cycle Laser Pulses", *Phys. Rev. Lett.*, vol. 93, pp. 183202-1–4.

Andersson, Y., D. C. Langreth, and B. I. Lundqvist, 1996, "van der Waals Interactions in Density-Functional Theory", *Phys. Rev. Lett.*, vol. 76, pp. 102–105.

Ando, T., 1977a, "Inter-subband optical-absorption in space-charge layers on semiconductor surfaces", *Z. Phys. B*, vol. 26, pp. 263–272.

Ando, T., 1977b, "Inter-subband optical transitions in a surface space-charge layer", *Solid State Commun.*, vol. 21, pp. 133–136.

Ando, K., and M. Santer, 2003, "Mixed quantum-classical Liouville molecular dynamics without momentum jump", *J. Chem. Phys.*, vol. 118, pp. 10399–10406.

Andrae, K., P.-G. Reinhard, and E. Suraud, 2002, "Theoretical exploration of pump and probe in medium-sized Na clusters", *J. Phys. B: At. Mol. Opt. Phys.*, vol. 35, pp. 4203–4210.

Andrae, K., P.-G. Reinhard, and E. Suraud, 2004, "Crossed Beam Pump and Probe Dynamics in Metal Clusters", *Phys. Rev. Lett.*, vol. 92, pp. 173402-1–4.

Andreoni, W., 1990, "The Car-Parrinello method and its application to microclusters", in *The Chemical Physics of Atomic and Molecular Clusters*, edited by G. Scoles, pp. 159–168 (North-Holland, Amsterdam).

Andreu, R., Garín, and J. Orduna, 2001, "Electronic absorption spectra of closed and open-shell tetrathiafulvalenes: the first time-dependent density-functional study", *Tetrahedron*, vol. 57, pp. 7883–7892.

Andruniow, T., M. Pawlikowski, and M. Z. Zgierski, 2000, "Density Functional Study of Absorption and Resonance Raman Spectra of Pyromellitic Diahydride (PMDA) Anion", *J. Phys. Chem. A*, vol. 104, pp. 845–851.

Andruniow, T., P. M. Kozlowski, and M. Z. Zgierski, 2001, "Theoretical analysis of electronic absorption spectra of vitamin B_{12} models", *J. Chem. Phys.*, vol. 115, pp. 7522–7533.

Angyan, J. G., I. C. Gerber, A. Savin, J. Toulouse, 2005, "van der Waals forces in density functional theory: perturbational long-range electron interaction corrections", *Phys. Rev. A*, vol. 72, pp. 012510-1–9.

Anni, M., F. Della Sala, M. Raganato, E. Fabiano, S. Lattante, R. Cingolani, G. Gigli, G. Barbarella, L. Favaretto, and A. Görling, 2005, "Nonradiative Relaxation in Thiophene-S,S-dioxide Derivatives: The Role of the Environment", *J. Phys. Chem. B*, vol. 109, pp. 6004–6011.

Appel, H., E. K. U. Gross, and K. Burke, 2003, "Excitations in Time-Dependent Density-Functional Theory", *Phys. Rev. Lett.*, vol. 90, pp. 043005-1–4.

Aquino, A. J. A., H. Lischka, and C. Hättig, 2005, "Excited-state intramolecular proton transfer: A survey of TDDFT and RI-CC2 excited-state potential energy surface", *J. Phys. Chem. A*, vol. 109, pp. 3201–3208.

Arenal, R., O. Stephan, M. Kociak, D. Taverna, A. Loiseau, and C. Colliex, 2005, "Electron Energy Loss Spectroscopy Measurement of the Optical Gaps on

Individual Boron Nitride Single-Walled and Multiwalled Nanotubes", *Phys. Rev. Lett.*, vol. 95, pp. 127601-1–4.

Arnaud, B., and M. Alouani, 2001, "Local-field and excitonic effects in the calculated optical properties of semiconductors from first-principles", *Phys. Rev. B*, vol. 63, pp. 085208-1–14.

Aryasetiawan, F., and O. Gunnarsson, 1998, "The *GW* method", *Rep. Prog. Phys.*, vol. 61, pp. 237–312.

Aryasetiawan, F., T. Miyake, and K. Terakura, 2002, "Total Energy Method from Many-Body Formulation", *Phys. Rev. Lett.*, vol. 88, pp. 166401-1–4.

Asbury, J. B., E. Hao, Y. Wang, H. N. Ghosh, and T. Lian, 2001, "Ultrafast Electron Transfer Dynamics from Molecular Adsorbates to Semiconductor Nanocrystalline Thin Films", *J. Phys. Chem. B*, vol. 105, pp. 4545–4557.

Asgari, R., M. Polini, B. Davoudi, and M. P. Tosi, 2003, "Correlation energy of a two-dimensional electron gas from static and dynamic exchange-correlation kernels" *Phys. Rev. B*, vol. 68, pp. 235116-1–6.

Ashcroft, N. W., and N. D. Mermin, 1976, *Solid State Physics* (Saunders College Publishing, New York).

Ashoori, R. C., 1996, "Electrons in artificial atoms", *Nature* (London), vol. 379, pp. 413–419.

Atkins, P. W., and R. S. Friedman, 1997, *Molecular Quantum Mechanics*, 3rd edition (Oxford University Press, New York).

Aulbur, W. G., L. Jönsson, and J. W. Wilkins, 2000, "Quasiparticle calculations in solids", *Solid State Physics*, vol. 54, 1–218.

Autschbach, J., F. R. Jorge, and T. Ziegler, 2003, "Density functional calculations on electronic circular dichroism spectra of chiral cobalt(III) complexes", *Inorg. Chem.*, vol. 42, pp. 2867–2877.

Badaeva, E. A., T. V. Timofeeva, A. M. Masunov, and S. Tretiak, 2005, "Role of donor-acceptor strengths and separation on the two-photon absorption response of cytotoxic dyes: A TD-DFT study", *J. Phys. Chem. A*, vol. 109, pp. 7276–7284.

Baer, R., and R. Gould, 2001, "A method for *ab initio* nonlinear electron-density evolution", *J. Chem. Phys.*, vol. 114, pp. 3385–3392.

Baer, R., D. Neuhauser, P. R. Ždánská, and N. Moiseyev, 2003, "Ionization and high-order harmonic generation in aligned benzene by a short intense circularly polarized laser pulse", *Phys. Rev. A*, vol. 68, pp. 043406-1–8.

Baer, R., T. Seideman, S. Ilani, and D. Neuhauser, 2004, "*Ab initio study* of the alternating current impedance of a molecular junction", *J. Chem. Phys.*, vol. 120, pp. 3387–3396.

Baer, R., Y. Kurzweil, and L. S. Cederbaum, 2005, "Time-dependent density functional theory for nonadiabatic processes", *Isr. J. Chem.*, vol. 45, pp. 161–170.

Baerends, E. J., D. E. Ellis, and P. Ros, 1973, "Self-consistent molecular Hartree-Fock-Slater calculations I. The computational procedure", *Chem. Phys.*, vol. 2, pp. 41–51.

Baerends, E. J., G. Ricciardi, A. Rosa, S. J. A. van Gisbergen, 2002, "A DFT/TDDFT Interpretetation of the Ground and Excited States of Porphyrin and Porphyrazine Complexes", *Coord. Chem. Rev.*, vol. 230, pp. 5–27.

Baev, A., O. Rubio-Pons, F. Gel'mukhanov, and H. Ågren, 2004, "Optical Limiting Properties of Zinc- and Platinum-Based Organometallic Compounds", *J. Phys. Chem. A*, vol. 108, pp. 7406–7416.

Baggins, B., 1973, *There and Back Again* (Ballantine Books, New York).

Baguenard, B., J. C. Pinar, C. Bordas, and M. Broyer, 2001, "Photoelectron imaging spectroscopy of small tungsten clusters: Direct observation of thermionic emission", *Phys. Rev. A*, vol. 63, pp. 023204-1–13.

Baierle, R. J., M. J. Caldas, E. Molinari, and S. Ossicini, 1997, "Inter-subband optical transitions in a surface space-charge layer", *Solid State Commun.*, vol. 102, pp. 545–549.

Bandrauk, A. D., and H. Shen, 1993, "Exponential split operator methods for solving coupled time-dependent Schrödinger equations" *J. Chem. Phys.*, vol. 99, pp. 1185–1193.

Bandrauk, A. D., 1994a, *Molecules in Laser Fields* (M. Dekker, New York).

Bandrauk, A. D., 1994b, "Molecular multiphoton transitions. Computational spectroscopy for perturbative and non-perturbative regimes", *Int. Rev. Phys. Chem.*, vol. 13, pp. 123–162.

Bandrauk, A. D., and H. Shen, 1994c, "Exponential operator methods for coupled time-dependent nonlinear Schroedinger equations.", *J. Phys. A: Math. Gen.*, vol. 27, pp. 7147–7155.

Bandrauk, A. D., and H. Yu, 1998, "High-order harmonic generation at long range in intense laser pulses", *J. Phys. B: At. Mol. Opt. Phys.*, vol. 31, pp. 4243–4255.

Bandrauk, A. D. and H. Yu, 1999, "High-order harmonic generation by one- and two-electron molecular ions with intense laser pulses", *Phys. Rev. A*, vol. 59, pp. 539–548.

Bandrauk, A. D., and S. Chelkowski, 2001, "Dynamic Imaging of Nuclear Wave Functions with Ultrashort UV Laser Pulses", *Phys. Rev. Lett.*, vol. 87, pp. 273004-1–4.

Bandrauk, A. D., and H. Kono, 2003a, "Molecules in intense laser fields: nonlinear multiphoton spectroscopy and near-femtosecond to sub-femtosecond (attosecond) dynamics", in *Advances in Multiphoton Processes and Spectroscopy* vol. 15, edited by S. H. Lin, A. A. Villaeys, and Y. Fujimura, pp. 147–214 (World Scientific, Singapore).

Bandrauk, A. D., S. Chelkowski, and I. Kawata, 2003b, "Molecular above-threshold-ionization spectra: The effect of moving nuclei", *Phys. Rev. A*, vol. 67, pp. 013407-1–13.

Bandrauk, A. D., and H. Z. Lu, 2005a, "Laser-induced electron recollision in H_2 and electron correlation", *Phys. Rev. A*, vol. 72, pp. 023408-1–7.

Bandrauk, A. D., and H. Z. Lu, 2005b, "Electron correlation and double ionization of a 1D H_2 in an intense laser field", to appear in *J. Mod. Opt.*.

Bark, T., A. von Zelewsky, D. Rappoport, M. Neuburger, S. Schaffner, J. Lacour, and J. Jodry, 2004, "Synthesis and Stereochemical Properties of Chiral Square Complexes of Iron(II)", *Chem.-Eur. J.*, vol. 10, pp. 4839–4845.

Barker, D., and M. Sprik, 2003, "Molecular dynamics study of electron gas models for liquid water", *Mol. Phys.*, vol. 101, pp. 1183–1198.

Barone, V., A. Palma, and N. Sanna, 2003, "Toward a reliable computational support to the spectroscopic characterization of excited state intramolecular proton transfer: [2,2'-bipyridine]-3,3'-diol as a test case", *Chem. Phys. Lett.*, vol. 381, pp. 451–457.

Baroni, S., P. Giannozzi, and A. Testa, 1987, "Green's-function approach to linear response in solids", *Phys. Rev. Lett.*, vol. 58, pp. 1861–1864.

Baroni, S., S. de Gironcoli, A. Dal Corso, and P. Giannozzi, 2001, "Phonons and related crystal properties from density-functional perturbation theory", *Rev. Mod. Phys.*, vol. 73, pp. 515–562.

Barranco, M., L. Colletti, E. Lipparini, A. Emperador, M. Pi, and L. Serra, 2000, "Wave-vector dependence of spin and density multipole excitations in quantum dots", *Phys. Rev. B*, vol. 61, pp. 8289–8297.

Bartholomew, G., M. Rumi, S. Pond, J. Perry, S. Tretiak, and G. Bazan, 2004, "Two-Photon Absorption in Three-Dimensional Chromophores Based on [2.2]-Paracyclophane", *J. Am. Chem. Soc.*, vol. 126, pp. 11529–11542.

Bartkowiak, W., and P. Lipkowski, 2005, "Hydrogen-bond effects on the electronic absorption spectrum and evaluation of nonlinear optical properties of an aminobenzodifuranone derivative that exhibits the largest positive solvatochromism", *J. Mol. Model.*, vol. 11, pp. 317–322.

Bartolotti, L. J., and R. G. Parr, 1980, "The concept of pressure in density functional theory", *J. Chem. Phys.*, vol. 72, pp. 1593–1596.

Bashkin, S., and J. O. Stoner Jr., 1975, *Atomic Energy-Level and Grotrian Diagrams*, vol. I and II (North-Holland, Amsterdam).

Bastard, G., 1988, *Wave Mechanics applied to Semiconductor Heterostructures* (Halsted Press, New York).

Batista, E., and R. Martin, 2005, "On the Excited States Involved in the Luminescent Probe $[Ru(bpy)_2dppz]^{2+}$", *J. Phys. Chem. A*, vol. 109, pp. 3128–3133.

Bauer, D., 1997, "Two-dimensional, two-electron model atom in a laser pulse: Exact treatment, single-active-electron analysis, time-dependent density-functional theory, classical calculations, and nonsequential ionization", *Phys. Rev. A*, vol. 56, pp. 3028–3039.

Bauer, D., and F. Ceccherini, 2001, "Time-dependent density functional theory applied to nonsequential multiple ionization of Ne at 800 nm", *Optics Express*, vol. 8, pp. 377–382.

Bauernschmitt, R., and R. Ahlrichs, 1996a, "Treatment of electronic excitations within the adiabatic approximation of time-dependent density functional theory", *Chem. Phys. Lett*, vol. 256, pp. 454–464.

Bauernschmitt, R., and R. Ahlrichs, 1996b, "Stability analysis for solutions of the closed shell Kohn-Sham equation", *J. Chem. Phys.*, vol. 104, pp. 9047–9052.

Bauernschmitt, R., and R. Ahlrichs, 1997, "Calculation of excitation energies within time dependent density functional theory using auxiliary basis set expansions", *Chem. Phys. Lett.*, vol. 264, pp. 573–578.

Bauernschmitt, R., R. Ahlrichs, F. H. Hennrich, and M. M. Kappes, 1998, "Experiment versus Time Dependent Density Functional Theory Prediction of Fullerene Electronic Absorption", *J. Am. Chem. Soc.*, vol. 120, pp. 5052–5059.

Bauschlicher, C. W., 2005, "Time-dependent density functional theory for polycyclic aromatic hydrocarbon anions: What is the best approach", *Chem. Phys. Lett.*, vol. 409, pp. 235–239.

Bawagan, A. O., C. E. Brion, E. R. Davidson, C. Boyle, and R. F. Frey, 1988, "The valence orbital momentum distributions and binding energy spectra of H_2CO: A comparison of electron momentum spectroscopy and quantum chemical calculations using near-Hartree-Fock quality and correlated wavefunctions", *Chem. Phys.*, vol. 128, 439–455.

Baym, G., and L. P. Kadanoff, 1961, "Conservation Laws and Correlation Functions", *Phys. Rev.*, vol. 124, pp. 287–299.

Baym, G., 1962, "Self-Consistent Approximations in Many-Body Systems", *Phys. Rev.*, vol. 127, pp. 1391–1401.

Beck, D. E., 1984, "Self-consistent calculation of the electronic structure of small jellium spheres" *Sol. St. Comm.*, vol. 49, pp. 381–385.

Beck, T. L., 2000, "Real-space mesh techniques in density-functional theory", *Rev. Mod. Phys.*, vol. 72, pp. 1041–1080.

Becke, A. D., 1988a, "Correlation energy of an inhomogeneous electron gas: A coordinate-space model", *J. Chem. Phys.*, vol. 88, pp. 1053–1062.

Becke, A. D., 1988b, "Density-functional exchange-energy approximation with correct asymptotic behavior", *Phys. Rev. A*, vol. 38, pp. 3098–3100.

Becke, A. D., 1993a, "A new mixing of Hartree-Fock and local density-functional theories", *J. Chem. Phys.*, vol. 98, pp. 1372–1377.

Becke, A. D., 1993b, "Density-functional thermochemistry. III. The role of exact exchange", *J. Chem. Phys.*, vol. 98, pp. 5648–5652.

Becker, A., and F. H. M. Faisal, 1996, "Mechanism of laser-induced double ionization of helium", *J. Phys. B: At. Mol. Opt. Phys.*, vol. 29, pp. L197–L202.

Becker, W., F. Grasbon, R. Kopold, D. B. Milošević, G. G. Paulus, and H. Walther, 2002, "Above-threshold ionization: from classical features to quantum effects", *Adv. Atom. Mol. Opt. Phys.*, vol. 48, pp. 35–98.

Becker, A., and F. H. M. Faisal, 2005, "Intense-field many-body *S*-matrix theory", *J. Phys. B: At. Mol. Opt. Phys.*, vol. 38, pp. R1–R56.

Beenken, W. L. D., and H. Lischka, 2005, "Spectral broadening and diffusion by torsional motion in biphenyl", *J. Chem. Phys.*, vol. 123, pp. 144311-1–9.

Belelli, P., D. Damiani, N. Castellani, 2005, "DFT theoretical studies of UV-Vis spectra and solvent effects in olefin polymerization catalysts", *Chem. Phys. Lett.*, vol. 401, pp. 515–521.

Benedict, L. X., E. L. Shirley, and R. B. Bohn, 1998a, "Optical Absorption of Insulators and the Electron-Hole Interaction: An *Ab Initio* Calculation", *Phys. Rev. Lett.*, vol. 80, pp. 4514–4517.

Benedict, L. X., E. L. Shirley, and R. B. Bohn, 1998b, "Theory of optical absorption in diamond, Si, Ge, and GaAs", *Phys. Rev. B*, vol. 57, pp. R9385–R9387.

Benedict, L. X., N. G. Chopra, M. L. Cohen, A. Zettl, S. G. Louie, and V. H. Crespi, 1998c, "Microscopic determination of the interlayer binding energy in graphite", *Chem. Phys. Lett.*, vol. 286, pp. 490–496.

Benedict, L. X., A. Puzder, A. J. Williamson, J. C. Grossman, G. Galli, J. E. Klepeis, J.-Y. Raty, and O. Pankratov, 2003, "Calculation of optical absorption spectra of hydrogenated Si clusters: Bethe-Salpeter equation versus time-dependent local-density approximation", *Phys. Rev. B*, vol. 68, pp. 085310-1–8.

Berger, J. A., P. L. de Boeij, and R. van Leeuwen, 2005, "Analysis of the viscoelastic coefficients in the Vignale-Kohn functional: The cases of one- and three-dimensional polyacetylene", *Phys. Rev. B*, vol. 71, pp. 155104-1–13.

Berger, J. A., P. L. de Boeij, and R. van Leeuwen, 2006, "Analysis of the Vignale-Kohn current functional: the dielectric function of silicon", *to be published.*

Berman, O., and S. Mukamel, 2003, "Quasiparticle density-matrix representation of nonlinear time-dependent density-functional response functions", *Phys. Rev. A*, vol. 67, pp. 042503-1–13.

Bernu, B., D. M. Ceperley, and W. A. Lester Jr., 1990, "The calculation of excited states with quantum Monte Carlo. II. Vibrational excited states", *J. Chem. Phys.*, vol. 93, pp. 552–561; *ibid*, vol. 95, p. 7782(E) (1991).

Bertsch, G. F., and S. Das Gupta, 1988, "A guide to microscopic models for intermediate energy heavy ion collisions", *Phys. Rep.*, vol. 160, pp. 189–233.

Bertsch, G. F., A. Rubio, and K. Yabana, 2001, "Comparision of direct and Fourier space techniques in time-dependent density functional theory", physics/0003090.

Bertsch, G. F., J.-I. Iwata, A. Rubio, and K. Yabana, 2000, "Real-space, real-time method for the dielectric function", *Phys. Rev. B*, vol. 62, pp. 7998–8002.

Bhatia, A. K., 2002, "Electron-He$^+$ elastic scattering", *Phys. Rev. A*, vol. 66, pp. 064702-1–3.

Billeter, S. R., and A. Curioni, 2005, "Calculation of nonadiabatic couplings in density-functional theory", *J. Chem. Phys.*, vol. 122, pp. 034105-1–7.

Binggeli, N., J. L. Martins, and J. R. Chelikowsky, 1992, "Simulation of Si clusters via Langevin molecular dynamics with quantum forces", *Phys. Rev. Lett.*, vol. 68, pp. 2956–2959.

Blaizot, J.-P., and G. Ripka, 1986, *Quantum Theory of Finite Systems* (MIT Press, Cambridge, Massachusetts).

Blandin, A., A. Nourtier, and D. W. Hone, 1976, "Localized time-dependent perturbations in metals: formalism and simple examples", *J. Phys. (Paris)* vol. 37, pp. 369–378.

Blanes, S., F. Casas, and J. Ros, 2000, "Improved high order integrators based on Magnus expansion", *BIT*, vol. 40, pp. 434–450.

Blase, X., A. Rubio, M. L. Cohen, and S. G. Louie, 1995, "Mixed-space formalism for the dielectric response in periodic systems", *Phys. Rev. B*, vol. 52, pp. R2225–R2228.

Bloss, W. L., 1989, "Effects of Hartree, exchange, and correlation energy on intersubband transitions", *J. Appl. Phys.*, vol. 66, pp. 3639–3642.

Blum, V., G. Lauritsch, J. A. Maruhn, and P.-G. Reinhard, 1992, "Comparison of coordinate-space techniques in nuclear mean-field calculations", *J. Comput. Phys.*, vol. 100, pp. 364–376.

Bobbert, P. A., and W. van Haeringen, 1994, "Lowest-order vertex-correction contribution to the direct gap of silicon", *Phys. Rev. B*, vol. 49, pp. 10326–10331.

Böhm, H. M., S. Conti, and M. P. Tosi, 1996, "Plasmon dispersion and dynamic exchange – correlation potentials from two-pair excitations in degenerate plasmas", *J. Phys. Condens. Matter*, vol. 8, pp. 781–797.

Boisbourdain, V., R. Taïeb, and V. Véniard, 2005, unpublished.

Bonačić-Koutecký, V., P. Fantucci, and J. Koutecký, 1989, "Ab initio configuration-interaction study of the photoelectron spectra of small sodium cluster anions", *J. Chem. Phys.*, vol. 91, pp. 3794–3795.

Bonačić-Koutecký, V., P. Fantucci, and J. Koutecký, 1991, "Quantum chemistry of small clusters of elements of groups Ia, Ib, and IIa: fundamental concepts, predictions, and interpretation of experiments", *Chem. Rev.*, vol. 91, pp. 1035–1108.

Bostrom, M., and B. E. Sernelius, 2000, "Fractional van der Waals interaction between thin metallic films", *Phys. Rev. B*, vol. 61, pp. 2204–2210.

Botti, S., F. Sottile, N. Vast, V. Olevano, L. Reining, A. Rubio, G. Onida, R. Del Sole, and R. W. Godby, 2004, "Long-range contribution to the exchange-correlation kernel of time-dependent density functional theory", *Phys. Rev. B*, vol. 69, pp. 155112-1–14.

Botti, S., A. Fourreau, F. Nguyen, Y.O. Renault, F. Sottile, and L. Reining, 2005, "Energy dependence of the exchange-correlation kernel of time-dependent density functional theory: A simple model for solids", *Phys. Rev. B*, vol. 72, pp. 125201-1–9.

Boulet, P., H. Chermette, C. Daul, F. Gilardoni, F. Rogemond, J. Weber, and G. Zuber, 2001, "Absorption Spectra of Several Metal Complexes Revisited by the Time-Dependent Density-Functional Theory-Response Theory Formalism", *J. Phys. Chem. A*, vol. 105, pp. 885–894.

Brabec, T., and F. Krausz, 2000, "Intense few-cycle laser fields: Frontiers of nonlinear optics", *Rev. Mod. Phys.*, vol. 72, pp. 545–591.

Brack, M., 1999, "Multipole vibrations of small alkali-metal spheres in a semiclassical description", *Phys. Rev. B*, vol. 39, pp. 3533–3542.

Brack, M., 1993, "The physics of simple metal clusters: self-consistent jellium model and semiclassical approaches", *Rev. Mod. Phys.*, vol. 65, pp. 677–732.

Brandbyge, M., J.-L. Mozos, P. Ordejón, J. Taylor, and K. Stokbro, 2002, "Density-functional method for nonequilibrium electron transport", *Phys. Rev. B*, vol. 65, pp. 165401-1–17.

Bransden, B.H., and C.J. Joachain, 1983, *Physics of Atoms and Molecules* (Longman, New York).

Breidbach, J., and L.S. Cederbaum, 2005, "Universal Attosecond Response to the Removal of an Electron", *Phys. Rev. Lett.*, vol. 94, pp. 033901-1–4.

Brejc, K., T.K. Sixma, P.A. Kitts, S.R. Kain, R.Y. Tsien, M. Ormö, and S.J. Remington, 1997, "Structural basis for dual excitation and photoisomerization of the *Aequorea victoria* green fluorescent protein", *Proc. Natl. Acad. Sci. USA*, vol. 94, pp. 2306–2311.

Brey, L., N.F. Johnson, and B.I. Halperin, 1989, "Optical and magneto-optical absorption in parabolic quantum wells", *Phys. Rev. B*, vol. 40, pp. 10647–10649.

Briggs, E.L., D.J. Sullivan, and J. Bernholc, 1995, "Large-scale electronic-structure calculations with multigrid acceleration", *Phys. Rev. B*, vol. 52, pp. R5471–R5474.

Brocławik, E., and T. Borowski, 2001, "Time-dependent DFT study on electronic states of vanadium and molybdenum oxide molecules", *Chem. Phys. Lett.*, vol. 339, pp. 433–437.

Brooks, B.R., R.E. Bruccoleri, B.D. Olafson, D.J. States, S. Swaminathan, and M. Karplus, 1983, "CHARMM: A program for macromolecular energy, minimization, and dynamics calculations", *J. Comput. Chem.*, vol. 4, 187–217.

Brooks, A., M. Vlasie, R. Banerjee, and T. Brunold, 2004, "Spectroscopic and Computational Studies on the Adenosylcobalamin-Dependent Methylmalonyl-CoA Mutase: Evaluation of Enzymatic Contributions to Co-C Bond Activation in the Co^{3+} Ground State" *J. Am. Chem. Soc.*, vol. 126, pp. 8167–8180.

Bruna, P.J., M.R.J. Hachey, and F. Grein, 1998, "The electronic structure of the H_2CO^+ radical and higher Rydberg states of H_2CO", *Mol. Phys.*, vol. 94, pp. 917–928.

Bruneval, F., F. Sottile, V. Olevano, R. Del Sole, and L. Reining, 2005, "Many-Body Perturbation Theory Using the Density-Functional Concept: Beyond the GW Approximation", *Phys. Rev. Lett.*, vol. 94, pp. 186402-1–4.

Brunger, A.T., and M. Karplus, 1988, "Polar hydrogen positions in proteins: empirical energy function placement and neutron diffraction comparison", *Proteins*, vol. 4, pp. 148–156.

Bublitz, G., B. King, and S. Boxer, 1998, "Electronic Structure of the Chromophore in Green Fluorescent Protein (GFP)", *J. Am. Chem. Soc.*, vol. 120, pp. 9370–9371.

Bucksbaum, P. H., R. R. Freeman, M. Bashkansky, and T. J. McIlrath, 1987, "Role of the ponderomotive potential in above-threshold ionization", *J. Opt. Soc. Am. B*, vol. 4, pp. 760–764.

Buenker, R. J., S. D. Peyerimhoff, and W. Butscher, "Applicability of the multi-reference double-excitation CI (MRD-CI) method to the calculation of electronic wavefunctions and comparison with related techniques", *Mol. Phys.*, vol. 35, pp. 771–791.

Bulliard, C., M. Allan, G. Wirtz, E. Haselbach, K. A. Zachariasse, N. Detzer, and S. Grimme, 1999, "Electron Energy Loss and DFT/SCI Study of the Singlet and Triplet Excited States of Aminobenzonitriles and Benzoquinuclidines: Role of the Amino Group Twist Angle", *J. Phys. Chem. A*, vol. 103, pp. 7766–7772.

Bunker, P. R., and P. Jensen, 1998, "Molecular Symmetry and Spectroscopy" (NRC Research Press, Ottawa).

Burke, P. G., P. Francken, and C. J. Joachain, 1990, "R-matrix-Floquet theory of multiphoton processes", *Europhys. Lett.*, vol. 13, pp. 617–622.

Burke, P. G., P. Francken, and C. J. Joachain, 1991, "R-matrix-Floquet theory of multiphoton processes", *J. Phys. B: At. Mol. Opt. Phys.*, vol. 24, pp. 761–790.

Burke, P. G., 1994, "Electron atom scattering theory and calculations", *Adv. Atom. Mol. Opt. Phys.*, vol. 32, pp. 39–55.

Burke, K., J. P. Perdew, and M. Ernzerhof, 1998a, "Why semi-local functionals work: Accuracy of the on-top pair density and importance of system averaging", *J. Chem. Phys.*, vol. 109, pp. 3760–3771.

Burke, K., and E. K. U. Gross, 1998b, "A guided tour of time-dependent density-functional theory" in *Density Functionals: Theory and Applications*, edited by D. Joubert, pp. 116–146 (Springer, Berlin).

Burke, K., M. Petersilka, and E. K. U. Gross, 2002, "A hybrid functional for the exchange-correlation kernel in time-dependent density functional theory", in *Recent advances in density functional methods,* vol. III, edited by P. Fantucci and A. Bencini, pp. 67–79 (World Scientific Press).

Burke, K., J. Werschnik, and E. K. U. Gross, 2005a, "Time-dependent density functional theory: Past, present, and future", *J. Chem. Phys.*, vol. 123, pp. 062206-1–9.

Burke, K., R. Car, and R. Gebauer, 2005b, "Density Functional Theory of the Electrical Conductivity of Molecular Devices", *Phys. Rev. Lett.*, vol. 94, pp. 146803-1–4.

Burke, K., M. Koentopp, and F. Evers, 2005, "Zero-bias molecular electronics: Exchange-correlation corrections to Landauer's formula", cond-mat/0502385.

Burnett, K., V. C. Reed, and P. L. Knight, 1993, "Atoms in ultra-intense laser fields", *J. Phys. B: At. Mol. Opt. Phys.*, vol. 26, pp. 561–598.

Butriy, O., R. van Leeuwen, and P. L. de Boeij, 2005, "Linear response in multi-component density-functional theory", unpublished.

Cai, Z.-L., K. Sendt, and J. R. Reimers, 2002a, "Failure of density-functional theory and time-dependent density-functional theory for large extended π systems", *J. Chem. Phys.*, vol. 117, pp. 5543–5549.

Cai, Z.-L., and J. R. Reimers, 2002b, "The First Singlet (n,π^*) and (π,π^*) Excited States of the Hydrogen-Bonded Complex between Water and Pyridine", *J. Phys. Chem. A*, vol. 106, pp. 8769–8778.

Cai, Z.-L., and J. R. Reimers, 2005, "First Singlet (n,π^*) Excited State of Hydrogen-Bonded Complexes between Water and Pyrimidine", *J. Phys. Chem. A*, vol. 109, pp. 1576–1586.

Caliebe, W. A., J. A. Soininen, E. L. Shirley, C.-C. Kao, and K. Hämäläinen, 2001, "Dynamic Structure Factor of Diamond and LiF Measured Using Inelastic X-Ray Scattering", *Phys. Rev. Lett.*, vol. 84, pp. 3907–3910.

Callen, H. B., and T. A. Welton, 1951, "Irreversibility and Generalized Noise", *Phys. Rev.*, vol. 83, pp. 34–40.

Calvayrac, F., P.-G. Reinhard, and E. Suraud, 1995, "Nonlinear plasmon response in highly excited metallic clusters", *Phys. Rev. B*, vol. 52, pp. R17056–R17059.

Calvayrac, F., P.-G. Reinhard, and E. Suraud, 1997, "Spectral signals from electronic dynamics in sodium clusters", *Ann. Physics*, vol. 255, pp. 125–162.

Calvayrac, F., P.-G. Reinhard, and E. Suraud, 1998, "Coulomb explosion of an Na_{13} cluster in a diabatic electron-ion dynamical picture", *J. Phys. B: At. Mol. Opt. Phys.*, vol. 31, pp. 5023–5030.

Calvayrac, F., P.-G. Reinhard, E. Suraud, and C. A. Ullrich, 2000, "Nonlinear electron dynamics in metal clusters", *Phys. Rep.*, vol. 337, pp. 493–578.

Calzolari, A., N. Marzari, I. Souza, and M. B. Nardelli, 2004, "Ab initio transport properties of nanostructures from maximally localized Wannier functions", *Phys. Rev. B*, vol. 69, pp. 035108-1–10.

Campbell, E. E. B., K. Hansen, K. Hoffmann, G. Korn, M. Tchaplyguine, M. Wittmann, and I. V. Hertel, 2000, "From Above Threshold Ionization to Statistical Electron Emission: The Laser Pulse-Duration Dependence of C_{60} Photoelectron Spectra", *Phys. Rev. Lett.*, vol. 84, pp. 2128–2131.

Capitani, J. F., R. F. Nalewajski, and R. G. Parr, 2000, "Non-Born-Oppenheimer density functional theory of molecular systems", *J. Chem. Phys.*, vol. 76, pp. 568–573.

Car, R., and M. Parrinello, 1985, "Unified Approach for Molecular Dynamics and Density-Functional Theory", *Phys. Rev. Lett.*, vol. 55, pp. 2471–2471.

Casadesús, R., M. Moreno, and J. M. Lluch, 2003, "A theoretical study of the ground and first excited singlet state proton transfer reaction in isolated 7-azaindole-water complexes", *Chem. Phys.*, vol. 290, pp. 319–336.

Casado, J., L. L. Miller, K. R. Mann, T. M. Pappenfus, Y. Kanemitsu, E. Orti, P. M. Viruela, R. Pou-Amerigo, V. Hernandez, and J. T. Lopez Navarette, 2002, "Combined Spectroelectrochemical and Theoretical Study of a Vinylene-Bridged Sexithiophene Cooligomer: Analysis of the π-Electron Delocalization and of the Electronic Defects Generated upon Doping", *J. Phys. Chem. B*, vol. 106, pp. 3872–3881.

Casida, M. E., 1995a, "Time-dependent density functional response theory for molecules", in *Recent Advances in Density Functional Methods*, edited by D. E. Chong, vol. 1 of *Recent Advances in Computational Chemistry*, pp. 155–192 (World Scientific, Singapore).

Casida, M. E., 1995b, "Generalization of the optimized-effective-potential model to include electron correlation: A variational derivation of the Sham-Schlüter equation for the exact exchange-correlation potential", *Phys. Rev. A*, vol. 51, 2005–2013.

Casida, M. E., 1996, "Time-dependent density functional response theory of molecular systems: Theory, computational methods, and functionals", in *Recent Developments and Application of Modern Density Functional Theory*, edited by J. M. Seminario, pp. 391–439 (Elsevier, Amsterdam).

Casida, M. E., C. Jamorski, K. C. Casida, and D. R. Salahub, 1998a, "Molecular excitation energies to high-lying bound states from time-dependent density-functional response theory: Characterization and correction of the time-dependent local density approximation ionization threshold", *J. Chem. Phys.*, vol. 108, pp. 4439–4449.

Casida, M. E., K. C. Casida, and D. R. Salahub, 1998b, "Excited-state potential energy curves from time-dependent density-functional theory: A cross section of formaldehyde's $^1A_1$ manifold", *Int. J. Quantum Chem.*, vol. 70, pp. 933–941.

Casida, M. E., and D. R. Salahub, 2000a, "Asymptotic correction approach to improving approximate exchange-correlation potentials: Time-dependent density-functional theory calculations of molecular excitation spectra", *J. Chem. Phys.*, vol. 113, pp. 8918–8935.

Casida, M. E., F. Gutierrez, J. Guan, F.-X. Cadea, D. Salahub, and J.-P. Duadey, 2000b, "Charge-transfer correction for improved time-dependent local density approximation excited-state potential energy curves: Analysis within the two-level model with illustration for H2 and LiH", *J. Chem. Phys.*, vol. 113, pp. 7062–7071.

Casida, M. E., 2002, "Jacob's ladder for time-dependent density-functional theory: Some rungs on the way to photochemical heaven", in *Accurate Description of Low-Lying Molecular States and Potential Energy Surfaces*, edited by M. Hoffmann and K. Dyall (ACS Press: Washington, D.C., 2002), pp. 199–220.

Casida, M. E., and T. A. Wesolowski, 2004, "Generalization of the Kohn-Sham equations with constrained electron density formalism and its time-dependent response theory formulation", *Int. J. Quantum Chem.*, vol. 96, pp. 577–588.

Casida, M., 2005, "Propagator corrections to adiabatic time-dependent density-functional theory linear response theory", *J. Chem. Phys.*, vol. 122, pp. 054111-1–9.

Castro, A., M. A. L. Marques, J. A. Alonso, G. F. Bertsch, K. Yabana, and A. Rubio, 2002, "Can optical spectroscopy directly elucidate the ground state of C_{20}?", *J. Chem. Phys.*, vol. 116, pp. 1930–1934.

Castro, A., A. Rubio, and M. J. Stott, 2003, "Solution of Poisson's equation for finite systems using plane-wave methods", *Can. J. Phys.*, vol. 81, pp. 1151–1164.

Castro, A., M. A. L. Marques, and A. Rubio, 2004a, "Propagators for the time-dependent Kohn-Sham equations", *J. Chem. Phys.*, vol. 121, pp. 3425–3433.

Castro, A., M. A. L. Marques, J. A. Alonso, and A. Rubio, 2004b, "Optical Properties of Nanostructures from Time-Dependent Density-Functional Theory", *J. of Comput. Theor. Nanoscience*, vol. 1, pp. 231–255.

Catalan, J., 2003, "On the inversion of the 1 B_u and 2 A_g electronic states in α,ω-diphenylpolyenes", *J. Chem. Phys.*, vol. 119, pp. 1373–1385.

Catalan, J., J. L. G. de Paz, 2004a, "On the ordering of the first two excited electronic states in all-*trans* linear polyenes", *J. Chem. Phys.*, vol. 120, pp. 1864–1872.

Catalan, J., P. Perez, J. C. del Valle, J. L. G. de Paz, and M. Kasha, 2004b, "H-bonded N-heterocyclic base-pair phototautomerizational potential barrier and mechanism: The 7-azaindole dimer", *Proc. Natl. Acad. Sci.*, vol. 101, pp. 419–422.

Catalan, J., and J. L. G. de Paz, 2005, "Study of 7-azaindole in its first four singlet states", *J. Chem. Phys.*, vol. 122, pp. 244320-1–7.

Cave, R. J., and E. W. Castner, 2002a, "Time-Dependent Density-Functional Theory Investigation of the Ground and Excited States of Coumarins 102, 152, 153, and 343", *J. Phys. Chem. A*, vol. 106, pp. 12117–12123.

Cave, R. J., K. Burke, and E. W. Castner, 2002b, "Theoretical Investigation of the Ground and Excited States of Coumarin 151 and Coumarin 120", *J. Phys. Chem. A*, vol. 106, pp. 9294–9305.

Cave, R. J., F. Zhang, N. T. Maitra, and K. Burke, 2004, "A dressed TDDFT treatment of the 2^1A_g states of butadiene and hexatriene", *Chem. Phys. Lett.*, vol. 389, pp. 39–42.

Cavillot, V., and B. Champagne, 2002, "Time-dependent density-functional theory simulation of UV/visible absorption spectra of zirconocene catalysts", *Chem. Phys. Lett.*, vol. 354, pp. 449–457.

Cavillot, V., and B. Champagne, 2005, "Simulation of UV/visible absorption spectra of (α-diimine)nickel(II) catalysts by time-dependent density-functional theory", *Int. J. Quantum Chem.*, vol. 101, pp. 840–848.

Ceccherini, F., D. Bauer, and P. Mulser, 2000, "Electron correlation *versus* stabilization of atoms in intense laser pulses", *Laser and Particle Beams*, vol. 18, pp. 449–454.

Ceperley, D. M., and B. J. Alder, 1980, "Ground State of the Electron Gas by a Stochastic Method", *Phys. Rev. Lett.*, vol. 45, pp. 566–569.

Champagne, B., D. H. Mosley, M. Vračko, and J.-M. André, 1995, "Electron-correlation effects on the static longitudinal polarizability of polymeric chains. II. Bond-length-alternation effects", *Phys. Rev. A*, vol. 52, pp. 1039–1053.

Champagne, B., E. A. Perpète, S. J. A. van Gisbergen, E. J. Baerends, J. G. Snijders, C. Soubra-Ghaoui, K. Robins, and B. Kirtman, 1998, "Assessment of conventional density-functional schemes for computing the polarizabilities and hyperpolarizabilities of conjugated oligomers: An ab initio investigation of polyacetylene chains", *J. Chem. Phys.*, vol. 109, pp. 10489–10498.

Champagne, B., E. A. Perpète, D. Jacquemin, S. J. A. van Gisbergen, E. J. Baerends, C. Soubra-Ghaoui, K. A. Robins, and B. Kirtman, 2000, "Assessment of Conventional Density-Functional Schemes for Computing the Dipole Moment and (Hyper)polarizabilities of Push-Pull π-Conjugated Systems", *J. Phys. Chem. A*, vol. 104, pp. 4755–4763.

Chang, E. K., M. Rohlfing, and S. G. Louie, 2000, " Excitons and Optical Properties of α-Quartz" *Phys. Rev. Lett.*, vol. 85, pp. 2613–2616.

Charlier, J.-C., X. Gonze, and J.-P. Michenaud, 1994, "Graphite interplanar bonding: electronic delocalization and van der Waals interaction", *Europhys. Lett.*, vol. 28, pp. 403–408.

Chattoraj, M., B. King, and G. Bublitz, S. Boxer, 1996, "Ultra-fast excited state dynamics in green fluorescent protein: Multiple states and proton transfer", *Proc. Natl. Acad. Sci. USA*, vol. 93, pp. 8362–8367.

Chelikowsky, J. R., and M. L. Cohen, 1992, "Pseudopotentials for Semiconductors", in *Handbook of Semiconductors,* 2nd edition, edited by T. S. Moss and P. T. Landsberg, pp. 219–256 (Elsevier, Amsterdam).

Chelikowsky, J. R., N. Troullier, and Y. Saad, 1994a, "Finite-difference-pseudopotential method: Electronic structure calculations without a basis", *Phys. Rev. Lett.*, vol. 72, pp. 1240–1243.

Chelikowsky, J. R., N. Troullier, K. Wu, and Y. Saad, 1994b, "Higher-order finite-difference pseudopotential method: An application to diatomic molecules", *Phys. Rev. B*, vol. 50, pp. 11355–11364.

Chelikowsky, J. R., L. Kronik, I. Vasiliev, M. Jain, and Y. Saad, 2003a, "Using Real Space Pseudopotentials for the Electronic Structure Problem", in *Handbook of Numerical Methods, Volume X: Computational Chemistry*, edited by P. G. Ciarlct and C. Le Bris, pp. 613–638 (Elsevier, Amsterdam).

Chelikowsky, J. R., L. Kronik, and I. Vasiliev, 2003b, "Time-dependent density-functional calculations for the optical spectra of molecules, clusters, and nanocrystals", *J. Phys.: Condens. Mat.*, vol. 15, pp. R1517–R1547.

Chelkowski, S., T. Zuo, and A. D. Bandrauk, 1992, "Ionization rates of H_2^+ in an intense laser field by numerical integration of the time-dependent Schrödinger equation", *Phys. Rev. A*, vol. 46, pp. R5342–R5345.

Chelkowski, S., and A. D. Bandrauk, 1995, "Two-step Coulomb explosions of diatoms in intense laser fields", *J. Phys. B: At. Mol. Opt. Phys.*, vol. 28, pp. L723–L731.

Chen, J., J. B. Krieger, Y. Li, and G. J. Iafrate, 1996, "Kohn-Sham calculations with self-interaction-corrected local-spin-density exchange-correlation energy functional for atomic systems", *Phys. Rev. A*, vol. 54, pp. 3939–3947.

Chen, R., and H. Guo, 1999, "The Chebyshev propagator for quantum systems", *Comput. Phys. Commun.*, vol. 119, pp. 19–31.

Chen, D. M., X. Liu, T. J. He, and F. C. Liu, 2003, "Density-functional theory investigation of porphyrin diacid: electronic absorption spectrum and conformational inversion", *Chem. Phys.*, vol. 289, pp. 397–407.

Chernyak, V., N. Wang, and S. Mukamel, 1995, "Four-Wave Mixing and Luminescence of Confined Excitons in Molecular Aggregates and Nanostructures. Many-Body Green Function Approach", *Physics Rep.*, vol. 263, pp. 213–309.

Chernyak, V., and S. Mukamel, 1996, "Size-consistent quasiparticle representation of nonlinear optical susceptibilities in many-electron systems", *J. Chem. Phys.*, vol. 104, pp. 444–459.

Chernyak, V., M. F. Schulz, S. Mukamel, S. Tretiak, and E. V. Tsiper, 2000, "Krylov-space algorithms for time-dependent Hartree-Fock and density-functional computations", *J. Chem. Phys.*, vol. 113, pp. 36–43.

Chayes, J. T., L. Chayes, and M. B. Ruskai, 1985, "Density-Functional Approach to Quantum Lattice Systems", *J. Stat. Phys.*, vol. 38, pp. 497–518.

Chin, S. L., and P. A. Golovinski, 1995, "High harmonic generation in the multiphoton regime: correlation with polarizability", *J. Phys. B: At. Mol. Opt. Phys.*, vol. 28, pp. 55–63.

Choi, H. J., J. S. Ihm, S. G. Louie, M. L. Cohen, 2000, "Defects, Quasibound States, and Quantum Conductance in Metallic Carbon Nanotubes", *Phys. Rev. Lett.*, vol. 84, pp. 2917–2920.

Chong, D. P., O. V. Gritsenko, and E. J. Baerends, 2002, "Interpretation of the Kohn-Sham orbital energies as approximate vertical ionization potentials", *J. Chem. Phys.*, vol. 116, pp. 1760–1772.

Choquet-Bruhat, Y., C. DeWitt-Morette, and M. Dillard-Bleick, 1991, *Analysis, Manifolds and Physics, Part I:Basics*, Section IIC (North-Holland, Amsterdam).

Christiansen, O., H. Koch, and P. Jørgensen, 1995, "The second-order approximate coupled cluster singles and doubles model CC2", *Chem. Phys. Lett.*, vol. 243, pp. 409–418.

Christov, I. P., M. M. Murnane, and H. C. Kapteyn, 1997, "High-Harmonic generation of Attosecond pulses in the "single-cycle" regime" *Phys. Rev. Lett.*, vol. 78, pp. 1251–1254.

Chu, X., and S.-I. Chu, 2001a, "Self-interaction-free time-dependent density-functional theory for molecular processes in strong fields: High-order harmonic generation of H2 in intense laser fields", *Phys. Rev. A*, vol. 63, pp. 023411-1–10.

Chu, X., and S.-I. Chu, 2001b, "Time-dependent density-functional theory for molecular processes in strong fields: Study of multiphoton processes and dynamical response of individual valence electrons of N2 in intense laser fields", *Phys. Rev. A*, vol. 64, pp. 063404-1–9.

Chu, S.-I, and D. A. Telnov, 2004a, "Beyond the Floquet theorem: generalized Floquet formalisms and quasienergy methods for atomic and molecular multiphoton processes in intense laser fields" *Physics Rep.*, vol. 390, pp. 1–131.

Chu, X., and S.-I Chu, 2004b, "Role of the electronic structure and multielectron responses in ionization mechanisms of diatomic molecules in intense short-pulse lasers: An all-electron *ab initio* study", *Phys. Rev. A*, vol. 70, pp. 061402-1–4(R).

Cianci, M., P. J. Rizkallah, A. Olczak, J. Raftery, N. E. Chayen, P. F. Zagalsky, and J. R. Helliwell, 2002, "The molecular basis of the coloration mechanism in lobster shell: β-Crustacyanin at 3.2-Åresolution", *Proc. Natl. Acad. Sci. USA*, vol. 99, pp. 9795–9800.

Cini, M., 1980, "Time-dependent approach to electron transport through junctions: General theory and simple applications", *Phys. Rev. B*, vol. 22, pp. 5887–5899.

Ciofini, I., P. Laine, F. Bedioui, and C. Adamo, 2004, "Photoinduced Intramolecular Electron Transfer in Ruthenium and Osmium Polyads: Insights from Theory", *J. Am. Chem. Soc.*, vol. 126, pp. 10763–10777.

Čížek, J., and J. Paldus, 1967, "Stability Conditions of the Hartree-Fock Equations for Atomic and Molecular Systems. Application to the Pi-Electron Model of Cyclic Polyenes", *J. Chem. Phys.*, vol. 47, pp. 3976–3985.

Clenshaw, C. W., 1955, "A note on the summation of Chebyshev series", *MTAC*, vol. 9, pp. 118–120.

Coe, B., L. Jones, J. Harris, B. Brunschwig, I. Asselberghs, K. Clays, A. Persoons, J. Garin, and J. Orduna, 2004a, "Synthesis and Spectroscopic and Quadratic Nonlinear Optical Properties of Extended Dipolar Complexes with Ruthenium(II) Ammine Electron Donor and N-Methylpyridium Acceptor Groups.", *J. Am. Chem. Soc.*, vol. 126, pp. 3880–3882.

Coe, B., J. Harris, B. Brunschwig, J. Garin, J. Orduna, S. Coles, and M. Hursthouse, 2004b, "Contrasting Linear and Quadratic Nonlinear Optical Behavior of Dipolar Pyridinium Chromophores with 4-(Dimethylamino)phenyl or Ruthenium(II) Ammine Electron Donor Groups", *J. Am. Chem. Soc.*, vol. 126, pp. 10418–10427.

Cohen-Tannoudji, C., J. Dupont-Roc, and G. Grynberg, 1992, *Atom-photon interactions: basic processes and applications* (Wiley, New York).

Cohen, M. L., and J. R. Chelikowsky, *Electronic Structure and Optical Properties of Semiconductors*, 2nd edition (Springer-Verlag, Berlin).

Cohen, A. E., and S. Mukamel, 2003, "Resonant Enhancement and Dissipation in Nonequilibrium van der Waals Forces", *Phys. Rev. Lett.*, vol. 91, pp. 233202-1–4.

Cohen, A. E., and S. Mukamel, 2003, "A Mechanical Force Accompanies Fluorescence Resonance Energy Transfer (FRET)", *J. Phys. Chem. A*, vol. 107, pp. 3633–3638.

Cohen, A. E., and S. Mukamel, 2005, "Superoperator approach to electrodynamics of intermolecular forces and nonlinear optics", unpublished.

Coleman, A. J., and V.I. Yukalov, 2000, *Reduced Density Matrices Coulson Challenge* (Springer-Verlag, Berlin).

Colwell, S. M., N. C. Handy, and A. M. Lee, 1996, "Determination of frequency-dependent polarizabilities using current density-functional theory", *Phys. Rev. A*, vol. 53, pp. 1316–1322.

Conti, S., R. Nifosì, and M. P. Tosi, 1997, "The exchange - correlation potential for current-density-functional theory of frequency-dependent linear response", *J. Phys. Condens. Matter*, vol. 9, pp. L475–L482.

Conti, S., and G. Vignale, 1999, "Elasticity of an electron liquid", *Phys. Rev. B*, vol. 60, pp. 7966–7980.

Cordova, F., L. Joubert Doriol, A. Ipatov, M.E. Casida, and A. Vela, 2006, "How Well Do Time-Dependent Density-Functional Theory and Related Methods Describe the Woodward-Hoffmann CC Ring-Opening of Oxirane?", unpublished.

Corkum, P. B., 1993, "Plasma perspective on strong field multiphoton ionization", *Phys. Rev. Lett.*, vol. 71, pp. 1994–1997.

Corkum, P. B., C. Ellert, M. Mehendale, P. Dietrich, S. Hankin, S. Aseyev, D. Rayner, and D. Villeneuve, 1999, "Molecular science with strong laser fields", *Faraday Discuss.*, vol. 113, pp. 47–59.

Cormier, E., and P. Lambropoulos, 1997, "Above-threshold ionization spectrum of hydrogen using B-spline functions", *J. Phys. B: At. Mol. Opt. Phys.*, vol. 30, pp. 77–91.

Corradini, M., R. Del Sole, G. Onida, and M. Palummo, 1998, "Analytical expressions for the local-field factor $G(\boldsymbol{q})$ and the exchange-correlation kernel $K_{\text{xc}}(\boldsymbol{r})$ of the homogeneous electron gas", *Phys. Rev. B*, vol. 57, pp. 14569–14571.

Cossi, M., and V. Barone, 2001, "Time-dependent density-functional theory for molecules in liquid solutions", *J. Chem. Phys.*, vol. 115, pp. 4708–4717.

Cramariuc, O., T. Hukka, and T. Rantala, 2004, "Time-Dependent Density-Functional Calculations on the Electronic Absorption Spectra of an Asymmetric Meso-Substituted Porphyrin and Its Zinc Complex", *J. Phys. Chem. A*, vol. 108, pp. 9435–9441.

Creemers, T. M. H., A. J. Lock, V. Subramaniam, T. M. Jovin and S. Völker, 1999, "Three photoconvertible forms of green fluorescent protein identified by spectral hole-burning", *Nat. Struc. Biol.*, vol. 6, pp. 557–560; *ibid*, vol. 6, p. 706(E) (1999).

Creemers, T. M. H., A. J. Lock, V. Subramaniam, T. M. Jovin, and S. Völker, 2000, "Photophysics and optical switching in green fluorescent protein mutants", *Proc. Natl. Acad. Sci. USA*, vol. 97, pp. 2974–2978.

Crespo, A., A. G. Turjanski, and D. A. Estrin, 2002, "Electronic spectra of indolyl radicals: a time-dependent DFT study", *Chem. Phys. Lett.*, vol. 365, pp. 15–21.

D'Agosta, R., and G. Vignale, 2005a, "On the non-V-representability of currents in time-dependent many particle systems", *Phys. Rev. B*, vol. 71, pp. 245103-1–6.

D'Agosta, R. and G. Vignale, (2005b), "Relaxation in time-dependent density-functional theory", in press.

Dahlen, N. E., and R. van Leeuwen, 2001, "Double ionization of a two-electron system in the time-dependent extended Hartree-Fock approximation", *Phys. Rev. A*, vol. 64, pp. 023405-1–7.

Dahlen, N. E., 2002, "Effect of Electron Correlation on the Two-Particle Dynamics of a Helium Atom in a Strong Laser Pulse", *Int. J. Mod. Phys. B*, vol. 16, pp. 415–452.

Dahlen, N. E., R. van Leeuwen, and U. von Barth, 2005, "Variational energy functionals of the Green function tested on molecules", *Int. J. Quan. Chem.*, vol. 101, pp. 512–519.

Dahlen, N. E., and R. van Leeuwen, 2005b, "Self-consistent solution of the Dyson equation for atoms and molecules within a conserving approximation", *J. Chem. Phys.*, vol. 122, pp. 164102-1–8.

Dal Corso, A., F. Mauri, and A. Rubio, 1996, "Density-functional theory of the nonlinear optical susceptibility: Application to cubic semiconductors", *Phys. Rev. B*, vol. 53, pp. 15638–15642.

Dalgarno, A., and J. T. Lewis, 1955, *Proc. Roy. Soc.* (London), vol. A233, p. 70.

Daniel, C., 2004, "Electronic Spectroscopy and Photoreactivity of Transition Metal Complexes: Quantum Chemistry and Wave Packet Dynamics", in *Transition metal and rare earth compounds III*, vol. 241 of *Topics in Current Chemistry*, pp. 119–165 (Springer, Berlin).

Danielewicz, P., 1984, "Quantum theory of nonequilibrium processes, I", *Ann. Physics*, vol. 152, pp. 239–304.

Danielsson, J., J. Ulicny, and A. Laaksonen, 2001, "A TD-DFT Study of the Photochemistry of Urocanic Acid in Biologically Relevant Ionic, Rotameric, and Protomeric Forms", *J. Am. Chem. Soc.*, vol. 123, pp. 9817–9821.

Datta, A., and S. Pal, 2005, "Effects of conjugation length and donor-acceptor functionalization on the nonlinear optical properties of organic push-pull molecules using density-functional theory", *J. Mol. Struct. (Theochem)*, vol. 715, pp. 59–64.

Daul, C., 1994, "Density-functional theory applied to the excited states of coordination compounds", *Int. J. Quant. Chem.*, vol. 52, pp. 867–877.

Davies, J. H., 1998, *The Physics of Low-Dimensional Semiconductors* (Cambridge University Press, Cambridge).

Day, P., K. Nguyen, and R. Pachter, 2005, "TDDFT Study of One- and Two-Photon Absorption Properties: Donor-π-Acceptor Chromophores", *J. Phys. Chem. B*, vol. 109, pp. 1803–1814.

De Angelis, F., S. Fantacci, and A. Selloni, 2004a, "Time-dependent density-functional theory study of the absorption spectrum of [Ru(4,4'-COOH-2,2'-bpy)$_2$(NCS)$_2$] in water solution: influence of the pH", *Chem. Phys. Lett.*, vol. 389, pp. 204–208.

De Angelis, F., A. Tilocca, and A. Selloni, 2004b, "Time-Dependent DFT Study of $[Fe(CN)_6]^{4-}$ Sensitization of TiO_2 Nanoparticles", *J. Am. Chem. Soc.*, vol. 126, pp. 15024–15025.

de Boeij, P. L., F. Kootstra, J. A. Berger, R. van Leeuwen, and J. G. Snijders, 2001, "Current density-functional theory for optical spectra: A polarization functional", *J. Chem. Phys.*, vol. 115, pp. 1995–1999.

de Boer, M. P., J. H. Hoogenraad, R. B. Vrijen, L. D. Noordam, and H. G. Muller, 1993, "Indications of high-intensity adiabatic stabilization in neon", *Phys. Rev. Lett.*, vol. 71, pp. 3263–3266.

de Boer, M. P., J. H. Hoogenraad, R. B. Vrijen, R. C. Constantinescu, L. D. Noordam, and H. G. Muller, 1994, "Adiabatic stabilization against photoionization: An experimental study", *Phys. Rev. A*, vol. 50, pp. 4085–4098.

de Heer, W. A., 1993, "The physics of simple metal clusters: experimental aspects and simple models", *Rev. Mod. Phys.*, vol. 65, pp. 611–676.

Deb, B. M., and S. K. Ghosh, 1982, "Schrödinger fluid dynamics of many-electron systems in a time-dependent density-functional framework", *J. Chem. Phys.*, vol. 77, pp. 342–348.

Del Sole, R., L. Reining, and R. W. Godby, 1994, "GWΓ approximation for electron self-energies in semiconductors and insulators", *Phys. Rev. B*, vol. 49, pp. 8024–8028.

Del Sole, R., G. Adragna, V. Olevano, and L. Reining, 2003, "Long-range behavior and frequency dependence of exchange-correlation kernels in solids", *Phys. Rev. B*, vol. 67, pp. 045207-1–5.

Delaere, D., M. T. Nguyen, and L. G. Vanquickenborne, 2003, "Structure-Property Relationships in Phosphole-Containing π-Conjugated Systems: A Quantum Chemical Study", *J. Phys. Chem. A*, vol. 107, pp. 838–846.

Delerue, C., G. Allan, and M. Lannoo, 1993, "Theoretical aspects of the luminescence of porous silicon", *Phys. Rev. B*, vol. 48, pp. 11024–11036.

Della Sala, F., and A. Görling, 2001a "Efficient localized Hartree-Fock methods as effective exact-exchange Kohn-Sham methods for molecules", *J. Chem. Phys.*, vol. 115, pp. 5718–5732.

Della Sala, F., H. H. Heinze, and A. Görling, 2001b, "Excitation energies of terthiophene and its dioxide derivative: a first-principles study", *Chem. Phys. Lett.*, vol. 339, pp. 343–350.

Della Sala, F., and A. Görling, 2002a, "Asymptotic Behavior of the Kohn-Sham Exchange Potential", *Phys. Rev. Lett.* vol. 89, pp. 033003-1–4.

Della Sala, F., and A. Görling, 2002b, "The asymptotic region of the Kohn-Sham exchange potential in molecules", *J. Chem. Phys.*, vol. 116, pp. 5374–5388.

Della Sala, F., and A. Görling, 2003a, "Excitation energies of molecules by time-dependent density-functional theory based on effective exact exchange Kohn-Sham potentials", *Int. J. Quantum Chem.*, vol. 91, pp. 131–138.

Della Sala, F., and A. Görling, 2003b, "Open-shell localized Hartree-Fock approach for an efficient effective exact-exchange Kohn-Sham treatment of open-shell atoms and molecules", *J. Chem. Phys.*, vol. 118, pp. 10439–10454.

Della Sala, F., M. F. Raganato, M. Anni, R. Cingolani, M. Weimer, A. Görling, L. Favaretto, G. Barbarello, and G. Gigli, 2003c, "Optical properties of functionalized thiophenes: a theoretical and experimental study", *Synth. Met.*, vol. 139, pp. 897–899.

Delley, B., and E. F. Steigmeier, 1993, "Quantum confinement in Si nanocrystals", *Phys. Rev. B*, vol. 47, pp. 1397–1400.

Delone, N. B., and V. P. Krainov, 2000, *Multiphoton Processes in Atoms* (Springer, Berlin).

Demel, T., D. Heitmann, P. Grambow, and K. Ploog, 1990, "Nonlocal dynamic response and level crossings in quantum-dot structures", *Phys. Rev. Lett.*, vol. 64, pp. 788–791.

deMon2k, A. M. Köster, P. Calaminici, M. E. Casida, R. Flores, G. Geudtner, A. Goursot, T. Heine, A. Ipatov, F. Janetzko, S. Patchkovskii, J. U. Reveles, A. Vela, and D. R. Salahub, Version 1.8, The deMon Developers (2005).

Derosa, P. A., and J. M. Seminario, 2001, "Electron Transport through Single Molecules: Scattering Treatment Using Density-Functional and Green Function Theories", *J. Phys. Chem. B*, vol. 105, pp. 471–481.

Dewar, M., E. Zoebisch, E. Healy, J. Stewart, 1985, "Development and use of quantum mechanical molecular models. 76. AM1: a new general purpose quantum mechanical molecular model", *J. Am. Chem. Soc.*, vol. 107, pp. 3902–3909.

Dhara, A. K., and S. K. Ghosh, 1987, "Density-functional theory for time-dependent systems", *Phys. Rev. A*, vol. 35, pp. 442–444.

Di Ventra, M., S. T. Pantelides, and N. D. Lang, 2000, "First-Principles Calculation of Transport Properties of a Molecular Device", *Phys. Rev. Lett.*, vol. 84, pp. 979–982.

Di Ventra, M., and T. N. Todorov, 2004, "Transport in nanoscale systems: the microcanonical versus grand-canonical picture", *J. Phys. C*, vol. 16, pp. 8025–8034.

Diedrich, C., and S. Grimme, 2003, "Systematic Investigation of Modern Quantum Chemical Methods to Predict Electronic Circular Dichroism Spectra", *J. Phys. Chem. A*, vol. 107, pp. 2524–2539.

Dierksen, M., and S. Grimme, 2004, "Density-functional calculations of the vibronic structure of electronic absorption spectra", *J. Chem. Phys.*, vol. 120, pp. 3544–3554.

Dierksen, M., and S. Grimme, 2005, "An efficient approach for the calculation of Franck-Condon integrals of large molecules", *J. Chem. Phys.*, vol. 122, pp. 244101-1–9.

Dietl, C., E. Papastathopoulos, P. Niklaus, R. Improta, F. Santoro, and G. Gerber, 2005, "Femtosecond photoelectron spectroscopy of *trans*-stilbene above the reaction barrier", *Chem. Phys.*, vol. 310, pp. 201–211.

Dinh, P. M., P.-G. Reinhard, and E. Suraud, 2005, submitted.

Dinte, B. P., 2004, "Novel constraints in the search for a van der Waals energy functional", Ph.D. thesis, Griffith University.

Dion, M., H. Rydberg, E. Schröder, D. C. Langreth, and B. I. Lundqvist, 2004, "van der Waals Density-Functional for General Geometries", *Phys. Rev. Lett.*, vol. 92, pp. 246401-1–4.

Dion, M., and K. Burke, 2005, "Coordinate scaling in time-dependent current density-functional theory", *Phys. Rev. A*, vol. 72, pp. 020502-1–4.

Ditmire, T., J. Zweiback, V. P. Yanovsky, T. E. Cowan, G. Hays, and K. B. Wharton, 1999, "Nuclear fusion from explosions of femtosecond laser-heated deuterium clusters", *Nature* (London), vol. 398, pp. 489–492.

Dmitrenko, O., W. Reischl, R. Bach, and J. Spanget-Larsen, 2004, "TD-DFT Computational Insight into the Origin of Wavelength-Dependent E/Z Photoisomerization of Urocanic Acid", *J. Phys. Chem. A*, vol. 108, pp. 5662–5669.

Dobson, J. F., 1992, "Electron-gas boundary properties in non-neutral jellium (wide-parabolic-quantum-well) systems", *Phys. Rev. B*, vol. 46, pp. 10163–10172.

Dobson, J. F., 1994a, "Harmonic-Potential Theorem: Implications for Approximate Many-Body Theories", *Phys. Rev. Lett.*, vol. 73, pp. 2244–2247.

Dobson, J. F., 1994b, "Quasi-Local Approximation for a van der Waals Energy Functional", in *Topics in Condensed Matter Physics*, edited by M. P. Das, Chap. 7 (Nova, New York); cond-mat/0311371.

Dobson, J. F., B. P. Dinte, 1996, "Constraint Satisfaction in Local and Gradient Susceptibility Approximations: Application to a van der Waals Density-Functional", *Phys. Rev. Lett.*, vol. 76, pp. 1780–1783.

Dobson, J. F., M. Bünner, and E. K. U. Gross, 1997, "Time-Dependent Density-Functional Theory beyond Linear Response: An Exchange-Correlation Potential with Memory", *Phys. Rev. Lett.*, vol. 79, pp. 1905–1908.

Dobson, J. F., B. Dinte, and J. Wang, 1998, "van der Waals Functionals via Local Approximations for Susceptibilities", in *Electronic Density-Functional Theory: Recent Progress and New Directions*, edited by J. F. Dobson, G. Vignale and M. P. Das, p. 261 (Plenum, New York).

Dobson, J. F., and J. Wang, 1999, "Successful Test of a Seamless van der Waals Density-Functional", *Phys. Rev. Lett.*, vol. 82, pp. 2123–2126.

Dobson, J. F., and J. Wang, 2000a, "Energy-optimized local exchange-correlation kernel for the electron gas: Application to van der Waals forces", *Phys. Rev. B*, vol. 62, pp. 10038–10045.

Dobson, J. F., B. P. Dinte, J. Wang, and T. Gould, 2000b, "A novel constraint for the simplified description of dispersion forces", *Australian J. Phys.*, vol. 53, pp. 575–596.

Dobson, J. F., K. McLennan, A. Rubio, J. Wang, T. Gould, H. M. Le, and B. P. Dinte, 2001, "Prediction of dispersion forces: is there a problem?" *Australian J. Chem.*, vol. 54, pp. 513–527.

Dobson, J. F., J. Wang, and T. Gould, 2002, "Correlation energies of inhomogeneous many-electron systems", *Phys. Rev. B*, vol. 66, pp. 081108-1–4(R).

Dobson, J. F., and J. Wang, 2004, "Testing the local density approximation with energy-versus-separation curves of jellium slab pairs", *Phys. Rev. B*, vol. 69, pp. 235104-1–9.

Dobson, J. F., J. Wang, B. P. Dinte, K. McLennan, and H. M. Le, 2005a, "Soft Cohesive Forces", *Int. J. Quant. Chem.*, vol. 101, pp. 579–598.

Dobson, J. F., and A. Rubio, 2005b, unpublished.

Dobson, J. F., and A. Rubio, 2005c, "Nonuniversality of the dispersion interaction: analytic benchmarks for van der Waals energy functionals", cond-mat/0502422.

Dobson J. F., A. White, and A. Rubio, 2006, "Asymptotics of the dispersion interaction: analytic benchmarks for van der Waals energy functionals", *Phys. Rev. Lett.*, vol. 96, pp. 073201-1–4.

Domps, A., P.-G. Reinhard, and E. Suraud, 1998, "Theoretical Estimation of the Importance of Two-Electron Collisions for Relaxation in Metal Clusters", *Phys. Rev. Lett.*, vol. 81, pp. 5524–5527.

Dörner, R., T. Weber, M. Weckenbrock, A. Staudte, M. Hattass, R. Moshammer, J. Ullrich, and H. Schmidt-Böcking, 2002, "Multiple ionization in strong laser fields", *Adv. Atom. Mol. Opt. Phys.*, vol. 48, pp. 1–34.

Drake, G. W. F. (ed.), 1996, in *Atomic, Molecular, and Optical Physics Handbook*, (AIP Press, Woodbury, New York).

Dreizler, R. M., and E. K. U. Gross, 1990, *Density-Functional Theory: An Approach to the Quantum Many-Body Problem* (Springer, Berlin).

Drescher, M., M. Hentschel, R. Kienberger, G. Tempea, C. Spielmann, G. A. Reider, P. B. Corkum, and F. Krausz, 2001, "X-ray Pulses Approaching the Attosecond Frontier", *Science*, vol. 291, pp. 1923–1927.

Drescher, M., M. Hentschel, R. Kienberger, M. Uiberacker, V. Yakovlev, A. Scrinzi, T. Westerwalbesloh, U. Kleineberg, U. Heinzmann, and F. Krausz, 2002, "Time-resolved atomic inner-shell spectroscopy", *Nature* (London), vol. 419, pp. 803–807.

Dreuw, A., B. D. Dunietz, and M. Head-Gordon, 2002, "Characterization of the Relevant Excited States in the Photodissociation of CO-Ligated Hemoglobin and Myoglobin", *J. Am. Chem. Soc.*, vol. 124, pp. 12070–12071.

Dreuw, A., G. R. Fleming, and M. Head-Gordon, 2003a, "Chlorophyll fluorescence quenching by xanthophylls", *Phys. Chem. Chem. Phys.*, vol. 5, pp. 3247–3256.

Dreuw, A., J. L. Weisman, and M. Head-Gordon, 2003b, "Long-range charge-transfer excited states in time-dependent density-functional theory require non-local exchange", *J. Chem. Phys.*, vol. 119, pp. 2943–2946.

Dreuw, A., G. R. Fleming, and M. Head-Gordon, 2003c, "Charge-Transfer State as a Possible Signature of a Zeaxanthin-Chlorophyll Dimer in the Non-photochemical Quenching Process in Green Plants", *J. Phys. Chem. B*, vol. 107, pp. 6500–6503.

Dreuw, A., and M. Head-Gordon, 2004, "Failure of Time-Dependent Density-Functional Theory for Long-Range Charge-Transfer Excited States: The Zincbacteriochlorin-Bacteriochlorin and Bacteriochlorophyll-Spheroidene Complexes", *J. Am. Chem. Soc.*, vol. 126, pp. 4007–4016.

Duan, X., X. Li, R. He, and X. Cheng, 2005, "Time-dependent density-functional theory study on intramolecular charge transfer and solvent effect of dimethylaminobenzophenone", *J. Chem. Phys.*, vol. 122, pp. 084314-1–9.

Dubrovin, B. A., A. T. Fomenko, and S. P. Novikov, 1984, *Modern Geometry – Methods and Applications*, vol. 1 (Springer, New York).

Dundas, D., and J. M. Rost, 2005, "Molecular effects in the ionization of N_2, O_2, and F_2 by intense laser fields", *Phys. Rev. A*, vol. 71, pp. 013421-1–8.

Dunlap, B. I., J. W. D. Connolly, and J. R. Sabin, 1979, "On some approximations in applications of $X\alpha$ theory", *J. Chem. Phys.*, vol. 71, pp. 3396–3402.

Dunietz, B. D., A. Dreuw, and M. Head-Gordon, 2003, "Initial Steps of the Photodissociation of the CO Ligated Heme Group", *J. Phys. Chem. B*, vol. 107, pp. 5623–5629.

Dunning Jr., T. H., 1989, "Gaussian basis sets for use in correlated molecular calculations. I. The atoms boron through neon and hydrogen", *J. Chem. Phys.*, vol. 90, pp. 1007–1023.

Durbeej, B., and L. A. Eriksson, 2000, "Thermodynamics of the Photoenzymic Repair Mechanism Studied by Density-Functional Theory", *J. Am. Chem. Soc.*, vol. 122, pp. 10126–10132.

Durbeej, B., and L. A. Eriksson, 2002, "Reaction mechanism of thymine dimer formation in DNA induced by UV light", *J. Photochem. Photobiol. A*, vol. 152, pp. 95–101.

Durbeej, B., and L. A. Eriksson, 2003, "On the bathochromic shift of the absorption by astaxanthin in crustacyanin: a quantum chemical study", *Chem. Phys. Lett.*, vol. 375, pp. 30–38.

Dyer, J., W. J. Blau, C. G. Coates, C. M. Creely, J. D. Gavey, M. W. George, D. C. Grills, S. Hudson, J. M. Kelly, P. Matousek, J. J. McGarvey, J. McMaster, A. W. Parker, M. Towrie, and J. A. Weinstein, 2003, "The photophysics of fac-$[Re(CO)_3(dppz)(py)]^+$ in CH_3CN: a comparative picosecond flash photolysis, transient infrared, transient resonance Raman and density-functional theoretical study", *Photochem. Photobiol. Sci.*, vol. 2, pp. 542–554.

Dzyaloshinskii, I. E., E. M. Lifshitz, and L. P. Pitaevskii, 1961, "The general theory of van Der Waals forces", *Adv. Phys.*, vol. 10, pp. 165–209.

Eberly, J. H., R. Grobe, C. K. Law, and Q. Su, 1992, "Numerical experiments in strong and super-strong fields", in *Atoms in Intense Laser Fields*, edited by M. Gavrila, pp. 301–334 (Academic Press, Boston).

Eberly, J. H., and K. C. Kulander, 1993, "Atomic Stabilization by Super-Intense Lasers", *Science*, vol. 262, pp. 1229–1233.

Eckart, C., 1935, "Some studies concerning rotating axes and polyatomic molecules", *Phys. Rev.*, vol. 47, pp. 552–558.

Eguiluz, A. G., 1985, "Self-consistent static-density-response function of a metal surface in density-functional theory", *Phys. Rev. B*, vol. 31, pp. 3303–3314.

Eguiluz, A. G., M. Heinrichsmeier, A. Fleszar, and W. Hanke, 1992, "First-principles evaluation of the surface barrier for a Kohn-Sham electron at a metal surface", *Phys. Rev. Lett.*, vol. 68, pp. 1359–1362.

Ehrenfest, P., 1927, "Bemerkungen über die angenäherte Gültigkeit der klassischen Mechanik", *Z. Phys.*, vol. 45, pp. 455–457.

Ehrler, O. T., J. M. Weber, F. Furche, and M. M. Kappes, 2003, "Photoelectron Spectroscopy of C_{84} Dianions", *Phys. Rev. Lett.*, vol. 91, pp. 113006-1–4.

Ehrler, O., F. Furche, J. Weber, and M. Kappes, 2005, "Photoelectron spectroscopy of fullerene dianions C_{76}^{2-}, C_{78}^{2-}, and C_{84}^{2-}", *J. Chem. Phys.*, vol. 122, pp. 094321-1–8.

Eichkorn, K., O. Treutler, H. Öhm, M. Häser, and R. Ahlrichs, 1995, "Auxiliary basis sets to approximate Coulomb potentials", *Chem. Phys. Lett.*, vol. 242, pp. 652–660.

Eichkorn, K., F. Weigend, O. Treutler, and R. Ahlrichs, 1997, "Auxiliary basis sets for main row atoms and transition metals and their use to approximate Coulomb potentials", *Theor. Chem. Acc.*, vol. 97, pp. 119–124.

Ekardt, W., 1984, "Dynamical Polarizability of Small Metal Particles: Self-Consistent Spherical Jellium Background Model", *Phys. Rev. Lett.*, vol. 52, pp. 1925–1928.

Elmaci, N., and E. Yurtsever, 2002, "Thermochromism in Oligothiophenes: The Role of the Internal Rotation", *J. Phys. Chem. A*, vol. 106, pp. 11981–11986.

Emperador, A., M. Barranco, E. Lipparini, M. Pi, and L. Serra, 1999, "Density-functional calculations of magnetoplasmons in quantum rings", *Phys. Rev. B*, vol. 59, pp. 15301–15307.

Engel, E., and R. M. Dreizler, 1999, "From explicit to implicit density-functionals", *J. Comput. Chem.*, vol. 20, pp. 31–50.

Engel, E., A. Höck, and R. M. Dreizler, 2000, "van der Waals bonds in density-functional theory", *Phys. Rev. A*, vol. 61, pp. 032502-1–5.

Engel, E., and F. Bonetti, 2001, "Implicit Density-Functionals for the Exchange-Correlation Energy: Description of Van Der Waals Bonds", *Int. J. Mod. Phys. B*, vol. 15, pp. 1703–1713.

Eremina, E., X. Liu, H. Rottke, W. Sandner, M. G. Schätzel, A. Dreischuh, G. G. Paulus, H. Walther, R. Moshammer, and J. Ullrich, 2004, "Influence of Molecular Structure on Double Ionization of N_2 and O_2 by High Intensity Ultrashort Laser Pulses", *Phys. Rev. Lett.*, vol. 92, pp. 173001-1–4.

Erhard, S., and E. K. U. Gross, 1997, "High harmonic generation in hydrogen and helium atoms subject to one- and two-color laser pulses", in *Multiphoton Processes 1996*, edited by P. Lambropoulos and H. Walther, pp. 37–46 (IOP Publishing, Bristol).

Ernzerhof, M., K. Burke, and J. P. Perdew, 1996, "Long-range asymptotic behavior of ground-state wave functions, one-matrices, and pair densities", *J. Chem. Phys.*, vol. 105, pp. 2798–2803.

Evers, F., F. Weigend, and M. Koentopp, 2004, "Conductance of molecular wires and transport calculations based on density-functional theory", *Phys. Rev. B*, vol. 69, pp. 235411-1–9.

Fabiano, E., F. Della Sala, R. Cingolani, M. Weimer, and A. Gorling, 2005, "Theoretical Study of Singlet and Triplet Excitation Energies in Oligothiophenes", *J. Phys. Chem. A*, vol. 109, pp. 3078–3085.

Facco Bonetti, A., E. Engel, R. N. Schmid, and R. M. Dreizler, 2001, "Investigation of the Correlation Potential from Kohn-Sham Perturbation Theory", *Phys. Rev. Lett.*, vol. 86, pp. 2241–2244.

Falkovskaia, E., V. G. Pivovarenko, and J. C. del Valle, 2002, "Observation of a single proton transfer fluorescence in a biaxially symmetric dihydroxy diflavonol", *Chem. Phys. Lett.*, vol. 352, pp. 415–420.

Falkovskaia, E., V. G. Pivovarenko, and J. C. Del Valle, 2003, "Interplay between Intra- and Intermolecular Excited-State Single- and Double-Proton-Transfer Processes in the Biaxially Symmetric Molecule 3,7-Dihydroxy-4H,6H-pyrano[3,2-*g*]-chromene-4,6-dione", *J. Phys. Chem. A*, vol. 107, pp. 3316–3325.

Fantacci, S., F. De Angelis, J. Wang, S. Bernhard, and A. Selloni, 2004a, "A Combined Computational and Experimental Study of Polynuclear Ru-TPPZ Complexes: Insight into the Electronic and Optical Properties of Coordination Polymers", *J. Am. Chem. Soc.*, vol. 126, pp. 9715–9723.

Fantacci, S., F. De Angelis, A. Sgamellotti, and N. Re, 2004b, "A TDDFT study of the ruthenium(II) polyazaaromatic complex $[Ru(dppz)(phen)_2]^{2+}$ in solution", *Chem. Phys. Lett.*, vol. 396, pp. 43–48.

Fattebert, J.-L., and J. Bernholc, 2000, "Towards grid-based $O(N)$ density-functional theory methods: Optimized nonorthogonal orbitals and multigrid acceleration", *Phys. Rev. B*, vol. 62, pp. 1713–1722.

Feit, M. D., and J. A. Fleck, 1982a, "Solution of the Schrödinger equation by a spectral method II: Vibrational energy levels of triatomic molecules", *J. Chem. Phys.*, vol. 78, pp. 301–308.

Feit, M. D., J. A. Fleck, and A. Steiger, 1982b, "Solution of the Schrödinger equation by a spectral method", *J. Comput. Phys.*, vol. 47, pp. 412–433.

Ferconi, M., and G. Vignale, 1994, "Current-density-functional theory of quantum dots in a magnetic field", *Phys. Rev. B*, vol. 50, pp. 14722–14725.

Feret, L., E. Suraud, F. Calvayrac, and P.-G. Reinhard, 1996, "On the electron dynamics in Na_9^+ metal clusters: a Vlasov approach", *J. Phys. B: At. Mol. Opt. Phys.*, vol. 29, pp. 4477–4491.

Fermi, E., and E. Amaldi, 1934, "Le orbite s degli elementi", *Accad. Ital. Rome*, vol. 6, pp. 117–149.

Ferray, M., A. L'Huillier, X. F. Li, L. A. Lompré, G. Mainfray, and C. Manus, 1988, "Multiple-harmonic conversion of 1064 nm radiation in rare gases", *J. Phys. B: At. Mol. Opt. Phys.*, vol. 21, pp. L31–L35.

Fetter, A. L., and J. D. Walecka, 1971, *Quantum Theory of Many-Particle Systems* (McGraw-Hill, New York).

Feuerstein, B., and U. Thumm, 2003, "Fragmentation of H_2^+ in strong 800-nm laser pulses: Initial-vibrational-state dependence", *Phys. Rev. A*, vol. 67, pp. 043405-1–8.

Feynman, R. P., 1939, "Forces in Molecules", *Phys. Rev.*, vol. 56, pp. 340–343.

Field, M. J., P. A. Bash, and M. Karplus, 1990, "A combined quantum mechanical and molecular mechanical potential for molecular dynamics simulations", *J. Comput. Chem.*, vol. 11, pp. 700–733.

Filippi, C., C. J. Umrigar, and X. Gonze, 1997, "Excitation energies from density-functional perturbation theory", *J. Chem. Phys.*, vol. 107, pp. 9994–10002.

Fiolhais, C., F. Nogueira, and M. A. L. Marques (eds.), 2003, *A Primer in Density-Functional Theory* (Springer, Berlin).

Fittinghoff, D. N., P. R. Bolton, B. Chang, and K. C. Kulander, 1992, "Observation of nonsequential double ionization of helium with optical tunneling", *Phys. Rev. Lett.*, vol. 69, pp. 2642–2645.

Fittinghoff, D. N., P. R. Bolton, B. Chang, and K. C. Kulander, 1994, "Polarization dependence of tunneling ionization of helium and neon by 120-fs pulses at 614 nm", *Phys. Rev. A*, vol. 49, pp. 2174–2177.

Foley, E. T., A. F. Kam, J. W. Lyding, and Ph. Avouris, 1998, "Cryogenic UHV-STM Study of Hydrogen and Deuterium Desorption from Si(100)", *Phys. Rev. Lett.*, vol. 80, 1336–1339.

Fornberg, B., and D. M. Sloan, 1994, "A review of pseudospectral methods for solving differential equations", *Acta Numerica*, vol. 94, pp. 203–268.

Francl, M. M., W. J. Pietro, W. J. Hehre, J. S. Binkley, M. S. Gordon, D. J. DeFrees, and J. A. Pople, 1982, "Self-consistent molecular orbital methods. XXIII. A polarization-type basis set for second-row elements", *J. Chem. Phys.*, vol. 77, pp. 3654–3665.

Fratiloiu, S., L. Candeias, F. Grozema, J. Wildeman, and L. Siebbeles, 2004, "VIS/NIR Absorption Spectra of Positively Charged Oligo (phenylenevinylene)s and Comparison with Time-Dependent Density-Functional Theory Calculations", *J. Phys. Chem. B*, vol. 108, pp. 19967–19975.

Frediani, L., Z. Rinkevicius, and H. Ågren, 2005, "Two-photon absorption in solution by means of time-dependent density-functional theory and the polarizable continuum model", *J. Chem. Phys.*, vol. 122, pp. 244104-1–12.

Freeman, R. R., and P. H. Bucksbaum, 1991, "Investigations of above-threshold ionization using subpicosecond laser pulses", *J. Phys. B: At. Mol. Opt. Phys.*, vol. 24, pp. 325–347.

Fricke, M., A. Lorke, J. P. Kotthaus, G. Medeiros-Ribeiro, and P. M. Petroff, 1996, "Shell structure and electron-electron interaction in self-assembled InAs quantum dots", *Europhys. Lett.*, vol. 36, pp. 197–202.

Friedrich, H., 1991, *Theoretical Atomic Physics* (Springer, New York).

Frydel, D., W. Terilla, and K. Burke, 2000, "Adiabatic connection from accurate wave-function calculations", *J. Chem. Phys.*, vol. 112, pp. 5292–5297.

Fuchs, M., and X. Gonze, 2002, "Accurate density-functionals: Approaches using the adiabatic-connection fluctuation-dissipation theorem", *Phys. Rev. B*, vol. 65, pp. 235109-1–4.

Fuchs, M., K. Burke, Y. M. Niquet, and X. Gonze, 2003, "Comment on "Total Energy Method from Many-Body Formulation"", *Phys. Rev. Lett*, vol. 90, p. 189701.

Fuchs, M., Y.-M. Niquet, X. Gonze, and K. Burke, 2005, "Describing static correlation in bond dissociation by Kohn-Sham density-functional theory", *J. Chem. Phys.*, vol. 122, pp. 094116-1–13.

Fuchs, M., Y. M. Niquet, K. Burke, and X. Gonze, 2005, "Describing static correlation in bond dissociation by Kohn-Sham density-functional theory", *J. Chem. Phys.*, vol. 122, pp. 094116-1–13.

Full, J., L. Gonzalez, and C. Daniel, 2001, "A CASSCF/CASPT2 and TD-DFT Study of the Low-Lying Excited States of η_5-CpMn(CO)$_3$", *J. Phys. Chem. A*, vol. 105, pp. 184–189.

Furche, F., 2000a, "Dichtefunktionalmethoden für elektronisch angeregte Moleküle. Theorie-Implementierung-Anwendung", Ph.D. thesis, Universität Karlsruhe.

Furche, F., R. Ahlrichs, C. Wachsmann, E. Weber, A. Sobanski, F. Vögtle, and S. Grimme, 2000b, "Circular Dichroism of Helicenes Investigated by Time-Dependent Density-Functional Theory", *J. Am. Chem. Soc.*, vol. 122, pp. 1717–1724.

Furche, F., 2001a, "On the density matrix based approach to time-dependent density-functional response theory", *J. Chem. Phys.* vol. 114, pp. 5982–5992.

Furche, F., and R. Ahlrichs, 2001b, "Fullerene C_{80}: Are there still more isomers?", *J. Chem. Phys.*, vol. 114, pp. 10362–10367.

Furche, F., 2001c, "Molecular tests of the random phase approximation to the exchange-correlation energy functional", *Phys. Rev. B*, vol. 64, 195120-1–8.

Furche, F., and R. Ahlrichs, 2002a, "Adiabatic time-dependent density-functional methods for excited state properties", *J. Chem. Phys.*, vol. 117, pp. 7433–7447; *ibid*, vol. 121, p. 12772(E) (2004).

Furche, F., and R. Ahlrichs, 2002b, "Absolute Configuration of D_2-Symmetric Fullerene C_{84}", *J. Am. Chem. Soc.*, vol. 124, pp. 3804–3805.

Furche, F., and K. Burke, 2005a, "Time-dependent density-functional theory in quantum chemistry", in *Annual Reports in Computational Chemistry*, vol. 1, edited by D. Spellmeyer, pp. 19–30 (Elsevier, Amsterdam).

Furche, F., and T. Van Voorhis, T, 2005b, "Fluctuation-dissipation theorem density-functional theory", *J. Chem. Phys.*, vol. 122, pp. 164106–164115.

Furche, F., D. Rappoport, 2005c, "Density-functional methods for excited states: equilibrium structure and excited spectra", in *Computational Photochemistry*, edited by M. Olivucci, in press pp. 93–128 (Elsevier, Amsterdam).

Gabrielsson, A., S. Zalis, P. Matousek, M. Towrie, and A. Vlcek, 2004, "Ultrafast Photochemical Dissociation of an Equatorial CO Ligand from trans(X,X)-[Ru(X)$_2$(CO)$_2$(bpy)] (X = Cl, Br, I): A Picosecond Time-Resolved Infrared Spectroscopic and DFT Computational Study", *Inorg. Chem.*, vol. 43, pp. 7380–7388.

Garoufalis, C. S., A. D. Zdetsis, and S. Grimme, 2001, "High Level *Ab Initio* Calculations of the Optical Gap of Small Silicon Quantum Dots", *Phys. Rev. Lett.*, vol. 87, pp. 276402-1–4.

Garraway, B. M., and K.-A. Suominen, 1995, "Wave-packet dynamics: new physics and chemistry in femto-time", *Rep. Prog. Phys.*, vol. 58, pp. 365–419.

Gaudoin, R., and K. Burke, 2004, "Lack of Hohenberg-Kohn Theorem for Excited States", *Phys. Rev. Lett.*, vol. 93, pp. 173001-1–4.

GAUSSIAN 03, Revision C.02, M. J. Frisch, G. W. Trucks, H. B. Schlegel, G. E. Scuseria, M. A. Robb, J. R. Cheeseman, J. A. Montgomery, Jr., T. Vreven, K. N. Kudin, J. C. Burant, J. M. Millam, S. S. Iyengar, J. Tomasi, V. Barone, B. Mennucci, M. Cossi, G. Scalmani, N. Rega, G. A. Petersson, H. Nakatsuji, M. Hada, M. Ehara, K. Toyota, R. Fukuda, J. Hasegawa, M. Ishida, T. Nakajima, Y. Honda, O. Kitao, H. Nakai, M. Klene, X. Li, J. E. Knox, H. P. Hratchian,

J. B. Cross, V. Bakken, C. Adamo, J. Jaramillo, R. Gomperts, R. E. Stratmann, O. Yazyev, A. J. Austin, R. Cammi, C. Pomelli, J. W. Ochterski, P. Y. Ayala, K. Morokuma, G. A. Voth, P. Salvador, J. J. Dannenberg, V. G. Zakrzewski, S. Dapprich, A. D. Daniels, M. C. Strain, O. Farkas, D. K. Malick, A. D. Rabuck, K. Raghavachari, J. B. Foresman, J. V. Ortiz, Q. Cui, A. G. Baboul, S. Clifford, J. Cioslowski, B. B. Stefanov, G. Liu, A. Liashenko, P. Piskorz, I. Komaromi, R. L. Martin, D. J. Fox, T. Keith, M. A. Al-Laham, C. Y. Peng, A. Nanayakkara, M. Challacombe, P. M. W. Gill, B. Johnson, W. Chen, M. W. Wong, C. Gonzalez, and J. A. Pople, Gaussian, Inc., Wallingford CT, 2004.

Gavrila, M. (ed.), 1992, *Atoms in Intense Laser Fields* (Academic Press, Boston).

Gavrila, M., 2002, "Atomic stabilization in superintense laser fields", *J. Phys. B: At. Mol. Opt. Phys.*, vol. 35, pp. R147–R193.

Gebauer, R., and R. Car, 2004a, "Current in Open Quantum Systems", *Phys. Rev. Lett.*, vol. 93, pp. 160404-1–4.

Gebauer, R., and R. Car, 2004b, "Kinetic theory of quantum transport at the nanoscale", *Phys. Rev. B*, vol. 70, pp. 125324-1–5.

Gebauer, R., S. Piccinin, and R. Car, 2005, "Quantum Collision Current in Electronic Circuits", *Chem.Phys.*, vol 6, pp. 1727–1730.

Ghizdavu, L., O. Lentzen, S. Schumm, A. Brodkorb, C. Moucheron, and A. Kirsch-De Mesmaeker, 2003, "Synthesis and Characterization of Optically Active and Racemic Forms of Cyclometalated Rh(III) Complexes. An Experimental and Theoretical Emission Study", *Inorg. Chem.*, vol. 42, pp. 1935–1944.

Ghosh, S. K., and A. K. Dhara, "Density-functional theory of many-electron systems subjected to time-dependent electric and magnetic fields", *Phys. Rev. A*, vol. 38, pp. 1149–1158.

Giannozzi, P., and S. Baroni, 1994, "Vibrational and dielectric properties of C_60 from density-functional perturbation theory", *J. Chem. Phys.*, vol. 100, pp. 8537–8539.

Giansiracusa, J., 2002, private communication.

Gidopoulos, N., 1998, "Kohn-Sham equations for multicomponent systems: The exchange and correlation energy functional", *Phys. Rev. B*, vol. 57, pp. 2146–2152.

Giglio, E., P.-G. Reinhard, and E. Suraud, 2003, "Angular distribution of emitted electrons in sodium clusters: A semiclassical approach", *Phys. Rev. A*, vol. 67, pp. 43202-1–6.

Girifalco, L. A., M. Hodak, and R. S. Lee, 2000, "Carbon nanotubes, buckyballs, ropes, and a universal graphitic potential", *Phys. Rev. B*, vol. 62, pp. 13104–13110.

Girifalco, L. A., and M. Hodak, 2002, "van der Waals binding energies in graphitic structures", *Phys. Rev. B*, vol. 65, pp. 125404-1–5.

Giuliani, G. F., and G. Vignale (eds.), 2005, *Quantum Theory of the Electron Liquid* (Cambridge University Press).

Giusti-Suzor, A., F. H. Mies, L. F. DiMauro, E. Charron, and B. Yang, 1995, "Dynamics of H_2^+ in intense laser fields", *J. Phys. B: At. Mol. Opt. Phys.*, vol. 28, pp. 309–L340.

Godby, R. W., M. Schlüter, and L. J. Sham, 1986, "Accurate Exchange-Correlation Potential for Silicon and Its Discontinuity on Addition of an Electron", *Phys. Rev. Lett.*, vol. 56, pp. 2415–2418.

Godby, R. W., M. Schlüter, and L. J. Sham, 1988, "Self-energy operators and exchange-correlation potentials in semiconductors", *Phys. Rev. B*, vol. 37, pp. 10159–10175.

Gonis, A., 1992, "Green Functions for Ordered and Disordered Systems", Studies in Mathematical Physics, Vol. 4, edited by E. van Groesen and E. M. de Jager, (Elsevier Science Publishers B.V., North-Holland).

Gonze, X., 1995a, "Adiabatic density-functional perturbation theory", *Phys. Rev. A*, vol. 52, pp. 1096–1114.

Gonze, X., Ph. Ghosez, and R. W. Godby, 1995b, "Density-Polarization Functional Theory of the Response of a Periodic Insulating Solid to an Electric Field", *Phys. Rev. Lett.*, vol. 74, pp. 4035–4038.

Gonze, X., Ph. Ghosez, and R. W. Godby, 1997a, "Long-wavelength behavior of the exchange-correlation kernel in the Kohn-Sham theory of periodic systems", *Phys. Rev. B*, vol. 56, pp. 12811–12817.

Gonze, X., Ph. Ghosez, and R. W. Godby, 1997b, "Density-Functional Theory of Polar Insulators", *Phys. Rev. Lett.*, vol. 78, pp. 294–297.

Gonze, X., and M. Scheffler, 1999, "Exchange and Correlation Kernels at the Resonance Frequency: Implications for Excitation Energies in Density-Functional Theory", *Phys. Rev. Lett.*, vol. 82, pp. 4416–4419.

Görling, A., and M. Levy, 1993a, "Correlation-energy functional and its high-density limit obtained from a coupling-constant perturbation expansion", *Phys. Rev. B*, vol. 47, pp. 13105–13113.

Görling, A., 1993b, "Symmetry in density-functional theory", *Phys. Rev. A*, vol. 47, pp. 2783–2799.

Görling, A., and M. Levy, 1994, "Exact Kohn-Sham scheme based on perturbation theory", *Phys. Rev. A*, vol. 50, pp. 196–204.

Görling, A., and M. Ernzerhof, 1995a, "Energy differences between Kohn-Sham and Hartree-Fock wave functions yielding the same electron density", *Phys. Rev. A*, vol. 51, pp. 4501–4513.

Görling, A., and M. Levy, 1995b, "DFT Ionization Formulas and a DFT Perturbation Theory for Exchange and Correlation", *Int. J. Quantum Chem. Symp.*, vol. 29, p. 93.

Görling, A., 1996, "Density-functional theory for excited states", *Phys. Rev. A*, vol. 54, pp. 3912–3915.

Görling, A., 1997, "Time-dependent Kohn-Sham formalism", *Phys. Rev. A*, vol. 55, pp. 2630–2639.

Görling, A., 1998a, "Exact exchange-correlation kernel for dynamic response properties and excitation energies in density-functional theory", *Phys. Rev. A*, vol. 57, pp. 3433–3436.

Görling, A., 1998b, "Exact exchange kernel for time-dependent density-functional theory", *Int. J. Quantum Chem.*, vol. 69, pp. 265–277.

Görling, A., 1999a, "New KS Method for Molecules Based on an Exchange Charge Density Generating the Exact Local KS Exchange Potential", *Phys. Rev. Lett.*, vol. 83, pp. 5459–5462.

Görling, A., H. H. Heinze, S. P. Ruzankin, M. Staufer, and N. Rösch, 1999b, "Density- and density-matrix-based coupled Kohn-Sham methods for dynamic polarizabilities and excitation energies of molecules", *J. Chem. Phys.*, vol. 110, pp. 2785–2799.

Görling, A., 1999c, "Density-functional theory beyond the Hohenberg-Kohn theorem", *Phys. Rev. A*, vol. 59, pp. 3359–3374.

Görling, A., 2000, "Proper treatment of symmetries and excited states in a computationally tractable Kohn-Sham method", *Phys. Rev. Lett.*, vol. 85, pp. 4229–4233.

Görling, A., 2005, "Orbital- and state-dependent functionals in density-functional theory", *J. Chem. Phys.*, vol. 123, pp. 062203-1–16.

Gorski, A., E. Vogel, J. L. Sessler, and J. Waluk, 2002, "Magnetic Circular Dichroism of Octaethylcorrphycene and Its Doubly Protonated and Deprotonated Forms", *J. Phys. Chem. A*, vol. 106, pp. 8139–8145.

Goto, H., N. Harada, J. Crassous, and F. Diederich, 1998, "Absolute configuration of chiral fullerenes and covalent derivatives from their calculated circular dichroism spectra", *J. Chem. Soc., Perkin Trans. 2*, pp. 1719–1723.

Goumans, T. P. M., A. W. Ehlers, M. C. van Hemert, A. Rosa, E. J. Baerends, and K. Lammertsma, 2003, "Photodissociation of the Phosphine-Substituted Transition Metal Carbonyl Complexes $Cr(CO)_5L$ and $Fe(CO)_4L$: A Theoretical Study", *J. Am. Chem. Soc.*, vol. 125, pp. 3558–3567.

Gouterman, M., 1961, "Spectra of porphyrins", *J. Mol. Spectr.*, vol. 6, pp. 138–163.

Gouterman, M., 1978, "Absorption Spectra and Electronic Structure of Porphyrins and Related Rings", in *Porphyrins*, vol. 3, edited by D. H. Dolphin, pp. 1–165 (Academic Press, New York).

Governale, M., 2002, "Quantum Dots with Rashba Spin-Orbit Coupling", *Phys. Rev. Lett.*, vol. 89, pp. 206802-1–4.

Grabo, T., T. Kreibich, S. Kurth, and E. K. U. Gross, 2000, "Orbital functionals in density-functional theory: the optimized effective potential method", in *Strong Coulomb Correlations in Electronic Structure: Beyond the Local Density Approximation*, edited by V.I. Anisimov, pp. 203–311 (Gordon & Breach, Amsterdam).

Grandi, A., F. A. Gianturco, and N. Sanna, 2004, "H^- Desorption from Uracil via Metastable Electron Capture", *Phys. Rev. Lett.*, vol. 93, pp. 048103-1–4.

Grasbon, F., G. G. Paulus, H. Walther, P. Villoresi, G. Sansone, S. Stagira, M. Nisoli, and S. De Silvestri, 2003, "Above-Threshold Ionization at the Few-Cycle Limit", *Phys. Rev. Lett.*, vol. 91, pp. 173003-1–4.

Greengard, L., and V. Rokhlin, 1987, "A fast algorithm for particle simulations", *J. Comput. Phys.*, vol. 73, pp. 325–348.

Griffiths, D. J., 1995, *Introduction to Quantum Mechanics* (Prentice Hall, Upper Saddle River).

Grimme, S., 1996, "Density-functional calculations with configuration interaction for the excited states of molecules", *Chem. Phys. Lett.*, vol. 259, pp. 128–137.

Grimme, S., I. Pischel, S. Laufenberg, and F. Vögtle, 1998, "Synthesis, structure, and chiroptical properties of the first 4-Oxa[7]Paracyclophane", *Chirality*, vol. 10, pp. 147–153.

Grimme, S., and M. Parac, 2003, "Substantial errors from time-dependent density-functional theory for the calculation of excited states of large (π) systems", *Chem. Phys. Chem.*, vol. 4, p. 292–295.

Grimme, S., 2004, "Calculation of the electronic spectra of large molecules", in *Reviews in Computational Chemistry*, vol. 20, edited by K. B. Lipkowitz and D. B. Boyd, pp. 153–218 (Wiley-VCH, New York).

Gritsenko, O. V., S. J. A. van Gisbergen, P. R. T. Schipper, and E. J. Baerends, 2000, "Origin of the field-counteracting term of the Kohn-Sham exchange-correlation potential of molecular chains in an electric field", *Phys. Rev. A*, vol. 62, pp. 012507-1–10.

Gritsenko, O. V., and E. J. Baerends, 2001, "Orbital structure of the Kohn-Sham exchange potential and exchange kernel and the field-counteracting potential for molecules in an electric field", *Phys. Rev. A*, vol. 64, pp. 042506-1–12.

Gritsenko, O. V., and E. J. Baerends, 2002, "The analog of Koopmans' theorem in spin-density-functional theory", *J. Chem. Phys.*, vol. 117, pp. 9154–9159.

Gritsenko, O. V., B. Braida, and E. J. Baerends, 2003, "Physical interpretation and evaluation of the Kohn-Sham and Dyson components of the $\epsilon - I$ relations between the Kohn-Sham orbital energies and the ionization potentials", *J. Chem. Phys.*, vol. 119, pp. 1937–1950.

Grobe, R., and J. H. Eberly, 1992, "Photoelectron spectra for a two-electron system in a strong laser field", *Phys. Rev. Lett.*, vol. 68, pp. 2905–2908.

Gross, E. K. U., and W. Kohn, 1985, "Local density-functional theory of frequency-dependent linear response", *Phys. Rev. Lett.*, vol. 55, pp. 2850–2852; *ibid*, vol. 57, p. 923(E) (1986).

Gross, E. K. U., and W. Kohn, 1990, "Time-Dependent Density-Functional Theory", *Adv. Quant. Chem.*, vol. 21, pp. 255–291.

Gross, E. K. U., C. A. Ullrich, and U. A. Gossmann, 1995a, "density-functional theory of time-dependent systems", in *Density-Functional Theory*, NATO ASI Series B, edited by R. Dreizler and E. K. U. Gross (Plenum, New York).

Gross, M., and C. Guet, 1995b, "Energy transfer in collisions of alkali-metal clusters and ions", *Z. Phys. D*, vol. 33, pp. 289–293.

Gross, E. K. U., J. Dobson, and M. Petersilka, 1996, "density-functional theory of time-dependent phenomena", in Density-Functional Theory II, edited by R. F. Nalewajski, vol. 181 of *Topics in Current Chemistry*, pp. 81–172 (Springer, Berlin).

Grossman, J. C., M. Rohlfing, L. Mitas, S. G. Louie, and M. L. Cohen, "High Accuracy Many-Body Calculational Approaches for Excitations in Molecules", *Phys. Rev. Lett.*, vol. 86, pp. 472–475.

Grüning, M., O. V. Gritsenko, and E. J. Baerends, 2002, "Exchange potential from the common energy denominator approximation for the Kohn-Sham Green's function: Application to (hyper)polarizabilities of molecular chains", *J. Chem. Phys.*, vol. 116, pp. 6435–6442.

Grüning, M., A. Marini, U. von Barth, and A. Rubio, 2005, "Variational many-body approaches to obtain the exchange-correlation kernel of extended systems", in preparation.

Guan, J., M.E. Casida, and D.R. Salahub, 2000, "Time-dependent density-functional theory investigation of excitation spectra of open-shell molecules", *J. Molec. Structure (Theochem)*, vol. 527, pp. 229–244.

Guillemoles, J. F., V. Barone, L. Joubert, and C. Adamo, 2002, "A Theoretical Investigation of the Ground and Excited States of Selected Ru and Os Polypyridyl Molecular Dyes", *J. Phys. Chem. A*, vol. 106, pp. 11354–11360.

Gunaratne, T., A. Gusev, X. Peng, A. Rosa, G. Ricciardi, E. Baerends, C. Rizzoli, M. Kenney, and M. Rodgers, 2005, "Photophysics of Octabutoxy Phthalocyaninato-Ni(II) in Toluene: Ultrafast Experiments and DFT/TDDFT Studies", *J. Phys. Chem. A*, vol. 109, pp. 2078–2089.

Gunnarsson, O., and B. I. Lundqvist, 1976, "Exchange and correlation in atoms, molecules, and solids by the spin-density-functional formalism", *Phys. Rev. B*, vol. 13, pp. 4274–4298.

Guo, C., and G. N. Gibson, 2001, "Ellipticity effects on single and double ionization of diatomic molecules in strong laser fields", *Phys. Rev. A*, vol. 63, pp. 40701-1–4.

Guo, G., K. Chu, D. Wang, and C. Duan, 2004, "Linear and nonlinear optical properties of carbon nanotubes from first-principles calculations", *Phys. Rev. B*, vol. 69, pp. 205416-1–11.

Gutierrez, F., J. Trzcionka, R. Deloncle, R. Poteau, and N. Chouini-Lalanne, 2005, "Absorption and solvatochromic properties of 2-methylisoindolin-1-one and related compounds: interplay between theory and experiments", *New J. Chem.*, vol. 29, pp. 570–578.

Gygi, F., 1995a, "*Ab initio* molecular dynamics in adaptive coordinates", *Phys. Rev. B*, vol. 51, pp. 11190–11193.

Gygi, F., and G. Galli, 1995b, "Real-space adaptive-coordinate electronic-structure calculations", *Phys. Rev. B*, vol. 52, pp. R2229–R2232.

Halasinski, T. M., J. L. Weisman, R. Ruiterkamp, T. J. Lee, F. Salama, and M. Head-Gordon, 2003, "Electronic Absorption Spectra of Neutral Perylene ($C_{20}H_{12}$), Terrylene ($C_{30}H_{16}$), and Quaterrylene ($C_{40}H_{20}$) and Their Positive and Negative Ions: Ne Matrix-Isolation Spectroscopy and Time-Dependent Density-Functional Theory Calculations", *J. Phys. Chem. A*, vol. 107, pp. 3660–3669.

Hamada, N., S. Sawada, and A. Oshiyama, 1992, "New one-dimensional conductors: Graphitic microtubules", *Phys. Rev. Lett.*, vol. 68, 1579–1581.

Hamel, S., M. E. Casida, and D. R. Salahub, 2001, "Assessment of the quality of orbital energies in resolution-of-the-identity Hartree-Fock calculations using deMon auxiliary basis sets", *J. Chem. Phys,*. vol. 114, pp. 7342–7350.

Han, Y., and S. Lee, 2004, "Time-dependent density-functional calculations of $S_0 - S_1$ transition energies of poly(p-phenylene vinylene)", *J. Chem. Phys.*, vol. 121, pp. 609–611.

Hanel, G., B. Gstir, S. Denifl, P. Scheier, M. Probst, B. Farizon, M. Farizon, E. Illenberger, and T. D. Märk, "Electron Attachment to Uracil: Effective Destruction at Subexcitation Energies", *Phys. Rev. Lett.*, vol. 90, pp. 188104-1–4.

Hangfeldt, A., and M. Gradzel, 1995, "Light-Induced Redox Reactions in Nanocrystalline Systems", *Chem. Rev.*, vol. 95, pp. 49–68.

Harbola, U., and S. Mukamel, 2004, "Intermolecular forces and nonbonded interactions: Superoperator nonlinear time-dependent density-functional-theory response approach", *Phys. Rev. A*, vol. 70, pp. 052506-1–15.

Hariharan, P. C., and J. A. Pople, 1973, "Influence of Polarization Functions on Molecular-Orbital Hydrogenation Energies", *Theoret. Chimica Acta*, vol. 28, pp. 213–222.

Harriman, J. E., 1981, "Orthonormal orbitals for the representation of an arbitrary density", *Phys. Rev. A*, vol. 24, pp. 680–682.

Harris, J., and A. Griffin, 1975, "Correlation energy and van der Waals interaction of coupled metal films", *Phys. Rev. B*, vol. 11, pp. 3669–3677.

Harumiya, K., H. Kono, Y. Fujimura, I. Kawata, and A. D. Bandrauk, 2002, "Intense laser-field ionization of H_2 enhanced by two-electron dynamics", *Phys. Rev. A*, vol. 66, pp. 043403-1–14.

Hasegawa, T., S. Mouri, Y. Yamada, and K. Tanaka, 2003, "Giant Photo-Induced Dielectricity in $SrTiO_3$", *J. Phys. Soc. Jpn.*, vol. 72, pp. 41–44.

Hasegawa, M., and K. Nishidate, 2004, "Semiempirical approach to the energetics of interlayer binding in graphite", *Phys. Rev. B*, vol. 70, pp. 205431-1–7.

Häser, M., and R. Ahlrichs, 1989, "Improvements of the direct SCF method", *J. Comput. Chem.*, vol. 10, pp. 104–111.

Hata, K., D. N. Futaba, K. Mizuno, T. Namai, M. Yumura, and S. Iijima, 2004, "Water-Assisted Highly Efficient Synthesis of Impurity-Free Single-Walled Carbon Nanotubes", *Science*, vol. 306, pp. 1362–1364.

Hättig, C., and F. Weigend, 2000, "CC2 excitation energy calculations on large molecules using the resolution of the identity approximation", *J. Chem. Phys.*, vol. 113, pp. 5154–5161.

Haug, H., and A.-P. Jauho, 1996, *Quantum Kinetics in Transport and Optics of Semiconductors* (Springer, Berlin).

Haupts, U., S. Maiti, P. Schwille, and W. W. Webb, 1998, "Dynamics of fluorescence fluctuations in green fluorescent protein observed by fluorescence correlation spectroscopy", *Proc. Natl. Acad. Sci. USA*, vol. 95, pp. 13573–13578.

Hedin, L., 1965, "New Method for Calculating the One-Particle Green's Function with Application to the Electron-Gas Problem", *Phys. Rev.*, vol. 139, pp. A796–A823.

Hedin, L., and S. Lundqvist, 1969, "Effects of the electron-electron and electron-phonon interactions on the one-electron states of solids", in *Solid State Physics*, vol. 23, edited by H. Ehrenreich, F. Seitz, and D. Turnbull, pp. 1–181 (Academic, New York).

Hedin, L., 1999, "On correlation effects in electron spectroscopies and the GW approximation", *J. Phys.: Condens. Matter*, vol. 11, pp. R489–R528.

Heim, R., and R. Tsien, 1996, "Engineering green fluorescent protein for improved brightness, longer wavelengths and fluorescence resonance energy transfer", *Curr. Biol.*, vol. 6, pp. 178–182.

Heinz, K., 1995, "LEED and DLEED as modern tools for quantitative surface structure determination", *Rep. Prog. Phys.*, vol. 58, pp. 637–704.

Heinze, H. H., A. Görling, and N. Rösch, 2000, "An efficient method for calculating molecular excitation energies by time-dependent density-functional theory", *J. Chem. Phys.*, vol. 113, pp. 2088–2099.

Heinze, H. H., F. Della Sala, and A. Görling, 2002, "Efficient methods to calculate dynamic hyperpolarizability tensors by time-dependent density-functional theory", *J. Chem. Phys.*, vol. 116, pp. 9624–9640.

Helm, M., 2000, "The basic physics of intersubband transitions", in *Intersubband Transitions in Quantum Wells I*, edited by H. C. Liu and F. Capasso, vol. 62 of *Semiconductors and Semimetals*, pp. 1–99 (Academic Press, San Diego).

Hentschel, M., R. Kienberger, C. Spielmann, G. A. Reider, N. Milosevic, T. Brabec, P. Corkum, U. Heinzmann, M. Drescher, and F. Krausz, 2001, "Attosecond metrology", *Nature* (London), vol. 414, pp. 509–513.

Hessler, P., J. Park, and K. Burke, 1999, "Several Theorems in Time-Dependent Density-Functional Theory", *Phys. Rev. Lett.*, vol. 82, pp. 378–381; *ibid* vol. 83, p. 5184(E) (1999).

Hessler, P., N. T. Maitra, and K. Burke, 2002, "Correlation in time-dependent density-functional theory", *J. Chem. Phys.*, vol. 117, pp. 72–81.

Hill, N. A., and K. B. Whaley, 1995, "Size Dependence of Excitons in Silicon Nanocrystals", *Phys. Rev. Lett.*, vol. 75, pp. 1130–1133.

Hindgren, M., and C.-O. Almbladh, 1997, "Improved local-field corrections to the G_0W approximation in jellium: Importance of consistency relations", *Phys. Rev. B*, vol. 56, pp. 12832–12839.

Hirata, S., and M. Head-Gordon, 1999a, "Time-dependent density-functional theory for radicals: An improved description of excited states with substantial double excitation character", *Chem. Phys. Lett.*, vol. 302, pp. 375–382.

Hirata, S., and M. Head-Gordon, 1999b, "Time-dependent density-functional theory within the Tamm-Dancoff approximation", *Chem. Phys. Lett.*, vol. 314, pp. 291–299.

Hirata, S., T. J. Lee, and M. Head-Gordon, 1999c, "Time-dependent density-functional study on the electronic excitation energies of polycyclic aromatic hydrocarbon radical cations of naphthalene, anthracene, pyrene, and perylene", *J. Chem. Phys.*, vol. 111, pp. 8904–8912.

Hirata, S., M. Head-Gordon, and R. J. Bartlett, 1999d, "Configuration interaction singles, time-dependent Hartree-Fock, and time-dependent density-functional theory for the electronic excited states of extended systems", *J. Chem. Phys.*, vol. 111, pp. 10774–10786.

Hirata, S., S. Ivanov, I. Grabowski, R. J. Bartlett, K. Burke, and J. D. Talman, 2001, "Can optimized effective potentials be determined uniquely?", *J. Chem. Phys.*, vol. 115, pp. 1635–1649.

Hirata, S., S. Invanov, I. Grabowski, and R. Bartlett, 2002, "Time-dependent density-functional theory employing optimized effective potentials", *J. Chem. Phys.*, vol. 116, pp. 6468–6481.

Hirata, S., M. Head-Gordon, J. Szczepanski, and M. Vala, 2003, "Time-Dependent Density-Functional Study of the Electronic Excited States of Polycyclic Aromatic Hydrocarbon Radical Ions", *J. Phys. Chem. A*, vol. 107, pp. 4940–4951.

Hirose, K., and N. S. Wingreen, 1999, "Spin-density-functional theory of circular and elliptical quantum dots", *Phys. Rev. B*, vol. 59, pp. 4604–4607.

Hirose, K., Y. Meir, and N. S. Wingreen, 2004, "Time-dependent density-functional theory of excitation energies of closed-shell quantum dots", *Physica E*, vol. 22, pp. 486–489.

Hochbruck, M., and C. Lubich, 1997, "On Krylov Subspace Approximations to the Matrix Exponential Operator", *SIAM J. Numer. Anal.*, vol. 34, pp. 1911–1925.

Hochbruck, M., C. Lubich, and H. Selhofer, 1998, "Exponential Integrators for Large Systems of Differential Equations", *SIAM J. Sci. Comput.*, vol. 19, pp. 1552–1574.

Hochbruck, M., and C. Lubich, 1999, "Exponential Integrators for Quantum-Classical Molecular Dynamics", *BIT Num. Math.*, vol. 39, pp. 620–645.

Hochbruck, M., and C. Lubich, 2003, "On Magnus Integrators for Time-Dependent Schrödinger Equations", *SIAM J. Numer. Anal.*, vol. 41, pp. 945–963.

Hohenberg, P., and W. Kohn, 1964, "Inhomogeneous Electron Gas", *Phys. Rev.*, vol. 136, pp. B864–B871.

Holas, A., and R. Balawender, 2002, "Maitra-Burke example of initial-state dependence in time-dependent density-functional theory", *Phys. Rev. A*, vol. 65, pp. 034502-1–4.

Hsu, C. P., P. J. Walla, M. Head-Gordon, and G. R. Fleming, 2001, "The Role of the S_1 State of Carotenoids in Photosynthetic Energy Transfer: The Light-Harvesting Complex II of Purple Bacteria", *J. Phys. Chem. B*, vol. 105, pp. 11016–11025.

Hult, E., H. Ridburg, B. I. Lundqvist, and D. C. Langreth, 1999, "Unified treatment of asymptotic van der Waals forces", *Phys. Rev. B*, vol. 59, pp. 4708–4713.

Hummel, P., J. Oxgaard, W. Goddard, and H. Gray, 2005, "Ligand-Field Excited States of Metal Hexacarbonyls", *Inorg. Chem.*, vol. 44, pp. 2454–2458.

Hunt, K. L. C., 1983, "Nonlocal polarizability densities and van der Waals interactions", *J. Chem. Phys.*, vol. 78, pp. 6149–6155.

Hutchison, G. R., M. A. Ratner, and T. J. Marks, 2002, "Accurate Prediction of Band Gaps in Neutral Heterocyclic Conjugated Polymers", *J. Phys. Chem. A*, vol. 106, pp. 10596–10605.

Hutchison, G. R., Y. J. Zhao, B. Delley, A. J. Freeman, M. A. Ratner, I. J. Marks, 2003, "Electronic structure of conducting polymers: Limitations of oligomer extrapolation approximations and effects of heteroatoms", *Phys. Rev. B*, vol. 68, pp. 035204-1–13.

Hutter, J., 2003, "Excited state nuclear forces from the Tamm-Dancoff approximation to time-dependent density-functional theory within the plane wave basis set framework", *J. Chem. Phys.*, vol. 118, pp. 3928–3934.

Hybertsen, M. S., and S. G. Louie, 1985, "First-Principles Theory of Quasiparticles: Calculation of Band Gaps in Semiconductors and Insulators", *Phys. Rev. Lett.*, vol. 55, pp. 1418–1421.

Hybertsen, M. S., and S. G. Louie, 1986, "Electron correlation in semiconductors and insulators: Band gaps and quasiparticle energies", *Phys. Rev. B*, vol. 34, pp. 5390–5413.

Hybertsen, M. S., and S. G. Louie, 1987, "*Ab initio* static dielectric matrices from the density-functional approach. I. Formulation and application to semiconductors and insulators", *Phys. Rev. B*, vol. 35, pp. 5585–5601.

Igami, M., T. Nakanishi, and T. Ando, 1999, "Conductance of Carbon Nanotubes with a Vacancy", *J. Phys. Soc. Jpn.*, vol. 68, pp. 716–719.

Iijima, S., 1991, "Helical microtubules of graphitic carbon", *Nature* (London), vol. 354, pp. 56–58.

Il'iechev, Y. V., and J. D. Simon, 2003, "Building Blocks of Eumelanin: Relative Stability and Excitation Energies of Tautomers of 5,6-Dihydroxyindole and 5,6-Indolequinone", *J. Phys. Chem. B*, vol. 107, pp. 7162–7171.

Improta, R., F. Santoro, C. Dietl, E. Papastathopoulos, and G. Gerber, 2004, "Time dependent DFT investigation on the two lowest ^1B_u states of the *trans* isomer of stilbene and stiff-stilbenes", *Chem. Phys. Lett.*, vol. 387, pp. 509–516.

Improta, R., and V. Barone, 2005, "Absorption and fluorescence spectra of uracil in the gas phase and in aqueous solution: A TD-DFT quantum mechanical study", *J. Am. Chem. Soc.*, vol. 126, pp. 14320–14321.

Imry, Y., 2002, *Introduction to Mesoscopic Physics* (Oxford University Press, Oxford).

Infante, I., and F. Lelj, "Role of methyl substitution on the spectroscopic properties of porphyrazines. A TDDFT study using pure and hybrid functionals on porphyrazine and its octamethyl derivative", *Chem. Phys. Lett.*, vol. 367, pp. 308–318.

Ipatov, A., F. Cordova, and M.E. Casida, 2006, "Excited-State Spin-Contamination in Time-Dependent Density-Functional Theory for Molecules with Open-Shell Ground States", unpublished.

Itatani, J., J. Levesque, D. Zeidler, H. Niikura, H. Pépin, J. C. Kieffer, P. B. Corkum, and D. M. Villeneuve, 2004, "Tomographic imaging of molecular orbitals", *Nature* (London), vol. 432, pp. 867–871.

Itoh, U., Y. Toyoshima, H. Onuki, N. Washida, and T. Ibuki, 1986, "Vacuum ultraviolet absorption cross sections of SiH_4, GeH_4, Si_2H_6, and Si_3H_8", *J. Chem. Phys.*, vol. 85, pp. 4867–4872.

Ivanov, S., S. Hirata, and R. J. Bartlett, 1999, "Exact Exchange Treatment for Molecules in Finite-Basis-Set Kohn-Sham Theory", *Phys. Rev. Lett.*, vol. 83, pp. 5455–5458.

Iwamoto, N., and E. K. U. Gross, 1987, "Correlation effects on the third-frequency-moment sum rule of electron liquids", *Phys. Rev. B*, vol. 35, pp. 3003–3004.

Iwata, J.-I., K. Yabana, and G. F. Bertsch, 2000, "Dynamic Hyperpolarizability Calculation without Basis Functions", *Nonlinear Optics*, vol. 26, pp. 9–16.

Jacak, L., P. Hawrylak, and A. Wójs, 1998, *Quantum Dots* (Springer, Berlin).

Jackiw, R., and A. Kerman, 1979, "Time-dependent variation principle and the effective action", *Phys. Lett. A*, vol. 71, pp. 158–162.

Jacobson, K. I., and R. E. Jacobson, 1976, *Imaging Systems* (John Wiley & Sons, New York).

Jamorski, C., M. E. Casida, and D. R. Salahub, 1996, "Dynamic polarizabilities and excitation spectra from a molecular implementation of time-dependent density-functional response theory: N_2 as a case study", *J. Chem. Phys.*, vol. 104, pp. 5134–5147.

Jamorski, C., J. B. Foresman, C. Thilgen, and H.-P. Lüthi, 2002a, "Assessment of time-dependent density-functional theory for the calculation of critical features in the absorption spectra of a series of aromatic donor-acceptor systems", *J. Chem. Phys.*, vol. 116, pp. 8761–8771.

Jamorski, C., and H.-P. Lüthi, 2002b, "Time-dependent density-functional theory investigation of the formation of the charge transfer excited state for a series of aromatic donor-acceptor systems. Part I", *J. Chem. Phys.*, vol. 117, pp. 4146–4156.

Jamorski, C., and H.-P. Lüthi, 2002c, "Time-dependent density-functional theory investigation of the formation of the charge transfer excited state for a series of aromatic donor-acceptor systems. Part II", *J. Chem. Phys.*, vol. 117, pp. 4157–4167.

Jamorski, C., and H.-P. Lüthi, 2003a, "Time-Dependent Density-Functional Theory (TDDFT) Study of the Excited Charge-Transfer State Formation of a Series of Aromatic Donor-Acceptor Systems", *J. Am. Chem. Soc.*, vol. 125, pp. 252–264.

Jamorski, C., and H.-P. Lüthi, 2003b, "A time-dependent density-functional theory investigation of the fluorescence behavior of related cyano and di-cyano isomers of 4-(*N*,*N*-dimethylamino) benzonitrile", *Chem. Phys. Lett.*, vol. 368, pp. 561–567.

Jamorski, C., and H.-P. Lüthi, 2003c, "Rational classification of a series of aromatic donor-acceptor systems within the twisting intramolecular charge transfer model, a time-dependent density-functional theory investigation", *J. Chem. Phys.*, vol. 119, pp. 12852–12865.

Jamorski, C., and M. Casida, 2004, "Time-Dependent Density-Functional Theory Investigation of the Fluorescence Behavior as a Function of Alkyl Chain Size for the 4-(N,N-Dimethylamino)benzonitrile-like Donor-Acceptor Systems 4-(N,N-Diethylamino)benzonitrile and 4-(N,N-Diisopropylamino)benzonitrile", *J. Phys. Chem. B*, vol. 108, pp. 7132–7141.

Jaramillo, J., and G. E. Scuseria, 2000, "Assessment of the Van Voorhis-Scuseria exchange-correlation functional for predicting excitation energies using time-dependent density-functional theory", *Theor. Chem. Acc.*, vol. 105, pp. 62–67.

Jauho, A.-P., N. S. Wingreen, and Y. Meir, "Time-dependent transport in interacting and noninteracting resonant-tunneling systems", *Phys. Rev. B*, vol. 50, pp. 5528–5544.

Javanainen, J., J. H., Eberly, and Q., Su, 1988, "Numerical simulations of multiphoton ionization and above-threshold electron spectra", *Phys. Rev. A*, vol. 38, pp. 3430–3446.

Jaworska, M., G. Kazibut, and P. Lodowski, 2003a, "Electronic Spectrum of Cobalt-Free Corrins Calculated by TDDFT Method", *J. Phys. Chem. A*, vol. 107, pp. 1339–1347.

Jaworska, M., and P. Lodowski, 2003b, "Electronic spectrum of Co-corrin calculated with the TDDFT method", *J. Mol. Struct. (Theochem)*, vol. 631, pp. 209–223.

Jaworska, M., W. Macyk, and Z. Stasicka, 2004, "Structure, spectroscopy and photochemistry of the $[M(\eta^5\text{-}C_5H_5)(CO)_2]_2$ complexes (M = Fe, Ru)", in *Optical spectra and chemical bonding in inorganic compounds*, vol. 106 of *Structure and Bonding*, pp. 153–172 (Springer, Berlin).

Jensen, K. L., and F. A. Buot, 1991, "Numerical simulation of intrinsic bistability and high-frequency current oscillations in resonant tunneling structures", *Phys. Rev. Lett.*, vol. 66, pp. 1078–1081.

Jensen, L., and P. T. van Duijnen, 2005, "The first hyperpolarizability of p-nitroaniline in 1,4-dioxane: A quantum mechanical/molecular mechanics study", *J. Chem. Phys.*, vol. 123, pp. 074307-1–7.

Jeziorski, B., R. Moszynski, and K. Szalewicz, 1994, "Perturbation Theory Approach to Intermolecular Potential Energy Surfaces of Van der Waals Complexes", *Chem. Rev.*, vol. 94, pp. 1887–1930.

Joachain, C. J., M. Dörr, and N. J. Kylstra, 2000, "High Intensity Laser-Atom Physics", *Adv. Atom. Mol. Opt. Phys.*, vol. 42, pp. 225–286.

Joester, D., E. Walter, M. Losson, R. Pugin, H. P. Merkle, and F. Diederich, 2003, "Amphiphilic Dendrimers: Novel Self-Assembling Vectors for Efficient Gene Delivery", *Angew. Chem.*, vol. 115, pp. 1524–1528.

Jogai, B., 1991, "Effect of many-body corrections on intersubband optical transitions in GaAs–Al_xGa_{1-x}As multiple quantum wells", *J. Vac. Sci. Technol. B*, vol. 9, pp. 2473–2478.

Johnson, B. G., P. M. W. Gill, and J. A. Pople, 1993, "The performance of a family of density-functional methods", *J. Chem. Phys.*, vol. 98, pp. 5612–5626.

Jorge, F. E., J. Autschbach, and T. Ziegler, 2003, "On the Origin of the Optical Activity in the d-d Transition Region of Tris-Bidentate Co(III) and Rh(III) Complexes", *Inorg. Chem.*, vol. 42, pp. 8902–8910.

Jorge, F., J. Autschbach, and T. Ziegler, 2005, "On the Origin of Optical Activity in Tris-diamine Complexes of Co(III) and Rh(III): A Simple Model Based on Time-Dependent Density Function Theory", *J. Am. Chem. Soc.*, vol. 127, pp. 975–985.

Jorgensen, W. L., D. S. Maxwell, J. Tirado-Rives, "Development and Testing of the OPLS All-Atom Force Field on Conformational Energetics and Properties of Organic Liquids", *J. Am. Chem. Soc.*, vol. 118, pp. 11225–11236.

Jung, J., P. Garcia-Gonzalez, J. F. Dobson, and R. W. Godby, 2004, "Effects beyond the random-phase approximation in calculating the interaction between metal films", *Phys. Rev. B*, vol. 70, pp. 205107-1–11.

Kadanoff, L. P., and G. Baym, 1962, *Quantum Statistical Mechanics* (Benjamin, New York).

Kamenev, A., and W. Kohn, "Landauer conductance without two chemical potentials", *Phys. Rev. B*, vol. 63, pp. 155304-1–11.

Kane, E. O., 1957, "Band structure of indium antimonide", *J. Phys. Chem. Solids*, vol. 1, pp. 249–261.

Karplus, M., and J. A. McCammon, 2002, "Molecular dynamics simulations of biomolecules", *Nature Struct. Biol.*, vol. 9, pp. 646–653; *ibid*, vol. 9, p. 788(E) (2002).

Katan, C., F. Terenziani, O. Mongin, M. Werts, L. Porres, T. Pons, J. Mertz, S. Tretiak, and M. Blanchard-Desce, 2005, "Effects of (Multi)branching of Dipolar Chromophores on Photophysical Properties and Two-Photon Absorption", *J. Phys. Chem. A*, vol. 109, pp. 3024–3037.

Katriel, J., and E. R. Davidson, 1980, "Asymptotic Behavior of Atomic and Molecular Wave Functions", *Proc. Natl. Acad. Sci. USA*, vol. 77, pp. 4403–4406.

Kawata, I., H. Kono, Y. Fujimura, and A. D. Bandrauk, 2001, "Intense-laser-field-enhanced ionization of two-electron molecules: Role of ionic states as doorway states", *Phys. Rev. A*, vol. 62, pp. 031401-1–4.

Kawata, I., H. Kono, and A. D. Bandrauk, 2001, "Mechanism of enhanced ionization of linear H_3^+ in intense laser fields", *Phys. Rev. A*, vol. 64, pp. 043411-1–15.

Ke, S., H. U. Baranger, and W. Yang, 2004, "Molecular Conductance: Chemical Trends of Anchoring Groups", *J. Am. Chem. Soc.*, vol. 126, pp. 15897–15904.

Keldysh, L. V., 1965, "Diagram technique for nonequilibrium processes", *Sov. Phys. JETP*, vol. 20, pp. 1018–1026.

Kienberger, R., M. Hentschel, M. Uiberacker, C. Spielmann, M. Kitzler, A. Scrinzi, M. Wieland, T. Westerwalbesloh, U. Kleineberg, U. Heinzmann, M. Drescher, and F. Krausz, 2002, "Steering Attosecond Electron Wave Packets with Light", *Science*, vol. 297, pp. 1144–1148.

Kienberger, R., E. Goulielmakis, M. Uiberacker, A. Baltuska, V. Yakovlev, F. Bammer, A. Scrinzi, T. Westerwalbesloh, U. Kleineberg, U. Heinzmann, M. Drescher, and F. Krausz, 2004, "Atomic transient recorder", *Nature* (London), vol. 427, pp. 817–821.

Kijak, M., A. Zielinska, C. Chamchoumis, J. Herbich, R. Thummel, and J. Waluk, 2004, "Conformational equilibria and photoinduced tautomerization in 2-(2'-pyridyl)pyrrole", *Chem. Phys. Lett.*, vol. 400, pp. 279–285.

Kim, Y.-H., and A. Görling, 2002a, "Excitonic Optical Spectrum of Semiconductors Obtained by Time-Dependent Density-Functional Theory with the Exact-Exchange Kernel", *Phys. Rev. Lett.*, vol. 89, pp. 096402-1–4.

Kim, Y. H., and A. Görling, 2002b, "Exact Kohn-Sham exchange kernel for insulators and its long-wavelength behavior", *Phys. Rev. B*, vol. 66, pp. 035114-1–6.

Kirtman, B., J. L. Toto, K.A. Robins, and M. Hasan, 1995, "Ab initio finite oligomer method for nonlinear optical properties of conjugated polymers. Hartree-Fock static longitudinal hyperpolarizability of polyacetylene", *J. Chem. Phys.*, vol. 102, pp. 5350–5356.

Kitzler, M., J. Zanghellini, C. Jungreuthmayer, M. Smits, A. Scrinzi, and T. Brabec, 2004, "Ionization dynamics of extended multielectron systems", *Phys. Rev. A*, vol. 70, pp. 041401-1–4(R).

Kjellberg, P., Z. He, and T. Pullerits, 2003, "Bacteriochlorophyll in Electric Field", *J. Phys. Chem. B*, vol. 107, pp. 13737–13742.

Klarsfeld, S., and J. A. Oteo, 1989, "Recursive generation of higher-order terms in the Magnus expansion", *Phys. Rev. A*, vol. 39, pp. 3270–3273.

Kobko, N., A. Masunov, and S. Tretiak, 2004, "Calculations of the third-order nonlinear optical responses in push-pull chromophores with a time-dependent density-functional theory", *Chem. Phys. Lett.*, vol. 392, pp. 444–451.

Koch, W., M. C. Holthausen, 2001, *A Chemist's Guide to Density-Functional Theory* (Wiley-VCH, Weinheim).

Kohn, W., 1961 "Cyclotron Resonance and de Haas-van Alphen Oscillations of an Interacting Electron Gas", *Phys. Rev.*, vol. 123, pp. 1242–1244.

Kohn, W., L. J. Sham, 1965, "Self-Consistent Equations Including Exchange and Correlation Effects", *Phys. Rev.*, vol. 140, pp. A1133-A1138.

Kohn, W., 1983a, "v-Representability and Density-Functional Theory", *Phys. Rev. Lett.*, vol. 51, pp. 1596–1598.

Kohn, W., and P. Vashishta, 1983b, "General Density-Functional Theory", in *Theory of the Inhomogeneous Electron Gas*, edited by S. Lundqvist and N. H. March, pp. 79–184 (Plenum Press, New York).

Kohn, W., Y. Meir and D. E. Makarov, 1998, "Van der Waals Energies in Density-Functional Theory", *Phys. Rev. Lett.*, vol. 80, pp. 4153–4156.

Köhn, A., and C. Hättig, 2004, "On the Nature of the Low-Lying Singlet States of 4-(Dimethyl-amino)benzonitrile", *J. Am. Chem. Soc.*, vol. 126, pp. 7399–7410.

Kondo, K., T. Tamida, Y. Nabekawa, and S. Watanabe, 1994, "High-order harmonic generation and ionization using ultrashort KrF and Ti: sapphire lasers", *Phys. Rev. A*, vol. 49, pp. 3881–3889.

Kootstra, F., P. L. de Boeij, and J. G. Snijders, 2000a, "Efficient real-space approach to time-dependent density-functional theory for the dielectric response of nonmetallic crystals", *J. Chem. Phys.*, vol. 112, pp. 6517–6531.

Kootstra, F., P. L. de Boeij, and J. G. Snijders, 2000b, "Application of time-dependent density-functional theory to the dielectric function of various nonmetallic crystals", *Phys. Rev. B*, vol. 62, pp. 7071–7083.

Kornberg, M. A., and P. Lambropoulos, 1999, "Photoelectron energy spectrum in "direct" two-photon double ionization of helium", *J. Phys. B*, vol. 32, pp. L603–L613.

Kosloff, R., 1988, "Time-dependent quantum-mechanical methods for molecular dynamics", *J. Phys. Chem.*, vol. 92, pp. 2087–2100.

Kotani, T., 1995, "Exact Exchange Potential Band-Structure Calculations by the Linear Muffin-Tin Orbital Atomic-Sphere Approximation Method for Si, Ge, C, and MnO", *Phys. Rev. Lett.*, vol. 74, pp. 2989–2992.

Kotani, T., 1998, "An optimized-effective-potential method for solids with exact exchange and random-phase approximation correlation", *J. Phys. Cond. Matt.*, vol. 10, pp. 9241–9261.

Krause, J. L., K. J. Schafer, and K.C. Kulander, 1991, "Optical harmonic generation in atomic and molecular hydrogen", *Chem. Phys. Lett.*, vol. 178, pp. 573–578.

Kreibich, T., 2000, "Multicomponent Density-Functional Theory for Molecules in Strong Laser Fields", Ph.D. Thesis, Universität Würzburg, Shaker-Verlag.

Kreibich, T., and E. K. U. Gross, 2001a, "Multicomponent Density-Functional Theory for Electrons and Nuclei", *Phys. Rev. Lett.*, vol. 86, pp. 2984–2987.

Kreibich, T., M. Lein, V. Engel, and E.K.U. Gross, 2001b, "Even-Harmonic Generation due to Beyond-Born-Oppenheimer Dynamics", *Phys. Rev. Lett.*, vol. 87, pp. 103901-1–4.

Kreibich, T., N. I. Gidopoulos, R. van Leeuwen, and E.K.U. Gross, 2003, "Towards time-dependent density-functional theory for molecules in strong laser pulses", in *Progress in Theoretical Chemistry and Physics*, vol. 14, pp. 69–78 (Springer, Berlin).

Kreibich, T., R. van Leeuwen, and E.K.U. Gross, 2004, "Time-dependent variational approach to molecules in strong laser fields", *Chem. Phys.*, vol. 304, pp. 183–202.

Kreibich, T., R. van Leeuwen, and E.K.U. Gross, 2005, "Multicomponent Density-Functional Theory for Electrons and Nuclei", unpublished.

Kreibig, U., and M. Vollmer, 1993, *Optical Properties of Metal Clusters*, vol. 25 of *Springer Series in Materials Science*, (Springer, Berlin).

Kresse, G., J. Paier, R. Hirsch, M. Marsman, and J. Gerber, 2005, in preparation.

Krieger, J. B., Y. Li, and G. J. Iafrate, 1992a, "Systematic approximations to the optimized effective potential: Application to orbital-density-functional theory", *Phys. Rev. A*, vol. 46, pp. 5453–5458.

Krieger, J. B., Y. Li, and G. J. Iafrate, 1992b, "Construction and application of an accurate local spin-polarized Kohn-Sham potential with integer discontinuity: Exchange-only theory", *Phys. Rev. A*, vol. 45, pp. 101–126.

Kroemer, H., "Nobel Lecture: Quasielectric fields and band offsets: teaching electrons new tricks", *Rev. Mod. Phys.*, vol. 73, pp. 783–793.

Krstić, P. S., D.J Dean, X.G. Zhang, D. Keffer, Y.S. Leng, P.T. Cummings, and J. C. Wells, 2003, "Computational chemistry for molecular electronics", *Comp. Mat. Sci.*, vol. 28, pp. 321–341.

Kruit, P., J. Kimman, H. G. Muller, and M. J. van der Wiel, 1983, "Electron spectra from multiphoton ionization of xenon at 1064, 532, and 355 nm", *Phys. Rev. A*, vol. 28, pp. 248–255.

Ku, W., A. G. Eguiluz, 2002, "Band-Gap Problem in Semiconductors Revisited: Effects of Core States and Many-Body Self-Consistency", *Phys. Rev. Lett.*, vol. 89, pp. 126401-1–4.

Kubo, R., 1957, "Statistical-Mechanical Theory of Irreversible Processes. I. General Theory and Simple Applications to Magnetic and Conduction Problems", *J. Phys. Soc. Jpn.*, vol. 12, pp. 570–586.

Kulander, K.C., 1987, "Multiphoton ionization of hydrogen: A time-dependent theory", *Phys. Rev. A*, vol. 35, pp. 445–447.

Kulander, K.C., 1998, "Time-dependent theory of multiphoton ionization of xenon", *Phys. Rev. A*, vol. 38, pp. 778–787.

Kulander, K.C, K. J. Schafer, and J. L. Krause, 1991a, "Dynamic stabilization of hydrogen in an intense, high-frequency, pulsed laser field", *Phys. Rev. Lett.*, vol. 66, pp. 2601–2604.

Kulander, K.C, K. J. Schafer, and J. L. Krause, 1991b, "Single active electron calculations of multiphoton processes in krypton", *Int. J. Quant. Chem. Symp.*, vol. 25, pp. 415–429.

Kulander, K.C, K. J. Schafer, and J. L. Krause, 1992a, "Time-dependent studies of multiphoton processes", in *Atoms in Intense Laser Fields*, edited by M. Gavrila, pp. 247–300 (Academic Press, Boston).

Krause, J. L., K.J. Schafer, and K. C. Kulander, 1992b, "Calculation of photoemission from atoms subject to intense laser fields", *Phys. Rev. A*, vol. 45, pp. 4998–5010.

Kulander, K.C., K.J. Schafer, and J. L. Krause, 1993, "Dynamics of short-pulse excitation, ionization and harmonic conversion", in *Super-Intense Laser-Atom Physics*, edited by B. Piraux, A. L'Huillier, and K. Rzazewski, NATO ASI Series B316, pp. 95–110 (Plenum Press, New York).

Kümmel, S., M. Brack, and P.-G. Reinhard, 1999, "Structure and optic response of the Na_9^+ and Na_{55}^+ clusters", *Euro. Phys. J.D*, vol. 9, pp. 149–152.

Kümmel, S., and J.P. Perdew, 2003, "Simple Iterative Construction of the Optimized Effective Potential for Orbital Functionals, Including Exact Exchange", *Phys. Rev. Lett.*, vol. 90, pp. 043004-1–4.

Kümmel, S., L. Kronik, and J. P. Perdew, 2004, "Electrical Response of Molecular Chains from Density-Functional Theory", *Phys. Rev. Lett.*, vol. 93, pp. 213002-1–4.

Kunert, T., and R. Schmidt, "Excitation and Fragmentation Mechanisms in Ion-Fullerene Collisions", *Phys. Rev. Lett.*, vol. 86, pp. 5258–5261.

Kurth, S., and J. P. Perdew, 1999, "Density-functional correction of random-phase-approximation correlation with results for jellium surface energies", *Phys. Rev. B*, vol. 59, pp. 10461–10468.

Kurth, S., G. Stefanucci, C.-O. Almbladh, A. Rubio, and E.K.U. Gross, 2005, "Time-dependent quantum transport: A practical scheme using density-functional theory", *Phys. Rev. B*, vol. 72, pp. 035308-1–13

Kurzweil, Y., and R. Baer, 2004, "Time-dependent exchange-correlation current density-functionals with memory", *J. Chem. Phys.*, vol. 121, pp. 8731–8741.

Kwon, O., and M. L. McKee, 2000, "Theoretical Calculations of Band Gaps in the Aromatic Structures of Polythieno[3,4-*b*]benzene and Polythieno[3,4-*b*]pyrazine", *J. Phys. Chem. A*, vol. 104, pp. 7106–7112.

Lafon, R., J. L. Chaloupka, B. Sheehy, P. M. Paul, P. Agostini, K.C. Kulander, and L. F. DiMauro, 2001, "Electron Energy Spectra from Intense Laser Double Ionization of Helium", *Phys. Rev. Lett.*, vol. 86, pp. 2762–2765.

Lahiri, A., J. Ulicny, A. Laaksonen, 2004, "Theoretical Analysis of the Excited State Properties of Wybutine: A Natural Probe for Transfer RNA Dynamics", *Int. J. Mol. Sci.*, vol. 5, pp. 75–83.

Lambropoulos, P., P. Maragakis, and J. Zhang, 1998, "Two-electron atoms in strong fields", *Phys. Rep.*, vol. 305, pp. 203–293.

Landau, L. D., and E. M. Lifshitz, 1969, *Statistical Physics* (Addison-Wesley, Reading, Massachusetts).

Landau, L. D., and E. M. Lifshitz, 1987, *Mechanics of Fluids*, vol. 6 of *Course of Theoretical Physics*, 2nd edition (Pergamon, New York).

Lang, N. D., 1995, "Resistance of atomic wires", *Phys. Rev B*, vol. 52, pp. 5335–5342.

Langbein, D., 1974, *Theory of Van der Waals Attraction, Springer Tracts in Modern Physics* (Springer-Verlag, Berlin).

Langreth, D. C., and J. P. Perdew, 1975, "The exchange-correlation energy of a metallic surface", *Solid State Commun.*, vol. 17, pp. 1425–1429.

Langreth, D. C., 1976, "Linear and Non-Linear Response Theory with Applications", in *Linear and Nonlinear Electron Transport in Solids*, edited by J.T. Devreese and E. van Doren, pp. 3–32 (Plenum, New York).

Langreth, D. C., J. P. Perdew, 1997, "Exchange-correlation energy of a metallic surface: Wave-vector analysis", *Phys. Rev. B*, vol. 15, pp. 2884–2901.

Langreth, D., M. Dion, H. Rydberg, E. Schröder, P. Hyldgaard, B.I. Lundqvist, 2005, "Van der Waals density-functional theory with applications", *Int. J. Quantum Chem.*, vol. 101, pp. 599–610.

Lappas, D., and R. van Leeuwen, 1998, "Electron correlation effects in the double ionization of He", *J. Phys. B: At. Mol. Opt. Phys.*, vol. 31, pp. L249–L256.

Larochelle, S., A. Talebpour, and S. L. Chin, 1998, "Non-sequential multiple ionization of rare gas atoms in a Ti: Sapphire laser field", *J. Phys. B: At. Mol. Opt. Phys.*, vol. 31, pp. 1201–1214.

Lautenschlager, P., M. Garriga, L. Viña, and M. Cardona, 1987. "Temperature dependence of the dielectric function and interband critical points in silicon", *Phys. Rev. B*, vol. 36, pp. 4821–4830.

Lebon, F., G. Longhi, F. Gangemi, S. Abbate, J. Priess, M. Juza, C. Bazzini, T. Caronna, and A. Mele, 2004, "Chiroptical Properties of Some Monoazapentahelicenes", *J. Phys. Chem. A*, vol. 108, pp. 11752–11761.

Lee, C., W. Yang, and R. G. Parr, 1988, "Development of the Colle-Salvetti correlation-energy formula into a functional of the electron density", *Phys. Rev. B*, vol. 37, pp. 785–789.

Légaré, F., I. V. Litvinyuk, P. W. Dooley, F. Quéré, A. D. Bandrauk, D. M. Villeneuve, and P. B. Corkum, 2003, "Time-Resolved Double Ionization with Few Cycle Laser Pulses", *Phys. Rev. Lett.*, vol. 91, pp. 093002-1–4.

Legrand, C, E. Suraud, and P.-G. Reinhard, 2002, "Comparison of self-interaction-corrections for metal clusters", *J. Phys. B: At. Mol. Opt. Phys.*, vol. 35, pp. 1115–1128.

Le Guennic, B., W. Hieringer, A. Görling, and J. Autschbach, 2005, "Density-functional calculation of the electronic circular dichroism spectra of the transition metal complexes $[M(phen)_3]^{2+}$ (M = Fe, Ru, Os)", *J. Phys. Chem. A*, vol. 108, pp. 11752–11761.

Lein, M., J. F. Dobson, and E. K. U. Gross, 1999, "Toward the description of Van der Waals interactions within density-functional theory", *J. Comput. Chem.*, vol. 20, pp. 12–22.

Lein, M., E. K. U. Gross, and V. Engel, 2000a, "Intense-Field Double Ionization of Helium: Identifying the Mechanism", *Phys. Rev. Lett.*, vol. 85, pp. 4707–4710.

Lein, M., E. K. U. Gross, and J. P. Perdew, 2000b, "Electron correlation energies from scaled exchange-correlation kernels: Importance of spatial versus temporal nonlocality", *Phys. Rev. B*, vol. 61, pp. 13431–13437.

Lein, M., and S. Kümmel, 2005, "Exact Time-Dependent Exchange-Correlation Potentials for Strong-Field Electron Dynamics", *Phys. Rev. Lett.*, vol. 94, pp. 143003-1–4.

Leising, G., 1988, "Anisotropy of the optical constants of pure and metallic polyacetylene", *Phys. Rev. B*, vol. 38, pp. 10313–10322.

Levine, Z. H., and D. C. Allan, 1989, "Linear optical response in silicon and germanium including self-energy effects", *Phys. Rev. Lett.*, vol. 63, pp. 1719–1722.

Levy, M., J. P. Perdew, V. Sahni, 1984, "Exact differential equation for the density and ionization energy of a many-particle system", *Phys. Rev. A*, vol. 30, pp. 2745–2748.

Lewenstein, M., P. Balcou, M. Yu. Ivanov, A. L'Huillier, and P. Corkum, 1994, "Theory of high-harmonic generation by low-frequency laser fields", *Phys. Rev. A*, vol. 49, pp. 2117–2132.

L'Huillier, A., L. A. Lompré, G. Mainfray, and C. Manus, 1983, "Multiply charged ions induced by multiphoton absorption in rare gases at 0.53 μm", *Phys. Rev. A*, vol. 27, pp. 2503–2512.

L'Huillier, A., L.A. Lompré, G. Mainfray, and C. Manus, 1992, "High-order harmonic generation in rare gases", in *Atoms in Intense Laser Fields*, edited by M. Gavrila, pp. 139–206 (Academic Press, Boston).

L'Huillier, A., and P. Balcou, 1993, "High-order harmonic generation in rare gases with a 1-ps 1053-nm laser", *Phys. Rev. Lett.*, vol. 70, pp. 774–777.

L'Huillier, A., 2002, "Atoms in strong laser fields", *Europhysics News*, vol. 33, no. 6.

Li, T., and P. Tong, 1985, "Hohenberg-Kohn theorem for time-dependent ensembles", *Phys. Rev. A*, vol. 31, pp. 1950–1951.

Li, T., and P. Tong, 1986, "Time-dependent density-functional theory for multicomponent systems", *Phys. Rev. A*, vol. 34, pp. 529–532.

Li, X. F., A. L'Huillier, M. Ferray, L.A. Lompré, and G. Mainfray, 1989, "Multiple-harmonic generation in rare gases at high laser intensity", *Phys. Rev. A*, vol. 39, pp. 5751–5761.

Li, X.-D., W.-D. Cheng, D.-S. Wu, H. Zhang, Y.-J. Gong, and Y.-Z. Lan, 2003, "Theoretical studies on photophysical properties of fullerene and its two derivatives (C_{60}, $C_{60}COOCH_2$, $C_{60}COOHCH_3$)", *Chem. Phys. Lett.*, vol. 380, pp. 480–485.

Li, X., W. Cheng, D. Wu, Y. Lan, H. Zhang, Y. Gong, F. Li, and J. Shen, 2004, "Modeling of configurations and third-order nonlinear optical properties of C_{36} and $C_{34}X_2$ (X = B,N)", *J. Chem. Phys.*, vol. 121, pp. 5885–5892.

Liang, Y., S. Augst, S. L. Chin, Y. Beaudoin, and M. Chaker, 1994, "High harmonic generation in atomic and diatomic molecular gases using intense picosecond laser pulses - a comparison", *J. Phys. B: At. Mol. Opt. Phys.*, vol. 27, pp. 5119–5130.

Liao, Y., B. Eichinger, K. Firestone, M. Haller, J. Luo, W. Kaminsky, J. Benedict, P. Reid, A. Jen, L. Dalton, and B. Robinson, 2005, "Systematic Study of the Structure-Property Relationship of a Series of Ferrocenyl Nonlinear Optical Chromophores", *J. Am. Chem. Soc.*, vol. 127, pp. 2758–2766.

Lieb, E. H., 1983, "Density-functionals for Coulomb systems", *Int. J. Quant. Chem.*, vol. 24, pp. 243–277.

Lieb, E. H., 1985, "Density-Functionals For Coulomb Systems", in *Density-Functional Methods in Physics*, edited by R. M. Dreizler, and J. da Providência, vol. 123 of *NATO ASI Series; Series B: Physics*, pp. 31–80 (Plenum Press, New York).

Lieb, E. H., J. P. Solovej, and J. Yngvason, 1992, "Heavy atoms in the strong magnetic field of a neutron star", *Phys. Rev. Lett.*, vol. 69, pp. 749–752.

Lipparini, E., and L. Serra, 1998, "Spin-wave excitations in quantum dots", *Phys. Rev. B*, vol. 57, pp. R6830–R6833.

Lipparini, E., M. Barranco, A. Emperador, M. Pi, and L. Serra, 1999, "Transverse dipole spin modes in quantum dots", *Phys. Rev. B*, vol. 60, pp. 8734–8742.

Lipparini, E., L. Serra, and A. Puente, 2002, "Magnetic dipole and electric quadrupole responses of elliptic quantum dots in magnetic fields", *Eur. Phys. J.B*, vol. 27, pp. 409–415.

Littlejohn, R. G., and M. Reinsch, "Gauge fields in the separation of rotations and internal motions in the n-body problem", *Rev. Mod. Phys.*, vol. 69, pp. 213–276.

Liu, H. C., and F. Capasso (eds.), 2000, *Intersubband Transitions in Quantum Wells I* and *II*, vols. 62 and 66 of *Semiconductors and Semimetals* (Academic Press, San Diego).

Liu, X. J., J.K. Feng, A.M. Ren, and X. Zhou, 2003, "Theoretical studies of the spectra and two-photon absorption cross sections for porphyrin and carbaporphyrins", *Chem. Phys. Lett.*, vol. 373, pp. 197–206.

Liu, X., H. Rottke, E. Eremina, W. Sandner, E. Goulielmakis, K. O. Keeffe, M. Lezius, F. Krausz, F. Lindner, M.G. Schätzel, G.G. Paulus, and H. Walther, 2004, "Nonsequential Double Ionization at the Single-Optical-Cycle Limit", *Phys. Rev. Lett.*, vol. 93, pp. 263001-1–4.

Liyanage, P., R. de Silva, and K. de Silva, 2003, "Nonlinear optical (NLO) properties of novel organometallic complexes: high accuracy density-functional theory (DFT) calculations", *J. Mol. Struct. (Theochem)*, vol. 639, pp. 195–201.

Llano, J., J. Raber, and L.A. Eriksson, 2003, "Theoretical study of phototoxic reactions of psoralens", *J. Photochem. Photobiol. A*, vol. 154, pp. 235–243.

Longuet-Higgins, H. C., 1965, "Spiers memorial lecture. Intermolecular Forces", Discussions of the Faraday Society, vol. 40, pp. 7–18.

Loridant-Bernard, D., S. Mezière, M. Constant, N. Dupuy, B. Sombret, and J. Chevallier, 1998, "Infrared study of light-induced reactivation of neutralized dopants in hydrogenated *n*-type GaAs doped with silicon", *Appl. Phys. Lett.*, vol. 73, pp. 644–646.

Louck, J. D., 1976, "Derivation of the Molecular Vibration-Rotation Hamiltonian from the Schrödinger equation for the Molecular Model", *J. Mol. Spectr.*, vol. 61, pp. 107–137.

Louisell, W. H., 1973, *Quantum Statistical Properties of Radiation* (Wiley, New York).

Löwdin, P.-O., and P. K. Mukherjee, 1972, "Some comments on the time-dependent variation principle", *Chem. Phys. Lett.*, vol. 14, pp. 1–7.

Lubich, C., 2002, "Integrators for Quantum Dynamics: A Numerical Analyst's Brief Review", in *Quantum simulations of complex many-body systems: From theory to algorithms*, edited by J. Grotendorst, D. Marx, and A. Muramatsu, vol. 10 of *NIC Series*, pp. 459–466, (John von Neumann Institute for Computing, Jülich).

Lucchese, R. R., and V. McKoy, 1980, "Application of the Schwinger variational principle to electron-ion scattering in the static-exchange approximation", *Phys. Rev. A*, vol. 21, pp. 112–123.

Luttinger, J. M., and J. C. Ward, "Ground-State Energy of a Many-Fermion System. II", *Phys. Rev.*, vol. 118, pp. 1417–1427.

Ma, J., S. Li, and Y. Jiang, 2002, "A Time-Dependent DFT Study on Band Gaps and Effective Conjugation Lengths of Polyacetylene, Polyphenylene, Polypentafulvene, Polycyclopentadiene, Polypyrrole, Polyfuran, Polysilole, Polyphosphole, and Polythiophene", *Macromol.*, vol. 35, pp. 1109–1115.

Macklin, J. J., J. D. Kmetec, and C. L. Gordon III, 1993, "High-order harmonic generation using intense femtosecond pulses", *Phys. Rev. Lett.*, vol. 70, pp. 766–769.

Macleod, N., P. Butz, J. Simons, G. Grant, C. Baker, and G. Tranter, "Electronic Circular Dichroism Spectroscopy of 1-(R)-Phenylethanol: The "Sector Rule" Revisited and an Exploration of Solvent Effects", *Isr. J. Chem.*, vol. 44, pp. 27–36.

Magnus, W., "On the exponential solution of differential equations for a linear operator", *Commun. Pure Appl. Math.*, vol. 7, pp. 649–673.

Magyar, R. J., A. Fleszar, and E.K.U. Gross, 2004, "Exact-exchange density-functional calculations for noble-gas solids", *Phys. Rev. B*, vol. 69, pp. 045111-1–7.

Magyar, R. J., and K. Burke, 2004, "Density-functional theory in one dimension for contact-interacting fermions", *Phys. Rev. A*, vol. 70, pp. 032508-1–8.

Magyar, R., S. Tretiak, Y. Gao, H. Wang, and A. Shreve, 2005, "A joint theoretical and experimental study of phenylene-acetylene molecular wires", *Chem. Phys. Lett.*, vol. 401, pp. 149–156.

Mahan, G. E., 1980, "Modified Sternheimer equation for polarizability", *Phys. Rev. A*, vol. 22, pp. 1780–1785.

Mahan, G. D., and B. E. Sernelious, "Electron-electron interactions and the bandwidth of metals", *Phys. Rev. Lett.*, vol. 62, pp. 2718–2720.

Mahan, G.E., and K. R. Subbaswamy, 1990, *Local Density Theory of Polarizability* (Plenum, New York).

Mahanty, J., and B. W. Ninham, 1976, *Dispersion Forces* (Academic Press, London).

Mainfray, G., and C. Manus, 1991, "Multiphoton ionization of atoms", *Rep. Prog. Phys.*, vol. 54, pp. 1333–1372.

Maitra, N. T., and K. Burke, 2001, "Demonstration of initial-state dependence in time-dependent density-functional theory", *Phys. Rev. A*, vol. 63, pp. 042501-1–7; *ibid* vol. 64, p. 039901(E) (2001).

Maitra, N. T., and K. Burke, 2002a, "On the Floquet formulation of time-dependent density-functional theory", *Chem. Phys. Lett.*, vol. 359, pp. 237–240.

Maitra, N. T., K. Burke, and C. Woodward, 2002b, "Memory in Time-Dependent Density-Functional Theory", *Phys. Rev. Lett.*, vol. 89, pp. 023002-1–4.

Maitra, N. T., K. Burke, H. Appel, E. K. U. Gross, and R. van Leeuwen, 2002c, "Ten topical questions in time-dependent density-functional theory", in *Reviews in Modern Quantum Chemistry, A celebration of the contributions of Robert Parr*, edited by K. D. Sen, pp. 1186–1225 (World Scientific, Singapore).

Maitra, T. N., I. Souza, and K. Budrke, 2003a, "Current-density-functional theory of the response of solids", *Phys. Rev. B*, vol. 68, pp. 045109-1–5.

Maitra, N. T., A. Wasserman, and K. Burke, 2003b, "What is time-dependent density-functional theory? Successes and Challenges", in *Electron Correlations and Materials Properties 2*, edited by A. Gonis, N. Kioussis, and M. Ciftan, pp. 285–298 (Kluwer Academic/Plenum Publishers).

Maitra, N., F. Zhang, R. Cave, and K. Burke, 2004, "Double excitations within time-dependent density-functional theory linear response", *J. Chem. Phys.*, vol. 120, pp. 5932–5937.

Maitra, N. T., 2005a, "Memory formulas for perturbations in time-dependent density-functional theory", *Int. J. Quant. Chem.*, vol. 102, pp. 573–581.

Maitra, N. T., 2005b, "Undoing static correlation: Long-range charge transfer in time-dependent density-functional theory", *J. Chem. Phys.*, vol. 122, pp. 234104-1–6.

Major, D. T., and B. Fischer, 2003, "Theoretical Study of the pH-Dependent Photophysics Of N1,N^6-Ethenoadenine and N3,N^4-Ethenocytosine", *J. Phys. Chem. A*, vol. 107, pp. 8923–8931.

Maksym, P. A., and T. Chakraborty, 1990, "Quantum dots in a magnetic field: Role of electron-electron interactions", *Phys. Rev. Lett.*, vol. 65, pp. 108–111.

Malloci, G., G. Mulas, and C. Joblin, 2004, "Electronic absorption spectra of PAHs up to vacuum UV: Towards a detailed model of interstellar PAH photophysics", *Astron. Astrophys.*, vol. 426, pp. 105–117.

Marangos, J. P., 2004, "Molecules in a strong laser field", in *Atoms and Plasmas in Super-Intense Laser Fields*, edited by D. Batani, C. J. Joachain, and S. Martellucci, SIF Conference Proceedings, vol. 88, pp. 213–243 (Società Italiana di Fisica, Bologna).

Marini, A., R. Del Sole, 2003a, "Dynamical Excitonic Effects in Metals and Semiconductors", *Phys. Rev. Lett.*, vol. 91, 176402-1–4.

Marini, A., R. Del Sole, and A. Rubio, 2003b, "Bound Excitons in Time-Dependent Density-Functional Theory: Optical and Energy-Loss Spectra", *Phys. Rev. Lett.*, vol. 91, 256402-1–4.

Marini, A., A. Rubio, 2004, "Electron linewidths of wide-gap insulators: Excitonic effects in LiF", *Phys. Rev. B*, vol. 70, 081103-1–4(R).

Marini, A., P. García-González, A. Rubio, 2006, "First-principle description of correlation effects in layered materials", *Phys. Rev. Lett.*, vol. 96, pp. 136404-1–3.

Markevitch, A. N., S. M. Smith, D.A. Romanov, H. B. Schlegel, M. Yu. Ivanov, and R. J. Levis, 2003, "Nonadiabatic dynamics of polyatomic molecules and ions in strong laser fields", *Phys. Rev. A*, vol. 68, pp. 011402-1–4(R)

Markevitch, A. N., D. A. Romanov, S. M. Smith, H. B. Schlegel, M. Yu. Ivanov, and R. J. Levis, 2004, "Sequential nonadiabatic excitation of large molecules and ions driven by strong laser fields", *Phys. Rev. A*, vol. 69, pp. 013401-1–13.

Marmorkos, I. K., and S. Das Sarma, 1993, "Interacting intersubband excitations in parabolic semiconductor quantum wells", *Phys. Rev. B*, vol. 48, pp. 1544–1561.

Marques, M.A.L., A. Castro, and A. Rubio, 2001, "Assessment of exchange-correlation functionals for the calculation of dynamical properties of small clusters in time-dependent density-functional theory", *J. Chem. Phys.*, vol. 115, pp. 3006–3014.

Marques, M.A.L., X. Lopez, D. Varsano, A. Castro, and A. Rubio, 2003a, "Time-Dependent Density-Functional Approach for Biological Chromophores: The Case of the Green Fluorescent Protein", *Phys. Rev. Lett.*, vol. 90, pp. 258101-1–4.

Marques, M.A.L., A. Castro, G. F. Bertsch, and A. Rubio, 2003b, "octopus: a first-principles tool for excited electron-ion dynamics", *Comput. Phys. Commun.*, vol. 151, pp. 60–78.

Marques, M.A.L., and E. K. U. Gross, 2004, "Time-Dependent Density-Functional Theory" *Annu. Rev. Phys. Chem.*, vol. 55, pp. 427–455.

Martin, P. C., and J. Schwinger, 1959, "Theory of Many-Particle Systems. I", *Phys. Rev.*, vol. 115, pp. 1342–1373.

Martin, R. M., and G. Ortiz, 1997a, "Functional theory of extended Coulomb systems", *Phys. Rev. B*, vol. 56, pp. 1124–1140.

Martin, J.D.D., and J. W. Hepburn, 1997b, "Electric Field Induced Dissociation of Molecules in Rydberg-like Highly Vibrationally Excited Ion-Pair States", *Phys. Rev. Lett.*, vol. 79, pp. 3154–3157.

Maruyama, S., R. Kojima, Y. Miyauchi, S. Chiashi, and M. Kohno, 2002, "Low-temperature synthesis of high-purity single-walled carbon nanotubes from alcohol", *Chem. Phys. Lett.*, vol. 360, pp. 229–234.

Masson, W. P. (ed.), 1964, *Physical Acoustic*, vol. I (Academic Press, New York).

Masunov, A. M., and S. Tretiak, 2004, "Prediction of two-photon absorption properties for organic chromophores using time-dependent density-functional theory", *J. Phys. Chem. B*, vol. 108, pp. 899–907.

Mattuck, R.D., 1967, *A Guide to Feynman Diagrams in the Many-Body Problem* (McGraw-Hill, London).

Maurice, D., and M. Head-Gordon, 1995, "Configuration interaction with single substitutions for excited states of open-shell molecules", *Int. J. Quant. Chem. Symp.*, vol. 29, pp. 361–370.

May, V., and O. Kühn, 2004, *Charge and Energy Transfer Dynamics in Molecular Systems*, 2nd edition (Wiley-VCH, Weinheim).

Mazzoni, M. S. C., H. Chacham, P. Ordejón, D. Sánchez-Portal, J. M. Soler, and E. Artacho, 1999, "Energetics of the oxidation and opening of a carbon nanotube", *Phys. Rev. B*, vol. 60, pp. R2208–R2211.

McCammon, J. A., and S. C. Harvey, 1987, *Dynamics of Proteins and Nucleic Acids* (Cambridge University Press, Cambridge).

McHugh, K. M., J.G. Eaton, G.H. Lee, H.W. Sarkas, L.H. Kidder, J.T. Snodgrass, M.R. Manaa, and K.H. Bowen, 1989, "Photoelectron spectra of the alkali metal cluster anions: $Na^-_{n=2-5}$, $K^-_{n=2-7}$, $Rb^-_{n=2-3}$, and $Cs^-_{n=2-3}$", *J. Chem. Phys.*, vol. 91, pp. 3792–3793.

McLachlan, A. D., 1963a, *Proc. Roy. Soc.*, vol. 271, p. 387.

McLachlan, A. D., 1963b, *Proc. Roy. Soc.*, vol. 274, p. 80.

McPherson, A., G. Gibson, H. Jara, U. Johann, T. S. Luk, I. A. McIntyre, K. Boyer, and C. K. Rhodes, 1987, "Studies of multiphoton production of vacuum-ultraviolet radiation in the rare gases", *J. Opt. Soc. Am. B*, vol. 4, pp. 595–601.

McWeeny, R., 1989, *Methods of Molecular Quantum Mechanics*, 2nd edition (Academic Press, London).

Mearns, D., and W. Kohn, 1987, "Frequency-dependent v-representability in density-functional theory", *Phys. Rev. A*, vol. 35, pp. 4796–4799.

Mennucci, B., A. Toniolo, and J. Tomasi, 2001, "Theoretical Study of the Photophysics of Adenine in Solution: Tautomerism, Deactivation Mechanisms, and Comparison with the 2-Aminopurine Fluorescent Isomer", *J. Phys. Chem. A*, vol. 105, pp. 4749–4757.

Mermin, N. D., 1965, "Thermal Properties of the Inhomogeneous Electron Gas", *Phys. Rev.*, vol. 137, pp. A1441–A1443.

Mermin, N. D., 1970, "Lindhard Dielectric Function in the Relaxation-Time Approximation", *Phys. Rev. B*, vol. 1, pp. 2362–2363.

Meyer, H., 2002, "The Molecular Hamiltonian", *Annu. Rev. Phys. Chem.*, vol. 53, pp. 141–172.

Mikhailova, T. Y., and V. I. Pupyshev, 1999, "Symmetric approximations for the evolution operator", *Phys. Lett. A*, vol. 257, pp. 1–6.

Milfeld, K. F., and R. E. Wyatt, 1983, "Study, extension, and application of Floquet theory for quantum molecular systems in an oscillating field", *Phys. Rev. A*, vol. 27, pp. 72–94.

Miller, R. J. D., G. McLendon, A. Nozik, W. Schmickler, F. Willig, 1995, *Surface electron-transfer processes* (VCH Publishers, New York).

Milton, K. A., 2001, *The Casimir Effect: physical manifestations of zero-point energy* (World Scientific, Singapore).

Mintmire, J. W., B. I. Dunlap, and C. T. White, 1992, "Are fullerene tubules metallic?", *Phys. Rev. Lett.*, vol. 68, pp. 631–634.

Misquitta, A. J., B. Jeziroski, and K. Szalewicz, 2003, "Dispersion Energy from Density-Functional Theory Description of Monomers", *Phys. Rev. Lett.*, vol. 91, pp. 033201-1–4.

Miyake, T., F. Aryasetiawan, T. Kotani, M. van Schilfgaarde, M. Usuda, K. Terakura, 2002, "Total energy of solids: An exchange and random-phase approximation correlation study", *Phys. Rev. B*, vol. 66, pp. 245103-1–4.

Miyamoto, Y., O. Sugino, and Y. Mochizuki, 1999, "Real-time electron-ion dynamics for photoinduced reactivation of hydrogen-passivated donors in GaAs", *Appl. Phys. Lett.*, vol. 75, pp. 2915–2917.

Miyamoto, Y., and O. Sugino, 2000, "First-principles electron-ion dynamics of excited systems: H-terminated Si(111) surfaces", *Phys. Rev. B*, vol. 62, pp. 2039–2044.

Miyamoto, Y., 2001, "Anti-bonding driving caused by electron emission: halogen desorption from Si surfaces", *Solid State Comm.*, vol. 117, pp. 727–732.

Miyamoto, Y., S. Berber, M. Yoon, A. Rubio, and D. Tománek, 2002, "Onset of nanotube decay under extreme thermal and electronic excitations", *Physica B*, vol. 323, pp. 78–85.

Miyamoto, Y., S. Berber, M. Yoon, A. Rubio, and D. Tománek, 2004a, "Can photo excitations heal defects in carbon nanotubes?", *Chem. Phys. Lett.*, vol. 392, pp. 209–213.

Miyamoto, Y., A. Rubio, S. Berber, M. Yoon, and D. Tománek, 2004b, "Spectroscopic characterization of Stone-Wales defects in nanotubes", *Phys. Rev. B*, vol. 69, pp. 121413-1–4(R).

Miyamoto, Y., A. Rubio, and D. Tománek, 2004c, "Photodesorption of oxygen from carbon nanotubes", *Phys. Rev. B*, vol. 70, pp. 233408-1–4.

Miyazaki, K., and H. Sakai, 1992, "High-order harmonic generation in rare gases with intense subpicosecond dye laser pulses", *J. Phys. B: At. Mol. Opt. Phys.*, vol. 25, pp. L83–L89.

Mizuta, H., and C.J. Goodings, 1991, "Transient quantum transport simulation based on the statistical density matrix", *J. Phys.: Condens. Matter*, vol. 3, pp. 3739–3756.

Moler, C., and C. van Loan, 2003, "Nineteen Dubious Ways to Compute the Exponential of a Matrix, Twenty-Five Years Later", *SIAM Review*, vol. 45, pp. 3–49.

Monat, J. E., J. H. Rodriguez, and J. K. McCusker, 2002, "Ground- and Excited-State Electronic Structures of the Solar Cell Sensitizer Bis(4,4'-dicarboxylato-2,2'-bipyridine)bis(isothiocyanato)ruthenium(II)", *J. Phys. Chem. A*, vol. 106, pp. 7399–7406.

Montag, B., Th. Hirschmann, J. Meyer, P.-G. Reinhard, and M. Brack, 1995a, "Shape isomerism in sodium clusters with $10 \leq Z \leq 44$: Jellium model with quadrupole, octupole, and hexadecapole deformations", *Phys. Rev. B*, vol. 52, pp. 4775–4778.

Montag, B., and P.-G. Reinhard, 1995b, "Ionic structure and global deformation of axially symmetric simple metal clusters", *Z. Phys. D*, vol. 33, pp. 265–279.

Moore, J. H., and N. D. Spencer (eds.), 2001, *Encyclopedia of Chemical Physics and Physical Chemistry*, vol. I-III (Institute of Physics).

Mori-Sanchez, P., Q. Wu, and W. T. Yang, 2003, "Accurate polymer polarizabilities with exact exchange density-functional theory", *J. Chem. Phys.*, vol. 119, pp. 11001–11004.

Moroni, S., D. M. Ceperley, and G. Senatore, 1995, "Static Response and Local Field Factor of the Electron Gas", *Phys. Rev. Lett.*, vol. 75, pp. 689–692.

Morrison, M.A., 1983, "The Physics of Low-energy Electron-Molecule Collisions: A Guide for the Perplexed and the Uninitiated", *Aust. J. Phys.*, vol. 36, p. 239

Moshammer, R., B. Feuerstein, W. Schmitt, A. Dorn, C.D. Schröter, J. Ullrich, H. Rottke, C. Trump, M. Wittmann, G. Korn, K. Hoffmann, and W. Sandner, 2000, "Momentum Distributions of Ne^{n+} Ions Created by an Intense Ultrashort Laser Pulse", *Phys. Rev. Lett.*, vol. 84, pp. 447–450.

Moyer, C., 2004, "Numerov extension of transparent boundary conditions for the Schrödinger equation in one dimension", *Am. J. Phys.*, vol. 72, pp. 351–358.

Mukamel, S., 1995, *Principles of Nonlinear Optical Spectroscopy*, (Oxford University Press, New York).

Mukamel, S., 2003, "Superoperator representation of nonlinear response: Unifying quantum field and mode coupling theories", *Phys. Rev. E*, vol. 68, pp. 021111-1–14.

Mukamel, S., 2005, "Generalized time-dependent density-functional-theory response functions for spontaneous density fluctuations and response: Resolving the causality paradox in real time", *Phys. Rev. A*, vol. 71, pp. 024503-1–4.

Mulliken, R. S., 1939, "Intensities of Electronic Transitions in Molecular Spectra II. Charge-Transfer Spectra", *J. Chem. Phys.*, vol. 7, pp. 20–34.

Mundt, M., and S. Kümmel, 2005, "Derivative discontinuities in time-dependent density-functional theory", *Phys. Rev. Lett.*, vol. 95, pp. 203004-1–4.

Nakata, A., T. Baba, H. Takahashi, and H. Nakai, 2003, "Theoretical study on the excited states of psoralen compounds bonded to a thymine residue", *J. Comput. Chem.*, vol. 25, pp. 179–188.

Nakatsuji, H., 2000, "Structure of the exact wave function", *J. Chem. Phys.*, vol. 113, pp. 2949–2956.

Negele, J. W., and H. Orland, 1998, *Quantum Many Particle Systems* (Addison-Wesley, New York).

Neiss, C., P. Saalfrank, M. Parac, and S. Grimme, 2003, "Quantum Chemical Calculation of Excited States of Flavin-Related Molecules", *J. Phys. Chem. A*, vol. 107, pp. 140–147.

Neugebauer, J., M.J́. Louwerse, E. J. Baerends, and T. A. Wesolowski, 2005, "The merits of the frozen-density embedding scheme to model solvatochromic shifts", *J. Chem. Phys.*, vol. 122, pp. 094115-1–13.

Nguyen, K. A., and R. Pachter, 2001, "Ground state electronic structures and spectra of zinc complexes of porphyrin, tetraazaporphyrin, tetrabenzoporphyrin, and phthalocyanine: A density-functional theory study", *J. Chem. Phys.*, vol. 114, pp. 10757–10767.

Nguyen, K. A., J. Kennel, and R. Pachter, 2002a, "A density-functional theory study of phosphorescence and triplet-triplet absorption for nonlinear absorption chromophores", *J. Chem. Phys.*, vol. 117, pp. 7128–7136.

Nguyen, K. A., P. N. Day, R. Pachter, S. Tretiak, V. Chernyak, and S. Mukamel, 2002b, "Analysis of Absorption Spectra of Zinc Porphyrin, Zinc meso-Tetraphenylporphyrin, and Halogenated Derivatives", *J. Phys. Chem. A*, vol. 106, pp. 10285–10293.

Nguyen, K. A., and R. Pachter, 2003, "Jahn-Teller triplet excited state structures and spectra of zinc complexes of porphyrin and phthalocyanine: A density-functional theory study", *J. Chem. Phys.*, vol. 118, pp. 5802–5810.

Nguyen, H. S., A. D. Bandrauk, and C. A. Ullrich, 2004, "Asymmetry of above-threshold ionization of metal clusters in two-color laser fields: A time-dependent density-functional study", *Phys. Rev. A*, vol. 69, pp. 063415-1–8.

Niehaus, T. A., S. Suhai, F. Della Sala, P. Lugli, M. Elstner, G. Seifert, and T. Frauenheim, 2001, "Tight-binding approach to time-dependent density-functional response theory", *Phys. Rev. B*, vol. 63, pp. 085108-1–9.

Niehaus, T. A., D. Heringer, B. Torralva, and Th. Frauenheim, 2005, "Importance of electronic self-consistency in the TDDFT based treatment of nonadiabatic molecular dynamics", Eur. Phys. J. D, vol. 35, pp. 467-477.

Nielsen, S. B., A. Lapierre, J. U. Andersen, U. V. Pedersen, S. Tomita and L. H. Andersen, "Absorption Spectrum of the Green Fluorescent Protein Chromophore Anion *In Vacuo*", *Phys. Rev. Lett.*, vol. 87, pp. 228102-1–4.

Nielsen, S. B, and T. I. Solling, 2005, "Are conical intersections responsible for the ultrafast processes of adenine, protonated adenine, and the corresponding nucleosides?", *ChemPhysChem.*, vol. 6, pp. 1276–1281.

Nifosì, R., S. Conti, and M. P. Tosi, 1998, "Dynamic exchange-correlation potentials for the electron gas in dimensionality $D = 3$ and $D = 2$", *Phys. Rev. B*, vol. 58, pp. 12758–12769.

Niquet, Y. M., M. Fuchs, and X. Gonze, 2003a, "Comment on "Investigation" of the Correlation Potential from Kohn-Sham Perturbation Theory", *Phys. Rev. Lett.*, vol. 90, pp. 219301-1–4.

Niquet, Y. M., M. Fuchs, and X. Gonze, 2003b, "Asymptotic behavior of the exchange-correlation potentials from the linear-response Sham-Schlüter equation", *J. Chem. Phys.*, vol. 118, pp. 9504–9518.

Niquet, Y. M., M. Fuchs, and X. Gonze, 2003c, "Exchange-correlation potentials in the adiabatic connection fluctuation-dissipation framework", *Phys. Rev. A*, vol. 68, pp. 032507-1–13.

Niquet, Y. M., and X. Gonze, 2004, "Band-gap energy in the random-phase approximation to density-functional theory", *Phys. Rev. B*, vol. 70, pp. 245115-1–12.

Niquet, Y. M., M. Fuchs, and X. Gonze, 2005, "Avoiding asymptotic divergence of the potential from orbital- and energy-dependent exchange-correlation functionals", *Int. J. Quantum. Chem.*, vol. 101, pp. 635–644.

Nitzan, A., and M. A. Ratner, 2003, "Electron Transport in Molecular Wire Junctions", *Science*, vol. 300, pp. 1384–1389.

Nobusada, K., and K. Yabana, 2004, "High-order harmonic generation from silver clusters: Laser-frequency dependence and the screening effect of *d* electrons" *Phys. Rev. A*, vol. 70, pp. 043411-1–7.

Nooijen, M., 2000, "Can the Eigenstates of a Many-Body Hamiltonian Be Represented Exactly Using a General Two-Body Cluster Expansion?", *Phys. Rev. Lett.*, vol. 84, pp. 2108–2111.

Nozières, P., and D. Pines, 1958, "Correlation Energy of a Free Electron Gas", *Phys. Rev.*, vol. 111, pp. 442–454.

Nozières, P., *Theory of Interacting Fermi Systems* (Benjamin, New-York).

Nozières, P., and D. Pines, 1999, *The Theory of Quantum Liquids*, Chap. 4 (Perseus Books, Cambridge, Massachusetts).

Nunes, R. W., and X. Gonze, 2001, "Berry-phase treatment of the homogeneous electric field perturbation in insulators", *Phys. Rev. B*, vol. 63, pp. 155107-1–22.

Odelius, M., B. Kirchner, and J. Hutter, 2004, "s-Tetrazine in Aqueous Solution: A Density-Functional Study of Hydrogen Bonding and Electronic Excitations", *J. Phys. Chem. A*, vol. 108, pp. 2044–2052.

Odifreddi, P., 2004, *Il diavolo in cattedra* (Einaudi Tascabili, Torino).

Öğüt, S., J. R. Chelikowsky, and S. G. Louie, 1997, "Quantum Confinement and Optical Gaps in Si Nanocrystals", *Phys. Rev. Lett.*, vol. 79, pp. 1770–1773.

Olevano, V., and L. Reining, 2001, "Excitonic Effects on the Silicon Plasmon Resonance", *Phys. Rev. Lett.*, vol. 86, pp. 5962–5965.

Olsen, J., H. J. A. Jensen, and P. Jørgensen, 1988, "Solution of the large matrix equations which occur in response theory", *J. Comput. Phys.*, vol. 74, pp. 265–282.

Onida, G., L. Reining, R. W. Godby, R. Del Sole and W. Andreoni, 1995, "*Ab Initio* Calculations of the Quasiparticle and Absorption Spectra of Clusters: The Sodium Tetramer", *Phys. Rev. Lett.*, vol. 75, pp. 818–821.

Onida, G., L. Reining, and A. Rubio, 2002, "Electronic excitations: density-functional versus many-body Green's-function approaches", *Rev. Mod. Phys.*, vol. 74, pp. 601-1–59.

Ortiz, G., I. Souza, and R. M. Martin, 1998, "Exchange-Correlation Hole in Polarized Insulators: Implications for the Microscopic Functional Theory of Dielectrics", *Phys. Rev. Lett.*, vol. 80, pp. 353–356.

Osinga, V. P., S. J. A. van Gisbergen, J. G. Snijders, and E. J. Baerends, 1997, "Density-functional results for isotropic and anisotropic multipole polarizabilities and C_6, C_7, and C_8 Van der Waals dispersion coefficients for molecules", *J. Chem. Phys.*, vol. 106, pp. 5091–5101.

Pacheco, J. M., and W. Ekardt, 1992a, "A new formulation of the dynamical response of many-electron systems and the photoelectron cross-section of small metal clusters", *Z. f. Physik D*, vol. 24, p. 65.

Pacheco, J. M., and W. Ekardt, 1992b, "Response of finite many-electron systems beyond the time-dependent LDA: application to small metal clusters", Annalen der Physik vol. 1, p. 254.

Pacheco, J. M., and J. P. P. Ramalho, 1997, "First-Principles Determination of the Dispersion Interaction between Fullerenes and Their Intermolecular Potential", *Phys. Rev. Lett.*, vol. 79, pp. 3873–3876.

Parac, M., and S. Grimme, 2002, "Comparison of Multireference Möller-Plesset Theory and Time-Dependent Methods for the Calculation of Vertical Excitation Energies of Molecules", *J. Phys. Chem. A*, vol. 106, pp. 6844–6850.

Parac, M., and S. Grimme, 2003, "A TDDFT study of the lowest excitation energies of polycyclic aromatic hydrocarbons", *Chem. Phys.*, vol. 292, pp. 11–21.

Park, T. J., and J. C. Light, 1986, "Unitary quantum time evolution by iterative Lanczos reduction", *J. Chem. Phys.*, vol. 85, pp. 5870–5876.

Park, C-H., C. D. Spataru, and S. G. Louie, 2005, "Excitons and Many-Electron Effects in the Optical Response of Single-Walled Boron Nitride Nanotubes", cond-mat/0508705.

Parker, J., K. T. Taylor, C. W. Clark, and S. Blodgett-Ford, 1996, "Intense-field multiphoton ionization of a two-electron atom", *J. Phys. B: At. Mol. Opt. Phys.*, vol. 29, pp. L33–L42.

Parker, J. S., E. S. Smyth, and K.T. Taylor, 1998, "Intense-field multiphoton ionization of helium", *J. Phys. B: At. Mol. Opt. Phys.* , vol. 31, pp. L571–L578.

Parker, J, L. R. Moore, D. Dundas, and K. T. Taylor, 2000, "Double ionization of helium at 390 nm", *J. Phys. B: At. Mol. Opt. Phys.*, vol. 33, pp. L691-L698.

Parker, J. S., L. R. Moore, K. J. Meharg, D. Dundas, and K. T. Taylor, 2001, "Double-electron above threshold ionization of helium", *J. Phys. B: At. Mol. Opt. Phys.*, vol. 34, pp. L69–L78.

Parusel, A. B. J., G. Köhler, and S. Grimme, 1998, "Density-Functional Study of Excited Charge Transfer State Formation in 4-(*N*,*N*-Dimethylamino)benzonitrile", *J. Phys. Chem. A*, vol. 102, pp. 6297–6306.

Parusel, A. B. J., and A. Ghosh, 2000a, "Density-Functional Theory Based Configuration Interaction Calculations on the Electronic Spectra of Free-Base Porphyrin, Chlorin, Bacteriochlorin, and *cis*- and *trans*-Isobacteriochlorin", *J. Phys. Chem. A*, vol. 104, pp. 2504–2507.

Parusel, A. B. J., T. Wondimagegn, and A. Ghosh, 2000b, "Do Nonplanar Porphyrins Have Red-Shifted Electronic Spectra? A DFT/SCI Study and Reinvestigation of a Recent Proposal", *J. Am. Chem. Soc.*, vol. 122, pp. 6371–6374.

Parusel, A. B. J., 2000c, "A DFT/MRCI study on the excited state charge transfer states of N-pyrrolobenzene, N-pyrrolobenzonitrile and 4-N,N-dimethylaminobenzonitrile", *Phys. Chem.* , vol. 2, pp. 5545-5552.

Parusel, A. B. J., and S. Grimme, 2001a, "DFT/MRCI calculations on the excited states of porphyrin, hydroporphyrins, tetrazaporphyrins and metalloporphyrins", *J. Porph. Phthal.*, vol. 5, pp. 225–232.

Parusel, A. B. J., and G. Köhler, 2001b, "Influence of the alkyl chain length on the excited-state properties of 4-dialkyl-benzonitriles. A theoretical DFT/MRCI study", *Int. J. Quant. Chem.*, vol. 84, pp. 149–156.

Parusel, A. B. J., 2001c, "Excited state intramolecular charge transfer in N,N-heterocyclic-4-aminobenzonitriles: a DFT study", *Chem. Phys. Lett.*, vol. 340, pp. 531–537.

Parusel, A. B. J., W. Rettig, and W. Sudholt, 2002, "A Comparative Theoretical Study on DMABN: Significance of Excited State Optimized Geometries and Direct Comparison of Methodologies", *J. Phys. Chem. A*, vol. 106, pp. 804–815.

Pask, J. E., B. M. Klein, P. A. Sterne, and C. Y. Fong, 2001, "Finite-element methods in electronic-structure theory", *Comput. Phys. Commun.*, vol. 135, pp. 1–34.

Pasquarello, A., and A. Quattropani, 1993, "Application of variational techniques to time-dependent perturbation theory", *Phys. Rev. B*, vol. 48, pp. 5090–5094.

Patton, D. C., and M. R. Pederson, 1997a, "Application of the generalized-gradient approximation to rare-gas dimers", *Phys. Rev. A*, vol. 56, pp. R2495–R2498.

Patton, D. C., D. V. Porezag, and M. R. Pederson, 1997b, "Simplified generalized-gradient approximation and anharmonicity: Benchmark calculations on molecules", *Phys. Rev. B*, vol. 55, pp. 7454–7459.

Paul, P. M., E. S. Toma, P. Breger, G. Mullot, F. Augé, P. Balcou, H. G. Muller, and P. Agostini, 2001, "Observation of a Train of Attosecond Pulses from High Harmonic Generation", *Science*, vol. 292, pp. 1689–1692.

Paulus, G. G., W. Nicklich, H. Xu, P. Lambropoulos, and H. Walther, 1994, "Plateau in above threshold ionization spectra", *Phys. Rev. Lett.*, vol. 72, pp. 2851–2854.

Paulus, G. G., F. Grasbon, H. Walther, R. Kopold, and W. Becker, 2001, "Channel-closing-induced resonances in the above-threshold ionization plateau", *Phys. Rev. A*, vol. 64, pp. 021401-1–4(R).

Payne, M. C., M. P. Teter, D. C. Allan, T. A. Arias, and J. D. Joannopoulos, 1992, "Iterative minimization techniques for *ab initio* total-energy calculations: molecular dynamics and conjugate gradients", *Rev. Mod. Phys.*, vol. 64, pp. 1045–1097.

Pearlman, D. A., D. A. Case, J. W. Caldwell, W. R. Ross, T. Cheatham III, S. DeBolt, D. Ferguson, G. Seibel, P. Kollman, 1995, "AMBER, a package of computer programs for applying molecular mechanics, normal mode analysis, molecular dynamics and free energy calculations to simulate the structural and energetic properties of molecules", *Comput. Phys. Commun.*, vol. 91, pp. 1–41.

Pearton, S. J., W. C. Dautremont-Smith, J. Chevallier, C. W. Tu, and K. D. Cummings, 1986, "Hydrogenation of shallow-donor levels in GaAs", *J. Appl. Phys.*, vol. 59, pp. 2821–2827.

Perdew, J. P., and A. Zunger, 1981 "Self-interaction correction to density-functional approximations for many-electron systems", *Phys. Rev. B*, vol. 23, pp. 5048–5079.

Perdew, J. P., R. G. Parr, M. Levy, and J. L. Balduz Jr., 1982, "Density-Functional Theory for Fractional Particle Number: Derivative Discontinuities of the Energy", *Phys. Rev. Lett.*, vol. 49, pp. 1691–1694.

Perdew, J. P., and M. Levy, 1983, "Physical Content of the Exact Kohn-Sham Orbital Energies: Band Gaps and Derivative Discontinuities", *Phys. Rev. Lett.*, vol. 51, pp. 1884–1887.

Perdew, J. P., 1986, "Density-functional approximation for the correlation energy of the inhomogeneous electron gas", *Phys. Rev. B*, vol. 33, pp. 8822–8824.

Perdew, J. P., and Y. Wang, 1992a, "Accurate and simple analytic representation of the electron-gas correlation energy", *Phys. Rev. B*, vol. 45, pp. 13244–13249.

Perdew, J. P., and Y. Wang, 1992b, "Pair-distribution function and its coupling-constant average for the spin-polarized electron gas", *Phys. Rev. B*, vol. 46, pp. 12947–12954; *ibid*, vol. 56, p. 7018(E) (1997).

Perdew, J. P., A. Savin, and K. Burke, 1995, "Escaping the spin-symmetry dilemma through a pair-density reinterpretation of spin-density-functional theory", *Phys. Rev. A*, vol. 51, pp. 4531–4541.

Perdew, J. P., K. Burke, and Y. Wang, 1996a, "Generalized gradient approximation for the exchange-correlation hole of a many-electron system", *Phys. Rev. B*, vol. 54, pp. 16533–16539; *ibid* vol. 57, p. 14999(E) (1998).

Perdew, J. P., K. Burke, and M. Ernzerhof, 1996b, "Generalized Gradient Approximation Made Simple", *Phys. Rev. Lett.*, vol. 77, pp. 3865–3868; *ibid*, vol. 78, p. 1396(E) (1997).

Perdew, J. P., M. Ernzerhof, and K. Burke, 1996c, "Rationale for mixing exact exchange with density-functional approximations", *J. Chem. Phys.*, vol. 105, pp. 9982–9985.

Perdew, J. P., and S. Kurth, 2003, "Density-Functionals for Non-relativistic Coulomb Systems in the New Century", in *A Primer in Density-Functional Theory*, edited by C. Fiolhais, F. Nogueira, and M. A. L. Marques, pp. 1–55 (Springer, Berlin).

Perdew, J. P., A. Ruzsinszky, J. Tao, V. N. Staroverov, G. E. Scuseria, and G. I. Csonka, 2005, "Prescription for the design and selection of density-functional approximations: More constraint satisfaction and fewer fits", submitted to *J. Chem. Phys.*

Perez-Jorda, J. M., and A. D. Becke, 1995, "A density-functional study of Van der Waals forces: rare gas diatomics", *Chem. Phys. Lett.*, vol. 233, pp. 134–137.

Perez-Jorda, J. M., E. San-Fabian, and A. J. Perez-Jimenez, 1999, "Density-functional study of Van der Waals forces on rare-gas diatomics: Hartree-Fock exchange", *J. Chem. Phys.*, vol. 110, pp. 1916–1920.

Perry, M. D., A. Szöke, O. L. Landen, and E. M. Campbell, 1988, "Nonresonant multiphoton ionization of noble gases: Theory and experiment", *Phys. Rev. Lett.*, vol. 60, pp. 1270–1273.

Persico, V., M. Carotenuto, and A. Peluso, 2004, "The Photophysics of Free-Base Hemiporphyrazine: A Theoretical Study", *J. Phys. Chem. A*, vol. 108, pp. 3926–3931.

Petersilka, M., U. J. Gossmann, and E. K. U. Gross, 1996a, "Excitation Energies from Time-Dependent Density-Functional Theory", *Phys. Rev. Lett.*, vol. 76, pp. 1212–1215.

Petersilka, M., and E. K. U. Gross, 1996b, "Spin-multiplet energies from time-dependent density-functional theory", *Int. J. Quant. Chem. Symp.*, vol. 30, pp. 1393–1407.

Petersilka, M., U. J. Gossman, and E. K. U. Gross, 1998, "Time-dependent optimized effective potential in the linear response regime", in *Electronic Density-Functional Theory: recent progress and new directions*, edited by J. F. Dobson, G. Vignale, and M. Das, pp. 177–197 (Plenum, New York).

Petersilka, M., and E. K. U. Gross, 1996, "Strong-Field Double Ionization of Helium: A Density-Functional Perspective", *Laser Phys.*, vol. 9, pp. 105–114.

Petit, L., C. Adamo, and N. Russo, 2005, "Absorption spectra of first-row transition metal complexes of bacteriochlorins: A theoretical analysis", *J. Phys. Chem. B*, vol. 109, pp. 12214–12221.

Peuckert, V., 1978, "A new approximation method for electron systems", *J. Phys. C*, vol. 11, pp. 4945–4956.

Pfannkuche, D., V. Gudmundsson, and P. A. Maksym, "Comparison of a Hartree, a Hartree-Fock, and an exact treatment of quantum-dot helium", *Phys. Rev. B*, vol. 47, pp. 2244–2250.

Pi, M., M. Barranco, A. Emperador, E. Lipparini, and L. Serra, 1998, "Current-density-functional approach to large quantum dots in intense magnetic fields", *Phys. Rev. B*, vol. 57, pp. 14783–14792.

Piccinin, S., 2006, Ph.D. thesis, Princeton University; and S. Piccinin, R. Gebauer, R. Car, in preparation.

Pickett, W., 1989, "Pseudopotential methods in condensed matter applications", *Comput. Phys. Rep.*, vol. 9, pp. 115–197.

Piecuch, P., K. Kowalski, P. D. Fang, and K. Jedziniak, "Exactness of Two-Body Cluster Expansions in Many-Body Quantum Theory", *Phys. Rev. Lett.*, vol. 90, pp. 113001-1–4.

Pinczuk, A., S. Schmitt-Rink, G. Danan, J. P. Valladares, L. N. Pfeiffer, and K. W. West, 1989, "Large exchange interactions in the electron gas of GaAs quantum wells", *Phys. Rev. Lett.*, vol. 63, pp. 1633–1636.

Pindzola, M. S., D. C. Griffin, and C. Bottcher, 1991, "Validity of time-dependent Hartree-Fock theory for the multiphoton ionization of atoms", *Phys. Rev. Lett.*, vol. 66, pp. 2305–2307.

Pindzola, M. S., P. Gavras, and T. W. Gorczyca, 1995, "Time-dependent unrestricted Hartree-Fock theory for the multiphoton ionization of atoms", *Phys. Rev. A*, vol. 51, pp. 3999-4004.

Pindzola, M. S., R. Robicheaux, and P. Gavras, 1997, "Double multiphoton ionization of a model atom", *Phys. Rev. A*, vol. 55, pp. 1307–1313.

Pines, D., and P. Nozières, 1966, *The Theory of Quantum Liquids*, vol. 1 (Benjamin, New York).

Piraux, B., A. L'Huillier, and K. Rzazewski (eds.), 1993, *Super-Intense Laser-Atom Physics*, NATO ASI Series B316 (Plenum Press, New York).

Pisignano, D., F. Della Sala, L. Persano, G. Gigli, R. Cingolani, G. Barbarella, and L. Favaretto, "Oligomer molecules: first-principles investigation of the optical properties and applications to luminescent devices", *Physica A*, vol. 339, pp. 106–111.

Pitarke, J. M., and A. G. Eguiluz, 1998, "Surface energy of a bounded electron gas: Analysis of the accuracy of the local-density approximation via *ab initio* self-consistent-field calculations", *Phys. Rev. B*, vol. 57, pp. 6329–6332.

Pitarke, J. M., and A. G. Eguiluz, 2001, "Jellium surface energy beyond the local-density approximation: Self-consistent-field calculations", *Phys. Rev. B*, vol. 63, pp. 045116-1–11.

Pitarke, J. M., and J. P. Perdew, 2003, "Metal surface energy: Persistent cancellation of short-range correlation effects beyond the random phase approximation", *Phys. Rev. B*, vol. 67, pp. 045101-1–5.

Platt, J. R., 1949, "Classification of Spectra of Cata-Condensed Hydrocarbons", *J. Chem. Phys.*, vol. 17, pp. 484–495.

Pogantsch, A., G. Heimel, and E. Zojer, 2002, "Quantitative prediction of optical excitations in conjugated organic oligomers: A density-functional theory study", *J. Chem. Phys.*, vol. 117, pp. 5921–5928.

Pohl, A., P.-G. Reinhard, and E. Suraud, 2000, "Towards Single-Particle Spectroscopy of Small Metal Clusters", *Phys. Rev. Lett.*, vol. 84, pp. 5090–5093

Pohl, A., P.-G. Reinhard, and E. Suraud, 2001, "Influence of intermediate states on photoelectron spectra", *J. Phys. B: At. Mol. Opt. Phys.*, vol. 34, pp. 4969–4981.

Pohl, A., P.-G. Reinhard, and E. Suraud, 2004, "Angular distribution of electrons emitted from Na clusters", *Phys. Rev. A*, vol. 70, pp. 023202-1–7.

Pollet, R., A. Savin, T. Leininger, and H. Stoll, 2002, "Combining multideterminantal wave functions with density-functionals to handle near-degeneracy in atoms and molecules", *J. Chem. Phys.*, vol. 116, pp. 1250–1258.

Polson, M., M. Ravaglia, S. Fracasso, M. Garavelli, and F. Scandola, 2005, "Iridium Cyclometalated Complexes with Axial Symmetry: Time-Dependent Density Functional Theory Investigation of *trans*-Bis-Cyclometalated Complexes Containing the Tridentate Ligand 2,6-Diphenylpyridine", *Inorg. Chem.*, vol. 44, pp. 1282–1289.

Pont, M., N. R. Walet, M. Gavrila, and C. W. McCurdy, 1998, "Dichotomy of the Hydrogen Atom in Superintense, High-Frequency Laser Fields", *Phys. Rev. Lett.*, vol. 61, pp. 939–942.

Pont, M., N. R. Walet, and M. Gavrila, 1990, "Radiative distortion of the hydrogen atom in superintense, high-frequency fields of linear polarization", *Phys. Rev. A*, vol. 41, pp. 477–494.

Pou-Amérigo, P. M. Viruela, R. Viruela, M. Rubio, and E. Orti, 2002, "Electronic spectra of tetrathiafulvalene and its radical cation: analysis of the performance of the time-dependent DFT approach", *Chem. Phys. Lett.*, vol. 352, pp. 491–498.

Prieto, J., F. Arbeloa, V. Martinez, and L. Arbeloa, 2004, "Theoretical study of the ground and excited electronic states of pyrromethene 546 laser dye and related compounds", *Chem. Phys.*, vol. 296, pp. 13–22.

Proot, J. P., C. Delerue, and G. Allan, 1992, "Electronic structure and optical properties of silicon crystallites: Application to porous silicon", *Appl. Phys. Lett.*, vol. 61, pp. 1948–1950.

Protopapas, M., C. H. Keitel, and P. L. Knight, 1997, "Atomic physics with super-high intensity lasers", *Rep. Prog. Phys.*, vol. 60, pp. 389–486.

Pryor, C., M.-E. Pistol, and L. Samuelson, 1997, "Electronic structure of strained InP/Ga_0.51In_0.49P quantum dots", *Phys. Rev. B*, vol. 56, pp. 10404–10411.

Pryor, C., 1998, "Eight-band calculations of strained InAs/GaAs quantum dots compared with one-, four-, and six-band approximations", *Phys. Rev. B*, vol. 57, pp. 7190–7195.

Puente, A., and L. Serra, 1999, "Oscillation Modes of Two-Dimensional Nanostructures within the Time-Dependent Local-Spin-Density Approximation", *Phys. Rev. Lett.*, vol. 83, pp. 3266–3269.

Puente, A., and L. Serra, 2001, "Ground state and far-infrared absorption of two-electron rings in a magnetic field", *Phys. Rev. B*, vol. 63, pp. 125334-1–9.

Puff, R. D., and N. S. Gillis, 1968, "Fluctuations and transport properties of many-particle systems", *Ann. Physics*, vol. 46, pp. 364–397.

Pulay, P., 1980, "Convergence acceleration of iterative sequences. the case of scf iteration", *Chem. Phys. Lett.*, vol. 73, pp. 393–398.

Pulay, P., 1982, "Improved SCF convergence acceleration", *J. Comput. Chem.*, vol. 3, pp. 556–560.

Pulci, O., A. Marini, M. Palummo, and R. Del Sole, 2005, in preparation.

Puschnig, P., and C. Ambrosch-Draxl, 2002, "Suppression of Electron-Hole Correlations in 3D Polymer Materials", *Phys. Rev. Lett.*, vol. 89, pp. 056405-1–4.

Puzder, A., A. J. Williamson, F. A. Reboredo, and G. Galli, 2003, "Suppression of Electron-Hole Correlations in 3D Polymer Materials", *Phys. Rev. Lett.*, vol. 91, pp. 157405-1–4.

Qian, Z., and G. Vignale, 2002, "Dynamical exchange-correlation potentials for an electron liquid", *Phys. Rev. B*, vol. 65, pp. 235121-1–12.

Qian, Z., and G. Vignale, 2003, "Dynamical exchange-correlation potentials for an electron liquid", *Phys. Rev. B*, vol. 68, pp. 195113-1–14.

Qteish A., A. I. Al-Sharif, M. Fuchs, M. Scheffler, S. Boeck, and J. Neugebauer, 2005, "Exact-exchange calculations of the electronic structure of ALN, GaN and InN", *Comp. Phys. Comm.*, vol. 169, pp. 28–31.

Radziszewski, J. G., M. Gil, A. Gorski, J. Spanget-Larsen, J. Waluk, and B. Mroz, 2000, "Electronic states of the phenoxyl radical", *J. Chem. Phys.*, vol. 115, pp. 9733–9738.

Raganato, M., V. Vitale, F. Della Sala, M. Anni, R. Cingolani, G. Gigli, L. Favaretto, G. Barbarella, M. Weimer, and A. Görling, 2004, "The effects of oxygenation on the optical properties of dimethyl-dithienothiophenes: Comparison between experiments and first-principles calculations", *J. Chem. Phys.*, vol. 121, pp. 3784–3791.

Raghavachari, K., D. Ricci, and G. Pacchioni, 2002, "Optical properties of point defects in SiO_2 from time-dependent density-functional theory", *J. Chem. Phys.*, vol. 116, pp. 825–831.

Rajagopal, A. K., 1996, "Time-dependent variational principle and the effective action in density-functional theory and Berry's phase", *Phys. Rev. A*, vol. 54, pp. 3916–3922.

Rapcewicz, K., and N. W. Ashcroft, 1991, "Fluctuation attraction in condensed matter: A nonlocal functional approach", *Phys. Rev. B*, vol. 44, pp. 4032–4035.

Rappoport, D., and F. Furche, 2004, "Photoinduced Intramolecular Charge Transfer in 4-(Dimethyl)aminobenzonitrile - A Theoretical Perspective", *J. Am. Chem. Soc.*, vol. 126, pp. 1277–1284.

Rappoport, D., and F. Furche, 2005, "Analytical time-dependent density-functional derivative methods within the RI-J approximation, an approach to excited states of large molecules", *J. Chem. Phys.*, vol. 122, pp. 064105-1–8.

Rassolov, V. A., J. A. Pople, M. A. Ratner, and T. L. Windus, 1998, "6-31G* basis set for atoms K through Zn", *J. Chem. Phys.*, vol. 109, pp. 1223–1229.

Ray, P., 2004, "Remarkable solvent effects on first hyperpolarizabilities of zwitterionic merocyanine dyes: ab initio TD-DFT/PCM approach", *Chem. Phys. Lett.*, vol. 395, pp. 269–273.

Reboredo, F. A., A. Franceschetti, and A. Zunger, 2000, "Dark excitons due to direct Coulomb interactions in silicon quantum dots", *Phys. Rev. B*, vol. 61, pp. 13073–13087.

Reimann, S. M., and M. Manninen, 2002, "Electronic structure of quantum dots", *Rev. Mod. Phys.*, vol. 74, pp. 1283-1–60.

Reinhard, P.-G., and R. Y. Cusson, 1982, "A comparative study of Hartree-Fock iteration techniques", *Nucl. Phys. A*, vol. 378, pp. 418–442.

Reinhard, P.-G., and E. Suraud, 1998, "Field amplification in Na clusters", *Euro. Phys. J.D*, vol. 3, pp. 175–178.

Reinhard, P.-G., F. Calvayrac, C. Kohl, S. Kümmel, E. Suraud, C. A. Ullrich, and M. Brack, 1999, "Frequencies, times, and forces in the dynamics of Na clusters", *Euro. Phys. J. D*, vol. 9, pp. 111–117.

Reinhard, P.-G., and E. Suraud, 2001, "Dynamics of Na clusters in picosecond laser pulses", *Appl. Phys. B*, vol. 73, pp. 401–406.

Reinhard, P.-G., and E. Suraud, 2002, "DFT studies of ionic vibrations in Na clusters", *Eur. Phys. J. D*, vol. 21, pp. 315–322.

Reinhard, P.-G., and E. Suraud, 2003, *Introduction to Cluster Dynamics* (Wiley, New York).

Reining, L., V. Olevano, A. Rubio, and G. Onida, 2002, "Excitonic Effects in Solids Described by Time-Dependent Density-Functional Theory", *Phys. Rev. Lett.*, vol. 88, pp. 066404-1–4.

Reis, H., M. Makowska-Janusika, and M. Papadopoulos, 2004, "Nonlinear Optical Susceptibilities of Poled Guest-Host Systems: A Computational Approach", *J. Phys. Chem. B*, vol. 108, pp. 8931–8940.

Resta, R., 1994, "Macroscopic polarization in crystalline dielectrics: the geometric phase approach", *Rev. Mod. Phys.*, vol. 66, pp. 899–915.

Reuter, N., A. Dejaegere, B. Maigret, and M. Karplus, 2000, "Frontier Bonds in QM/MM Methods: A Comparison of Different Approaches", *J. Phys. Chem. A*, vol. 104, pp. 1720–1735.

Ricciardi, G., A. Rosa, S. J. A. van Gisbergen, and E. J. Baerends, 2000, "A Density-Functional Study of the Optical Spectra and Nonlinear Optical Properties of Heteroleptic Tetrapyrrole Sandwich Complexes: The Porphyrinato-Porphyrazinato-Zirconium(IV) Complex as a Case Study", *J. Phys. Chem. A*, vol. 104, pp. 635–643.

Ricciardi, G., A. Rosa, and E. J. Baerends, 2001, "Ground and Excited States of Zinc Phthalocyanine Studied by Density-Functional Methods", *J. Phys. Chem. A*, vol. 105, pp. 5242–5254.

Ricciardi, G., A. Rosa, E. J. Baerends, and S. J. A. van Gisbergen, 2002, "Electronic Structure, Chemical Bond, and Optical Spectra of Metal Bis(porphyrin) Complexes: A DFT/TDDFT Study of the Bis(porphyrin)M(IV) (M = Zr, Ce, Th) Series", *J. Am. Chem. Soc.*, vol. 124, pp. 12319–12334.

Rice, S. A., and M. Zhao, 2000, *Optical Control of Molecular Dynamics*, (John Wiley and Sons).

Richardson, C. F., and N. W. Ashcroft, 1994, "Dynamical local-field factors and effective interactions in the three-dimensional electron liquid", *Phys. Rev. B*, vol. 50, pp. 8170–8181.

Ridley, J., M. Zerner, 1973, "Intermediate neglect of differential overlap (INDO) technique for spectroscopy. Pyrrole and the azines" *Theor. Chim. Acta*, vol. 32, pp. 111–134.

Ring, P., and P. Schuck, 1980, *The Nuclear Many Body Problem* (Springer-Verlag, New York).

Rinke, P., A. Qteish, J. Neugebauer, C. Freysoldt, and M. Scheffler, 2005, "Combining GW calculations with exact-exchange density-functional theory: An analysis of valence-band photoemission for compound semiconductors", *New J. Phys.*, vol. 7, pp. 126-1–35.

Rinkevicius, Z., I. Tunell, P. Salek, O. Vahtras, and H. Ågren, 2003, "Restricted density-functional theory of linear time-dependent properties in open-shell molecules" *J. Chem. Phys.*, vol. 119, pp. 34–46.

Rogers, C. L., and A. M. Rappe, 2002, "Geometric formulation of quantum stress fields", *Phys. Rev. B*, vol. 65, pp. 224117-1–8.

Rogers, J. E., K. A. Nguyen, D. C. Hufnagle, D. G. McLean, W. Su, K. M. Gossett, A. R. Burke, S. A. Vinogradov, R. Pachter, and P. A. Fleitz, 2003, "Observation and Interpretation of Annulated Porphyrins: Studies on the Photophysical Properties of *meso*-Tetraphenylmetalloporphyrins", *J. Phys. Chem. A*, vol. 107, pp. 11331–11339.

Rohlfing, M., and S. G. Louie, 1998a, "Electron-Hole Excitations in Semiconductors and Insulators", *Phys. Rev. Lett.* vol. 81, pp. 2312–2315.

Rohlfing, M., and S. G. Louie, 1998b, "Excitonic Effects and the Optical Absorption Spectrum of Hydrogenated Si Clusters", *Phys. Rev. Lett.*, vol. 80, pp. 3320–3323.

Rohlfing, M., and S. G. Louie, 1999a, "Optical Excitations in Conjugated Polymers", *Phys. Rev. Lett.*, vol. 82, pp. 1959–1962.

Rohlfing, M., and S. G. Louie, 1999b, "Excitons and Optical Spectrum of the Si(111)-(21) Surface", *Phys. Rev. Lett.*, vol. 83, pp. 856–859.

Rohlfing, M., and S. G. Louie, 2000a, "Electron-hole excitations and optical spectra from first principles", *Phys. Rev. B*, vol. 62, pp. 4927–4944.

Rohlfing, M., and S. G. Louie, 2000b, "Electron-Hole Excitations in Semiconductors and Insulators", *Phys. Rev. Lett.*, vol. 81, pp. 2312–2315.

Rohra S., E. Engel, and A. Görling, 2005, "Exact-Exchange Kohn-Sham formalism applied to one-dimensional periodic electronic systems", *Phys. Rev. B*, in press; cond-mat/0512299.

Rohrig, U. F., I. Frank, J. Hutter, A. Laio, J. VandeVondele, and U. Rothlisberger, 2003, "QM/MM Car-Parrinello Molecular Dynamics Study of the Solvent Effects on the Ground State and on the First Excited Singlet State of Acetone in Water", *ChemPhysChem* vol. 4, pp. 1177–1182.

Rojas, H. N., R. W. Godby, and R. J. Needs, 195, "Space-Time Method for *Ab Initio* Calculations of Self-Energies and Dielectric Response Functions of Solids", *Phys. Rev. Lett.*, vol. 74, pp. 1827–1830.

Romaniello, P., M. C. Aragoni, M. Arca, T. Cassano, C. Denotti, F. A. Devillanova, F. Esaia, F. Lelj, V. Lippolis, and R. Tommasi, 2003a, "Ground and Excited States of [M(H_2timdt)$_2$] Neutral Dithiolenes (M = Ni, Pd, Pt; H_2timdt = Monoanion of Imidazolidine-2,4,5-trithione): Description within TDDFT and Scalar Relativistic (ZORA) Approaches", *J. Phys. Chem. A*, vol. 107, pp. 9679–9687.

Romaniello, P., and F. Lelj, 2003b, "Optical non-linear properties of the [MXY] neutral mixed-ligand dithiolenes (M = Ni, Pd, Pt; X = R_2timdt, dmit, mnt; Y = R_2timdt, dmit, mnt; X $\neq$ Y). The role of coordinated metal, substituents and of high lying excited states", *Chem. Phys. Lett.*, vol. 372, pp. 51–58.

Romaniello, P., and P. L. de Boeij, 2005, "Time-dependent current-density-functional theory for the metallic response of solids", *Phys. Rev. B*, vol. 71, pp. 155108-1–17.

Rosa, A., E. J. Baerends, S. J. A. van Gisbergen, E. van Lenthe, J. A. Groeneveld, and J. G. Snijders, 1999, "Electronic Spectra of M(CO)$_6$ (M = Cr, Mo, W) Revisited by a Relativistic TDDFT Approach", *J. Am. Chem. Soc.*, vol. 121, pp. 10356–10365.

Rosa, A., G. Ricciardi, E. J. Baerends, S. J. A. van, and Gisbergen, 2001, "The Optical Spectra of NiP, NiPz, NiTBP, and NiPc: Electronic Effects of Meso-tetraaza Substitution and Tetrabenzo Annulation", *J. Phys. Chem. A*, vol. 105, pp. 3311–3327.

Rosa, A., G. Ricciardi, E. J. Baerends, A. Romeo, and L. M. Scolar, 2003, "Effects of Porphyrin Core Saddling, meso-Phenyl Twisting, and Counterions on the Optical Properties of *meso*-Tetraphenylporphyrin Diacids: The [H_4TPP](X)$_2$ (X = F, Cl, Br, I) Series as a Case Study", *J. Phys. Chem. A*, vol. 107, pp. 11468–11482.

Rosa, A., G. Ricciardi, O. Gritsenko, E. Baerends, 2004, "Excitation Energies of Metal Complexes with Time-dependent Density-Functional Theory", in *Principles and applications of density-functional theory in inorganic chemistry I*, vol. 112 of *Structure and Bonding*, pp. 49–115 (Springer, Berlin).

Rosenthal, C. M., 1971, "Solution of the Delta Function Model for Heliumlike Ions", *J. Chem. Phys.* vol. 55, pp. 2474–2483.

Rotenberg B., R. Taïeb, V. Véniard, and A. Maquet, 2002, "H_2^+ in intense laser field pulses: ionization versus dissociation within moving nucleus simulations", *J. Phys. B: At. Mol. Opt. Phys.*, vol. 35, pp. L397–L402.

Rubio, A., J. A. Alonso, J. M. Lopez, and M. J. Stott, 1993, "Surface plasmon excitations in C_{60}, C_{60}K and C_{60}H clusters", *Physica B*, vol. 183, pp. 247–263.

Rubio, A., J. A. Alonso, X. Blase, L. C. Balbás, and S. G. Louie, 1996, "Ab-initio Photoabsorption Spectra and Structures of Small Semiconductor and Metal Clusters", *Phys. Rev. Lett.*, vol. 77, pp. 247–250; *Phys. Rev. Lett.*, vol. 77, pp. 5442(E) (1996).

Ruiz, C., L. Plaja, R. Taïeb, V. Véniard and A. Maquet, 2005, "Quantum and semi-classical simulations in intense laser-H_2^+ interactions", to be published.

Runge, E., and E. K. U. Gross, 1984, "Density-Functional Theory for Time-Dependent Systems", *Phys. Rev. Lett.*, vol. 52, pp. 997–1000.

Runge, E., E. K. U. Gross, and O. Heinonen, 1991, *Many-particle theory* (Adam Hilger, Bristol).

Ryan, J. C., 1991, "Collective excitations in a spin-polarized quasi-two-dimensional electron gas", *Phys. Rev. B*, vol. 43, pp. 4499–4502.

Rydberg, H., B. I. Lundqvist, D. C. Langreth, and M. Dion, 2000, "Tractable non-local correlation density-functionals for flat surfaces and slabs", *Phys. Rev. B.*, vol. 62, pp. 6997–7006.

Rydberg, H., N. Jacobson, P. Hyldgaard, S. Simak, B. I. Lundqvist, and D. C. Langreth, 2003a, "Hard Numbers on Soft Matter", *Surf. Sci.*, vol. 532, pp. 606.

Rydberg, H., M. Dion, N. Jacobson, E. Schröder, P. Hyldgaard, S. I. Simak, D. C. Langreth, and B.I. Lundqvist, 2003b, "Van der Waals Density-Functional for Layered Structures", *Phys. Rev. Lett.*, vol. 91, pp. 126402-1–4.

Ryeng, H., and A. Ghosh, 2002, "Do Nonplanar Distortions of Porphyrins Bring about Strongly Red-Shifted Electronic Spectra? Controversy, Consensus, New Developments, and Relevance to Chelatases", *J. Am. Chem. Soc.*, vol. 124, pp. 8099–8103.

Saad, Y., 1992, "Analysis of Some Krylov Subspace Approximations to the Matrix Exponential Operator", *SIAM J. Numer. Anal.*, vol. 29, pp. 209–228.

Saalmann, U., and R. Schmidt, 1996, "Non-adiabatic quantum molecular dynamics: basic formalism and case study", *Z. Phys. D*, vol. 38, pp. 153–163.

Sahni, V., J. Gruenebaum, and J. P. Perdew, 1982, "Study of the density-gradient expansion for the exchange energy", *Phys. Rev. B*, vol. 26, pp. 4371–4377.

Sai, N., M. Zwolak, G. Vignale, and M. Di Ventra, 2005, "Dynamical corrections to the DFT-LDA electron conductance in nanoscale systems", *Phys. Rev. Lett.*, vol. 94, pp. 186810-1–4.

Saito, R., M. Fujita, G. Dresselhaus, and M. S. Dresselhaus, 1992, "Electronic structure of graphene tubules based on C_60", *Phys. Rev. B*, vol. 46, pp. 1804–1811.

Sakai, H., J. J. Larsen, I. Wendt-Larsen, J. Olesen, P. B. Corkum, and H. Stapelfeldt, 2003, "Nonsequential double ionization of D_2 molecules with intense 20-fs pulses", *Phys. Rev. A*, vol. 67, pp. 063404-1–4.

Sałek, P., O. Vahtras, J. D. Guo, Y. Luo, T. Helgaker, and H. Ågren, 2003, "Calculations of two-photon absorption cross sections by means of density-functional theory", *Chem. Phys. Lett.*, vol. 374, pp. 446–452.

Salières, P., A. L'Huillier, P. Antoine, and M. Lewenstein, 1999, "Study of the Spatial and Temporal Coherence of High Order Harmonics", *Adv. Atom. Mol. Opt. Phys.*, vol. 41, pp. 83–142.

Santer, M., U. Manthe, and G. Stock, 2001, "Quantum-classical Liouville description of multidimensional nonadiabatic molecular dynamics", *J. Chem. Phys.*, vol. 114, pp. 2001–2012.

Sarukura, N., K. Hata, T. Adachi, R. Nodomi, M. Watanabe, and S. Watanabe, 1991, "Coherent soft-x-ray generation by the harmonics of an ultrahigh-power KrF laser", *Phys. Rev. A*, vol. 43, pp. 1669–1672.

Saunders, V. R., and J. H. van Lenthe, 1983, "The direct CI method. A detailed analysis", *Mol. Phys.*, vol. 48, pp. 923–954.

Savin, A., 1988, "A combined density-functional and configuration interaction method", *Int. J. Quantum Chem.*, vol. S22, pp. 59.

Savin, A., 1991, "Correlation contributions from density-functionals", in *Density-Functional Methods in Chemistry*, edited by J. K. Labanowski and J. W. Andzelm, pp. 213 (Springer, Berlin).

Schäfer, A., H. Horn, and R. Ahlrichs, 1992, "Fully optimized contracted Gaussian basis sets for atoms Li to Kr", *J. Chem. Phys.*, vol. 97, pp. 2571–2577.

Schäfer, A., C. Huber, and R. Ahlrichs, 1994, "Fully optimized contracted Gaussian basis sets of triple zeta valence quality for atoms Li to Kr", *J. Chem. Phys.*, vol. 100, pp. 5829–5835.

Schinke, R., 1993, "Photodissociation Dynamics" (Cambridge University Press, Cambridge).

Schipper, P. R. T., O. V. Gritsenko, and E. J. Baerends, 1997, "Kohn-Sham potentials corresponding to Slater and Gaussian basis set densities", *Theor. Chem. Acc.*, vol. 98, pp. 16–24.

Schnürer, M., C. Spielmann, P. Wobrauschek, C. Streli, N. H. Burnett, C. Kan, K. Ferencz, R. Koppitsch, Z. Cheng, T. Brabec, and F. Krausz, 1998, "Coherent 0.5-keV X-Ray Emission from Helium Driven by a Sub-10-fs Laser", *Phys. Rev. Lett.*, vol. 80, pp. 3236–3239.

Schöne, W. D., and A. G. Eguiluz, 1998, "Self-Consistent Calculations of Quasiparticle States in Metals and Semiconductors", *Phys. Rev. Lett.*, vol. 81, pp. 1662–1665.

Schroder, E., and P. Hyldegaard, 2003, "The Van der Waals interactions of concentric nanotubes", *Surf. Sci.*, vol. 532-535, pp. 880.

Schüller, C., G. Biese, K. Keller, C. Steinebach, D. Heitmann, P. Grambow, and K. Eberl, 1996, "Single-particle excitations and many-particle interactions in quantum wires and dots", *Phys. Rev. B*, vol. 54, pp. R17304–R17307.

Schultz, T., J. Quenneville, B. Levine, A. Toniolo, T. J. Martinez, S. Lochbrunner, M. Schmitt, J. P. Shaffer, M. Z. Zgierski, and A. Stolow, 2003, "Mechanism and Dynamics of Azobenzene Photoisomerization", *J. Am. Chem. Soc.*, vol. 125, pp. 8098–8099.

Seaton, M. J., 1958, "The Quantum Defect Method", *Mon. Not. R. Astron. Soc.*, vol. 118, pp. 504–518.

Seibert, E., J. B. A. Ross, and R. Osman, 2002, "Quantum mechanical investigation of the electronic structure and spectral properties of 6,8-dimethylisoxanthopterin", *Int. J. Quant. Chem.*, vol. 88, pp. 28–33.

Seideman, T., M. Yu. Ivanov, and P. B. Corkum, 1995, "Role of Electron Localization in Intense-Field Molecular Ionization", *Phys. Rev. Lett.*, vol. 75, pp. 2819–2822.

Seidl, M., J. P. Perdew, and S. Kurth, 2000, "Simulation of All-Order Density-Functional Perturbation Theory, Using the Second Order and the Strong-Correlation Limit", *Phys. Rev. Lett.*, vol. 84, pp. 5070–5073.

Sergi, A., M. Gruening, M. Ferrario, and F. Buda, 2001, "Density-Functional Study of the Photoactive Yellow Protein's Chromophore", *J. Phys. Chem. B*, vol. 105, pp. 4386–4391.

Serra, L., and E. Lipparini, 1997, "Spin response of unpolarized quantum dots", *Europhys. Lett.*, vol. 40, pp. 667–672.

Serra, L., M. Barranco, A. Emperador, M. Pi, and E. Lipparini, 1999a, "Spin and density longitudinal response of quantum dots in the time-dependent local-spin-density approximation", *Phys. Rev. B*, vol. 59, pp. 15290–15300.

Serra, L., M. Pi, A. Emperador, M. Barranco, and E. Lipparini, 1999b, "Longitudinal modes of quantum dots in magnetic fields", *Eur. Phys. J. D*, vol. 9, pp. 643–646.

Serra, L., A. Puente, and E. Lipparini, 1999c, "Orbital current mode in elliptical quantum dots", *Phys. Rev. B*, vol. 60, pp. R13966–R13969.

Serra, L., A. Puente, and E. Lipparini, 2003a, "Breathing modes of 2-D quantum dots with elliptical shape in magnetic fields", *Int. J. Quant. Chem.*, vol. 91, pp. 483–489.

Serra, L., M. Valín-Rodríguez, and A. Puente, 2003b, "Spin-orbit coupling and the far infrared response of quantum dots in magnetic fields", *Surf. Sci.*, vol. 532–535, pp. 576–581.

Shaginyan, V. R., 1993, "Construction of the exact exchange potential of density-functional theory", *Phys. Rev. A*, vol. 47, pp. 1507–1509.

Sham, L. J., and T. M. Rice, 1966, "Many-Particle Derivation of the Effective-Mass Equation for the Wannier Exciton", *Phys. Rev.*, vol. 144, pp. 708–714.

Sham, L. J., and M. Schlüter, 1983, "Density-Functional Theory of the Energy Gap", *Phys. Rev. Lett.*, vol. 51, 1888–1891.

Sham, L. J., 1985, "Exchange and correlation in density-functional theory", *Phys. Rev. B*, vol. 32, pp. 3876–3882.

Sham, L. J., and M. Schlüter, 1985, "Density-functional theory of the band gap", *Phys. Rev. B*, vol. 32, pp. 3883–3889.

Shao, Y., M. Head-Gordon, and A. I. Krylov, 2003, "The spinflip approach within time-dependent density-functional theory: Theory and applications to diradicals" *J. Chem. Phys.*, vol. 118, pp. 4807–4818.

Sharma, S., J. K. Dewhurst, and C. Ambrosh-Draxl, 2005, "All-Electron Exact Exchange Treatment of Semiconductors: Effect of Core-Valence Interaction on Band-Gap and *d*-Band Position", *Phys. Rev. Lett.*, vol. 95, pp. 136402-1–4.

Sharp, R. T., and G. K. Horton, 1953, "A Variational Approach to the Unipotential Many-Electron Problem", *Phys. Rev.*, vol. 90, pp. 317.

Shelnutt, J. A., X.-Z. Song, J.-G. Ma, S.-L. Jia, W. Jentzen, and C. J. Medforth, 1998, "Nonplanar porphyrins and their significance in proteins", *Chem. Soc. Rev.*, vol. 27, pp. 31–42.

Shen, T.-C., C. Wang, G. C. Abeln, J. R. Tucker, J. W. Lyding, Ph. Avouris, and R. E. Walkup, 1995, "Atomic-Scale Desorption Through Electronic and Vibrational Excitation Mechanisms", *Science*, 268, pp. 1590–1592.

Sheng, Y., J. Leszczynski, A. Garcia, R. Rosario, D. Gust, and J. Springer, 2004, "Comprehensive Theoretical Study of the Conversion Reactions of Spiropyrans: Substituent and Solvent Effects", *J. Phys. Chem. B*, vol. 108, pp. 16233–16243.

Sherwood, P., 2000, "Hybrid quantum mechanics/molecular mechanics approaches", in *Modern Methods and Algorithms of Quantum Chemistry*, pp. 285–305, edited by J. Grotendorst, (John von Neumann Institute for Computing, Forschungszentrum Jülich)

Shirley, E. L., 1996, "Self-consistent GW and higher-order calculations of electron states in metals", *Phys. Rev. B*, vol. 54, pp. 7758–7764.

Shukla, M. K., and J. Leszczynski, 2002, "Interaction of Water Molecules with Cytosine Tautomers: An Excited-State Quantum Chemical Investigation", *J. Phys. Chem. A*, vol. 106, pp. 11338–11346.

Shukla, M., and J. Leszczynski, 2004, "TDDFT investigation on nucleic acid bases: Comparison with experiments and standard approach", *J. Comput. Chem.*, vol. 25, pp. 768–778.

Shukla, M. K., and J. Leszczynski, 2005a, "Time-dependent density-functional theory (TD-DFT) study of the excited state proton transfer in hypoxanthine", *Int. J. Quantum Chem.*, vol. 105, pp. 387–395.

Shukla, M. K., and J. Leszczynski, 2005b, "Excited state proton transfer in guanine in the gas phase and in water solution: A theoretical study", *J. Phys. Chem. A*, vol. 109, pp. 7775–7780.

Siddique, Z., Y. Yamamoto, T. Ohno, and K. Nozaki, 2003, "Structure-Dependent Photophysical Properties of Singlet and Triplet Metal-to-Ligand Charge Transfer States in Copper(I) Bis(diimine) Compounds", *Inorg. Chem.*, vol. 42, pp. 6366–6378.

Sikorska, E., I. Khmelinskii, J. Koput, J. Bourdelande, and M. Sikorski, 2004a, "Electronic structure of isoalloxazines in their ground and excited states", *J. Mol. Struct. (Theochem)*, vol. 697, pp. 137–141.

Sikorska, E., I. Khmelinskii, W. Prukala, S. Williams, M. Patel, D. Worrall, J. Bourdelande, J. Koput, and M. Sikorski, 2004b, "Spectroscopy and Photophysics of Lumiflavins and Lumichromes", *J. Phys. Chem. A*, vol. 108, pp. 1501–1508.

Sikorska, E., I. Khmelinskii, A. Komasa, J. Koput, L. F. V. Ferreira, J. R. Herance, J. L. Bourdelande, S. L. Williams, D. R. Worrall, M. Insinska-Rak, and M. Sikorski, 2005, "Spectroscopy and photophysics of flavin related compounds: Riboflavin and iso-(6,7)-riboflavin", *Chem. Phys.*, vol. 314, pp. 239–247.

Sikorski, C., and U. Merkt, 1989, "Spectroscopy of electronic states in InSb quantum dots", *Phys. Rev. Lett.*, vol. 62, pp. 2164–2167.

Sinicropi, A., T. Andruniow, N. Ferr, R. Basosi, and M. Olivucci, 2005, "Properties of the Emitting State of the Green Fluorescent Protein Resolved at the CASPT2//CASSCF/CHARMM Level", *J. Am. Chem. Soc.*, vol. 127, pp. 11534–11535.

Singwi, K. S., M. P. Tosi, R. H. Land, and A. Sjölander, 1968, "Electron Correlations at Metallic Densities", *Phys. Rev.*, vol. 176, pp. 589–599.

Slipchenko, L. V., and A. I. Krylov, 2003, "Electronic structure of the trimethylenemethane diradical in its ground and electronically excited states: Bonding, equilibrium geometries, and vibrational frequencies" *J. Chem. Phys.*, vol. 118, pp. 6874–6883.

Smirnov, V. I., and M.A. Lebedov, 1968, *Functions of a Complex Variable* (ILIFFE Books, London).

Smith, G., 1978, *Numerical Solutions of Partial Differential Equation: Finite Difference Methods*, 2nd edition (Oxford, New York).

Smith, N., 2001, *Physics Today*, vol. 54. pp. 29–35.

Sobolewski, A. L., and W. Domcke, 1999, "*Ab initio* potential-energy functions for excited state intramolecular proton transfer: a comparative study of o-hydroxybenzaldehyde, salicylic acid and 7-hydroxy-1-indanone", *Phys. Chem. Chem. Phys.*, vol. 1, pp. 3065–3072.

Sobolewski, A. L., and W. Domcke, 2002, "On the mechanism of nonradiative decay of DNA bases: ab initio and TDDFT results for the excited states of 9H-adenine", *Eur. Phys. J.D*, vol. 20, pp. 369–374.

Sobolewski, A., and W. Domcke, 2004, "Intramolecular Hydrogen Bonding in the $S_1(\pi\pi^*)$ Excited State of Anthranilic Acid and Salicylic Acid: TDDFT Calculation of Excited-State Geometries and Infrared Spectra", *J. Phys. Chem. A*, vol. 108, pp. 10917–10922.

Song, K. S., and R. T. Williams, 1996, *Self-Trapped Excitons*, 2nd edition (Springer, Berlin).

Sottile, F., V. Olevano, and L. Reining, 2003, "Parameter-Free Calculation of Response Functions in Time-Dependent Density-Functional Theory", *Phys. Rev. Lett.*, vol. 91, pp. 056402-1–4.

Sottile, F., et al., 2005, unpublished.
Spataru, C. D., S. Ismail-Beigi, L. X. Benedict, and S. G. Louie, 2004, "Excitonic Effects and Optical Spectra of Single-Walled Carbon Nanotubes", *Phys. Rev. Lett.*, vol. 92, pp. 077402-1–4.
Spielfiedel, A., and N. C. Handy, 1999, "Potential energy curves for PO, calculated using DFT and MRCI methodology", *Chem. Phys.*, vol. 1, pp. 2401–2409.
Spielmann, C., N. H. Burnett, S. Sartania, R. Koppitsch, M. Schnürer, C. Kan, M. Lenzner, W. Wobrauschek, and F. Krausz, 1997, "Generation of Coherent X-rays in the Water Window Using 5-Femtosecond Laser Pulses", *Science*, vol. 278, pp. 661–664.
Srivastava, G. P., and D. Weaire, 1987, "The theory of the cohesive energies of solids", *Adv. Phys.*, vol. 36, pp. 463–517.
Städele, M., J. A. Majewski, P. Vogl, and A. Görling, 1997, "Exact Kohn-Sham Exchange Potential in Semiconductors", *Phys. Rev. Lett.*, vol. 79, pp. 2089–2092.
Städele, M., M. Moukara, J. A. Majewski, P. Vogl and A. Görling, 1999, "Exact exchange Kohn-Sham formalism applied to semiconductors", *Phys. Rev. B*, vol. 59, pp. 10031–10043.
Stapelfeldt, H., and T. Seideman, 2003, "Colloquium: Aligning molecules with strong laser pulses", *Rev. Mod. Phys.*, vol. 75, pp. 543-1–15.
Stefanucci G., and C.-O. Almbladh, 2004a, "Time-dependent partition-free approach in resonant tunneling systems", *Phys. Rev. B*, vol. 69, pp. 195318-1–17.
Stefanucci, G., and C.-O. Almbladh, 2004b, "Time-dependent quantum transport: An exact formulation based on TDDFT", *Europhys. Lett.*, vol. 67, pp. 14–20.
Steffens, O., M. Suhrke, and U. Rössler, 1998, "Spontaneously broken time-reversal symmetry in quantum dots", *Europhys. Lett.*, vol. 44, pp. 222–228.
Stephens, P., D. McCann, F. Devlin, J. Cheeseman, and M. Frisch, 2004, "Determination of the Absolute Configuration of [3_2](1,4)Barrelenophanedicarbonitrile Using Concerted Time-Dependent Density-Functional Theory Calculations of Optical Rotation and Electronic Circular Dichroism", *J. Am. Chem. Soc.*, vol. 126, pp. 7514–7521.
Stich, T. A., A. J. Brooks, N. R. Buan, and T. C. Brunold, 2003, "Spectroscopic and Computational Studies of Co^{3+}-Corrinoids: Spectral and Electronic Properties of the B_{12} Cofactors and Biologically Relevant Precursors", *J. Am. Chem. Soc.*, vol. 125, pp. 5897–5914.
Stich, T., N. Buan, and T. Brunold, 2004, "Spectroscopic and Computational Studies of Co^{2+}Corrinoids: Spectral and Electronic Properties of the Biologically Relevant Base-On and Base-Off Forms of Co^{2+}Cobalamin", *J. Am. Chem. Soc.*, vol. 126, pp. 9735–9749.
Stoll, H., and A. Savin, 1985, "density-functionals for correlation energies of atoms and molecules", in *Density-Functional Methods in Physics*, edited by R. M. Dreizler and J. da Providencia, pp. 177 (Plenum, New York).
Stone, A. J., 1996, *The Theory of Intermolecular Forces* (Clarendon Press, Oxford).
Stott, M. J., and E. Zaremba, 1980, "Linear-response theory within the density-functional formalism: Application to atomic polarizabilities", *Phys. Rev. A*, vol. 21, pp. 12–23; *ibid*, vol. 22, pp. 2293(E) (1980).
Stoyanov, S. R., J. M. Villegas, and D. P. Rillema, 2003, "Time-Dependent Density-Functional Theory Study of the Spectroscopic Properties Related to Aggregation in the Platinum(II) Biphenyl Dicarbonyl Complex", *Inorg. Chem.*, vol. 42, pp. 7852–7860.

Stratmann, R. E., G. E. Scuseria, and M. J. Frisch, 1998, "An efficient implementation of time-dependent density-functional theory for the calculation of excitation energies of large molecules", *J. Chem. Phys.*, vol. 109, pp. 8218–8224.

Strenz, R., U. Bockelmann, F. Hirler, G. Abstreiter, G. Böhm, and G. Weimann, 1994, "Single-Particle Excitations in Quasi-Zero- and Quasi-One-Dimensional Electron Systems", *Phys. Rev. Lett.*, vol. 73, pp. 3022–3025.

Strinati, G., 1988, "Application of the Green's functions methods to the study of the optical properties of semiconductors", *Rivista del Nuovo Cimento*, vol. 11, pp. 1–86.

Stubner, R., I. V. Tokatly, and O. Pankratov, 2004, "Excitonic effects in time-dependent density-functional theory: An analytically solvable model", *Phys. Rev. B*, vol. 70, pp. 245119-1–12.

Sturm, K., and A. Gusarov, 2000, "Dynamical correlations in the electron gas", *Phys. Rev. B*, vol. 62, pp. 16474–16491.

Su, Q., and J. H. Eberly, 1991, "Model atom for multiphoton physics", *Phys. Rev. A*, vol. 44, pp. 5997–6008.

Sugino, O., and Y. Miyamoto, 1999, "Density-functional approach to electron dynamics: Stable simulation under a self-consistent field", *Phys. Rev. B*, vol. 59, pp. 2579–2586; *ibid*, vol. 66, 89901(E) (2002).

Sullivan, K. F., and S. A. Kay, 1999, *Green Fluorescent Proteins* (Academic Press, San Diego).

Sulpizi, M., P. Carloni, J. Hutter, and U. Röthlisberger, 2003, "A hybrid TDDFT/MM investigation of the optical properties of aminocoumarins in water and acetonitrile solution", *Phys. Chem. Chem. Phys.*, vol. 5, pp. 4798–4805.

Sun, Y., K. Zhao, C. Wang, Y. Luo, Y. Yan, X. Tao, and M. Jiang, 2004, "Theoretical studies on nonlinear optical properties of two newly synthesized compounds: PVPHC and DPVPA", *Chem. Phys. Lett.*, vol. 394, pp. 176–181.

Sundholm, D., 1999, "Density-functional theory calculations of the visible spectrum of chlorophyll *a*", *Chem. Phys. Lett.*, vol. 302, pp. 480–484.

Sundholm, D., 2000a, "Density-functional theory study of the electronic absorption spectrum of Mg-porphyrin and Mg-etioporphyrin-I", *Chem. Phys. Lett.*, vol. 317, pp. 392–399.

Sundholm, D., 2000b, "Comparison of the electronic excitation spectra of chlorophyll *a* and pheophytin *a* calculated at density-functional theory level", *Chem. Phys. Lett.*, vol. 317, pp. 545–552.

Sundholm, D., 2000c, "Interpretation of the electronic absorption spectrum of free-base porphin using time-dependent density-functional theory", *Phys. Chem. Chem. Phys.*, vol. 2, pp. 2275–2281.

Sundholm, D., 2003, "A density-functional-theory study of bacteriochlorophyll *b*", *Chem. Phys.*, vol. 5, pp. 4265–4271.

Sutcliffe, B., 2000, "The decoupling of electronic and nuclear motions in the isolated molecule Schrödinger Hamiltonian", *Adv. Chem. Phys.*, vol. 114, pp. 1–121.

Suzuki, M., 1992, "General Nonsymmetric Higher-Order Decomposition of Exponential Operators and Symplectic Integrators", *J. Phys. Soc. Jpn.*, vol. 61, pp. 3015–3019.

Suzuki, M., and T. Yamauchi, 1993, "Convergence of unitary and complex decompositions of exponential operators", *J. Math. Phys.*, vol. 34, pp. 4892–4897.

Suzuki, T., S. Minemoto, T. Kanai, and H. Sakai, 2004, "Optimal Control of Multiphoton Ionization Processes in Aligned I_2 Molecules with Time-Dependent Polarization Pulses", *Phys. Rev. Lett.*, vol. 92, pp. 133005-1–4.

Szabo, A., and N. S. Ostlund, 1989, *Modern Quantum Chemistry, Introduction to Advanced Electronic Structure Theory* (McGraw-Hill, New York).

Szemik-Hojniak, A., I. Deperasinska, W. Buma, G. Balkowski, A. Pozharskii, N. Vistorobskii, and X. Allonas, 2005, "The asymmetric nature of charge transfer states of the cyano-substituted proton sponge", *Chem. Phys. Lett.*, vol. 401, pp. 189–195.

Szydlowska, I., A. Kyrychenko, A. Gorski, J. Waluk, and J. Herbich, 2003a, "Excited states of 4-dimethylaminopyridines: Magnetic circular dichroism and computational studies", *Photochem. Photobiol. Sci.*, vol. 2, pp. 187–194.

Szydlowska, I., A. Kyrychenko, J. Nowacki, and J. Herbich, 2003b, "Photoinduced intramolecular electron transfer in 4-dimethylaminopyridines", *Phys. Chem. Chem. Phys.*, vol. 5, pp. 1032–1038.

Takesada, M., T. Yagi, M. Itoh, and S. Koshihara, 2003, "A Gigantic Photoinduced Dielectric Constant of Quantum Paraelectric Perovskite Oxides Observed under a Weak DC Electric Field", *J. Phys. Soc. Jpn.*, vol. 72, pp. 37–40.

Tal-Ezer, H., and R. Kosloff, 1984, "An accurate and efficient scheme for propagating the time dependent Schrödinger equation", *J. Chem. Phys.*, vol. 81, pp. 3967–3971.

Talman, J. D., and W. F. Shadwick, 1976, "Optimized effective atomic central potential", *Phys. Rev. A*, vol. 14, pp. 36–40.

Tamuliene, J., M. Balevicius, and A. Tamulis, 2004, "How Has the Bridge Fragment Chosen to Design Charge Transfer Molecular Device?", *Struct. Chem.*, vol. 15, pp. 579–585.

Tamulis, A., J. Tamuliene, and V. Tamulis, 2003, "Quantum Mechanical Design of Light Driven Molecular Logical Machines and Elements of Molecular Quantum Computer", in *Molecular Electronics: Bio-sensors and Bio-computers*, vol. 96 of *NATO Sci. Ser. II*, pp. 1–27.

Tamulis, A., J. Tamuliene, V. Tamulis, and A. Ziriakoviene, 2004, "Quantum Mechanical Design of Molecular Computers Elements Suitable for Self-Assembling to Quantum Computing Living Systems", in *Self Formation Theory and Applications*, vol. 97-98 of *Solid State Phenomena*, pp. 173–179 (Trans Tech Publications, Zürich-Uetikon).

Tanaka, K., K. Okahara, M. Okada, and T. Yamabe, 1992, "Electronic properties of bucky-tube model", *Chem. Phys. Lett.*, vol. 191, pp. 469–472.

Tang, X., H. Rudolph, and P. Lambropoulos, 1991, "Nonperturbative time-dependent theory of helium in a strong laser field", *Phys. Rev. A*, vol. 44, pp. R6994–R6997.

Tao, J., J. P. Perdew, V. N. Staroverov, and G. E. Scuseria, 2003, "Climbing the Density-Functional Ladder: Nonempirical Meta-Generalized Gradient Approximation Designed for Molecules and Solids", *Phys. Rev. Lett.*, vol. 91, pp. 146401-1–4.

Tatarczyk, K., A. Schindlmayr, and M. Scheffler, 2001, "Exchange-correlation kernels for excited states in solids", *Phys. Rev. B*, vol. 63, pp. 235106–235112.

Taut, M., 1993, "Two electrons in an external oscillator potential: Particular analytic solutions of a Coulomb correlation problem", *Phys. Rev. A*, vol. 48, pp. 3561–3566.

Taylor, P. R., 1985, "Symmetrization of operator matrix elements", *Int. J. Quantum Chem.*, vol. 27, pp. 89–96.

Thompson, M. J., D. Bashford, L. Noodleman, and E. D. Getzoff, 2003, "Photoisomerization and Proton Transfer in Photoactive Yellow Protein", *J. Am. Chem. Soc.*, vol. 125, pp. 8186–8194.

Thompson, K., and N. Miyake, 2005, "Properties of a New Fluorescent Cytosine Analogue, Pyrrolocytosine", *J. Phys. Chem. B*, vol. 109, pp. 6012–6019.

Thouless, D. J., 1961, *The Quantum Mechanics of Many-Body Systems* (Academic Press, New York).

Tiago, M., S. Ismail-Beigi, and S. Louie, 2005, "Photoisomerization of azobenzene from first-principles constrained density-functional calculations", *J. Chem. Phys.*, vol. 122, pp. 094311-1–7.

Tobita, M., S. Hirata, and R. J. Bartlett, 2001, "A crystalline orbital study of polydiacetylenes", *J. Chem. Phys.*, vol. 114, pp. 9130–9141.

Toher, C., A. Filippetti, S. Sanvito, and K. Burke, 2005, "Self-interaction errors in density-functional calculations of electronic transport", *submitted.*

Tokatly, I. V., and O. Pankratov, 2001, "Many-Body Diagrammatic Expansion in a Kohn-Sham Basis: Implications for Time-Dependent Density-Functional Theory of Excited States", *Phys. Rev. Lett.*, vol. 86, pp. 2078–2081.

Tokatly, I. V., R. Stubner, and O. Pankratov, 2002, "Many-body diagrammatic expansion for the exchange-correlation kernel in time-dependent density-functional theory", *Phys. Rev. B*, vol. 65, pp. 113107-1–4.

Tokatly, I. V., and O. Pankratov, 2003, "Local exchange-correlation vector potential with memory in time-dependent density-functional theory: The generalized hydrodynamics approach", *Phys. Rev. B*, vol. 67, pp. 201103-1–4(R).

Tokatly, I. V., 2005a, "Quantum many-body dynamics in a Lagrangian frame: I. Equations of motion and conservation laws", *Phys. Rev. B*, vol. 71, pp. 165104-1–13.

Tokatly, I. V., 2005b, "Quantum many-body dynamics in a Lagrangian frame: II. Geometric formulation of time-dependent density-functional theory", *Phys. Rev. B*, vol. 71, pp. 165105-1–17.

Tomic, K., J. Tatchen, and C. M. Marian, 2005, "Quantum chemical investigation of the electronic spectra of the keto, enol, and keto-imine tautomers of cytosine", *J. Phys. Chem. A*, vol. 109, pp. 8410–8418.

Tong, X. M., and S.-I Chu, 1997, "Theoretical study of multiple high-order harmonic generation by intense ultrashort pulsed laser fields: A new generalized pseudospectral time-dependent method", *Chem. Phys.*, vol. 217, pp. 119–130.

Tong, X. M., and S.-I Chu, 1998, "Time-dependent density-functional theory for strong-field multiphoton processes: Application to the study of the role of dynamical electron correlation in multiple high-order harmonic generation", *Phys. Rev. A*, vol. 57, pp. 452–461.

Tong, X. M., and S.-I Chu, 2001, "Multiphoton ionization and high-order harmonic generation of He, Ne, and Ar atoms in intense pulsed laser fields: Self-interaction-free time-dependent density-functional theoretical approach", *Phys. Rev. A*, vol. 64, pp. 013417-1–8.

Torres, L., R. Gelabert, M. Moreno, and J. M. Lluch, 2003, "Fast hydrogen elimination from the $[Ru(PH_3)_3(CO)(H)_2]$ complex in the first singlet excited states. A quantum dynamics study", *Chem. Phys.*, vol. 286, pp. 149–163.

Toto, T. T., J. L. Toto, C. P. de Melo, M. Hasan, and B. Kirtman, 1995, "Ab initio finite oligomer method for nonlinear optical properties of conjugated polymers. Effect of electron correlation on the static longitudinal hyperpolarizability of polyacetylene" *Chem. Phys. Lett.*, vol. 244, pp. 59–64.

Tournus, F., J.-C. Charlier, and P. Mélinon, 2005, "Mutual orientation of two C_{60} molecules: An *ab initio* study", *J. Chem. Phys.* vol. 122, pp. 094315-1–9.

Toyota, S., T. Shimasaki, N. Tanifuji, and K. Wakamatsu, 2003, "Experimental and theoretical investigations of absolute stereochemistry and chiroptical properties of enantiopure 2,2'-substituted 9,9'-bianthryls", *Tetrahedron: Asymmetry*, vol. 14, pp. 1623–1629.

Toyota, S., T. Shimasaki, T. Ueda, N. Tanifuji, and K. Wakamatsu, 2004, "Absolute Stereochemisty and Chiroptical Properties of 3,3'-Bis(methoxycarbonyl)-9,9'-bianthryl", *Bull. Chem. Soc. Jpn.*, vol. 77, pp. 2065–2070.

Tozer, D. J., R. D. Amos, N. C. Handy, B. O. Roos, and L. Serrano-Andres, 1999, "Does density-functional theory contribute to the understanding of excited states of unsaturated organic compounds?" *Mol. Phys.*, vol. 97, pp. 859–868.

Tozer, D. J., 2003, "Relationship between long-range charge-transfer excitation energy error and integer discontinuity in Kohn-Sham theory", *J. Chem. Phys.*, vol. 119, pp. 12697–12699.

Tretiak, S., and S. Mukamel, 2002, " Density Matrix Analysis and Simulation of Electronic Excitations in Conjugated and Aggregated Molecules", *Chem. Rev.*, vol. 102, pp. 3171–3212.

Tretiak, S., and Chernyak V., 2003, "Resonant nonlinear polarizabilities in the time-dependent density-functional theory" *J. Chem. Phys.*, vol. 119, pp. 8809–8823.

Tretiak, S., K. Igumenshchev, and V. Chernyak, 2005, "Exciton sizes of conducting polymers predicted by time-dependent density-functional theory", *Phys. Rev. B*, vol. 71, pp. 033201-1–4.

Treutler, O, and R. Ahlrichs, 1995, "Efficient molecular numerical integration schemes", *J. Chem. Phys.*, vol. 102, pp. 346–354.

Troullier, N., and J. L. Martins, 1991, "Efficient pseudopotentials for plane-wave calculations", *Phys. Rev. B*, vol. 43, pp. 1993–2006.

Truong, T. N., J. J. Tanner, P. Bala, and J. A. McCammon, 1992, "A comparative study of time dependent quantum mechanical wave packet evolution methods", *J. Chem. Phys.* vol. 96, pp. 2077–2084.

Tsolakidis, A., and E. Kaxiras, 2005, "A TDDFT Study of the Optical Response of DNA Bases, Base Pairs, and Their Tautomers in the Gas Phase", *J. Phys. Chem. A*, vol. 109, pp. 2373–2380.

Tuckerman, M. R., and M. Parrinello, 1994a, "Integrating the Car-Parrinello equations. I. Basic integration techniques", *J. Chem. Phys.*, vol. 101, pp. 1302–1315.

Tuckerman, M. R., and M. Parrinello, 1994b, "Integrating the Car-Parrinello equations. II. Multiple time scale techniques", *J. Chem. Phys.*, vol. 101, pp. 1316–1329.

Tully, J. C., and R. K. Preston, 1971, "Trajectory Surface Hopping Approach to Nonadiabatic Molecular Collisions: The Reaction of H^+ with D_2", *J. Chem. Phys.*, vol. 55, pp. 562–572.

Tully, J. C., 1990, "Molecular dynamics with electronic transitions", *J. Chem. Phys.*, vol. 93, pp. 1061–1071.

Ullrich, C. A., U. J. Gossmann and E. K. U. Gross, 1995a, "Time-dependent optimized effective potential", *Phys. Rev. Lett.*, vol. 74, pp. 872–875.

Ullrich, C. A., U. J. Gossmann, and E. K. U. Gross, 1995b, "Density-functional approach to atoms in strong laser pulses", *Ber. Bunsenges. Phys. Chem.*, vol. 99, pp. 488–497.

Ullrich, C. A., S. Erhard, and E.K.U. Gross, 1996, "Density-Functional Approach to Atoms in Strong Laser Pulses", in *Super Intense Laser Atom Physics IV*, edited by H. G. Muller and M. V. Fedorov, Nato ASI 3/13, pp. 267–284 (Kluwer, Dorderecht).

Ullrich, C. A., and E. K. U. Gross, "Many-electron atoms in strong femtosecond laser pulses: A density-functional study", *Comments At. Mol. Phys.*, vol. 33, pp. 211–236.

Ullrich, C. A., and G. Vignale, 1998, "Collective intersubband transitions in quantum wells: A comparative density-functional study", *Phys. Rev. B*, vol. 58, pp. 15756–15765.

Ullrich, C. A., and G. Vignale, 2000a, "Collective charge-density excitations of noncircular quantum dots in a magnetic field", *Phys. Rev. B*, vol. 61, pp. 2729–2736.

Ullrich, C. A., 2000b, "Time-dependent Kohn-Sham approach to multiple ionization", *J. Mol. Struct. (Theochem)*, vol. 501–502, pp. 315–325.

Ullrich, C. A., P.–G. Reinhard, and E. Suraud, 2000c, "Simplified implementation of self-interaction correction in sodium clusters", *Phys. Rev. A*, vol. 62, pp. 053202-1–9.

Ullrich, C. A., and G. Vignale, 2001, "Theory of the Linewidth of Intersubband Plasmons in Quantum Wells", *Phys. Rev. Lett.*, vol. 87, pp. 037402-1–4.

Ullrich, C. A., and W. Kohn, 2002a "Degeneracy in Density-Functional Theory: Topology in the v and n Spaces", *Phys. Rev. Lett.*, vol. 89, pp. 156401-1–4.

Ullrich, C. A., and G. Vignale, 2002b, "Time-dependent current-density-functional theory for the linear response of weakly disordered systems", *Phys. Rev. B*, vol. 65, pp. 245102-1–19.

Ullrich, C. A., and M. E. Flatté, 2002c, "Intersubband spin-density excitations in quantum wells with Rashba spin splitting", *Phys. Rev. B*, vol. 66, pp. 205305-1–10.

Ullrich, C. A., and M. E. Flatté, 2003, "Anisotropic splitting of intersubband spin plasmons in quantum wells with bulk and structural inversion asymmetry", *Phys. Rev. B*, vol. 68, pp. 235310-1–8.

Ullrich, C. A., and K. Burke, 2004, "Excitation energies from time-dependent density-functional theory beyond the adiabatic approximation", *J. Chem. Phys.*, vol. 121, pp. 28–35.

Ullrich, C. A., and Tokatly, I. V., 2006, "Nonadiabatic electron dynamics in time-dependent density-functional theory", *Phys. Rev. B*, vol. 73, pp 235102-1-15

Ummels, R. T. M., P. A. Bobbert and W. van Haeringen, 1994, "First-order corrections to random-phase approximation *GW* calculations in silicon and diamond", *Phys. Rev. B*, vol. 57, 11962–11973.

Umrigar, C. J., and X. Gonze, 1994, "Accurate exchange-correlation potentials and total-energy components for the helium isoelectronic series", *Phys. Rev. A*, vol. 50, pp. 3827–3837.

Valín-Rodríguez, M., A. Puente, and L. Serra, 2000, "Collective oscillations in quantum rings: A broken symmetry case", *Eur. Phys. J. D*, vol. 12, pp. 493–498.

Valín-Rodríguez, M., A. Puente, and L. Serra, 2001a, "Far-infrared response of quantum-dot molecules", *Eur. Phys. J. D* vol. 16, pp. 387–390.

Valín-Rodríguez, M., A. Puente, and L. Serra, 2001b, "Far-infrared absorption in triangular and square quantum dots: Characterization of corner and side modes", *Phys. Rev. B*, vol. 64, pp. 205307-1–5.

Valín-Rodríguez, M., A. Puente, and L. Serra, 2002, "Role of spin-orbit coupling in the far-infrared absorption of lateral semiconductor dots", *Phys. Rev. B* vol. 66, pp. 045317-1–5.

van Caillie, C., and R. D. Amos, 1999, "Geometric derivatives of excitation energies using SCF and DFT", *Chem. Phys. Lett.*, vol. 308, pp. 249–255.

van Caillie, C., and R. D. Amos, 2000, "Geometric derivatives of density-functional theory excitation energies using gradient-corrected functionals", *Chem. Phys. Lett.*, vol. 317, pp. 159–164.

van Druten, N. J., R. Constantinescu, J. M. Schins, H. Nieuwenhuize, and H. G. Muller, 1997, "Adiabatic stabilization: Observation of the surviving population", *Phys. Rev. A*, vol. 55, pp. 622–629.

van Faassen, M., P. L. de Boeij, R. van Leeuwen, J. A. Berger, and J. G. Snijders, 2002, "Ultranonlocality in Time-Dependent Current-Density-Functional Theory: Application to Conjugated Polymers", *Phys. Rev. Lett.*, vol. 88, pp. 186401-1–4.

van Faassen, M., P. L. de Boeij, F. Kootstra, J. A. Berger, R. van Leeuwen, and J. G. Snijders, 2003a, "Application of time-dependent current-density-functional theory to nonlocal exchange-correlation effects in polymers", *J. Chem. Phys.*, vol. 118, pp. 1044–1053.

van Faassen, M., and P. L. de Boij, 2003b, "Excitation energies for a benchmark set of molecules obtained within time-dependent current-density-functional theory using the Vignale-Kohn functional", *J. Chem. Phys.*, vol. 120, pp. 8353–8363.

van Faassen, M., and P. L. de Boeij, 2004, "Excitation energies of π-conjugated oligomers within time-dependent current-density-functional theory". *J. Chem. Phys.*, vol. 121, pp. 10707–10714.

van Gisbergen, S. J. A., J. G. Snijders, and E. J. Baerends, 1995, "A density-functional theory study of frequency-dependent polarizabilities and Van der Waals dispersion coefficients for polyatomic molecules", *J. Chem. Phys.*, vol. 103, pp. 9347–9354.

van Gisbergen, S. J. A., J. G. Snijders, and E. J. Baerends, 1997, "Time-dependent Density-Functional Results for the Dynamic Hyperpolarizability of C_{60}", *Phys. Rev. Lett.*, vol. 78, pp. 3097–3100.

van Gisbergen, S. J. A., F. Kootstra, P. R. T. Schipper, O. V. Gritsenko, J. G. Snijders, and E. J. Baerends, 1998, "Density-functional-theory response-property calculations with accurate exchange-correlation potentials", *Phys. Rev. A*, vol. 57, pp. 2556–2571.

van Gisbergen, S. J. A., J. A. Groeneveld, A. Rosa, J. G. Snijders, and E. J. Baerends, 1999a, "Excitation Energies for Transition Metal Compounds from Time-Dependent Density-Functional Theory. Applications to MnO_4^-, $Ni(CO)_4$, and $Mn_2(CO)_{10}$", *J. Phys. Chem. A*, vol. 103, pp. 6835–6844.

van Gisbergen, S. J. A., P. R. T. Schipper, O. V. Gritsenko, E. J. Baerends, J. G. Snijders, B. Champagne, and B. Kirtman, 1999b, "Electric Field Dependence of the Exchange-Correlation Potential in Molecular Chains", *Phys. Rev. Lett.*, vol. 83, pp. 694–697.

van Gisbergen, S. J. A., A. Rosa, G. Ricciardi, and E. J. Baerends, 1999c, "Time-dependent density-functional calculations on the electronic absorption spectrum of free base porphin", *J. Chem. Phys.*, vol. 111, pp. 2499–2506.

van Holde, K. E., W. C. Johnson, P. S. Ho (eds.), 1998, *Principles of Physical Biochemistry*, (Prentice Hall, New Jersey).

van Leeuwen, R., and E. J. Baerends, 1994, "Exchange-correlation potential with correct asymptotic behavior", *Phys. Rev. A*, vol. 49, pp. 2421–2431.

van Leeuwen, R., 1996, "The Sham-Schlüter Equation in Time-Dependent Density-Functional Theory", *Phys. Rev. Lett.*, vol. 76, pp. 3610–3613.

van Leeuwen, R., 1998, "Causality and Symmetry in Time-Dependent Density-Functional Theory", *Phys. Rev. Lett.*, vol. 80, pp. 1280–1283.

van Leeuwen, R., 1999, "Mapping from Densities to Potentials in Time-Dependent Density-Functional Theory", *Phys. Rev. Lett.* vol. 82, pp. 3863–3866.

van Leeuwen, R., 2001, "Key Concepts in Time-Dependent Density-Functional Theory", *Int. J. Mod. Phys. B*, vol. 15, pp. 1969–2024.

van Leeuwen, R., 2003, "Density-Functional Approach to the Many-Body Problem: Key Concepts and Exact Functionals", *Adv. Quant. Chem.*, vol. 43, pp. 25–94.

van Leeuwen, R., and N. E. Dahlen, 2004a, "Conserving approximations in nonequilibrium Green function and density-functional theory", in *The electron liquid model in condensed matter physics*, edited by G. F. Giuliani and G. Vignale, pp. 169–188 (IOS Press, Amsterdam).

van Leeuwen, R., 2004b, "First-principles approach to the electron-phonon interaction", *Phys. Rev. B*, vol. 69, pp. 115110-1–20; ibid 199901(E) (2004).

van Voorhis, T., and M. Head-Gordon, 2001, "Two-body coupled cluster expansions", *J. Chem. Phys.*, vol. 115, pp. 5033–5040.

Varsano, D., A. Marini, and A. Rubio, 2005, "Polarisability and optical absorption of polymers and one dimensional chains", to be submitted.

Vasiliev, I., S. Öğüt, and J. R. Chelikowsky, 1999, "*Ab Initio* Excitation Spectra and Collective Electronic Response in Atoms and Clusters", *Phys. Rev. Lett.*, vol. 82, pp. 1919–1922.

Vasiliev, I., S. Öğüt, and J. R. Chelikowsky, 2001, "*Ab Initio* Absorption Spectra and Optical Gaps in Nanocrystalline Silicon", *Phys. Rev. Lett.*, vol. 86, pp. 1813–1816.

Vasiliev, I., S. Öğüt, and J. R. Chelikowsky, 2002a, "First-principles density-functional calculations for optical spectra of clusters and nanocrystals", *Phys. Rev. B*, vol. 65, pp. 115416-1–18.

Vasiliev, I., J. R. Chelikowsky, and R. M. Martin, 2002b, "Surface oxidation effects on the optical properties of silicon nanocrystals", *Phys. Rev. B*, vol. 65, pp. 121302-1–4(R).

Vasiliev, I., and R. Martin, 2002c, "Optical Properties of Hydrogenated Silicon Clusters with Reconstructed Surfaces", *Phys. Stat. Sol. (b)*, vol. 233, pp. 5–9.

Vasiliev, I., and R. M. Martin, 2004, "Time-dependent density-functional calculations with asymptotically correct exchange-correlation potentials", *Phys. Rev. A*, vol. 69, pp. 052508-1–10.

Vaswani, H. M., C. P. Hsu, M. Head-Gordon, and G. R. Fleming, 2003, "Quantum Chemical Evidence for an Intramolecular Charge-Transfer State in the Carotenoid Peridinin of Peridinin-Chlorophyll-Protein", *J. Phys. Chem. B*, vol. 107, pp. 7940–7946.

Vendrell, O., R. Gelabert, M. Moreno, and J. Lluch, "Photoinduced proton transfer from the green fluorescent protein chromophore to a water molecule: analysis of the transfer coordinate", *Chem. Phys. Lett.*, vol. 396, pp. 202–207.

Véniard, V., R. Taïeb, and A. Maquet, 2002, "Atomic clusters submitted to an intense short laser pulse: A density-functional approach", *Phys. Rev. A*, vol. 65, pp. 013202-1–7.

Véniard, V., R. Taïeb and A. Maquet, 2003, "Photoionization of atoms using time-dependent density-functional theory", *Laser Phys.*, vol. 13, pp. 465–474.

Verlet, L., 1967, "Computer "Experiments" on Classical Fluids. I. Thermodynamical Properties of Lennard-Jones Molecules", *Phys. Rev.*, vol. 159, pp. 98–103.

Veseth, L., 2001, "Molecular excitation energies computed with Kohn-Sham orbitals and exact exchange potentials", *J. Chem. Phys.*, vol. 114, pp. 8789–8795.

Vignale, G., and M. Rasolt, 1987, "Density-functional theory in strong magnetic fields", *Phys. Rev. Lett.*, vol. 59, pp. 2360–2363.

Vignale, G., and M. Rasolt, 1988, "Current- and spin-density-functional theory for inhomogeneous electronic systems in strong magnetic fields", *Phys. Rev. B*, vol. 37, pp. 10685–10696.

Vignale, G., 1995a, "Center of Mass and Relative Motion in Time Dependent Density-Functional Theory", *Phys. Rev. Lett.*, vol. 74, pp. 3233–3236.

Vignale, G., 1995b, "Sum rule for the linear density response of a driven electronic system", *Phys. Lett. A*, vol. 209, pp. 206–210.

Vignale, G., and W. Kohn, 1996, "Current-Dependent Exchange-Correlation Potential for Dynamical Linear Response Theory", *Phys. Rev. Lett.*, vol. 77, pp. 2037–2040.

Vignale, G., C. A. Ullrich, and S. Conti, 1997, "Time-Dependent Density-Functional Theory Beyond the Adiabatic Local Density Approximation", *Phys. Rev. Lett.*, vol. 79, pp. 4878–4881.

Vignale, G., and W. Kohn, 1998, "Current-density-functional theory of linear response to time-dependent electromagnetic fields", in *Electronic Density-Functional Theory: Recent Progress and New Directions*, edited by J. F. Dobson, G. Vignale, and M. P. Das, pp. 199–216 (Plenum Press, New York).

Vignale, G., 2004, "Mapping from current densities to vector potentials in time-dependent current density-functional theory", *Phys. Rev. B*, vol. 70, pp. 201102-1–4(R).

Villeneuve, D., M. Yu. Ivanov, and P. B. Corkum, "Enhanced ionization of diatomic molecules in strong laser fields: A classical model", *Phys. Rev. A*, vol. 54, pp. 736–741.

Villars, F. M. H., and G. Cooper, 1970, "Unified Theory of Nuclear Rotations", *Ann. Physics*, vol. 56, pp. 224–258.

Vitale, V., F. Della Sala, and R. Cingolani, 2004, "First-principles time-dependent density-functional theory study of functionalized benzo[b]thiophenes", *Phys. Stat. Solid. (c)*, vol. 1, pp. 555–559.

Vitale, V., F. Della Sala and A. Görling, 2005, "Open-shell localized Hartree-Fock method based on the generalized adiabatic connection Kohn-Sham formalism for a self-consistent treatment of excited states", *J. Chem. Phys.*, vol. 122, pp. 244102-1–17.

von Barth, U., and L. Hedin, 1972, "A local exchange-correlation potential for the spin polarized case: I", *J. Phys. C*, vol. 5, pp. 1629–1642.

von Barth, U., N. E. Dahlen, R. van Leeuwen, and G. Stefanucci, 2005, "Conserving Approximations in Time-Dependent Density-Functional Theory", *Phys. Rev. B*, vol. 72, pp. 235109-1–10.

von Lilienfeld, O. A., I. Tavernelli, U. Rothlisberger, and D. Sebastiani, 2004, "Optimization of Effective Atom Centered Potentials for London Dispersion Forces in Density-Functional Theory", *Phys. Rev. Lett.*, vol. 93, pp. 153004-1–4.

Vosko, S. H., L. Wilk, and M. Nusair, "Accurate spin-dependent electron liquid correlation energies for local spin density calculations: a critical analysis", *Can. J. Phys.*, vol. 58, pp. 1200–1211.

Voskoboynikov, O., C. P. Lee, and O. Tretyak, 2001, "Spin-orbit splitting in semiconductor quantum dots with a parabolic confinement potential", *Phys. Rev. B*, vol. 63, pp. 165306-1–6.

Wachter, R., B. King, R. Heim, K. Kallio, R. Tsien, S. Boxer, and S. Remington, 1997, "Crystal Structure and Photodynamic Behavior of the Blue Emission Variant Y66H/Y145F of Green Fluorescent Protein", *Biochemistry*, vol. 36, pp. 9759–9765.

Wagner, M., 1991, "Expansions of nonequilibrium Green's functions", *Phys. Rev. B*, 44, pp. 6104–6117.

Wahlström, C.-G., J. Larsson, A. Persson, T. Starczewski, S. Svanberg, P. Salières, P. Balcou, and A. L'Huillier, 1993, "High-order harmonic generation in rare gases with an intense short-pulse laser", *Phys. Rev. A*, vol. 48, pp. 4709–4720.

Walker, B., B. Sheehy, L. F. DiMauro, P. Agostini, K. J. Schafer, and K. C. Kulander, 1994, "Precision Measurement of Strong Field Double Ionization of Helium", *Phys. Rev. Lett.*, vol. 73, pp. 1227–1230.

Walsh, T. R., 2005, "Exact exchange and Wilson-Levy correlation: a pragmatic device for studying complex weakly-bonded systems", *Phys. Chem. Chem. Phys.*, vol. 7, pp. 443–451.

Wan, J., Y. Ren, J. Wu, and X. Xu, 2004, "Time-Dependent Density-Functional Theory Investigation of Electronic Excited States of Tetraoxaporphyrin Dication and Porphycene", *J. Phys. Chem. A*, vol. 108, pp. 9453–9460.

Wang, C. R. C., S. Pollack, D. Cameron, and M. M. Kappes, 1990a, "Optical absorption spectroscopy of sodium clusters as measured by collinear molecular beam photodepletion", *J. Chem. Phys.*, vol. 93, pp. 3787–3801.

Wang, C. R. C., S. Pollack, and M. M. Kappes, 1990b, "Molecular excited states versus collective electronic oscillations: Optical absorption proves of Na_4 and Na_8", *Chem. Phys. Lett.*, vol. 166, pp. 26–31.

Wang, C. R. C., S. Pollack, T. A. Dahlseid, G. M. Koretsky, and M. M. Kappes, 1992, "Photodepletion probes of Na_5, Na_6, and Na_7. Molecular dimensionality transition (2D→3D)?", *J. Chem. Phys.*, vol. 96, pp. 7931–7937.

Wang, L. W., and A. Zunger, 1993, "Solving Schrödinger's equation around a desired energy: Application to silicon quantum dots", *J. Chem. Phys.*, vol. 100, pp. 2394–2397.

Wang, L. W., and A. Zunger, 1994, "Electronic Structure Pseudopotential Calculations of Large (apprx. 1000 Atoms) Si Quantum Dots", *J. Phys. Chem.*, vol. 98, pp. 2158–2165.

Wang, E., and U. Heinz, 2002, "Generalized fluctuation-dissipation theorem for nonlinear response functions", *Phys. Rev. D*, vol. 66, pp. 025008-1–17.

Wang, X., L. Chen, A. Endou, M. Kubo, and A. Miyamoto, 2003, "A study on the excitations of ligand-to-metal charge transfer in complexes Cp_2MCl_2 ($Cp=\pi$-

C_5H_5, M=Ti, Zr, Hf) by density functional theory", *J. Organomet. Chem.*, vol. 678, p. 156

Wang, F., and T. Ziegler, 2004, "Time-dependent density-functional theory based on a noncollinear formulation of the exchange-correlation potential", *J. Chem. Phys.*, vol. 121, pp. 12191–12196.

Wang, B., H. Liao, W. Chen, Y. Chou, J. Yeh, and J. Chang, 2005a, "Theoretical investigation of electro-luminescent properties in red emission DCM, DCJ, RED and DAD derivatives", *J. Mol. Struct. (Theochem)*, vol. 716, pp. 19–25.

Wang, X., C. Lv, M. Koyama, M. Kubo, and A. Miyamoto, 2005b, "A theoretical investigation of the photo-induced intramolecular charge transfer excitation of cuprous (I) bis-phenanthroline by density-functional theory", *J. Organomet. Chem.*, vol. 690, pp. 187–192.

Wanko, M., M. Garavelli, F. Bernardi, A. Niehaus, T. Frauenheim, and M. Elstner, 2004, "A global investigation of excited state surfaces within time-dependent density-functional response theory", *J. Chem. Phys.*, vol. 120, pp. 1674–1692.

Warburton, R. J., C. Gauer, A. Wixforth, J. P. Kotthaus, B. Brar, and H. Kroemer, 1996, "Intersubband resonances in InAs/AlSb quantum wells: Selection rules, matrix elements, and the depolarization field", *Phys. Rev. B*, vol. 53, pp. 7903–7910.

Wasbotten, I. H., T. Wondimagegn, and A. Ghosh, 2002, " Electronic Absorption, Resonance Raman, and Electrochemical Studies of Planar and Saddled Copper(III) *meso*-Triarylcorroles. Highly Substituent-Sensitive Soret Bands as a Distinctive Feature of High-Valent Transition Metal Corroles", *J. Am. Chem. Soc.*, vol. 124, pp. 8104–8116.

Wasserman, A., N. T. Maitra, and K. Burke, 2003, "Accurate Rydberg Excitations from the Local Density Approximation", *Phys. Rev. Lett.*, vol. 91, pp. 263001-1–4.

Wasserman, A., 2005a, "Scattering States from Time-dependent Density-Functional Theory", PhD Thesis, Rutgers University.

Wasserman, A., N. T. Maitra, and K. Burke, 2005b, "Continuum states from Time-dependent Density-Functional Theory", *J. Chem. Phys.*, vol. 122, pp. 144103-1–5.

Watanabe, N., and M. Tsukada, 2002, "Efficient method for simulating quantum electron dynamics under the time-dependent Kohn-Sham equation", *Phys. Rev. E*, vol. 65, pp. 036705-1–6.

Watson, J. B., A. Sanpera, D. G. Lappas, P. L. Knight, and K. Burnett, 1997, "Nonsequential Double Ionization of Helium", *Phys. Rev. Lett.*, vol 78, pp. 1884–1887.

Weber, T., M. Weckenbrock, A. Staudte, L. Spielberger, O. Jagutzki, V. Mergel, F. Afaneh, G. Urbasch, M. Vollmer, H. Giessen, and R. Dörner, 2000, "Recoil-Ion Momentum Distributions for Single and Double Ionization of Helium in Strong Laser Fields", *Phys. Rev. Lett.*, vol. 84, pp. 443–446.

Weigend, F., M. Häser, H. Patzelt, and R. Ahlrichs, 1998, "RI-MP2: Optimized auxiliary basis sets and demonstration of efficiency", *Chem. Phys. Lett.*, vol. 294, pp. 143–152.

Weigend, F., F. Furche, and R. Ahlrichs, 2003, "Gaussian basis sets of quadruple zeta quality for atoms H-Kr", *J. Chem. Phys.*, vol. 119, pp. 12753–12762.

Weigend, F., and R. Ahlrichs, 2005, "Balanced basis sets of split valence, triple zeta valence and quadruple zeta valence quality for H to Rn: Design and assessment of accuracy", *Phys. Chem. Chem. Phys.*, vol. 7, pp. 3297–3305.

Weimer, M., W. Hieringer, F. Della Sala, and A. Görling, 2005, "Electronic and optical properties of functionalized carbon chains with the localized Hartree-Fock and conventional Kohn-Sham methods", *Chem. Phys.*, vol. 309, pp. 77–87.

Weisman, J. L., T. J. Lee, and M. Head-Gordon, 2001a, "Electronic spectra and ionization potentials of a stable class of closed shell polycyclic aromatic hydrocarbon cations", *Spectrochim. Acta A*, vol. 57, pp. 931–945.

Weisman, J. L., and M. Head-Gordon, 2001b, "Origin of Substituent Effects in the Absorption Spectra of Peroxy Radicals: Time Dependent Density-Functional Theory Calculations", *J. Am. Chem. Soc.*, vol. 123, pp. 11686–11694.

Weisman, J. L., T. J. Lee, F. Salama, and M. Head-Gordon, 2003, "Time-dependent Density-Functional Theory Calculations of Large Compact Polycyclic Aromatic Hydrocarbon Cations: Implications for the Diffuse Interstellar Bands", *Astrophys. J.*, vol. 587, pp. 256–261.

Weiss, H., R. Ahlrichs, and M. Häser, 1993, "A direct algorithm for self-consistent-field linear response theory and application to C_{60}: Excitation energies, oscillator strengths, and frequency-dependent polarizabilities", *J. Chem. Phys.*, vol. 99, pp. 1262–1270.

Wertsching, A. K., A. S. Koch, and S. G. DiMagno, 2001, "On the Negligible Impact of Ruffling on the Electronic Spectra of Porphine, Tetramethylporphyrin, and Perfluoroalkylporphyrins", *J. Am. Chem. Soc.*, vol. 123, pp. 3932–3939.

Wesolowski, T. A., and A. Warshel, 1993, "Frozen density-functional approach for ab-initio calculations of solvated molecules", *J. Phys. Chem.*, vol. 97, pp. 8050–8053.

White, A., 2003, "Dispersion forces in quasi-one-dimensional conductors", Honours Thesis, Griffith University, Australia. See also White, A. and Dobson, J. F., 2006, submitted.

Whitem, C. A., and M. Head-Gordon, 1994, "Derivation and efficient implementation of the fast multipole method", *J. Chem. Phys.*, vol. 101, pp. 6593–6605.

Wickenhauser, M., J. Burgdörfer, F. Krausz, and M. Drescher, 2005, "Time Resolved Fano Resonances", *Phys. Rev. Lett.*, vol. 94, pp. 023002-1–4.

Wijewardane, H. O., and C. A. Ullrich, 2004, "Coherent control of intersubband optical bistability in quantum wells", *Appl. Phys. Lett.*, vol. 84, pp. 3984–3986.

Wijewardane, H. O., and C. A. Ullrich, 2005, "Time-dependent Kohn-Sham theory with memory", *Phys. Rev. Lett.*, vol. 95, pp. 086401-1–4.

Williams, J. B., M. S. Sherwin, K. D. Maranowski, and A. C. Gossard, 2001, "Dissipation of Intersubband Plasmons in Wide Quantum Wells", *Phys. Rev. Lett.*, vol. 87, pp. 037401-1–4.

Williamson. A. J., J. C. Grossman, R. Q. Hood, A. Puzder, and G. Galli, 2002, "Quantum Monte Carlo Calculations of Nanostructure Optical Gaps: Application to Silicon Quantum Dots", *Phys. Rev. Lett.*, vol. 89, pp. 196803-1–4.

Wilson, L. C., and M. Levy, 1990, "Nonlocal Wigner-like correlation-energy density-functional through coordinate scaling", *Phys. Rev. B*, vol. 41, pp. 12930–12932.

Wilson, L. C., and S. Ivanov, 1998, "A Correlation-Energy Functional For Addition to the Hartree-Fock Energy", in *Electronic Density-Functional Theory: Recent Progress and New Directions*, edited by J. F. Dobson, G. Vignale, and M. P. Das, p. 133 (Plenum, New York).

Wilson, A. K., D. E. Woon, K. A. Peterson, and T. H. Dunning, Jr., 1999, "Gaussian basis sets for use in correlated molecular calculations. IX. The atoms gallium through krypton", *J. Chem. Phys.*, vol. 110, pp. 7667–7676.

Wingreen, N. S., A-P. Jauho and Y. Meir, 1993, "Time-dependent transport through a mesoscopic structure", *Phys. Rev. B*, vol. 48, pp. 8487–8490.

Winstead, C., and V. McKoy, 1996, "Electron Scattering by Small Molecules", in *Advances in Chemical Physics*, edited by I. Prigogine and S. A. Rice, vol. XCVI, pp. 103–190.

Wirtz, L., A. Marini, and A. Rubio, 2006, "Excitons in boron nitride nanotubes: dimensionality effects", *Phys. Rev. Lett.*, vol. 96, pp. 126104-1–4.

Woon, D. E., and T. H. Dunning, Jr., 1993, "Gaussian basis sets for use in correlated molecular calculations. III. The atoms aluminium through argon", *J. Chem. Phys.*, vol. 98, pp. 1358–1371.

Wrigge, G., M. A. Hoffmann, and B. von Issendorff, 2002, "Photoelectron spectroscopy of sodium clusters: Direct observation of the electronic shell structure", *Phys. Rev. A*, vol. 65, pp. 063201-1–5.

Wu, Q., and W. Yang, 2002, "Empirical correction to density-functional theory for van der Waals interactions", *J. Chem. Phys.*, vol. 116, pp. 515–524.

Wu, D., W. Cheng, X. Li, Y. Lan, D. Chen, Y. Zhang, H. Zhang, and Y. Gong, 2004, "First Principles Treatment of Configuration Optimizations, Excited-State Properties, and Dynamic Third-Order Polarizabilities of Chloro-Metal Phthalocyanines MPcCl (M = Al, Ga, In)", *J. Phys. Chem. A*, vol. 108, pp. 1837–1843.

Wunderlich, C., E. Kobler, H. Figger, and T. W. Hänsch, 1997, "Light-induced molecular potentials", *Phys. Rev. Lett.*, vol. 78, pp. 2333–2336.

Xiao, X., B. Xu, and N. J. Tao, 2004, "Measurement of Single Molecule Conductance: Benzenedithiol and Benzenedimethanethiol", *Nano Lett.*, vol. 4, pp. 267–271.

Xie, R., G. Bryant, G. Sun, T. Kar, Z. Chen, V. Smith, Y. Araki, N. Tagmatarchis, H. Shinohara, and O. Ito, 2004a, "Tuning spectral properties of fullerenes by substitutional doping", *Phys. Rev. B*, vol. 69, pp. 201403-1–4.

Xie, R., G. Bryant, G. Sun, M. Nicklaus, D. Heringer, T. Frauenheim, M. Manaa, V. Smith, Y. Araki, O. Ito, 2004b, "Excitations, optical absorption spectra, and optical excitonic gaps of heterofullerenes. I. C_{60}, $C_{59}N^+$, and $C_{48}N_{12}$: Theory and experiment", *J. Chem. Phys.*, vol. 120, pp. 5133–5147.

Xu, H., X. Tang, and P. Lambropoulos, 1992, "Nonperturbative theory of harmonic generation in helium under a high-intensity laser field: The role of intermediate resonance and of the ion", *Phys. Rev. A*, vol. 46, pp. R2225–R2228.

Xu, H., X. Tang, and P. Lambropoulos, 1993, "Atoms in a High-Intensity Laser Field of Short Duration", *Laser Phys.*, vol. 3, pp. 759–766.

Xue, Y., S. Datta, and M. A. Ratner, 2002, "First-principles based matrix Green's function approach to molecular electronic devices: general formalism", *Chem. Phys.*, vol. 281, pp. 151–170.

Yabana, K., and G. F. Bertsch, 1996, "Time-dependent local-density approximation in real time", *Phys. Rev. B*, vol. 54, pp. 4484–4487.

Yabana, K., and G. F. Bertsch, 1997, "Optical response of small carbon clusters", *Z. Phys. D*, vol. 42, pp. 219–225.

Yabana, K., and G. F. Bertsch, 1999a, "Time-dependent local-density approximation in real time: Application to conjugated molecules", *Int. J. Quantum Chem.*, vol. 75, pp. 55–66.

Yabana, K., and G. F. Bertsch, 1999b, "Application of the time-dependent local density approximation to optical activity", *Phys. Rev. A*, vol. 60, pp. 1271–1279.

Yabana, K., and G. F. Bertsch, 1999c, "Optical response of small silver clusters", *Phys. Rev. A*, vol. 60, pp. 3809–3814.

Yamaguchi, Y., S. Yokoyama, S. Mashiko, 2002a, "Strong coupling of the single excitations in the Q-like bands of phenylene-linked free-base and zinc bacteriochlorin dimers: A time-dependent density-functional theory study" *J. Chem. Phys.*, vol. 116, pp. 6541–6548.

Yamaguchi, Y., 2002b, "Time-dependent density-functional calculations of fully π-conjugated zinc oligoporphyrins", *J. Chem. Phys.*, vol. 117, pp. 9688–9694.

Yamaguchi, Y., 2002c, "Excited states of analogue models of M-bacteriochlorophylls (M = Mg, Zn) in the photosynthetic reaction center: a time-dependent density-functional theory study", *J. Porph. Phthal.*, vol. 6, pp. 617–625.

Yamaguchi, Y., 2004, "Theoretical prediction of electronic structures of fully π-conjugated zinc oligoporphyrins with curved surface structures" *J. Chem. Phys.*, vol. 120, pp. 7963–7970.

Yamakawa, K., Y. Akahane, Y. Fukuda, M. Aoyama, N. Inoue, H. Ueda, and T. Utsumi, 2004, "Many-Electron Dynamics of a Xe Atom in Strong and Super-strong Laser Fields", *Phys. Rev. Lett.*, vol. 92, pp. 123001-1–4.

Yan, Z., J. P. Perdew, S. Kurth, 2000, "density-functional for short-range correlation: Accuracy of the random-phase approximation for isoelectronic energy changes", *Phys. Rev. B*, vol. 61, pp. 16430–16439.

Yang, F., L. H. Moss, and G. N. Phillips Jr., 1996, "The molecular structure of green fluorescent protein", *Nature Biotech.*, vol. 14, pp. 1246–1251.

Yang, W., and Q. Wu, 2002, "Direct Method for Optimized Effective Potentials in Density-Functional Theory", *Phys. Rev. Lett.*, vol. 89, pp. 143002-1–4.

Yip, S. K., 1991, "Magneto-optical absorption by electrons in the presence of parabolic confinement potentials", *Phys. Rev. B*, vol. 43, pp. 1707–1718.

Yokozawa, A., and Y. Miyamoto, 2000, "Hydrogen dynamics in SiO2 triggered by electronic excitations", *J. Appl. Phys.*, vol. 88, pp. 4542–4546.

Yu, H., and A. D. Bandrauk, 1995, "Three-dimensional Cartesian finite element method for the time dependent Schrödinger equation of molecules in laser fields", *J. Chem. Phys.*, vol. 102, pp. 1257–1265.

Yudin, G. L., and M. Yu. Ivanov, 2001, "Physics of correlated double ionization of atoms in intense laser fields: Quasistatic tunneling limit", *Phys. Rev. A*, vol. 63, pp. 033404-1–14.

Yurtsever, M., G. Sonmez, A. S. Sarac, 2003, "Time dependent density-functional theory calculations for the electronic excitations of pyrrole-acrylamide copolymers", *Synth. Met.*, vol. 135–136, pp. 463–464.

Zacharia, R., H. Ulbricht, and T. Hertel, 2004, "Interlayer cohesive energy of graphite from thermal desorption of polyaromatic hydrocarbons", *Phys. Rev. B*, vol. 69, 155406-1–7.

Zahradník, R., M. Srnec, and Z. Havlas, 2005, "Electronic spectra of conjugated polyynes, cumulenes and related systems: A theoretical study", *Collect. Czech. Chem. Commun.*, vol. 70, pp. 559–578.

Zakrzewski, J., J. Delaire, C. Daniel, and I. Cote-Bruand, 2004, "$W(CO)_5$-pyridine π-acceptor complexes: theoretical calculations and a laser photolysis study", *New J. Chem.*, vol. 28, pp. 1514–1519.

Zalis, S., I. Farrell, and A. Vlcek, 2003, "The Involvement of Metal-to-CO Charge Transfer and Ligand-Field Excited States in the Spectroscopy and Photochemistry of Mixed-Ligand Metal Carbonyls. A Theoretical and Spectroscopic Study of [$W(CO)_4$(1,2-Ethylenediamine)] and [$W(CO)_4$(N, N'-Bis-alkyl-1,4-diazabutadiene)]", *J. Am. Chem. Soc.*, vol. 125, pp. 4580–4592.

Zalis, S., N. Ben Amor, and C. Daniel, 2004, "Influence of the Halogen Ligand on the Near-UV-Visible Spectrum of [Ru(X)(Me)$(CO)_2$(-diimine)] (X = Cl, I; α-Diimine = Me-DAB, iPr-DAB; DAB = 1,4-Diaza-1,3-butadiene): An ab Initio and TD-DFT Analysis", *Inorg. Chem.*, vol. 43, pp. 7978–7985.

Zanghellini, J., M. Kitzler, T. Brabec, and A. Scrinzi, 2004, "Testing the multi-configuration time-dependent Hartree-Fock method", *J. Phys. B: At. Mol. Opt. Phys.*, vol. 37, pp. 763–773.

Zanghellini, J., M. Kitzler, Z. Zhang, and T. Brabec, 2005, "Multi-electron dynamics in strong laser fields", *J. Mod. Opt.*, vol. 52, pp. 479–488.

Zangwill, A., and P. Soven, 1980a, "Density-functional approach to local-field effects in finite systems: Photoabsorption in the rare gases", *Phys. Rev. A*, vol. 21, pp. 1561–1572.

Zangwill, A., and P. Soven, 1980b, "Resonant Photoemission in Barium and Cerium", *Phys. Rev. Lett.*, vol. 45, pp. 204–207.

Zangwill, A., and P. Soven, 1981, "Resonant two-electron excitation in copper", *Phys. Rev. B*, vol. 24, pp. 4121–4127.

Zaremba, E., and W. Kohn, 1976, "Van der Waals interaction between an atom and a solid surface", *Phys. Rev. B*, vol. 13, pp. 2270–2285.

Zewail, A. H., 1994, *Femtochemistry*, vol. I & II (World Scientific, Singapore).

Zgierski, M. Z., 2003, "Cu(I)-2,9-dimethyl-1,10-phenanthroline: density-functional study of the structure, vibrational force-field, and excited electronic states", *J. Chem. Phys.*, vol. 118, pp. 4045–4051.

Zhang, Y., W. Pan, and W. Yang, 1997, "Describing van der Waals Interaction in diatomic molecules with generalized gradient approximations: The role of the exchange functional", *J. Chem. Phys.*, vol. 107, pp. 7921–7925.

Zhang, F., and K. Burke, 2004a, "Adiabatic connection for near degenerate excited states" *Phys. Rev. A*, vol. 69, pp. 052510-1–6.

Zhang, C., Z. Cao, H. Wu, and Q. Zhang, 2004b, "Linear and nonlinear feature of electronic excitation energy in carbon chains $HC_{2n+1}H$ and $HC_{2n}H$", *Int. J. Quantum Chem.*, vol. 98, pp. 299–308.

Zhou, J., J. Peatross, M. M. Murnane, and H. C. Kapteyn, 1996, "Enhanced High-Harmonic Generation Using 25 fs Laser Pulses", *Phys. Rev. Lett.*, vol. 76, pp. 752–755.

Zhou, X., A. Ren, and J. Feng, 2005, "Theoretical studies on the ground states in M(terpyridine)$_2^{2+}$ and M(n-butyl-phenylterpyridine)$_2^{2+}$ (M = Fe, Ru, Os) and excited states in Ru(terpyridine)$_2^{2+}$ using density-functional theory", *J. Organomet. Chem.*, vol. 690, pp. 338–347.

Zhu, Z., Y. Wang, and Y. Lu, 2003, "Time-Dependent Density-Functional Theory Study on Polyazopyrrole and Polyazothiophene", *Macromol.*, vol. 36, pp. 9585–9593.

Ziegler, T., A. Rauk, and E.J. Baerends, 1977, "On the calculation of multiplet energies by the Hartree-Fock-Slater method", *Theor. Chim. Acta*, vol. 43, 261 – 271.

Zimmer, M., 2002, "Green Fluorescent Protein (GFP): Applications, Structure, and Related Photophysical Behavior", *Chem. Rev.*, vol. 102, pp. 759–781.

Zubarev, D. N., 1974, *Nonequilibrium Statistical Thermodinamics* (Consultants Bureau, New York).

Zumbach, G., N. A. Modine, and E. Kaxiras, 1996, "Adaptive coordinate, real-space electronic structure calculations on parallel computers", *Solid State Commun.*, vol. 99, pp. 57–61.

Zuo, T., and A. D. Bandrauk, 1995, "Charge-resonance-enhanced ionization of diatomic molecular ions by intense lasers", *Phys. Rev. A*, vol. 52, pp. R2511–R2514.

Zuo, T., A. D. Bandrauk, and P. B. Corkum, 1996a, "Laser-induced electron diffraction: a new tool for probing ultrafast molecular dynamics", *Chem. Phys. Lett.*, vol. 259, pp. 313–320.

Zuo, T., and A. D. Bandrauk, 1996b, "Phase control of molecular ionization: H_2^+ and H_3^{2+} in intense two-color laser fields", *Phys. Rev. A*, vol. 54, pp. 3254–3260.

Zuo, T., H. Yu, and A. D. Bandrauk, 1996c, "Molecules in intense laser fields: enhanced ionization in a one-dimensional model of H_2", *Phys. Rev. A*, vol. 54, pp. 3290–3298.

Index

Lecture Notes in Physics

For information about earlier volumes please contact your bookseller or Springer LNP Online archive: springerlink.com

Vol.660: A. N. Gorban, I. V. Karlin, Invariant Manifolds for Physical and Chemical Kinetics

Vol.661: N. Akhmediev, A. Ankiewicz (Eds.) Dissipative Solitons

Vol.662: U. Carow-Watamura, Y. Maeda, S. Watamura (Eds.), Quantum Field Theory and Noncommutative Geometry

Vol.663: A. Kalloniatis, D. Leinweber, A. Williams (Eds.), Lattice Hadron Physics

Vol.664: R. Wielebinski, R. Beck (Eds.), Cosmic Magnetic Fields

Vol.665: V. Martinez (Ed.), Data Analysis in Cosmology

Vol.666: D. Britz, Digital Simulation in Electrochemistry

Vol.667: W. D. Heiss (Ed.), Quantum Dots: a Doorway to Nanoscale Physics

Vol.668: H. Ocampo, S. Paycha, A. Vargas (Eds.), Geometric and Topological Methods for Quantum Field Theory

Vol.669: G. Amelino-Camelia, J. Kowalski-Glikman (Eds.), Planck Scale Effects in Astrophysics and Cosmology

Vol.670: A. Dinklage, G. Marx, T. Klinger, L. Schweikhard (Eds.), Plasma Physics

Vol.671: J.-R. Chazottes, B. Fernandez (Eds.), Dynamics of Coupled Map Lattices and of Related Spatially Extended Systems

Vol.672: R. Kh. Zeytounian, Topics in Hyposonic Flow Theory

Vol.673: C. Bona, C. Palenzula-Luque, Elements of Numerical Relativity

Vol.674: A. G. Hunt, Percolation Theory for Flow in Porous Media

Vol.675: M. Kröger, Models for Polymeric and Anisotropic Liquids

Vol.676: I. Galanakis, P. H. Dederichs (Eds.), Half-metallic Alloys

Vol.677: A. Loiseau, P. Launois, P. Petit, S. Roche, J.-P. Salvetat (Eds.), Understanding Carbon Nanotubes

Vol.678: M. Donath, W. Nolting (Eds.), Local-Moment Ferromagnets

Vol.679: A. Das, B. K. Chakrabarti (Eds.), Quantum Annealing and Related Optimization Methods

Vol.680: G. Cuniberti, G. Fagas, K. Richter (Eds.), Introducing Molecular Electronics

Vol.681: A. Llor, Statistical Hydrodynamic Models for Developed Mixing Instability Flows

Vol.682: J. Souchay (Ed.), Dynamics of Extended Celestial Bodies and Rings

Vol.683: R. Dvorak, F. Freistetter, J. Kurths (Eds.), Chaos and Stability in Planetary Systems

Vol.684: J. Dolinšek, M. Vilfan, S. Žumer (Eds.), Novel NMR and EPR Techniques

Vol.685: C. Klein, O. Richter, Ernst Equation and Riemann Surfaces

Vol.686: A. D. Yaghjian, Relativistic Dynamics of a Charged Sphere

Vol.687: J. W. LaBelle, R. A. Treumann (Eds.), Geospace Electromagnetic Waves and Radiation

Vol.688: M. C. Miguel, J. M. Rubi (Eds.), Jamming, Yielding, and Irreversible Deformation in Condensed Matter

Vol.689: W. Pötz, J. Fabian, U. Hohenester (Eds.), Quantum Coherence

Vol.690: J. Asch, A. Joye (Eds.), Mathematical Physics of Quantum Mechanics

Vol.691: S. S. Abdullaev, Construction of Mappings for Hamiltonian Systems and Their Applications

Vol.692: J. Frauendiener, D. J. W. Giulini, V. Perlick (Eds.), Analytical and Numerical Approaches to Mathematical Relativity

Vol.693: D. Alloin, R. Johnson, P. Lira (Eds.), Physics of Active Galactic Nuclei at all Scales

Vol.694: H. Schwoerer, J. Magill, B. Beleites (Eds.), Lasers and Nuclei

Vol.695: J. Dereziński, H. Siedentop (Eds.), Large Coulomb Systems

Vol.696: K.-S. Choi, J. E. Kim, Quarks and Leptons From Orbifolded Superstring

Vol.697: E. Beaurepaire, H. Bulou, F. Scheurer, J.-P. Kappler (Eds.), Magnetism: A Synchrotron Radiation Approach

Vol.698: S. Bellucci (Ed.), Supersymmetric Mechanics – Vol. 1

Vol.699: J.-P. Rozelot (Ed.), Solar and Heliospheric Origins of Space Weather Phenomena

Vol.700: J. Al-Khalili, E. Roeckl (Eds.), The Euroschool Lectures on Physics with Exotic Beams, Vol. II

Vol.701: S. Bellucci, S. Ferrara, A. Marrani, Supersymmetric Mechanics – Vol. 2

Vol.702: J. Ehlers, C. Lämmerzahl, Special Relativity

Vol.703: M. Ferrario, G. Ciccotti, K. Binder (Eds.), Computer Simulations in Condensed Matter Systems Volume 1

Vol.704: M. Ferrario, G. Ciccotti, K. Binder (Eds.), Computer Simulations in Condensed Matter Systems Volume 2

Vol.705: P. Bhattacharyya, B.K. Chakrabarti (Eds.), Modelling Critical and Catastrophic Phenomena in Geoscience

Vol.706: M.A.L. Marques, C.A. Ullrich, F. Nogueira, A. Rubio, K. Burke, E.K.U. Gross (Eds.), Time-Dependent Density Functional Theory